T0136656

CACTI
OF EASTERN BRAZIL

Facheiroa squamosa (Photo: Nigel Taylor)

Cereus jamacaru subsp. *jamacaru* (Mandacaru). The largest specimen ever seen, growing at Morro do Chapéu, Bahia, 2002. Photograph by Marlon Machado.

CACTI
OF EASTERN BRAZIL

Nigel Taylor and Daniela Zappi

Published by
The Royal Botanic Gardens, Kew

PLANTS PEOPLE
POSSIBILITIES

© The Board of Trustees of the Royal Botanic Gardens, Kew 2004

All rights reserved. No part of this publication may be reproduced,
stored in a retrieval system, or transmitted, in any form, or by any
means, electronic, mechanical, photocopying, recording or otherwise,
without written permission of the publisher unless in accordance with
the provisions of the Copyright Designs and Patents Act 1988.

First published in 2004 by
Royal Botanic Gardens, Kew
Richmond, Surrey, TW9 3AB, UK
www.kew.org

ISBN 1 84246 056 0

Production editor: Ruth Linklater
Typesetting and page layout: Christine Beard

Design by Media Resources,
Information Services Department,
Royal Botanic Gardens, Kew.

Printed in Italy by L.E.G.O.

For information or to purchase all Kew titles please visit
www.kewbooks.com or email publishing@kew.org

CONTENTS

TEXT TABLES, MAPS, HALFTONE ILLUSTRATIONS AND COLOUR PLATES

HALFTONE ILLUSTRATIONS

COLOUR PLATES

PREFACE

'*The day has passed delightfully. Delight itself, however, is a weak term to express the feelings of a naturalist who, for the first time, has wandered himself in a Brazilian forest. The novelty of the plants, the beauty of flowers, but above all the general luxuriance of the vegetation, filled me with admiration.*' It was not by accident that Charles Darwin wrote this remark in his journal of researches on February 23rd in 1823, when he entered a tropical forest in Eastern Brazil. And it was not by accident that one of the most significant international agreements, the Convention on Biological Diversity (CBD), was initiated in Rio de Janeiro in 1992.

It is generally accepted that Eastern Brazil is one of the global centres of biodiversity. The region comprises some 2 million square kilometres of tropical Rain Forest (*Mata atlântica*), dry woodlands (*Caatinga*), savannas (*Campos cerrados*) and highland vegetation (*Campos rupestres*) with altogether at least 20,000 species of flowering plants, a high percentage of these being endemics.

With cacti we associate the dry regions of the Sonoran, Chihuahuan or Atacama Deserts, but, surprisingly, the species-rich forests and highlands of Eastern Brazil harbour around 10% of the species of the Cactus Family. Remarkable and fantastic plants of completely differing life forms: large foliage trees (*Pereskia*), lianas (*Hylocereus*) and a wealth of epiphytes within the canopy, like the Christmas Cacti (*Schlumbergera*) and their relatives (*Rhipsalis*). I think the most surprising fact is the occurrence of highly succulent, columnar (eg. *Coleocephalocereus*) or even globular (*Melocactus ernestii*) cacti on rock outcrops and inselbergs in the humid rain forests of the *Mata atlântica*. Even the Sugar-loaf (Pão de Açucar) and adjacent inselbergs around Rio de Janeiro harbour endemic species. The present work, however, focuses on the much less studied regions to the north of Rio, which the ecotourist is only just beginning to discover.

Early explorers could document little of Eastern Brazil's cactus riches due to the difficulties these plants present for the preservation and description of living and museum specimens, not to mention the inaccessibility of so many of the most biodiverse cactus habitats prior to the mid-twentieth century. Most taxa now known were discovered in the past 100 years, the majority being named and classified by horticulturists, whose activities have rather distorted the number of botanically justifiable entities and their geographical range. Even so, many genuinely distinct taxa do have rather restricted ranges and in such a large area numerous, targeted field excursions are essential for a proper understanding and characterisation of species and adequate documentation of distribution and ecological data. Such activities have an overriding purpose in acquiring the evidence and arguments for the conservation of these remarkable examples of our natural inheritance — a richness which is slipping away day-by-day as man's reckless progress continues unabated.

Nigel Taylor is the Curator of the living collections at Kew, and his wife, Daniela Zappi, Assistant Keeper of the Kew Herbarium. I admired Nigel's broad knowledge in various plant groups when I first met him at the Herbarium of the Royal Botanic Gardens in the late 1970s. This led to a very fruitful co-operation on one of our mutual favourite groups, epiphytic cacti, and to wonderful field trips in North and South America. The same broad interests were later shared by Daniela — both are truly not typical 'herbarium' botanists. They have experience of cultivating the plants they have chosen to study and through this have built up extensive personal networks in the amateur and academic communities that serve the cactus hobby. They have also devoted considerable time to teaching and enthusing a significant number of Brazilian university students at undergraduate and postgraduate levels. Such contacts are essential not only for promoting awareness and conservation, but also for filling the inevitable gaps in knowledge of the plants in the field — gaps that exist even after a study of 15 years and many more excursions in the wilds of Brazil. Their work has been one of critical inventory, attempting to bring order to a complex group that, prior to the 1990s, was poorly understood in terms of taxonomic relationships, phytogeography and conservation status.

The 'Cacti of Eastern Brazil' is a monographic treatment of more than 130 species, with keys, maps and marvellous coloured photographic illustrations of the living plants, mostly in habitat, and all carefully documented. It includes a wealth of ecological data and chapters on their biogeography and conservation. A long-existing gap in our knowledge of one of the most fascinating biomes of the world has been filled.

Wilhelm Barthlott
Director, Botanisches Institut
und Botanischer Garten der
Universität Bonn, Germany

1. INTRODUCTION

1.1 SIGNIFICANCE OF THE AREA DELIMITED AS 'EASTERN BRAZIL'

In the biodiverse neotropics the cactus family, with more than 1300 species (Hunt 1999a: 160), represents the second in order of size amongst higher plant groups that are endemic or almost so (Bromeliaceae are in first place). In the Americas the Cactaceae have 4 major geographical centres of diversity (Taylor in Oldfield 1997: 18–19), of which the most significant is Mexico and the south-western USA. Baseline floristic-monographic treatments of cacti from this North American region have been published in the last 25 years (cf. Hunt 1992b). The second centre of diversity is within the Andean chain, Peru and Bolivia being especially rich. Here the taxonomic complexities of the family are too great and currently too little understood (cf. Taylor *et al.* in Oldfield 1997: 111). The third centre in order of importance is Eastern Brazil — a large but discrete region cut off from other areas of cactus diversity by broad environmental zones that are ecologically unsuitable for most members of this primarily dryland family. The last monographic-floristic treatment of the Cactaceae in Brazil was published 114 years ago (Schumann 1890) and accounted for less than 10% of the 130+ native species now recorded from Eastern Brazil, most of which are endemic, making this area a priority for taxonomic inventory and conservation.

As defined here 'Eastern Brazil' includes the habitats of all cacti native to the vegetation known as *caatinga*, its ecotones with the Atlantic Forest (*agreste*) and all but 9 species endemic to *campo rupestre* and other, geographically associated rupicolous formations, such as *mata calcária*, within the contiguous *cerrados* (for vegetation details, see Chapter 3.2). The 9 taxa excluded are from the adjacent parts of the Brazilian states of Goiás, Tocantins and Minas Gerais★, but expanding the study area westwards would mean including elements of two more Brazilian Regions ('Grandes Regiões'), ie. Central-western and Northern Brazil, and for relatively little gain, since these are without significant cactus floras. Expanding further southwards would increase the number of epiphytic Rhipsalideae to be treated, but as noted below, these are now better understood and less deserving of study than the cacti of the Brazilian drylands. The taxonomic survey presented here deals with c. 1000 botanical names applied to cactus taxa found in Eastern Brazil, more than 800 of these being treated as synonyms.

★ Namely, *Pilosocereus albisummus, P. flexibilispinus, P. vilaboensis, P. diersianus* (Zappi 1994), *Cereus bicolor,* C. sp. nov. (Tocantins), C. *pierre-braunianus* (Goiás), *Arrojadoa* sp. nov. (Goiás) and *Micranthocereus estevesii* (see Appendix II).

1.2 BACKGROUND TO THE CACTI OF EASTERN BRAZIL PROJECT

The present study has its origins in a field excursion to Eastern Brazil led by Dr Ray Harley (Herbarium, Royal Botanic Gardens, Kew), which departed the U.K. at the close of September 1988. This excursion, in which the authors were invited to participate as Cactaceae specialists with interests in Brazilian taxa (Taylor 1980, 1981, 1982, Zappi 1989), was part of a long-term collaborative project to study the flora of the East Brazilian Highlands and especially the *campos rupestres* of the Chapada Diamantina, Bahia (cf. Stannard 1995). This involved RBG Kew, the Instituto de Biociências of the Universidade de São Paulo (São Paulo state) and the herbarium of the Centro de Pesquisas de Cacau (CEPEC), Ilhéus (Bahia state). It was carried out under the terms of a *convênio* (agreement) and collecting permit issued by the Brazilian authorities (Conselho Nacional de Pesquisas Científicas e Tecnológicas, CNPq and Instituto Brasileiro do Meio Ambiente, IBAMA). Whilst the excursion focused on the highland areas, including visits to various localities in the state of Minas Gerais as well as the Pico das Almas in Bahia, participating specialists were allowed time and resources to make field collections of their chosen plant families in other vegetation types. Thus, the authors were able to study and sample some of the Cactaceae characteristic of the *caatinga* and *cerrado* biomes (Zappi had begun field studies on cacti in the *campos rupestres* the previous year for her Master's thesis, Table 1.1). Prior to Taylor's return to the U.K., in February 1989, there was also an opportunity to study members of the family in the Atlantic Forest (*Mata atlântica*) of São Paulo state.

Having gained an insight into the ecology and variation patterns of Brazilian cacti in habitat, the authors collaborated in an initial field excursion towards a taxonomic revision of the genus *Pilosocereus* Byles & G. D. Rowley, the subject selected for Zappi's doctoral thesis (Zappi 1992, 1994). This also greatly furthered progress with a long-running project to revise the genus *Melocactus* Link & Otto in Central and South America (Taylor 1991a), whose centre of diversity is in Bahia, which was extensively travelled by the authors during 3 weeks in July 1989. Following this rewarding experience, approval was gained from the Keeper of the Herbarium at Kew, Prof. Gren Lucas, to begin a regional taxonomic monograph of the family, to be entitled 'Cacti of Eastern Brazil'. The decision to take this project forward was based on three factors: (1) the perceived high levels of endemism of Cactaceae in this area, (2) the apparent lack of knowledge about their taxonomy and distribution, and (3) the established series of collaborations and floristic treatments already realized by RBG Kew in the North-eastern region of Brazil (eg. Harley & Mayo 1980, Renvoize 1983, Harley & Simmons 1986, Lewis 1987).

However, whereas the above-cited treatments had focused on Bahia state, or particular sites in its mountainous interior, the 'Cacti of Eastern Brazil' project, with a more modest number of taxa, could afford to be more ambitious and cover a wider area. Initial field studies suggested this should be the most species-rich parts of North-eastern Brazil and the adjacent state of Minas Gerais from the South-eastern Region, which holds many endemic taxa besides being the southern limit of range of many others. Ultimately the area chosen expanded further to encompass NE Brazil in its entirety and the northern half of South-eastern Brazil, limited at c. 22°S and 46°W. Hence the convenient term, 'Eastern Brazil', an area amounting to some 2 million km² (Map 1, page 69).

From 1990 further field excursions were planned and executed and opportunities taken to collaborate with students and their professors, besides giving papers and mini-courses at Brazilian botanical congresses and universities. In the interests of gaining a better knowledge of the largely epiphytic cactus flora of the Atlantic Forest — a vegetation type that has been all but comprehensively modified within Eastern Brazil — additional field excursions within the contiguous South-eastern region of Brazil, in the states of Rio de Janeiro and São Paulo, were also carried out. These contributed to a precursory treatment of the tribe Rhipsalideae (Barthlott & Taylor 1995), whose species found in Eastern Brazil would otherwise have been difficult to interpret, since their centre of greatest diversity is outside the area dealt with here.

Subsequently the authors began developing ideas on the phytogeography and conservation status of the cacti of Eastern Brazil (Taylor 1991c, Taylor & Zappi 1992a, Taylor *et al.* in Oldfield 1997: 111–124, 143–144), which have become particular foci in the work presented here, backed up by the detailed taxonomic inventory in Chapter 5. Finally, a significant part of this book is based on studies made for our respective doctoral theses (Zappi 1992, Taylor 2000).

1.3 STUDY METHODS AND TAXONOMIC CONCEPTS

In order to build a comprehensive understanding and documentation of morphology, ecology and geographic distribution, field and herbarium studies and observations of living materials in cultivation were made. The methods employed have already been described in Taylor (1991a: 2–4) and Zappi (1994: 11) and have encompassed surveys of seed-morphology using Scanning Electron Microscope facilities at RBG Kew. Field excursions are detailed in Table 1.1, below. Field studies are especially important for Cactaceae, because the difficulties that face the collector in their preparation for the herbarium and for subsequent identification have meant that the family has tended to be ignored or avoided by professional botanists and has in general remained poorly represented in museum collections (see Chapter 2). Nevertheless, specimens at the following 58 herbaria were studied and augmented in some cases with materials collected for the project★ (institutions are listed by their standard *Index Herbariorum* codes, those indicated in **bold** typeface being the more important for Cactaceae in Eastern Brazil, some of the remainder being significant at the state level): ALCB, ASE, B†, BAH, BHCB, BHMG, BM, BONN†, BR†, [C — list by D. R. Hunt consulted], CEN, **CEPEC**, CESJ, E†, EAN, ESA†, F, GUA, HB, HNT, **HRB**, **HRCB**†, **HUEFS**, IAN, **IPA**, JPB, **K**†, L, M†, MAC, MBM, MBML, MEXU, MG, MO, **NY**†, P (loan only), **PEUFR**, R, **RB**†, RSA, S, SI, **SP**, **SPF**, SPSF, **U**†, UB, UEC, UFG, UFMT†, UFP, UPS, **US**, W,

---

★Duplicates of specimens have been deposited in one local herbarium in each state of collection and at SPF or HRCB.

WAG, **ZSS**†, Herb. Fundação Zoo-Botânica de Belo Horizonte† and Herb. Univ. Federal de Espírito Santo (the dagger symbol, †, indicates that associated living collections were also examined). Most of the above-cited institutions have been visited in person, so that all of their materials could be studied, this in view of the need to have as comprehensive a view of taxon distribution as possible. Another major source of distributional data has been the authors' many contacts, both in Brazil and amongst the cactus hobbyist community in the U.K., Germany and elsewhere. The extensive literature on Cactaceae has also been consulted and published records and details of type localities were particularly helpful in completing the discussion and distribution maps for taxa from beyond Eastern Brazil, as detailed in Chapter 3 (Maps 7–14). Much of this bibliography is difficult to access for those in Brazil who would like to study the family and thus it is hoped that the majority of key literature references are cited in Chapter 6, which should serve as a guide to what is available in European libraries. However, it should be noted that recent publications by Russian botanist Prof. A. Doweld, in which he has 'atomised' the Cactaceae, have not been considered. They add nothing to our understanding of the family and further increase its overburdened nomenclature.

Distribution maps and calculations of 'extent of occurrence' for phytogeographic and conservation purposes have been produced digitally with the assistance of Kew's Geographical Information Systems Unit (GIS Unit, Herbarium), employing ArcView software and a convexhull algorithm run on a networked desktop computer with a 21 inch, high resolution screen. During the production of maps it became obvious that over-reliance on latitude and longitude coordinates determined by collectors and their assistants prior to the advent of global positioning systems equipment (GPS) can introduce disturbingly great errors. Such records, therefore, have been localized using other label data aided by various GIS overlays, eg. road and river systems. Obviously erroneous coordinate label data have been excised from the records cited in Chapter 5. GIS overlays have also been checked against a diverse range of available printed maps acquired in Brazil during the course of this study (besides those held in the Library at RBG Kew) and also compared with actual routes travelled on the ground, revealing significant discrepancies in some areas.

Generic concepts. Unlike specific concepts, where many taxonomists can agree that it is possible to recognize well-defined entities in nature whose circumscription is often relatively unambiguous, concepts that group species into genera are much more liable to differ in breadth. Cactaceae systematics is a classic exemplar of this, 'with the archetypal lumpers and splitters traditionally engaged in regular and unwearying conflict' (Hunt 1999b: 3). The key point, however, is that for the purpose of communicating evolutionary relationship the group defined should have a single common ancestor and include all taxa derived from that ancestor, ie. it is monophyletic. Thus, in an earlier, morphologically-based, cladistic treatment of the genera of Cereeae (the largest cactus tribe in Eastern Brazil) we stated that 'the operational taxonomic units to be employed . . . as far as our knowledge permits . . . are monophyletic' (Taylor & Zappi 1989: 15). This is still very much the principle embodied here, so that the genera *Melocactus* Link & Otto

Table 1.1

Field excursions directly related to the project, involving the authors (1987–2003).

Dates	Areas visited	Main purpose of excursion	Collaborator(s) (*Index Herbariorum* Acronyms)
1987–88	Minas Gerais (Serra do Espinhaço), various excursions/localities	Master's thesis research (Zappi only)	J. Prado *et al.* (SPF)
Oct. 1988–Feb. 1989	Minas Gerais, Bahia, coastal São Paulo	Collecting material for above, plus 'Flora of the Pico das Almas' and related projects	A.M. Giulietti & N.L. de Menezes (SPF), R.M. Harley *et al.* (K)
July 1989	Bahia, N Minas Gerais	Collecting material for Zappi's PhD revision of *Pilosocereus*	
Jan.–Feb. 1990	Cent. & SW Ceará, SE Piauí, E Maranhão, N Bahia, NW & cent. Pernambuco (by public bus)	As above	
May 1990	Coast of E Rio de Janeiro & S Espírito Santo	Cactaceae course at Univ. Federal, Vitória, ES	O.J. Pereira, Univ. Fed. ES J.R. Pirani *et al.* (SPF)
Sep. 1990	Cent.-N Minas Gerais	Collecting material for 'Cacti of E Brazil' project (Zappi only)	
Dec. 1990	NE & SE Minas Gerais, W Espírito Santo	As above (Taylor/Zappi)	
Jan.–Feb. 1991	Bahia, E Goiás, N Minas Gerais, Sergipe, Alagoas, Pernambuco	As above	U. Eggli (ZSS)
July 1991	S Minas Gerais	As above (Zappi only)	Univ. Fed. Juiz de Fora
April 1992	N & E Bahia	Noting/photographing additional distribution records (Taylor/Zappi)	
Feb. 1995	E Bahia, E Alagoas, cent.-N Pernambuco, E & S Paraíba, N Piauí, NW Ceará	As above	BA: A.M. Carvalho (CEPEC) AL: R. Lyra-Lemos (MAC) PE/PB: M.F. Agra & E.A. Rocha (JPB) PI/CE: J.B. da Silva (MG)

Aug. 1998	NE Bahia (by public bus), E Sergipe	Taylor, to determine identity of *Pilocereus rupicola* Werderm. (Serra da Itabaiana, Sergipe)	M.L. Santos, Aracaju
Nov.–Dec. 1999	N Rio de Janeiro, Espírito Santo, E Pernambuco	Zappi only, for Rubiaceae (RJ/ES) and Cactaceae Course, Univ. Federal Pernambuco	RJ/ES: L. Aona (UEC) PE: E.A. Rocha (UFPE)
March–April 2000	W Pernambuco, N & W Paraíba, S Rio Grande do Norte, S Ceará, SE Piauí, N Bahia	Noting/photographing additional distribution records and testing phytogeographical hypotheses for Taylor's PhD thesis (Taylor only)	E.A. Rocha, Univ. Estadual Santa Cruz, Bahia & P. Griffiths (K)
Aug. 2001	coastal Paraíba to SE Bahia (Zappi), cent.-S Minas Gerais (Taylor)	Conservation assessments and additional geographical records	PB–BA: J. Jardim (CEPEC) MG: M. Pimentel M. & J. Ordones Rego, Fundação Zoo-Bot., BH, MG
Nov. 2001	E Bahia	Cactaceae Course, Univ. Federal Feira de Santana (Zappi only)	J. Jardim *et al.* (HUEFS)
Aug. 2002	N & E Bahia	Conservation assessments and additional records & photographs (Taylor only)	M. Machado plus staff & students from HUEFS; R. Augusto, Cruz das Almas
April 2003	SE Bahia	Additional records and photos of new taxa (Taylor/Zappi)	M. Machado (HUEFS) and friends from Vitória da Conquista, incl. UESB
Aug. 2003	S Minas Gerais	Serra do Ibitipoca etc. for photos/records	M. Hjertson (UPS)

and *Coleocephalocereus* Backeb. are each regarded as monophyletic (they are held to be sister taxa on the basis of the synapomorphy of fruits expressed from the cephalium). *Melocactus* has autapomorphies of a terminal cephalium, seeds with relatively few testa-cells and pollen with simple tectal perforations (Taylor & Zappi 1989, Taylor 1991a: 17–18), while its potentially paraphyletic sister taxon is now believed to be monophyletic on the basis of having fruits dehiscent by means of a small basal pore. Recently, the monophyly of both genera has been supported in a preliminary analysis of the chloroplast *rpl*16 gene sequence (Soffiatti unpubl.).

Independent cladistic analyses of Rhipsalideae DC., based on morphological characters (Taylor unpubl. [1996]) and molecular data (Wallace unpubl. [1996]), when taken together, suggest that in comparison to the plesiomorphic, mesotonically branched *Lepismium* Pfeiff. (Barthlott 1987), the more derived E Brazilian genera of the tribe are each potentially monophyletic, as follows: (1) based on acrotonic branching, freely disarticulating old/diseased stem-segments and expanded flowers with ± colourless perianth-segments (*Rhipsalis* Gaertn.); (2) these same vegetative states combined with strictly determinate stem-segments (ie. complete absence of indeterminate extension shoots) and ± erect (*vs* pendent) habit, perianth highly coloured (*Hatiora* Britton & Rose, *sens. str.* [3 spp.]); and (3) determinate stem-segments combined with the presence of a well-developed flower-tube, highly coloured perianth and stamens inserted in 2 series (*Schlumbergera* Lem.). The floral characters of *Hatiora* and *Schlumbergera* are here interpreted as independent adaptations towards pollination by hummingbirds. However, in Wallace's molecular analysis, *Hatiora* is paired with *Rhipsalis pachyptera*, indicating *Rhipsalis* as paraphyletic, but the morphological evidence clearly places this species (and all other members of *R.* subg. *Phyllarthrorhipsalis*, see Chapter 5) together with the remainder of *Rhipsalis*. Defined on this basis *Rhipsalis, Hatiora (sens. str.)* and *Schlumbergera*, are easily recognized and may well be monophyletic, but there are also theoretically less clear cut, yet practical considerations that flavour some of the generic circumscriptions presently adopted in other tribes of the family.

Thus, various genera are recognised here on the basis of 'obvious morphological character[s]' (Judd *et al.* 1999: 29) that are *assumed* to be autapomorphies, but cannot be proved to be such, at least partly because of the lack of molecular phylogenies that would help determine the polarities of the character states employed, eg. characters defining taxa in the major tribes Cereeae and Trichocereeae (see Chapter 2.7 for explanations). In other cases it may not even be wise to state that such obvious morphological characters can be assumed to be autapomorphies, but their significance is that the taxa they define cannot be allied with any other group and seem isolated taxonomically. This latter situation describes, and is used to justify, the recognition of *Pseudoacanthocereus* F. Ritter, *Leocereus* Britton & Rose (as a monotype) and *Uebelmannia* Buining (see Table 2.1), each of which possesses a unique suite of characters and whose inclusion in any other genus — and it is hard to decide which this might be (!) — would greatly increase the risk of creating at least a paraphyletic, if not a polyphyletic assemblage. These uncertainties stem from a lack of sufficient characters that can be safely employed in phylogenetic analysis (Taylor & Zappi 1989: 14), since the highly specialized and reduced nature of cactus morphology effectively limits the features that can be used, many being strongly linked to environmental factors and homoplasious. This said, a comprehensive investigation into anatomy (as begun by Mauseth 1996) and stem chemistry might well add valuable data sets for Brazilian and other South American taxa, as it has for the North American Pachycereeae (Gibson & Horak 1978). In the short term, molecular data look to be those most likely to contribute to a better understanding of generic relationships in Brazilian cacti and various teams of researchers, including Brazilians (eg. Dr P. Soffiatti, USP, São Paulo), are currently seeking new sources of informative gene sequences.

In a couple of other cases it is not so much a lack of knowledge, as the needs of practical convenience, that have shaped decisions on the circumscriptions adopted, it being arguably better to be aware of a problem than to use ignorance as one's excuse — the situation described in the previous paragraph. For example, based on molecular data (Wallace 1995: 9, Butterworth unpubl.) the species comprising *Pereskia* Mill. fall into 4 distinct clades, whose relationship to one another at present cannot be further resolved. And, a more serious problem is that the genus is potentially paraphyletic in respect of the Cactoideae (represented by *Leptocereus* in Wallace's analysis, l.c.). It could, in theory, be broken up and in that case the Brazilian taxa would end up being distributed between 3 different genera (see Chapter 5), not that this would be either helpful for identification purposes or more informative than their recognition as subgenera or sections. However, although the genus is implicitly circumscribed here to include all its named species, the element that makes it paraphyletic in respect of the Cactoideae is *P. lychnidiflora* DC., a native of Central America, not Brazil. A not dissimilar situation is presented by the treatment of *Cereus* Mill. adopted here. Two sympatric subgenera are recognized, Subg. *Cereus* and Subg. *Mirabella* (F. Ritter) N. P. Taylor (1991b, 1992a), but the latter lacks the potential autapomorphy for the genus, which is the early-deciduous spent perianth (inclusion of Subg. *Mirabella* has required *Cereus* to be keyed out twice in the 'Key to genera', Chapter 5). However, on other characters there is no reason for its exclusion from *Cereus* and it seems particularly close in habit, flowers and fruit to another allopatric-vicariant element known as *C.* subg. *Ebneria* (see Chapter 3 and Map 8), which has the early-deciduous perianth. In these circumstances the authors prefer to await molecular evidence that may help determine whether the persistent perianth in Subg. *Mirabella* represents the plesiomorphic state or an apomorphic reversal. The same remark must apply to another case where, in contrast, two taxa have been kept apart. This concerns *Pilosocereus* Byles & G. D. Rowley and *Micranthocereus* Backeb. (see Chapter 5 for details), where there is an unclear relationship and fragile distinctions between two taxa that have never been combined hitherto. The much larger and more widespread genus is *Pilosocereus*, which has the younger of the two generic names involved, and in the interests of nomenclatural stability would need to be conserved, if they were lumped as a single genus.

Besides the already-mentioned *Leocereus*, the status of two other monotypic genera treated below merits particular justification. It could be argued that a monotype is an admission of failure on the part of the taxonomist/systematist, since the otherwise conveniently informative binomial system of nomenclature in this case may not indicate the relationship of the species involved. Thus, the names of *Espostoopsis dybowskii* and *Brasiliopuntia brasiliensis* do not give an immediate clue as to their relationship, although the well-informed will get a steer from the generic etymologies. In fact, their names do indeed refer to the putative relationship of *Espostoopsis* and *Brasiliopuntia*, which are potentially close to *Espostoa* Britton & Rose and *Opuntia* Mill., respectively. In the case of *Espostoopsis* its inclusion in *Espostoa* (tribe Trichocereeae) would be premature on current evidence and require an expansion of the latter's circumscription, which would make for difficulties in writing a simple key to South American columnar genera. Alternatively, the various unusual features or combinations of characters that *Espostoopsis dybowskii* displays argue against its inclusion in any of the genera of tribe Cereeae (Taylor & Zappi 1989), where its similarities could be due to convergence. On the basis of robust evidence from gene sequences and seed anatomy (Wallace & Dickie 2002, Stuppy 2002) *Brasiliopuntia* belongs in the same clade as *Opuntia* Mill. and shares some obvious morphological similarities, eg. combination of flattened stem-segments and flowers with spreading yellow perianth-segments. However, the previously ill-defined 'dustbin genus' *Opuntia* can now be most conveniently recognized in a restricted sense on the basis of the autapomorphic character state of pollen with a reticulate exine, a unique feature within the centrosperms, according to Wilhelm Barthlott (pers. comm.). This circumscription is further supported in more practical terms, since all species to be included in *Opuntia sens. str.* can be readily identified by flowers possessing sensitive stamens, which excludes *Brasiliopuntia* and most other genera in the *Opuntia* clade (Taylor *et al.* 2002). In relation to most of these the habit, dimorphic stems and pollen characters of *Brasiliopuntia* are clearly strong autapomorphies, while its inclusion in any of them would likely result in the creation of a paraphyletic group or at least one which was difficult to key out and comprehend. (The somewhat similar Caribbean genus, *Consolea* Lem., differs in having seeds with very distinct funicular envelope anatomy and flowers with numerous small ovules, and is regarded as convergent.)

Specific concepts and use of subspecies. Judd *et al.* (1999: 128–130, Table 6.4) describe more than seven different species concepts and admit that 'there is no consensus about species concepts in plants'. Earlier they note that 'The ascendency of phylogeny as an organising principle in systematics motivated a phylogenetic species concept'. This has various potential interpretations, but we like the definition advanced by Nixon & Wheeler (1990) of a *phylogenetic* species as the 'smallest aggregation of populations (sexual) or lineages (asexual) diagnosable by a unique combination of character states in comparable individuals'. If such combinations of characters are fixed, diagnosable *and testable* by phylogenetic analysis, then we have a better method of recognizing species than that which has, in practice, been applied here — ie. essentially the *phenetic* species concept, based on 'the overall similarity of members of a species, which are separated from other species by a gap in variation' (Judd *et al.,* l.c.). For reasons explained in the preceding paragraphs it is difficult to employ phylogenetic methods for testing species concepts in Brazilian cacti and where it has been attempted (eg. Zappi 1994: 29) the robustness of the clades identified could easily be questioned. Such techniques were also attempted in the investigative stages of a revision of *Melocactus* (Taylor 1991a), but the phylogenies deduced for the species level treatments were not published because of their inherent weaknesses. A sufficient number of characters that can be confidently polarized into plesiomorphic and apomorphic states is necessary and this has proved difficult enough at the generic level (Taylor & Zappi 1989), let alone at the rank of species, without access to molecular data. The size of the treatment realized here (130 native species) has unfortunately prevented the application of molecular techniques and phylogenetic methodology in a uniform way at this level. Furthermore, this will only become a feasible and meaningful activity when the *generic* relationships and, therefore, the significance of the characters and their states, can be clarified in the tribes Cereeae and Trichocereeae.

In employing a phenetic species concept this treatment has, however, taken into account, wherever possible, the reproductive strategy and inferred breeding system of the plants, as observed in nature and in cultivation (where the authors have studied a considerable diversity of taxa over many years). Understanding reproductive strategy is important if similarities or differences and suites of linked characters that relate to pollination and dispersal syndromes are not to be over-valued. For example, it seems that some Brazilian cacti are probably able to take advantage of different pollinators by quite minor changes to floral morphology and timing of anthesis, eg. *Micranthocereus (Austrocephalocereus) purpureus* (Taylor & Zappi 1989: 22) and *Pilosocereus glaucochrous* (Zappi 1994: 78). Little is known at the cytological level about the breeding systems of cacti, but circumstantial evidence strongly implies that the majority of Brazilian taxa (as is true of cacti in general) are self-incompatible and outbreeders (cf. Ross 1981). The actual mechanism has recently been investigated in two genera, namely *Schlumbergera* (O'Leary & Boyle 1998) and *Echinopsis* (Boyle & Idnurm 2001), but its operation in the rest of the family is presently unknown. However, exceptions to the obligate outbreeder status of most cacti are known or suspected. Many, but not all, *Melocactus* species appear to be self-compatible (Taylor 1991a: 16, Nassar *et al.* 2002), and there is evidence that a couple of Brazilian taxa are cleistogamous (see *M. lanssensianus* and other geographically disjunct look-alikes of uncertain status discussed in Chapter 5). The species definition that has been maintained in this latter circumstance is somewhat different, since there can be no gene exchange between such lineages and thus it is possible that a series of very narrowly defined taxa could be recognized if this phenomenon should prove to be more widespread than current field knowledge suggests. Going down that road suggests abandoning a pragmatic phenetic species concept in favour of the largely discredited *biological* species concept (Judd *et al.*, l.c.) and for the time being the temptation to recognize additional narrowly defined taxa has been resisted.

The very few chromosome counts so far published for E Brazilian cacti, eg. in *Pereskia* (Leuenberger 1986) and Rhipsalideae (Barthlott 1976), indicate that nearly all are diploids, although the authors are aware of a recently published study documenting a predominance of tetraploids in *Melocactus* (Assis *et al.* 2003). Clearly, there is much work to be done here and studies of North American members of the family have painted a more interesting picture of polyploidy in members of the Cactoideae (especially the 'HPE clade') and Opuntioideae (Pinkava 2002).

Field studies of cacti conducted in Brazil and elsewhere indicate that related taxa, and especially sister taxa, are only rarely sympatric and suggest that speciation has probably occurred by allopatric means in a majority of cases. When species belonging to the same genus occur together they can hybridize (eg. *Tacinga, Cipocereus, Arrojadoa* and *Melocactus*), but this is not the norm and fewer hybrids have been recorded in the large genera *Pilosocereus* and *Rhipsalis*, whose species are frequently sympatric (cf. Zappi 1994: 35, Barthlott & Taylor 1995, Taylor 1999). However, even if inability to interbreed is a good criterion for defining some species, its obvious limitation is the converse situation, since more distantly related taxa hybridize on occasion and can produce viable offspring, eg. *Pilosocereus pentaedrophorus* × *Micranthocereus purpureus*, known from two sites at the eastern edge of the Chapada Diamantina, Bahia. Likewise, no one would seriously wish to suggest that *Tacinga inamoena* and *T. palmadora*, or for that matter *Arrojadoa rhodantha* and *A. penicillata*, should be lumped together as more broadly defined species because they hybridize at some (and certainly not all) sites of sympatry. In these cases the individual species here recognized can each be separated on suites of mutually exclusive characters and the *Tacinga* species would deserve being classified in different sections were the genus not so small as to make this of limited value.

Leaving aside theoretical concepts it is worth recording the authors' experience that it has been a relatively straightforward task to delimit species amongst the cacti of Eastern Brazil. Straightforward, at least once the necessary fieldwork has been completed, because most initial uncertainties have revolved around regional or local variation, in the case of a number of widespread taxa, and the geographical area to be surveyed has been substantial. The initial difficulties have been created through the use of typological species concepts by amateur 'cactus hunters', whose desire to discover and publish something new has been greater than that to understand the overall patterns of variation (see Chapter 2.5). Put another way, their approach to defining and naming taxa has not been synthetic, but driven only by a search for differences. While species have often been more broadly defined here than in previous treatments, their regional variation is now better understood and many of the more widespread taxa have been subdivided into subspecies.

In nature, many such cactus species, eg. *Arrojadoa rhodantha* and *Pilosocereus pachycladus*, are comprised of numerous geographically sequential races, none of which is sufficiently distinct to be worthy of being named (though many have). In some cases, however, these races can grouped on a geographical basis into subspecific taxa sharing mutually exclusive similarities. Thus, the races of *P. pachycladus* are presently grouped into two subspecies defined on numbers of stem ribs correlated with amount of areolar wool. At their points of contact in northern Bahia they are scarcely distinguishable, whereas towards their margins it is likely that additional subspecies will be required once further field studies have been made. A somewhat different situation where the rank of subspecies has been employed is exemplified by *P. fulvilanatus*. This is not as wide-ranging as its aforementioned relative and much less variable, but it includes a rather disjunct population that has begun to evolve some constant differences suggestive of incipient speciation (*P. fulvilanatus* subsp. *rosae*). Here similarities greatly outweigh differences, but geographical separation makes its recognition as a subspecies appropriate and informative.

The most problematical taxa encountered in this study have been in the 3 largest genera, two of which have already been monographed (ie. *Melocactus* and *Pilosocereus*). Here, there remain doubts or controversies about the circumscriptions of the *M. oreas* / *M. bahiensis*, *P. pachycladus*, *P. machrisii* / *P. aurisetus* and *Rhipsalis burchellii* complexes, but it is anticipated that these can be resolved by more focused field studies.

Example of an herbarium specimen from the 1988 expedition to Bahia (*Coleocephalocereus goebelianus*) (K).

2. HISTORY OF DISCOVERY, NAMING AND CLASSIFICATION

2.1 PROBLEMS OF THE '*HORTUS SICCUS*'

The history of the discovery and naming of Cactaceae from Eastern Brazil, as defined here, is detailed for each taxon in Appendix I and can be resolved into 4 major periods. First, however, it is necessary to make some remarks about this subject in more general terms.

Being for the most part highly succulent, normally unpleasantly spiny, yet unusual to look at, cacti have always been worthy of study and comment, but never easy to preserve by conventional methods, nor to draw accurately. Early botanical expeditions to Brazil undoubtedly encountered many cacti, but while even the most illustrious authorities, such as VON MARTIUS, frequently refer to them in their field notebooks and scientific publications (Martius 1832, 1841), remarkably few actually succeeded in bringing significant numbers of preserved specimens back to their herbaria and museums. Many, however, attempted and were evidently successful in sending live material back for cultivation in Europe, eg. *Melocactus violaceus* subsp. *violaceus* (*M. depressus*)★, sent back to Britain from Pernambuco by George GARDNER (Hooker 1838). Once introduced to the stove house, their naming and description soon followed, but frequently the living specimens were never subsequently preserved, either because they were considered too precious as items of horticultural interest, or as simply 'incapable of an *hortus siccus*' (Bradley 1716–1727). Thus, until relatively recent times detailed knowledge of the majority of species has been hampered by a profound lack of both museum specimens and reliable literature, and this to an extent much greater than with other, more 'conventional' plants.

As noted already, the last and only comprehensive study that focused on the Cactaceae from this vast region of Eastern Brazil was made by Karl Moritz SCHUMANN, as part of Martius's monumental *Flora brasiliensis* (Schumann 1890). For the reasons just explained his treatment is rather limited in its coverage, since many conspicuous species that were accessible to, and very probably noticed by 17th and 19th Century collectors in Brazil are either missing or not distinguished, for lack of adequate material, eg. the caatinga's ubiquitous Xique-xique (*Pilosocereus gounellei*). Unluckily for Schumann, the 20th Century's golden age of cactus discovery in Brazil was still to come, as the following notes and Appendix I hopefully make clear. In fact, Schumann (1890) accounted for only 13 of the 162 native taxa at the ranks of species and subspecies (heterotypic) now accepted from the area covered here.

★ Botanical authorities for E Brazilian cactus names are not cited here but can be found in Chapter 5.

2.2 COUNT JOHAN MAURITS IN NORTH-EASTERN BRAZIL: 1637–1644

Our knowledge of Brazilian cacti begins with the Dutch occupation of North-eastern Brazil, during the years 1630–1654. This included a seven-year period, commencing in 1637, under the command of Governor General Johan MAURITS of Nassau-Siegen, whose company included two accomplished scientists, Willem Pies (latinized to the more familiar PISO) and Georg MARCGRAF (also written Marcgrave, Markgraf etc.), and the talented artists, Albert ECKHOUT and Frans POST. Whitehead & Boeseman (1989) have brought together a substantial body of information on the work of the Maurits team, amongst which six cactus species can be confidently identified from the illustrations they cite and reproduce: *Tacinga palmadora, Brasiliopuntia brasiliensis, Cereus fernambucensis, C. jamacaru, Pilosocereus gounellei* and *Melocactus violaceus* (see Appendix I). They also illustrate what may be *Harrisia adscendens* (l.c., tt. 89b & 99b) and refer to a *Rhipsalis* species that remains to be identified (l.c., 84). Although the Dutch were mostly restricted to the coastal regions of the Nordeste, ranging from Salvador (Bahia), to São Luis (Maranhão), maps drawn by Marcgraf indicate that they travelled far inland, via the Rio São Francisco. This explains how they encountered *caatinga* species, such as *Cereus jamacaru, Pilosocereus gounellei, Tacinga palmadora* and, perhaps, *Harrisia adscendens*.

While other kinds of plants illustrated during the Dutch occupation were described in the 18th Century by Linnaeus and his contemporaries, no cacti appear to have been named from this source until 1814, when Willdenow published *Cactus brasiliensis* (= *Brasiliopuntia brasiliensis*), followed in 1828 by *Cereus jamacaru* DC., both of these based on illustrations in Piso (1648). The latter is now a conserved name, since the illustration specifically cited by De Candolle is identifiable as *C. fernambucensis*, but subsequent authors consistently employed the name for the much larger *caatinga* species we know today (Taylor & Zappi 1992c).

2.3 19TH CENTURY COLLECTORS

After the Dutch natural historians left North-eastern Brazil about 170 years passed before records of additional cactus species were made from the region covered here. With the Brazilian capital and principal port of entry now situated at Rio de Janeiro, which is outside the geographical limits of this study, many of the humid forest and coastal sand-dune species were first recorded from around that city long before they turned up further to the north. Certainly, parts of Rio de Janeiro were intensively botanized in the 19th Century, as was the southern half of Minas Gerais (Urban 1906). However, during the period 1815–1860, only 6 additional species and 2 heterotypic subspecies of cacti were collected in the area treated here and 4 of these — *Pereskia grandifolia, Rhipsalis lindbergiana, Arthrocereus glaziovii, Discocactus placentiformis* — were found in northern Rio de Janeiro, Minas Gerais and Espírito Santo, in areas relatively close to the then capital. The others — *Rhipsalis baccifera* subsp. *hileiabaiana, Pilosocereus pentaedrophorus, Melocactus oreas, M. violaceus* subsp. *margaritaceus* — were discovered in the state of Bahia, close to, or within a few days' journey of the state capital and principal sea port of Salvador. The *Pilosocereus* was obtained by a French collector, MOREL, who could also have been the discoverer and source of *Melocactus violaceus* subsp. *margaritaceus*, which was originally described as *M. ellemeetii* in 1857–58. The collectors of *Discocactus placentiformis* and *Melocactus oreas* are unknown, although the type of one of the contemporary synonyms of the former was said to have been obtained by RIEDEL (see Chapter 5, for details of synonyms). Four of the remaining taxa mentioned above were gathered as herbarium specimens by other famous European collectors from this period, including Prince Maximilian of WIED-NEUWIED, SAINT-HILAIRE, SELLO and BLANCHET (see Urban 1906, for details).

Two collectors, who were active during the remaining period up to the end of the century, are worthy of particular mention: the Dane, WARMING, who spent 6 years at Lagoa Santa in Minas Gerais during the 1860s, and the Frenchman, GLAZIOU, who sent collections from the same state to Schumann prior to 1890. These individuals are notable for reasons besides their relative productivity. Warming's collections from the gallery forests, mountains and Bambuí limestone outcrops around Lagoa Santa were mostly lost or spoilt (Zappi 1994: 80–82), but he subsequently discussed and illustrated the plants (Warming 1908), so that it is possible to be sure about their identity (here using modern nomenclature): *Hylocereus setaceus*, *Lepismium warmingianum*, *Cereus jamacaru* subsp. *calcirupicola* and *Pilosocereus floccosus*. Glaziou probably can be credited with the discovery of *Arthrocereus melanurus* and *Cereus hildmannianus*, but his claims to have personally collected *Tacinga (Opuntia) inamoena* and *T. braunii* (the latter described and named more than a century later) are almost certainly not genuine. The *Tacinga* species are indeed found in Minas Gerais, but not in the southern parts of the state that Glaziou is known to have visited and thus it is fairly clear that here we have a further instance of his mis-appropriation of material from another and as yet unidentified collector (cf. Wurdack 1970). There is also uncertainty arising between differences in the stated origins of taxa 'collected' by Glaziou and described by Schumann (1890), who sometimes gave Rio de Janeiro as the source of plants that were subsequently stated by the collector to have come from Minas Gerais (Glaziou 1909).

2.4 THE GOLDEN AGE OF CACTUS DISCOVERY: 1900–1950

From the beginning of the 20th Century until around 1930 six field collectors made major contributions to our knowledge of the Cactaceae of Eastern Brazil. Between 1906 and 1907 the German, Ernst ULE, made a series of long expeditions into the interior of Piauí and Bahia, partly in search of plants that would provide a source of rubber (Zappi 1994: 135). This was the first time that any serious attempt to collect the cacti of the *caatinga* had been made and so nearly everything that Ule found was new to science. Most of the novelties were named by Gürke at Berlin-Dahlem from 1906 onwards, the great cactologist-monographer, Schumann, having died in 1903. The following key *caatinga* species are amongst the many significant discoveries made by Ule: *Pereskia bahiensis*, *Stephanocereus leucostele*, *Arrojadoa penicillata*, *A. rhodantha*, *Pilosocereus catingicola*, *P.*

pachycladus (the earliest definite record, but named much later), *P. piauhyensis, Melocactus ernestii, Harrisia adscendens* (see 2.2 above), *Facheiroa ulei* and *F. squamosa.* He also made the first collection of the most widespread of the Chapada Diamantina's *campo rupestre* cacti, *Micranthocereus purpureus.*

The Swiss-german emigré, Leo ZEHNTNER, who is known to have established a cactus garden in the *Horto Florestal* (forestry station) at Juazeiro in northernmost Bahia, remains a little-researched figure in the history of Brazilian plant collectors, yet there is no doubting his importance in relation to Cactaceae. As a collector he was active during the period 1912–1920 and seems to have complemented the activities of Ule, visiting a number of Bahian sites the latter did not reach. He is justly commemorated in the generic names of the cereoids, *Zehntnerella* (= *Facheiroa*) and *Leocereus,* both East Brazilian endemics, and in the epithets of many other Brazilian cacti named by Britton & Rose (1919–1923). This eponymy reflected the considerable assistance he gave to the American collectors, Rose & Russell, as discussed below, who made field studies in 1915 in connection with the great monograph of the family by the above-cited authors. At this period Zehntner must have relied extensively on access to habitats in Bahia by means of the fluvial highway of the Rio São Francisco and via the few railway systems that had already been constructed. Amongst the taxa he is believed to have collected for the first time are: *Pereskia stenantha, Quiabentia zehntneri, Tacinga funalis, Pseudoacanthocereus brasiliensis, Pilosocereus gounellei* subsp. *zehntneri, Micranthocereus flaviflorus, Coleocephalocereus goebelianus, Leocereus bahiensis, Discocactus zehntneri* and *D. bahiensis.* Another little-known collector from the earliest part of this century was DYBOWSKI, the discoverer of the remarkable *Espostoopsis dybowskii,* Plate 68.

Ule and Zehntner between them are the most important discoverers of cacti from the North-east Region of Brazil until quite recent times. Others amongst their contemporaries, however, improved our knowledge of distribution, while also finding a few novelties. Phillip von LUETZELBURG, another phytographer of Germanic origin, made extensive journeys over much of North-eastern Brazil and also through Espírito Santo state during the period 1913–1933, while in the employ of the Brazilian government. However, his 3-volume work, *Estudo Botânico do Nordéste,* published by the Inspetoria Federal de Obras Contra as Secas (Luetzelburg 1925–1926), is more important as a source of information and illustrations than are the few cactus collections that survived these expeditions. Unfortunately, as Werdermann (1933) noted, much of Luetzelburg's cactus material was lost, and the present study has revealed that the part which survived suffers from many confusions of labelling, rendering his data unreliable and his published lists of species ± unverifiable. His certain discoveries include *Cereus fernambucensis* subsp. *sericifer, Stephanocereus luetzelburgii* (the remarkable bottle-cactus of Bahia, Plate 33), *Pilosocereus brasiliensis* subsp. *ruschianus, P. chrysostele* and *Coleocephalocereus pluricostatus,* but only two of these were described from his material.

More important as collectors were the two aforementioned North Americans, J. N. ROSE and his assistant P. G. RUSSELL, who spent a couple of months collecting in Bahia, via the railway system, in 1915. Apart from their obviously successful collaboration with Leo Zehntner and the use made of his living collections at Juazeiro, they obtained numerous gatherings in northern Bahia and made a significant contribution at the limits of the *caatinga* with the *Mata atlântica,* in eastern Bahia, utilizing railway lines that formerly connected the cities of Juazeiro and Jequié with Salvador. Amongst their discoveries were *Tacinga werneri, Rhipsalis russellii, Cereus albicaulis, Melocactus bahiensis* and *M. zehntneri,* but their extensive herbarium records, preserved at US, NY, K and elsewhere, are perhaps more important for documenting the distribution of already known taxa in parts of the region which have since suffered much habitat destruction (eg. at Jaguaquara ['Toca da Onça']).

During the period 1930–1950 a greater number of collectors made mostly small but worthy additions to our knowledge of the cacti of Eastern Brazil. By far the most important of these was the German cactus specialist, Erich WERDERMANN, from the Berlin-Dahlem Botanical Garden & Museum. He planned and executed a very challenging expedition during 1932, which was reported the following year in his entertaining botanical travelogue entitled *Brasilien und seine Säulenkakteen,* an English translation appearing 9 years later (Werdermann 1933, 1942). With colleagues from Berlin his journey began in Pernambuco, at the coast, then working inland via the railroad, subsequently spending a considerable period in Bahia (Salvador, São Félix on the Rio Paraguaçu, Morro do Chapéu, Tremedal, Caitité), before passing through Minas Gerais (Diamantina, Serras do Cipó, Caraça and Curral) *en route* to São Paulo. Much of the journey was made by Ford Zeppel car, but the condition of roads in the Brazilian interior left much to be desired and his account tells of the many stops they made to effect repairs. Werdermann made numerous collections, including living specimens and some in spirit, but sadly much of his material was subsequently destroyed at Berlin during the second World War (cf. Leuenberger 1978). Amongst the taxa he discovered are *Cipocereus minensis, Pilosocereus tuberculatus, P. glaucochrous, Micranthocereus polyanthus* and *Melocactus salvadorensis.* He also collected and described a significant number of entities that have passed into synonymy, besides creating some confusions of identity that survive even to this day (eg. he used the name *Melocactus bahiensis* for the much more widespread *M. zehntneri*).

Other collectors who made notable contributions contemporary with that of Werdermann included resident Brazilian botanists as well as foreign explorers, eg. Bento PICKEL (especially at the easternmost limits of the *caatinga-agreste* in Pernambuco★), MARKGRAF *et al.* (Minas Gerais, discovered *Brasilicereus markgrafii*), MELLO-BARRETO (Minas Gerais, *Uebelmannia gummifera*), HERINGER (Minas Gerais), CUTLER (Ceará), DROUET (Ceará, *Pilosocereus flavipulvinatus*), BRADE (Minas Gerais and Espírito Santo, *Cipocereus bradei, C. minensis* subsp. *leiocarpus, Arthrocereus rondonianus*), DUARTE (Minas Gerais, *Pilosocereus densiareolatus*) and PINTO (Bahia, *Pereskia aureiflora*).

★ Pickel collected many specimens at the adjacent localities of 'Tapera' (Mun. Moreno) and 'São Bento' (Mun. São Lourenço da Mata), which are situated in an area where dry vegetation penetrates the otherwise broad band of humid Atlantic Forest in this state.

Some key individuals in the history of discovery and description of the cacti of Eastern Brazil: **1.** J.E.B. Warming (K), **2.** A.F.M. Glaziou (K), **3.** E. Ule (B), **4.** E. Werdermann (B), **5.** W. Uebelmann (UE). **6.** K.M. Schumann (B), author of the last monograph of Brazilian cacti in 1890.

2.5 MODERN COLLECTORS: POST 1950

In modern times, cactus discovery in Eastern Brazil began with the arrival of the German, Friedrich RITTER, in 1959 (Eggli *et al.* 1995), and was enhanced in the 1960s by the German-Brazilian, Leopoldo HORST, the latter's Swiss nurseryman sponsor, Werner UEBELMANN, and their Dutch collaborator, Albert BUINING (Uebelmann 1996). These collectors were largely motivated by the horticultural trade and cactus hobbyist interests, especially in Europe, where new discoveries were eagerly sought. However, although many so-called *'spec. nov.'* were collected, introduced to cultivation and swiftly named, the number of genuine first discoveries of taxonomically 'good' species was far less than it seems, for two reasons. First, many of the taxa that these and subsequent cactus plant hunters claimed as new discoveries had already been collected much earlier by professional botanists and deposited in herbaria, which such 'cactophiles' did not usually consult (see Appendix I). Secondly, the taxonomic concepts they employed ignored regional variation and resulted in a plethora of weakly defined microspecies, many of which have since been dumped into synonymy or down-graded in rank by the studies of the present authors (Taylor 1980, 1981, 1982, 1991a; Taylor & Zappi 1990, 1991, 1997; Taylor in Hunt 1992b, 1999; Zappi 1994). Nevertheless, Ritter can lay claim to having discovered at least 21 distinct taxa between 1959 and 1971, although he was not always the first person to publish these (Ritter 1979): *Tacinga saxatilis, Rhipsalis floccosa* subsp. *oreophila (R. monteazulensis), Cereus mirabella (Mirabella minensis), Cipocereus crassisepalus, C. pusilliflorus, Arrojadoa dinae, Pilosocereus floccosus* subsp. *quadricostatus, P. magnificus, P. aurisetus* subsp. *aurilanatus, P. multicostatus, Micranthocereus albicephalus, Coleocephalocereus buxbaumianus* subsp. *flavisetus, C. fluminensis* subsp. *decumbens, C. aureus, Melocactus ernestii* subsp. *longicarpus, M. bahiensis* subsp. *amethystinus, M. levitestatus, M. concinnus, M. violaceus* subsp. *ritteri (M. macrodiscus* var. *minor), Facheiroa cephaliomelana, Arthrocereus melanurus* subsp. *odorus.*

Horst and his associates, Uebelmann and Buining, were even more successful than Ritter, since they were evidently able to devote more time to exploration of the remote Brazilian interior (the *'sertão'*) and this often by means of suitable vehicles, on roads that were steadily improving. Over a period of some 30 years, starting in the mid-1960s, a steady stream of novelties was reported in journals and catalogues back in Europe and the USA, and plant material was distributed to avid cactophiles via Uebelmann's nursery. No less than 27 distinct new taxa can be attributed to the efforts of their explorations, although, as already indicated, the number they actually claimed as new was probably in terms of hundreds (a measure of this can be gained from the many names listed for relevant Brazilian genera in Eggli & Taylor 1991). The following were discovered between the years 1966 and 1982:– *Tacinga inamoena* subsp. HU 336 (new — awaiting description), *Arrojadoa dinae* subsp. *eriocaulis, Pilosocereus pentaedrophorus* subsp. *robustus, P. fulvilanatus* (2 subspp.), *P. aureispinus, P. parvus, Micranthocereus violaciflorus, M. auriazureus, M. dolichospermaticus, Coleocephalocereus buxbaumianus* subsp. *buxbaumianus, C. purpureus, Melocactus conoideus, M. deinacanthus, M. azureus, M. ferreophilus, M. pachyacanthus, M. lanssensianus, M. glaucescens, Discocactus zehntneri* subsp. *boomianus, D. heptacanthus* subsp.

catingicola, D. pseudoinsignis, D. horstii, Uebelmannia buiningii, U. pectinifera (3 subspp.). The achievement and significance of these collectors is reflected in the fact that many of the above novelties are extremely localized in habitat and now figure amongst the threatened taxa discussed in Chapter 4.

Other collectors that have been active during this period include professional botanists, such as our Kew colleague David HUNT, who collected cacti in Pernambuco in 1966. The Argentinian cactus specialist, Alberto CASTELLANOS, made frequent visits to Eastern Brazil while resident in Rio de Janeiro and collected cacti on a number of occasions. In 1968 he collected material of a *Pilosocereus, P. azulensis*, which was only recognised as new quite recently, Plate 40 (Taylor & Zappi 1997). He is known to have made other herbarium collections from the core area, but it is believed that an important part of his material was destroyed in a fire (M. Vianna [GUA], pers. comm.). Around the mid 1960s Paulo MARTINS collected Rhipsalideae at the behest of Prof. F. Brieger (Universidade Federal de Brasília) and discovered *Rhipsalis paradoxa* subsp. *septentrionalis* in Bahia, which was not described until 1995, Plate 16 (Barthlott & Taylor 1995). Dárdano de ANDRADE-LIMA, Raymond HARLEY (discoverer of *Melocactus oreas* subsp. *cremnophilus*), Leopoldo KRIEGER (discoverer of *Arthrocereus melanurus* subsp. *magnus*) and Gustavo MARTINELLI are amongst a handful of mostly Brazilian botanists that included cacti in their general field collections for the herbarium during the 1970s. In particular Andrade-Lima deserves praise for his many collections in areas that have since undergone considerable habitat modification, and for his various helpful publications on Brazilian cacti (cf. Prance & Mori 1982).

During the 1980s and 90s there has been much more field activity, both by Brazilian botanists and their European collaborators, as well as hobbyist cactophiles, but this has resulted in fewer discoveries of genuinely new taxa. Some of those that have been found are extreme rarities, narrow endemics or plants from more or less inaccessible habitats. Nonetheless, a steady stream of novel plants has been reported in the specialist cactus & succulent literature. In 1981, a team of botanists from universities in São Paulo state, including Antônio FURLAN and Inês CORDEIRO, discovered a peculiar new *Arrojadoa, A. bahiensis*, restricted to nearly vertical cliffs in the Chapada Diamantina, Bahia (Taylor & Zappi 1996). In the same year another remarkable cactus, *Melocactus paucispinus*, which imitates the unrelated genus *Discocactus*, was discovered in the same region by the German cactophiles, HEIMEN *et al.*, Plate 63. In 1984, Brazilian Eddie ESTEVES PEREIRA found two new subspecies at a locality in southern Bahia (*Tacinga saxatilis* subsp. *estevesii* and *Facheiroa cephaliomelana* subsp. *estevesii*), while the following year VAN HEEK and VAN CRIEKINGE discovered the very rare *Micranthocereus streckeri* in central Bahia. In 1986, Prof. Werner RAUH and Roberto KAUTSKY collected a series of little-known or undescribed Rhipsalideae from the region of Domingos Martins in Espírito Santo state: *Rhipsalis pacheco-leonis* subsp. *catenulata, R. cereoides, R. sulcata, R. clavata, R. pilocarpa* and *Schlumbergera kautskyi*. Their work was soon after complemented by field studies conducted by Prof. Wilhelm BARTHLOTT and Dieter SUPTHUT, and thereafter by the former's student, Stefan POREMBSKI, from the University of Bonn, with surveys of inselberg vegetation. Meanwhile, in 1987, Countess Beatrix ORSSICH obtained the extraordinary, red-flowered *Rhipsalis hoelleri* from Espírito Santo, Plate 22 (Barthlott & Taylor 1995). Also in 1987, one of the present authors, Daniela ZAPPI, and Brazilian colleague, Vera SCATENA, found a most unusual new cereoid cactus, *Cipocereus laniflorus*, on exposed rocks of the Serra do Caraça, in central-southern Minas Gerais, Plate 29. Then, the following year the present authors discovered the 'Critically Endangered' *Melocactus pachyacanthus* subsp. *viridis* in central-northern Bahia (Taylor 1991a).

Since the last-mentioned there have been no significant discoveries until very recently (apart from presumed hybrids and other plants of dubious status). However, in the period 2000–2002 three quite remarkable 'good' new taxa have come to light. The first is *Pilosocereus bohlei* Hofacker — see Chapter 5, under *Pilosocereus* 'Insufficiently known taxa'. The second and third await further study prior to their description, but are briefly mentioned in Chapter 5 under *Cereus jamacaru* and *Arrojadoa rhodantha*. Both are discoveries made by staff from the Universidade Estadual do Sudoeste da Bahia, at Vitória da Conquista, and communicated to the authors through the ever enthusiastic Marlon MACHADO. Besides these, during the past 15 years, extensive field studies focused on Cactaceae have been carried out for this book and have involved numerous collaborators (see Chapter 6.2), especially Urs EGGLI, Emerson ROCHA and, latterly, Marlon Machado. Others, from the University of Feira de Santana, including Luciano Paganucci de QUEIROZ and Jomar JARDIM, have made significant herbarium collections in the Nordeste and, unlike many botanists, have not been afraid of gathering substantial numbers of cacti, greatly extending our knowledge of geographical distribution and improving our estimations of current levels of threats to habitats and taxa.

As this text was being finalized a review focusing on the last 25 years of Brazilian cactus discovery was published (Braun & Esteves Pereira 2002). This review contains large amounts of literature references, colour photographs and opinion, but is generally short on useful data, especially precise localities, although its authors emphasize their many years of exploration in the field. In particular, students of the family should be wary of the distributional information offered (eg. l.c. 79), which is mostly not backed up by cited herbarium vouchers or photographs, and in part contradicts the distribution maps presented. In the introduction (l.c. 8) it is noted that there is now a greater amount of material in herbaria, presumably for study, but little evidence is presented of this resource having been utilized except, of course, for the deposition of obligatory nomenclatural types. In addition, the application of Brazilian vegetation terminology is sometimes quite bizarre, especially with reference to *carrasco* ['charasco'], *cerradão, agreste* and *caatinga* — a tradition established much earlier by Ritter and Buining.

2.6 TAXONOMIC HISTORY OF THE CACTACEAE FROM EASTERN BRAZIL: 1890–1979

In terms of monographic and synoptic taxonomic treatments, the following names are particularly relevant in relation to the cacti of Brazil, amongst botanists and field collectors that published prior to the 1980s: Schumann, Britton & Rose, Berger, Werdermann, Backeberg, Buxbaum, Hunt and Ritter. The first treatment that merits mention is that by

SCHUMANN (1897–98), the *Gesamtbeschreibung der Kakteen*, which followed closely on the heels of his accounts for Martius's *Flora brasiliensis* and Engler & Prantl's *Das Pflanzenfamilien* (Schumann 1890, 1894). Schumann's classification was, by present standards, extremely conservative at generic level, where he recognized the following 'hold-all' genera: *Cereus* Mill. (including a variety of Brazilian columnar-cereoid and scandent species, plus some rhipsaloids now referred to *Schlumbergera* Lem., but excluding part of the modern *Pilosocereus* Byles & G. D. Rowley as *Pilocereus* Lem.), *Rhipsalis* Gaertn. (for a diversity of epiphytic taxa belonging to tribe Rhipsalideae DC., but also including some now referred to tribe Hylocereeae Buxb.), *Echinocactus* Link & Otto (for all the low-growing globular forms like *Discocactus* Pfeiff., but not *Melocactus*) and *Opuntia* Mill., this last being used in the traditional broad sense that has persisted until quite recent times. In addition he accepted *Melocactus* Link & Otto and *Pereskia* Mill. in their current senses, *Hariota* DC. (*nom. illeg.* = *Hatiora* Britton & Rose), *Pilocereus* Lem. (in a sense excluding its type), *Phyllocactus* Link (correctly *Epiphyllum* Haw.), *Epiphyllum* (for part of what is now *Schlumbergera*) and *Zygocactus* K. Schum. (= *Schlumbergera*). Schumann classified the genera of Cactaceae into 3 subfamilies, Pereskioideae, Opuntioideae and Cereoideae (correctly Cactoideae), an arrangement which has remained little changed until very recently (*Maihuenia* (Weber) K. Schum., which he placed in Pereskioideae, is now placed in its own subfamily).

In their great 4-volume work, *The Cactaceae*, BRITTON & ROSE (1919–1923) recognized Schumann's 3 subfamilies as tribes and otherwise radically changed the classification of the family at generic level, liberally dividing his hold-all genera and describing many new ones to account for the numerous discoveries made in the early years of the 20th Century. In total they recognised 124 genera for the family. A good part of the changes they made to the abundant cereoid species was considerably influenced by Alwin BERGER's detailed subgeneric rearrangement of *Cereus* (Berger 1905), some of whose subgeneric names and/or taxa they upgraded to generic status. Thus, Schumann's concept of *Cereus* became restricted to the group immediately close to the type species, *C. hexagonus* (L.) Mill. However, *Pilocereus* was included in *Cephalocereus* Pfeiff., whose type (*Cactus senilis* Haw.) and Brazilian taxa (ie. *Pilosocereus*) are nowadays placed in different tribes (Pachycereeae and Cereeae, respectively). The scandent cereoids including Brazilian species were separated into the new genera, *Hylocereus* and *Mediocactus*. The then known E Brazilian globular cacti were placed in the reinstated *Discocactus*, but allied with *Melocactus*, for which Britton & Rose dug up the abandoned name *Cactus* L. These were the only genera included in their subtribe Cactinae (*Melocactus* is now seen as the most derived element in tribe Cereeae, while *Discocactus* is referred to Trichocereeae, Taylor & Zappi 1989). Britton & Rose also split up the E Brazilian elements in Schumann's concept of *Rhipsalis*, recognizing *Lepismium* Pfeiff., *Erythrorhipsalis* A. Berger (1920) and *Epiphyllanthus* A. Berger (1905), as well as *Hatiora* Britton & Rose. New genera were created for various taxa that had been described in *Cereus* and *Cephalocereus* shortly after Schumann's death, namely, *Arrojadoa, Leocereus, Facheiroa* and *Zehntnerella*. New opuntioid genera were also created for recently discovered taxa: *Quiabentia* (for a leafy species initially described as a *Pereskia*) and *Tacinga* (for a curious, scandent species).

While some commentators welcomed these new generic names, others, especially in Germany, where Schumann's view still held sway, rejected some or all, or even hedged their bets. For example, Berger (1929) ambiguously treated many of Britton & Rose's genera simultaneously as subgenera and genera in his handbook, *Kakteen*.

WERDERMANN (1933, 1942), in dealing with the cereoid cacti of Eastern Brazil, adopted a compromise position, recognising some of the segregates employed by Britton & Rose, but sinking others. For example, *Trichocereus* (Berger) Riccob. he accepted for some of the Brazilian taxa subsequently realigned by others in *Arthrocereus* A. Berger (1929), but Britton & Rose's *Arrojadoa* went into the synonymy of *Cephalocereus*, from which he separated *Pilocereus* auctt. non Lemaire. In the latter genus he placed the curious *Cereus luetzelburgii,* and also *C. leucostele,* for which Berger (1926) had recently created the then monotypic *Stephanocereus*. Another of Berger's splits he did not accept was *Brasiliopuntia* A. Berger (1926), which he referred back to *Opuntia*, although he seems to have been prepared to accept *Tacinga* Britton & Rose.

If Britton & Rose started a trend towards the splitting of genera, then the German cactus nurseryman, Curt BACKEBERG (1938, 1958–62), went many stages further, his 6-volume monograph promulgating a more than 10-fold increase in genera over Schumann's treatment (220 *vs* 21)! His poorly researched innovations, typological species concepts, disregard for taxonomic and nomenclatural conventions and unsatisfactory suprageneric classification (Barthlott 1988, Hunt 1991: 152) resulted in little less than a state of taxonomic and nomenclatural chaos, which probably frightened most professional botanists away from serious study of the family until the mid-1960s. In 1938 he established 3 very poorly defined genera, which have fortuitously turned out to be worthy of recognition, now that the family is better understood in Eastern Brazil: *Brasilicereus, Micranthocereus* and *Coleocephalocereus*. At the same time he published *Austrocephalocereus*, which modern authors now agree should be subsumed in *Micranthocereus*, according to the type species he cited for the former (*Cephalocereus purpureus* Gürke), although it is very doubtful whether he was using that binomial in its correct sense! He also accepted nearly all previous generic splits involving E Brazilian species. His great adversary and critic was the cactus evolutionary morphologist, Franz BUXBAUM, who supported the description of perhaps the most distinct of all the newer genera to emanate from Eastern Brazil — *Uebelmannia* Buining (1967). On reflection, it is surprising that Backeberg made no attempt to name this group himself, having previously described its earliest-known species (*Parodia gummifera* Backeb. & Voll) in 1950.

David HUNT (1967), in John Hutchinson's *The Genera of the Flowering Plants* (vol. 2), provided a comprehensive botanical account of the Cactaceae down to generic level. He adopted a more conservative attitude than both Britton & Rose and Backeberg and classified genera within a modified version of the former monographers' tribal/subtribal scheme. In total 84 genera were recognised for the family and their arrangement was partly influenced by the morphological-phylogenetical studies of Buxbaum (see below). Notable in Hunt's treatment was his use of *Cephalocereus* in a very broad sense to include both North American and Brazilian columnar taxa bearing cephalioid structures, ie. *Arrojadoa,*

Stephanocereus, Pilosocereus, Micranthocereus, Austrocephalocereus and *Coleocephalocereus.* The only Brazilian exception to this was *Facheiroa,* which was subsumed into the Andean genus *Espostoa* Britton & Rose, a move earlier proposed by Buxbaum (1959). Similarly, *Rhipsalis* was used in a more inclusive sense than either Schumann or Britton & Rose, but *Schlumbergera* was circumscribed in the manner accepted today. Both *Leocereus* and *Arthrocereus* were recognised, but the latter, besides Brazilian taxa, included a divergent Argentinian species now placed in *Echinopsis* (*E. mirabilis* Speg.).

The aforementioned Friedrich RITTER based his classification scheme on very careful observation of the plants in the field, which knowledge he had gained during many years of exploration (Eggli *et al.* 1995). In 1959 he was the first to recognize the generic status of what later was named *Uebelmannia* Buining (1967), for which his manuscript name was '*Gummocactus*' (cf. Ritter 1979). Later, in a paper on Brazilian cephalium-bearing cacti (Ritter 1968) he independently recognised the distinctness of the rare Bahian endemic, *Cereus dybowskii,* which he named *Gerocephalus dybowskii,* this, unfortunately, a few weeks after Franz Buxbaum had published the priorable generic name, *Espostoopsis,* for the same plant (Buxbaum 1968). He was also the first among modern authors to recognize that the globular-stemmed genus, *Melocactus,* was closely related to the cereoid *Coleocephalocereus* (Ritter, l.c., Taylor 1991a: 17). Subsequently, when writing up his many years' results of cactus study in South America, he published a series of new genera reflecting his excellent knowledge of the Cactaceae (Ritter 1979). Of these, *Pseudoacanthocereus* and *Cipocereus* are recognized today, but *Floribunda* and *Mirabella* have passed into the synonymy of *Cipocereus* and *Cereus,* respectively (Zappi & Taylor 1991, Taylor 1991b, 1992a). Although Ritter lacked a proper training in plant systematics, his generic concepts were certainly more realistic than those of either Backeberg or Buxbaum. This is evidenced by his treatment of *Pilosocereus* as including *Pseudopilocereus* Buxb., of *Micranthocereus* including *Austrocephalocereus,* and of *Coleocephalocereus* including *Buiningia* Buxb., all of which are circumscriptions followed in standard treatments today (eg. Hunt 1992b, 1999a, Barthlott & Hunt 1993, Zappi 1994).

While Buxbaum described and named a new genus and a few new subgenera amongst the cacti of Eastern Brazil, as well as rearranging some others, the major contribution he made to the study of the Cactaceae as a whole was the evolution of a tribal classification based on a comprehensive understanding of morphology (Buxbaum 1950–1954). His system was first presented in 1958 with further notes and adjustments appearing over the next 17 years (Buxbaum 1958, 1959, 1962, 1968a & b, 1975; Endler & Buxbaum 1974). His studies, terminology and tribal nomenclature still form the basis for current schemes of classification, as described in the following section (Chap. 2.7). Towards the close of the period under discussion a new tool for investigating micro-morphology and especially that of pollen and seeds, was becoming important for the study of cacti — the Scanning Electron Microscope (Leuenberger 1976, Barthlott & Voit 1979). Influenced by data derived from these sources, the last treatment of the Cactaceae to appear prior to the 1980s recognised a little over 100 genera (Barthlott 1977, 1979).

2.7 CACTACEAE SYSTEMATICS 1980–2003 AND THE IOS CONSENSUS INITIATIVES

An unusual feature of Cactaceae systematics over the past two decades is the degree to which its proponents have been organised. Since 1950, the International Organization for Succulent Plant Study (IOS) has encompassed a significant group of cactus specialists (as well as those with wider interests in succulent plants). A notable exception to this was the wayward Curt Backeberg, whose major detractors were mostly IOS members! A study of the names of those in attendance at successive biennial IOS congresses highlights the opportunities that have existed for those interested in 'cactology' to exchange ideas (Supthut 1999). Thus, it was perhaps not surprising that a proposal by IOS Secretary, David Hunt, to establish a Cactaceae Working Party, which would aim towards a 'consensus classification', was made at the Frankfurt-am-Main congress of the organization in 1984 (Anderson 1999). This consensus approach was partly driven by the need of two of its major protagonists, David Hunt and Wilhelm Barthlott, to complete treatments of the family for major reference works on vascular plants (ie. Walters *et al.* 1989: 202–301, Barthlott & Hunt 1993). Such treatments were at risk of being hampered by the singular lack of orthodox taxonomic revisions of individual genera and the widely contrasting approaches adopted in modern 'standard works' on the family, such as those by the 'splitters', Curt Backeberg and Friedrich Ritter (see Chapter 2.6) and 'lumper', Lyman Benson. The latter's circumscription of *Cereus* Mill. was as broad as that of its 18th Century author, Phillip Miller, taking in elements placed in at least 4 of Buxbaum's tribes (Benson 1982).

In these circumstances, and recognizing that a considerable body of knowledge existed amongst the IOS membership, but was fragmentary in nature, the drawing together of a Working Party at annual meetings could bring benefits. This process was agreed in 1984 and meetings have been held at least annually since then, facilitated by the decision to hold 'inter-congress' meetings of IOS between its biennial congresses, starting in 1985. Following a second meeting of the Working Party a draft list of genera was published in 1986 (Hunt & Taylor 1986). This recognized a total of 86 genera in 3 categories: 'unanimously accepted' by the majority of Working Party participants (51 genera), 'less than unanimously accepted' by at least a third of participants (26 genera) and the remainder (9 genera) to be retained out of nomenclatural expediency or because they were considered '*incertae sedis*'. This first 'consensus' list was relatively conservative, but as the Working Party grew, and with it the sources of useful information, there was a gradual inflation of genera, so that the next list published (Hunt & Taylor 1990) recognized a total of 93, once again in 3 categories: 55 accepted by at least 80% of participants, 21 accepted by c. 60% and 17 for nomenclatural reasons or as *incertae sedis* and accepted by 35–60%. The genera accepted in the second list were determined by a postal ballot, which attracted responses from more than 20 specialists. Proposed changes to the previous list were justified in a series of short printed notes and accompanied by the sometimes contrasting views of particular participants, the present authors included (Hunt & Taylor 1990: 98–104). Further thoughts and views were subsequently aired by consensus participants in a paper printed the following year (Hunt & Taylor 1991) and opportunities were taken to publish new names for use in the first edition of the CITES Cactaceae Checklist (Hunt 1992b), which closely followed the IOS consensus list, but accepted a further 8 genera in the third category, bringing the total for the family to 101. This Checklist went down to the level of species and consequently drew attention to taxa whose generic placement was particularly controversial. A series of short papers by members of the Working Party, as well as by others not necessarily connected with the IOS, has continued to appear (eg. Hunt & Taylor 1992 and in the newsletters, *Cactaceae Consensus Initiatives* nos. 1–8, 1996–99, *Cactaceae Systematics Initiatives* nos. 9–15, 2000–03). Besides these shorter commentaries, more substantial monographic, synoptic and cladistic treatments of larger or complex groups by members of the IOS Working Party have begun to appear, those with relevance to Brazil including *Pereskia* (Leuenberger 1986), *Facheiroa* (Braun & Esteves Pereira 1986–89), tribe Cereeae (Taylor & Zappi 1989), *Melocactus* (Taylor 1991a), *Pilosocereus* (Zappi 1994), tribe Rhipsalideae (Barthlott & Taylor 1995) and *Uebelmannia* (Nyffeler 1998). More recently, a second edition of the CITES Checklist has appeared (Hunt 1999a), drawing heavily on the above-cited treatments and accepting 108 genera, with their included taxa listed down to the level of subspecies.

Since 1992, as with other groups of organisms, our understanding of Cactaceae systematics has become heavily influenced by molecular data, in the form of phylogenies derived from analysis of gene sequence variation (DNA/RNA), and this has been further supported by anatomical, phytochemical, pollen and seed micro-morphological information, which the molecular data are beginning to help interpret (Stuppy & Huber 1991, Wallace 1995, Wallace & Cota 1996, Maffei *et al.* 1997, Santos *et al.* 1997, Barthlott & Hunt 2000, Stuppy 2002, Taylor *et al.* 2002, Wallace & Dickie 2002, Wallace & Gibson 2002). The ways in which the IOS consensus process and subsequent studies have contributed to the classification of the 30 cactus genera here recognized for Eastern Brazil are listed in Table 2.1.

Relationships are clearly defined on molecular and other evidence at the subfamily level (ie. Pereskioideae, Opuntioideae and Cactoideae, see Table 2.2; Barthlott & Hunt 1993, Wallace 1995, Wallace & Cota 1996, Wallace & Dickie 2002, Wallace & Gibson 2002, Stuppy 2002, Nyffeler 2002), but in the complex Cactoideae, comprising > 100 genera, the situation is less clear, though there appear to be c. 7 major clades of uneven size.

Unfortunately, some of the molecular research already conducted is yet to be formally published, although regular presentations and updates have been provided by Robert Wallace and his students (Iowa State Univ., USA) at IOS and other meetings, and most recently by him and also Reto Nyffeler (Harvard Univ. Herbaria, USA; cf. Nyffeler 1999, 2000a–c) at the 26th IOS congress in Zürich, March 2000.

These investigations have demonstrated that the Brazilian Rhipsalideae DC. are a monophyletic group (Nyffeler 2000a, 2002) and in Eastern Brazil recognition of 4 genera based on morphological characters is well supported by the *trn*L-F intergenic spacer and *rpl*16 intron molecular markers (*fide* Wallace). As already argued by Barthlott (1988), this strongly contradicts Buxbaum's tribal arrangement (Buxbaum 1958 etc., cf. Leuenberger

1976), where the slender-stemmed, small-flowered South American rhipsaloid epiphytes were included within the Hylocereeae, which are larger-growing epiphytes and climbers with a centre of diversity in Central America.

In-so-far as they have been sampled, the remaining E Brazilian Cactoideae divide into two groups of Buxbaumian tribes (based on the *rpl*16 intron and *rbc*L markers, Wallace & Gibson 2002, and *trn*K intron, including *mat*K, Nyffeler 2000a, 2001, 2002): the Hylocereeae-Pachycereeae-Echinocereeae clade (HPE, also includes Leptocereeae *pro parte* and *Corryocactus* Britton & Rose) and the Browningieae-Cereeae-Trichocereeae (BCT) clade. In the Taxonomic Inventory in Chapter 5 the Rhipsalideae are positioned after the HPE clade and before the BCT clade, since more recently announced molecular evidence suggests that the Rhipsalideae (excluding *Pfeiffera, sens. lat.*) are basal to the latter (Nyffeler 2000c). The HPE clade includes the large-flowered, robust climbers and epiphytes belonging to the Hylocereeae (*Hylocereus, Epiphyllum* etc.) to which should probably be added the similar if enigmatic *Pseudoacanthocereus*, whose position awaits confirmation with molecular evidence. In an earlier molecular analysis the Hylocereeae appeared as the most basal element in Cactoideae (Wallace 1995: Fig. 11), but that position has now been taken by *Calymmanthium* F. Ritter, an aberrant Peruvian cereoid.

The BCT clade is robustly monophyletic, characterised by a substantial deletion of some 300 nucleotide base-pairs from the *rpl*16 intron, but its further resolution into tribes awaits the completion of molecular systematics projects currently in progress. The clade represents 16 out the total of 30 cactus genera of Eastern Brazil. However, preliminary molecular studies by Soffiatti (unpubl.) have raised doubts about the current circumscription of tribe Cereeae (8–9 genera). In contrast, evidence from a phytochemical source suggests that the Cereeae could be a monophyletic component within BCT, since the cuticular *n*-alkanes of *Cereus, Pilosocereus* and *Melocactus*, representatives of markedly different parts of the tribe (Taylor & Zappi 1989: Figs 2–5), are reported to be much more similar to each other than to those of other tribes (Maffei *et al.* 1997). *Brasilicereus* is provisionally included in Cereeae here, but has aberrant, scaly flowers and potentially might represent the Browningieae component of BCT. The presumed Trichocereeae representatives are 7 genera, of which *Leocereus, Espostoopsis* and *Uebelmannia* are each aberrant in different ways and their inclusion in the tribe is provisional at present. Indeed, the last-named appears closer to Brazilian endemic genera of Cereeae than to Trichocereeae, according to recent gene sequence studies (Soffiatti unpubl.). Unlike the naked-flowered Cereeae, the Trichocereeae are characterised by flowers ± clothed in hair-spines and/or sometimes spines or bract-scales, although *Espostoopsis* has almost naked flowers (and may be convergent with the Cereeae).

The remaining major clades of Cactoideae recognizable on molecular evidence, that are absent from Eastern Brazil, are the Cacteae (N Hemisphere), Notocacteae *sens. str.* (S South America), *Calymmanthium* (Peru) and *Copiapoa* Britton & Rose (Chile), the latter two lacking tribal names at present. A comparison of Buxbaum's higher level classification of Cactaceae with that implied by modern molecular phylogenies is offered in Table 2.3, together with brief notes on the circumscription of his groupings.

Table 2.1

Status of genera of Cactaceae found in Eastern Brazil according to the IOS consensus process (Hunt & Taylor 1986, 1990), CITES Checklist editions (Hunt 1992b, 1999a) and their present treatment, with brief justifications and notes.

(See further discussion under individual generic treatments in Chapter 5.3–5.5)

E Brazilian genus (* = introduced)	IOS Consensus status	Comments/justification
PERESKIOIDEAE K. Schum.:		
1. *Pereskia* Mill. 1754	Unanimously accepted since 1986	Earliest generic name in subfam., but genus potentially paraphyletic as circumscribed here
OPUNTIOIDEAE K. Schum.:		
2. *Quiabentia* Britton & Rose 1923	Sunk into *Pereskiopsis* 1986 & 1990; recognised by Hunt (1992b, 1999a)	Seed anatomy (Stuppy 2002) supports separation from *Pereskiopsis*
3. *Tacinga* Britton & Rose 1919	Unanimously accepted 1986 & 1990, but only in original circumscription *(sens. str.):* 2 spp.	Recognized in an expanded sense here for 6 spp., based on seed anatomy and floral similarities (Taylor *et al.* 2002)
4. *Brasiliopuntia* (K. Schum.) A. Berger 1926	Sunk into *Opuntia /* 1986 & 1990	Recognized here based on seed- and pollen-morphology (lacks reticulate exine of *Opuntia sens. str.*) and unique, autapomorphic habit and behaviour (Taylor *et al.* 2002)
*5. *Nopalea* Salm-Dyck 1850	Sunk into *Opuntia /* 1986 & 1990	Recognized here based on pollen-morphology (lacks reticulate exine of *Opuntia sens. str.*) and floral differences (Taylor *et al.* 2002)
6. *Opuntia* Mill. 1754	Unanimously accepted since 1986 with *sens. lat.* circumscription	Earliest generic name in subfamily, here recognized *sens. str.* for taxa with unique autapomorphic reticulate pollen exine and sensitive stamens
CACTOIDEAE (HPE clade) — Hylocereeae:		
7. *Hylocereus* Britton & Rose 1909	Unanimously accepted 1986 & 1990, employing circumscription adopted by Hunt (1967)	Circumscription expanded here to include *Selenicereus* sect. *Salmdyckia* D. R. Hunt, based on trigonous stem-morphology and taxa with transitional floral/fruit characters, supported by gene sequence data (Wallace unpubl.)

*8. *Selenicereus* Britton & Rose 1909	Unanimously accepted 1986 & 1990 with its circumscription adjusted in 1990 in line with Hunt (1989)	Circumscription adjusted here, as indicated above
9. *Epiphyllum* Haw. 1812	Unanimously accepted since 1986	Earliest generic name in tribe

CACTOIDEAE (HPE clade?) — Echinocereeae(?):

10. *Pseudoacanthocereus* F. Ritter 1979	Included in *Acanthocereus* in 1986 & 1990, and by Hunt (1992b) with a note about its probable distinctness. Recognised by Taylor *et al.* (1992) and by Hunt (1999a)	Fruit- and seed-morphology (Taylor *et al.* 1992) unique within the HPE clade and sister group as yet unidentified

CACTOIDEAE — Rhipsalideae:

11. *Lepismium* Pfeiff. 1835	Accepted to include only the Brazilian type species for reasons of nomenclatural expediency in 1986 then in an expanded sense by a minority of IOS WP members in 1990; accepted by Hunt (1992b, 1999a)	Circumscription employed here defined on the basis of plesiomorphic stem- and fruit-morphology character states, as detailed in Barthlott (1987) and Barthlott & Taylor (1995). Brazilian elements shown to be a monophyletic group on the basis of gene sequence data (Wallace unpubl.), but genus becomes paraphyletic if some Andean taxa are included (Nyffeler 2000c)
12. *Rhipsalis* Gaertn.	Unanimously accepted 1986–90	Earliest generic name in the tribe; present generic circumscription is supported by phylogenetic analyses
13. *Hatiora* Britton & Rose 1915	Unanimously accepted 1986, then by only a minority in 1990; subsequently accepted by Hunt (1992b, 1999a)	Circumscription employed here defined on the basis of stem- and floral-morphology, as described by Barthlott (1987) and Barthlott & Taylor (1995). E Brazilian type species shown to be part of a monophyletic group (*Hatiora* subg. *Hatiora*) on the basis of phylogenetic analyses, but genus becomes paraphyletic if taxa referable to *Rhipsalidopsis* B. & R. are included
14. *Schlumbergera* Lem. 1858	Unanimously accepted since 1986 following expanded circumscription employed in revision by Hunt (1969)	Recognition of this genus is supported by gene sequence data (Wallace unpubl.) and floral morphology: presence of true perianth tube

CACTOIDEAE (BCT clade) — Cereeae:

15. *Brasilicereus* Backeb. 1938	Included in *Monvillea* Britton & Rose 1920 in 1986, but accepted by a minority of IOS WP members in 1990, following recommendations in Cereeae paper by Taylor & Zappi (1989); accepted by Hunt (1992b, 1999a)	Defined on the basis of a unique combination of vegetative and floral features within the BCT clade (cereoid habit, short-tubed flowers with glabrous but scaly pericarpel, stamens inserted in 2 series and indehiscent fruits bearing brownish, not blackish, perianth remains). Strongly supported as distinct in a recent gene sequence analysis (Soffiatti unpubl.)
16. *Cereus* Mill. 1754	Unanimously accepted since 1986	Earliest generic name in the tribe
17. *Cipocereus* F. Ritter 1979	Lumped with *Pilosocereus* in 1986, then accepted (incl. *Floribunda* F. Ritter) by a minority of IOS WP members in 1990, following recommendations in Cereeae paper by Taylor & Zappi (1989); accepted by Hunt (1992b, 1999a)	Distinguished from other Cereeae by its blue-waxy, globose, indehiscent fruits with translucent funicular pulp and persistent, erect, blackened perianth remains in combination with small seeds. Shown to be monophyletic in a recent gene sequence analysis (Soffiatti unpubl.)
18. *Stephanocereus* A. Berger 1926	Accepted to include only the type species for reasons of nomenclatural expediency in 1986, then in an expanded sense by a minority of IOS WP members in 1990; accepted by Hunt (1992b, 1999a)	Present circumscription reflects presumed homology between the two included species in respect of juvenile to adult developmental stages and reproductive structures (cephalia, flowers & fruit). It differs from other Cereeae in these same characteristics (Taylor & Zappi 1989)
19. *Arrojadoa* Britton & Rose 1920	Unanimously accepted since 1986	Earlier generic name than its sister group, *Stephanocereus*, having derived pollen characters (Taylor & Zappi 1989)
20. *Pilosocereus* Byles & G. D. Rowley 1957 [*Pilocereus* K. Schum. 1894 *nom. illeg.*]	Unanimously accepted 1986 (incl. *Cipocereus*) and then with present circumscription in 1990	Differentiated from other Cereeae by its depressed-globose, dehiscent fruits (Taylor & Zappi 1989, Zappi 1994)
21. *Micranthocereus* Backeb. 1938	Sunk into *Arrojadoa* in 1986, but accepted by a majority of IOS WP members in 1990, following papers by Taylor & Zappi (1989) and Braun & Esteves Pereira (1990)	Since 1990 its circumscription has followed that of Ritter (1968, 1979) and Taylor & Zappi (1989), including *Austrocephalocereus* Backeberg 1938, which the 1986 IOS report accepted as distinct. Probably very close to *Pilosocereus*, but with small, hummingbird-syndrome flowers and indehiscent fruit

22. *Coleocephalocereus* Backeb. 1938	Accepted by a majority of IOS WP members in 1986, then by fewer votes in 1990, following the suggestion that it might be paraphyletic in respect of *Melocactus* by Taylor & Zappi (1989). Accepted by Hunt (1992b, 1999a)	The presence of fruits dehiscent by a small basal pore is supported by molecular evidence (Cowan & Soffiatti unpubl.) as a sound autapomorphy, distinguishing the genus from both its potential sister group, *Melocactus*, and other Cereeae (cf. Taylor 1991a: 18)
23. *Melocactus* Link & Otto 1828	Unanimously accepted since 1986	Highly derived within Cereeae, with autapomorphic non-chlorophyllous terminal cephalium and distinctive fruit, pollen and seeds (Taylor & Zappi 1989, Taylor 1991a)

CACTOIDEAE (BCT clade) — Trichocereeae:

24. *Harrisia* Britton & Rose 1908	Unanimously accepted since 1986 in the broad sense including *Eriocereus* Riccob. (1909)	Distinctive for its seed-anatomy (Barthlott & Hunt 2000: 27, pl. 23–24); circumscription supported by gene sequence data (Wallace 1997)
25. *Leocereus* Britton & Rose 1920	Accepted less than unanimously by IOS WP members in 1986 (when it was hinted that *Arthrocereus* A. Berger might be included), then fully accepted as a monotypic entity in 1990, following commentary by Taylor (Hunt & Taylor 1990: 100) and data in Taylor & Zappi (1990)	Fruit- and seed-morphology unique within the BCT clade and sister group as yet unidentified. Position awaits confirmation based on molecular data
26. *Facheiroa* Britton & Rose 1920	Accepted less than unanimously by IOS WP members in 1986 & 1990; accepted by Hunt (1992b, 1999a)	Sister group within tribe at present uncertain and its position remains to be elucidated with molecular data
27. *Espostoopsis* Buxb. 1968	Maintained as a synonym of *Austrocephalocereus* in 1986 (which had been Buxbaum's later view, cf. Leuenberger 1976), then recognized by a minority of IOS WP members in 1990, following recommendations by Taylor & Zappi (1989); accepted by Hunt (1992b, 1999a)	In stem and habit characters this plant strongly resembles the Andean genus, *Espostoa* Britton & Rose, but has naked flowers and polycolpate pollen. Its placement in Trichocereeae, rather than Cereeae, is provisional and based on the strong suspicion that its floral features are merely convergent with Cereeae, where it appears to lack any obvious relatives
28. *Arthrocereus* A. Berger 1929	Tentatively referred to *Leocereus* in 1986, then accepted by a minority of IOS WP members in 1990, following debate reported in Hunt & Taylor 1990: 99–100; accepted by Hunt (1992b, 1999a)	Close to the very large and complex genus, *Echinopsis* Zucc., but differing in its pollen (Leuenberger 1976) and and indehiscent fruits. Molecular evidence is needed to confirm its generic status or otherwise

29. *Discocactus* Pfeiff. 1837	Unanimously accepted since 1986	Assumed to be the sister group of the much larger *Gymnocalycium* Pfeiff. ex Mittler 1844, but distinguished by having 12–15-colpate (vs 3-colpate) pollen (Leuenberger 1976) and the presence of a terminal cephalium — a parallelism with *Melocactus*
30. *Uebelmannia* Buining 1967	Accepted less than unanimously by IOS WP members in 1986, then unanimously in 1990; accepted by Hunt (1992b, 1999a)	Strongly supported as monophyletic in a recent gene sequence analysis (Soffiatti unpubl.), which places it close to *Cipocereus* (cf. Cereeae above)

Table 2.2

Characters defining the subfamilies of Cactaceae represented in Eastern Brazil.

Subfamily	Characters / molecular markers
PERESKIOIDEAE K. Schum. (*Pereskia* only)	Woody, with essentially non-succulent stems bearing broad, functional, scarcely succulent leaves and un-barbed spines; seeds of 'common ancestral centrospermous type' (Barthlott & Hunt 1993); pollen 3–15-colpate. Has no shared cpDNA restriction site changes with the putatively basal Maihuenioideae (*Maihuenia* only), but may be paraphyletic in respect of Cactoideae (Wallace 1995: 9)
OPUNTIOIDEAE K. Schum.	Stems succulent, at least at first, bearing cylindrical to awl-shaped or rarely flattened, succulent and often caducous leaves and two distinct kinds of barbed spines; seeds covered in a bony aril formed from the funicle (unique in the centrosperms); pollen mostly 12–18-porate. Has autapomorphic deletion in the plastid gene accD, ORF 512, from the large single copy portion of cpDNA (Wallace & Gibson 2002)
CACTOIDEAE	Stems succulent, leaves reduced to minute ephemeral scales or usually entirely absent; spines un-barbed; seeds various, lacking an aril or this a corky appendage at the hilum only; pollen mostly 3(–12)-colpate. Has autapomorphic c. 700 base-pair deletion from the chloroplast encoded gene *rpo*C1 intron (Wallace & Cota 1996)

Table 2.3

Buxbaum's tribal arrangement of Cactaceae-Cactoideae (after Leuenberger 1976) evaluated by modern molecular data

(Wallace, pers. comm. etc.). For terminology, see Chapter 7.1.

Tribes/subtribes of Buxbaum, principal included genera [* = native in E Brazil] and their characteristics	Status based on molecular data (gene sequence variation)	Comments
LEPTOCEREEAE Buxb. *Leptocereus, Calymmanthium, Armatocereus, Neoraimondia, Samaipaticereus* etc. Erect columnar-cereoid taxa with few high stem ribs and shortly tubular flowers bearing areoles and spines	Buxbaum's circumscription is not supported by molecular data and these genera should be dispersed amongst the tribes that follow, except for *Calymmanthium*, which probably merits its own tribe	Buxbaum was grouping these taxa on the basis of what are now seen to be symplesiomorphies
HYLOCEREEAE Buxb.: 5 subtribes *Nyctocereus, Peniocereus, Acanthocereus, Harrisia*★, *Aporocactus, Selenicereus, Deamia, Hylocereus*★, *Epiphyllum*★, *Disocactus, Pseudorhipsalis, Pfeiffera, Hatiora*★, *Schlumbergera*★, *Rhipsalis*★, *Lepismium*★ etc. Suberect shrubs, lianas and epiphytes (both erect and pendent) with few-ribbed cylindric or flattened stems, very diverse flowers and ± smooth seedcoats (except *Harrisia*)	Molecular evidence requires the removal of the small-flowered, S Hemisphere epiphytes as the Rhipsalideae DC., *sens. str.* (excl. *Pseudorhipsalis*, and also *Pfeiffera sens. lat.*; Nyffeler 2000c). *Acanthocereus, Leptocereus* (see above), *Deamia* and the Hylocereeae *sens. str.* are basal to the Echinocereeae (Pachycereeae) elements in the HPE clade. *Harrisia* belongs in Trichocereeae	As circumscribed by Buxbaum, the Hylocereeae should have been called the Rhipsalideae DC., which is an older tribal name
PACHYCEREEAE Buxb.: 4 subtribes *Pterocereus, Escontria, Pachycereus, Stenocereus, Carnegiea, Neobuxbaumia, Cephalocereus, Myrtillocactus* etc. Erect columnar-cereoid taxa with woody tissues organized into discrete rods and diverse flowers with spiny, bristly, scaly or woolly pericarpels	These form a significant part of the HPE clade, to which must be added the relevant parts of Buxbaum's Leptocereeae and Hylocereeae, and the Echinocereeae (*q.v.*)	Buxbaum included *Pilosocereus* p.p. (tribe Cereeae) in *Cephalocereus*
BROWNINGIEAE Buxb. *Browningia, Castellanosia* etc. Erect columnar-cereoid taxa with densely scaly tubular flowers	Part of *Browningia sens. lat.* belongs to the BCT clade, but recent reports suggest that *Castellanosia* is not related (Nyffeler 2002)	

CEREEAE Salm-Dyck
Jasminocereus, Stetsonia, Praecereus, Cereus,
Pseudopilocereus [Pilosocereus p.p.]*,
Stephanocereus*, Coleocephalocereus*,
Brasilicereus**
Erect columnar-cereoid taxa, bearing
± naked, tubular, mostly nocturnal flowers
and ± naked fruit

With the exception of *Jasminocereus*, which belongs with *Armatocereus* in the HPE clade, the remainder included here by Buxbaum, where sampled, are BCT clade members and those having a naked pericarpel are presently referred to Cereeae	*Cereus*, *Praecereus* and *Brasilicereus* may not be close to the E Brazilian taxa included here according to *rpl16* gene sequences (Soffiatti unpubl.)	

TRICHOCEREEAE Buxb.: 4 subtribes
Echinopsis, Haageocereus, Espostoa, Facheiroa,
Austrocephalocereus (incl. Espostoopsis)*,
Leocereus*, Arthrocereus*, Cleistocactus,
Oreocereus, Matucana, Micranthocereus*,
Arrojadoa*, Rebutia* etc.
Erect columnar-cereoid to low-growing
taxa, bearing long-tubed flowers with
bract-scales and/or spines/hair-spines

All elements included here by Buxbaum belong in the BCT clade, but some taxa are here referred to the Cereeae on the basis of floral morphology, and other BCT-Trichocereeae were misplaced in the Notocacteae. *Samaipaticereus* (Leptocereeae) and *Harrisia* (Hylocereeae *sensu* Buxb.) belong here (Wallace 1997)

The position of *Leocereus* remains to be clarified on the basis of molecular markers

NOTOCACTEAE Buxb.: 5 subtribes
*Corryocactus, Austrocactus, Eriosyce,
Eulychnia, Copiapoa, Parodia, Frailea,
Uebelmannia*, Astrophytum, Gymnocalycium,
Sulcorebutia [= Rebutia p.p., see above],
Discocactus*, Melocactus** etc.
Mostly globular-stemmed, ribbed or
tuberculate cacti, bearing diurnal short-
tubed flowers with mostly scaly and/or
bristly and woolly pericarpels; many
unique seedcoat characters

Notocacteae in its strictest sense (cf. Nyffeler 2001; ie. *Parodia sens. lat.* and *Eriosyce*) is strongly supported by molecular data, but the tribe *sensu* Buxbaum is otherwise a complete mixture, including elements of at least 4 other major lines, eg. the BCT clade, the HPE clade (*Eulychnia, Austrocactus, Corryocactus*), *Copiapoa* & Cacteae (*Astrophytum*)

The Notocacteae as previously defined were recognizable on the basis of floral characters (woolly pericarpel with bristles above) in combination with unique seedcoat morphology

ECHINOCEREEAE Buxb.
Bergerocactus, Echinocereus.
Low-growing, shortly cereoid cacti
bearing diurnal flowers with areolate
-spiny pericarpels

Not recognizable as a discrete entity on molecular data, being an integral part of the large HPE clade and close to *Stenocereus* (Pachycereeae)

Echinocereeae will likely prove to be the oldest name for the HPE clade

CACTEAE: 4 subtribes
*Echinocactus, Sclerocactus, Thelocactus,
Turbinicarpus, Lophophora, Strombocactus,
Ariocarpus, Ferocactus, Escobaria,
Mammillaria, Coryphantha* etc.
Mostly globular-stemmed, ribbed or
tuberculate cacti, bearing diurnal
short-tubed flowers with naked,
scaly and/or woolly pericarpels

Well-defined major clade strongly supported by molecular and other data when the North American *Astrophytum* is included from Buxbaum's Notocacteae

The majority of taxa are characterized by a suite of apomorphic seedcoat characters that sets them apart from all other cacti

3. PHYTOGEOGRAPHY

3.1 OVERVIEW

Following detailed study of the systematics, habitats and range of the Cactaceae native in Eastern Brazil it becomes obvious that there are some more or less well-defined distribution patterns into which the majority of taxa can be classified. These patterns generally correspond with geographical, climatic, edaphic and other ecological phenomena and thus may offer help in defining phytogeographical regions and give clues about past vegetational history (Harley 1988, Prado & Gibbs 1993). Such studies are also important from the standpoint of conservation, since they indicate the minimum number, range and diversity of areas and habitats that need to be considered for protection (for priorities, see Chapter 4: Table 4.3).

Perhaps the most interesting and potentially informative aspect of this study is to use the areas defined below as a model to compare with other floristically important plant groups found in Eastern Brazil, especially those with ecological preferences similar to Cactaceae, such as Araceae (eg. *Philodendron, Anthurium*) and Bromeliaceae (eg. *Dyckia, Encholirium*), both of which are frequently lithophytic. Another group including *caatinga* species, which is sometimes compared in the discussion below, is *Caesalpinia* (Lewis 1998), while the Hyptidinae (Lamiaceae) include a number of *campo rupestre* taxa with ranges similar to those of Cactaceae (cf. Harley 1988). However, at present, opportunities for making such comparisons are few, because, if they are to be meaningful, it is necessary to have a rather detailed data set on the group concerned. This means monographic studies, to enable the careful determination of species limits, relationships, geographical range and reproductive-dispersal strategies, all requiring extensive field knowledge. This is perhaps the first study of its kind to focus on a taxonomic group of reasonable size and ecological complexity. Hopefully, there will be future opportunities to evaluate its phytogeographical significance in relation to other families found in Eastern Brazil, especially those mentioned above.

This chapter considers all named, fully accepted taxa of Cactaceae that are native of the North-eastern region ('Nordeste') of Brazil, and of its South-eastern region ('Sudeste') north of c. 22°S and east of c. 46°W, as indicated on Map 1. As defined, the area chosen includes all Brazilian Cactaceae endemic to the great *caatingas* dominion (Andrade-Lima 1981) and all but two of those endemic to the *campos rupestres,* the vegetation type that characterizes the East Brazilian Highlands (Harley in Stannard 1995: 25), whose western part, with few cacti (W Minas Gerais & Goiás), is excluded (see Map 1). Three-quarters or 76% (123) of the 162 native species and heterotypic subspecies of cacti found in Eastern Brazil are endemic and we believe this justifies its choice for study.

While the whole of this geographical area has been considered, initial field and herbarium investigations suggested focusing on a somewhat smaller core area delimited at c. 7°S and c. 46°W, ie. including south-eastern Piauí, the southernmost tip of Ceará, the southern half of Paraíba, then southwards to 22°S, in which there is a much greater diversity of taxa. *Cereus insularis* (Fernando de Noronha), is the only endemic from the Nordeste that does not enter this core area. Table 3.1 indicates taxa ranging beyond the core area and those endemic to it, the latter being two-thirds (107 taxa) of the overall total. A significant part of the territories of the northernmost states of the Brazilian Nordeste, comprising Maranhão, Piauí, Ceará, Rio Grande do Norte and Paraíba, has a relatively poor cactus flora, none having more than 16 native taxa. Only Minas Gerais and Bahia have significant numbers of state endemic taxa. Possible explanations for these differing levels of diversity will be discussed below.

As stated above, the core geographical area defined within Eastern Brazil and particularly focused on in this study has been recognized primarily for its remarkable endemism of Cactaceae genera and species, especially those of the *caatinga* of the North-eastern states and associated highland *campos rupestres*. Both vegetation types extend into the northern part of the SE Brazilian state of Minas Gerais, where the highest concentration of Cactaceae taxa is found. Habitats including a significant representation of cacti also reach into the Atlantic Forest (*Mata atlântica*) of eastern Minas Gerais and the neighbouring state of Espírito Santo, where a substantial area drained by the Rio Doce and adjacent river systems receives less than 1000–1250 mm of rain per year (see map 'isoietas anuais 1914–1938' in Azevedo 1972; Nimer 1973: 40, fig. 18). In this extensive core area representatives of the family appear to have evolved for long periods in isolation, so that many distinctive taxa without parallel elsewhere have arisen. It is also true that various major cactus genera, widespread in the neotropics, are absent from this region. These include *Selenicereus* (NB. *S.* subg. *Salmdyckia* is here referred to *Hylocereus*), *Disocactus, Pseudorhipsalis, Echinopsis* and *Cleistocactus*, besides others mentioned below, which, unlike the above, are replaced in Eastern Brazil by obvious, closely related vicariants. Genera from the widespread, species-rich, South American Notocacteae are likewise absent. Other significant neotropical genera are represented by only a single native species in each case, eg. *Opuntia (sens. str.), Hylocereus, Epiphyllum* and *Harrisia*. In the case of *Opuntia*, the single species, *O. monacantha*, is only marginally represented, being a rarity at the northern limits of its range.

It is perhaps not difficult to see why this should be so, since today the region is effectively cut off from other centres of cactus diversity by very broad zones of habitat unfavourable to most members of the family. To the north-west of the dry region of North-eastern Brazil is the nearly constantly humid Amazonian region, with its rainforests and included savanna formations, where only few specialized cacti are able to compete with other plants (see Appendix II). Amongst these is the epiphyte, *Rhipsalis baccifera* subsp. *baccifera*, whose range southwards into Eastern Brazil stops at c. 8°S in the coastal forests of Pernambuco (around Recife), indicating the southern limit of Amazonian floristic influence in terms of Cactaceae. To the west, in Central-western Brazil, are the extensive fire-swept *cerrados*, which are avoided by most cactus genera

Table 3.1

Distribution of species and subspecies of Cactaceae native in Eastern Brazil by state.

Key: † = taxon endemic to the total area included in this table; ‡ = taxon endemic to core area (northern limit 7°S); § = 'single-site endemic' (definition in Chap. 3.3); + = definitely recorded as native; ± = record requiring confirmation or of dubious identity; +? = possibly native or cultivated; **c** = cultivated; **FN** = Fernando de Noronha (PE) only; MG¹/RJ² = these states included only east of 46°W and north of c. 22°S. For locations of states/key to 2-letter codes, see caption to Map 1.

TAXON / STATE	MA	PI	CE	RN	PB	PE	AL	SE	BA	MG¹	ES	RJ²
Pereskia aculeata	+					+	+	+	+	+	+	+
P. grandifolia subsp. *grandifolia*	+?		+			+			+	+	+	+
P. grandifolia subsp. *violacea* ‡									c	+	+	
P. bahiensis ‡									+			
P. stenantha ‡									+	+		
P. aureiflora ‡									+	+		
Quiabentia zehntneri ‡									+	+		
Tacinga funalis ‡		±				+			+			
T. braunii ‡										+		
T. werneri ‡									+	+		
T. palmadora †				+	+	+	+	+	+			
T. saxatilis subsp. *saxatilis*									+	+		
T. saxatilis subsp. *estevesii* ‡§									+			
T. inamoena †	+	+	+	+	+	+	+	+	+	+		
Brasiliopuntia brasiliensis					+	+	+	+	+	+	+	+
Opuntia monacantha								c	c	+	+	
Hylocereus setaceus	±	+	+			+	+	+	+	+	+	+
Epiphyllum phyllanthus	+	+	+		+	+	+	+	+	+	+	+
Pseudoacanthocereus brasil. ‡									+	+		
Lepismium houlletianum										+		
L. warmingianum										+	+	
L. cruciforme						+			+	+	+	
Rhipsalis russellii									+	+	+	
R. elliptica										+		
R. oblonga									+		+	
R. crispata					+				+			
R. floccosa subsp. *floccosa*					+			+	+		+	
R. floccosa subsp. *oreophila* ‡									+	+		
R. floccosa subsp. *pulvinigera*										+	+	
R. paradoxa subsp. *septentrion.* ‡						+			+	+	+	
R. pacheco-leonis subsp. *catenul.*											+	
R. cereoides											+	
R. sulcata ‡§											+	
R. lindbergiana						+		+	+	+	+	
R. teres										+	+	
R. baccifera subsp. *baccifera*	+		+		+	+						
R. bacc. subsp. *hileiabaiana* ‡									+			

	1	2	3	4	5	6	7	8	9	10	11	12
R. pulchra										+		
R. burchellii										±	+	
R. juengeri										+		
R. clavata										±	+	+
R. cereuscula						+			+	+		
R. pilocarpa										+	+	
R. hoelleri ‡§										+		
Hatiora salicornioides									+	+	+	+
H. cylindrica									+		±	
Schlumbergera kautskyi ‡										+		
S. microsphaerica										+	±	
S. opuntioides										+		
Brasilicereus phaeacanthus ‡									+	+		
B. markgrafii ‡										+		
Cereus mirabella	+								+	+		
C. albicaulis †		+	+			+	±		+	±		
C. fernambucensis subsp. fern.				+	+	+	+	+	+		+	+
C. fernambuc. subsp. sericifer										+	+	+
C. insularis †						FN						
C. jamacaru subsp. jamacaru †?	+	+	+	+	+	+	+	+	+	+	c	c
C. jamacaru subsp. calcirupicola‡									+	+		
C. hildmannianus										+		
Cipocereus laniflorus ‡§										+		
Cipocereus crassisepalus ‡										+		
C. bradei ‡										+		
C. minensis subsp. leiocarpus ‡										+		
C. minensis subsp. minensis ‡										+		
C. pusilliflorus ‡§										+		
Stephanocereus leucostele ‡									+			
S. luetzelburgii ‡									+			
Arrojadoa bahiensis ‡									+			
A. dinae subsp. dinae ‡									+	+		
A. dinae subsp. eriocaulis ‡									+	+		
A. penicillata ‡									+	+		
A. rhodantha ‡		+				+			+	+		
Pilosocereus tuberculatus ‡						+		+	+			
P. gounellei subsp. gounellei †	+	+	+	+	+	+	+	+	+			
P. gounellei subsp. zehntneri ‡									+	+		
P. catingicola subsp. catingicola ‡									+			
P. catingic. subsp. salvadorensis †			±	+	+	+	+	+	+			
P. azulensis ‡§?										+		
P. arrabidae									+		+	+
P. brasiliensis subsp. brasiliensis											+	
P. brasiliensis subsp. ruschianus ‡									+	+	+	
P. flavipulvinatus †	+	+	+									
P. pentaedrophorus subsp. pent. ‡						+		+	+			
P. pentaedroph. subsp. robustus ‡									+	+		
P. glaucochrous ‡									+			
P. floccosus subsp. floccosus ‡										+		

Taxon	1	2	3	4	5	6	7	8	9	10	11
P. flocc. subsp. *quadricostatus* ‡									+		
P. fulvilanatus subsp. *fulvil.* ‡									+		
P. fulvilanatus subsp. *rosae* ‡§									+		
P. pachycladus subsp. *pachycl.* ‡								+	+		
P. pachycl. subsp. *pernambuco.* †	+	+	+	+	+	+		+			
P. magnificus ‡									+		
P. machrisii	+							+			
P. aurisetus subsp. *aurisetus* ‡									+		
P. aurisetus subsp. *aurilanatus* ‡									+		
P. aureispinus ‡								+			
P. multicostatus ‡									+		
P. piauhyensis †	+	+	+								
P. chrysostele †		+	+	+	+						
P. densiareolatus ‡								+	+		
Micranthocer. violaciflorus ‡									+		
M. albicephalus ‡								+	+		
M. purpureus ‡								+			
M. auriazureus ‡§									+		
M. streckeri ‡§								+			
M. polyanthus ‡								+			
M. flaviflorus ‡								+			
M. dolichospermaticus ‡								+	+		
Coleocephalocer. buxb. subsp. *b* ‡									+	+	
C. buxbaum. subsp. *flavisetus* ‡									+		
C. fluminensis subsp. *fluminensis*									+	+	+
C. fluminensis subsp. *decumbens* ‡									+		
C. pluricostatus ‡									+	+	
C. goebelianus ‡								+	+		
C. aureus ‡									+		
C. purpureus ‡§									+		
Melocactus oreas subsp. *oreas* ‡								+			
M. oreas subsp. *cremnophilus* ‡								+			
M. ernestii subsp. *ernestii* †		±	±	+	+	±	+	+	+		
M. ernestii subsp. *longicarpus* ‡								+			
M. bahiensis subsp. *bahiensis* ‡					+			+			
M. bahiensis subsp. *amethystinus* ‡								+			
M. conoideus ‡§								+			
M. deinacanthus ‡§								+			
M. levitestatus ‡								+	+		
M. azureus ‡								+			
M. ferreophilus ‡								+			
M. pachyacanthus subsp. *pachy.* ‡								+			
M. pachyacanthus subsp. *viridis* ‡								+			
M. salvadorensis ‡								+			
M. zehntneri †	+	+	+	+	+	+	+	+			
M. lanssensianus ‡§?				±	+						
M. glaucescens ‡								+			
M. concinnus ‡								+	+		
M. paucispinus ‡								+			

	MA	PI	CE	RN	PB	PE	AL	SE	BA	MG[1]	ES	RJ[2]
M. violaceus subsp. *violaceus*				+	+	+		+	+	+	+	
M. violaceus subsp. *ritteri* ‡									+			
M. violac. subsp. *margaritaceus* ‡							+	+	+			
Harrisia adscendens ‡		±	+			+	+	+	+			
Leocereus bahiensis ‡	±	+							+		+	
Facheiroa ulei ‡									+			
F. cephaliomelana subsp. *c.* ‡									+		+	
F. cephaliom. subsp. *estevesii* ‡§									+			
F. squamosa ‡		+				+			+			
Espostoopsis dybowskii ‡									+			
Arthrocereus melanurus subsp. *me.*										+		
A. melanurus subsp. *magnus* ‡§										+		
A. melanurus subsp. *odorus* ‡										+		
A. rondonianus ‡										+		
A. glaziovii ‡										+		
Discocactus zehntneri subsp. *z.* ‡									+			
D. zehntneri subsp. *boomianus* ‡									+			
D. bahiensis †		+	+			±			+			
D. heptacanthus subsp. *catingic.*		+							+	+		
D. placentiformis ‡										+		
D. pseudoinsignis ‡										+		
D. horstii ‡§										+		
Uebelmannia buiningii ‡										+		
U. gummifera ‡										+		
U. pectinifera subsp. *pectinifera* ‡										+		
U. pectinifera subsp. *flavispina* ‡										+		
U. pectinifera subsp. *horrida* ‡§										+		
STATE	MA	PI	CE	RN	PB	PE	AL	SE	BA	MG[1]	ES	RJ[2]
STATE ENDEMICS	0	0	0	0	0	1?	0	0	30	35	3	0
TAXA SCORED +	8	16	15	11	15	33	14	19	95	97	34	11
SPECIES SCORED +	8	16	15	11	15	33	14	19	79	80	30	11

(except *Discocactus* in gravelly areas), although included rock outcrops and gallery forests do provide some suitable habitats (eg. for *Rhipsalis russellii, Pilosocereus machrisii* and allies, *Cereus* spp. and *Arthrocereus spinosissimus*) and have probably permitted a limited amount of migration to and from the *caatingas, Mata atlântica* and *campos rupestres* in the past (cf. Prado & Gibbs 1993). To the south the diversity of non-epiphytic cacti abruptly decreases as humidity increases, and these are replaced by numerous epiphytic species from tribe Rhipsalideae (Barthlott & Taylor 1995) until the grasslands (or *campos*) of Rio Grande do Sul are reached. Here the diversity of non-epiphytic cacti suddenly increases again; these, however, are globular species from tribe Notocacteae. Such ecological barriers in recent times must have severely limited migration of cactus species and genera to and from neighbouring regions of high diversity, such as the Caribbean, the Andes and the part of south-eastern South America where the Notocacteae are most abundant.

While Table 3.1 indicates which taxa are endemic to Eastern Brazil and the core area, it is also worthwhile to summarize endemism in terms of the major taxonomic groups represented:

Pereskioideae. *Pereskia*, nowadays the only genus included in this subfamily, is comprehensively represented in Eastern Brazil in terms of diversity, since elements of the 3 major groups of species recognized by Leuenberger (1986) are present and give a total of 5 species, with 3 endemic to the core area. The endemic species (*P. bahiensis, P. stenantha* and *P. aureiflora*) are restricted to the *caatinga-agreste*, while *P. aculeata* and *P. grandifolia* (and its endemic subsp. *violacea,* which may be specifically distinct) are *Mata atlântica* species.

Opuntioideae. In terms of species, the 4 rather distinct genera of Opuntioideae native of Eastern Brazil are very small, with only 9 species in total (6 belonging to *Tacinga, sens. lat.*, an endemic of the NE and SE regions of Brazil). At first sight this seems hard to explain, since elsewhere this subfamily is generally species-rich in the dry zones of the Americas, eg. in the Andes, Mexico and south-western United States, where they invariably constitute a major component of cactus floras. So why have so few taxa evolved in Eastern Brazil? The answer may lie in the origin of the subfamily. It includes taxa resembling the genus *Maihuenia* (Maihuenioideae, cf. Wallace 1995), a strong candidate for the plesiomorphic sister group of Opuntioideae and a clear parallel to some members of the Portulacaceae, a part of which is the sister group of the Cactaceae (Hershkovitz & Zimmer 1997, Savolainen *et al.* 2000: Fig. 2). *Maihuenia* is a genus from temperate South America (Patagonia etc.) and the Opuntioideae in major part are also represented by plants of cooler mountainous regions (Andes) or higher plateau lands (northern Mexico etc.), perhaps suggesting that they are generally less well adapted to life in the constantly warm, lowland parts of the tropics. In the drier parts of such tropical regions the Opuntioideae have, however, evolved some highly specialized and derived forms (derived when compared to a supposed *Maihuenia*-like ancestor), such as *Tacinga* (including 2 spp. of cactus lianas, E Brazil), *Brasiliopuntia* (1 treelike sp., neotropics south of the River Amazon), *Consolea* (c. 8 treelike spp., E Caribbean) and *Pereskiopsis* (6 semi-scandent or treelike spp., trop. Mexico & Cent. America). *Quiabentia* (2 treelike spp., South American Chaco region & E Brazil) is not included in this list, since at least its western species comes from a region that experiences frost in winter (Map 13) and its Brazilian counterpart is confined to the more seasonal south-western *caatingas*. *Tacinga, sens. lat.* (6 spp.), extends considerably beyond the E Brazilian core area into the northernmost *caatingas* (*T. inamoena*), and slightly beyond it to the west, in Minas Gerais, on edaphically dry limestone outcrops (*T. saxatilis*). It completely replaces *Opuntia, sens. str.,* in most of Eastern Brazil, but is lacking at the southern margins of the core area where *O. monacantha* reaches its northern limit.

Cactoideae. Out of the 8 tribes and 3 or more other major clades of Cactoideae currently recognised (Wallace 1995, Wallace & Gibson 2002, Nyffeler 2000a–c; see Chapter 2.7) nearly all E Brazilian cactus genera of this relationship fall into tribes Cereeae

(9 genera), Trichocereeae (7 genera) and Rhipsalideae (4 genera), with only *Hylocereus* (1 sp.) and *Epiphyllum* (1 sp.) in the Hylocereeae and *Pseudoacanthocereus* (1 sp.) doubtfully in Echinocereeae, *sens. lat.* (incl. Leptocereeae & Pachycereeae). Therefore, Eastern Brazil is apparently rather lacking in diversity at the tribal level or equivalent. However, at generic level the situation is different. Cereeae has its major centre of diversity in Eastern Brazil (Taylor & Zappi 1989) and many of its genera and subgenera are endemic: *Cipocereus* (5 spp.), *Brasilicereus* (2 spp.), *Stephanocereus* (2 spp.), *Micranthocereus* subg. *Micranthocereus* (4 spp.), *M.* subg. *Austrocephalocereus* (3 spp.), *Coleocephalocereus* subg. *Buiningia* (2 spp.) and *C.* subg. *Simplex* (1 sp.). *Micranthocereus* (9 spp.), *Arrojadoa* (4–5 spp.) and *Coleocephalocereus* (6 spp.) are almost endemic, the first with only *M. estevesii* (Subg. *Siccobaccatus*) located outside the core area on limestone outcrops between southernmost Tocantins and NW Minas Gerais, the second with a recently discovered taxon in NE Goiás. The Trichocereeae, while represented by relatively few species, are also diverse at generic level, with 4 out of 7 genera being endemic: *Uebelmannia* (3 spp.), *Facheiroa* (3 spp.), *Espostoopsis* (1 sp.) and *Leocereus* (1 sp.). The stronghold of this tribe in Eastern Brazil at species/subspecies level is in the South-eastern *campos rupestres* and adjacent *cerrados*, where they total nearly half of the Cactaceae. All 4 genera of Rhipsalideae *sens. str.* are found in Eastern Brazil, but none is endemic and few species and subspecies are endemic, since the *Mata atlântica* they mostly inhabit extends well beyond the limits of the core area to the south and south-west. Excepting *Rhipsalis baccifera*, their range northward terminates at the northern border of Pernambuco.

3.2 DEFINITION OF VEGETATION TYPES

In most cases, the terminology used in this book to identify types of vegetation in Eastern Brazil follows that employed by Prado (1991) and Harley in Stannard (1995: 7–11, 13–27, 37–40, 47–50, 52–66), which contain more detailed treatments including numerous valuable references. Azevedo (1972), Cavalcanti Bernardes (1951), Domingues (1973), King (1956) and Nimer (1973) have been consulted as sources of edaphic, geological and climatic data.

Mata atlântica (Atlantic Forest, Plate 1.1, Map 2). In its broadest sense this term covers a wide range of evergreen and semi-evergreen forest formations, beginning at the coast, from the high tide mark, and stretching far inland until, in NE Brazil, the deciduous *caatinga* is reached. It also reappears, in fragmented form, at the base of the eastern side of the Chapada Diamantina in Bahia and as riverine forest along the Rio São Francisco (Brandão 2000: 75). In SE Brazil such vegetation continues much further inland, on to the *planalto central*, and there once formed extensive tracts of semi-humid '*mata de planalto*', which is considered a part of the Atlantic Forest system by many Brazilian botanists. In some areas the forest includes extensive outcrops of smooth gneissic rocks or larger inselbergs and these represent an important habitat for rupicolous cacti in such areas, especially from south-eastern Bahia (Floresta Azul) southwards and south-westwards to Rio de Janeiro, where, for example, *Pilosocereus brasiliensis* and

Coleocephalocereus subg. *Coleocephalocereus* are characteristic rupicolous taxa (Maps 36 & 41). Although normally thought of as a very humid environment, this is by no means uniformly the case, since rainfall can vary from more than 2000 mm down to less than 800 mm per annum and often includes a more or less well-marked dry season. The cactus flora varies considerably with levels and frequency of precipitation and, as a consequence, many taxa show markedly disjunct distributions. Minimum temperatures vary in line with latitude and altitude, but are probably never less than 0°C anywhere in the regions considered in this study, and are never <12°C, from Rio Grande do Norte to coastal S Bahia (Nimer 1973: 42).

In the drier or rocky and less dense phases of the *Mata atlântica* large, terrestrial (ie. non-epiphytic) cacti are able to compete with other forest species, such as in the coastal forests around the border regions of Alagoas, Sergipe and Bahia (eg. *Cereus jamacaru* — typically a *caatinga* species), in eastern Minas Gerais and central Espírito Santo (eg. *Pereskia grandifolia* subsp. *violacea*) and in the lowlands of southern Espírito Santo and north-eastern Rio de Janeiro (eg. *Brasiliopuntia brasiliensis*). At or near the coast the forest is termed *restinga* and even in more humid areas frequently includes edaphically dry sand-dunes in which an open or denser scrub, or taller forest supports a variety of non-epiphytic cacti. However, the dense, multilayered forest in areas where there is rain every month of the year includes only cactus epiphytes, but even these can be absent or rare in some phases of this ecosystem, eg. the 'mata de tabuleiro' found on elevated Tertiary Barreiras sediments in many coastal areas of NE Brazil. Further landwards the 'brejo' forests (*mata de brejo*) represent isolated patches of more humid, evergreen vegetation on mountain ridges surrounded by seasonally dry Atlantic Forest or even *caatinga* (Map 6), eg. the Pico do Jabre (Mun. Maturéia, Paraíba; Rocha & Agra 2002). These benefit from lower average temperatures, greater cloud cover and overnight dews that compensate for rain during dry periods (Rodal *et al.* 1998). Besides *Melocactus ernestii,* which is common on exposed rocks, they are often rich in epiphytes, with a greater variety of Rhipsalideae than is found generally in *Mata atlântica*. Finally, in eastern parts of the Chapada Diamantina (Bahia) and Serra do Espinhaço (N Minas Gerais) there occur isolated montane cloud forests ('mata de neblina' at 1200–1800 m), which, at the highest elevations, are extremely humid. These are treated as part of the *campo rupestre* mosaic here (see below), because their flora lacks various of the more widespread cactus epiphytes characteristic of forests further east and even includes an endemic subspecies (*Rhipsalis floccosa* subsp. *oreophila*).

Connecting the extremes of perhumid coastal rainforest and dry inland *caatinga* is the ecotonal type known as *agreste* (here used to include *Mata de cipó*). Like much of the *Mata atlântica* (Myers *et al.* 2000), most of this has been destroyed eliminating an important habitat for certain cacti, but the remaining fragments in SE Bahia contain some of the world's largest cacti, including 20–25 metre tall *Brasiliopuntia* and the massive *Cereus* sp. (5c) — see Chapter 5.

Caatinga (Plate 1.2, Map 3). The low forest or semi-open thorny thicket vegetation most prevalent in NE Brazil, in which the great majority of species is drought-deciduous. There is a pronounced dry season, which can be of unpredictable duration (2–12+

months), the total annual rainfall being less than 1000 mm and sometimes less than 500 mm. Soils vary from very shallow and stony to deep and sandy, but many of the cacti are restricted to rock outcrops of various kinds, including gneiss/granite, sandstone and limestone (± flat expanses of rock are called '*lajedos*'). The minimum temperature throughout the area never drops below 8°C and the maximum can reach 40°C (Nimer 1973: 49). This vegetation is normally encountered at less than 750 metres altitude, occurring at sea level at its northern limits in coastal Ceará, but occasionally from 750–1100 metres ('*caatinga de altitude*'), such as around Morro do Chapéu, Bahia, c. 50 km west of Seabra (Mun. Ibitiara, BA) and on the west slope of the Serra Geral, east of Monte Azul (northern Minas Gerais). The southern limit of true *caatinga* is uncertain (although cacti may help in defining it, see below), but in the Rio São Francisco valley islands of taller, dry forest (deciduous or semi-deciduous *mata seca*) mostly on limestone outcrops ('*mata calcária*') and associated deep soils extend far to the south of the main *caatinga* area. At its southern and western limits the *caatinga* merges into floristically somewhat different formations termed *cerradão*, which are ecotonal with the *cerrado*. Andrade-Lima (1981) divides *caatinga* vegetation into 12 different types, only some having a diversity of Cactaceae, while others are quite devoid of the family.

Cerrado (Map 4). This is the Brazilian savanna, which occupies a huge area in the Central-Western and South-eastern Regions, occurring in only a few, mostly small patches in the North-east Region where precipitation and soil conditions dictate. It is a kind of grassland, but usually includes an open, ± evergreen layer of trees with curious contorted trunks and a continuous cover of mixed, herbaceous and woody ground flora. The plants show many adaptations against high insolation and fire: sclerophyllous foliage, xylopodia, thick bark etc. Usually around 1500 mm or more of rain falls each year, but there is a regular dry season during winter when fires sometimes sweep through. The soils are very deep and strongly leached of nutrients, often with a very low pH. Cacti are rare or quite absent, except in open areas on some deep sandy or gravelly substrates, or restricted to rock outcrops protected from the fire, such as those on limestone supporting dry forest ('*mata calcária*'). In Eastern Brazil *cerrado* vegetation, as described here, is usually found below 750 metres altitude. When occurring above this elevation it is generally included with the following type, although some localities at higher altitudes of certain specialized taxa are marked on Map 4.

Campo rupestre (Plates 1.3 & 1.4, Map 5). This is often described as a mosaic of different vegetation types, which reflects the intimate mixture of different topographies, substrates and microclimates, resulting from the juxtaposition of mountain ridges, grassy or marshy valleys, plateaux with ± open vegetation ('*gerais*'), bare rocks, sand and gravel deposits and sharp diurnal fluctuations in temperature, humidity and light, all combined with elements of the 3 preceding vegetation types (hence the terms '*mata de neblina*', '*caatinga de altitude*' and '*cerrado de altitude*'; also '*carrasco*' — sometimes used for the *caatinga/ campo rupestre/cerrado* ecotone). However, *campo rupestre* is most commonly and intricately associated with *cerrado*, so that the distinction between the two is often difficult to make and, therefore, in Area No. 4, below, the two are treated together. Rainfall is rather variable, but frequently well in excess of 1000 mm per annum and accompanied by heavy dews and mist. Maximum temperatures are lower than in the other vegetation types mentioned above and winter minima can descend to between 0°C and -4°C at the highest elevations (Nimer 1973: 46). Cacti are mainly restricted to the rock outcrops and areas of deep sands and gravels, from c. 750–1950 m, examples near the upper altitudinal limits being the Pico das Almas, BA (*Arrojadoa bahiensis*) and Pico do Itambé, MG (*Cipocereus minensis*), although vegetation of this nature can descend to c. 400 metres (eg. near Andaraí, Bahia: *Micranthocereus purpureus*). Such vegetation is commonest along the mountain backbone of Eastern Brazil, formed by the Serra do Espinhaço and Chapada Diamantina (Serra da Moeda, MG, northwards to Jacobina, BA), but also occurs in smaller pockets further to the east on more isolated ridges, such as at Rui Barbosa and Monte Santo (BA), and further south in the Serra da Mantiqueira (Serras do Lenheiro and Ibitipoca, MG).

3.3   ANALYSIS OF DISTRIBUTION PATTERNS

The phytogeographical Areas and their subdivisions elaborated below are largely defined by the distribution patterns of cacti indigenous to Eastern Brazil. The lines delimiting Areas 2–4 on Map 6 reflect the recorded native occurrences of the taxa described and documented in Chapter 5. Gaps in the authors' field knowledge, and where herbarium records are also lacking, are indicated on Map 6*, so that any extrapolated lines and other assumptions can be identified. No less than 95% of the taxa (including c. 9% thought to be 'single-site' endemics, ie. those known from only a single locality or group of closely adjacent localities) have distributions that show a strong correlation with the major vegetation types and topographic features, falling into one of the 3 major categories described below (Area Nos 2–4).

Leaving aside the first distribution category (No. 1, below), the other major categories recognised, ie. Area Nos 2, 3 & 4, are as delimited on Map 6, with minor exceptions, ie. where disjunct occurrences of one vegetation type are found inside the area of another or as rare disjunct records of particular taxa, these being noted in the discussion under each Area heading below. In terms of vegetation types, Area No. 2 broadly corresponds with the *Mata atlântica* (Atlantic Forest) and its constituent sub-types, such as *restinga* (at the coast) and *mata de brejo* (on higher ground well inland). Further west, Area No. 3 corresponds to the *caatinga* in its entirety, which also surrounds and includes the *campos rupestres/cerrados* of the northern part of the East Brazilian Highlands and disjunct islands

---

*By state, from north to south, areas deserving of further investigation in the field are as follows. Maranhão: northernmost (coastal dunes) and southernmost limits (towards Piauí); SW Piauí; Ceará (most of state except NW & S); northern Rio Grande do Norte; SE Pernambuco; N & W Alagoas (especially around Mata Grande); southernmost coastal Sergipe; Bahia: NE (Raso da Catarina), cent.–W (the entire region N & S of road BR 242, including land draining into the Rio São Francisco from its E side, to the borders with Minas Gerais, Goiás and Piauí) & SE inselbergs (BA/MG/ES border region); Minas Gerais: NW, NE (drained by Rios Jequitinhonha & Mucuri), E (Caratinga eastwards) & SW (towards 46°W); and Espírito Santo: NW (inselbergs) & S (at > 600 m for Rhipsalideae).

of dry forest on limestone outcrops beyond its vaguely defined southern limits. Area No. 4 comprises the western *cerrados* and the *campos rupestres* of the south-eastern part of the East Brazilian Highlands, which are bounded by the *Mata atlântica* (Area No. 2) on their eastern and southern slopes. At the borders between these adjacent Areas species assigned to one or other may sometimes be sympatric over a small distance, eg. upon gneissic rocks south of Padre Paraíso, Minas Gerais (Nos 2 & 3) and in the municípios of Grão Mogol and Bocaiúva, MG (Nos 3 & 4). Apart from these, the zones of overlap, sometimes ecotonal in character, are generally not very extensive or do not appear to be so today due to their widespread destruction, as is the case, for example, with the *agrestes*, which lie between the *caatinga* and *Mata atlântica*. However, in a few places such overlaps can be detected, eg. in the region south and east of Vitória da Conquista, Bahia, where there are occasional records of *caatinga-agreste* taxa (*Brasilicereus phaeacanthus, Pereskia bahiensis,* Maps 27, 17) to the east of sites of typical *Mata atlântica* species (*Lepismium cruciforme, Rhipsalis baccifera* subsp. *hileiabaiana,* Maps 21, 24). At their southern limits Areas 2–4 can no longer be distinguished geographically, differing only as ecological concepts (Map 6). This means that species referred to No. 2 or to No. 3 may sometimes grow in very close proximity to *campo rupestre* taxa from No. 4. From the above it should be noted that while each of the Areas 2, 3 and 4 broadly corresponds to a major vegetation type, namely the *Mata atlântica, caatinga* and *cerrado,* the *campos rupestres* are divided between Areas 3 & 4. This reflects the strong divergence in cactus genera and species between the northern and south-eastern parts of the East Brazilian Highlands, a feature already noted in relation to studies of *Eriope* by Harley (1988). These points are discussed under the four Areas, below.

As will shortly become clear, within each Area it is possible to define further patterns or distribution categories. These are summarized below with lists of their constituent taxa and they are also employed in short-hand form at the beginning of the statement describing the range and ecology of each taxon in the taxonomic inventory (Chapter 5). These patterns/subcategories are not all mutually exclusive, ie. they overlap, especially in the areas of highest diversity, eg. in the Southern *caatingas*: (3a:iii:c), (3b:iii), (3d:ii).

AREA 1. This is Eastern Brazil in its entirety, ie. for taxa that are widespread and/or non-specific, or disjunct between different vegetation/Area categories (9 taxa; 5.5% of total).

(1a). *Very widespread/non-specific in Areas 2–4* (2 taxa): **Hylocereus setaceus, Epiphyllum phyllanthus.**

(1b). *Ranging between Area No. 2 and 3 or 4, in restinga/Mata atlântica/mata de brejo/mata de neblina/campo rupestre/caatinga/agreste etc.* (7 taxa): **Brasiliopuntia brasiliensis, Rhipsalis lindbergiana, R. baccifera** subsp. **hileiabaiana, R. russellii, Hatiora salicornioides, Pilosocereus catingicola** subsp. **salvadorensis, Melocactus violaceus** subsp. **violaceus.**

Under this Area heading are listed the taxa which cannot be fitted into any of the more specific vegetation/area categories or patterns outlined below (Nos 2–4). Only one is endemic to Eastern Brazil, and the remainder includes a few taxa very widely distributed in the neotropics as well as those ranging into other parts of Brazil but not beyond. That most of these are epiphytes confirms the assessment made in Ibisch *et al.* (1996), in relation to Peru, that epiphyte taxa are generally more wide-ranging than terrestrial taxa (see also Area No. 2).

The two taxa in Category (1a) are wide-ranging neotropical species of broad ecological tolerance, being strongly xerophytic epiphytes and lithophytes (*Epiphyllum phyllanthus* can even be found in the cactus-poor *cerrado*). They belong to tribe Hylocereeae, which phylogenetic research based on gene sequences (Wallace 1995 & pers. comm.) has shown to be amongst the most basal branches in the Cactoideae or its HPE clade. It is also possible that these species themselves are rather old, as is suggested by their very extensive distributions in tropical America, which in the case of *Hylocereus setaceus* could represent the combined trans-Amazonian and Andean historical migration route hypothesized by Prado (1991).

In Category (1b) *Rhipsalis baccifera* subsp. *hileiabaiana* and *Hatiora salicornioides* are taxa mostly occurring in areas of relatively high rainfall in the *Mata atlântica/mata de brejo* of Area No. 2 and in *campos rupestres* and montane cloud forests of the East Brazilian Highlands further west (Chapada Diamantina, Bahia, Area No. 3). Strictly speaking, *H. salicornioides* occurs in Areas 2–4, but in Area No. 2 is restricted to its southern part, so hardly qualifies for inclusion in the previous Category, (1a). The disjunctions of such plants, restricted to habitats with very high humidity, is striking and is unlikely to be an artefact of poor collecting or habitat destruction. In Eastern Brazil *Rhipsalis russellii* is known from central-E/SE Bahia and two apparently disjunct collections from central/southern Minas Gerais (as an epiphyte) and southern Espírito Santo. From there it ranges into Paraná, Goiás and Mato Grosso, where its habitat details are unknown at present. *R. lindbergiana* has most of its distribution within Eastern Brazil, ranging from Pernambuco to the border region between eastern São Paulo and Rio de Janeiro. It is found in the *restinga/Mata atlântica* on the coast, in *mata de brejo* on the Serra Negra (Pernambuco), and more rarely upon rocks or in semi-humid forests within or surrounding the East Brazilian Highlands, as well as in patches of more humid forest within the *caatinga* area east of the Chapada Diamantina (Bahia) and in *mata seca* west of the Serra do Espinhaço, Minas Gerais.

Melocactus violaceus subsp. *violaceus* (Map 46) is commonly found at the coast (*restinga*) and up to 35 km inland, in open or low-shrubby vegetation on sands, at c. 0–150 metres altitude, from Rio de Janeiro to Rio Grande do Norte. However, it has also been collected in north-eastern Minas Gerais, in a very sandy phase of '*cerrado de altitude*' at 1100 metres elevation (Taylor 1991a: 56). This is an example of the links between the montane and coastal floras of Eastern Brazil, such as have been reported for other plant groups (Giulietti & Pirani 1988: 47, 53, 60; Harley 1988: 100).

Brasiliopuntia (Map 18) is a most unusual cactus, inhabiting semi-humid, high forest where it forms a tree with a well-developed trunk. It ranges from dry *restinga* forest at the coast into the *agrestes* and *caatinga* (especially along seasonal watercourses), from north-western Paraíba to NE Minas Gerais, and also occurs in *brejo* forest (central Pernambuco, western Alagoas and Bahia), reappearing in north-western Bahia (at Formosa do Rio Preto) and then on limestone outcrops amongst *mata seca* in central Minas Gerais (Area No. 3).

Pilosocereus catingicola subsp. *salvadorensis* appears to be the only cactus that is characteristic of both *restinga* and true *caatinga* (Map 35). While its range is primarily littoral, from Salvador (Bahia) northwards along the coast erratically to Rio Grande do Norte (and perhaps Ceará), it is also a significant component in *caatinga* and similar vegetation from the southern foot of the Serra Negra (PE) and the Raso da Catarina (BA) eastwards, through northern Sergipe and southern Alagoas. The clue to this unusual distribution pattern is the corridor provided by the rather dry climate characteristic of the lower parts of the Rio São Francisco's drainage, which links the inland *caatinga* with the coastal *restinga*.

AREA 2. Humid/subhumid, evergreen/semi-evergreen forest, including *Mata atlântica, restinga, mata de neblina, mata de brejo, capão de mato, agreste* in part etc. (Map 6, east of line C–D; summarized distribution in Map 3) (42 taxa; 26% of total).

(2a) *Widespread taxa and those from SE Brazil or beyond with disjunct occurrences in brejo forests of NE Brazil* (9 taxa): **Pereskia aculeata, P. grandifolia** subsp. **grandifolia, Lepismium cruciforme, Rhipsalis crispata, R. floccosa** subsp. **floccosa, R. paradoxa** subsp. **septentrionalis, R. cereuscula, Hatiora cylindrica, Cereus fernambucensis** subsp. **fernambucensis.**

(2b) *Ranging mainly to the north of the core area* (2 taxa): **Rhipsalis baccifera** subsp. **baccifera, Cereus insularis.**

(2c) *Restingas etc. between Alagoas and Salvador, Bahia*: **Melocactus violaceus** subsp. **margaritaceus.**

(2d) *Southern subhumid and perhumid forest (SE Bahia southwards)* (30 taxa):
(i) *Taxa characteristic of less humid, seasonally dry habitats, especially gneiss/granite inselbergs below 900 metres altitude* (8 taxa): **Pereskia grandifolia** subsp. **violacea, Cereus fernambucensis** subsp. **sericifer, Pilosocereus brasiliensis** subsp. **ruschianus, Coleocephalocereus** subg. **Coleocephalocereus** (**C. fluminensis** [2 subspecies], **C. buxbaumianus** [2 subspp.], **C. pluricostatus**).
(ii) *Taxa from more humid habitats* (22 taxa): **Opuntia monacantha, Lepismium houlletianum, L. warmingianum, Rhipsalis elliptica, R. oblonga, R. floccosa** subsp. **pulvinigera, R. pacheco-leonis** subsp. **catenulata, R. cereoides, R. sulcata, R. teres, R. pulchra, R. burchellii, R. juengeri, R. clavata, R. pilocarpa, R. hoelleri, Schlumbergera kautskyi, S. microsphaerica, S. opuntioides, Cereus hildmannianus, Pilosocereus arrabidae, P. brasiliensis** subsp. **brasiliensis.**

As might be expected, the high proportion of epiphytic taxa in this category means that there are fewer endemics in this part of the core area of Eastern Brazil in comparison with Areas 3 & 4 (*caatingas/campos rupestres*). Nevertheless, in Eastern Brazil this vegetation type is characterised by some distinctive and widespread taxa (eg. the endemic *Rhipsalis paradoxa* subsp. *septentrionalis*, Map 23) and there are some interesting disjunctions of non-endemic taxa. In its northern parts, north of the Rio Paraguaçu (BA), the *Mata atlântica* becomes a generally rather narrow coastal band, of which the more humid phases sometimes reappear further inland on higher ground as isolated stands of *mata de brejo*. These are located at or near the limit with Area No. 3 (*caatingas*), or actually within it, west of the line on Map 6, but are assigned to Area No. 2.

It should also be re-emphasized that Area No. 2 includes an extensive zone receiving less than 1000–1250 mm of rain/year (see map 'isoietas anuais 1914–1938' in Azevedo 1972; Nimer 1973: 40, fig. 18), located between eastern Minas Gerais, central & western Espírito Santo (Rio Doce drainage) and northern Rio de Janeiro, at less than 900 metres altitude. Although this south-east region includes all 5 representatives of *Coleocephalocereus* subg. *Coleocephalocereus*, whose other two subgenera are restricted to the *caatingas* (Area No. 3), its referal to the *Mata atlântica* is supported, in terms of Cactaceae at least, by the widespread presence of *Pereskia aculeata* (Map 2), a species characteristic of this vegetation and its ecotone with the *caatinga* (the *agreste*), but one rarely found in the *caatinga* proper. This SE area also lacks any of the taxa typical of Area No. 3, such as *Tacinga, Pseudoacanthocereus, Cereus jamacaru* etc., and instead is characterized by vicariant subspecies of widespread *Mata atlântica* species, such as *Pereskia grandifolia* subsp. *violacea* and *Cereus fernambucensis* subsp. *sericifer*. However, its lower rainfall appears to interrupt the ranges of various species characteristic of more humid phases of the *Mata atlântica* (eg. *Lepismium cruciforme, Rhipsalis floccosa, R. oblonga, R. russellii*) and separates closely related taxa such as *R. baccifera* subsp. *hileiabaiana* and *R. teres* (Map 24). An area of high diversity and endemism of Rhipsalideae is thus restricted to much wetter habitats in southern Espírito Santo (ie. Serra do Caparaó, Domingos Martins, Santa Teresa), which is effectively isolated from the stronghold of this group further southwest, in the mountains of Rio de Janeiro etc. Such habitat disjunction can explain the morphological divergence of the northern and southern taxa in the widespread species, *Rhipsalis paradoxa*, the southernmost record for subsp. *septentrionalis* (at Domingos Martins, ES) being separated by some 250 km from the nearest of subsp. *paradoxa* (Macaé de Cima, Rio de Janeiro).

Lepismium cruciforme, Rhipsalis crispata and *R. cereuscula* (Category 2a) are Rhipsalideae whose range in NE Brazil is mainly restricted to montane *brejo* forest and *agreste* (Maps 21, 22 & 25), where all three reach their north-eastern limits of distribution in Pernambuco. They are mostly of more frequent occurrence in SE Brazil and two range much further in a south-westerly direction, *R. cereuscula* attaining Argentina and central Bolivia (La Paz). Of these, *R. crispata* is an interesting example of disjunction between subhumid *restinga* of eastern Rio de Janeiro (from near the state capital to Cabo Frio), forests of the *planalto* of inner São Paulo and *brejo* forest rising out of the *caatinga-agreste* in eastern Bahia and Pernambuco (Caruaru). It is also recorded from Santa Catarina in Southern Brazil.

As noted already, in terms of Cactaceae, *Rhipsalis baccifera* subsp. *baccifera* (Category 2b) potentially marks the limits of Amazonian floristic influence, or past expansions of its flora (cf. Oliveira-Filho & Ratter 1995: 144). Significantly, it is the only member of the large tribe Rhipsalideae recorded from the northern third of Eastern Brazil. *Cereus insularis* is an endemic of the Atlantic archipelago of Fernando de Noronha and seems taxonomically close to *C. fernambucensis* subsp. *fernambucensis* from (2a).

Melocactus violaceus subsp. *margaritaceus,* Category (2c), is the cactus representative amongst other endemics known from the extensive coastal sand-dunes that begin north of Salvador, Bahia (cf. Harley 1988: 100–101, Taylor 1991a). It also occurs in *restinga*-like vegetation at up to 400 metres elevation on the coastal slope of the Serra da Itabaiana, Sergipe (Map 46).

In Category (2d:i) the 5 taxa of *Coleocephalocereus* subg. *Coleocephalocereus* and *Cereus fernambucensis* subsp. *sericifer* have extensive distributions in the drainage basins of the Rios Mucuri (MG/BA), Doce (MG/ES) and Paraíba do Sul (MG/RJ), with *Co. buxbaumianus* subsp. *flavisetus* reaching westwards to the watershed between the Rio Grande and Rio São Francisco (MG): Maps 29 & 41. These six, plus the more northerly-ranging *Pilosocereus brasiliensis* subsp. *ruschianus* (Map 36), are characteristic inhabitants of this region's abundant, smooth, gneiss/granite inselbergs, which do not retain much water or permit the accumulation of soil and are otherwise home mainly to certain bromeliads and specialized bulbs, besides annual herbs and more wide-ranging cacti (eg. *Pereskia aculeata, Hylocereus setaceus*). *Pereskia grandifolia* subsp. *violacea* is an endemic plant with a similar distribution area, but not always associated with inselbergs and occasionally recorded above 900 metres altitude at the western limits of the Area (Map 17). It may be specifically distinct and represents the basal element of the P. GRANDIFOLIA Group within Eastern Brazil (Wallace 1995), from which the *caatinga* taxa belonging to this Group have been derived.

Of those included in Category (2d:ii), 10 out of the total of 18 Rhipsalideae are taxa with the major part of their range outside of Eastern Brazil. Only 3 in this category are endemic and appear to have very restricted ranges, in southern Espírito Santo. This tribe has a considerable number of species endemic to the region comprising E São Paulo, SE Minas Gerais and Rio de Janeiro, where its centre of diversity is located. The non-endemic *Rhipsalis oblonga* (Map 22) appears to be restricted to the 'Hiléia Baiana' or Bahian phase of the Atlantic forest, but this could reflect lack of collecting, eg. in Espírito Santo (where there is one doubtful record from Luetzelburg). Its occurrence seems to be linked to regions of very high rainfall (ie. 1750 mm/year or more). It reappears to the west of the city of Rio de Janeiro, reaching coastal São Paulo (Ilha São Sebastião) and represents an ecological vicariant of *R. crispata,* see (2a). *Pilosocereus arrabidae* and *P. brasiliensis* subsp. *brasiliensis* (Maps 35 & 36) are *restinga* taxa almost endemic to the area, reaching southwards to the environs of Rio de Janeiro. *Opuntia monacantha*, a widely distributed species from SE South America and at its northern limit in Eastern Brazil, has a mainly littoral range in SE & S Brazil (from S Espírito Santo south-westwards), but has also been recorded from dunes near Diamantina and Pedra Menina (Mun. Rio Vermelho), central-eastern Minas Gerais, at c. 1000 metres altitude (Map 18). Its occurrence in the latter is comparable with records from inland sites in São Paulo state (eg. near Piracicaba, and historically at Lorena and, even, Congonhas — nowadays engulfed by the city of São Paulo).

The most diverse tribe of Cactaceae in this major area are the epiphytic Rhipsalideae, but all of these appear to have a southern or south-western origin. Thus, the Cactaceae of the *Mata atlântica* can be said to show little direct floristic affinity with those of the Amazonian rainforest.

AREA 3. *Caatingas* & Northern *campos rupestres* (+ included *cerrados*)/*caatinga-agreste*/*mata seca* (Map 6, north of line A–D; summarized distribution of *caatinga* elements in Map 3) (83 taxa; c. 51% of total).

(3a) *Widespread from east to west etc.* (23 taxa):

   (i) *Widespread east to west and north to south* (5 taxa): **Tacinga inamoena, Cereus albicaulis, C. jamacaru** subsp. **jamacaru, Pilosocereus gounellei** subsp. **gounellei, Melocactus zehntneri.**

   (ii) *Northern caatingas (S Piauí, N Bahia, Alagoas & Pernambuco northwards)* (5 taxa): **Pilosocereus flavipulvinatus, P. pachycladus** subsp. **pernambucoensis, P. piauhyensis, P. chrysostele, Discocactus bahiensis.**

   (iii) *Central-Southern caatingas, from S Piauí, S Ceará, S Paraíba southwards* (13 taxa):

      (a) *Wide-ranging in the central-southern caatingas* (7 taxa): **Tacinga funalis, Arrojadoa penicillata, A. rhodantha, Pilosocereus tuberculatus, P. pachycladus** subsp. **pachycladus, Harrisia adscendens, Leocereus bahiensis**.

      (b) *Bahian caatingas, including and surrounding the Chapada Diamantina* (2 taxa): **Pereskia bahiensis, Stephanocereus leucostele.**

      (c) *Southern caatingas (Cent./S Bahia & N Minas Gerais)* (4 taxa): **Pereskia aureiflora, Brasilicereus phaeacanthus, Coleocephalocereus goebelianus, Melocactus ernestii** subsp. **longicarpus.**

(3b) *Caatingas (and mata seca on southern limestone outcrops) in the middle and upper drainage of the Rio São Francisco (cent. Minas Gerais to W & cent.-N Bahia/SW Pernambuco) and adjacent SE Piauí* (20 taxa):

   (i) *Widespread in the caatingas of the middle part of the Rio São Francisco drainage* (2 taxa): **Pilosocereus gounellei** subsp. **zehntneri, Facheiroa squamosa.**

   (ii) *Caatingas of central-northern Bahia:* **Melocactus azureus, M. ferreophilus, M. pachyacanthus** (2 subspp.), **Facheiroa ulei, Discocactus zehntneri** subsp. **zehntneri.**

   (iii) *Southern caatingas and other seasonally dry forests (on islands of limestone) of the Rio São Francisco/Rio das Velhas drainage (western & central-southern Bahia to central Minas Gerais)* (12 taxa): **Pereskia stenantha, Quiabentia zehntneri, Tacinga saxatilis** (2 subspp.), **Cereus jamacaru** subsp. **calcirupicola, Pilosocereus floccosus** subsp. **floccosus, P. densiareolatus, Micranthocereus dolichospermaticus, Melocactus deinacanthus, M. levitestatus, Facheiroa cephaliomelana** (2 subspp.).

(3c) *Northern campos rupestres etc., East Brazilian Highlands (Chapada Diamantina and northern part of Serra do Espinhaço, BA/MG)* (17 taxa):

   (i) *Chapada Diamantina & Northern Serra do Espinhaço (BA/MG)* (3 taxa): **Rhipsalis floccosa** subsp. **oreophila, Arrojadoa dinae** subsp. **eriocaulis, Melocactus paucispinus.**

   (ii) *Chapada Diamantina (BA)* (9 taxa): **Stephanocereus luetzelburgii, Arrojadoa bahiensis, Pilosocereus glaucochrous, Micranthocereus purpureus, M. flaviflorus, M. streckeri, Melocactus oreas** subsp. **cremnophilus, M. glaucescens, Discocactus zehntneri** subsp. **boomianus.**

(iii) *Northern Serra do Espinhaço (Bocaiúva MG – Caetité BA)* (5 taxa): **Arrojadoa dinae** subsp. **dinae, Cipocereus pusilliflorus, Micranthocereus violaciflorus, M. albicephalus, M. polyanthus.**

(3d) *Eastern caatingas-agrestes/campos rupestres (Minas Gerais & Bahia, from the Serra do Espinhaço & Chapada Diamantina, eastwards & north-eastwards to Pernambuco and sometimes beyond)* (23 taxa):

(i:a) *Widespread eastern taxa and those restricted to E/NE Bahia and Paraíba/Pernambuco* (10 taxa): **Tacinga palmadora, Pseudoacanthocereus brasiliensis, Pilosocereus catingicola** subsp. **catingicola, P. pentaedrophorus** subsp. **pentaedrophorus, Melocactus ernestii** subsp. **ernestii, M. bahiensis** subsp. **bahiensis, M. lanssensianus, M. concinnus, M. violaceus** subsp. **ritteri, Espostoopsis dybowskii.**

(i:b) *Restricted to the E & SE caatingas-agrestes/and associated campos rupestres from the Rio Paraguaçu drainage (BA) southwards* (6 taxa): **Tacinga werneri, Pilosocereus pentaedrophorus** subsp. **robustus, Melocactus oreas** subsp. **oreas, M. bahiensis** subsp. **amethystinus, M. conoideus, M. salvadorensis.**

(ii) *SE caatingas-agrestes of NE Minas Gerais (Rio Jequitinhonha drainage and watersheds with Rio Pardo and Rio Mucuri)* (7 taxa): **Tacinga braunii, Pilosocereus azulensis, P. floccosus** subsp. **quadricostatus, P. magnificus, P. multicostatus, Coleocephalocereus** subg. **Buiningia** (**C. aureus** & **C. purpureus**).

This major area category includes all that considered here as part of the *caatingas* dominion in Eastern Brazil (cf. Andrade-Lima 1981), plus the following:– those areas referred to *caatinga* to the west of the Rio São Francisco (Andrade-Lima 1975); the areas of *campo rupestre* and *cerrado* from the northern part of the East Brazilian Highlands (ie. the northern parts of the Serra do Espinhaço and Chapada Diamantina, and their extensions eastwards); ecotones at the southern limits of the *caatinga* and *mata seca* in northern Minas Gerais; and part of the *agreste* or ecotonal vegetation between the *caatinga* and *Mata atlântica* in the east. (Much of the *agreste* has now been so severely altered or destroyed that its cactus flora is mainly to be inferred from a few herbarium records and scrappy extant remnants of what was once an extensive vegetation zone.) In addition, to the west and south of the *caatinga* proper, in western Bahia, south-western Piauí and central Minas Gerais (Rio São Francisco drainage), a few taxa characteristic of this area occasionally appear in disjunct dry habitats, eg. *Melocactus zehntneri, Leocereus bahiensis, Pilosocereus flavipulvinatus, P. floccosus* and *Cereus jamacaru* subsp. *calcirupicola*. These are mainly rock outcrops located inside Area No. 4, but represent disjunct extensions from Area No. 3. The same applies to occurrences of *Cereus jamacaru* and *Pilosocereus pentaedrophorus* in the *Mata atlântica* to the east (Area No. 2). A cactus-based approximation to the *caatingas* area is offered in Map 3.

Area No. 3, as delimited here, makes good sense on the basis of endemism of Cactaceae, as the following list hopefully makes clear. At generic level the endemics are: *Facheiroa* (3 spp.), *Stephanocereus* (2 spp.), *Espostoopsis* (1 sp.) and *Leocereus* (1 sp.), with *Arrojadoa* (4–5 spp.) almost endemic. There are also significant infrageneric taxa in *Pilosocereus, Micranthocereus, Coleocephalocereus* and *Melocactus. Facheiroa* and *Espostoopsis* are typical *caatinga* elements. However, the diverse species of *Arrojadoa* and *Stephanocereus* are equally divided between the *caatingas* and Northern *campos rupestres* and this pair of genera is believed to have common ancestry with *Cipocereus*, a more southerly ranging, endemic genus from Minas Gerais, which is associated with *campo rupestre* and appears to be relictual (Taylor & Zappi 1996). This suggests an autochthonous E Brazilian source for some of the most characteristic elements of the *caatinga*'s cactus flora, such as *A. penicillata, A. rhodantha* and *S. leucostele*, and this may also be assumed to be the case with the majority of species in the largest *caatinga* genera, *Pilosocereus* and *Melocactus* (Zappi 1994, Taylor 1991a). *Micranthocereus* subg. *Austrocephalocereus* is a Northern *campo rupestre* endemic.

The southernmost records of cactus species that can be considered as typical elements of the *Caatinga*/Northern *campo rupestre* flora are, west to east:– *Pilosocereus pachycladus* (near Francisco Dumont, MG), *Melocactus concinnus* (Peixe Cru, Mun. Turmalina, MG — Map 46), *M. bahiensis* (Mercês, Diamantina, MG — Map 43) and *Brasilicereus phaeacanthus* (near Padre Paraíso, MG — Map 27). However, this is not intended to suggest that any of these species are actually found in *caatinga* vegetation at these southern sites. The southern limit of *caatinga* vegetation proper is uncertain and depends on one's definition, but, in the Rio São Francisco valley, deciduous thorn forest including cacti, such as *Pereskia stenantha* (Map 17), *Pilosocereus pachycladus* and *Arrojadoa rhodantha* (Map 33), extends southwards at least as far as the municípios of Varzelândia, Janaúba and Porteirinha, MG (c. 15°40'S). South of there cacti interpreted as belonging to the *caatingas* are found only on edaphically dry, exposed rock outcrops, mostly of Precambrian Bambuí limestone and gneiss/granite, and are represented by taxa from categories (3a) & (3b). The southernmost occurrences on such outcrops are those of *Tacinga saxatilis, Melocactus levitestatus, Cereus jamacaru* subsp. *calcirupicola* and *Pilosocereus floccosus* subsp. *floccosus*, the latter two (Maps 29 & 36) inhabiting *mata seca* as far south as the region of Lagoa Santa (c. 19°40'S). *Arrojadoa dinae*, from the Northern *campo rupestre* flora, has its southern limit at c. 17°30'S. Further east, crossing the Serra do Espinhaço, the south–easternmost *caatingas* are isolated in the middle section of the Rio Jequitinhonha valley between Mun. Jacinto (16°10'S) and Mun. Araçuaí (16°50'S), as determined by the ranges of species representative of Categories (3a) and (3d:ii). The Cactaceae characteristic of *campo rupestre* vegetation and endemic to Area No. 3 are about 45% fewer than those endemic to the South-eastern *campos rupestres* (Area No. 4c below).

In Category (3a) all the taxa listed are primarily *caatinga* elements. They represent less than a third (28%) of the taxa included in Area 3, the remainder falling into three parallel geographical and/or ecological categories: (3b), (3c), (3d).

The widespread taxa included in (3a:i) are amongst the cacti that are most characteristic of the *caatinga* and *agreste* of North-eastern Brazil, with *Tacinga inamoena* also ascending into the *campos rupestres*. *Cereus jamacaru* ranges further than any of the others included here, reaching western Maranhão, entering drier phases of the Atlantic forest in NE Brazil and penetrating the area of the SE *campos rupestres* as far as the region of Diamantina, MG. The distribution of the endemic and much less frequent, *C. albicaulis*, is linked to sandy habitats. It occurs mainly in *caatinga* associated with the Serra do Espinhaço and Chapada Diamantina in the southern half of its range, but expands northwards into SE Piauí, W & NE Bahia, S Pernambuco etc., finally reaching the Serra da Ibiapaba (N Piauí & NW Ceará). Its sister species is *C. mirabella*, with a complementary range west and south (Map 28), and is one of the few cacti restricted to the *cerrado*, being a rarity limited to eastern parts of this biome.

The five taxa characteristic of the Northern *caatingas* (Category 3a:ii) are amongst a small minority with a significant part of their ranges north of the core area of Eastern Brazil (ie. north of 7°S). This area corresponds to some degree with the eco-region ['ecorregião'] termed as 'Depressão Sertaneja Setentrional' in Velloso *et al.* (2002). The most characteristic species of this area, *Pilosocereus chrysostele* and *P. piauhyensis,* are relatives of the southern endemic, *P. multicostatus,* see (3d:ii) below. *P. flavipulvinatus,* which reaches beyond the limit of the *caatingas* as far as Carolina (W Maranhão), is the sister taxon of the *P. pentaedrophorus/P. glaucochrous* species pair from (3c:ii) & (3d:i). A legume whose range corresponds to Category (3a:ii) is *Caesalpinia gardneriana* Benth. (Lewis 1998: 129).

The Central-Southern *caatingas* Category, (3a:iii:a), corresponds with the eco-region termed as 'Depressão Sertaneja Meridional' (Velloso *et al.,* l.c.). Excepting *Leocereus bahiensis* and *Pilosocereus pachycladus* subsp. *pachycladus,* which have more western distributions beginning in the Chapada Diamantina (Maps 47 & 37), included taxa are restricted to the *caatinga* proper and, together with those listed for Category (3a:i/ii), are the most important cactus species of the *caatingas* generally. Although falling reasonably within the widespread category, in detail the distribution of *Leocereus* is unlike that of any other and is hard to characterize in terms of its prefered vegetation type, being a plant of ecotones between mostly higher altitude *caatinga, cerrado* and the margins of *campo rupestre,* but is generally associated with arenitic rocks and sands. *Pilosocereus tuberculatus* is perhaps an example of an erratically distributed relict species (see Chapter 5), whose range (Map 34) fits neatly inside Köppen's semi-arid climatic zone (Cavalcanti Bernardes 1951). Like *Cereus albicaulis,* it is usually found on dunes or light sandy substrates (referred to the 'Cipó' soil series) and both species appear to be characteristic of a distinct type of *caatinga* identified as 'No. 5' by Andrade-Lima (1981: 159). A legume with a distribution pattern not dissimilar to that of *Pilosocereus tuberculatus* is *Caesalpinia microphylla* G. Don, while the more widespread *C. laxiflora* Tul. matches that of *Tacinga funalis* (Lewis 1998: 140, 155).

Both species categorized as (3a:iii:b) are widespread and typical elements of the Bahian *caatingas* located within and around all sides of the East Brazilian Highlands, including the Chapada de Maracás. On present knowledge, they do not range southwards into Minas Gerais or, apparently, northwards into Pernambuco or Piauí. They also avoid the drier NW and NE parts of Bahia.

The four species from the Southern *caatingas,* (3a:iii:c), are typical elements on, or associated with, gneiss/granite outcrops (inselbergs) or derived substrates in the southernmost *caatingas,* outside of the semi-arid climatic zone as defined by Köppen (Cavalcanti Bernardes 1951). *Coleocephalocereus goebelianus* and *Melocactus ernestii* subsp. *longicarpus* are absent from NE Minas Gerais (east of the Serra do Espinhaço), where they are replaced by sister taxa, see (3d:ii/iii). *Brasilicereus phaeacanthus* has its much rarer sister species in the South-eastern *campos rupestres* (4c:ii:a).

All bar one of the taxa listed for the Rio São Francisco drainage, (3b), are endemic to this region (11 restricted to Bahia), giving it a very characteristic cactus flora and suggesting that it may have been a refugium for such drought-adapted plants during past periods of greater humidity and/or cooler conditions. It should be noted that all 3 species of *Facheiroa* are endemic here, with one in each of the subdivisions (3b:i–iii), *F. squamosa* straying slightly outside of the Rio São Francisco drainage basin in southern Piauí and central Bahia (Mun. Rio de Contas). *Pereskia stenantha* and *Pilosocereus gounellei* subsp. *zehntneri* replace their respective allies, *Pe. bahiensis* and *Pil. gounellei* subsp. *gounellei*, which have extensive ranges to the north and east, Category (3a). Both these endemics are relatively widespread, but many of the others are restricted either to central-northern or S Bahia (south or east of the curving course of the Rio São Francisco), or to western Bahia and northern Minas Gerais on both sides of the São Francisco on *Bambuí* limestone, eg. *Pilosocereus densiareolatus* and *Melocactus levitestatus* (cf. Andrade-Lima 1977). Most of the endemics are found only on one of various rock types, eg. the widespread limestone (numerous taxa, including two endemic species-groups in *Melocactus*) or gneiss/granite (*Melocactus deinacanthus*, a taxonomically isolated species), but some, eg. *Quiabentia zehntneri* and *Pilosocereus gounellei* subsp. *zehntneri*, are less specific. West of the river on soils derived from limestone the cactus flora is rather different, since most of the widespread *caatinga* species (from 3a:i/ii/iii) are lacking, the chief exceptions being *Cereus jamacaru*, *Pilosocereus pachycladus* and *Arrojadoa rhodantha*, none of which is common. A legume whose distribution conforms to (3b) is *Caesalpinia pluviosa* DC. var. *sanfranciscana* G. P. Lewis (1998: 150).

Within (3b) 3 subcategories are recognized, (ii/iii) conforming to the limits between Köppen's semi-arid and hot/humid climates (Cavalcanti Bernardes 1951). The range of both widespread taxa (3b:i) is somewhat disjunct between their northern and southern occurrences (Maps 34 & 48), which probably in part reflects a lack of suitable habitat, there being extensive intervening areas of *cerrado* and marshy sand-dunes (Tricart 1985: 209–211). *Facheiroa squamosa* (3b:i) is restricted to crystalline rocks or sandstones, whereas its south-western congener, *F. cephaliomelana* (3b:iii), occurs only on limestone, and a third, more poorly understood species, *F. ulei*, is found on non-calcareous rocks in central-northern Bahia (3b:ii). The *Melocactus* taxa included in (3b:ii) represent the M. AZUREUS Group, an endemic of this area, and are restricted to outcrops of limestone in valleys draining northwards (Rios Verde, Jacaré & Salitre) and dissecting the Chapada Diamantina area (cf. Category 3c), but they clearly belong to (3b), being absent from limestone further to the east. These and other taxa included here are threatened with extinction due to extensive modification of their habitat by agriculture and formerly by the great Represa de Sobradinho dam-lake.

Most of the species included in (3b:iii) are rock-dwellers. The two *Melocactus* each represent a monotypic species-group endemic to this area — *M. deinacanthus* being restricted to a few outcrops of gneiss, *M. levitestatus* to *Bambuí* limestone, but more widespread. The most widespread taxon is *Cereus jamacaru* subsp. *calcirupicola*, which is found in the middle and upper drainage of the Rio São Francisco, in both the south-western *caatingas* and on comparable limestone islands, in *cerrado*, further south, in central

Minas Gerais (Map 29). It occasionally occurs on substrates other than limestone, straying into the South-eastern *campos rupestres,* in the region of the Serra do Cabral. In the southern part of its range it is often found in association with *Pilosocereus floccosus* subsp. *floccosus* (Map 36), which occurs in proximity to *campo rupestre* vegetation (Zappi 1989), but is actually a plant of *mata seca.*

In relation to Category (3c) it should be noted that the term Chapada Diamantina as employed here extends this mountain area somewhat farther north than the definition implied in Bandeira (1995), including parts of the municípios of Sento Sé, Umburanas, Campo Formoso and Jacobina, whose highlands are ± continuous with those to the south. All of the taxa included are endemic to Eastern Brazil and none has a close link with an extra-Brazilian cactus flora. Represented are some morphologically rather unusual cacti, such as *Stephanocereus luetzelburgii, Arrojadoa dinae* and *Melocactus paucispinus*, which have evolved specialized habit forms in keeping with their environment and/or pollinators. Despite the availability of many suitable habitats, it is curious that the genus *Discocactus* is absent from all but the northernmost part of this area (having the strongest *caatinga* influence), which is in stark contrast to the ecologically similar Category (4c) below, where the genus is well-represented. The taxa included here display the full spectrum from widespread within either one of the two subdivisions recognized (eg. *Stephanocereus luetzelburgii, Micranthocereus purpureus, Arrojadoa dinae* subsp. *dinae*), to those known from single localities (eg. *Micranthocereus streckeri, Cipocereus pusilliflorus*), but only 3 are known from both subdivisions (3c:i) and each of these is infrequent or disjunct. Notable species are *Pilosocereus glaucochrous*, restricted to dense '*caatinga/cerrado de altitude*', *Rhipsalis floccosa* subsp. *oreophila*, found on rocks and as an epiphyte in pockets of perhumid cloud forest (*mata de neblina*), and *Melocactus glaucescens* and *Discocactus zehntneri* subsp. *boomianus*, plants of northern habitats associated with sparse *caatinga* on sand and gravel. The remainder comprise typical *campo rupestre* elements (including '*cerrado de altitude*'). *Arrojadoa dinae* has a recently discovered, probable sister-species in NE Goiás (Map 32).

As can be seen from the above list for Category (3c:ii), the Chapada Diamantina has a significant number of endemic Cactaceae, although it should be noted that most of these are restricted to its eastern segment. They are nearly all allopatric or ecological-vicariant sister taxa of species from the adjacent *caatingas* at lower elevations, such as *Pilosocereus pentaedrophorus* and *Melocactus oreas* subsp. *oreas* (3d:i:a/b), or from the *campos rupestres* further south (Serra do Espinhaço, S Bahia & Minas Gerais). The southern portion of this mountain region, (3c:iii), is separated from the broader ranges to its north and south by areas of only moderate elevation and is itself generally lower, and as a consequence probably drier, being surrounded by *caatinga* on all sides. The flora exists as a complex patchwork of small areas of *campo rupestre* and *cerrado*, and their ecotones with *caatinga*, the latter ascending high up the west-facing slopes. Of the five taxa listed here *Arrojadoa dinae* is the most widespread and ecologically most tolerant, being found on very sandy *cerrado* as well as on more stony *campo rupestre* and *caatinga* ecotone substrates (Map 32). *Cipocereus pusilliflorus* is the only member of its genus found outside the South-eastern *campos rupestres* of Area category (4c) and is related to *C. minensis* and *C. bradei* from that area (Map 30).

While the distribution patterns and phytogeographical affinities of *campo rupestre* cacti are closely matched by those described for *Eriope* in Harley (1988), there is one key difference that concerns Category (3c:iii). Harley (l.c. 83, 93), on the basis of one of his species, *E. salviifolia* (Benth.) Harley (l.c. 87), would include the highlands between Caetité (BA), Vitória da Conquista (BA) and Grão Mogol (MG) within the South-eastern *campos rupestres* of Minas Gerais, whereas on many cactus distributions and even generic affinities they are considered to be part of the Northern highlands here (eg. see *Melocactus concinnus*, Map 46). Nevertheless, it is clear from both cactus and *Eriope* distributions that the Chapada Diamantina (BA) and South-eastern *campos rupestres* (4c, below) are discrete entities, the latter having their northern limit near Grão Mogol.

All except two of the taxa listed for Category (3d) are currently thought to be endemic to Eastern Brazil and comparison with Category (3b) suggests that the East Brazilian Highlands have been an important barrier isolating the cactus flora of the Rio São Francisco drainage from that further east. Thus, east of the Chapada Diamantina we have *Pseudoacanthocereus* and *Espostoopsis*, while the genera *Facheiroa* and *Discocactus*, characteristic of the Rio São Francisco valley, are lacking. Some of the species in (3d) are associated with the now much depleted part of the *caatinga* biome that forms the *Mata atlântica* ecotone (the *agreste*), especially *Pseudoacanthocereus brasiliensis* and *Pilosocereus pentaedrophorus*, while others, also found partly associated with such transitional vegetation, are characteristic of the granitic/gneissic outcrops found in many parts of this region, eg. *Tacinga werneri* and *Melocactus ernestii*. Four of the taxa included in (3d:i) — *Tacinga palmadora*, *Pilosocereus pentaedrophorus* subsp. *pentaedrophorus*, *Melocactus ernestii* and *M. bahiensis* subsp. *bahiensis* — have extensive distributions, extending northwards as far as Pernambuco or beyond. In contrast, *Melocactus violaceus* subsp. *ritteri* is found on isolated areas of quartz sand/gravel on the eastern slope of the Chapada Diamantina at Jacobina and further south at Rui Barbosa (BA). *M. lanssensianus* is a local, cleistogamous endemic of uncertain status from the region of Garanhuns (PE), which can be associated with other, very similar plants from elsewhere in that state and from neighbouring Paraíba. The distribution of the Bahian endemic, *Espostoopsis,* is markedly disjunct (2 areas c. 400 km apart) and is presumably indicative of its relict status (Map 48). *Pilosocereus pentaedrophorus* ranges only within Köppen's hot/humid zone, avoiding the semi-arid climate (cf. Cavalcanti Bernardes 1951). It has recently been reported from the coastal *restinga* vegetation in NE Bahia (N of Salvador) and also occurs in *Mata atlântica* clothing the eastern slopes of the Chapada Diamantina (Map 36). A non-cactus with an eastern distribution matching Category (3d) is *Caesalpinia pyramidalis* Tul. var. *pyramidalis* (Lewis 1998: 129).

Tacinga werneri, Pilosocereus pentaedrophorus subsp. *robustus* and *Melocactus bahiensis* subsp. *amethystinus,* (3d:i:b), are southern relatives of taxa in Category (3d:i:a). *M. salvadorensis* replaces the more widespread *M. zehntneri* (3a:i) in southern and eastern Bahia, except for a small area to the east of Brumado where they are sympatric (Map 45).

The cactus flora of the relatively small region categorized as (3d:ii) is extremely interesting, not only because of the variety of unusual cactus endemics it has, but also for the absence (or only marginal presence) of certain widespread *caatinga-agreste* Cactaceae, while other 'indicator' species, with which they are normally associated, are present and common, especially those from Categories (3a) & (3d:i:a/b). In many cases these absent or marginally present cacti are replaced by vicariant sister taxa or species from the same infrageneric group, eg. *Tacinga funalis* by *T. braunii*, *Pilosocereus pachycladus* by *P. magnificus*, *P. catingicola* by *P. azulensis*, and *Coleocephalocereus* subg. *Simplex (C. goebelianus)* by *C.* subg. *Buiningia*. An instance of a more disjunct distribution pattern is that of the species group to which *Pilosocereus multicostatus* belongs (Map 39), its nearest relatives being *P. piauhyensis* and *P. chrysostele* from the distant Northern *caatingas*. While there can be no doubt that parts of the middle Rio Jequitinhonha valley have typical *caatinga* vegetation, the aforementioned species composition and endemism of Cactaceae also suggest that this region has been somewhat isolated from the main *caatingas* area further north and west during a substantial period in its history. However, there is a small zone of contact between the two *caatinga* areas, near the borders of the Municípios of Taiobeiras and Águas Vermelhas (see the 'Biomas' map for Minas Gerais in Costa *et al*. 1998: 21), and, interestingly, it is here that the otherwise allopatric *Pilosocereus pentaedrophorus* (3d:i:a/b) and its relative *P. floccosus* subsp. *quadricostatus* (Map 36) occur together. The Rio Jequitinhonha valley deserves to be a key area for the attention of conservationists.

In summary, the ranges of most (60 out of 83, ie. c. 72%) of the cactus taxa from Area No. 3 can be characterized in terms of three topographical/ecological subdivisions, running in parallel from SSW to NNE, and broadly corresponding to the major river and mountain systems. These are: (3b) the middle and upper parts of the Rio São Francisco drainage, with 20 taxa, the majority *caatinga* elements; (3c) the *campos rupestres* etc. on primarily crystalline rocks and sandstones of the East Brazilian Highlands (Chapada Diamantina & northern Serra do Espinhaço), with 17 taxa; and (3d) the complex of *caatingas-agrestes, campos rupestres* etc. for taxa ranging eastwards and north-eastwards from within or to the east of the latter mountain system, with 23 taxa (Table 3.2).

The southern part of the *Caatingas*/Northern *campos rupestres* area is particularly rich in species and there are instances of high levels of sympatry. For example, climbing eastwards into the Serra Geral east of Monte Azul (MG) it is possible, including epiphytes, to find at least 16 cactus species over a distance of less than 2 kilometres, and in southern Bahia the *caatinga* may often have 10 or more sympatric cactus species. However, from the limits of Bahia northwards the diversity of Cactaceae diminishes rapidly and it is significant that none of the 3 *caatinga* species of the widespread neotropical genus *Pereskia* appears to range outside of Bahia and northern Minas Gerais. There are no cacti endemic to the extensive northern *caatingas* in Paraíba, Rio Grande do Norte, Ceará and northern Piauí and only 4 species can be said to have the majority of their ranges north of 7°S (3a:ii). This perhaps suggests that either the northern *caatingas* are younger than the dry areas further south, or that they have experienced stronger forces of extinction in the past, or that they lack the diversity of habitats and refugia created by the combination of the Rio São Francisco valley, the East Brazilian Highlands and more humid ecotonal areas connected with the Atlantic Forest. It is certainly curious that some rather widespread *caatinga* species, in Bahia and northern Minas Gerais, with distribution patterns around or between the East Brazilian Highlands and the *agrestes*, have not spread further north,

Table 3.2

Geographical subdivision of the *caatingas* dominion based on the distribution of Cactaceae

Caatinga area	Endemic/characteristic Cactaceae
(3a:ii). Northern *caatingas*: MA, PI, CE, RN, PB, PE, AL, northern BA	*Pilosocereus flavipulvinatus, P. pachycladus* subsp. *pernambucoensis, P. piauhyensis, P. chrysostele, Discocactus bahiensis*
(3a:iii). Central-southern *caatingas*: southern PI & PB, southernmost CE, PE, AL, BA, SE, MG	*Pereskia aureiflora, P. bahiensis, Tacinga funalis, Brasilicereus phaeacanthus, Stephanocereus leucostele, Arrojadoa penicillata, A. rhodantha, Pilosocereus tuberculatus, Coleocephalocereus goebelianus, Melocactus ernestii* subsp. *longicarpus, Harrisia adscendens*
(3b). *Caatingas* of the Rio São Francisco drainage (MG, BA, PE) and adjacent south-eastern PI	*Pereskia stenantha, Quiabentia zehntneri, Tacinga saxatilis, Pilosocereus gounellei* subsp. *zehntneri, P. densiareolatus, Micranthocereus dolichospermaticus, Melocactus deinacanthus, M. levitestatus, M. azureus, M. ferreophilus, M. pachyacanthus, Discocactus zehntneri* subsp. *zehntneri, Facheiroa* (3 spp.)
(3d:i). Eastern *caatingas-agrestes*: CE [part] & RN southwards to central BA & north-eastern MG (from the E Brazilian Highlands eastwards)	*Tacinga werneri, T. palmadora, Pseudoacanthocereus brasiliensis, Pilosocereus pentaedrophorus, P. catingicola* subsp. *catingicola, Melocactus oreas* subsp. *oreas, M. ernestii* subsp. *ernestii, M. salvadorensis, M. concinnus, Espostoopsis dybowskii*
(3d:ii). South-eastern *caatingas-agrestes*: endemics of north-eastern Minas Gerais	*Tacinga braunii, Pilosocereus azulensis, P. floccosus* subsp. *quadricostatus, P. magnificus, P. multicostatus, Coleocephalocereus* subg. *Buiningia* (2 spp.)

examples being *Pereskia bahiensis* and *Pseudoacanthocereus brasiliensis*. Has the drier zone along the Rio São Francisco valley (and elsewhere in NE Bahia), or the great river itself, halted their expansion northwards?

AREA 4. *Cerrados* and *South-eastern campos rupestres* (Map 6, west of line A–B & south of line B–C) (28 taxa; 17% of the total).

(4a) *Widespread (cerrados):* **Cereus mirabella.**

(4b) *Western cerrados (including those immediately east of the Rio São Francisco in central-southern Bahia)* (3 taxa): **Pilosocereus machrisii, P. aureispinus, Discocactus heptacanthus** subsp. **catingicola**.

(4c) *South-eastern campos rupestres and associated sandy/gravelly cerrados, Minas Gerais* (Map 6, south of line B–C) (24 taxa):
   (i) *Widespread taxa* (3 taxa): **Cipocereus minensis** subsp. **leiocarpus, Pilosocereus aurisetus** subsp. **aurisetus, Discocactus placentiformis.**

(ii) *Northern part of area* (15 taxa):

(a) *Municípios of Grão Mogol, Botumirim & Cristália (Rio Jequitinhonha drainage)* (5 taxa): **Pilosocereus fulvilanatus** subsp. **fulvilanatus, Discocactus pseudoinsignis, D. horstii, Micranthocereus auriazureus, Brasilicereus markgrafii.**

(b) *Serra do Cabral and western slopes of Serra do Espinhaço (Rio São Francisco drainage)* (5 taxa): **Cipocereus bradei, Pilosocereus fulvilanatus** subsp. **rosae, P. aurisetus** subsp. **aurilanatus, Arthrocereus rondonianus, Uebelmannia pectinifera** subsp. **horrida.**

(c) *Município Diamantina & E to the Serra Negra / Serra do Ambrósio* (5 taxa): **Cipocereus crassisepalus, Uebelmannia** (3 spp. + 1 heterotypic subsp.).

(iii) *Southern part: Município Diamantina south- and south-westwards* (6 taxa): **Cipocereus laniflorus, C. minensis** subsp. **minensis, Arthrocereus glaziovii, A. melanurus** (3 subspp.).

As in the case of the *Caatingas* and Northern *campos rupestres* (Area No. 3), this region is well-defined in terms of endemic cactus taxa, including the genus *Uebelmannia* (3 spp. & 2 heterotypic subspp.), an ecologically highly specialized group with no close relatives and apparently relictual. Also notable is *Cipocereus*, a peculiar member of tribe Cereeae, which is represented by 4 endemic species plus one heterotypic subspecies. A characteristic non-endemic genus found only in Area No. 4 within Eastern Brazil is *Arthrocereus* (3 spp. + 2 subspp.). Excluding taxa from the widespread Area Category No. 1, the genera of cacti that occur in common with Area No. 3 are all represented by different species, although in some cases the southern species are actually sisters or probable sister taxa of those from further north or east, eg. *Brasilicereus phaeacanthus* (3a:iii:c) & *B. markgrafii* (4c:ii:a) and *Cereus albicaulis* (3a:iii:a) & *C. mirabella* (4a). Table 3.3 summarizes the principal differences between the Northern and South-eastern *campos rupestres* in terms of Cactaceae (excluding a few very narrow endemics).

Table 3.3

Cactaceae characteristic of the Northern and South-eastern *campo rupestre* vegetation

(† = taxon also found in other included vegetation types — especially *caatinga* — of Area No. 3).

Campo rupestre areas within E Brazil	Characteristic taxa of Cactaceae
Area 3. Northern *campos rupestres* (and included *cerrados*): Bahia & northern Minas Gerais	*Stephanocereus* subg. *Lagenopsis, Arrojadoa* p.p. (2 spp.), *Pilosocereus glaucochrous†, P. pachycladus* subsp. *pachycladus†, Melocactus paucispinus, M. oreas* subsp. *cremnophilus, M. violaceus* subsp. *ritteri, Micranthocereus* subg. *Austrocephalocereus* (3 spp.), *M. polyanthus, M. flaviflorus, Discocactus zehntneri* subsp. *boomianus*
Area 4. South-eastern *campos rupestres* (and included *cerrados*): Minas Gerais	*Uebelmannia, Cipocereus* p.p. (4 spp.), *Arthrocereus* p.p. (3 spp.), *Brasilicereus markgrafii, Pilosocereus aurisetus, P. fulvilanatus, Micranthocereus auriazureus, Discocactus placentiformis, D. pseudoinsignis*

The geography of cactus taxa within this southern area is very far from being random and a series of well-defined patterns of distribution and endemism can be recognized. Two exceptions to the line delimiting this Area on Map 6 should be noted, both in (4b). The first is provided by a narrow zone of *cerrado* located on the east side of the Rio São Francisco in central-southern Bahia, where *Pilosocereus aureispinus* is endemic (its relatives in the P. AURISETUS Group are all *Cerrado*/South-eastern *campo rupestre* taxa). The second is a similarly located disjunct record of *Discocactus heptacanthus* subsp. *catingicola*.

The only widespread *cerrado* taxon (4a), *Cereus mirabella,* on present knowledge has a markedly disjunct distribution, including the Rio Doce/Rio Jequitinhonha watershed (W of Água Boa, MG — the *locus classicus*), western & northern Minas Gerais, western Bahia and western Maranhão (near Carolina). It is assumed to be a relict species, the major part of whose range appears to be within Eastern Brazil, mostly in sandy phases of the *cerrado* (Map 28). *C. albicaulis*, its sister species, has a parallel range eastwards in *caatinga*.

Turning to Category (4b), *Pilosocereus aureispinus,* from east of the Rio São Francisco, is the sister species of *P. vilaboensis* from Goiás (Zappi 1994: 126–129), while the other two species are wide-ranging in Central-western Brazil and eastern Paraguay. The variable *P. machrisii*, which is found only on rock outcrops, is the sister species of *P. aurisetus*, Category (4c:i). In spite of it having received the epithet *catingicola*, the so-called subspecies of *Discocactus heptacanthus* is usually found in the *cerrados* and ecotonal areas.

Nearly all of the taxa listed for Category (4c) are endemic to the core area of Eastern Brazil and none of the species restricted to this type of habitat is common to that of the comparable (3c), Chapada Diamantina and N Serra do Espinhaço, where there are only 17 taxa, although some genera are shared. The absence of various non-endemic genera and species groups characteristic of adjacent areas, including *Tacinga, Harrisia, Leocereus, Coleocephalocereus* and *Facheiroa*, is particularly worthy of note. From the conservation standpoint it is also important to point out that (4c) has a significant number of 'single-site' endemics (6 out of its 24 taxa).

Of those taxa that are categorized as widely distributed in the South-eastern *campos rupestres* (4c:i), *Discocactus placentiformis* and *Pilosocereus aurisetus* subsp. *aurisetus* have somewhat differentiated regional forms in each of the included subdivisions (4c:ii:a–c/4c:iii). These taxa have been formally recognised by Braun & Esteves Pereira (see Chapter 5), but are not as clearly distinct as other regional variants given such recognition in this treatment.

Subdivision (4c:ii:a) is at the border with Area No. 3, and the above taxa can be found sympatric with, or in close proximity to *Tacinga inamoena, Arrojadoa dinae, Melocactus bahiensis* and *M. concinnus*, which are characteristic members of the *Caatingas*/Northern *campos rupestres* flora. *Pilosocereus fulvilanatus,* whose northern relatives are all *caatinga* taxa, is the sister species of *P. ulei*, from the dry coastal forest at Cabo Frio, eastern Rio de Janeiro (Zappi 1994). *Brasilicereus markgrafii* is considered to be a plesiomorphic relict. All the taxa listed for (4c:ii:b) have sister taxa elsewhere in (4c). *Cipocereus crassisepalus* and the species from *Uebelmannia* subg. *Uebelmannia* endemic to (4c:ii:c) are restricted to the abundant deposits of quartz sands and gravels found eastwards from Diamantina. In relation to the taxa listed for (4c:ii:c) & (4c:iii) it is worth noting the statement by Giulietti & Pirani (1988: 65), that Hensold regards the most primitive forms of

Paepalanthus subg. *Xeractis* (Eriocaulaceae) as being from the southern and eastern Serra do Espinhaço. This observation can also be applied to *Cipocereus*, whose taxa with most plesiomorphies are *C. laniflorus* (Serra do Caraça), *C. minensis* subsp. *minensis* (Serra do Cipó southwards) and *C. crassisepalus* (E of Diamantina).

The relationships of the East Brazilian Highlands' cactus flora (including the *campos rupestres, sensu stricto*, plus '*cerrado de altitude*' and montane cloudforest) with other vegetation zones are quite varied, although most do not extend much beyond the core area. They include significant links with the *caatinga* flora, contrary to what is stated in

Table 3.4

Principal links between the E Brazilian Highlands and other areas (based on Cactaceae).

Highland area/substrate or vegetation	Other area	Linking taxon/taxa
Widespread/humid montane forest & rocks	*Mata atlântica/ mata de planalto*	*Rhipsalis russellii, R. floccosa, sens. lat.*
E drainage of highlands from Itamarandiba (MG) to Jacobina (BA)/rocks, sands & gravels	Eastern *caatingas-agrestes*	*Pilosocereus catingicola* subsp. *catingicola, Melocactus oreas, sens. lat., M. ernestii* subsp. *ernestii, M. bahiensis, M. concinnus*
E edge of Highlands and disjunct serras: Serra da Areia (MG), Rui Barbosa & Jacobina (BA), Serra da Itabaiana (SE)/quartz sand & gravel	coastal *restinga* from RN to RJ	*Melocactus violaceus, sens. lat.*
From Grão Mogol (MG) northwards/arenitic rocks, quartz sands & gravels	SW *caatingas* & planalto central, on limestone	*Micranthocereus*
N Chapada Diamantina, region of Morro do Chapéu northwards/arenitic rocks, sands & gravels	*caatingas* of Rio São Francisco	*Discocactus zehntneri, sens. lat.*
Chapada Diamantina & N Serra do Espinhaço/sands & gravels	Central-southern *caatingas*	*Arrojadoa*
Chapada Diamantina/perhumid montane forest	*Mata atlântica* (BA, ES, MG)	*Rhipsalis baccifera* subsp. *hileiabaiana, Hatiora salicornioides*
SE *campos rupestres,* between Grão Mogol and Augusto de Lima/arenitic rocks	coastal rocks, Cabo Frio, RJ	*Pilosocereus fulvilanatus/P. ulei* species pair
SE *campos rupestres,* Diamantina & Serra do Cabral to Serra do Cipó/rocks & gravels	Cent.-western *cerrados,* on diverse rocks	PILOSOCEREUS AURISETUS Group
SE *campos rupestres,* Serra do Cabral (MG) southwards/crystalline rocks	Chapada dos Guimarães, MT	*Arthrocereus*

general terms by Giulietti *et al.* in Davis *et al.* (1997: 400), but cacti may represent an exception because of their common preference for rupicolous habitats, which exist in both ecosystems. These links are summarized in Table 3.4.

Taking together all the species of Cactaceae confined to *campo rupestre* and associated vegetation (eg. 'cerrado de altitude', 'mata de neblina'), ie. most taxa from categories (3c) and (4c), plus those from (3d:i), there is a total of 42 taxa, or more than one quarter (26%) of the cacti of Eastern Brazil, all but one of these being endemic to the core area. These include the endemic genera *Cipocereus* (5 spp.) and *Uebelmannia* (3 spp.), and the subgenera *Micranthocereus* subg. *Micranthocereus* (4 spp.), *M.* subg. *Austrocephalocereus* (3 spp.) and *Stephanocereus* subg. *Lagenopsis* (1 sp.). Making the same analysis for species found mainly in the *caatinga-agreste*, we have a total of 58 taxa or 36%, of which 50 (c. 31%) are endemic to the core area of Eastern Brazil. Thus, the number of endemic Cactaceae restricted to either *campo rupestre* (26%) or *caatinga-agreste* (31%) in Eastern Brazil is not remarkably different and accounts for c. 57% of the total. However, it should not be forgotten that there are some important taxa that occur in both vegetation types (eg. *Tacinga inamoena*, *Melocactus bahiensis*, *Pilosocereus glaucochrous*), as well as a few that have an even wider ecological tolerance. The taxa found in neither *campo rupestre* nor *caatinga-agreste*, are from either the more humid forests, their included rock outcrops and coastal sand-dunes (Area No. 2), or the *cerrados,* and amount to c. 26% of the total, only 7% being endemic to the core area.

To summarize, these figures indicate that, although one might have expected the great majority of Cactaceae of Eastern Brazil to be from the extensive *caatingas*, in point of fact less than half occur there, and the representation of cactus taxa in the *campos rupestres* appears to be nearly as important, yet the area they occupy is relatively much smaller. Taxa that can be regarded as the best overall markers of each vegetation type include, for the *caatinga-agrestes*: *Brasilicereus phaeacanthus*, *Pilosocereus gounellei* subsp. *gounellei* and *P. pachycladus, sens. lat.* (see Map 3, noting that the last-named occasionally strays into the *campos rupestres*); for the *campos rupestres*: *Arthrocereus*, *Cipocereus* and *Micranthocereus* subg. *Micranthocereus* & subg. *Austrocephalocereus* (Map 5); for the Atlantic Forest, *sensu lato*, including *restinga*, riverine vegetation and eastern slopes of the Chapada Diamantina etc.: *Pereskia aculeata* and *P. grandifolia* (Map 2); and for the *cerrados* (including 'cerrado de altitude'): *Cereus mirabella*, *Melocactus paucispinus*, *Discocactus heptacanthus* and *D. placentiformis* (Map 4).

3.4 PHYTOGEOGRAPHICAL LINKS AND PALAEOCLIMATES

Amongst the endemic cactus genera and subgenera there are some which have obvious sister taxa or close relatives beyond the bounds of Eastern Brazil. For example, the genus *Brasilicereus* Backeb., with a pair of allopatric species (see Area Nos 3 & 4, above), is judged to be the sister group of the widespread *Praecereus* Buxb. (N South America to N Argentina; Hunt 1999a), which it replaces in Eastern Brazil (Taylor 1992a, 1997a & b): Map 7. *Praecereus* ranges westwards and northwards from São Paulo and Paraná, Brazil, through NE and NW Argentina, Paraguay, Bolivia, Peru, Ecuador, Colombia and

Venezuela. This range is complemented by that of *Brasilicereus* in Eastern Brazil (Minas Gerais & Bahia) and, when these are taken together, they conform closely to the extended Pleistocene Arc distribution pattern described by Prado & Gibbs (1993) for dry seasonal woodlands in South America. Another vicariant pair of taxa comparable with the *Praecereus/Brasilicereus* case is that of *Cereus* subg. *Mirabella* (2 spp.), which replaces *C.* subg. *Ebneria* in Eastern Brazil (the latter comprises 6–7 species, ranging through Central-western Brazil, Bolivia, Paraguay and Argentina; Taylor 1992a): Map 8. A different and much more disjunct example of vicariance is that of *Espostoopsis* and its presumed relative, the central Andean genus, *Espostoa* — Map 9. Interestingly, this is matched by the almost identical example of *Hyptis* sect. *Leucocephala,* discussed by Harley (1988: 110). The sister relationships proposed for the three Cactaceae cases identified above await confirmation with molecular data.

Below generic rank the high level of endemism that characterizes the cacti of Eastern Brazil means that most have no close relatives outside of this region. However, some notable exceptions provide clear phytogeographical links with other, more or less remote cactus floras. The most important of these involve the Caribbean, northern South America and Amazonia. Thus, the PILOSOCEREUS PENTAEDROPHORUS Group (8 spp., 5 in E Brazil, 3 endemic to the core area) includes one markedly disjunct species in Roraima and the Guianas (*P. oligolepis;* Zappi 1994), while the MELOCACTUS VIOLACEUS Group (10 spp., 7 in E Brazil, 5 endemic to the core area) includes the geographically isolated *M. smithii* and *M. neryi* in the Amazonas-Orinoco drainage region and *M. matanzanus* in Cuba (Taylor 1991a). A gene sequence phylogeny (Wallace 1997: 11) suggests that the geographically isolated *caatinga* endemic, *Harrisia adscendens,* links members of the basal *H.* subg. *Eriocereus* (E Bolivia, Paraguay, Argentina & Uruguay) with the Caribbean taxa of Subg. *Harrisia,* indicating their path of radiation: Map 10. This would seem to have been via an eastern route, rather than a western Andean route, as is suggested by the *Praecereus/Brasilicereus* example given in the previous paragraph. Equally interesting links are those between the *caatinga-agreste* of Eastern Brazil and similar dry habitats in northern South America (Colombia & Venezuela): *Pereskia aureiflora* and *P. guamacho* (Wallace 1995: 9); *Pseudoacanthocereus brasiliensis* and *P. sicariguensis* (Taylor *et al.* 1992). These vicariant species-pairs are disjunct across the currently humid Amazonas-Orinoco region: Map 11. A geographically similar case, but not so disjunct, involving a pair of species, or perhaps a single widespread species with regional forms, is that of the E Brazilian *Cereus jamacaru* and the Venezuelan/Amazonian *C. hexagonus.* Links with the cactus flora of SE Bolivia, Argentina, Paraguay and immediately adjacent parts of Central-western and Southern Brazil are provided by an E Brazilian group of 3 *Pereskia* spp. (*P. grandifolia, P. bahiensis, P. stenantha*) and *Quiabentia zehntneri,* which have south-western counterparts in the taxa comprising *P. nemorosa/P. sacharosa* and *Q. verticillata* (a species ranging in and around the western part of the Chaco). *Quiabentia* (Map 13) is the only clear link between Chaco and *caatinga* (cf. Prado 1991), while the *Pereskia* species-group is another which conforms to the Pleistocene Arc distribution pattern described by Prado & Gibbs (1993): Map 12. Likewise, the monotypic *Brasiliopuntia,* a wide-ranging species, distributed through Eastern Brazil, eastern Paraguay, the Misiones nucleus (Argentina) and the eastern Andes of Peru and Bolivia, is another clear example of this Pleistocene pattern. *Arthrocereus* is a genus of four species, with 3 native to Eastern Brazil (South-eastern *campos rupestres*) and one disjunct in the Chapada dos Guimarães, Mato Grosso (*A. spinosissimus*): Map 13. Further examples of such 'western' connections include *Pilosocereus machrisii* and *Discocactus heptacanthus, sens. lat.*, from eastern Paraguay and Central-western & Northern Brazil etc., both ranging into the western part of Eastern Brazil and having endemic sister taxa in Minas Gerais and/or Bahia: *P. aurisetus* and *D. placentiformis/D. bahiensis* etc. Other such links involve, for example, *Rhipsalis cereuscula*, which ranges from the *Mata atlântica* of NE & SE Brazil via S Brazil and Paraguay to the west as far as the east Andean Yungas of Bolivia. The widespread and almost endemic *R. lindbergiana* is morphologically most similar to *R. baccifera* subsp. *shaferi*, which ranges between São Paulo state (Campinas) and the eastern Andes. Two *Lepismium* species, *L. cruciforme* (NE to S Brazil etc.) and *L. warmingianum* (SE & S Brazil etc.), have sister taxa in the eastern Andes of Bolivia (in the Yungas) and north-western Argentina — *L. incachacanum* and *L. lorentzianum*, respectively: Map 14. *Pereskia aculeata*, from dry habitats in the *Mata atlântica* in Eastern Brazil, has a bimodal distribution north and south of the Amazon basin (Leuenberger 1986: 62), mainly on the eastern side of the neotropics. Its allies are all Andean (Wallace 1995: Fig. 10).

Cactus phytogeography lends support to hypotheses on historical plant migration routes involving dry seasonal forest formations in South America (Pennington *et al.* 2000). Amongst the origins and routes discussed by Prado (1991: Fig. 8.3) for the flora of the *caatingas*, the following are supported by examples from the Cactaceae of the *caatinga-agrestes* and adjacent drier phases of the *Mata atlântica* of Eastern Brazil: (i) the Caribbean islands and coastal regions of northern South America ('Guajira province'), via a trans-Amazonian route; (ii) the same region and/or dry valleys of the central Andes, via a western route including the 'Pleistocene Arc'; (iii) N South America or beyond, via a 'pincers movement' (ie. via both trans-Amazonian and Andean routes); (iv) foothills of the eastern Andes (including the 'Piedmont' and W Chaco) and Misiones nucleus, via the 'Pleistocene Arc'; (v) the Atlantic Forests of Brazil; and (vi) the *cerrados* of Central-western Brazil. Some of the clearest examples supporting routes (i)–(vi) are summarized below in Table 3.5. It should be noted that here it is not necessarily the origin of the *caatinga* flora that is being elucidated, rather that the *caatinga* is one of the nodes for migration routes, which may either have brought taxa *to it,* or *from it,* to other dry areas. Indeed, molecular phylogenies for the PERESKIA GRANDIFOLIA Group and *Coleocephalocereus* indicate radiations in opposite directions along the same route — the former from the *Mata atlântica* (2d:i) into the Southern *caatingas* (3a:iii/3b:iii), the latter in reverse (Wallace 1995, Cowan & Soffiatti unpubl.). And, as noted above, another molecular phylogeny suggests that *Harrisia* both entered the *caatinga* (from the eastern Andes etc.) and radiated from it (into the drier parts of the eastern Caribbean). A further point to note is that Prado's routes should not been seen as mutually exclusive, since, for example, (i), (ii) and (iv) are essentially subsets of (iii). Nevertheless, for ease of comparison his list has been followed in the table below.

In addition to his discussion of migration routes for species of dry seasonal forests in South America, Prado (1991: 232–239) considered palaeoclimates and particularly the Pleistocene fluctuations, when the *caatingas* and similar dry forests appear to have

Table 3.5

Cactus evidence in support of historical migration routes to and from the *caatingas* etc., as proposed by Prado (1991).

Region/migration route (Map refs)	Vicariant taxa/ranges (habitat/country codes [+ Brazilian state codes]: E Brazilian vegetation types)	Notes
(i) Caribbean & 'Guajira province', via Amazônia (see Maps 10 & 11)	(1) *Pereskia guamacho* (CO,VE) & allies (Caribbean): *P. aureiflora* (S caatingas-agrestes) (2) *Pseudoacanthocereus sicariguensis* (CO,VE): *P. brasiliensis* (E caatingas-agrestes) (3) *Cereus hexagonus* (CO,VE,GY,GF,SR,BR [RR,PA]): *C. jamacaru* (caatingas-agrestes) (4) *Harrisia* subg. *Harrisia* (Caribbean): *H. adscendens* (caatingas)	(1) & (3) were examples given in Prado, l.c.; (1) & (4) are now confirmed by molecular evidence. NB. *Pseudoacanthocereus* has only 2 species
(ii) Ibid., via dry cent. Andean valleys and the 'Pleistocene Arc' (see Map 7)	(1) *Praecereus* (CO,VE,EC,PE,BO,PY,AR,BR [MS,SP,PR]): *Brasilicereus* (S caatingas-agrestes & adjacent campos rupestres)	Relationship awaits confirmation by molecular data
(iii) N South America and beyond, via both (i) and (ii) above	(1) *Hylocereus setaceus* complex (MX,BE,CO,EC, PE,BO,PY,AR,GY,SR,BR[RR,PA,MS,RJ, SP,PR]/caatingas & Mata atlântica)	See Chapter 5.5 for notes on this complex
(iv) Via the 'Pleistocene Arc', from the floristic nodes it connects (see Maps 8, 10, 12 & 13)	(1) *Pereskia sacharosa* & *P. nemorosa* (BO,AR,PY,BR [MT,MS,RS]: remainder of P. GRANDIFOLIA Group (Mata atlântica & S caatingas) (2) *Quiabentia verticillata* (BO,PY,AR): *Q. zehntneri* (SW caatingas) (3) *Brasiliopuntia brasiliensis* (PE,BO,AR,PY/caatingas-agrestes & dry phases of Mata atlântica) (4) *Cereus* subg. *Ebneria* (BO,AR,PY,BR[MT,MS, RS]): *C.* subg. *Mirabella* (caatingas & E Brazilian cerrados) (5) *Harrisia* subg. *Eriocereus* (BO,AR,PY,UY,BR [MS]): *H. adscendens* (caatingas)	Molecular evidence has the western *Harrisia* spp. and the andean pereskias as basal or potentially basal within their respective groups (Wallace 1995, 1997). NB. *Quiabentia* has only 2 species
(v) Brazilian Atlantic Forest (see Maps 17, 35 & 41)	(1) *Pereskia grandifolia* (Mata atlântica, BR[CE,PE, BA,MG,ES,RJ,SP,SC]): *P. bahiensis* & *P. stenantha* (Bahian & SW caatingas) (2) *Pilosocereus catingicola* subsp. *salvadorensis* (restinga/ caatinga BR[RN,PB,PE,AL,SE,BA]): *P. catingicola* subsp. *catingicola* (Bahian caatingas-agrestes) (3) *Coleocephalocereus* subg. *Simplex* (S caatingas [BA, MG]): *C.* subg. *Coleocephalocereus* (Mata atlântica)	Molecular evidence indicates *Pereskia grandifolia* subsp. *violacea* and *C. goebelianus* as basal taxa in each case (Wallace 1995, Cowan & Soffiatti unpubl.)
(vi) Central-western Brazilian cerrados (Map 50)	(1) *Discocactus heptacanthus, sens. lat.* (PY,BO,BR[MS, MT,GO,TO,PI,BA,MG]): *D. bahiensis* & *D. zehntneri* (N caatingas & adjacent N campos rupestres)	See Chapter 5.5

expanded (12–18,000 years BP). He presents evidence that the northern *caatinga* area expanded westwards into Maranhão and thence northward, forming a corridor along a route including the Monte Alegre/Faro area (region of Santarém, Pará) and north of the Amazon River to the Roraima/Guyana area, potentially linking ultimately with the dry areas in present day northern South America ('Guajira province'). Such a route seems plausible on the basis of the known distribution of the as yet inadequately resolved *Cereus jamacaru/C. hexagonus* taxonomic complex, *C. jamacaru* subsp. *jamacaru* ranging north-westwards at least as far as the limits of Eastern Brazil (see Map 29). Equally significant is *Pilosocereus flavipulvinatus*, which also ranges to the western border of Maranhão, reaching Carolina (see Map 36). Furthermore, Zappi (1994: 69) suggests that this species provides a link to the related *P. oligolepis*, known only from the Roraima/Guyana region and the northernmost member of the diverse and largely endemic E Brazilian P. PENTAEDROPHORUS Species Group. Do the present ranges of *P. flavipulvinatus* and *P. oligolepis* represent remnants of a Pleistocene expansion of the caatingas? If so, then it is possible that the presence of *Tacinga inamoena, Melocactus zehntneri* and *Leocereus bahiensis* on rock outcrops in the *cerrados* of westernmost Bahia, of *Pilosocereus pentaedrophorus* in *restinga* north of Salvador, Bahia and of *P. ulei* at Cabo Frio, Rio de Janeiro (Ab'Sáber 1974, Araújo in Davis *et al.* 1997: 373) are likewise remnants representative of former *caatinga* expansions (Maps 19, 45, 47 & 36). Similarly, the disjunct occurrences of *Tacinga funalis, Arrojadoa penicillata* and *Harrisia adscendens* in sand-dunes west of the Rio São Francisco (NW Bahia) could represent historical range extensions during the Pleistocene fluctuations, when the barrier presented by the river was removed as it dried up in its middle section (Tricart 1985: 210). These are taxa whose current range is mainly in the eastern sector of the *caatingas* and are not likely to have been long-distance dispersed into NW Bahia, to judge from their presumed dispersal strategies (see below).

3.5 REPRODUCTIVE AND DISPERSAL STRATEGIES

While distribution patterns may be heavily influenced by geographic, climatic, edaphic and temporal factors, the range of a taxon will also be dependent, to a greater or lesser extent, on its reproductive ability and dispersal strategies. As far as can be determined, all of the cactus taxa native of Eastern Brazil reproduce by means of seeds, only very few also employing vegetative means. The chief examples of the latter are in the Opuntioideae, where the jointed stem-segments are frequently capable of being detached, transported and then forming roots upon contact with the ground. A few of the epiphytic and scrambling taxa in tribes Rhipsalideae and Hylocereeae may also indulge in a limited amount of vegetative propagation, but this is unlikely to spread the plant much beyond adjacent branches of the tree in which it originally established itself. Table 3.6, page 133, lists taxa in systematic order noting, where possible, the observed, reported or presumed dispersal vectors, principal habitat type(s) and a categorization of geographical range within specified habitat type(s) in Eastern Brazil into either 'widespread', 'restricted' or 'single site'. Chapter 5 should be consulted for data on size, morphology and dehiscence of fruit and seed.

Amongst the range of observed or probable dispersal strategies demonstrated by the cacti of Eastern Brazil, the rarest appear to be those involving wind or water — *Micranthocereus dolichospermaticus* (W. Barthlott, pers. comm.), *Pilosocereus gounellei* (Zappi 1994) and *Discocactus bahiensis*. However, it is likely that in each case zoochory is still part of the initial stages of dispersal, enabling the seeds to escape from the fruit and/or wool of the cephalium or stem apex. Wind dispersal of the seeds of *M. dolichospermaticus* from its elevated cephalium is a strategy matched by associated species of the bromeliaceous genus *Encholirium*, both taxa needing to disperse their seeds across deeply fissured, karstic limestone outcrops. However, it may be doubted whether this strategy alone is capable of achieving dispersal over longer distances. *Pilosocereus gounellei* is very widely distributed, perhaps because its fruits are attractive to bats and other vectors (large wasps have been observed flying away bearing funicular pulp with seeds attached), but its buoyant seeds may be adapted for local dispersal during flash-floods, which affect the impermeable, flat rocky substrates the species tends to frequent. The same strategy for local dispersal may apply to the much rarer *Discocactus bahiensis*, which has been observed on a low-lying river flood plain subject to occasional inundation. It could also play a role in other species of this genus.

Bats, birds, lizards and ants are almost certainly the commonest seed vectors for E Brazilian cacti, with non-flying mammals being of lesser significance and linked to species with larger fruits and seeds. The most interesting amongst the latter vectors is that suggested by the behaviour of 3 species of *Pereskia* (the P. GRANDIFOLIA Group), *Brasiliopuntia brasiliensis, Opuntia monacantha* and *Pseudoacanthocereus brasiliensis*. In these, the fruits seem to ripen only once they have fallen to the ground, turning yellow or reddish and then smelling strongly of pineapple, just like the ripe infructescences of ground-dwelling species of *Bromelia*, with which they are often associated. Here it is hypothesized that such taxa are (or were formerly) dispersed by peccaries ('caitité', 'caititu', 'queixada' etc.), within whose historic range they are included. This could explain the wide distribution achieved by *Brasiliopuntia, Pseudoacanthocereus* and the 5 species of the PERESKIA GRANDIFOLIA Group across South America, via the Amazon basin, *cerrado* and/or dry seasonal forest environments. In the much altered *caatingas* of the present day, where the peccary is either extinct or very rare, the fallen fruits of *P. bahiensis* and *P. stenantha* are eaten by man's cattle and good crops of pereskia seedlings can be observed germinating amongst recently deposited cow-pats. Two climbing or epiphytic taxa of Hylocereeae (*Hylocereus* & *Epiphyllum*), wide-ranging in the neotropics, produce large succulent fruit that may be of interest to monkeys, which have been observed feeding in forest where *E. phyllanthus* was fruiting (bats are also possible vectors here).

Map 1. The Caribbean and South America, with Brazil, indicating its states, the core study area (box delimited at 7 & 22°S and 46°W), Rio São Francisco, East Brazilian Highlands [▲] and Atlantic Archipelago of Fernando de Noronha, the home of *Cereus insularis* [⊙]. The overall area treated in this study includes the states of Maranhão (MA), Piauí (PI), Ceará (CE), Rio Grande do Norte (RN), Paraíba (PB), Pernambuco (PE), Alagoas (AL), Sergipe (SE) and Bahia (BA), which comprise the NE Brazil region, and parts of both Minas Gerais (MG) & Rio de Janeiro (RJ) and Espírito Santo (ES) from SE Brazil. Note that in the following maps (Nos 2–51) a symbol may sometimes represent two or more closely situated localities, as cited in the text of Chapter 5, and the position of some symbols has been slightly adjusted where different taxa are coincident at a given locality.

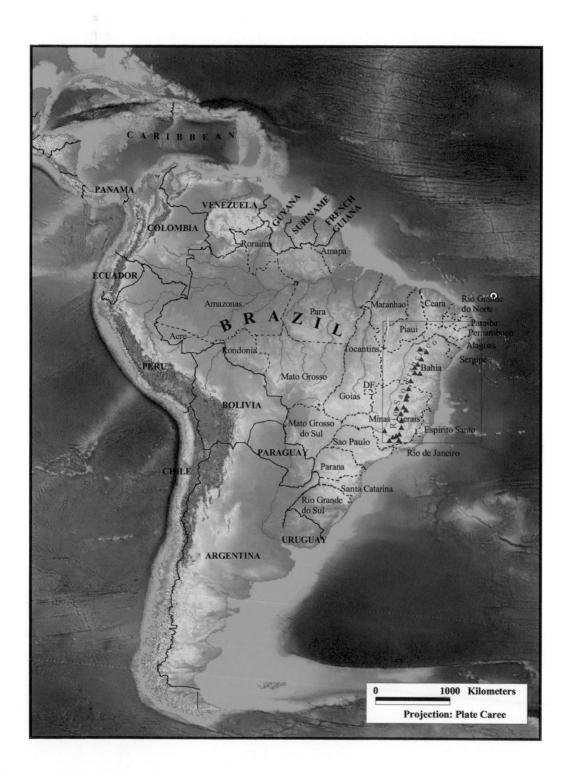

0 1000 Kilometers

Projection: Plate Caree

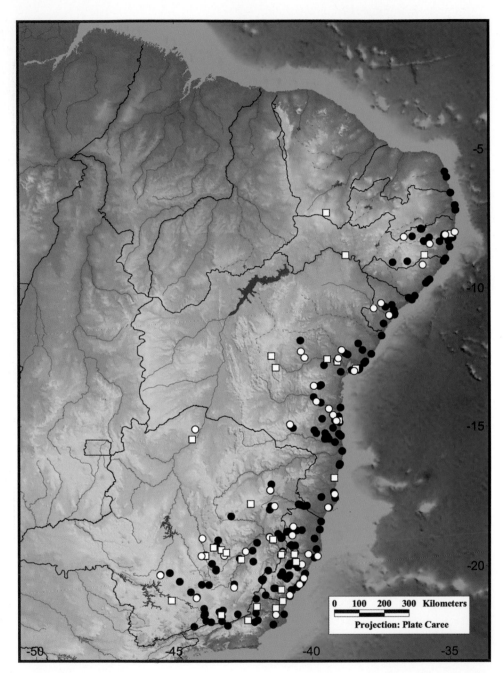

Map 2. Distribution records in Eastern Brazil of all cactus taxa restricted to *Mata atlântica, sens. lat.*, distinguishing the marker species *Pereskia aculeata* [○] and *P. grandifolia, sens. lat.* [□]. NB. Only records within the study area, ie. north of 22°S are displayed (this vegetation extends much further south).

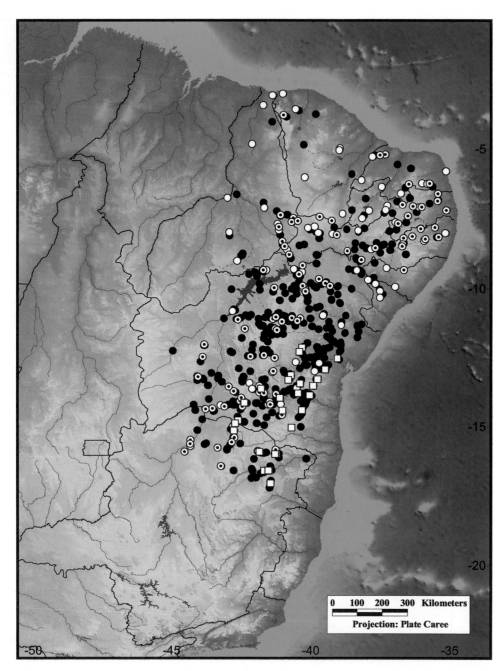

Map 3. Distribution records of taxa typical of *caatinga* and *caatinga-agreste* vegetation, distinguishing *Brasilicereus phaeacanthus* [□], *Pilosocereus gounellei* subsp. *gounellei* [○] and *P. pachycladus, sens. lat.* [⊙] (omitting a southern *campo rupestre* site) and indicating the approximate limits of the *caatinga* biome (excluding south–western *mata seca*).

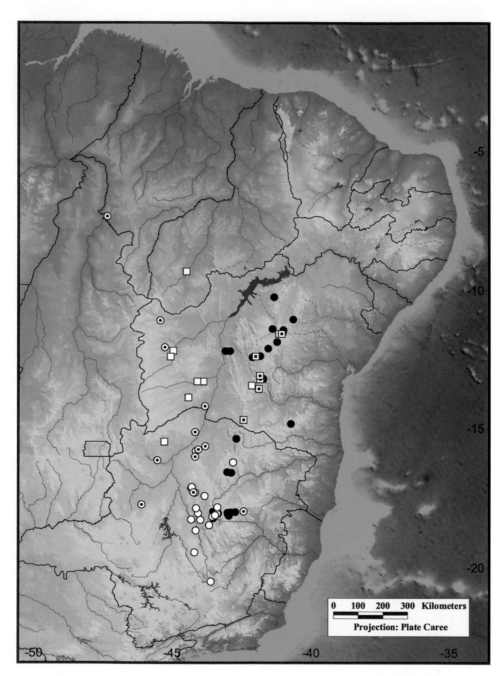

Map 4. Distribution records in Eastern Brazil of taxa restricted to, or with specialised adaptations for *cerrado* (including that within *campo rupestre*), distinguishing the marker species *Cereus mirabella* [⊙], *Melocactus paucispinus* [▣], *Discocactus heptacanthus* [□] and *D. placentiformis* [○].

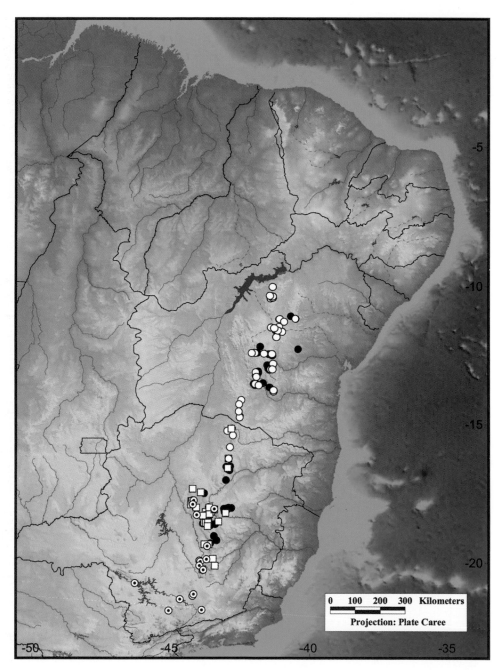

Map 5. Distribution records in Eastern Brazil of all taxa restricted to *campo rupestre* (excluding pure *cerrado* elements), distinguishing *Arthrocereus* [⊙], *Cipocereus* [□] and *Micranthocereus* [○] (subgenera *Micranthocereus* & *Austrocephalocereus*).

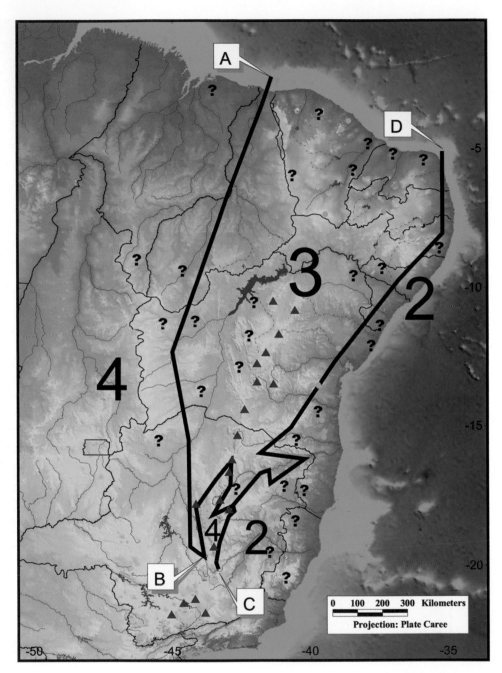

Map 6. Eastern Brazil, showing the principal phytogeographic Area categories (2 = *Mata atlântica*, 3 = *Caatingas* & Northern *campos rupestres*, 4 = *Cerrados* & South-eastern *campos rupestres*), 'brejo' forests [●], *campo rupestre* highlands [▲] and areas of potential interest that await thorough investigation [**?**]. Taxa defining lines A-B-C-D: *Pereskia stenantha, Quiabentia zehntneri, Tacinga braunii, T. inamoena, Rhipsalis paradoxa* subsp. *septentrionalis, Brasilicereus phaeacanthus, Cereus jamacaru* subsp. *calcirupicola, Cipocereus laniflorus, Pilosocereus gounellei* subsp. *gounellei, P. floccosus, P. pachycladus* subsp. *pernambucoensis, P. aurisetus* subsp. *aurisetus, Melocactus bahiensis* subsp. *amethystinus, M. concinnus, Harrisia adscendens, Uebelmannia gummifera*.

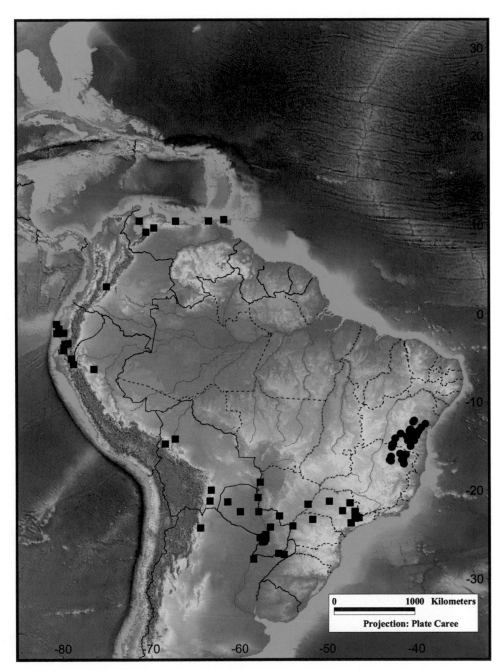

Map 7. Distribution of *Praecereus* [■] and *Brasilicereus* [●]. (South America)

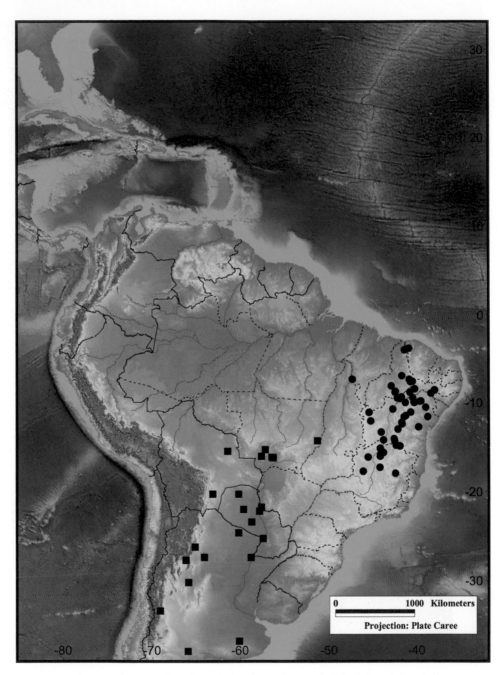

Map 8. Distribution of *Cereus* subg. *Ebneria* [■] and *C.* subg. *Mirabella* [●]. (South America)

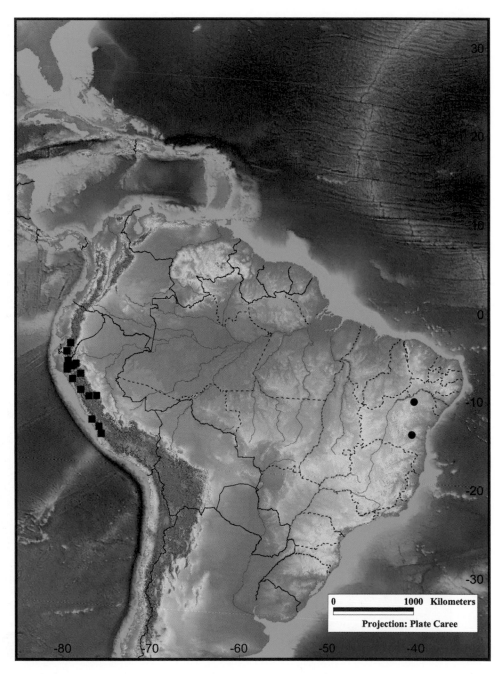

Map 9. Distribution of *Espostoa, sens. str.* [■] and *Espostoopsis* [●]. (South America)

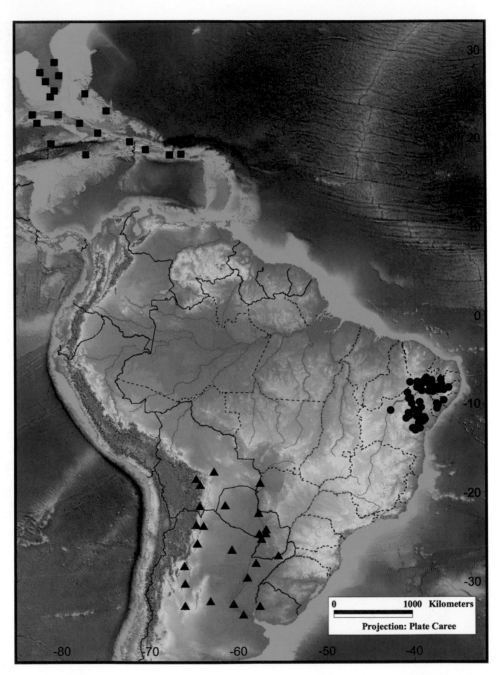

Map 10. Distribution of *Harrisia:* subg. *Eriocereus* [▲], *H. adscendens* [●] and subg. *Harrisia* [■]. (Caribbean & South America)

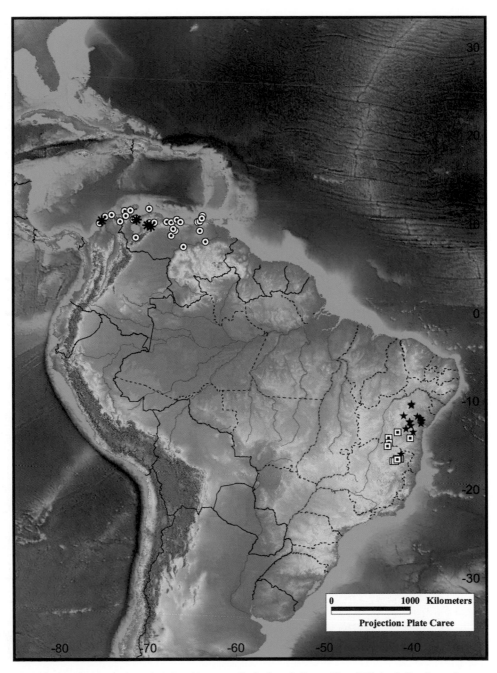

Map 11. Vicariant species-pairs: *Pereskia guamacho* [⊙] and *P. aureiflora* [▣]; and *Pseudoacanthocereus sicariguensis* [✳] and *P. brasiliensis* [★]. (South America)

Map 12. Distribution of the PERESKIA GRANDIFOLIA Group: *P. sacharosa* [●], *P. nemorosa* [▲], *P. grandifolia, sens. lat.* [■], *P. bahiensis* [○] and *P. stenantha* [□]. (South America)

Map 13. Distribution of the genus *Arthrocereus*: *A. spinosissimus* [⊙] and E Brazilian taxa [▣], see Map 49; and of the genus *Quiabentia*: *Q. verticillata* [✦] and *Q. zehntneri* [★]. (South America)

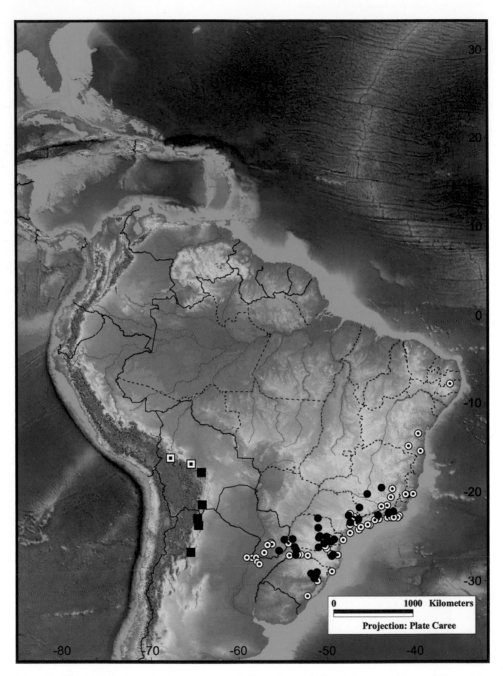

Map 14. Vicariant species–pairs in *Lepismium: L. lorentzianum* [■] and *L. warmingianum* [●]; and *L. incachacanum* [▣] and *L. cruciforme* [◉]. (South America)

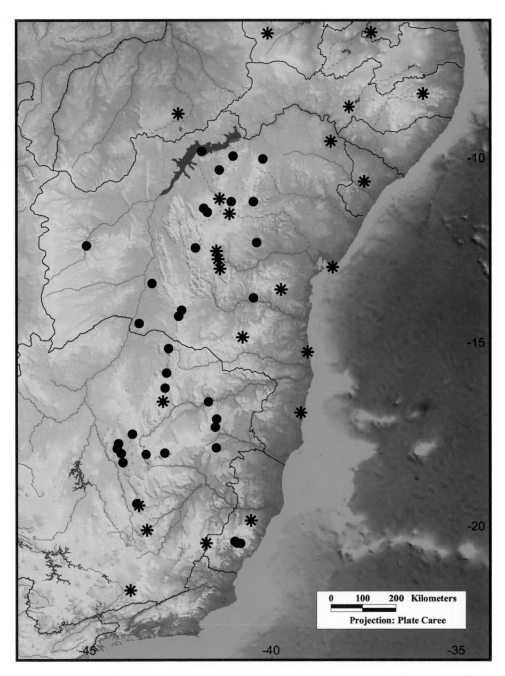

Map 15. Locations of protected areas [✳] including threatened cactus taxa compared with sites of rare and threatened taxa [●] lacking protected areas in Eastern Brazil (endemic taxa only).

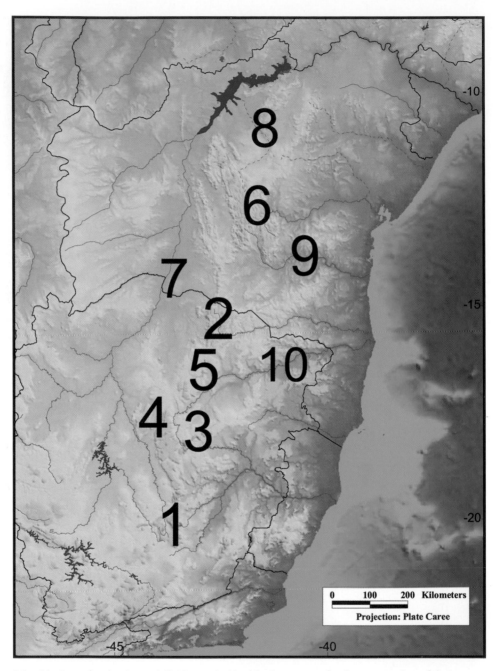

Map 16. Second order Area subdivisions prioritized by importance for the conservation of their cactus endemics, according to the summation of short-list values of taxa scoring 8 or more (see Table 4.3). *Key*. **1** (1st) = SE *Campos Rupestres* (SECR) – Southern part, **2** (2nd) = Northern *Campos Rupestres* (NCR) – N Serra do Espinhaço, **3** (3rd) = SECR – Diamantina (MG) eastwards, **4** = SECR – Serra do Cabral etc., **5** = SECR – Grão Mogol etc., **6** = NCR – Chapada Diamantina, Bahia, **7** = Rio São Francisco *Caatingas* (RSFC) – southern *caatinga*, **8** = RSFC – cent.-northern Bahia, **9** = Eastern *Caatingas-agrestes / Campos Rupestres*, **10** = SE *Caatingas-agrestes*, Minas Gerais.

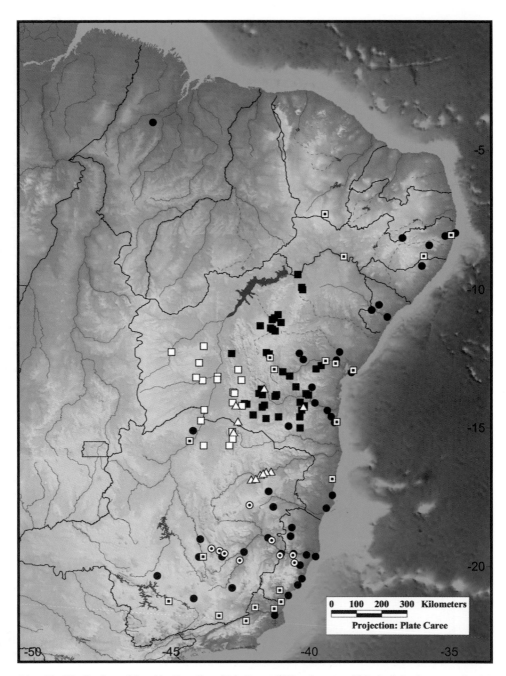

Map 17. Distribution of *Pereskia: P. aculeata* [●], *P. grandifolia* subsp. *grandifolia* [▣] (both non-endemic) & subsp. *violacea* [⊙], *P. bahiensis* [■], *P. stenantha* [□] and *P. aureiflora* [△] (all endemic).

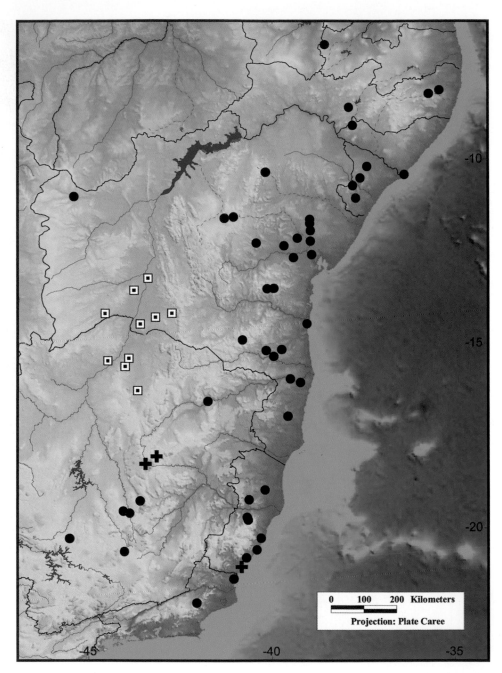

Map 18. Distribution of *Quiabentia zehntneri* [▣] (endemic), *Brasiliopuntia brasiliensis* [●] and *Opuntia monacantha* [✚] (Eastern Brazil only).

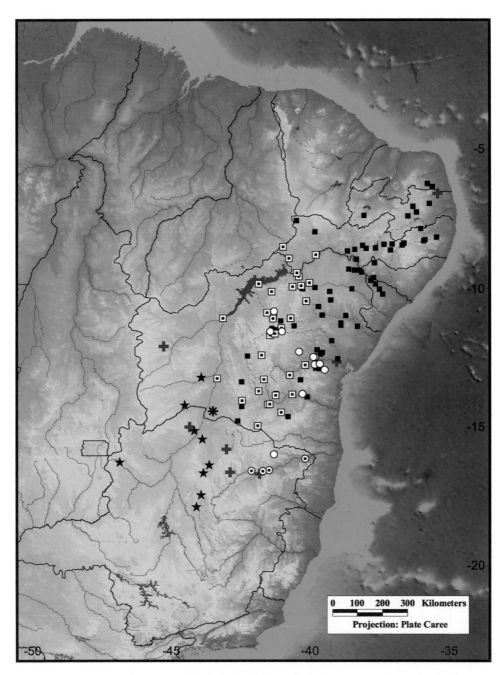

Map 19. Distribution of *Tacinga*: *T. funalis* [⊡], *T. braunii* [⊙], *T. werneri* [○], *T. palmadora* [■], *T. saxatilis* subsp. *saxatilis* [★] & subsp. *estevesii* [✳]; and *T. inamoena* [✚] — eastern-, southern- and westernmost records only.

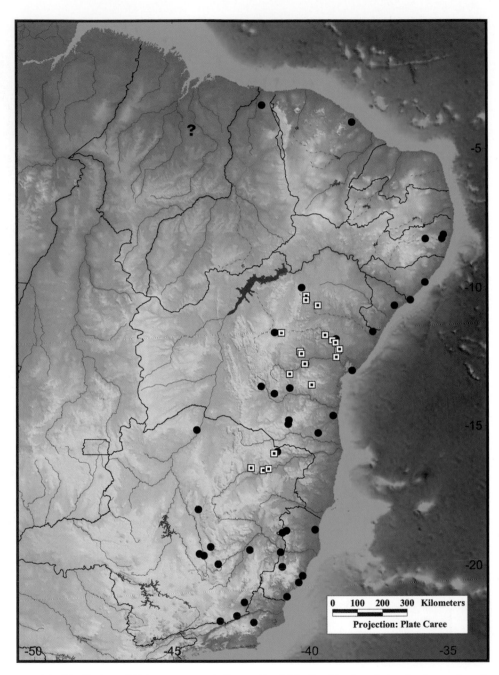

Map 20. Distribution of *Hylocereus setaceus* in Eastern Brazil [●; **?** = identity of record to be confirmed] and the endemic *Pseudoacanthocereus brasiliensis* [▣].

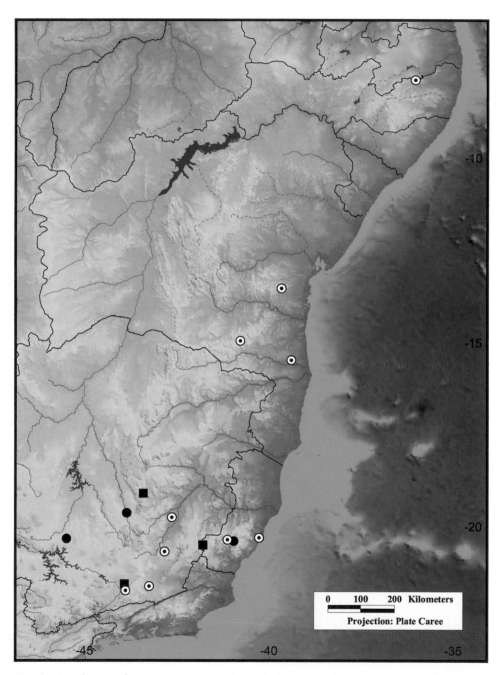

Map 21. Distribution of *Lepismium* in Eastern Brazil: *L. houlletianum* [■], *L. warmingianum* [●] and *L. cruciforme* [⊙].

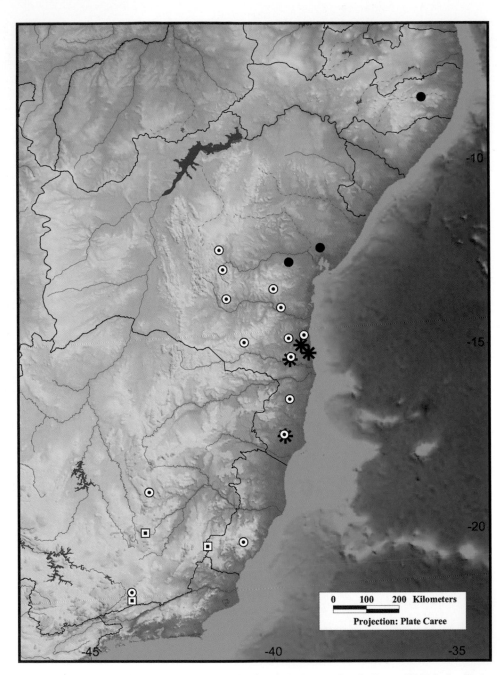

Map 22. Distribution of *Rhipsalis* subg. *Phyllarthrorhipsalis* in Eastern Brazil: *R. russellii* [⊙], *R. elliptica* [▣], *R. crispata* [●] and *R. oblonga* [✳].

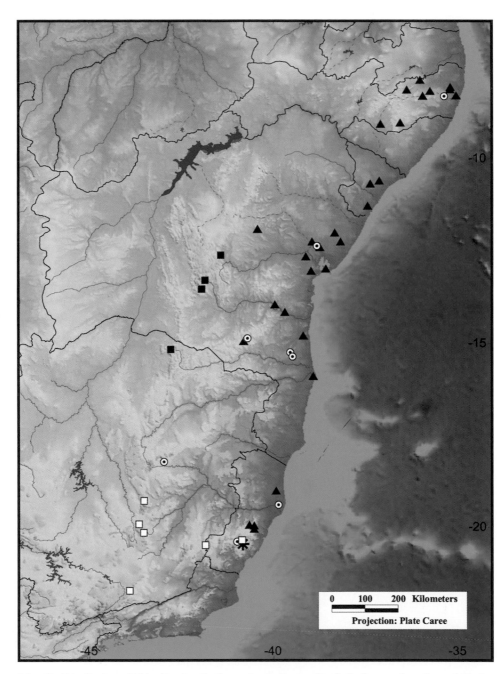

Map 23. Distribution of *Rhipsalis* subg. *Epallagogonium* in Eastern Brazil: *R. floccosa* subsp. *floccosa* [▲], the endemic subsp. *oreophila* [■] & subsp. *pulvinigera* [□], the endemic *R. paradoxa* subsp. *septentrionalis* [☉] and *R. pacheco-leonis* subsp. *catenulata*, *R. sulcata* & *R. cereoides* [all indicated by ✳].

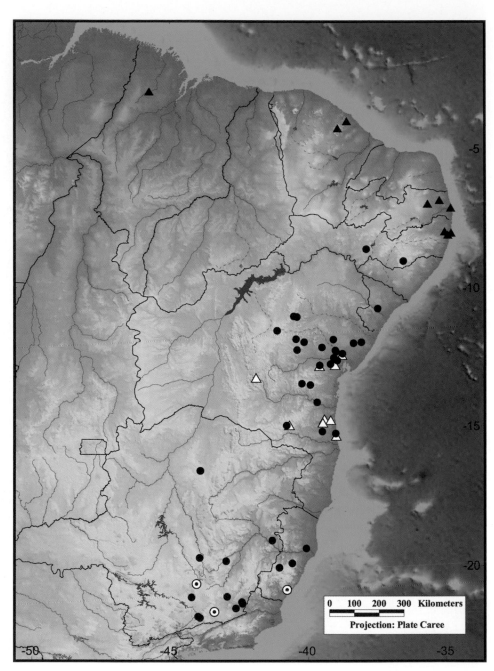

Map 24. Distribution of *Rhipsalis* subg. *Rhipsalis* in Eastern Brazil: *R. lindbergiana* [●], *R. baccifera* subsp. *baccifera* [▲] & the endemic subsp. *hileiabaiana* [△] and *R. teres* [⊙].

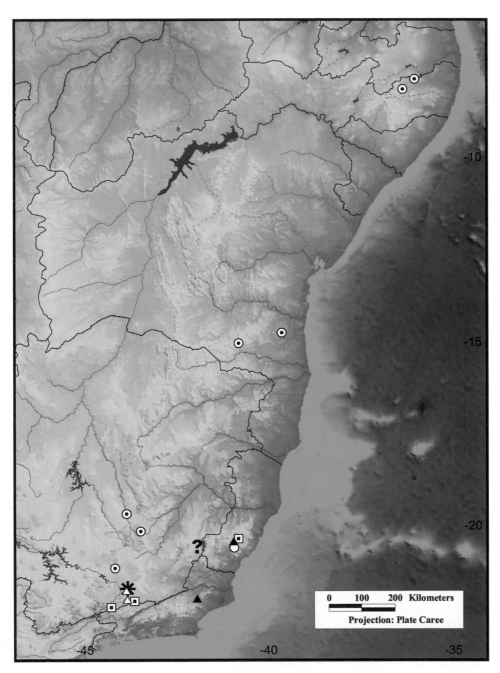

Map 25. Distribution of *Rhipsalis* subg. *Erythrorhipsalis* and *R.* subg. *Calamorhipsalis* in Eastern Brazil: *R. pulchra* [✱], *R. juengeri* [△], *R. clavata* [▲], *R.* indet. cf. *R. juengeri* / *R. clavata* [**?**], *R. cereuscula* [⊙], *R. pilocarpa* [▣] and *R. burchellii* & the endemic *R. hoelleri* [both indicated by ○].

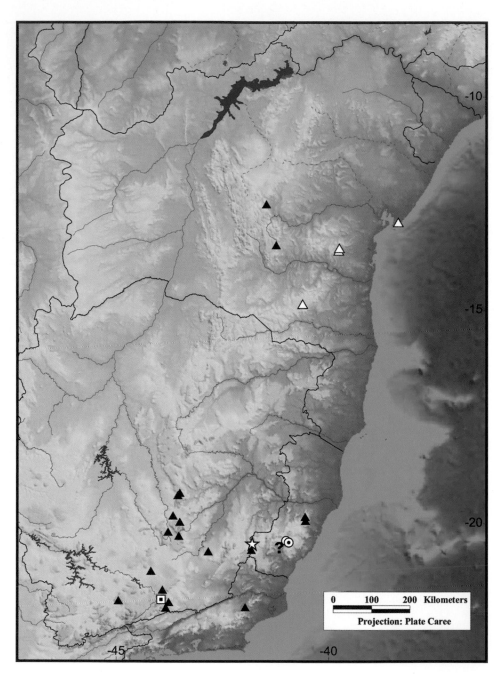

Map 26. Distribution of *Hatiora* and *Schlumbergera* in Eastern Brazil: *H. salicornioides* [▲], *H. cylindrica* [△; **?** = identity of record to be confirmed], the endemic *S. kautskyi* [⊙], *S. microsphaerica* [☆] and *S. opuntioides* [▣].

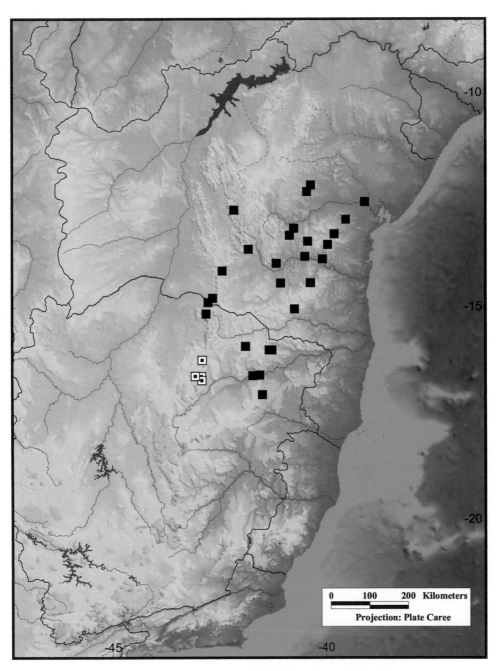

Map 27. Distribution of *Brasilicereus*: *B. phaeacanthus* [■] and *B. markgrafii* [▫].

Map 28. Distribution of *Cereus* subg. *Mirabella* in Eastern Brazil: *C. mirabella* [⊙ ;**?** = record to be confirmed] and the endemic *C. albicaulis* [▣].

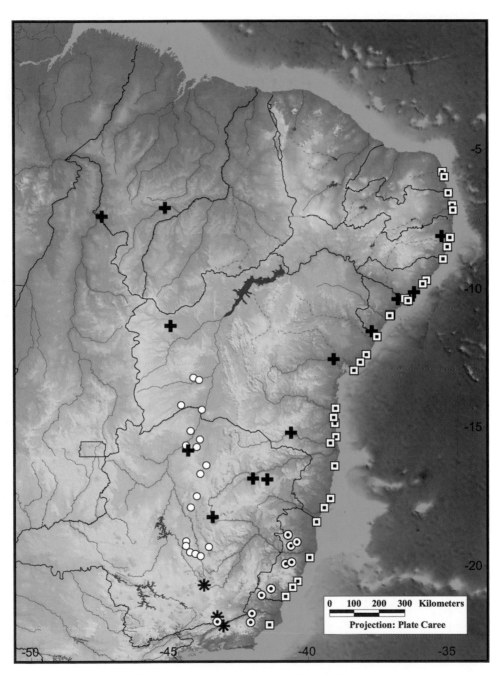

Map 29. Distribution of *Cereus* subg. *Cereus* in Eastern Brazil (excluding *C. insularis*, see Map 1): *C. fernambucensis* subsp. *fernambucensis* [▣] & subsp. *sericifer* [◉], *C. jamacaru* subsp. *jamacaru* [✚] — eastern, southern- and westernmost records only & subsp. *calcirupicola* [○] and *C. hildmannianus* [✳].

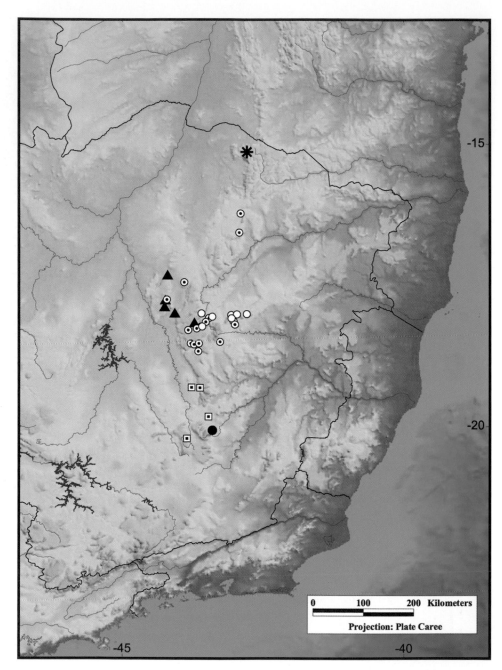

Map 30. Distribution of *Cipocereus: C. laniflorus* [●], *C. crassisepalus* [○], *C. bradei* [▲], *C. minensis* subsp. *leiocarpus* [☉] & subsp. *minensis* [▣] and *C. pusilliflorus* [✳]. (Minas Gerais)

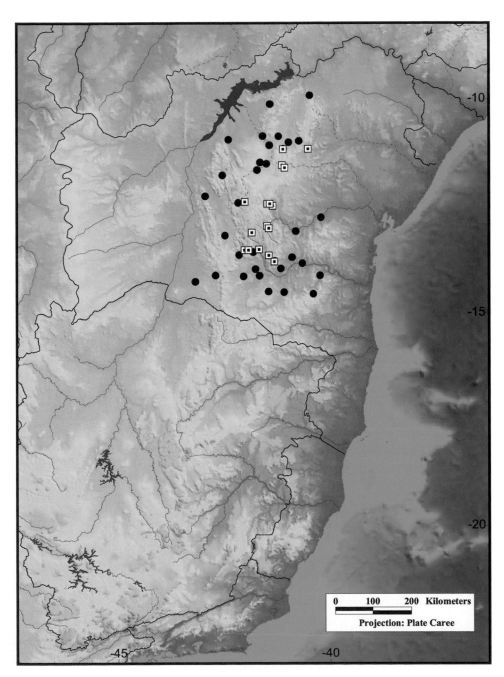

Map 31. Distribution of *Stephanocereus: S. leucostele* [●] and *S. luetzelburgii* [▣].

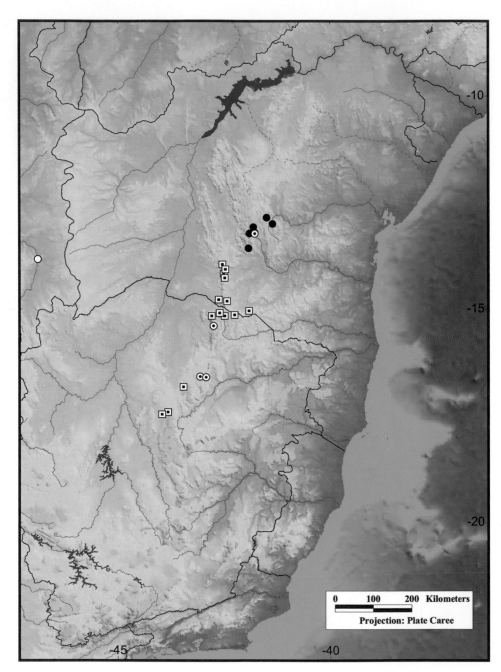

Map 32. Distribution of *Arrojadoa bahiensis* [●], *A. dinae* subsp. *dinae* [▣] & subsp. *eriocaulis* [◉] and *A.* cf. *dinae* (Goiás) [○].

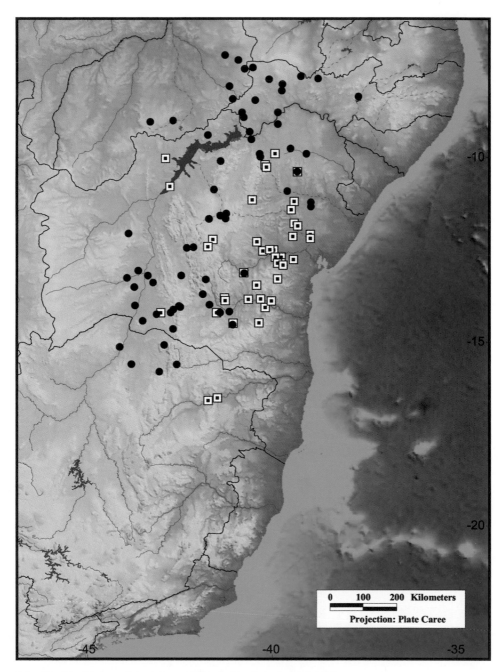

Map 33. Distribution of *Arrojadoa penicillata* [▣] and *A. rhodantha* [●].

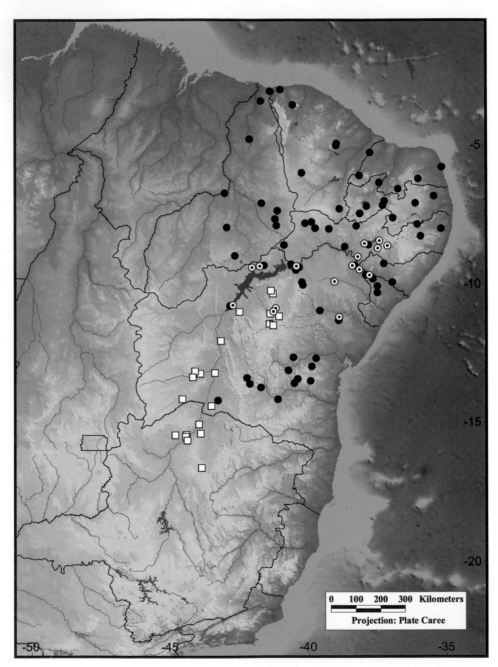

Map 34. Distribution of *Pilosocereus* subg. *Gounellea*: *P. tuberculatus* [⊙] and *P. gounellei* subsp. *gounellei* [●] & subsp. *zehntneri* [□].

Map 35. Distribution of the Pilosocereus arrabidae Species Group: *P. catingicola* subsp. *catingicola* [⊙]
& subsp. *salvadorensis* [▣], *P. azulensis* [★] and *P. arrabidae* [○].

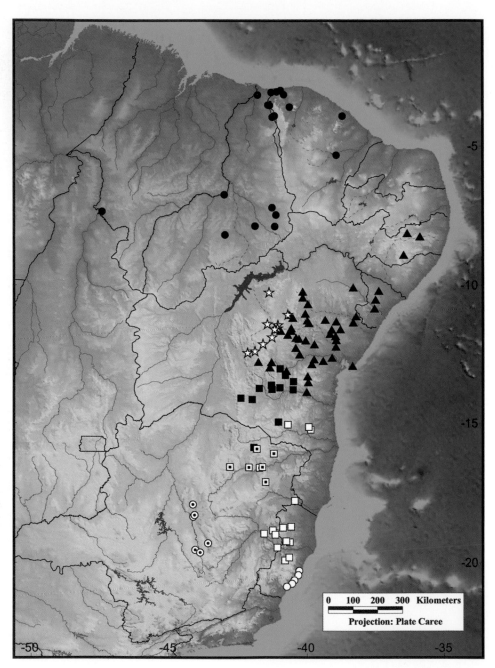

Map 36. Distribution of the PILOSOCEREUS PENTAEDROPHORUS Species Group: *P. brasiliensis* subsp. *brasiliensis* [O] & subsp. *ruschianus* [□], *P. flavipulvinatus* [●], *P. pentaedrophorus* subsp. *pentaedrophorus* [▲] & subsp. *robustus* [■], *P. glaucochrous* [☆] and *P. floccosus* subsp. *floccosus* [⊙] & subsp. *quadricostatus* [⊡].

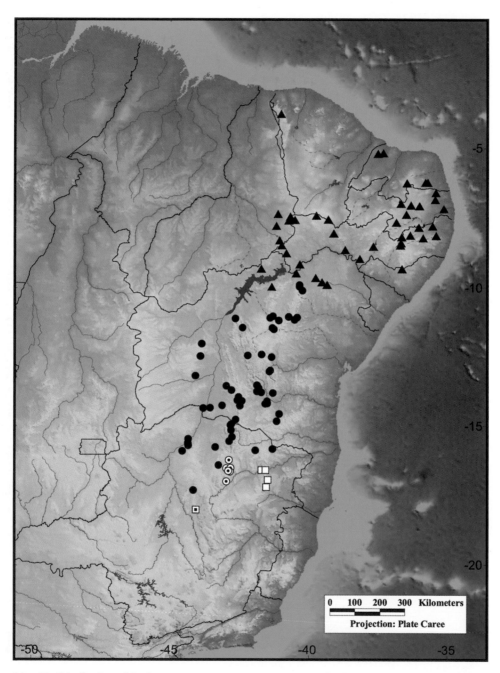

Map 37. Distribution of the PILOSOCEREUS ULEI Species Group: *P. fulvilanatus* subsp. *fulvilanatus* [⊙] & subsp. *rosae* [▣], *P. pachycladus* subsp. *pachycladus* [●] & subsp. *pernambucoensis* [▲] and *P. magnificus* [□].

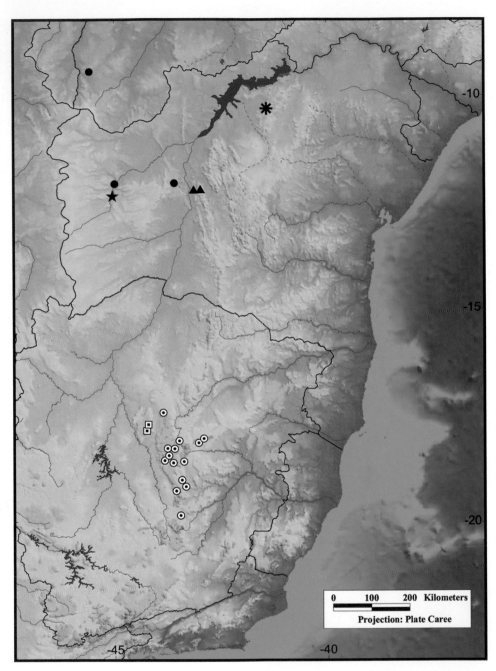

Map 38. Distribution of the PILOSOCEREUS AURISETUS Species Group in Eastern Brazil: *P. machrisii* [●], *P. parvus* [★] and the endemics *P. aurisetus* subsp. *aurisetus* [⊙] & subsp. *aurilanatus* [▣], *P. bohlei* [✳] and *P. aureispinus* [▲].

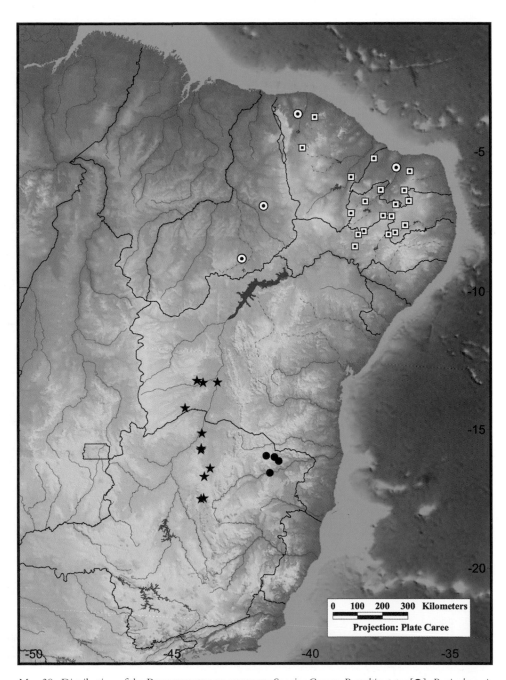

Map 39. Distribution of the PILOSOCEREUS PIAUHYENSIS Species Group: *P. multicostatus* [●], *P. piauhyensis* [⊙], *P. chrysostele* [▣] and *P. densiareolatus* [★].

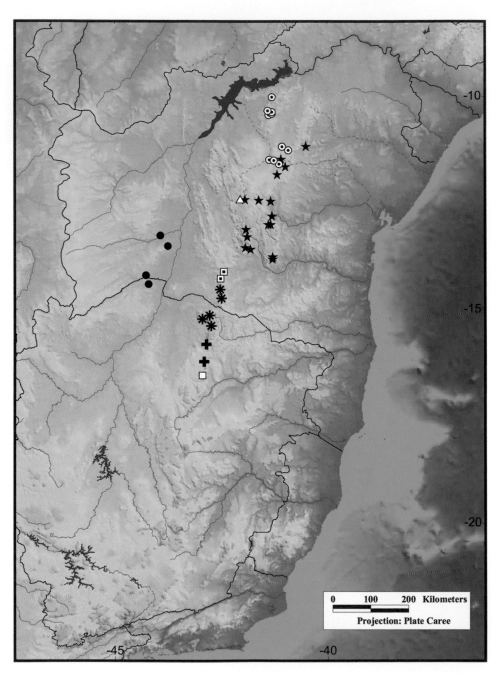

Map 40. Distribution of *Micranthocereus* in Eastern Brazil (all taxa treated here are endemic): *M. violaciflorus* [✚], *M. albicephalus* [✳], *M. purpureus* [★], *M. auriazureus* [□], *M. streckeri* [△], *M. polyanthus* [▣], *M. flaviflorus* [◉] and *M. dolichospermaticus* [●].

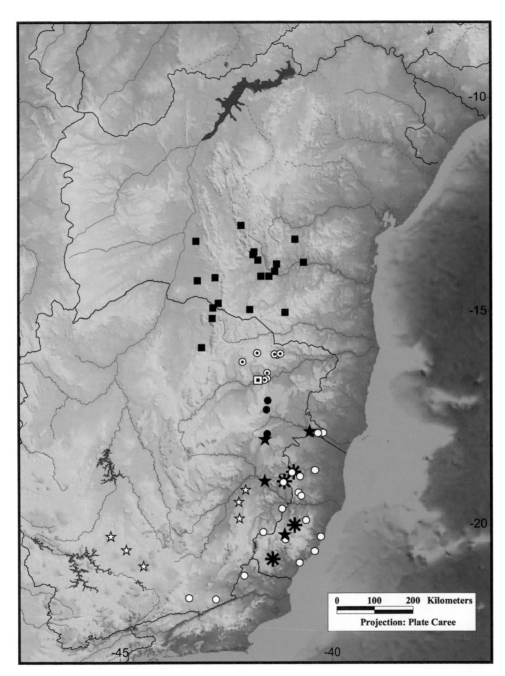

Map 41. Distribution of *Coleocephalocereus* in Eastern Brazil: *C. buxbaumianus* subsp. *buxbaumianus* [★] & subsp. *flavisetus* [☆], *C. fluminensis* subsp. *fluminensis* [○] & subsp. *decumbens* [●], *C. pluricostatus* [✳], *C. goebelianus* [■], *C. aureus* [◉] and *C. purpureus* [◨].

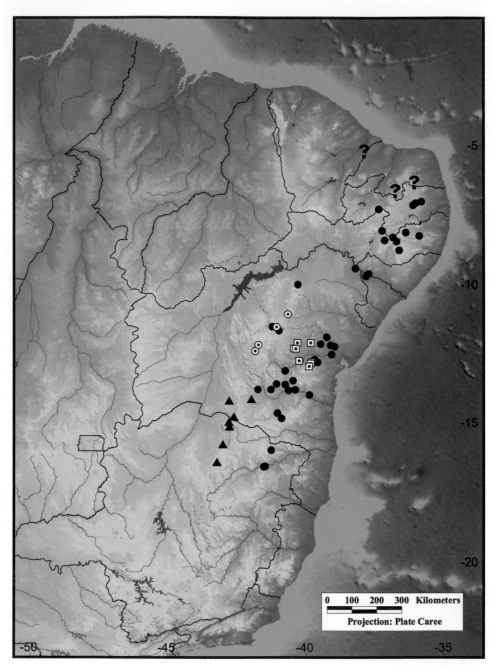

Map 42. Distribution of the MELOCACTUS OREAS Species Group (I): *M. oreas* subsp. *oreas* [▫] & subsp. *cremnophilus* [⊙] and *M. ernestii* subsp. *ernestii* [●; **?** = sites to be confirmed] & subsp. *longicarpus* [▲].

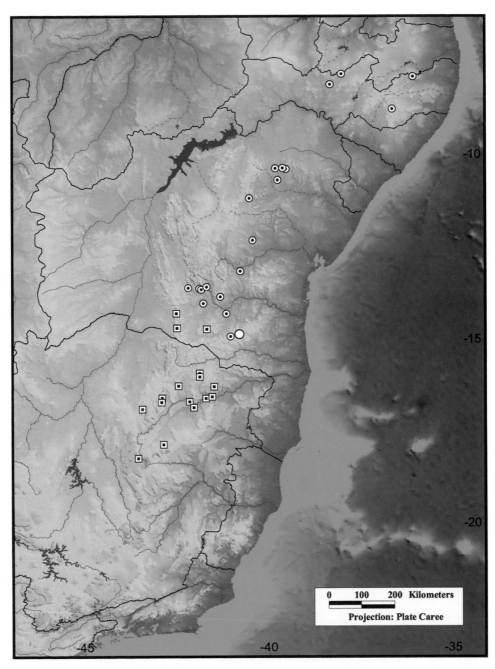

Map 43. Distribution of the MELOCACTUS OREAS Species Group (II): *M. bahiensis* subsp. *bahiensis* [⊙] & subsp. *amethystinus* [▣] and *M. conoideus* [O].

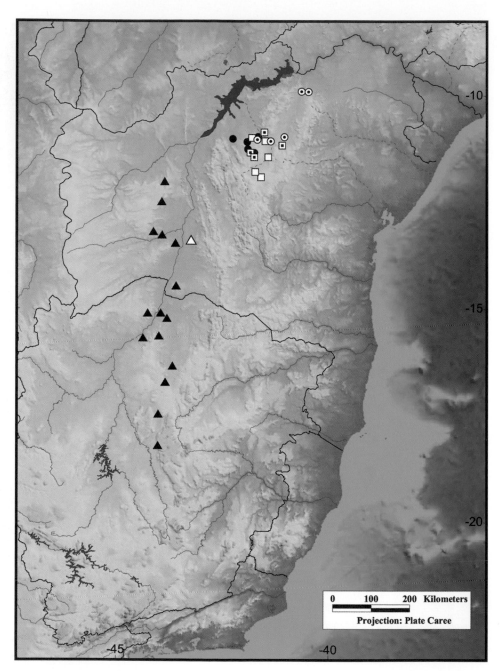

Map 44. Distribution of *Melocactus deinacanthus* [△], *M. levitestatus* [▲] and the M. AZUREUS Species Group: *M. azureus* [●], *M. ferreophilus* [□] and *M. pachyacanthus* subsp. *pachyacanthus* [⊙] & subsp. *viridis* [▣].

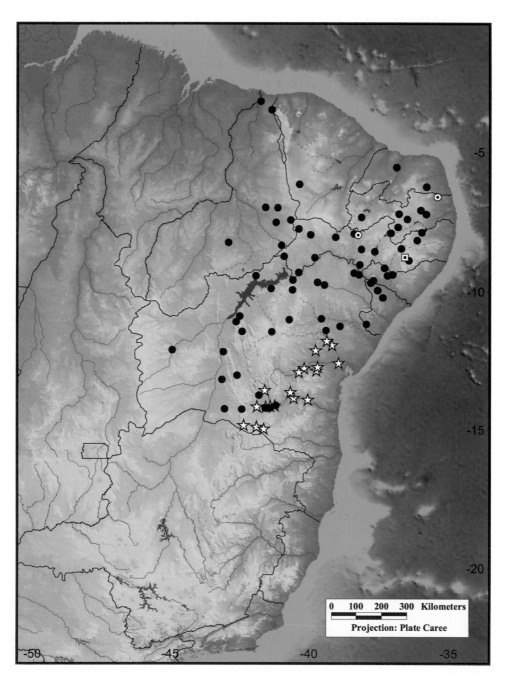

Map 45. Distribution of the MELOCACTUS VIOLACEUS Species Group (I): *M. salvadorensis* [☆], *M. zehntneri* [●] and *M. lanssensianus* [▫] ; ⊙ = records of comparable taxa].

Map 46. Distribution of the MELOCACTUS VIOLACEUS Species Group (II): *M. glaucescens* [✳], *M. concinnus* [○], *M. paucispinus* [△] and *M. violaceus* subsp. *violaceus* [▣], subsp. *ritteri* [◉] & subsp. *margaritaceus* [■].

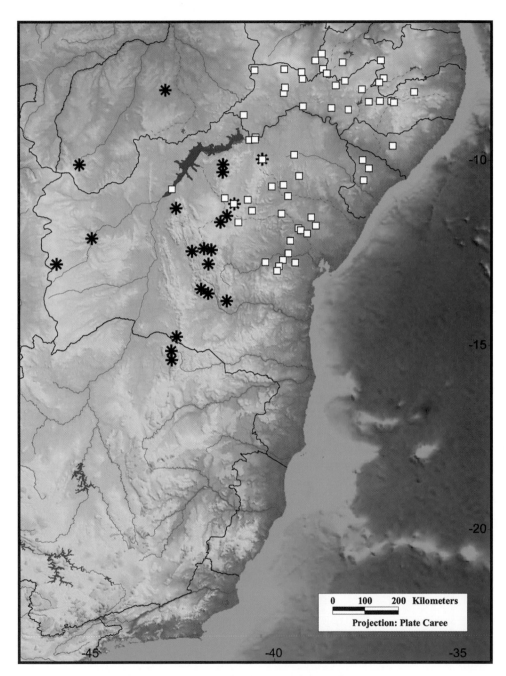

Map 47. Distribution of *Harrisia adscendens* [□] and *Leocereus bahiensis* [✱].

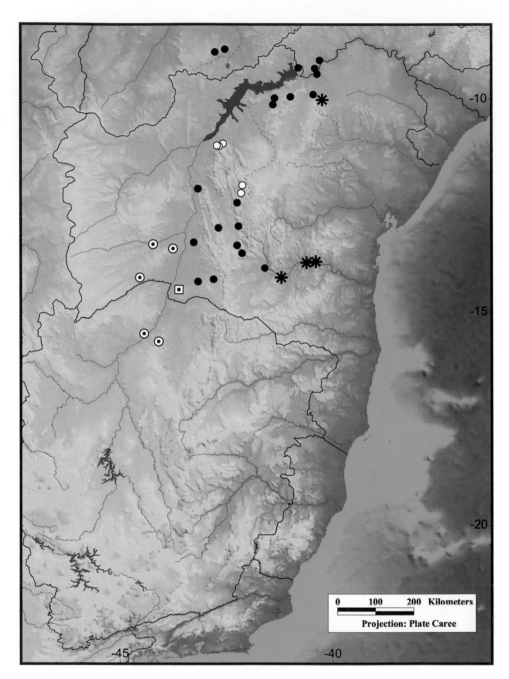

Map 48. Distribution of *Espostoopsis dybowskii* [✳] and *Facheiroa: F. ulei* [○], *F. cephaliomelana* subsp. *cephaliomelana* [⊙] & subsp. *estevesii* [▣] and *F. squamosa* [●]. Notable is the marked disjunction of both *Espostoopsis* and *Facheiroa squamosa* between northern and southern Bahia.

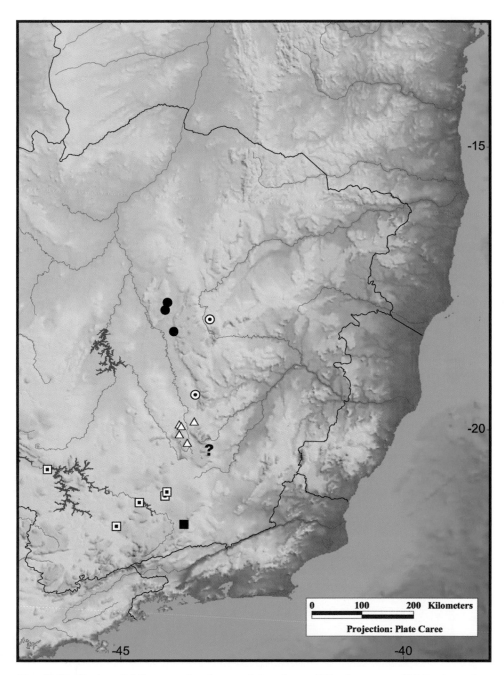

Map 49. Distribution of *Arthrocereus: A. melanurus* subsp. *melanurus* [▣], subsp. *magnus* [■] & subsp. *odorus* [⊙], *A. rondonianus* [●] and *A. glaziovii* [△; **?** = unconfirmed record]. (Minas Gerais)

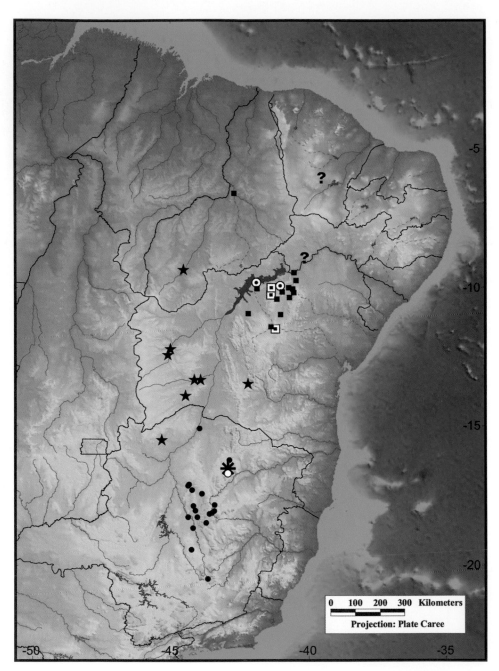

Map 50. Distribution of *Discocactus*: *D. zehntneri* subsp. *zehntneri* [⊙] & subsp. *boomianus* [⊡], *D. bahiensis* [■; **?** = records to be confirmed or localized], *D. heptacanthus* subsp. *catingicola* [★], *D. placentiformis* [●], *D. pseudoinsignis* [○] and *D. horstii* [✳].

Map 51. Distribution of *Uebelmannia: U. buiningii* [●], *U. gummifera* [○] and *U. pectinifera* subsp. *pectinifera* [■], subsp. *flavispina* [⊙] & subsp. *horrida* [▣]. Minas Gerais

Plate 1: examples of vegetation types including cacti. **1.1** *Mata atlântica. Lepismium cruciforme* and *Rhipsalis russellii,* in *agreste (mata de cipó),* SE Bahia, 2002 (MM). **1.2** *Caatinga. Melocactus pachyacanthus* on a limestone *lajedo, caatinga* forest behind, N Bahia, 2002 (NT). **1.3** *Cerrado.* Partly scorched *Melocactus conoideus* in '*cerrado de altitude*', SE Bahia, 2002 (MM). **1.4** *Campo rupestre.* Young *Pilosocereus pachycladus,* nr Brejinho das Ametistas, S Bahia, 1991 (NT).

Plate 2. 2.1 *Pereskia aculeata*. Cult., Univ. Heidelberg (WR). **2.2** *Ibid*. Cult. Les Cèdres (WR). **2.3** *P. grandifolia* subsp. *grandifolia*. SW Minas Gerais, *Harley et al*. 24802 (NT). **2.4** *P. grandifolia* subsp. *violacea*, with Daniela Zappi. E Minas Gerais, 2 km N of Penha do Norte, 1990 (NT).

Plate 3. **3.1** *Pereskia grandifolia* subsp. *violacea*. E Minas Gerais, *Taylor & Zappi* 773 (NT). **3.2** *Ibid*. Ibid. (NT). **3.3** *P. bahiensis*. Bahia, *Harley et al*. 25542 (NT). **3.4** *P. stenantha*. N Minas Gerais, *Harley et al*. 25523 (NT).

Plate 4. 4.1 *Pereskia aureiflora*. NE Minas Gerais, *Harley et al.* 25525 (DZ). **4.2** *Quiabentia zehntneri*. N Minas Gerais, Capitão Enéas, 2002 (GC). **4.3** *Tacinga funalis*. N Bahia, NE of Gruta dos Brejões, 2002 (NT). **4.4** *Ibid.* Bahia, W of Morro do Chapéu, 2002 (MM).

Plate 5. 5.1 *Tacinga funalis*, in fruit. Bahia, E of Seabra, 1989 (WB). **5.2** *T. braunii*. NE Minas Gerais, W of Itaobim, *Harley et al*. 25530 (NT). **5.3** *Ibid*. Loc. cit. (GC).

Plate 6. **6.1** *Tacinga werneri*, with open flower and buds. Bahia, S of Morro do Chapéu, 2002 (MM). **6.2** *Ibid.*, in fruit. NE Minas Gerais, *Taylor et al.* 1520 (NT). **6.3** *T. palmadora*. N Bahia, N of Brejão, 2002 (NT). **6.4** *T. saxatilis* subsp. *estevesii*. Type locality, *Horst & Uebelmann* HU 761, cult. A. Hofacker (AH).

Plate 7. **7.1** *Tacinga saxatilis* subsp. *saxatilis*. N Minas Gerais, Capitão Enéas, 2002 (GC). **7.2** *T. inamoena*. N Bahia, N of Brejão, 2002 (NT). **7.3** *Ibid.*, in fruit. Bahia, 1989 (NT).

Plate 8. 8.1 *Tacinga werneri* × *T. inamoena*. Bahia, W of Morro do Chapéu, 2002 (NT). **8.2** *Brasiliopuntia brasiliensis*. E Pernambuco, Caruaru, *E.A. Rocha et al.* (DZ). **8.3** *Ibid.,* seedling. Rio de Janeiro, Ilha Grande, cult. Kew, 1992 (K).

Plate 9. *Brasiliopuntia brasiliensis*, in fruit. Bahia, S of Vitória da Conquista, 2003 (NT).

Plate 10. 10.1 *Brasiliopuntia brasiliensis*. Rio de Janeiro, Ilha Grande, cult. Kew, flower, 1999 (K). **10.2** *Nopalea cochenillifera*. Pernambuco, Igarassu, 1999 (DZ). **10.3** *Opuntia monacantha*. Espírito Santo, *Zappi* 470 (DZ). **10.4** *Ibid*. Ibid. (DZ).

Plate 11. 11.1 *Opuntia dillenii*. Bahia, *Taylor et al.* 1350 (NT). **11.2** *O. ficus-indica*. N Bahia, Morro do Chapéu, Brejão, 2002 (NT). **11.3** *Hylocereus setaceus*, in *mata de encosta*. Piauí, *J.B. Fernandes da Silva* 365 (NT). **11.4** *Ibid*., fruit. Espírito Santo, *Taylor & Zappi* 779 (NT).

Plate 12. 12.1 *Hylocereus undatus*. Bahia, Morro do Chapéu, 2002 (NT). **12.2** *Epiphyllum phyllanthus*. Brazil, *Horst*, cult. Univ. Heidelberg (WR). **12.3** *Pseudoacanthocereus brasiliensis*. Bahia, *Taylor et al.* 1579 (NT). **12.4** *Ibid*., in fruit. Ibid. (NT).

Plate 13. **13.1** *Pseudoacanthocereus brasiliensis*. NE Minas Gerais, *Taylor & Zappi* 768 (NT). **13.2** *Lepismium houlletianum* f. *regnellii* (left) & f. *houlletianum*. Cult., Univ. Bonn (WB). **13.3** *L. warmingianum*, in fruit. Ibid. (WB). **13.4** *L. cruciforme*, 2 different forms. Ibid. (WB).

Table 3.6

Seed dispersal vectors of species and subspecies of Cactaceae in Eastern Brazil, with details of habitat and extent of range within that habitat type.

Actual observations of presumed vectors at work are indicated by (●). Some infraspecific taxa are not distinguished from their respective species, where their entries would otherwise be identical and repetitive (" = as above)

TAXON	DISPERSAL VECTOR	HABITAT	RANGE
Pereskia aculeata	mammal	Mata atlântica, sens. lat.	widespread
P. grandifolia subsp. *grandifolia*	ground-feeding mammal (peccary?)	"	"
P. grandifolia subsp. *violacea*	"	"	restricted
P. bahiensis	"	caatinga	"
P. stenantha	"	"	"
P. aureiflora	mammal	"	"
Quiabentia zehntneri	?	"	"
Tacinga funalis	?	"	"
T. braunii	?	"	"
T. (Opuntia) werneri	mammal	"	"
T. (Opuntia) palmadora	"	"	widespread
T. (Opuntia) saxatilis subsp. *saxatilis*	"	"	restricted
T. (Opuntia) saxatilis subsp. *estevesii*	"	"	single site
T. (Opuntia) inamoena	mammal (peccary?)	"	widespread
Brasiliopuntia (Opuntia) brasiliensis	ground-feeding mammal (peccary?)	diverse forests	"
Opuntia monacantha	"	Mata atlântica: dunes	restricted
Hylocereus setaceus	mammal (bat/primate?) & bird	diverse forest	widespread
Epiphyllum phyllanthus	mammal: monkey (●) & bat (?); bird?	"	"
Pseudoacanthocereus brasiliensis	ground-feeding mammal (peccary?)	caatinga	restricted
Lepismium houlletianum	bird & small monkeys	Mata atlântica	"
L. warmingianum	"	"	"
L. cruciforme	"	"	widespread
Rhipsalis russellii	"	Mata atlântica & campo rupestre	"
R. elliptica	"	Mata atlântica	restricted
R. oblonga	"	"	"
R. crispata	"	"	widespread
R. floccosa subsp. *floccosa*	"	"	"
R. floccosa subsp. *oreophila*	"	"	"
R. floccosa subsp. *pulvinigera*	"	"	restricted
R. paradoxa subsp. *septentrionalis*	"	"	widespread
R. pacheco-leonis subsp. *catenulata*	"	"	single site
R. cereoides	"	"	"
R. sulcata	"	"	"
R. lindbergiana	"	"	widespread
R. teres	"	"	restricted
R. baccifera subsp. *baccifera*	"	"	"
R. baccifera subsp. *hileiabaiana*	"	Mata atlântica/ de neblina	"

R. pulchra	"	"	"
R. burchellii	"	Mata atlântica	"
R. juengeri	"	"	"
R. clavata	"	"	"
R. cereuscula	"	"	widespread
R. pilocarpa	"	"	restricted
R. hoelleri	"	"	single site
Hatiora salicornioides	"	Mata atlântica & campo rupestre	widespread
H. cylindrica	bird and (?) small lizards	Mata atlântica	
Schlumbergera kautskyi	bird & small monkeys	"	restricted
S. microsphaerica	"	mata de neblina	"
S. opuntioides	"	"	"
Brasilicereus phaeacanthus	bat?	caatinga-agreste	restricted
B. markgrafii	"	campo rupestre	"
Cereus mirabella	mammal	cerrado	widespread
C. albicaulis	"	caatinga	"
C. fernambucensis subsp. *fernambuc.*	bat, other mammal & bird	restinga	"
C. fernambucensis subsp. *sericifer*	"	Mata atlântica	restricted
C. insularis	?	restinga	"
C. jamacaru subsp. *jamacaru*	bird (●)	caatinga-agreste	widespread
C. jamacaru subsp. *calcirupicola*	"	"	restricted
C. hildmannianus	bat & bird	Mata atlântica	"
Cipocereus laniflorus	"	campo rupestre	single site
C. crassisepalus	"	campo rupestre/ cerrado	restricted
C. bradei	"	campo rupestre	"
C. minensis subsp. *leiocarpus*	"	"	"
C. minensis subsp. *minensis*	"	"	"
C. pusilliflorus	"	"	single site
Stephanocereus leucostele	bat & ants	caatinga	restricted
S. luetzelburgii	bat	campo rupestre	"
Arrojadoa bahiensis	bird?	"	"
A. dinae subsp. *dinae*	bird & mammal?	campo rupestre/ cerrado	"
A. dinae subsp. *eriocaulis*	"	"	"
A. penicillata	"	caatinga	"
A. rhodantha	"	"	widespread
Pilosocereus tuberculatus	bat	"	restricted
P. gounellei subsp. *gounellei*	bat, insect (wasp ●) & water	"	widespread
P. gounellei subsp. *zehntneri*	"	"	restricted
P. catingicola subsp. *catingicola*	bird (●) & bat	"	restricted
P. catingicola subsp. *salvadorensis*	bat (Locatelli *et al.* 1997)	caatinga-agreste & restinga	restricted
P. azulensis	bat	caatinga-agreste	single site?
P. arrabidae	lizard (●), bat & bird?	restinga	restricted
P. brasiliensis subsp. *brasiliensis*	bat & bird?	"	"
P. brasiliensis subsp. *ruschianus*	"	Mata atlântica	"
P. flavipulvinatus	"	caatinga	restricted
P. pentaedrophorus subsp. *pentaedrophorus*	"	caatinga-agreste	widespread
P. pentaedrophorus subsp. *robustus*	"	"	restricted

P. glaucochrous	"	caatinga de altitude	widespread
P. floccosus subsp. *floccosus*	"	mata seca	restricted
P. floccosus subsp. *quadricostatus*	"	caatinga-agreste	"
P. fulvilanatus subsp. *fulvilanatus*	"	campo rupestre	"
P. fulvilanatus subsp. *rosae*	"	"	single site
P. pachycladus subsp. *pachycladus*	"	caatinga & campo rupestre	widespread
P. pachycl. subsp. *pernambucoensis*	"	caatinga-agreste	"
P. magnificus	"	"	restricted
P. machrisii	"	rocks in cerrado	"
P. aurisetus subsp. *aurisetus*	"	campo rupestre	"
P. aurisetus subsp. *aurilanatus*	"	"	"
P. aureispinus	bat, bird (?) & ant	rocks in cerrado	"
P. multicostatus	bat & bird?	caatinga-agreste	"
P. piauhyensis	"	caatinga	"
P. chrysostele	"	"	"
P. densiareolatus	"	"	"
Micranthocereus violaciflorus	ant?	campo rupestre	"
M. albicephalus	"	"	"
M. purpureus	bat & bird (●)	"	"
M. auriazureus	bird	"	single site
M. streckeri	"	"	"
M. polyanthus	"	"	restricted
M. flaviflorus	"	"	"
M. dolichospermaticus	wind	caatinga	"
Coleocephalocereus buxbaumianus, sens. lat.	ant (●) (also bird/bat?)	Mata atlântica	"
C. fluminensis subsp. *fluminensis*	(ant ●)	"	"
C. fluminensis subsp. *decumbens*	ant (also bird/bat?)	"	"
C. pluricostatus	"	"	"
C. goebelianus	(ant ●) "	caatinga	"
C. aureus	(ant ●) "	"	"
C. purpureus	ant (also bird/bat?)	"	single site
Melocactus oreas subsp. *oreas*	lizard (●) & bird	"	restricted
M. oreas subsp. *cremnophilus*	"	campo rupestre	"
M. ernestii subsp. *ernestii*	" (lizard ●)	caatinga	widespread
M. ernestii subsp. *longicarpus*	"	"	restricted
M. bahiensis subsp. *bahiensis*	"	caatinga & campo rupestre	restricted
M. bahiensis subsp. *amethystinus*	"	"	"
M. conoideus	"	cerrado de altitude	single site
M. deinacanthus	"	caatinga	"
M. levitestatus	"	"	restricted
M. azureus	"	"	"
M. ferreophilus	"	"	"
M. pachyacanthus, sens. lat.	"	"	"
M. salvadorensis	"	"	"
M. zehntneri	" (lizard ●)	"	widespread
M. lanssensianus, sens. lat.	"	"	restricted
M. glaucescens	"	campo rupestre	"

M. concinnus	"	caatinga & campo rupestre	"
M. paucispinus	"	cerrado de altitude	"
M. violaceus subsp. *violaceus*	" (lizard ●)	restinga	widespread
M. violaceus subsp. *ritteri*	"	campo rupestre	restricted
M. violaceus subsp. *margaritaceus*	"	restinga	"
Harrisia adscendens	bat, other mammal & bird	caatinga	widespread
Leocereus bahiensis	"	campo rupestre & caatinga	"
Facheiroa ulei	bat	caatinga	restricted
F. cephaliomelana subsp. *cephaliomelana*	"	"	"
F. cephaliomelana subsp. *estevesii*	"	"	single site
F. squamosa	"	"	restricted
Espostoopsis dybowskii	bat & bird	"	"
Arthrocereus melanurus subsp. *melanurus*	bat, other mammal & bird	campo rupestre	"
A. melanurus subsp. *magnus*	"	"	single site
A. melanurus subsp. *odorus*	"	"	restricted
A. rondonianus	"	"	"
A. glaziovii	"	"	"
Discocactus zehntneri subsp. *zehntneri*	ant	caatinga	restricted
D. zehntneri subsp. *boomianus*	"	campo rupestre	"
D. bahiensis	ant (●) & water	caatinga	"
D. heptacanthus subsp. *catingicola*	ant	cerrado	"
D. placentiformis	" (●)	campo rupestre/ cerrado	"
D. pseudoinsignis	"	cerrado	"
D. horstii	"	campo rupestre/ cerrado	single site
Uebelmannia buiningii	ant?	"	restricted
U. gummifera	"	"	"
U. pectinifera subsp. *pectinifera*	"	campo rupestre	"
U. pectinifera subsp. *flavispina*	"	"	"
U. pectinifera subsp. *horrida*	"	"	single site

Bats are assumed to be important seed vectors for the great majority of Brazilian columnar cacti with fleshy fruit (Taylor & Zappi 1989: 22; Locatelli *et al*. 1997) and this role is certainly well known in relation to the famous North American Sahuaro cactus, *Carnegiea gigantea* (Engelm.) Britton & Rose, and other cereoid cacti from the Northern Hemisphere (Dobat & Peikert-Holle 1985). It seems reasonable to assume that the distances fruit-eating bats regularly travel are not likely to be a limiting factor in the expansion of range for such columnar species, and the substantial distributions attained by genera such as *Pilosocereus* in the Americas presumably reflect this. Birds may play a similar role (eg. *Cereus jamacaru, Melocactus*) and are likely also to be important vectors for the many smaller-fruited, epiphytic Rhipsalideae (*Lepismium, Rhipsalis, Hatiora* & *Schlumbergera*), whose ranges vary from 'widespread' to 'single site' (Table 3.6). Some of these taxa have very sticky fruit pulp and are assumed to be dispersed by birds in a manner similar to mistletoes (*Viscum* spp.), but small monkeys ('sagüis') have also been reported eating their fruits in South-Eastern Brazil (herbarium label data).

Lizards have been frequently observed upon *Melocactus* plants, eating the watery fruit, and are assumed to be effective local dispersal agents (Taylor 1991a, Figueira *et al*. 1993, 1994). However, it is improbable that this genus has achieved its currently extensive, yet discontinuous, neotropical range by this means alone and dispersal over longer distances by birds is presumed to have occurred (Taylor 1991a). Lizards may also consume the fruits of other cacti that grow close to the ground or which are not so densely covered in sharp spines as to render them inaccessible (however, dead lizards impaled on the spines of *Melocactus* have been observed on more than one occasion!). These vectors were also observed eating *Pilosocereus arrabidae* fruits.

Ants are frequently attracted to the funicular pulp in which nearly all cactus seeds are embedded, but generally do not appear to penetrate the often leathery pericarp of the fruit on their own and are probably less effective with or unable to transport larger seeds. However, the small-seeded genera, *Coleocephalocereus* and *Discocactus,* have dehiscent fruits, the former opening by means of a basal pore, the latter by lateral fissures, and each type has been observed in the process of being raided by ants, which were carrying seeds away. Furthermore, *Discocactus bahiensis* plants have been observed growing in the centre of old ants' nests. Both genera are characterized by erratic occurrence, being abundant in some suitable habitats yet curiously absent from others, suggesting the possibility that longer distance dispersal is less effective and may rely on birds. It is interesting to note that while their fruits at first sight appear to differ markedly, *Coleocephalocereus goebelianus* and *Stephanocereus leucostele* have seeds of remarkably similar shape, yet are not closely related taxa. However, it turns out that both have fruits which open by means of a basal pore and are strongly attractive to ants (M. Machado, unpubl. observations for *S. leucostele*), this suggesting that the coincidence in seed morphology may be connected with dispersal vectors.

While it has been possible to make observations in habitat of a few taxa, and speculate on rather more, in relation to their dispersal vectors, there remain some species for which it is difficult to be sure how dispersal might be achieved. These include the opuntioids, *Quiabentia zehntneri, Tacinga funalis* and *T. braunii*, whose fruits are scarcely either conspicuously coloured, fragrant or juicy/fleshy at maturity. While *Quiabentia* may indulge in vegetative propagation on a local scale, this can hardly explain the present range of either the genus or its Brazilian species, unless this was once much more continuous than it is today, or was achieved over a long time scale. The same remark applies, to a lesser degree, to the above-mentioned *Tacinga* species, but careful studies of these, and of Brazilian cacti in general, are clearly desirable.

Study of Table 3.6 suggests that dispersal strategies do have a significance for the interpretation of distribution patterns, but that this is probably much less important than other, even if interrelated, climatic and edaphic factors. While, as noted occasionally above, the range and frequency of certain taxa may well be explained by such strategies, many closely related taxa, with presumably similar or identical dispersal vectors, differ markedly in the range they have attained, suggesting that other habitat factors, climatic history or the age of the taxon itself have come into play. This is assumed to be the case with the disjunct occurrences of *Harrisia adscendens*, *Arrojadoa penicillata* and *Tacinga funalis* to the west/north-west of the Rio São Francisco, none of these displaying dispersal strategies, such as very sticky fruit pulp enabling epizoochory, which could facilitate the extension of range over longer distances. The points made in this paragraph can also be applied to pollination vectors, which are briefly discussed under the heading of Conservation (Chapter 4). Pollinators are clearly of considerable importance, since extant knowledge suggests that the family is largely comprised of self-incompatible taxa (Ross 1981).

4. CONSERVATION

4.1 WHY CONSERVE THE CACTI OF EASTERN BRAZIL?

There are various arguments to be advanced in support of the conservation of the Cactaceae of Eastern Brazil. Perhaps the strongest, however, is the degree of biological uniqueness these plants represent in terms of endemic genera and species. This can be expressed both in terms of the family in Brazil and in the Americas as a whole, to which the Cactaceae is all but endemic (save only for *Rhipsalis baccifera*, which ranges from the neotropics into the palaeotropics). Brazil has a total of 37 native genera of Cactaceae (ie. c. 30% out of a New World total of c. 120), of which 28 or 75% are found in Eastern Brazil, 12 (32%) of these being Brazilian endemics native in Eastern Brazil. Of the 28 genera from this area, 8 (29%) are endemic to it and the remaining 4 Brazilian endemics have the major parts of their ranges and nearly all of their biological diversity within the area. The 28 native cactus genera of Eastern Brazil comprise 130 named, fully accepted species, of which 117 (90%) are endemic to Brazil and 88 (68%) are endemic to the area covered here. If heterotypic subspecies are added to the total for Eastern Brazil, then we have 162, of which 123 (76%) are endemic. Taking the family as a whole, 3 out of the 4 subfamilies occur in Eastern Brazil, and the largest of these, the Cactoideae, includes about 9 tribal groups. One such is Cereeae, whose 10 genera all occur in Brazil, 3 being endemic to the Eastern region (*Brasilicereus, Cipocereus & Stephanocereus*) and a further 5 having most of their diversity in the region. Another large tribe are the Trichocereeae, with 7 genera in Eastern Brazil, 4 of these being endemic to the region (*Leocereus, Facheiroa, Espostoopsis & Uebelmannia*). The taxonomic isolation, remarkable morphology and ecology of *Uebelmannia* make the strongest of biological arguments for its conservation and the preservation of its habitats.

While it may be legitimate to analyse the above cactus diversity in isolation, it is more environmentally relevant to consider the ways in which its loss might impact the overall ecology of the habitats the plants occupy and sometimes co-dominate. We know too little of the overall ecology of cacti in Eastern Brazil to be able to cite specific cases, but at least in parts of the *caatinga, campo rupestre,* drier phases of the *Mata atlântica* and coastal *restinga* there exist communities where the loss of cacti could certainly affect the survival of their pollinators and seed vectors, and *vice versa*. Perhaps the most vulnerable of higher animal groups in this respect are bats, which visit columnar cacti for nectar, pollen and fruit (Zappi 1994, Ruiz *et al.* 1997, Locatelli *et al.* 1997, Petit 1999). In some of these communities the cacti flower and fruit for much of the year, providing an ongoing food resource even when other vegetation is seasonally dormant or suffering from drought. Another group reliant on cactus nectar as an energy source, at least in some of the driest areas, are the hummingbirds, and in particular those which locate their territories amongst populations of *Melocactus, Arrojadoa* and *Tacinga*, which like some columnar cacti tend to flower for a significant part of the year (Taylor 1991a, Taylor & Zappi 1996, Raw 1996, Locatelli & Machado 1999a). In the case of *Melocactus*, the subsequent and regular production of juicy fruits is an important water resource for lizards (Figueira *et al.* 1993, 1994), which locally disperse the seeds. Besides these organisms, there are other birds, lepidopterans (Taylor 1991a, Locatelli *et al.* 1997, Locatelli & Machado 1999b), bees (Schlindwein & Wittmann 1997), ants and terrestrial mammals that interact with Brazilian cacti in various ways (see also Chapter 3.5), not to mention the important roles played by other plants. These last are many and various, ranging, for example, from the carnaúba palm (*Copernicia prunifera*) in flooded forests in northern Piauí and Ceará, upon which *Pilosocereus gounellei* is often epiphytic, to great trees in the Atlantic Forest supporting many epiphytes belonging to the tribe Rhipsalideae. And while birds may need cacti as a source of energy, they can also be a source of nesting material, the authors having seen birds collecting the cephalium wool of *Micranthocereus purpureus* for this purpose. The same is reported by Schulz & Machado (2000: 63), involving *Pilosocereus aurisetus* and a hummingbird.

The numerous locality records for cacti in Eastern Brazil assembled for the present study may in future enable assessments of the well-being of diverse vegetation types to be made and guide those who have the power to create reserves, as environments inevitably deteriorate further. Few, if any, cacti found in Eastern Brazil appear to increase significantly when habitat disturbance occurs (*Quiabentia zehntneri, Cereus jamacaru* and *Pilosocereus pachycladus* subsp. *pernambucoensis* being possible exceptions), but it is probable that a more complete knowledge of the ecology of their diverse habitats will permit the identification of cactus species assemblages indicative of primary vegetation. This would seem a real possibility in the case of certain Rhipsalideae, whose presence, diversity and abundance in the more humid parts of the Atlantic Forest should be a reliable indicator of its primary or secondary status, since these epiphytes seem only to occur on mostly large trees of considerable antiquity.

Another compelling argument in favour of cactus conservation in Eastern Brazil is their economic potential, if used sustainably, both at the local level and for international trade. They are already employed for a variety of purposes locally, whether as substitute livestock fodder in times of drought, for their delicious fruit, for making 'cactus candy' or planted to form living, impenetrable fences (especially *Pereskia bahiensis* and *Cereus jamacaru*). The frequent attributions of medicinal value made by locals in relation to cacti mostly lack any kind of scientific insight at present (cf. Agra 1996: 32), but in other countries these plants are frequently used in alternative medicine and some have been cultivated commercially for extraction of drugs, eg. *Selenicereus* (see Backeberg 1966: plate pp. 10–11). While they are already recognized for their potential in amenity horticulture in Brazil, where an established nursery industry exports large volumes to Europe, the USA and beyond, more could be made of *native* species from Eastern Brazil. Some are especially attractive as seedlings and certainly marketable, provided their production does not rely on a constant supply of seeds and plants from the wild, but employs artificial propagation. A further area of commerce that native cacti have a role to play in is that of the ever-increasing trend for 'eco-tourism'. There are enough cactophiles in the world to make it worthwhile for specialised tour companies to arrange bespoke holidays for aficionados to visit the more spectacular plants and habitats, eg. around Morro do Chapéu, BA, once again given that the object is to photograph, video and observe in nature, but not collect! Such tours have operated in Mexico for some years now and, if properly policed, can heighten public awareness of the value of nature conservation.

Cacti are also important to preserve for purely aesthetic and cultural reasons. The Xique-xique, Mandacaru and Palmatória (*Pilosocereus gounellei, Cereus jamacaru* & *Tacinga palmadora*, respectively) frequently appear on the signs of restaurants and bars in the Brazilian Nordeste, where they form a part of the folklore, as well as providing names for numerous villages and towns. Their vernacular nomenclature is extensive, interesting and sometimes quite entertaining for those with knowledge of Brazilian Portuguese and, as such, is listed in Chapters 5 & 7.3. Cacti are frequently planted as decorative symbols of the region in village and town squares and sometimes allowed to remain in the fields when other vegetation has been cleared during agricultural development. Last, but not least, it is hard to forget the Brazilian passer-by in a remote part of Minas Gerais, who upon noticing our interest in a large planted *Cereus* flowering by the roadside, warmly offered the comment, "É Mandacaru da Bahia! — Quer uma muda?" [... *Do you want a cutting?*].

4.2 STATUS OF THE ENVIRONMENT AND CONSERVATION OF CACTI IN E BRAZIL

The following background information has been updated from the relevant Brazilian elements in the chapter on South America in the IUCN Species Survival Commission's Cactus & Succulent Specialist Group Action Plan, by Taylor *et al.* (see Oldfield 1997). Since then an important official report recording the levels of habitat destruction in Brazil has also been published (Brasil 1998: 49 etc.).

In terms of East Brazilian cactus species, special conservation concerns include endemic taxa from the genera *Discocactus* (7 species, 5 endemic), *Uebelmannia* (endemic, 3 spp.) and *Melocactus* (*M. conoideus, M. deinacanthus, M. glaucescens, M. paucispinus*), which are 'Critically Endangered', 'Endangered' or 'Vulnerable' (*sensu* IUCN 2001) and, since 1992, have been placed in Appendix I of CITES (Convention on Trade in Endangered Species of fauna & flora) to afford them protection from the export trade (Taylor 1991c). Most of these 'Appendix I' plants and equally threatened taxa in the genera *Cipocereus, Arthrocereus* and *Espostoopsis* are known from only one or very few localities, where the populations number between less than ten to at most a few thousand individuals. *Discocactus placentiformis, D. horstii, D. pseudoinsignis, D. zehntneri* subsp. *boomianus, Melocactus glaucescens* (all except the first being known from between only 1 and 5 small areas each) and all *Uebelmannia* species (*U. buiningii* being 'Critically Endangered', cf. Braun & Esteves Pereira 1988) are threatened in part by trade, via regular collection of plants for seed production, or of seeds, for wholesale export in large quantities. *Discocactus bahiensis* and *Melocactus deinacanthus* (the latter currently known from only a single very small area) are also seriously threatened by agricultural development, and both the former and *D. zehntneri* subsp. *zehntneri* have had their ranges and numbers significantly reduced by inundation from the Represa de Sobradinho, a huge dam-lake created in the 1970s on the Rio São Francisco (Bahia/Pernambuco). Repeated commercial collecting was only partly responsible for the decline of *Melocactus conoideus* at its type locality above the expanding city of Vitória da Conquista, southern Bahia, a species that remains 'Critically Endangered' due to the extraction of the quartz gravel in which it grows (Taylor 1992b). On the positive side the 'Critically Endangered' *Melocactus glaucescens* has recently received increased protection by the creation of the Parque Estadual do Morro do Chapéu (BA), whose area includes most of the small remaining population at its *locus classicus* where, however, it still needs protection from the burning of vegetation by local farmers.

Caatinga. The driest zone of Eastern Brazil is the *caatinga* ('Area 3', see Chapter 3) and its ecotones with Atlantic Forest to the east (*agreste*), dry forests to the south (in Minas Gerais & Espírito Santo) and savannas (*cerrados*) to its west. It represents a severely disturbed ecosystem (Andrade-Lima 1981) that has been subject to forest clearance for agriculture and fuelwood over more than two centuries (Lleras in Davis *et al.* 1997: 395). Over half of the energy needs in North-eastern Brazil are met by plant biomass (Brasil 2000: 10). Desertification in its northern part has recently begun to accelerate at an alarming rate, as reports in the popular press testify and have been stimulating government action towards grand irrigation schemes (Gusmão 1999). Sadly, this seems to be driven more by an understandable desire to further exploit the land than to conserve its remaining biodiversity.

Nevertheless, many cacti have probably suffered less than most other plants as a consequence of their frequent occurrence on rock outcrops unsuitable for cultivation or livestock grazing. Thus, species of *Coleocephalocereus,* various *Pilosocereus* and some *Melocactus* (eg. *M. ernestii*), have significant populations in places dominated by gneiss/granite inselbergs, which are probably at less risk from habitat modification, unless situated near expanding towns or selected for quarrying.

Of those cacti that are not mainly restricted to rock outcrops, the least threatened are those which seem able to regenerate when their forest habitat is cut over. These include *Cereus jamacaru, Pereskia grandifolia* (Plate 2.4), *P. bahiensis* and *P. stenantha*, and all are also conserved by their use in the form of impenetrable livestock hedges and fences surrounding homesteads, both within and sometimes outside their natural ranges. A few very widely distributed endemic cacti, which inhabit little-utilised or sufficiently diverse habitats, are probably not at risk, even though their numbers may have dropped significantly, eg. *Facheiroa squamosa, Harrisia adscendens, Leocereus bahiensis, Tacinga inamoena, Pilosocereus gounellei* subsp. *gounellei* and *P. pachycladus, sens. lat.* However, other, mostly wide-ranging species that are mainly found growing in the soil of the *caatinga-agreste*, or on exposed rocks more or less level with the floor of the surrounding thorn forest, have suffered considerable reductions in their distributions and abundance through forest clearance. Endemic species affected in this way, whose ranges now appear to some degree fragmented, include *Arrojadoa penicillata, A. rhodantha, Brasilicereus phaeacanthus, Cereus albicaulis, Coleocephalocereus goebelianus, Melocactus salvadorensis, M. zehntneri, Pereskia aureiflora, Pseudoacanthocereus brasiliensis, Pilosocereus catingicola, sens. lat., P. floccosus* subsp. *quadricostatus, P. flavipulvinatus, P. glaucochrous, P. pentaedrophorus, sens. lat., Stephanocereus leucostele, Tacinga palmadora, T. braunii* and *T. funalis*. Although most of these are unlikely to become seriously threatened in the immediate future (some are already 'Vulnerable'), regular monitoring and protection is essential if they are not to become 'Endangered' in the longer term.

Of more urgent concern are taxa, such as *Melocactus azureus* and *M. pachyacanthus, sens. lat.*, which have smaller ranges and are restricted to local, low-lying outcrops of limestone (Rio São Francisco basin, Plate 1.2) whose vegetation gets destroyed when the surrounding *caatinga* forest is cleared for cultivation and grazing. These taxa are here assessed as 'Endangered' or 'Critically Endangered' on the basis of their known populations, but more field studies are needed in the remoter parts of northern Bahia, where additional and less disturbed habitats could exist, although as time goes on this seems less likely. Even if cacti found on raised rock outcrops within the *caatinga* are generally at less risk from agricultural development etc., some, and particularly those close to roads or human settlements, are at risk from the quarrying of stone for building materials. Those found only on limestone or gneiss/granite outcrops and probably most at risk are: *Coleocephalocereus aureus, Facheiroa cephaliomelana, Melocactus ferreophilus, M. levitestatus, Micranthocereus dolichospermaticus, Pilosocereus densiareolatus, P. floccosus, P. multicostatus, P. gounellei* subsp. *zehntneri* and *Tacinga saxatilis*.

Examples of the few and mostly relatively small protected areas (WCMC 1992, Brasil 1998: 66–91) within the vast '*Caatinga* dominion' (Andrade-Lima 1981) are, as follows: 'Parque Nacional Serra de Capivara' (includes *Pilosocereus piauhyensis*) and 'Parque Nacional Sete Cidades' (incl. *P. flavipulvinatus*; both parks are in Piauí state), 'Parque Nacional de Ubajara' (Ceará), 'Estação Ecológica de Seridó' (Rio Grande do Norte), 'Reserva Ecológica do Raso da Catarina' (NE Bahia), the 'Áreas de proteção ambiental (APA)' known as 'Serra de Baturité' (Ceará), 'Lago de Pedra do Cavalo' (Bahia — includes *Pseudoacanthocereus brasiliensis*) and 'Gruta dos Brejões/Vereda do Romão Gramacho' (BA — includes *Melocactus pachyacanthus* and *M. ferreophilus*), 'Estação Ecológica Federal de Aiuaba' (Ceará) and the 'Estação Experimental do IPA', Caruaru (PE). These can offer protection to only few species, since, unfortunately, there are currently no significant protected areas in the southern part of the *caatinga* zone in cent.-S Bahia and N Minas Gerais, where a very high species diversity and endemism is matched by a most disturbing level of habitat destruction (mainly for agriculture and charcoal production). One of the most important areas needing protection amongst the southern *caatinga-agrestes* is the middle section of the Rio Jequitinhonha valley (between the towns of Araçuaí and Jacinto) in north-eastern Minas Gerais, where a remarkably rich assortment of cacti exists, including many endemic and potentially threatened cactus species (Taylor & Zappi 1992a). Another promising site for protection, with a comprehensive range of southern *caatinga* cacti, including the rare *Espostoopsis dybowskii*, is situated to the east of the village of Porto Alegre, on the north bank of the Rio de Contas drainage, Mun. Maracás, Bahia. Other sites need to be identified for the conservation of taxa characteristic of the deep soils and 'Bambuí' limestone outcrops in the middle section of the valley of the São Francisco River (especially for columnar Cactaceae). One such would be the massive raised outcrop south of the town of Iuiú on the east bank of the river (SW Bahia), which has two very local endemics restricted to the rock itself (*Facheiroa cephaliomelana* subsp. *estevesii* and *Tacinga saxatilis* subsp. *estevesii*). Other sites should be found to the west of the river, where further endemics, such as the aforementioned *Micranthocereus dolichospermaticus* and *Facheiroa cephaliomelana* subsp. *cephaliomelana,* are located.

Campo rupestre & cerrado. The East Brazilian Highlands, with their mosaic of *campo rupestre* and *cerrado* vegetation (Giulietti & Pirani 1988, Zappi & Taylor 1994: 77), represent the least modified of the environments in Eastern Brazil. However, they have much greater concentrations of threatened species than the *caatingas-agrestes* just discussed, and many are of extremely local occurrence and therefore potentially at considerable risk. Widespread and mostly common, non-threatened exceptions include *Cipocereus minensis* subsp. *leiocarpus, Melocactus bahiensis, M. concinnus, Pilosocereus aurisetus* subsp. *aurisetus, Micranthocereus purpureus* and *Stephanocereus luetzelburgii,* the latter two endemic to the extensive uplands of the Chapada Diamantina, Bahia, and also found within its national park ('Parque Nacional da Chapada Diamantina', Mucugê–Lençóis).

Utilization of the *campos rupestres* is generally limited to cattle grazing, with associated burning to induce re-growth of edible pasture, and local extraction of some plants, eg. Eriocaulaceae (dried flower export trade — a serious conservation issue), orchids and *Vellozia* spp. (Harley in Stannard 1995: 35–37, Giulietti *et al.* in Davis *et al.* 1997: 403), and there is also limited disturbance caused by small-scale mining for gold and precious stones. In certain areas tourism, including eco-tourism, is becoming popular due to the spectacular scenery, and, as noted already, this can create either negative or positive environmental pressures, depending on how sustainably it is managed. Some parts where *cerrado* vegetation is more abundant are being cut over for the production of charcoal and later converted into *Eucalyptus* plantations, especially in Minas Gerais, where this activity is one of the factors threatening *Uebelmannia* species and *Cipocereus crassisepalus.* Burning the land for grazing and trampling by cattle does affect some native populations of cacti, but the regular collection of plants, and nowadays more especially of seed, of certain rare cacti may be cause for greater concern. In addition to some of the CITES Appendix I taxa noted above, the following *campo rupestre/cerrado* cacti are known from only one or two small populations, or at best have a very localized range which does not include any kind of designated protected area: *Arrojadoa dinae* (especially the rare variant, subsp. *eriocaulis*), *Arthrocereus rondonianus, Cipocereus bradei, C. crassisepalus, C. pusilliflorus, Melocactus violaceus* subsp. *ritteri, Micranthocereus albicephalus, M. polyanthus, M. streckeri, Pilosocereus aurisetus* subsp. *aurilanatus* and *P. fulvilanatus* subsp. *rosae.*

Similarly restricted taxa located within protected areas are rather few: *Arrojadoa bahiensis* (partly inside the 'Parque Nacional Chapada Diamantina' and the 'APA da Serra do Barbado', Bahia), *Brasilicereus markgrafii, Micranthocereus auriazureus, M. violaciflorus, Discocactus pseudoinsignis* & *D. horstii* (inside the 'Parque Estadual da Serra do Barão', Minas

Gerais), *Uebelmannia gummifera* (found within the 'Parque Estadual da Serra Negra', MG), *Cipocereus minensis* subsp. *minensis* (partly within the 'Parque Nacional da Serra do Cipó', MG), *C. laniflorus* (entirely within the 'Reserva Particular de Patrimônio Natural do Caraça', MG) and *Arthrocereus melanurus* subsp. *magnus* ('Parque Estadual de Ibitipoca', MG). If extended slightly to its west, the 'Parque Nacional Chapada Diamantina' would include a further population of the remarkable *Arrojadoa bahiensis*. The Serra da Piedade (Mun. Caeté, Minas Gerais) is not a designated protected area, but benefits from some protection as a site of religious significance, which has a population of *Arthrocereus glaziovii*, a specialized species generally restricted to rocks very rich in iron (known as *canga*). Many of its former habitats have disappeared through ore extraction. A peculiar and specialized cactus, found in the sandy *cerrados* bordering on the *caatinga* and *campo rupestre* zones, from western Maranhão to central-eastern Minas Gerais, is *Cereus mirabella*. It is widespread, but erratically occurring, and much of the habitat in the southern part of its range is being destroyed by charcoal producers. Its status needs to be monitored.

Locations within the *campos rupestres*, where new protected areas have been suggested to assist the conservation of the above-listed rarities (Taylor in Oldfield 1997, Costa *et al.* 1998), including the earlier discussed CITES Appendix I taxa, are: the southern end of the Serra Chapada and associated *cerrado* (27–28 km W of Seabra, BA), the quartzitic outcrops at Brejinho das Ametistas (S Bahia), the Serra Geral c. 12–15 km east of Monte Azul (Minas Gerais), the Serra Geral with white sand *cerrado* 12 km east of Mato Verde (MG), the Serra do Cabral (MG), the western slopes of the Serra de Minas east of Santa Bárbara (Mun. Augusto de Lima, MG) and one or more sites for *Uebelmannia pectinifera* in the vicinity of Diamantina (MG).

Mata atlântica. Of great concern, in general terms, is the Brazilian Atlantic Forest. This comprises the coastal rainforest (*Mata atlântica* in its strictest sense) and sandy littoral dunes (*restingas*) of North-eastern Brazil and their extensions southwards, where the former broadens and merges with the *planalto* forests of South-eastern Brazil. This more or less humid area, which has a high alpha-diversity of epiphytic cacti from the tribe Rhipsalideae (cf. Ibisch *et al.* 1996), is represented by only a small fraction of the original forest. Myers *et al.* (2000: Table 1) state that only 7.5% of the original primary forest remains. Rising above the forest or its remnants are the frequently encountered gneiss/granite inselbergs, which represent important habitats for rupicolous taxa. While these may appear safe, they are nevertheless affected when the surrounding forest is cut, because this tends to allow weedy species and especially alien grasses (eg. *'capim gordura'*) to occupy the niches that would be taken by seedling cacti.

The Atlantic Forest zone is home to some horticulturally and economically important genera, such as *Schlumbergera* (species of 'Christmas cactus') and *Hatiora*. A few very widespread or regionally common taxa, such as the epiphytic *H. salicornioides, Lepismium cruciforme, L. houlletianum, L. warmingianum, Rhipsalis floccosa, R. teres, R. elliptica* and *R. cereuscula*, and the non-epiphytes *Brasiliopuntia brasiliensis, Opuntia monacantha, Cereus fernambucensis, Pilosocereus arrabidae* and *P. brasiliensis* are here regarded as of 'Least Concern' or 'Near Threatened'. However, the remaining Brazilian endemic species are of conservation concern to varying degrees and the not infrequent but mostly small coastal protected areas may not be sufficient to halt their decline. For example, the wide-ranging, but erratically occurring *restinga* taxa, *Melocactus violaceus* subspp. *violaceus* & *margaritaceus*, and other coastal cacti, are threatened at many points in their ranges by ever-expanding tourism, urban and agricultural developments (sugarcane, coconut and pineapple). At least one extensive population of *M. violaceus* subsp. *margaritaceus* and *Pilosocereus catingicola* subsp. *salvadorensis* are, however, currently offered some protection in the 'Estação Ecológica da Serra de Itabaiana', Sergipe and 'Reserva Ecológica Dunas do Abaete', Bahia.

The flora of southern Espírito Santo is poorly understood and its habitats severely altered, but it can count a recently described species of Christmas Cactus, *Schlumbergera kautskyi* (known from only 2 small sites), and the remarkable, red-flowered *Rhipsalis hoelleri*, as yet not localized with certainty. Other species, represented by disjunct populations, are *Rhipsalis cereoides, R. pilocarpa* and *Schlumbergera microsphaerica* (the last within the boundaries of the 'Parque Nacional do Caparaó', which straddles the border with the state of Minas Gerais). Another protected area in this state, which merits investigation, is the 'Reserva Biológica Federal de Nova Lombardia' (Mun. Santa Teresa).

Destruction of the Atlantic Forest has been greatest in North-eastern Brazil, where very little remains (Wayt Thomas in Davis *et al.* 1997: 364) and, therefore, our knowledge of the flora is correspondingly fragmentary. To judge from the number of taxa known from very few, disjunct records, it is quite possible that epiphytic Cactaceae from here have become extinct before discovery and description. In Pernambuco remnants of this forest include those termed as *'brejos',* on higher land far away from the coast, where the watersheds are an important resource for the human populations living below them. Such forests have recently been studied and catalogued (Rodal *et al.* 1998, Sales *et al.* 1998) as part of an Anglo-Brazilian project, supported by the U.K. Government's Darwin Initiative and the Royal Botanic Gardens, Kew ('Plantas do Nordeste' programme). Great emphasis is being placed on the need to preserve these floristic refuges, which, *inter alia*, include disjunct northern populations of cactus epiphytes, such as *Lepismium cruciforme* and *Rhipsalis crispata*. A protected area including one of these *brejos* is the 'Reserva Biológica Federal da Serra Negra', Pernambuco, which includes a population of the monotypic *Brasiliopuntia brasiliensis*.

Further south, in coastal Bahia (up to 100 km inland), between the capital Salvador and Teixeira de Freitas, where annual rainfall is generally in excess of 1750 mm, there are occasional records of various species of Rhipsalideae, indicating a once rich centre of diversity, including *Hatiora cylindrica* ('Vulnerable'), *Rhipsalis crispata, R. paradoxa* subsp. *septentrionalis* ('Endangered'), *R. baccifera* subsp. *hileiabaiana* & *R. russellii* (both 'Vulnerable') and *R. oblonga*. With so little forest remaining, post 1971, when road BR 101 was completed, it seems reasonable to assume that all of these are threatened to a significant extent. Hopefully, some may benefit from protection in local reserves, such as the 'Reserva Biológica Federal de Una' (south of Ilhéus, BA) and 'Parque Nacional de Monte Pascoal' (N of Itamaraju, BA) — other, smaller protected areas are described by Wayt Thomas in Davis *et al.* (1997: 367). Table 4.1 lists *brejo* forests in NE Brazil (see Map

Table 4.1

Brejo forests in North-eastern Brazil with notable Cactaceae and their most recent collection dates.

Precise localities are given under materials cited for the above in Chapter 5.

Locality of brejo (state)	Cactaceae (date of most recent record), notes
Serra de Baturité (Ceará)	*Rhipsalis baccifera* subsp. *baccifera* (1979)
N slope of Chapada do Araripe (Ceará)	*Pereskia grandifolia* subsp. *grandifolia* (1971)
Areia (Paraíba)	*Rhipsalis baccifera* subsp. *baccifera* (1998)
Pico do Jabre (Paraíba)	*Melocactus ernestii* (1998) — see Rocha & Agra (2002)
Taquaritinga do Norte (Pernambuco)	*Lepismium cruciforme* & *Rhipsalis floccosa* (2001), *R. cereuscula* (1972)
Poção (Pernambuco)	*Pereskia aculeata* (1991)
Brejo da Madre de Deus (Pernambuco)	*Rhipsalis floccosa* (1980), *R. cereuscula* (1995), *Melocactus ernestii* (1993)
Caruaru (Pernambuco)	*Rhipsalis crispata* (1970), *R. floccosa* (1971) — these species records urgently need re-confirmation in habitat
Serra Negra, Inajá (Pernambuco)	*Brasiliopuntia brasiliensis* & *Rhipsalis lindbergiana* (1995)
Serra do Cumanati (Pernambuco)	*Rhipsalis floccosa* (1969) — this brejo may have been destroyed, since it does not figure in Rodal *et al.* (1998)
Mata Grande (Alagoas)	*Brasiliopuntia brasiliensis* (1974)
Serra de São José, Feira de Santana (Bahia)	*Pereskia aculeata, Brasiliopuntia brasiliensis, Rhipsalis floccosa, R. lindbergiana* & *Melocactus ernestii* (all 2001)
Rui Barbosa (Bahia)	*Pereskia aculeata* (1973), *Brasiliopuntia brasiliensis* (1978)
Monte Verde, Itaberaba (Bahia)	*Pereskia aculeata* (1973)
Serra da Jibóia, Santa Teresinha (Bahia)	*Rhipsalis baccifera* subsp. *hileiabaiana, R. crispata* & *Melocactus ernestii* (2002)
Três Braços, Ubaíra (Bahia)	*Lepismium cruciforme* & *Rhipsalis floccosa* (1994)
Jaguaquara (Bahia)	*Pereskia aculeata, Brasiliopuntia brasiliensis, Rhipsalis russellii* (all 1915) — brejo assumed to have been destroyed
Venceslau Guimarães (Bahia)	*Hatiora cylindrica* (1993) — in protected area?
Almadina/Floresta Azul (Bahia)	*Rhipsalis russellii* (before 1966), *R. baccifera* subsp. *hileiabaiana* (before 1966), *R. cereuscula* (1972)
Barra do Choça (Bahia)	*Rhipsalis floccosa, R. paradoxa* subsp. *septentrionalis, R. lindbergiana, Hatiora cylindrica* (all 2003)
Juçari/Jussari (Bahia)	*Rhipsalis paradoxa* subsp. *septentrionalis* (before 1966)
Camacã/Camacan (Bahia)	*Lepismium cruciforme* (2003), *Rhipsalis russellii* (1971), *R. oblonga* (1971), *R. paradoxa* subsp. *septentrionalis* (before 1979)

6) including notable cacti, with dates of most recent collections/observations. Unfortunately, it is likely that many of these are now much altered or destroyed. Those that survive merit regular monitoring (Rodal *et al.* 1998). Also part of the Brazilian Nordeste, is the Archipelago of Fernando de Noronha, a Federal Environment Protection area. These Atlantic islands are home to an endemic cactus, *Cereus insularis* (a close relative of the Brazilian coastal *C. fernambucensis*), which seems adequately protected at present.

The locations of reserves that potentially assist the conservation of threatened cacti in Eastern Brazil are indicated on Map 15, which also marks the sites of rare and threatened or narrowly endemic taxa that are currently without protected areas.

4.3 CRITERIA FOR THE IDENTIFICATION OF PRIORITY TAXA

The internationally agreed process for assessing the level of threat of extinction to living organisms is that established by IUCN, whose system aims, as far as is possible, to provide an objective means of determining Categories of Threat, or 'Red List Categories' (IUCN 2001). The principal categories employed in the IUCN system are 'Extinct (EX)', 'Extinct in the Wild (EW)', 'Critically Endangered (CR)', 'Endangered (EN)', 'Vulnerable (VU)' and 'Least Concern (LC)'. The series of standard criteria that lead to their determination should be stated when an assessment is published, in order that the category can be easily verified or reassessed in the future, as circumstances change. The Categories CR, EN & VU are defined by any one of 5 sets of criteria, which are of a consistent type throughout, but differ in degree for each Category.

It is a relief to be able to report that the present study has not identified any Cactaceae native of Eastern Brazil that belong in either the 'Extinct' or 'Extinct in the Wild' categories. However, there is no shortage of taxa whose assessments place them in the remaining categories. Besides the categories for threatened taxa listed above, there is also 'Near Threatened (NT)' for taxa that are close to qualifying as VU and 'Data Deficient (DD)' where there is inadequate information to make a direct or indirect assessment of risk of extinction based on geographical distribution and/or population status. This last subcategory applies to 10 out of the 162 taxa covered in the present study, while a further 3 present taxonomic difficulties and together with the few natural hybrids have not been assessed, being referred to as 'Not Evaluated (NE)'. The assessments made in this study are recorded under each taxon in the taxonomic inventory (Chapter 5) and are based on the total range of the taxon, whether or not it is endemic to Eastern Brazil as defined here. They were communicated to IUCN, via the Species Survival Commission's Cactus & Succulent Specialist Group, in August 2001 (some have been revised here).

Farjon & Page criteria. While the standard methodology embodied in the published IUCN system has been employed here, it has recently been noted by Farjon & Page (1999) that its application can result in rather long lists of equally threatened taxa. This suggests that further prioritisation is desirable if the limited resources currently applied to nature conservation are not to be spread too thinly. Farjon & Page (1999: 28) have devised a novel additional formula for achieving such prioritisation, which has proved eminently capable of being applied here. This formula calculates a score based on the Category of Threat, where arbitrarily assigned values of CR=4, EN=3, VU=2 and LC/NT=1, are multiplied by the sum of 3 other criteria, namely 'Phylogenetic Distinction', 'Ecological Importance' and 'Genetic Diversity', whose ranges of values are given below. The resulting total score effectively short-lists and prioritises between taxa with the same IUCN rating, as can be seen for the cacti of Eastern Brazil in Table 4.2. The Farjon & Page criteria are defined as follows:

Phylogenetic distinction **(PD).** This is a measure of the relative taxonomic isolation of an organism and recognises that not all taxa of the same rank are equal. Thus, if the taxon is representative (a) of a monotypic genus, it scores 4; (b) of a species or infraspecific rank of a small genus (2–5 species) or monotypic infrageneric rank within any genus, it scores 3; (c) of a species of a larger genus (>5 species★), it scores 2; and (d) of an infraspecific rank of a species of a larger genus, it scores 1.

Ecological importance **(EI).** This was devised with forest conifers in mind, but can be applied equally to cacti, which are also woody and frequently take the place, or are an important component of arborescent vegetation in dryland ecosystems. Here, if a taxon is co-dominant in a distinct vegetation type, it scores 2; if it is only a more minor constituent of the vegetation, it scores 1. (Farjon & Page, l.c., give a score of 3 to what they call 'keystone species of a biotic community', but none of the cacti treated here seems to fall into this category, although some of the arborescent cereoids arguably come close and those found elsewhere would definitely qualify, eg. some Mexican Pachycereeae.)

Genetic diversity **(GD).** This criterion distinguishes between taxa known to be unusually rich in regional diversity, often expressed in high levels of morphological variation or varied ecological adaptation, from those whose genetic diversity is less. This is the most subjective of the Farjon & Page criteria, at least if based only on morphological and ecological assessments, because it is known from modern studies of conservation genetics that taxa which appear to be variable may have low levels of measurable genetic diversity and *vice versa*. However, since it will be many years before the population genetics of the cacti of Eastern Brazil have been properly investigated in the laboratory, use of this criterion remains justifiable as the best option available. Taxa displaying genetic diversity within and/or between regional populations receive a score of 2 (eg. *Uebelmannia gummifera* and *Espostoopsis dybowskii*), less diverse taxa scoring 1 (eg. *Melocactus violaceus* subsp. *ritteri*) — see Chapter 5 for details in each case.

The short-list formula is, therefore, calculated as follows: IUCN category of threat (score 1–4) × (PD [score 1–4] + EI [score 1–2] + GD [score 1–2]); see Table 4.2 (below).

More sophisticated uses of phylogenetics and population genetics for assessing conservation priorities are discussed by Linder (1995), in relation to the Southern African orchid genus, *Herschelia*, but such methods require a considerably more detailed knowledge of relationships and biology/ecology than is currently available for most Brazilian cacti.

★ Farjon & Page, l.c., divide small and larger genera on the basis of '2–5' *versus* '>6' species, leaving genera with exactly 6 species in limbo. The authors are grateful to Aljos Farjon (pers. comm., 30.11.99) for clarifying that the stated '>6' should be read as >5 species.

Table 4.2

'League-table' of prioritised Red List categories for all taxa assessed as CR, EN or VU.

These have short-list scores ranging from 24 to 6. The 65 taxa below represent 40% of the 162 Cactaceae native to Eastern Brazil considered in this exercise (including 3 that were 'Not Evaluated'). The full range of scores was from 24 down to 3 and it is important to stress that many of the remaining 94 taxa rated as 'Data Deficient', 'Near Threatened' or 'Least Concern' also merit conservation action, including taxonomic field studies and monitoring. *Key.* † = Convention on International Trade in Endangered Species (CITES): trade in wild-collected CITES Appendix I taxa and their derivatives is prohibited, while those in App. II and artificially propagated App. I taxa may be traded under export/import licences issued by national management authorities.

Red List Categories prioritised by conservation short-list scores (see Chapter 5): priority order/taxon name/(Brazilian state code)	IUCN Red List Category [criteria]/ CITES App.†	Area code (Chapter 3.3)	References to existing and proposed protected areas and recommendations
CRITICALLY ENDANGERED (CR)			
24 1. *Melocactus deinacanthus* (BA)	CR [B1ab(iii) + 2ab(iii)]/App. I	(3b:iii)	**A cluster of small sites is recommended for protection here (see below).**
20 2. *Cipocereus pusilliflorus* (MG)	CR [D]/App. II	(3c:iii)	Taylor in Oldfield (1997): 144, prop. 105(e).
2. *Melocactus glaucescens* (BA)	CR [B1ab(iii) + 2ab(iii)]/App. I	(3c:ii)	Offered protection by the Parque Estadual do Morro do Chapéu. **Protected status for the new populations discovered in 2000–03 in Mun. Morro do Chapéu should be sought. Needs monitoring!**
2. *Uebelmannia buiningii* (MG)	CR [C2a]/App. I	(4c:ii:c)	Costa *et al.* (1998): 62, prop. C7; Taylor in Oldfield (1997): 144, prop. 105(l). **Monitoring essential.**
16 3. *Pilosocereus azulensis* (MG)	CR [D]/App. II	(3d:ii)	Taylor in Oldfield (1997): 143, prop. 105(a).
3. *Micranthocereus streckeri* (BA)	CR [B1ab(iii) + 2ab(iii); C2a(ii); D]/App. II	(3c:ii)	Taylor in Oldfield (1997): 144, prop. 105(c).
3. *Coleocephalocereus purpureus* (MG)	CR [C2b]/App. II	(3d:ii)	Taylor in Oldfield (1997): 143, prop. 105(a).

3. *Melocactus conoideus* (BA) 2ab(iii,iv,v)]/App. I	CR [B1ab(iii,iv,v) +	(3d:i:b)	**Adequate protection for a recently designated site is needed (see below).** Re-introduction is planned.
4. *Pilosocereus fulvilanatus* subsp. *rosae* (MG)	**12** CR [B1ab(iii) + 2ab(iii)]/App. II	(4c:ii:b)	Taylor in Oldfield (1997): 144, prop. 105(j).
4. *Melocactus pachyacanthus* subsp. *viridis* (BA) App. II	CR [B2ab(i,ii,iii, iv); C2a(ii); D]/	(3b:ii)	**Protected status for the populations discovered in 2002–03 in Mun. Morro do Chapéu should be sought (see below).**
4. *M. violaceus* subsp. *ritteri* (BA) App. II	CR [B2ab(iii,v)]/	(3d:i:a)	**Sites are recommended for protection here (see below).**

ENDANGERED (EN)

24 1. *Espostoopsis dybowskii* (BA) App. II	EN [B2ab(iii)]/	(3d:i:a)	Taylor in Oldfield (1997): 143, prop. 105(b). **A second site is recommended for protection here (see below).**
18 2. *Cipocereus minensis* subsp. *minensis* (MG)	EN [B2ab(iii)]/ App. II	(4c:iii)	North-eastern part of range included in the Parque Nacional Serra do Cipó.
2. *Arrojadoa dinae* subsp. *eriocaulis* (BA/MG)	EN [B2ab(iii)]/ App. II	(3c:iii)	Costa *et al.* (1998): 64, prop. N22; Taylor in Oldfield (1997): 144, props 105(f/h).
2. *Arthrocereus glaziovii* (MG)	EN [B1ab(i,ii,iii, iv,v) + 2ab(i,ii,iii, iv,v)]/App. II	(4c:iii)	Costa *et al.* (1998): 63, prop. C10.
15 3. *Brasilicereus markgrafii* (MG)	EN [B1ab(iii) + 2ab(iii)]/App. II	(4c:ii:a)	Protected in the Parque Estadual Serra do Barão, Grão Mogol.
3. *Cipocereus laniflorus* (MG)	EN [D]/App. II	(4c:iii)	Adequately protected in the RPPN do Caraça, Catas Altas.
3. *C. bradei* (MG)	EN [B1ab(iii) + 2ab(iii)]/App. II	(4c:ii:b)	Costa *et al.* (1998): 63, prop. C6; Taylor in Oldfield (1997): 144, prop. 105(i).
3. *Melocactus azureus* (BA)	EN [B1ab(i,ii,iii, iv) + 2ab(i,ii,iii, iv)]/App. II	(3b:ii)	Conservation *ex situ* may be the only viable option unless population discovered in 2002 can be adequately protected.
3. *Discocactus bahiensis* (BA/PE?/PI/CE)	EN [B2ab(i,ii,iii, iv,v)]/App. I	(3a:ii)	**A site is recommended for protection here (see below).**

ENDANGERED (EN) (continued)

12			
4. *Schlumbergera kautskyi* (ES)	EN [B2ab(iii)]/ App. II	(2d:ii)	**Sites are recommended for protection here (see below).**
4. *Pilosocereus aurisetus* subsp. *aurilanatus* (MG)	EN [B1ab(iii) + 2ab(iii)]/App. II	(4c:ii:b)	Costa *et al.* (1998): 63, prop. C6; Taylor in Oldfield (1997): 144, prop. 105(i).
4. *Micranthocereus auriazureus* (MG)	EN [B1ab(iii) + 2ab(iii)]/App. II	(4c:ii:a)	Protected in the Parque Estadual Serra do Barão, Grão Mogol.
4. *M. polyanthus* (BA)	EN [B2ab(iii)]/ App. II	(3c:iii)	Taylor in Oldfield (1997): 144, prop. 105(d).
4. *Coleocephalocereus fluminensis* subsp. *decumbens* (MG)	EN [B1ab(iii) + 2ab(iii)]/App. II	(2d:i)	Costa *et al.* (1998): 67, prop. E7?
4. *Melocactus ferreophilus* (BA)	EN (B1/2c–e)/ App. II	(3b:ii)	Protected in the APA Gruta dos Brejões.
4. *Melocactus pachyacanthus* subsp. *pachyacanthus* (BA)	EN [B1ab(i,ii,iii, iv)]/App. II	(3b:ii)	Protected in the APA Gruta dos Brejões, Plate 1.2.
4. *M. paucispinus* (BA)	EN [B2ab(v)]/ App. I	(3c:i)	Taylor in Oldfield (1997): 144, prop. 105(c). Now protected on a private reserve at Morro do Chapéu.
4. *Discocactus pseudoinsignis* (MG)	EN [B1ab(iii) + 2ab(iii)]/App. I	(4c:ii:a)	Protected in the Parque Estadual Serra do Barão, Grão Mogol.
4. *D. horstii* (MG)	EN [B1ab(iii) + 2ab(iii)]/App. I	(4c:ii:a)	Protected in the Parque Estadual Serra do Barão, Grão Mogol.
9			
5. *Rhipsalis paradoxa* subsp. *septentrionalis* (PE/BA/MG/ES)	EN [B2ab(iii)]/ App. II	(2a)	Costa *et al.* (1998): 67, prop. C12?

VULNERABLE (VU)

14			
1. *Facheiroa cephaliomelana* subsp. *cephaliomelana* (BA/MG)	VU [B1ab(iii)]/ App. II	(3b:iii)	Taylor in Oldfield (1997): 143–144, prop. 105(c)(ii).
12			
2. *Pseudoacanthocereus brasiliensis* (BA/MG)	VU [C1 + 2a(i)]/ App. II	(3d:i:a)	Taylor in Oldfield (1997): 143, prop. 105(a). May be offered some protection in the APA do Lago de Pedra do Cavalo, Bahia.
2. *Arrojadoa bahiensis* (BA)	VU [D2]/App. II	(3c:ii)	Probably not a high priority for further protection.
2. *A. dinae* subsp. *dinae* (BA/MG)	VU [B1ab(iii)]/ App. II	(3c:iii)	Taylor in Oldfield (1997): 144, prop. 105(e).

2. *Uebelmannia gummifera* (MG)	VU [B1ab(iii,v)]/ App. I	(4c:ii:c)	Some populations may be offered protection inside the Parque Estadual da Serra Negra.
2. *U. pectinifera* subsp. *pectinifera* (MG)	VU [B1ab(iii,v)]/ App. I	(4c:ii:c)	Costa *et al.* (1998): 62, prop. C7; Taylor in Oldfield (1997): 144, prop. 105(k). **Monitoring essential.**
2. *U. pectinifera* subsp. *flavispina* (MG)	VU [B1ab(iii,v)]/ App. I	(4c:ii:c)	Costa *et al.* (1998): 62, prop. C7; Taylor in Oldfield (1997): 144, prop. 105(k). **Monitoring essential.**
2. *U. pectinifera* subsp. *horrida* (MG)	VU [D2]/App. I	(4c:ii:b)	Costa *et al.* (1998): 62, prop. C7. **Monitoring essential.**
10 3. *Tacinga werneri* (BA/MG)	VU [B2ab(ii,iii,iv, v)]/App. II	(3d:i:b)	Taylor in Oldfield (1997): 143, prop. 105(a/b).
3. *Rhipsalis russellii* (BA/MG/ES)	VU [B2ab(iii,iv)]/ App. II	(1b)	Part of range incl. within Parque Nacional Chapada Diamantina (nr Mucugê, BA) and, possibly, Parque Nac. de Monte Pascoal etc. (S Bahia).
3. *R. pilocarpa* (MG/ES)	VU [B2ab(i,ii,iii, iv)]/App. II	(2d:ii)	**See site recommended for** *Schlumbergera kautskyi* **here.**
3. *Cipocereus crassisepalus* (MG)	VU [B1ab(iii) + 2ab(iii)]/App. II	(4c:ii:c)	Some populations may be offered protection inside the Parque Estadual da Serra Negra.
3. *Facheiroa cephaliomelana* subsp. *estevesii* (BA)	VU [D2]/App. II	(3b:iii)	Taylor in Oldfield (1997): 143, prop. 105(c): (i).
3. *Arthrocereus melanurus* subsp. *melanurus* (MG)	VU [B1ab(iii)]/ App. II	(4c:iii)	Costa *et al.* (1998): 64, prop. S9.
3. *A. melanurus* subsp. *odorus* (MG)	VU [D2]/App. II	(4c:iii)	Part of range incl. in Parque Nacional Serra do Cipó.
3. *A. rondonianus* (MG)	VU [D2]/App. II	(4c:ii:b)	Costa *et al.* (1998): 63, prop. C6; Taylor in Oldfield (1997): 144, prop. 105(i).
3. *Discocactus placentiformis* (MG)	VU [B2ab(iii,v)]/ App. I	(4c:i)	Costa *et al.* (1998): 62, prop. C7; Taylor in Oldfield (1997): 144, prop. 105(g, i–k).
8 4. *Pereskia aureiflora* (BA/MG)	VU [A2c + 3c]/ App. II	(3a:iii:c)	Taylor in Oldfield (1997): 143, prop. 105(a).
4. *Tacinga braunii* (MG)	VU [B1ab(iii) + 2ab(iii)]/App. II	(3d:ii)	Taylor in Oldfield (1997): 143, prop. 105(a).
4. *Rhipsalis crispata* (PE/BA)	VU [B2ab(iii)]/ App. II	(2a)	**Field studies are required to determine if it is still extant in PE and monitoring in BA.**

VULNERABLE (VU) score 8 (continued)

4. *R. cereoides* (ES)	VU [B2ab(iii)]/ App. II	(2d:ii)	**See site recommended for** *Schlumbergera kautskyi* **here.**
4. *Hatiora cylindrica* (BA/ES?)	[VU B2ab(iii)]/ App. II	(2a)	In theory protected at some of its Bahian localities, but these are likely to remain under threat.
4. *Cereus mirabella* (MA/BA/MG)	VU [B2ab(iii)]/ App. II	(4a)	**Known populations require regular monitoring.**
4. *Pilosocereus brasiliensis* subsp. *brasiliensis* (ES/RJ)	VU [B2ab(iii)]/ App. II	(2d:ii)	**Sites are recommended for protection here (see below).**
4. *P. fulvilanatus* subsp. *fulvilanatus* (MG)	VU [B1ab(iii)]/ App. II	(4c:ii:a)	Protected in the Parque Estadual Serra do Barão, Grão Mogol.
4. *Micranthocereus violaciflorus* (MG)	VU [D2]/App. II	(3c:iii)	Protected in the Parque Estadual Serra do Barão, Grão Mogol.
4. *Coleocephalocereus buxbaumianus* subsp. *flavisetus* (MG)	VU [B2ab(iii)]/ App. II	(2d:i)	**A site is recommended for protection here (see below).**
4. *Melocactus violaceus* subsp. *violaceus* (RN/PB/PE/SE/BA/MG/ES)	VU [A3c]/App. II	(1b)	**Recorded from within 3 small protected areas (PB/ ES/RJ), but others are needed to maintain genetic diversity.**
4. *M. violaceus* subsp. *margaritaceus* (AL/SE/BA)	VU [A3c]/App. II	(2c)	Protected in the Serra de Itabaiana reserve (IBAMA), Sergipe and Res. Ecol. Dunas de Abaete, Bahia. **Additional reserves needed.**
4. *Discocactus zehntneri* subsp. *boomianus* (BA)	VU [D2]/App. I	(3c:ii)	Offered protection by the Parque Estadual do Morro do Chapéu.
4. *D. heptacanthus* subsp. *catingicola* (PI/BA/MG)	VU [B2ab(iii)]/ App. I	(4b)	**Needs regular monitoring.**
6 5. *Tacinga saxatilis* subsp. *estevesii* (BA)	VU [D2]/App. II	(3b:iii)	Taylor in Oldfield (1997): 143, prop. 105(c): (i).
5. *Rhipsalis baccifera* subsp. *hileiabaiana* (BA)	VU [B2ab(i,ii,iii, iv)]/App. II	(1b)	**The possibility for its protection inside the area of CEPLAC (Ilhéus, BA) should be investigated.**
5. *Pilosocereus floccosus* subsp. *quadricostatus* (MG)	VU [B2ab(iii)]/ App. II	(3d:ii)	Taylor in Oldfield (1997): 143, prop. 105(a).

4.4 CONSERVATION HOTSPOTS

The above short-listing process can also be used to identify geographical areas of high priority for attention by conservationists, these nowadays commonly being referred to as 'hotspots' (Reid 1998, Myers *et al.* 2000). Here (Table 4.3), the short-list scores of the 65 threatened taxa in Table 4.2 have been summed according to the geographical areas they are characteristic of (Chapter 3), giving a prioritised list of these area subdivisions, which can be analysed at different levels. This indicates that of the First Order subdivisions, the South-eastern *campos rupestres* (4c) are by far the most important for attention by conservationists and land managers, having a score nearly twice that of any of the others. This subdivision is followed by 3 with broadly similar scores: 2nd the Northern *campos rupestres* (3c), 3rd the Eastern *caatingas-agrestes/campos rupestres* (3d) and 4th the *Caatingas* of the Rio São Francisco (3b). In 5th and last place is the Southern subhumid/humid forests and inselbergs (2d), which is part of the Atlantic Forest (Area 2), whose present state of destruction is so great that all remaining areas merit immediate preservation. Unfortunately, cactus diversity can only play a small role in the biological and other arguments in support of this obvious need.

Table 4.3

'League-table' of principal conservation area hot-spots based on a summation of the scores presented in Table 4.2. See also Map 16.

Priority hot-spots (sum of taxon scores from Table 4.2)	Endemics: no. of taxa × score
A). Second order Area subdivisions	
1 (71). SE *campos rupestres* (MG): Southern part (4c:iii)	2×18, 1×15, 2×10
2 (70). Northern *campos rupestres*: N Serra do Espinhaço (3c:iii)	1×20, 1×18, 2×12, 1×8
3 (66). SE *campos rupestres* (MG): Diamantina & E (4c:ii:c)	1×20, 3×12, 1×10
4 (61). SE *campos rupestres* (MG): Serra do Cabral etc. (4c:ii:b)	1×15, 3×12, 1×10
5 (59). SE *campos rupestres* (MG): Grão Mogol etc. (4c:ii:a)	1×15, 3×12, 1×8
6 (56). N *campos rupestres*: Chapada Diamantina, BA (3c:ii)	1×20, 1×16, 1×12, 1×8
7 (54). Rio São Francisco *caatingas*: southern *caatingas* (3b:iii)	1×24, 1×14, 1×10, 1×6
8 (51). Rio São Francisco *caatingas*: cent.-northern Bahia (3b:ii)	1×15, 3×12
9 (48). Eastern *caatingas-agrestes/campos rupestres* (3d:i:a)	1×24, 2×12
10 (46). South-eastern *caatingas-agrestes* (3d:ii)	2×16, 1×8, 1×6
B). First order Area subdivisions	
1 (267). South-eastern *campos rupestres*, Minas Gerais (4c)	1×20, 2×18, 3×15, 9×12, 5×10, 1×8
2 (138). Northern *campos rupestres*, Bahia & Minas Gerais (3c)	2×20, 1×18, 1×16, 4×12, 2×8
3 (120). Eastern *caatingas-agrestes/campos rupestres* (3d)	1×24, 3×16, 2×12, 1×10, 1×8, 1×6
4 (105). Rio São Francisco *caatingas*, Bahia & Minas Gerais (3b)	1×24, 1×15, 1×14, 3×12, 1×10, 1×6
5 (58). Southern subhumid/humid forest/inselbergs, MG/ES (2d)	2×12, 1×10, 3×8

At the level of Second Order subdivisions (Table 4.3: A), the top 5 are all contiguous *campo rupestre* areas, followed by the *campos rupestres* of the Chapada Diamantina (6th) and central-southern *caatingas* (7th–10th). Thus, in terms of conservation importance for cacti, the *campos rupestres* stand well in front of the *caatingas*, even though the threats to the former environments are generally of lesser significance (see Map 16). Nevertheless, all subdivisions include various taxa of key conservation importance.

4.5 PRIORITY ACTIONS RECOMMENDED

Since the late 1970s when eminent Brazilian botanists, such as Dárdano de Andrade-Lima (1981) and Nanuza de Menezes (many verbal presentations), drew attention to the levels of modification and threats being suffered by the *caatingas* and *campos rupestres*, only relatively few and mostly very small protected areas have been created in the more than one million square kilometres of land surface in which these major ecosystems are represented in Eastern Brazil. The various federal and state reserves offering protection to Cactaceae have been mentioned already, but many more are needed (eg. see Map 15), as has been indicated by recent symposia and studies conducted by international, governmental and non-governmental interests (Oldfield 1997, Costa *et al.* 1998). The most encouraging development of recent times is the evolution of various legal instruments that offer Brazilian municipalities and citizens opportunities for the setting aside for conservation reasons of natural areas they oversee or own (cf. Costa *et al.* 1998). In addition to these, Brazilian cactus enthusiasts, such as Marlon Machado, are persuading municipal authorities and private land owners to establish local protected areas and take pride in the rare taxa and unusual habitats that they were probably unaware of until very recently. Such developments are to be greatly applauded and, none more so, than the remarkable achievement represented by the publication of the comprehensive report, entitled 'Biodiversidade em Minas Gerais' (Costa *et al.* 1998). This places the collective knowledge of conservation priorities for the state of Minas Gerais on a level that all other Brazilian states from the area covered in this study need to emulate. It is only to be hoped that the recommendations in this ground breaking report will be actioned by the authorities empowered to do so, and especially, since the present study identifies by far the greatest concentration of conservation priority taxa as endemics of Minas Gerais (see Tables 3.1, 4.2 & 4.3, Maps 15 & 16). The same aspirations apply to the need for protected areas and associated actions detailed in the IUCN-SSC Cactus & Succulents Action Plan (Oldfield 1997). The following additional actions are recommended here:

• *Establishment of local reserves/protected areas* (A1 = taxon in CITES App. I):
1. For the highest scoring taxon in the 'Critically Endangered' category in Table 4.2, *Melocactus deinacanthus* (A1), known for certain from only a small cluster of sites within a few kilometres north and east of the type location at 'Morro da Barriguda, Juá', Mun. Bom Jesus da Lapa (at the border of Mun. Riacho de Santana), southern Bahia, to protect it from agricultural activities including depredation by livestock and collecting.

2. Similarly, for the highest priority taxon in the 'Endangered' category in Table 4.2, *Espostoopsis dybowskii*, in the Município of Jaguarari, northern Bahia, to complement that recommended for its morphologically distinct southern populations (Oldfield 1997: 143) and to protect it from urban expansion.

3. Designation as a reserve of the area known as 'Lagedo Bordado' and its surroundings (course of the Rio Salitre, N of Brejão, W of Icó, Mun. Morro do Chapéu, Bahia) for recently discovered, contiguous populations of the Critically Endangered *Melocactus pachyacanthus* subsp. *viridis* and *M. glaucescens* (A1). Also, expansion of the APA Gruta dos Brejões to include an adjacent population of the former taxon discovered in 2003.

4. A reserve for *Melocactus conoideus* (A1), above Mun. Vitória da Conquista, Bahia, has recently been designated, but needs better enforcement, since habitat destruction caused by gravel extraction and urban expansion continue. Establishment of a secure area will be essential if the re-introduction programme planned for this species by staff from the Univ. Estadual Sudoeste da Bahia is to succeed.

5. For *Discocactus bahiensis* (A1), in an area of former flood plain of the Rio São Francisco, near Rodeadouro, WSW of Juazeiro, northern Bahia. More than half of this population was recently destroyed by the construction of an embanked asphalt road, but healthy plants remain at the western edge of the former site dominated by bushes of jurema preta, *Mimosa tenuiflora*.

6. For *Schlumbergera kautskyi*, at mountain sites (inselbergs) in Mun. Domingos Martins (Pico da Pedra Azul) and Mun. Alfredo Chaves (São Bento de Urânia), Espírito Santo, from which region a range of other rare Rhipsalideae is also reported: *Rhipsalis pacheco-leonis* subsp. *catenulata*, *R. cereoides*, *R. sulcata*, *R. burchellii*, *R. pilocarpa* and, perhaps, *R. hoelleri*. This region is being developed as a high cost residential and tourism area, alongside the appearance of plantations of the dreaded *Eucalyptus*.

7. For *Melocactus paucispinus* (A1), at its sites around the Pico das Almas, Mun. Rio de Contas/Érico Cardoso (Água Quente), Chapada Diamantina, Bahia. (It is already receiving protection at Morro do Chapéu.) Inclusion of the Pico das Almas itself would also protect a population of the remarkable *Arrojadoa bahiensis*.

8. For *Melocactus violaceus* subsp. *ritteri*, at its two sites near to the towns of Jacobina and Rui Barbosa, Bahia, to protect it from urban expansion and general habitat disturbance, as well as from collectors.

9. For any surviving western populations of *Coleocephalocereus buxbaumianus* subsp. *flavisetus*, such as that recently found at Carmo da Mata, south-western Minas Gerais (see Chapter 5).

10. For *Pilosocereus brasiliensis* subsp. *brasiliensis* along the coast at Guarapari and southwards (Espírito Santo), which could also protect one or more populations of the Vulnerable *Melocactus violaceus* subsp. *violaceus*.

• *Establishment of* ex situ *gene banks to enable future re-introductions, if Extinction in the Wild should occur:*

1. Specifically for *Melocactus azureus*, whose known habitats are in imminent peril of destruction and hold wild populations that are highly fragmented or numbering only tens

of individuals. It is desirable that projects for the cold storage of seeds in one or more seed banks (Dickie *et al.* 1990, Yang 1999), supplemented by *ex situ* gene banks of living plants under controlled conditions of pollination (to avoid hybridization), are established. However, this should not detract from any efforts to conserve wild populations, if sufficiently intact habitats can be protected, such as that located by Aloísio Cardoso in 2002. 2. The above actions are equally applicable to many of the taxa discussed in this chapter, as well as others in Chapter 5, where it is difficult to identify secure protected areas, or where there is a severe risk of catastrophic reduction in genetic diversity. In the latter case, seed banking may provide an effective insurance, if collecting programmes are carefully designed.

• *Monitoring of wild populations*

Regular monitoring of the status in the wild of many of the taxa documented in Chapter 5 is essential, whether or not these are currently rated as threatened (eg. those listed in Table 4.1, above). Many taxa presently assessed as 'Least Concern' or 'Near Threatened' are likely to become 'Vulnerable' in due course and could become yet more seriously threatened shortly thereafter, if timely conservation actions are not triggered. In particular, populations of *Uebelmannia pectinifera* and *U. buiningii* need regular investigation to assess the ongoing impact of collecting for the horticultural trade. The latter species is already 'Critically Endangered' and the former could easily pass from 'Vulnerable' to 'Endangered' if a few more of its sites are ransacked. In addition to monitoring known populations there is also the need for field studies of those taxa that are apparently restricted to only one or two small sites in the hope that new populations can be located and protected before their habitats are irreversibly modified. Examples include a number of the taxa listed in Table 4.2: *Rhipsalis crispata, Melocactus deinacanthus, Cipocereus pusilliflorus, Micranthocereus streckeri, Pilosocereus azulensis, P. fulvilanatus* subsp. *rosae, Coleocephalocereus purpureus* and *C. fluminensis* subsp. *decumbens*. Discovery of one or two additional localities for any of these could make a considerable difference to our perception of their conservation status. While some may be truly restricted in range, we know that significant extensions are possible when similar habitats are located. Thus, only recently the range of the very rare *Micranthocereus violaciflorus* was considerably extended, from a single small site in the region of Grão Mogol to a new one at Serranópolis. Locating a third site for this taxon might well remove it from the 'Vulnerable' to the 'Least Concern' category of risk. Likewise, the discovery between 2000 and 2003 of four additional sites for the 'Critically Endangered' *Melocactus glaucescens,* some 40 and 80 km from its previously known area, increases our optimism that it can be saved.

5. TAXONOMIC INVENTORY

5.1 INTRODUCTORY NOTES

As its scope this account attempts to include all taxa of Cactaceae that are native, naturalized or commonly cultivated outdoors in North-eastern Brazil, and in South-eastern Brazil north of 22°S and east of 46°W (entries for non-native taxa are indicated by an asterisk [*] below). However, the focus from the start has been on the terrestrial, rather than epiphytic taxa, the former having a centre of high endemism in the dry environments comprising south-eastern Piauí, southernmost Ceará, the southern half of Paraíba, Pernambuco, Bahia, Alagoas, Sergipe, Espírito Santo, most of Minas Gerais and the northern half of Rio de Janeiro (Map 1, box). While epiphytic taxa native within this area are fully treated below, their main centre of diversity in SE Brazil, around 22–26°S, has been intentionally excluded, since a preliminary account of the all-important tribe Rhipsalideae was published quite recently (Barthlott & Taylor 1995).

Except where expressly limited, synonymy aims to be comprehensive for names based on Brazilian types, but all relevant botanical nomenclature is included in the 'Index to botanical names and epithets' (Chapter 7.2). Also included for some accepted names and synonyms are references to good illustrations. Vernacular names are listed from various sources, eg. Menezes (1949), and as reported to the authors by 'sertanejos'. The morphological descriptions do not aim in principle to account for all aspects of the plant, but are meant to be primarily diagnostic, supplementing details in the dichotomous keys to facilitate identification. In some cases they are comprehensive where the data have been available or when the differences between taxa are particularly subtle. Details mentioned in descriptions of genera, subgenera or other infrageneric categories, which should be assumed to apply to all subordinate taxa, are not always repeated in descriptions of the included species etc. Unqualified measurements refer to length (or height in the case of erect plants); those connected by a multiplication sign (×) refer to length followed by width/diameter and, occasionally, also thickness, in the case of laterally compressed structures. Dimensions of the whole plant or its stem are always given exclusive of the spines; and distances between structures that themselves have a size, eg. between areoles on a rib, are given from organ centre to centre. Specialized terms are defined in the Glossary (Chapter 7.1).

Besides nomenclatural and diagnostic/descriptive details, bibliographical, phytogeographical and ecological data (see Chapter 3) are provided, likewise distribution records and conservation assessments (IUCN 2001 Categories of Threat and conservation short-listing values, including Phylogenetic Distinction [PD], Ecological Importance [EI] and Genetic Diversity [GD] — see Chapter 4 and cf. Golding & Smith 2001). Where appropriate, commentary is also provided on some historical aspects, other taxonomic treatments, relationships, uses and the need for further studies.

All materials cited have been seen unless indicated by 'n.v.' (*non vidi*). The cited records are organized by state and *município* (abbreviated to 'Mun.'), and mostly arranged from north to south and west to east. A minimum of at least one record per *município* has been included and, in the case of many taxa, all available records are cited, since Brazilian Cactaceae remain in general very poorly collected. For three taxa that are widespread, frequent and taxonomically well-defined within Eastern Brazil — *Tacinga inamoena, Epiphyllum phyllanthus* and *Cereus jamacaru* subsp. *jamacaru* — only a few representative collections/observations or records marking the limits of distribution etc. have been cited, since their abundance and range are otherwise not in doubt. Apart from herbarium *exsiccata,* whose locations are indicated by the standard *Index Herbariorum* codes, other field observations by the authors (or from reliable sources, even if unsupported by vouchers) are mentioned in some cases; likewise published and unpublished illustrations and photographs of plants in localized habitats. A bracketed questionmark (?) preceding a record's entry indicates uncertainty about the collection's identity, whereas after the name of the *município* it indicates doubt about the correct municipal attribution of the locality cited.

5.2 ARTIFICIAL KEY TO GENERA

1. Actively growing stems bearing broad or awl-shaped leaves (*Pereskia* & Opuntioideae) 2
1. Actively growing stems leafless or with only minute scale-like leaf primordia (Cactoideae) 7

2. Spines not microscopically barbed, not becoming strongly attached if allowed to penetrate the skin; glochids lacking; seed with blackish testa visible . **1. Pereskia**
2. Spines microscopically barbed, very difficult to detach if allowed to penetrate the skin, or true spines lacking and glochids present (at least on older stem-segments or the trunk); seed encased in a pale and sometimes fibrous/hairy funicular envelope (Opuntioideae) 3

3. Leaves broad, like those of *Pereskia* (SW Bahia & Cent.-N Minas Gerais) . . . . **2. Quiabentia**
3. Leaves awl-shaped (widespread) 4

4. Tree to 4 m or more with dimorphic stems, comprising a cylindric indeterminate leader shoot and flattened determinate lateral segments, the ultimate thin and leaf-like; seed 8–10 mm . . . . **4. Brasiliopuntia**
4. Trees to < 4 m, or shrubs, subshrubs or lianas; stems not as above; seed to c. 5 mm 5

5. Stamens and perianth spreading, the former sensitive, closing around the style when touched, the latter patent but never strongly reflexed, at least partly yellow; pollen exine reticulate . . . . **6. Opuntia**

5. Stamens erect, clustered around the style, not sensitive; perianth erect,
spreading or strongly reflexed, greenish, deep pinkish, red or purplish,
or orange-yellow and plants not exceeding 50 cm; pollen exine not as above . . . 6

6. Stamens long-exserted and perianth-segments erect, not spreading; plants
spineless (introduced and cultivated) . . . .*5. **Nopalea**
6. Stamens not as above or perianth-segments spreading to strongly reflexed
and/or stems spiny . . . . **3. Tacinga**

7. Flower > 10 cm long, or 8–10 cm and tube bract-scales bearing hairs
(hair-spines) in their axils 8
7. Flowers < 8 cm long, or 8–10 cm and tube bract-scales naked in their axils
or lacking or minute 15

8. Fruit yellow, globose, surface weakly ribbed, > 5 cm diam., falling to the
ground, smelling of pineapple and spineless when ripe; seed light brown when
fresh, brown when old (N & E Bahia & NE Minas Gerais: *caatinga-agreste*) . . . .
. . . . **10. Pseudoacanthocereus**
8. Fruit not as above; seed blackish 9

9. Stems flat, trigonous or 3-winged 10
9. Stems with > 3 wings/ribs 13

10. Pericarpel and flower-tube bearing conspicuous spines/bristles or broad-based
bract-scales; perianth > 15 cm diam. 11
10. Pericarpel and flower-tube with inconspicuous, narrow-based bract-scales,
minute spines and/or trichomes only, or expanded perianth < 15 cm diam. . . . 12

11. Stems trigonous or 3-winged . . . . **7. Hylocereus**
11. Stems flattened . . . .*8. **Selenicereus**

12. Epiphyte or garden plant with mostly flattened stems . . . . **9. Epiphyllum**
12. Terrestrial with mostly trigonous or 3-winged stems . . . . **16. Cereus** p.p.

13. Perianth remains cleanly abscissing from young fruit or persistent and
strongly blackened . . . . **16. Cereus** p.p.
13. Perianth remains not as above (blackening only if attacked by fungi) 14

14. Fruit > 5 cm, dehiscent; seeds 3–4 mm (Bahia northwards: *caatinga*) . **24. Harrisia**
14. Fruit 1–5 cm, indehiscent; seeds 1–2 mm (Minas Gerais: *campo rupestre*) . . . .
. . . . **28. Arthrocereus**

15. Epiphytic, or epilithic on coastal rocks, in *Mata atlântica* or at altitudes of
>1500 m, with flattened or 3–5-winged/angled stem-segments or stems
<2 cm diam., often only slightly succulent; flowers < 3 cm long or if
larger then magenta and zygomorphic (tribe Rhipsalideae) 16
15. Terrestrial, or epilithic in *caatinga*/*campo rupestre* and/or stems and flowers not
as above 19

16. Flowers ± zygomorphic, tube to 8 mm long or more (S Espírito Santo & S Minas Gerais) . . . . **14. Schlumbergera**
16. Flowers actinomorphic, tube 0 or < 3 mm long 17

17. New stem-segments (excluding greatly elongated, usually basal, extension shoots) arising mostly 2 or more together from the apices of older segments (branching acrotonic), old and diseased stems separating from the plant at the joints between segments, deciduous 18
17. New stem-segments arising only at base or singly from the sides of older segments (branching basi-/mesotonic), not deciduous when old or diseased . . . . **11. Lepismium**

18. Flowers whitish or not strongly coloured, or developed laterally on stem-segments of ± indeterminate pendent growth . . . . **12. Rhipsalis**
18. Flowers bright yellow or orange, from composite areoles at the apex of ultimate or penultimate segments of strictly determinate ± erect growth . . . . **13. Hatiora**

19. Plant unbranched, never segmented, shortly columnar, ± globose or depressed-globose, with or without a non-chlorophyllous bristly-woolly terminal cephalium 20
19. Plant branched (at least basally), slender cylindric to tall columnar, or stems solitary with lateral cephalia, segmented with cephalia at the joints or bottle-shaped and the juvenile stem tapering into an elongate chlorophyllous terminal cephalium 23

20. Flowers yellowish, rarely greenish, diurnal (central Minas Gerais) . . . . **30. Uebelmannia**
20. Flowers not as above (widespread) 21

21. Flowers to 4 cm long, tubular, deep magenta-pink to red at least without, diurnal to crepuscular 22
21. Flowers > 4 cm long, salverform, whitish, nocturnal . . . . **29. Discocactus**

22. Flowers from a cephalium . . . . **23. Melocactus**
22. Flowers from the stem apex, cephalium lacking (Chapada Diamantina, Bahia: *A. bahiensis*) . . . . **19. Arrojadoa** p.p.

23. Pericarpel and tube clothed in conspicuous bract-scales and/or wool and bristle-spines 24
23. Pericarpel and tube ± naked, bract-scales lacking, minute or very widely spaced . . 27

24. Fruit with red or purplish pulp when ripe, spiny at first; stems to 2.5 cm diam. . . . . **25. Leocereus**
24. Fruit with white or translucent pulp, spineless, or if spiny then stem >2.5 cm diam. 25

25. Pericarpel and tube bearing areoles and bristle-spines; fruit covered in an intensely blue waxy bloom (Cent.-S Minas Gerais) . . . . **17. Cipocereus** p.p.

25. Pericarpel and tube with bract-scales but lacking bristle-spines; fruit not
    as above 26

26. Flowers shortly funnelform, at least 4 cm diam. at full anthesis; seed 2–3 mm . . . . . .
    . . . . **15. Brasilicereus**
26. Flowers tubular, to 2.5 cm diam. at full anthesis; seed to c. 1.5 mm . . **26. Facheiroa**

27. Fruit depressed-globose (rarely globose), 2–6 cm diam., bursting open
    laterally or apically due to pressure from the expanding funicular pulp;
    stems never regularly segmented (widespread) . . . . **20. Pilosocereus**
27. Fruit not as above, opening via a basal pore or < 2 cm diam.; stems various 28

28. Stems segmented, with bristly/woolly ring cephalia at the joints and apex
    of distal stem-segments 29
28. Stems not as above; cephalium, if present, lateral and ± continuous or
    terminal and chlorophyllous 30

29. Flowers green without, 8–10 cm long; fruit >3 cm diam. (Bahia, *caatinga*) . . . .
    . . . . **18. Stephanocereus** subg. **Stephanocereus**
29. Flowers deep pink to bright red without, < 4 cm long; fruit < 3 cm
    diam. (widespread) . . . . **19. Arrojadoa** p.p.

30. Mature plant bottle-shaped, the upper part narrowed into a terminal
    chlorophyllous cephalium (Bahia, Chapada Diamantina) . . . .
    . . . . **18. Stephanocereus** subg. **Lagenopsis**
30. Mature plant not as above 31

31. Flower-bearing areoles not differing markedly from those on purely
    vegetative stems and/or fruit covered in blue wax (Minas Gerais) . . . .
    . . . . **17. Cipocereus** p.p.
31. Flower-bearing areoles ± modified or comprising a lateral cephalium;
    fruit not as above 32

32. Fruit clavate, sometimes laterally compressed, > 11 mm diam., deep pink to
    red, with a small basal pore allowing ants to enter (plants of naked gneiss/
    granite inselbergs of SE Brazil; *C. goebelianus* also on other rocks and in
    stony soil of the *caatinga* in cent.-E to S Bahia) . . . . . . **22. Coleocephalocereus**
32. Fruit depressed-globose, globose or very shortly clavate, variously coloured
    or whitish, or < 11 mm diam., not opening at base 33

33. Stem tissues almost lacking mucilage; shrub branched at base and above,
    not glaucous; perianth-segments white inside and out (*caatinga* to the
    N & E of the Chapada Diamantina, Bahia) . . . . **27. Espostoopsis**
33. Stem tissues highly mucilaginous and at least the outermost perianth-
    segments coloured (Minas Gerais & Bahia: *campo rupestre*), or plant a
    glaucous column to > 3 m (W of Rio São Francisco, N Minas
    Gerais & W Bahia: *Bambuí* limestone) . . . . **21. Micranthocereus**

5.3 PERESKIOIDEAE K. Schum.

This subfamily comprises only one genus following the removal of *Maihuenia* (S Chile & S Argentina: Patagonia) to the Maihuenioideae Fearn (cf. Wallace 1995). The Pereskioideae are distinguished from the 2 broad-leaved genera of Opuntioideae by their unbarbed spines and unspecialized seeds lacking a funicular envelope (aril).

1. PERESKIA Mill.

Gard. dict. abr. ed. 4 (1754). Type: *Pereskia aculeata* Mill.
Including *Rhodocactus* Backeb. & F. M. Knuth (1936).
Literature: Leuenberger (1986), Wallace (1995: 9, Fig. 10).

Leafy and spiny trees, shrubs or scramblers, 1–20 m or more. Areoles in the axils of deciduous leaves, sparsely to densely tomentose, always with felt and sometimes with long hairs, producing spines and sometimes brachyblast leaves. Leaves alternate, broad, exstipulate, somewhat fleshy, deciduous. Spines solitary or clustered, or paired, claw-like and recurved (see species No. 1), sometimes lacking on the flowering stems but increasing in number on the trunk. Flowers solitary or in paniculate to cymose inflorescences, sometimes developing by proliferation from the flower receptacle; flowers 2–7 cm diam., perigynous or epigynous, receptacle smooth or with prominent podaria and areoles, bract-scales leafy, fleshy, green or sometimes tinged like the sepaloids; perianth multiseriate, segments free, sepaloids lacking axillary areoles and partly tinged like the petaloids, these spreading, rotate or forming a campanulate-urceolate corolla; stamens numerous; filaments shorter than the perianth; ovary superior to inferior, unilocular, with septal ridges at the ovary roof only, or rarely septate; stigma-lobes 3–20, erect to spreading. Fruit solitary or clustered-proliferous, pear-shaped, turbinate or globular, terete or angled, with a narrow to broad umbilicus and persistent or deciduous flower remnants or leafy bract-scales; pericarp mucilaginous; locule without funicular pulp (in the Brazilian species). Seeds few to numerous, obovate to lenticular-reniform, 1.8–7.5 mm, black, smooth, shiny; hilum whitish.

A genus widespread in the neotropics with 17 species, of which 5 are native to Eastern Brazil (3 species and one heterotypic subspecies are endemic). Two further species are reported from western and southern Brazil, *P. sacharosa* (Mato Grosso & Mato Grosso do Sul) and *P. nemorosa* (Rio Grande do Sul), and are related to Nos 2–4 from Eastern Brazil. The E Brazilian taxa are restricted to various phases of the *Mata atlântica, agreste* and Southern *caatingas*. Plates 2–4.1.

1. Scrambling or climbing plant, spines on vigorous shoots paired, recurved; perianth white or cream, flower strongly scented (diosmin) .... **1. aculeata**
1. Erect shrubs or trees, spines always straight, spreading, never paired-recurved; perianth not as above 2

2. Flower bright yellow; fruit globose, 1–3-seeded .... **5. aureiflora**
2. Flower pink, magenta, orange or red; fruit turbinate, with > 3 seeds 3

3. Leaves narrowly elliptic to obovate-lanceolate; lateral veins (7–)10–13; seeds (5–)6–7 mm (*Mata atlântica, sens. lat.*, including *brejo* forest) .... **2. grandifolia**
3. Leaves ovate to broadly elliptic-obovate; lateral veins 5–7; seeds 4–5.5 mm (*caatinga*) 4

4. Flowers with campanulate to urceolate perianth; flower-buds orange; perianth-segments reddish pink, erect, recurving at apex only, stamens and style enclosed .... **4. stenantha**

4. Flowers with rotate, widely opening perianth; flower-buds greenish or
   pinkish; perianth-segments pink or magenta-pink, spreading; stamens and
   style not enclosed . . . . **3. bahiensis**

Three types of sclereids are found in the genus (Leuenberger 1986: fig. 22) and each is
represented amongst the species treated here, dividing them into three groups as
indicated below:

PERESKIA ACULEATA Group (No. 1): sclereids fusiform-simple; stomata present
on stem; periderm formation early; brachyblast leaves 0.

1. Pereskia aculeata *Mill.*, Gard. dict. ed. 8 (1768). Lectotype (Leuenberger 1986):
Dillenius, Hort. eltham.: tab. 227, fig. 294 (1732).

Cactus pereskia L., Sp. pl. 1: 467 (1753). *Pereskia pereskia* (L.) H. Karst., Deut. Fl.: 888 (1882), nom. inval.
(Arts 23.4 & 32.1(b)). Type: as above.

VERNACULAR NAMES. Ora-pro-nobis, Azedinha, Lobolôbô, Espinho-de-Santo-Antônio, Espinho-preto,
Surucucú, Cipó-Santo.

Climbing shrub or liana, stems 3–10 m long; extension shoots scandent, sometimes ascending up trees
to heights of 20–30 metres and trunk becoming 10 cm thick near the ground. Areoles initially 2 mm
diam., with white long hairs, later accrescent, cushion-like, to 15 mm diam., producing claw-like,
geminate primary spines and secondary straight spines, but not brachyblast leaves. Leaves lanceolate to
oblong or ovate, 4.5–7.0(–11.0) × 1.5–5.0 cm, shortly petiolate, base cuneate, attenuate or rounded;
blade green, concolorous or purplish below; lateral veins 4–7, often inconspicuous. Inflorescence
terminal and lateral on long shoots, racemose to profusely paniculate, with up to 70 flowers of more.
Flowers 2.5–5.0 cm diam., white or creamy-white, strongly fragrant (diosmin); pedicels 5–15 mm;
receptacle cup-shaped to turbinate, with 6–15 hairy areoles on inconspicuous to prominent podaria;
bracts leaf-like, fleshy, lower bracts 9–12, spreading to recurved, upper bracts 3–8, erect or adpressed
in bud; outer perianth-segments 2–5, pale green to whitish; inner segments 6–11, obovate to
spathulate, to 2.5 cm, delicate, white; stamens 5–10 mm, filaments white, yellow or orange-red; ovary
not distinctly delimited from the style, which is 10–12(–19) × 1–2 mm, thickening above, stigma-lobes
4–7, erect or suberect. Infructescences large or fruits solitary, pedicellate, 1.5–2.5 cm, globular, yellow
to orange, pericarp fleshy, usually bracteate and spiny when immature, mostly naked when ripe; locule
with gelatinous tissue enclosing the seeds. Seeds 2–5, lenticular, 4.5–5.0 mm diam., laterally
compressed. Chromosome number: 2n = 22 (Leuenberger 1986: 43).

Humid/subhumid evergreen forest element: scrambling over vegetation and inselbergs of gneiss/granite
etc., *Mata atlântica,* including *restinga, mata de brejo* and *agreste (mata de cipó),* rarely in *'carrasco'* (near the
Rio São Francisco, cent.-N Minas Gerais) and *caatinga* (SE Bahia), Maranhão, eastern Pernambuco to
central-eastern and south-eastern Bahia, eastern and southern Minas Gerais and Espírito Santo, from sea
level to > 1000 m; Goiás, South-eastern and Southern Brazil; eastern Paraguay and Argentina, Mexico,
Central America, Caribbean and northern South America (once recorded from Peru). Maps 2 & 17.
MARANHÃO: Mun. Bom Jardim ['Santa Luzia'], 'Faz. Agripec. da VARIG, margem esquerda do Rio
Pindaré', 2 Apr. 1983, *M.G.A. Lobo et al.* 337 (MO).
PERNAMBUCO: Mun. Paulista ['Olinda'], 'restinga do Rio Doce', 5 Jan. 1948, *Leal* 41 (RB); Mun. São
Lourenço da Mata, Engenho São Bento, Mata do Camocim (*fide* Andrade-Lima 1966: 1454); Mun.
Recife, Bonji, 3 Oct. 1968, *I. Pontual* 68/878 (PEUFR); Mun. Poção, road to Jataúba, 21 Oct. 1991,
F.A.R. Santos 2 (HUEFS, PEUFR); Mun. Bonito, 1 km before limit with Camocim de São Félix, 10
Feb. 1967, *Andrade-Lima* 67-4944 (SP, IPA).

ALAGOAS: Mun. União dos Palmares, 'Faz. Santo Antônio, mata ao N da sede', 27 Nov. 1966, *Andrade-Lima* 66–4769 (IPA).

SERGIPE: Mun. Frei Paulo, 8 Apr. 1986, *G. Viana* 1386 (ASE 4249); Mun. Simão Dias, 5 Feb. 1975, *A.C. Barreto* s.n. (ASE 295); Mun. São Cristóvão, 'Faz. Rio Comprido, 5 km após entrada para Rita Cacete – S. Cristo', 25 Mar. 1982, *G. Viana* 384 (IPA, ASE).

BAHIA: Mun. Feira de Santana, Faz. Boa Vista, Serra de São José, 12°15'S, 38°58'W, 10 May 1984, *Noblick* 3182 (CEPEC, K, HUEFS); Barragem de Bananeiras, Rio Paraguaçu and Rio Jacuípe, 12°32'S, 39°5'W, Apr. 1980, *Iscardino et al.* in Grupo Pedra do Cavalo 4 (ALCB, HUEFS); Mun. Cruz das Almas, 30 May 1981, *G.C.P. Pinto* 83 (HRB); Mun. Salvador, 26 May 1915, *Rose & Russell* 19639 (NY, US); Mun. Rui Barbosa, Serra do Orobó, 1 July 1973, *G.C.P. Pinto* s.n. (ALCB); Mun. Itaberaba, Faz. Monte Verde, Nov. 1973, *A.L. Costa* (ALCB); Mun. Jaguaquara [Toca da Onça], 27–29 June 1915, *Rose & Russell* 20069 (NY, US); Mun. Aiquara, 2 km N of town, 13 May 1969, *Jesus* 375 & *T.S. dos Santos* 424 (CEPEC, K); Mun. Aurelino Leal, between Itajuípe and Ubaitaba, 12 km N of Banco Central, 24 Apr. 1965, *Belém & Magalhães* 877 (CEPEC); Mun. Uruçuca, 20 May 1994, *W.W. Thomas et al.* 10407 (CEPEC, MO, HUEFS); Mun. Ilhéus, 14°48'S, 39°10'W, 27 July 1995, *L.A. Mattos-Silva* 3153 (HUEFS); l.c., Km 22 from Ilhéus on road to Itabuna, CEPLAC, 25 Apr. 1983, *Leuenberger & Brito* 3103 (CEPEC, K, MO); l.c., *Zehntner* s.n. (US); Mun. Vitória da Conquista, 7 km NE on road BR 116, then 3.5 km W, *caatinga*, 16 April 2003, *Taylor & Zappi* (obs.); l.c. 12 km from town centre towards Itambé, 17 April 2003, *Taylor* (K, photos); Mun. (?), between Prado and Alcobaça, 16 Apr. 1967, *Lanna* 1390 & *Castellanos* 26413 (CEPEC, K, HB); Mun. Nova Viçosa, 10 Apr. 1984, *Hatschbach* 47785 (CEPEC, HRB, MBM).

MINAS GERAIS: Mun. Itacarambi, Vale do Peruaçu, near the entrance to Fazenda Terra Brava, 15°6'45"S, 44°15'36"W, *carrasco*, 14 Feb. 1998, *A. Salino & A. Gotschalg* 3997 (K, BHCB); Mun. Caraí, 12 km N of Catuji, 13 Dec. 1990, *Taylor & Zappi* 757 (K, HRCB, ZSS, BHCB); Mun. Teófilo Otoni, Pedro Versiani, Aug. 1958, *Mendes Magalhães* 17397 (IAN); Mun. Galiléia, 23 km SE from São Vítor, 15 Dec. 1990, *Taylor & Zappi* 772 (K, HRCB, ZSS, BHCB); Mun. Aimorés, Terreno do Varejão, 7 July 1997, *M.F. de Vasconcelos* s.n. (K, BHCB); Mun. Santana do Pirapama, 18 Feb. 1971, *Krieger* 10145 (CESJ); Mun. Pedro Leopoldo, 11 Feb. 1995, *M.L. Guedes* 5042 (ALCB); Mun. Caratinga, road BR 458, between Ipatinga and Iapu, 12 km from the BR 381, 13 Dec. 1990, *Taylor & Zappi* 753 (K, HRCB, ZSS, BHCB); Mun. Itabira, Cauê, 11 Feb. 1934, *Sampaio* 7069 (R); Mun. Arcos, Serra do Minério, July 1983, *E. Esteves Pereira* 170 (UFG); Mun. Viçosa, 19 Dec. 1929 & 27 June 1930, *Mexia* 4129, 4804 (K, NY, U, US, MO); Mun. São João del Rei, Apr. 1896, *Silveira* 943 (R).

ESPÍRITO SANTO: Mun. Nova Venécia, 3 km de Todos os Santos, 18°37'S, 40°43'W, 8 Sep. 1989, *H.Q.B. Fernandes* 2823 (MBML, HRCB); Mun. São Gabriel da Palha, 30 km from Barra de São Francisco on road to Águia Branca, 16 Dec. 1990, *Taylor & Zappi* 785 (K, HRCB, ZSS, MBML); Mun. Linhares, Reserva Florestal da CVRD, Est. Flamengo, 5 Apr. 1986, *Folli* 589 (Univ. Fed. ES); l.c., Reserva Biológica de Comboios, IBDF, Regência, near Fazenda Cata-Vento, 22 May 1989, *Folli* 916 (Univ. Fed. ES); l.c., road BR 101 20 km S of town, 7 Apr. 1984, *Hatschbach* 47706 (MBM); Mun. Colatina, 30 Apr. 1934, *Kuhlmann* 290 (RB, MO); Mun. Santa Teresa, Rio Cinco de Novembro, 22 Sep. 1988, *H.Q.B. Fernandes et al.* 2550 (MBML, HRCB), 13 Mar. 1980, *Weinberg* 502 (MBML); Mun. Fundão, Três Barras, 1 Aug. 1984, *Pizziolo* 220 (MBML, HRCB); Mun. Vila Velha, between Lagoa do Milho and Barra do Jucu, 20 July 1973, *Araújo* 369 (RB); l.c., Barra do Jucu, 15 Mar. 1981, *Weinberg* 499 (MBML); l.c., Interlagos II, 27 July 1982, *Weinberg* 262 (R); Mun. Guarapari, Setiba, 25 Apr. 1989, *O.J. Pereira* 1959 (Univ. Fed. ES), 1 May 1981, *Krieger & Souza* 19564 (CESJ); Mun. Itapemirim, *Bello* 472 (R); l.c., road ES 060, between Marataízes and Pontal, 10 May 1987, *H.C. de Lima et al.* 2917 (K, MO).

RIO DE JANEIRO: NE Rio de Janeiro, Mun. Campos, Apr. 1939, *Sampaio* s.n. (R).

CONSERVATION STATUS. Least Concern [LC] (1); PD=2, EI=1, GD=2. Short-list score (1×5) = 5.

This taxon is variable in leaf size and shape and in its inflorescence morphology, both within and between populations. Its leaves and those of the following (No. 2) are eaten as a potherb.

PERESKIA GRANDIFOLIA Group (Nos 2–4): sclereids fusiform-aggregated; stomata on stem present; periderm formation retarded; brachyblast leaves present. Fusiform-aggregated sclereids are also found in *P. nemorosa* and *P. sacharosa* (SE South America, see Map 12), which are indicated as potentially basal members of this Group in molecular phylogenies produced by R. Wallace (1995) and C. Butterworth (*in litt.*, 10 Feb. 2000) at Iowa State University, USA.

2. Pereskia grandifolia *Haw.*, Rev. pl. succ.: 85 (1819). Neotype (Leuenberger 1986): 'raised from seed collected by Bowie & Cunningham in 1816 in the neighbourhood of Rio de Janeiro', drawing by T. Duncanson, 11 June 1824 (K).

Cactus grandifolius (Haw.) Link, Enum. pl. hort. Berol. 2: 25 (1822). *Rhodocactus grandifolius* (Haw.) F. M. Knuth in Backeberg & Knuth, Kaktus-ABC: 97 (1935, publ. 1936).

Shrub or tree, 2–10 m, forming a trunk to 80 cm diam.; main shoots erect to arching. Areoles rounded, cushion-like, initially 3–7(–8) mm diam., later to 12 mm diam., producing straight spines, and 1–4 brachyblast leaves. Spines 0–8(–11) on twigs, fasciculate to spreading, 1–4 cm, increasing in number and size on older branches. Leaves elliptic or narrowly elliptic-oblong, narrowly ovate to obovate-lanceolate, (6–)9–26(–30) × (3–)4–6(–9) cm, petiole to 1 cm or more; base of lamina attenuate, apex acute to acuminate, sometimes recurved; lateral veins (7–)10–13 or more. Inflorescence terminal, densely cymose-paniculate by proliferation from the receptacle, 10–15- to sometimes 30(–50)-flowered. Flowers 3–6(–7) cm diam., rose-like, showy; pedicels 1–3 cm, stout; receptacle turbinate, with conspicuous furrows and podaria, areoles on the upper half only; bracts small or leaf-like, fleshy, lower bracts 4–7, spreading to ascending, upper bracts 2–5, adpressed in bud; outer perianth-segments 2–5, asymmetric; inner segments 5–12, obovate to spathulate, delicate, rose, pale or purple-pink, apex often emarginate; stamens 5–10 mm, filaments white; ovary distinctly inferior, with conical roof narrowing into the style, which is 8–12 mm, cylindric, stigma-lobes 5–8, suberect, white to pale pink. Infructescences large, usually pendent; fruit 5–10 × 3–7 cm, pyriform or turbinate, with conspicuous angles, green or reddish to yellowish, bracts usually deciduous when ripe. Seeds 20–60, obovate-elliptic, 5–7 × 4.5–5.0 mm, laterally compressed.

CONSERVATION STATUS. Least Concern [LC]; see subspecies below.

This species is divisible into two subspecies:

1. Receptacular bracts green, the lowermost ones rarely with recurved apices; outer perianth-segments greenish to pink, inner segments 15–32 mm; anthers golden yellow (NE & SE Brazil) . . . . **2a.** subsp. **grandifolia**
1. Receptacular bracts and outer perianth-segments purplish-pink to dark purplish, the lowermost bracts with recurved apices; inner perianth-segments 10–18 mm; anthers pale yellow (cent.-S Minas Gerais to W Espírito Santo, cultivated in N Minas Gerais & S Bahia) . . . . **2b.** subsp. **violacea**

2a. subsp. **grandifolia**

VERNACULAR NAMES. Sabonete, Ora-pro-nobis, Quiabento, Entrada-de-baile, Rosa-mole, Sem-vergonha.

Leaves thin, green, concolorous, shortly acuminate to acute. Inflorescence and receptacular bracts green, flat, rarely recurved; receptacular bracts shorter to longer than the inner perianth-segments, outer segments greenish-pink, inner segments 15–32 mm, pink to deep pink; anthers golden yellow. Fruit variable, pyriform to obtriangular, angled; often nearly as broad as long; ovary locule cup-shaped to broadly obovate-triangular. Chromosome number: 2n = 22 (Leuenberger 1986: 43).

Humid/subhumid evergreen forest element: perhaps native in *agreste* and *Mata atlântica* (including that on the E slopes of the Chapada Diamantina, *brejo* and riverine forest), c. 100–1140 m, southernmost Ceará, Pernambuco (native and cultivated, *fide* Andrade-Lima 1966: 1454), central-eastern and south-eastern Bahia to southern Espírito Santo and south-western Minas Gerais, but widely cultivated; perhaps also native in Rio de Janeiro and São Paulo; commonly introduced in the neotropics. Map 17.

MARANHÃO: Mun. Barra do Corda, [cult.?], 20 July 1909, *Arrojado Lisboa* 2467 (MG).

CEARÁ: S Ceará, Chapada do Araripe, 30 Aug. 1971, *Gifford & Fonseca* G 335 (E, K, photo).

PERNAMBUCO: Mun. Olinda, cult., 12 Feb. 1925, *Pickel* 518 (SP); Mun. São Lourenço da Mata, Engenho São Bento, cult., *Pickel* 1911 (IPA, *fide* Andrade-Lima 1966: 1454); Mun. Recife, Dois Irmãos, 3 Oct. 1968, *I. Pontual* 68/879 (PEUFR); Mun. São Benedito do Sul, 9 Feb. 1994, *A.M. Miranda* 1286 (PEUFR).

BAHIA: Mun. Rodelas, Volta do Rio, 27 Jan. 1987, *L.B. Silva & G.O. Mattos e Silva* 98 (HRB); Mun. Juazeiro, cult., 11 Feb. 1990, *Zappi* 221A (K, photos); Mun. Palmeiras, Morro do Pai Inácio, 12°27'37"S, 41°28'40"W, 1100–1140 m, *M. Alves et al.* in EBNN 1016 (PEUFR); Porto Castro Alves, vale dos Rios Paraguaçu e Jacuípe, 10 Dec. 1980, *Iscardino et al.* in Grupo Pedra do Cavalo 1011 (ALCB, HRB, BAH); Mun. Andaraí, Km 0 da rodovia que liga à BR 242, 12 Jan. 1983, *L.A. Mattos Silva et al.* 1592 (CEPEC, K); l.c., Rio Paraguaçu ao longo da rodovia para Mucugê, 12°50'S, 41°19'W, 19 Oct. 1997, *M. Alves et al.* 1116 (HRB); Mun. Cruz das Almas, May 1973, *G.C.P. Pinto* s.n. (ALCB), *G.C.P. Pinto* 29/81 (HRB); Mun. Salvador, 27 May 1915, *Rose & Russell* 19656 (NY, US); Mun. Vitória da Conquista, 20 km SE of Anajé, [cult.?], 13 Apr. 1983, *Leuenberger et al.* 3062 (CEPEC, K); Mun. Ilhéus, 2 Dec. 1982, *T.S. dos Santos* 3827 (HUEFS); Mun. Porto Seguro, Monte Pascoal, road to 'Caraíba' [Caraíva], 14 May 1971, *T.S. dos Santos* 1641 (CEPEC, K).

MINAS GERAIS: S Minas Gerais, Mun. Juiz de Fora, 18 Oct. 1969, *Krieger* 7528 (CESJ), l.c., near Paraíbuna, Apr. 1970, *Krieger* 8244 (SPF); SW Minas Gerais, Mun. Lavras, road BR 381, 8 km S of bridge over Rio Grande, 10 Oct. 1988, *Taylor & Zappi* in Harley 24802 (SPF, K).

ESPÍRITO SANTO: Mun. Barra de São Francisco, cult., 16 Dec. 1990, *Taylor & Zappi* 779A (K, photo); Mun. Cachoeiro do Itapemirim, *Bello* 653 (R); unlocalized [perhaps from Campos, Rio de Janeiro], 'Campo Novo', [1815/1816], [*Wied-*]*Neuwied* s.n. [ex Herb. Martius anno 1869] (MO).

RIO DE JANEIRO: Sertão de Cacimbas, on right embankment of Rio Itabapoana, 10 Oct. 1909, *Sampaio* 985 (R); between Morro do Coco and Campos, 8 Aug. 1964, *Trinta* 1051 (R); Monte Alegre, Feb. 1927, *Vidal* s.n. (R); Mun. Cantagalo, *Peckolt* 121 (W).

CONSERVATION STATUS. Least Concern [LC] (1), but perhaps Near Threatened, since, although of very wide range, much of its original habitat has been destroyed; PD=1, EI=1, GD=1. Short-list score (1×3) = 3. It has been taken into cultivation as a hedge plant in the region where it may be native.

The natural range of this taxon remains poorly known, probably through early destruction of its habitat and for the uncertainty as to its native status caused by its widespread introduction as a cultivated ornamental.

2b. subsp. **violacea** *(Leuenb.) N. P. Taylor & Zappi* in Cact. Consensus Initiatives 3: 7 (1997). Holotype: Brazil, Minas Gerais, Mun. Santana do Riacho, Serra do Cipó, 19°21'S, 43°36'W, 23 Sep. 1981, *F. C. F. Silva* 89 (HRB; B, iso.).

Pereskia grandifolia var. *violacea* Leuenb. in Mem. New York Bot. Gard. 41: 116, fig. 46 (1986).

VERNACULAR NAME. Ora-pro-nobis.

Leaves thickish, dark green above, often slightly discolorous and purple below, nearly always distinctly acuminate and with recurved apex. Inflorescence bracts all dark purplish to bright purplish pink, uppermost almost resembling open flowers from a distance; lower receptacular bracts thick, keeled, recurved, nearly as long as the inner perianth-segments, outer segments purplish to pink, inner segments 10–15(–18) mm, pale pink to purple-pink; anthers pale yellow. Fruit and ovary locule narrowly obtriangular, bracts purplish tinged with green or becoming nearly green. Chromosome number: 2n = 22 (Leuenberger 1986: 43).

Southern humid/subhumid forest element: drier phases of *Mata atlântica*, c. 50–1400 m, drainage of the Rio Doce, central-southern and eastern Minas Gerais to central Espírito Santo, apparently cultivated elsewhere. Endemic to the core area within South-eastern Brazil. Map 17.

BAHIA: Mun. Vitória da Conquista, cult., 13 Apr. 1983, *Leuenberger et al.* 3060 (CEPEC).

MINAS GERAIS: Mun. Januária, 4 Nov. 1978, *Krieger* 16130 (CESJ) — cult.?; Mun. Pedra Azul, near town, 16°1'S, 41°17'W, cult., 19 Oct. 1988, *Taylor & Zappi* in *Harley* 25218 (K, SPF); Mun. Itaobim, 8 km W of town, 6 km S of road Itaobim–Araçuaí, cult., 9 Apr. 1983, *Leuenberger & Martinelli* 3057 (CEPEC, RB); cent.-S & E Minas Gerais, Mun. Capelinha/Malacacheta, *Uebelmann* HU 226 (M); Mun. Conselheiro Pena, 16.7 km SE of Galiléia, N bank of Rio Doce, 15 Dec. 1990, *Taylor & Zappi* 773 (K, HRCB, ZSS, BHCB); ibid., 2 km N of Penha do Norte, regenerating from cut-over forest, 15 Dec. 1990, *Taylor & Zappi* (K, photo); Mun. Santana do Riacho, Serra do Cipó, 19°21'S, 43°36'W, 23 Sep. 1981, *F.C.F. Silva* 89 (RB), 25 Mar. 1991, *Pirani et al.* in CFSC 12042 (SPF, K); Mun. Itambé do Mato Dentro, 24 Jan. 1922, *G. Santos* s.n. (R); Mun. Aimorés, Terreno do Varejão, 14 Sep. 1997, *M.F. de Vasconcelos* s.n. (K, BHCB); Mun. Itabira, Cauê, 11 Feb. 1934, *Sampaio* 7068 (R); Mun. Marliéria, Parque Est. do Rio Doce, 19°46'S, 42°36'W, 27 Sep. 1975, *Heringer & Eiten* 15252 (UB, US).

ESPÍRITO SANTO: SW of Colatina, *Braun* 476 (B, photo); Mun. Santa Teresa, Rio Cinco de Novembro, cult. Mus. Bot. Mello Leitão, 17 Dec. 1990, *H.B.Q. Fernandez* in *Taylor & Zappi* 789 (HRCB, K, ZSS).

CONSERVATION STATUS. Data Deficient [DD] (1); PD=1, EI=1, GD=1. Short-list score (1×3) = 3. It has been taken into cultivation as a hedge plant in the region where it is native. Its native habitat continues to decline.

Even though the native distribution of subsp. *grandifolia* is poorly understood, there are no records of it as other than a cultivated plant within the extensive area in which subsp. *violacea* is found, thereby justifying recognition of the latter as a subspecies rather than a variety. Indeed, recent studies of plastid DNA gene sequences, conducted by Wallace (1995: 9) and Butterworth (unpubl.), have indicated that this taxon may be worthy of specific status and that it is the basal element amongst the East Brazilian taxa belonging to the P. GRANDIFOLIA Group. Whatever the true status of subsp. *grandifolia*, it seems reasonably certain that subsp. *violacea* is native within the area drained by the Rio Doce, where it has been observed regenerating from stumps remaining in recently cut primary forest (see above and Plate 2.4).

3. Pereskia bahiensis *Gürke* in Monatsschr. Kakt.-Kunde 18: 86 (1908). Type (Leuenberger 1986: 119): Brazil, Bahia, 'Calderão' [Caldeirão, Mun. Maracás], Oct. 1906, *Ule* 7050 (HBG, lecto.; L, lectopara.).

VERNACULAR NAMES. Quiabento (Quiá-bento), Inhabento, Jumbeba, Surucurú, Flor-de-cera, Espinho-de-Santo-Antônio, Entrada-de-baile, Ora-pro-nobis.

Shrub to small tree, 1–6 m, forming a trunk to 33 cm diam.; main shoots erect to arching, stout. Areoles rounded, cushion-like, initially 3–6(–8) mm diam., later to 20 mm diam., producing straight spines, and 1–4 brachyblast leaves. Spines 0–6 on twigs, fasciculate to spreading, 2–5 cm, increasing in number and size on older branches. Leaves elliptic or obovate, often narrowly obovate on main shoots, flat or folded upwards and recurved, (4–)5–12 × 2–7 cm, petiole short to indistinct, base of lamina attenuate, apex rounded to broadly acute, lateral veins (3–)4–7(–8). Inflorescence terminal, cymose-paniculate by proliferation from the receptacle, 2–12-flowered, dense to lax, or flowers sometimes solitary. Flower perigynous, 4–7 cm diam., rose-like, showy; pedicel 1 cm, stout; receptacle turbinate, with prominent podaria and 3–8 areoles mainly on the upper half; bracts sepal-like, fleshy, lower bracts 4–6, spreading, upper bracts 4–6, erect, green with purplish to nearly black margin, keeled; outer perianth-segments 2–5, asymmetric, greenish red to reddish purple, inner segments 7–9, 20–30 mm, obovate to spathulate, delicate, pink to reddish purple, pale at base, apex rounded-truncate to emarginate; stamens 5–10 mm, clustering around the style, filaments white; ovary half-inferior to inferior, with distinct septal ridges at the roof; style 7–15 mm, cylindric, stigma-lobes 5–10, suberect. Infructescences lax, forming a cymose-paniculate cluster of up to 12 fruits, often pendent, or fruit solitary; 3–6 × 3–5 cm, pyriform or turbinate, ± angled, green to yellow when ripe, scented like pineapple, bracts usually deciduous. Seeds > 30, obovate, 3.8–5.5 × 2.8–3.5 mm, laterally compressed. Chromosome number: 2n = 22 (Leuenberger 1986: 43).

Central-southern (Bahian) *caatinga* element: *caatinga* surrounding the Chapada Diamantina, planalto de Maracás, northern Serra do Espinhaço and Serra Geral (Bahia), 300–900 m, east and south of the Rio São Francisco. Endemic to Bahia. Map 17.

BAHIA: Mun. Juazeiro, Nov. 1913, *Zehntner* 743 (US); Mun. Jaguarari, 64 km SE of Juazeiro towards Senhor do Bonfim, 8 Jan. 1991, *Taylor et al.* 1377 (K, HRCB, ZSS, CEPEC); l.c., Barrinha, 7 June 1915, *Rose et al.* 19786 (NY, US), 25 Feb. 1968, *I. Pontual* 68-738 (IPA, PEUFR); Mun. Ourolândia, NE of Olho D'Água do Fagundo, 4 Aug. 2002, *Taylor* (obs.); Mun. São Gabriel/João Dourado, APA Gruta dos Brejões, 4 Aug. 2002, *Taylor* (obs.); Mun. Morro do Chapéu, NE of Gruta dos Brejões, 4 Aug. 2002, *Taylor* (obs.); l.c., W of Icó, N of Brejão (Formosa), 5 Aug. 2002, *Taylor* (obs.); l.c., 20 km W of Morro do Chapéu, 3 Aug. 2002, *Taylor* (obs.); l.c., Bambuí limestone, 20 Jan. 1984, *Fotius* 3707 (IPA); Mun. Irecê, 26 Oct. 1970, *A.L. Costa* s.n. (IPA), 7 Oct. 1980, *M.S.G. Ferreira* 118 (HRB, ALCB, BAH); Mun. América Dourada, Nova América, 17 May 1975, *Costa & Barroso* s.n. (ALCB); Mun. Oliveira dos Brejinhos, 123 km W of Seabra towards Ibotirama, 14 Jan. 1991, *Taylor et al.* 1420 (K, HRCB, ZSS, CEPEC); Mun. Iraquara, Faz. Boa Vista, 12°16'S, 41°38'W, 14 Mar. 1984, *O.A. Salgado* 365 (HRB, RB); Mun. Iraquara/Palmeiras, 16 km from Rio Preto to Riacho do Mel, 19 Mar. 1980, *Brazão* 165 (HRB); Mun. Santa Teresinha, 20 Nov. 1986, *L.P. de Queiroz* 1341 (HUEFS); Mun. Itaetê, 19 Oct. 1978, *Almeida* 28 (HRB, RB, F); Mun. Milagres, 2 km from the town, 15 Apr. 1995, *E. de Melo* 1204 (HUEFS); Mun. Marcionílio Souza, Machado Portella, 19–23 June 1915, *Rose & Russell* 19901 (NY, US); l.c., 18 Dec. 1912, *Zehntner* 677 (RB, US); Mun. Maracás, towards Tamburi, 24 Jan. 1965, *E. Pereira* 9719 *& G. Pabst* 8608 (HB); l.c., Km 26 on road from Maracás to Tamburi, 20 Apr. 1983, *Leuenberger et al.* 3086 (CEPEC, K); l.c., 24 Jan. 1965, *Brazão* 109 (HRB, RB); Mun. Rio de Contas, 19 June 1978, *Araújo* 47 (RB); l.c., caminho para Lagoa Nova, 13°48'S, 41°47'W, 5 Feb. 1997, *E. Saar et al.* in PCD 5106 (ALCB, K, CEPEC, UEFS, SPF); Mun. Livramento do Brumado, 11 km S of the town, 13°45'S, 41°49'W, 23 Nov. 1988, *Taylor & Zappi* in *Harley* 25542 (K, SPF, CEPEC); Mun. Ituaçu, 13°48'S, 41°16'W, 22 June 1987, *L.P. de Queiroz* 1649 (HUEFS); l.c., Morro da Mangabeira, 22 Dec. 1983, *E.P. Gouveia* 46/88 (ALCB); Mun. Jequié, 4 km N of town, 28 Mar. 1965, *M. Magalhães* s.n. (HB 37923); l.c., 11–17 km W of town towards Lafaiete Coutinho, 19 Nov. 1978, *Mori et al.* s.n. (CEPEC, K); Mun. Brumado, Serra das Éguas, 23 Sep. 1989 & 8 Feb. 1990, *A.M. Miranda* 41 & 120 (PEUFR); Mun. Jequié/Poções, *A.L. Costa* s.n. (ALCB, in spirit); Mun. Malhada de Pedras, Km 11 on road from Brumado to Caetité, 14 Apr. 1983, *Leuenberger et al.* 3063 (CEPEC, MO); Mun. Lagoa Real, 81 km W of Brumado towards Caetité, 2 Feb. 1991, *Taylor et al.* 1529 (K, HRCB, ZSS, CEPEC); Mun. Caetité, 14 km towards Brumado, 19 Feb. 1992, *A.M. de Carvalho et al.* 3761 (K, CEPEC); l.c., Brejinho das Ametistas, cult., 15 Apr. 1983, *Leuenberger et al.* 3071 (CEPEC, K); Mun. Manuel Vitorino, Serra da Pipoca, 15 Jan. 1984, *G.C.P. Pinto* 65/84 (HRB, CEPEC); Mun. Boa Nova, 20 km S of Manuel Vitorino, 19 Apr. 1983, *Leuenberger et al.* 3083 (CEPEC, K); Mun. Aracatu, road BR 262, 50 km from town towards Vitória da Conquista, 12 Dec. 1984, *G.P. Lewis et al.* in CFCR 6716 (SPF, SP, K); Mun. (?), Vitória da Conquista to Brumado, 31 Mar. 1966, *Castellanos* 25947 (HB); Mun. Cacule, 14°31'S, 42°12'W, 2001, *J. Jardim* 3221 (CEPEC); Mun. Poções, Faz. Boa Esperança, road BR 116, 9 km S of Poções, 5 Apr. 1988, *L.A. Mattos Silva et al.* 2317 (CEPEC, K); Mun. Jânio Quadros, 8 km SW of Maetinga, 19 Feb. 1992, *S.C. Sant'Ana et al.* 227 (K, CEPEC); Mun. Caatiba, Faz. Bom Jardim, 10 Feb. 1975, *A.L. Costa* s.n (ALCB).

CONSERVATION STATUS. Least Concern [LC] (1); PD=2, EI=1, GD=1. Short-list score (1×4) = 4. It has been taken into cultivation as a hedge plant in the region where it is native.

The very close relationship between this species and the following deserves further investigation. A plant encountered near the border of municípios Piatã and Boninal, Bahia (*Taylor & Zappi* in *Harley* 25598, SPF, CEPEC, K), well beyond the known range of *P. stenantha*, had somewhat intermediate flowers, and in the region of Caitité the two species seem to hybridize or intergrade. As already noted by Leuenberger (1986), there are scarcely any vegetative differences to separate them, although *P. stenantha* seems capable of producing much larger leaves (especially in western Bahia where it inhabits a region of higher rainfall).

4. Pereskia stenantha *F. Ritter*, Kakt. Südamer. 1: 21, figs 3–4 (1979). Holotype: Brazil, Bahia, Caitité, *Ritter* 1251 (U, not found). Lectotype (Leuenberger 1986: 123): Ritter, l.c., fig. 3 (1979).

VERNACULAR NAMES. As for *P. bahiensis*.

Shrub to small tree, 2–4(–6) m, branching mostly near the base, sometimes forming a trunk to 15 cm diam.; main shoots erect to arching. Areoles rounded, cushion-like, initially 2–5 mm diam., later to 15 mm diam., producing straight spines, and 1–3 brachyblast leaves. Spines 0–7 on twigs, fasciculate to spreading, 1–5 cm, increasing in number and size on older branches. Leaves obovate to elliptic, often folded upwards along the midrib, (5–)7–11(–15) × (2–)4–6(–9) cm, petiole 2–10 mm, base of lamina cuneate to attenuate, apex broadly acute to rounded, lateral veins 5–7. Inflorescence terminal, densely cymose-paniculate by proliferation from the receptacle, c. 15-flowered, or flowers sometimes solitary. Flowers perigynous to epigynous, 1–2 cm diam., urceolate-campanulate, opening very little; flower-buds orange-red; pedicels hardly distinguishable; receptacle turbinate, with 5–8 areoles, bracteate in the upper half, bracts sepal-like, fleshy, lower bracts 3–5(–7), spreading, upper bracts (1–)3–5(–6), erect, green to reddish, keeled, outer perianth-segments 4–5, erect, concave, red to orange-red, inner segments 6–8, 20–25 mm, linear lanceolate, erect, spreading at the tip only, pink to purplish pink, acute to broadly acute at apex; stamens 12–15 mm, surrounding the slightly longer style, filaments white; ovary half-inferior to inferior, with septal ridges at the roof; style 13–20 mm, thick at base, stigma-lobes 4–7, suberect to spreading. Infructescence a densely cymose-paniculate cluster of up to 15 fruits, erect or arching at maturity, or fruits solitary; fruit 3–7 × 2–6 cm, pyriform or turbinate, ± angled, otherwise like *P. bahiensis*. Seeds c. 30(–50) per fruit, obovoid, 4.5–5.5 × 3.2–3.6 mm, laterally compressed. Chromosome number: 2n = 22 (Leuenberger 1986: 43).

Southern Rio São Francisco *caatinga* element: *caatinga*, 450–750 m, valley of the Rio São Francisco, western and central-southern Bahia and central-northern Minas Gerais. Endemic to the core area of Eastern Brazil. Map 17.

BAHIA: W Bahia, Mun. Barreiras, estrada para São Desidério, 23 Dec. 1954, *G.A. Black* 54-17686 (IAN); Mun. Vanderlei, 19 July 1989, *Zappi* 148A (SPF); Mun. Brejolândia, 5 km N of Tabocas do Brejo Velho, 12°39'S, 44°2'W, *Harley et al.* 22011 (K, U, SPF, UEC); Mun. Santana, Porto Novo do Rio Corrente, 12 Nov. 1912, *Zehntner* 563 (US), ibid. in IFOCS 5048 (M); l.c., 28 km S of town towards Santa Maria da Vitória, 15 Jan. 1991, *Taylor et al.* 1423 (K, HRCB, ZSS, CEPEC); cent.-S Bahia, Mun. Macaúbas, Veredinha, 12°54'S, 42°38'W, 20 Mar. 1984, *D.A. Salgado & H.P. Bautista* 283 (HRB); Mun. Bom Jesus da Lapa, road to Caldeirão, Faz. Imbuzeiro da Onça, 13°9'S, 43°22'W, 19 Apr. 1980, *Harley* 21543 (K, SPF, CEPEC); l.c., 2 km E of town, 16 Apr. 1983, *Leuenberger et al.* 3076 (CEPEC, K); l.c., near Santo Antônio, Nov. 1912, *Zehntner* 631 (US); ibid. (?), 'Lapa', Nov. 1912, *Zehntner* in IFOCS 5047 (M); Mun. Botoporã, 50 km SE of Macaúbas towards Paramirim, 26 Aug. 1988, *Eggli* 1297 (ZSS); Mun. Bom Jesus da Lapa/Riacho de Santana, 28 Dec. 1965, *A. Duarte* 9531 & *E. Pereira* 10442 (HB, RB); Mun. Riacho de Santana, 10 km W of Igaporã, 18 Apr. 1983, *Leuenberger et al.* 3081 (CEPEC, K); Mun. Igaporã, 6 km W of town, 16 Apr. 1983, *Leuenberger et al.* 3072 (CEPEC, K, MO); Mun. Caetité, type locality of Ritter (1979); l.c., 23 km S of town towards Brejinho das Ametistas, cult., 2 Feb. 1991, *Taylor et al.* 1539 (K, HRCB, ZSS, CEPEC); Mun. Guanambi, 9.5 km from town towards Caetité, 25 July 1989, *Taylor* (obs.); Mun. Iuiú, 2 km S of town, 21 July 1989, *Taylor & Zappi* (obs.).

MINAS GERAIS: Mun. Montalvânia, S of the ferry crossing of the Rio Carinhanha, 1999, *Klaassen et al.* (photo); Mun. Manga, 1999, illustrated by Braun & Esteves Pereira (2002: 155, Abb. 169); Mun. Monte Azul, 5 km E of town, E of Vila Angical, 15°10'S, 42°49'W, 9 Nov. 1988, *Taylor & Zappi in Harley* 25523 (K, SPF); Mun. Mato Verde, 4 km E of town towards Rio Pardo de Minas, 30 Jan. 1991, *Taylor et al.* 1496 (K, HRCB, ZSS, BHCB); Mun. Varzelândia, 36 km ENE of town, 23 July 1989, *Taylor & Zappi* (obs.); Mun. Porteirinha, 32 km from Mato Verde, 30 Jan. 1991, *Taylor et al.* 1497 (K, HRCB, ZSS, BHCB).

CONSERVATION STATUS. Least Concern [LC] (1); PD=2, EI=1, GD=1. Short-list score (1×4) = 4. It has been taken into cultivation as a hedge plant in the region where it is native.

Almost indistinguishable from *P. bahiensis* (see above) when not in flower, but with a distinct range.

PERESKIA PORTULACIFOLIA Group (No. 5): stone cells present; stomata on stem 0; periderm formation early; brachyblast leaves present. Other members of this group are from northern South America (*P. guamacho*, sister species of *P. aureiflora*; see Map 11) and the eastern Caribbean islands (4 spp.).

5. Pereskia aureiflora *F. Ritter*, Kakt. Südamer. 1: 22, fig. 5 (1979). Holotype: Brazil, Minas Gerais, Itaobim, *Ritter* 1413 (U, not found). Lectotype (designated here): Ritter, l.c., fig. 5 (1979).

VERNACULAR NAMES. Facho, Ora-pro-nobis-da-mata.

Small tree or shrub, to 6 m, trunk to 20 cm diam.; branches erect to arching. Areoles rounded, cushion-like, initially 3 mm diam., later to 6 mm diam., producing straight spines, and 1–3 brachyblast leaves. Spines 0–3 on twigs, fasciculate to spreading, 1–3 cm, increasing in number and size on older branches. Leaves obovate to elliptic-suborbicular in shade, elliptic to lanceolate and thicker in sun, flat or somewhat boat-shaped, dimorphic, auxoblast leaves to 11 × 2–3 cm, brachyblast 4–10 × (1.5–)2.5–5.0 cm, petiole 2–4 mm, base of lamina broadly cuneate, apex acute, lateral veins 3–5(–7). Flowers solitary, yellow, terminal or on short lateral twigs 3–13 cm, c. 3–4 cm diam., opening widely, showy, epigynous; pedicels short; receptacle turbinate, with areoles, bracteate; bracts leafy, fleshy, green, sometimes reddish at margin, lower bracts 5–8, spreading, upper bracts 5–8, erect, keeled; outer perianth-segments c. 3, obovate, inner segments 10–12, 15–20 mm, obovate, spreading to reflexed, bright yellow, obtuse at apex; stamens 5–10 mm, filaments white; ovary half-inferior; style 10 mm, cylindric, stigma-lobes 4–7, erect. Fruit 10–15(–20) mm diam., globular, reddish green to brown or 'chocolate-purple' when ripe, bracts leafy, deciduous. Seeds c. 1–3, obovoid, (4–)5–5.5 × 4.2–5.1 mm, laterally compressed.

Southern *caatinga* (inselberg) element: in *caatinga/agreste*, especially in association with gneiss/granite inselbergs or derived substrates, 300–920 m, central-southern Bahia to central-northern and north-eastern Minas Gerais. Endemic to the core area of Eastern Brazil. Map 17.

BAHIA: Mun. Rio de Contas, 13 km E of town on road to Marcolino Moura, 13°35'S, 41°45'W, 25 Mar. 1977, *Harley* 19992 (K, B, U, SPF, UEC); Mun. (?), between Jequié and Poções, Oct. 1950, *G. Pinto*

628 (IAN); Mun. Guanambi, 9.5 km NE of town, granite/gneiss outcrop beside road to Caitité, 25 July 1989, *Zappi* 171 (SPF, HRCB); Mun. Urandi (?), Km 11, Saco da Onça – Urandi, 16 Oct. 1970, *D. Andrade-Lima* 70-6067 (IPA).

MINAS GERAIS: cent.-N Minas Gerais, Mun. Monte Azul, 7 km E of town, beyond Vila Angical, 920 m, 28 Jan. 1991, *Taylor et al.* 1470 (K, HRCB, ZSS, BHCB); NE Minas Gerais, Mun. Itaobim, 8 km W of town, 9 Apr. 1983, *Leuenberger & Martinelli* 3056 (CEPEC, RB); l.c., W end of town, 16°33'S, 41°30'W, 18 Nov. 1988, *Taylor & Zappi in Harley* 25525 (K, ZSS, B, SPF); l.c., 4 km E of town towards Jequitinhonha, 9 Mar. 1977, *Shepherd et al.* 4422 (UEC, MG); Mun. Itinga, 4 Mar. 1984, *Mattos-F. & Rizzini* 1622 (RB); l.c., 21 km from road BR 116 towards Itinga, 14 Dec. 1990, *Taylor & Zappi* 767 (K, HRCB, ZSS, BHCB); l.c., 15 km SW of town towards Araçuaí, Taquaral, 8 Apr. 1983, *Leuenberger & Martinelli* 3054 (CEPEC, K, NY, RB, MO); Mun. Virgem da Lapa/Araçuaí, 17 Dec. 1967, *A. Duarte* 10534 (RB); Mun. Araçuaí, 12 Nov. 1981, *G.C.P. Pinto* 398/81 (HRB, RB).

CONSERVATION STATUS. Vulnerable [VU A2c + 3c] (2); extent of occurrence = 60627 km²; PD=2, EI=1, GD=1. Short-list score (2×4) = 8. This species appears to be rare within its extent of occurrence, except in north-eastern Minas Gerais, in the Rio Jequitinhonha valley, where considerable habitat modification is taking place.

In southern Bahia and central-northern Minas Gerais *P. aureiflora* grows in close proximity to, or sympatric with, *P. stenantha* and/or *P. bahiensis*, but seems to be much rarer than either. However, it is the commonest pereskia in the middle part of the Rio Jequitinhonha drainage system, north-eastern Minas Gerais (Itaobim/Itinga), whence it was originally described, and where its above-mentioned congeners are absent. Its status as sister species to *P. guamacho* (N South America) was earlier suspected on purely morphological grounds, but has more recently been confirmed by gene sequence phylogenies obtained by Wallace (1995: Fig. 10) and Butterworth (unpubl.).

5.4   OPUNTIOIDEAE K. Schum.

Areoles bearing barbed spines and/or glochids; pericarpel scarcely differentiated from stem-segments; pollen mostly polyporate; seed enclosed in a pale, usually bony funicular envelope (aril).

Cladistic analyses employing DNA gene sequence data (Wallace & Dickie 2002) and seed anatomical, pollen, gross morphological data and pragmatic considerations (Stuppy & Huber 1991, Stuppy 2002; Taylor *et al.* 2002) indicate that c. 16 genera deserve recognition in this subfamily. *Austrocylindropuntia* Backeb. and its sister group *Cumulopuntia* F. Ritter (both Andean) appear to be basal amongst Opuntioideae. The first genus treated below is the only South American representative of a tribe of 4 genera otherwise native to Central & North America and the Caribbean, comprising *Pereskiopsis* Britton & Rose, *Quiabentia, Cylindropuntia* (Engelm.) F. M. Knuth and *Grusonia* F. Rchb. ex Britton & Rose (incl. *Corynopuntia*). The broad flattened leaves of *Quiabentia* and *Pereskiopsis* may be a synapomorphy (and both are decaploids; Pinkava 2002: 61, Table 1), since the *Austrocylindropuntia-Cumulopuntia* clade and the subfamily's plesiomorphic potential sister group, the Maihuenioideae, possess cylindric to awl-shaped leaves.

Tribe **CYLINDROPUNTIEAE** Doweld

2. QUIABENTIA Britton & Rose

Cact. 4: 252 (1923). Type: *Quiabentia zehntneri* (Britton & Rose) Britton & Rose.

A genus of only 2 species, the second being Q. *verticillata* (Vaupel) Vaupel, a sometimes treelike plant (2–15 m high), from the western Chaco and its periphery, in Argentina, Paraguay and Bolivia (Map 13). Its Brazilian counterpart, treated below, is of restricted and presumably relictual distribution, representing a marginal floristic element of the *caatinga*, since it occurs only near the south-western limits of this vegetation type. Plate 4.2

1. Quiabentia zehntneri *(Britton & Rose) Britton & Rose*, Cact. 4: 252 (1923). Type: Brazil, Bahia, Bom Jesus da Lapa, Rio São Francisco, 15–16 Nov. 1912, *Zehntner 630* (US, lecto. designated here; NY, lectopara.).

Pereskia zehntneri Britton & Rose, Cact. 1: 14 (1919).

VERNACULAR NAMES. Quiabento, Flor-de-cera, Espinho-de-Santo-Antônio, Cai-cai.

Shrubby, erect, 2–3 m, leafy when growing; branches, to 30 cm (or more), 1–2.5 cm diam., cylindric, the stoutest sometimes faintly ribbed, especially towards apex when dry, grey-green to grey, glaucous; areoles 20–40 mm apart, with white felt. Leaves 20–55 × 5–25 mm, to 3.5 mm thick, elliptic-ovate, orbicular or elliptic-lanceolate, pointed at both ends, concave above, gradually decurved towards apex, sessile, fleshy, stiff, green. Spines 7 or more per areole, the longest to 50 mm, white, reddish or golden, flexible, glochids few, spine-like and poorly differentiated from the spines apart from their position at the upper edge of the areole. Flowers apical or subapical, 7 × 7–8 cm; pericarpel narrowly turbinate to cylindric, to 40 mm, grey-green, with areoles and fleshy bracts; tube short, to 1 cm; perianth-segments to 40 mm, ovate or spathulate, retuse at apex, dark-pink, spreading; stamens exposed; stigma-lobes exserted. Fruit 4–7.5 cm long, narrowly turbinate, umbilicus not very deep, greenish to reddish or purple outside; funicular pulp translucent, whitish, placenta yellowish. Seeds many per fruit, to 4 mm diam.

Southern Rio São Francisco *caatinga* element: on ± naked or thinly wooded limestone (Bambuí) outcrops or gneiss/granite inselbergs amidst high *caatinga* forest, 450–750 m, both sides of the Rio São Francisco valley, west-cent./southern Bahia and cent.-northern Minas Gerais. Endemic to the core area of Eastern Brazil. Map 18.
BAHIA: Mun. Bom Jesus da Lapa, Morro da Lapa, 15 & 16 Nov. 1912, *Zehntner* 567 & 630 (US, RB), 28 Jan. 1958, *Mendes Magalhães* s.n. (IAN), 17 Apr. 1983, *Leuenberger et al.* 3078 (CEPEC, K); l.c., 13°15'S, 43°26'W, 15 Apr. 1980, *Harley et al.* 21398 (K, CEPEC); l.c., gruta do Bom Jesus, 10 June 1992, *A.M. de Carvalho et al.* 3964 (K, CEPEC); l.c., W of Rio São Francisco, reported by Andrade-Lima (1977: 191, fig. 5); Mun. Cocos, c. 5 km W of town, 2000, *L. Aona*, cult. UNICAMP, Aug. 2000, Zappi (obs.); Mun. Guanambi, 9.5 km NE of town on road to Caitité, gneiss/granite outcrop S of road, 25 July 1989, *Taylor & Zappi* (obs.); Mun. Palmas de Monte Alto, hill with TV tower, gneiss/granite rocks in *caatinga* above base of tower, 21 July 1989, *Zappi* 154 (SPF); Mun. Iuiú, 2 km S of town, 21 July 1989, *Zappi* 155A (SPF, fr.).
MINAS GERAIS: Mun. Januária, Brejo do Amparo, Morro Itapiraçaba, 11 Nov. 1989, *P.E. Nogueira Silva et al.* 125 (K); 10 km W of Januária, 15°30'S, 44°30'W, 20 Oct. 1972, *Ratter et al.* 2632 (UB); Mun. Itacarambi/Manga, 15°26'S, 43°55'W, Sep. 1999, *I. Ribeiro* (photos); Mun. Varzelândia, c. 10 km N of town, 'Serra São Felipe' 12 Aug. 1988, *Eggli* 1148 (ZSS); Mun. Capitão Enéas, limestone outcrop visible from road BR 122, June 2002, *M. Machado & G. Charles* (K, photos).

CONSERVATION STATUS. Least Concern [LC] (1); PD=3, EI=1, GD=1. Short-list score (1×5) = 5.

Rizzini & Mattos-Filho (1992) claim that this species is found in the 'região de Brumado–Condeúba, Bahia' and is there used as hedges. However, the plant commonly used for this purpose in this region and called 'quiabento' is *Pereskia bahiensis* and it may be doubted that *Quiabentia* is native there. Braun & Esteves Pereira (2002: 170) state that *Q. zehntneri* is exclusively found on limestone outcrops, which is not correct (see Bahian localities cited above).

Tribe OPUNTIEAE

The following 4 genera form a monophyletic lineage culminating in *Opuntia, sens. str.*, which is the most derived, possessing sensitive stamens and pollen with a reticulate exine. Other genera included in this lineage, but absent from Brazil, are *Miqueliopuntia* F. Ritter (N Chile), *Tunilla* D. R. Hunt & Iliff (southern Andes) and *Consolea* Lem. (E Caribbean). The following genus has a relictual distribution almost confined to Eastern Brazil, where its extensive range does not overlap at any point with that of *Opuntia sens. str.*:

3. TACINGA Britton & Rose

Cact. 1: 39 (1919). Type: *Tacinga funalis* Britton & Rose.
Literature: Taylor *et al.* (2002).

Shrubs, subshrubs and lianas, 0.2–5.0 m; branches not dimorphic, pith often ± chambered or quite hollow when > 1–2 years old; stem-segments cylindric or compressed, then orbicular, obovate to elliptic or rhomboid in outline; areoles borne on non-existent to very low tubercles in the axils of caducous leaves, with abundant glochids and felt, spines present or absent. Leaves minute, subulate, sessile, fleshy, early deciduous. Flowers solitary, from the margin or apex of the stem-segments; flowers epigynous, pericarpel globose, turbinate or elongate, sometimes deeply depressed and forming a tube at apex, with areoles subtended by leaf-like, fleshy, green or coloured bract-scales, often with active extra-floral nectaries before and during anthesis (attracting ants); perianth multiseriate, tube relatively short, outer perianth-segments short, stiff, erect, patent or strongly reflexed, coloured, fleshy, inner segments coloured, delicate, erect, somewhat spreading or strongly reflexed; stamens numerous, erect, at least at first, sometimes strongly exserted, *not sensitive*, those adjacent to the perianth represented by hairlike staminodes in species 1–3; *pollen exine not reticulate, very finely punctate/spinulate*. Fruit solitary or clustered-proliferous, globose, turbinate or elongate clavate with a very deep umbilicus, flower remnants deciduous; funicular pulp translucent or opaque and coloured, fibrous or almost lacking; placenta white, greenish or coloured. Seeds few, globular to reniform, to 5 mm diam., enclosed in a hard, bony funicular envelope.

As recently amplified, a genus of 6 species named as an anagram of *caatinga,* the dry thorn forest of Eastern Brazil, where it is frequent, also ascending into the included *campos rupestres* (see No. 6) and extending slightly west of the core area into north-western Minas Gerais on limestone outcrops (No. 5). Species Nos 1–4 are endemic to the core area of Eastern Brazil as defined here. Hybrids between Nos 2 & 3, 3 & 6, 4 & 6 and 5 & 6 are known or suspected and sometimes not uncommon, but are not keyed out below. Plates 4.3–8.1.

1. Perianth-segments erect and forming a tube or only spreading slightly at apex, deep
   magenta-pink to orange-red 2
1. Perianth-segments ± spreading to strongly reflexed and lying against the pericarpel,
   yellow, orange-reddish, green or purplish 4

2. Stamens exserted; style strongly thickened near base (cultivated) . . . . . ***Nopalea cochenillifera**
2. Stamens included; style not as above (eastern *caatingas*) 3

3. Fruit greenish to reddish or purple outside, to 3 cm, funicular pulp yellowish . . . . **4. palmadora**
3. Fruit greenish white, sometimes with faint pink shades, 4–5.5 cm, funicular pulp
   bright pink . . . . **3. werneri**

4. Plants low-growing, rarely exceeding 1 m, mostly subshrubs; stem-segments always
   flattened; spines 0 or fine and slender, to 15 mm (widespread, especially on rocks) 5
4. Plants taller, to 2 m or more, scandent, with at least the lower parts of stems perfectly
   cylindric; spines 0 (NE Minas Gerais to SW Pernambuco, *caatinga*) 6

5. Stem-segments spineless at maturity; areoles well spaced, 10–20 mm apart . . . . **6. inamoena**
5. Stem-segments with small, persistent spines; areoles congested, 1–14 mm apart . . . . . . **5. saxatilis**

6. All stem-segments perfectly cylindric; flowers green to purplish (S Bahia to SW
   Pernambuco) . . . . **1. funalis**
6. Ultimate stem-segments ± flattened; flowers green (Rio Jequitinhonha valley,
   NE Minas Gerais) . . . . **2. braunii**

1. Tacinga funalis *Britton & Rose*, Cact. 1: 39–40 (1919). Type (Taylor *et al.* 2002): Brazil, Bahia, Juazeiro, 1915, *Rose & Russell* 19723 (US, lecto.; NY, lectopara.).

? T. luetzelburgii Küpper ex Luetzelb., Estud. bot. Nordéste 3: 70, 111 (1926), nom. nud.

T. atropurpurea Werderm. in Notizbl. Bot. Gart. Berlin-Dahlem 12: 223–224 (1934). *T. funalis* var.
  atropurpurea (Werderm.) P.J. Braun & E. Esteves Pereira in Kakt. and. Sukk. 43(5): Karteiblatt 14, [2]
  (1992). *T. funalis* subsp. *atropurpurea* (Werderm.) P.J. Braun & E. Esteves Pereira in Succulenta 74:
  134 (1995). Type: Brazil, Bahia, south of Boa Nova and [Vitória da] Conquista, 600 m, May 1932,
  Werdermann 3999 (B†).

T. zehntneri Backeb. & Voll in Backeb., Blätt. Kakteenf. 1935(3): unpaged [p. 4] & suppl. [p. 6], with
  illus. (1935). Type: not cited. Lectotype (Taylor *et al.* 2002): Backeberg, l.c. (illus., flower at left).

T. atropurpurea var. *zehntnerioides* Backeb., Descr. Cact. Nov. [1:] 10 (1956, publ. 1957). Type: Brazil,
  Bahia (assumed not to have been preserved).

VERNACULAR NAMES. Rabo-de-rato, Rabo-de-gato, Cipó-de-espinho, Quipá-voador, Trança-perna.

Shrubby, decumbent or scandent, stems to 12 m long, hollow inside and scarcely succulent once > 1 year-old. Stem-segments to 30–100 cm or more, 0.8–1.5 cm diam., cylindric, grey-green to purplish or strongly reddish, often somewhat glaucous. Areoles 5–20 mm apart, prominent, felt white to brownish, with long woolly hairs at first, spineless, glochids abundant, brownish to grey. Flowers from near the apex of terminal segments, 57–80 × 20 mm, green to purplish; pericarpel narrowly turbinate, to 30–50 mm, grey-green, with areoles; outer perianth-segments 10, 5–15 mm, narrowly ovate, acute, spreading to reflexed; inner segments c. 7, to 20–40 mm, acute, revolute; stamens erect, exserted c. 17 mm or more, the outermost replaced by abundant hairlike staminodes; anthers sometimes reddish; style to 45 mm, stigma-lobes exserted beyond stamens. Fruit 40–50 × 20 mm, bottle-shaped, with deep umbilicus, greenish to reddish or purple outside; funicular pulp translucent, whitish, placenta yellowish. Seeds 3–5 per fruit, to 4 mm diam.

Central-southern *caatinga* element: in *caatinga*, 380–950 m, south-western Pernambuco (also vaguely reported from adjacent Piauí) and western and northern to southern Bahia. Endemic to the core area within North-eastern Brazil. Map 19.

PIAUÍ: *fide* Andrade-Lima (1989: 26), but unlocalized.

PERNAMBUCO: W Pernambuco, Mun. Afrânio, 20 Apr. 1971, *Heringer, Andrade-Lima et al.* 212 (IPA); Mun. Petrolina, 17 km SSE of Rajada on road BR 407, 8°55'S, 40°44'W, 6 Apr. 2000, *E.A. Rocha et al.* (K, photos); l.c., between the town and the lock ('eclusa'), 14 Aug. 1966, *Hunt* 6498 (see Taylor *et al.* 2002); Mun. Santa Maria da Boa Vista, 4 Aug. 1955, *Andrade-Lima* 55-2116 (IPA).

BAHIA: W Bahia, Mun. Barra, Luetzelburg (1925–26, 3: 70); N–S Bahia, Mun. Juazeiro, 1915, *Rose & Russell* 19723 (US, NY); l.c., near Carnaíba do Sertão, 8 Aug. 1964, *Castellanos* 25387 (GUA); Mun. Sento Sé, [*Zehntner* in] *Rose* 19734 (US), l.c., 10°7'S, 41°25'W, 11 July 2000, *G. Charles* (photos); Mun. Jaguarari ['Curaçá'], 9°48'S, 40°3'W, 19 Aug. 1983, *S.B. Silva & G.C.P. Pinto* 300 (HRB); l.c., Barrinha, 7–8 June 1915, *Rose & Russell* 19792 (K, US); l.c., 64 km S of Juazeiro on road to Senhor do Bonfim, between Maçaroca and Barrinha, 8 Jan. 1991, *Taylor et al.* 1378 (K, HRCB, ZSS, CEPEC); Mun. Campo Formoso, 32 km S of Junco, near Curral Velho, 9 Jan. 1991, *Taylor et al.* (obs.); Mun. Senhor do Bomfim ['Villa Nova'], Aug. 1912, *Zehntner* 306 (US); Mun. (?), 10°53'S, 41°35'W, 13 July 2000, *G. Charles* (photos); Mun. João Dourado & Mun. Morro do Chapéu, APA Gruta dos Brejões, 4 Aug. 2002, *Taylor & Machado* (K, photos); Mun. Jacobina, 27–28 km WNW towards Lajes, 12 Jan. 1991, *Taylor et al.* (obs.); l.c., c. 18 km S of town, 'Buraco do Possidônio', 2001, *M. Machado* (obs.); Mun. Cafarnaum, 26 Oct. 1965, *A.P. Duarte* 9252 (HB, RB); Mun. Seabra, Apr. 1978, *E. Esteves Pereira* 71 (UFG); Mun. Iaçu, Nov. 1987, *E. Esteves Pereira* 273 (UFG); Mun. Marcionílio Sousa, Machado Portella, 19 June 1915, *Rose* 19919 (US); Mun. Abaíra, road Piatã–Abaíra, 13°17'S, 41°42'W, 22 Sep. 1992, *W. Ganev* 1167 (K, HUEFS); Mun. Bom Jesus da Lapa, reported by Luetzelburg (1925–26, 3: 70); Mun. Livramento do Brumado, 12 km S of town, 23 Nov. 1988, *Taylor & Zappi* in *Harley* 25545 (SPF, K); Mun. Ituaçu (?), 13°52'S, 41°17'W, 9 Aug. 1979, *J.E.M. Brazão* 119 (HRB, RB); Mun. Maracás, Porto Alegre, *fide* Uebelmann (1996): HU 384; Mun. Caetité, hill with cross above town, 27 Aug. 1988, *Eggli* 1304 (ZSS); Mun. Brumado, 19 km from town on road BR 030 towards Sussuarana, 3 Feb. 1991, *Taylor et al.* (obs.); Mun. Aracatu, near border with Mun. Brumado, 25 km on road BR 030, 3 Feb. 1991, *Taylor et al.* 1543 (K, HRCB, ZSS, CEPEC); ibid., c. 20 km N of Anajé towards Sussuarana, 17 Aug. 1988, *Eggli* 1184 (ZSS); Mun. Cordeiros, 15°4.5'S, 41°54'W, Sep. 2003, *M. Machado* (obs.).

CONSERVATION STATUS. Least Concern [LC] (1); PD=2, EI=1, GD=2. Short-list score (1×5) = 5. Least Concern at present, but habitat destruction is continuing throughout its range. It is protected inside the A.P.A. Gruta dos Brejões, Bahia.

This species remains poorly known as fertile material, but appears to be variable. Its flowers may be either green or purple, the latter colour variant apparently being characteristic in the southern part of its range, but recorded northwards to at least Mun. Cafarnaum, central Bahia. These variants do not appear to exhibit obvious vegetative differences and the flowers of Backeberg's var. *zehntnerioides*, as described, seem to be of somewhat intermediate colour. More collections and observations during its late winter (August/September) flowering period are needed. Marlon Machado (pers. comm.) has observed hummingbirds visiting the flowers of this species near Morro do Chapéu, Bahia.

2. Tacinga braunii *E. Esteves Pereira* in Kakt. and. Sukk. 40: 134–135 (1989). Holotype: Brazil, Minas Gerais, surroundings of the Rio Jequitinhonha, *P. J. Braun* 864 (ZSS).

[*Opuntia rubescens sensu* K. Schum. in Martius, Fl. bras. 2(4): 306 (1890), quoad *Glaziou* 14865, non Salm-Dyck ex DC. (1828).]

[*Tacinga funalis sensu* Rizzini & Mattos-Filho, Contrib. Conhecim. Fl. NE. Minas Gerais Bahia Medit.: 38 (1992) non Britton & Rose.]

VERNACULAR NAMES. Rabo-de-rato, Cipó-de-espinhos, Rabo-de-espinhos.

Shrubby or semi-scandent with the ultimate stem-segments sometimes pendulous, to 6 m, partially hollow inside; primary stem cylindric, erect, subsequent stem-segments rounded at base, somewhat flattened towards apex, ultimate segments to 35 × 1.5–3.5 × 0.5–0.8 cm, rarely broader, ± strongly flattened, grey-green to purplish, glaucous; areoles to c. 23 mm apart, prominent, felt white, with long hairs when young; spines 0 (seen only in a plant thought to show introgression with *T. werneri*), glochids abundant, greyish. Flowers apical or subapical, c. 7 × 2 cm; pericarpel narrowly turbinate, to 40 mm, grey-green to purplish, bearing areoles; outer perianth-segments green, c. 10, 5–15 mm, narrowly ovate, acute, erect to ± reflexed; inner segments pale green, to 40 mm, acute, revolute; stamens erect, the outermost replaced by abundant hairlike staminodes; style to 45 mm, stigma-lobes exserted. Fruit 40 mm, narrowly urceolate, with a deep umbilicus, greenish; funicular pulp translucent, whitish, placenta yellowish. Seeds 4–5 per fruit, to 4 mm diam.

South-eastern *caatinga* (inselberg) element: on gneiss/granite outcrops/inselbergs in *caatinga-agreste*, 170–350 m, Rio Jequitinhonha valley, north-eastern Minas Gerais. Endemic to the Rio Jequitinhonha valley. Map 19.

MINAS GERAIS: Mun. Jacinto, near the Rio Jequitinhonha, 16°9'S, 40°13'W, 25 May 1979, *P. Vaillant* 78 (HRB, RB); Mun. Itaobim, 1 km W of town, 0.5 km N of Rio Jequitinhonha, 18 Nov. 1988, *Taylor & Zappi in Harley* 25530 (K, SPF); ibid., on road BR 357, 8 km W of town towards Itinga, 14 Dec. 1990, *Taylor & Zappi* 760 (K, HRCB, ZSS, BHCB); Mun. Itinga, on road BR 357, 2 km E of town towards Itaobim, 14 Dec. 1990, *Taylor & Zappi* 769 (K, HRCB, ZSS, BHCB); Mun. Coronel Murta, 5 km SE of town towards Araçuaí, 21 Feb. 1988, *Supthut* 8865 (ZSS); município unknown, surroundings of the Rio Jequitinhonha, July 1987, *E. Esteves Pereira* 264 (UFG), *P.J. Braun* 864 (ZSS); unlocalized, ['Serra de São José', according to Glaziou (1909: 327), but not from there, see below], '1883–1884', [*anon.* in] *Glaziou* 14865 (K, C, F, photo).

CONSERVATION STATUS. Vulnerable [VU B1ab(iii) + 2ab(iii)] (2); extent of occurrence = 1515 km^2; PD=2, EI=1, GD=1. Short-list score (2×4) = 8.

Although correctly named only in 1989, this distinctive species was in fact collected (see *Glaziou* 14865) before its better-known sister-species treated above. The early collection, made 120 years ago and claimed by Glaziou, was misidentified by Schumann (1890) in the *Flora brasiliensis* as *Opuntia rubescens* Salm-Dyck ex DC., which is the type of *Consolea* Lem., an opuntioid genus endemic to the Caribbean (the original description of *O. rubescens* erroneously suggested it might have come from Brazil, so Schumann can, perhaps, be forgiven). Glaziou himself should not be credited with the first collection, however, since his locality data are clearly false (the Serra de São José, where he is known to have collected other plants, is in perhumid southern Minas Gerais). It seems that, as in other well-documented cases (Wurdack 1970), Glaziou appears to have appropriated material from an anonymous collector and given it his own number. For a further example of this deplorable practice, see *T. inamoena,* below.

3. Tacinga werneri *(Eggli)* N. P. Taylor & *Stuppy* in Succ. Pl. Res. 6: 111 (2002). Holotype: Brazil, Bahia, Mun. Jequié, *N. P. Taylor et al.* 1555 (CEPEC; HRCB, K, ZSS, isos.).

Opuntia werneri Eggli in Bradleya 10: 90, with illus. (1992).

Shrubby, mostly < 1 m high; stem-segments flat, elongate-ellipsoid to broadly ellipsoid, 10–20 × 4–10 × 0.5–2.0 cm, dark green, those of seedlings and juvenile growth nearly orbicular; areoles c. 3 mm diam., on scarcely raised podaria, mostly 12–18 mm apart, often displaying indeterminate growth near base of plant; glochids dirty whitish, few at first; spines (0–)3–8 at first (some forms remaining almost spineless), 3–63 × 1.5 mm near base, ash-grey tipped yellow to yellowish brown, diverging. Flowers c. 35–50 mm long; perianth-segments ± erect to slightly expanded, to c. 7 mm long, bright red, outer series rather fleshy, darker red; stamens numerous, the outermost represented by staminodes (with rudimentary anthers or these entirely lacking), filaments orange-yellow to orange above; style narrowly obclaviform, yellowish; stigma-lobes 4, white; pericarpel 28 × 20 mm, green, areoles with numerous glochids but spineless. Fruit elongate ovoid, broadly beaked at apex, 40–55 × 25–35 mm, with a narrow, deeply sunken umbilicus, pale greenish white to white, pericarp thick, pale green; funicular pulp pink when ripe. Seed c. 3 × 3.5 mm, nearly spherical, funicular envelope pale ochre.

Eastern *caatinga* element: margins of gneiss/granite outcrops, on inselbergs and occasionally on calcareous soils in *caatinga* of the Chapada Diamantina, 100–950 m, drainage of the Rios Paraguaçu, de Contas and Jequitinhonha, eastern Bahia and north-eastern Minas Gerais. Endemic to the core area of Eastern Brazil. Map 19.

BAHIA: Mun. Morro do Chapéu, between 'Vermelho' and 'Mulungu da Gruta', 11°0'27"S, 41°20'36"W, Aug. 2003, *M. Machado* (obs.); l.c., 15–20 km W of town, Sep. 2002, *M. Machado* (K, photos); l.c., nr Ventura, Sep. 2002, *M. Machado* (K, photos); l.c., nr 'Buraco do Possidônio', Sep. 2002, *M. Machado* (K, photos); Mun. Rui Barbosa, near granite quarry, Aug. 1988, *Eggli 1259* (CEPEC, ZSS), 7 Feb. 1991, *Taylor et al. 1568* (CEPEC, HRCB, K, ZSS); Mun. Rafael Jambeiro, Km 30 of road BR 242, 11 May 1975, *A.L. Costa s.n.* (ALCB, in spirit); l.c., Km 33, Rio Coruja, 29 Sep. 1973, *A.L. Costa s.n.* (ALCB, in spirit); Mun. Iaçu, Morro do Coité, 12°45'S, 39°53'W, 28 Sep. 1997, *E. de Melo 2259* (HUEFS); Mun. Itatim, Morro da Torre, 12°43'S, 39°42'W, 9 Nov. 1996, *E. de Melo 1831* (HUEFS; K, photo); Mun. Elísio Medrado, S of Monte Cruzeiro, 25 June 1915, *Rose & Russell 20051* (US, NY); Mun. Jequié, 41.5 km E of Porto Alegre (Mun. Maracás) towards Jequié, 4 Feb. 1991, *Taylor et al. 1555* (CEPEC, HRCB, K, ZSS).

MINAS GERAIS: Mun. Pedra Azul, 8 km from town towards road BR 116, 18 Oct. 1988, *Taylor & Zappi* in *Harley 25187* (K, SPF), 1 Feb. 1991, *Taylor et al. 1520* (BHCB, HRCB, K, ZSS).

CONSERVATION STATUS. Vulnerable [VU B2ab(ii, iii, iv, v)] (2); extent of occurrence >20000 km², but area of occupancy estimated to be < 2000 km²; PD=2, EI=1, GD=2. Short-list score (2×5) = 10. This species has a local and rather disjunct distribution and is nowhere common. Its habitat at Rui Barbosa (BA) is being destroyed by mining operations.

The flowers and fruit of this species provide a clear link between *T. funalis*, *T. braunii* and the following species, with which it is frequently sympatric. It is likely that it has been under-recorded due to confusion with *T. palmadora*. The collections from Rui Barbosa cited above (*Eggli 1259*, *Taylor et al. 1568*) represent a variant population with rather weak or almost absent spination. As just noted, its long term future looks doubtful while granite quarrying activities continue.

Plants of probable hybrid origin, involving *T. werneri* and *T. inamoena*, have been observed by the authors near Pedra Azul, Minas Gerais and west of Morro do Chapéu, Bahia (eg. 15–20 km W of Morro do Chapéu, 5 Aug. 2002, *Taylor & Machado* [K, photos]). The following names, as suggested by Ritter's original illustration, could possibly refer to this hybrid, although *T. werneri* has not yet been found near Urandi: *Platyopuntia inamoena* f. *spinigera* F. Ritter, Kakt. Südamer. 1: 32 (1979). *Opuntia inamoena* f. *spinigera* (Ritter) P. J. Braun & E. Esteves Pereira in Cact. Succ. J. (US) 61: 272–273 (1989). Holotype: Bahia, Urandi, *Ritter 1252A* (U). If they do not belong here, then it is assumed that they will be identifiable with *T.* ×*quipa* (F. A. C. Weber) N. P. Taylor & Stuppy.

4. Tacinga palmadora *(Britton & Rose) N. P. Taylor & Stuppy* in Succ. Pl. Res. 6: 112 (2002). Lectotype (Krook & Mottram 2001: 94): Brazil, Mun. Jaguarari, Barrinha, 7 June 1915, *Rose & Russell* 19787 (NY; NY, US, isos.).

Opuntia palmadora Britton & Rose, Cact. 1: 202 (1919).

VERNACULAR NAMES. Palmatória, Palmatória-de-espinho, Palma-de-espinhos, Palma, Palminha, Quipá-de-espinho, Quipá, Palmatória-de-quipá, Rabo-de-onça.

Shrubby, to 2(–4) m tall, main stem flattened, transversely elliptic, sometimes with brownish bark, very spiny, the areoles showing considerable indeterminate growth. Branches erect, stem-segments 8–17 × 3–7 cm, elliptic to obovate, grey-green, glaucous. Areoles 10–20 mm apart, prominent, felt white; spines (2–)3–9, (7–)15–35 mm, pungent, golden at first, pale grey with age. Flowers from the apex of the terminal segments, 30 × 15–20 mm; pericarpel turbinate, to 20 mm, pale to grey-green, with orange, acute, fleshy bract-scales subtending bristles; outer perianth-segments 10 mm, narrowly lanceolate, deep red or pinkish orange, erect; innermost 15 mm, spathulate, orange, spreading only at apex; stamens included; stigma-lobes exserted. Fruit to 3 cm long, turbinate to obovoid, greenish to reddish or purple outside; funicular pulp translucent, whitish, placenta yellowish. Seeds 3–5 per fruit, to 5 mm diam.

Eastern *caatinga* element: in *caatinga, agreste* and *carrasco,* frequent on deep sandy substrates but not restricted to these, c. 200–1020 m, Rio Grande do Norte to southern Bahia (from the Chapada Diamantina/northern Serra do Espinhaço eastwards). Endemic to North-eastern Brazil. Map 19.

RIO GRANDE DO NORTE: Mun. São José do Campestre, 58 km N of Guarabira (PB) on road to Tangará, 6°21'S, 35°40'W, 1 Apr. 2000, *E.A. Rocha et al.* (K, photos); Mun. Tangará, between the town and Currais Novos, 6°14'S, 35°51'W, 1 Apr. 2000, *E.A. Rocha et al.* (K, photos).

PARAÍBA: Mun. Areia, 'no agreste', 22 Sep. 1956, *J.C. Moraes* s.n. (SPSF 5008); Mun. Soledade, see Eiten (1983: figs 67 & 68); Mun. Itaporanga, Serra Água Branca, 7–10 Jan. 1994, *M.F. Agra et al.* 2560 (JPB); Mun. Campina Grande, Boa Vista, 27–29 Apr. 1994, *M.F. Agra et al.* 3125 (JPB); Mun. São João do Cariri, road BR 412, 27–29 Apr. 1994, *M.F. Agra et al.* 3122, 3123, 3124 (JPB).

PERNAMBUCO: Mun. Araripina, Dec. 1963, *Ritter* 1253 (SGO 122051, *fide* Eggli *et al.* 1995: 505); Mun. Ouricuri, 26 km ESE towards Parnamirim, 7°58'S, 39°52'W, 16 July 1962, *G. Eiten & L.T. Eiten* 4929 (SP, US); ibid., Faz. Tabuleiro, July 1984, *G. Costa-Lima* 21 (HRB, IPA); Mun. Sertânia, Moderna, 8 Apr. 2000, *E.A. Rocha et al.* (obs.); l.c., 13 km W of Arcoverde on road from Custódia, 12 Feb. 1991, *Taylor & Zappi* (obs.); Mun. Pesqueira, 5 Oct. 1934, *Pickel* 3661 (IPA); Mun. Caruaru, 24 Nov. 1991, *F.A.R. Santos* 11 (HUEFS, PEUFR), l.c., 9 km NE of town, 8°14'21"S, 35°54'52"W, Dec. 1999, *E.A. Rocha et al.* (K, photos); Mun. Taquaritinga do Norte, Gravatá do Ibiapina, 8 Aug. 2001, *Zappi* (obs.); Mun. Gravatá, Russinha, 13 Oct. 1934, *Pickel* 3656 (IPA 4486); Mun. Buíque, Fazenda Laranjeiras, 10 July 1995, *K. Andrade & L. Figueiredo* 126 (K, PEUFR); Mun. Alagoinha, Sítio Cajueiro, Sep. 1991, *J. Hamburgo Alves & R. Silva* s.n. (PEUFR); l.c., Sítio Lagoa Seca, 5.5 km from road PE 217, 13 Feb. 1993, *F.A.R. Santos* 74, 75 (ALCB, HUEFS, PEUFR); Mun. Belém de São Francisco, near border with Mun. Floresta, 8°40'S, 38°44'W, 8 Apr. 2000, *E.A. Rocha et al.* (K, photos); Mun. Floresta, 8°36'S, 38°27'W, 8 Apr. 2000, *E.A. Rocha et al.* (K, photos); l.c., N of the Serra Negra, 8°33'S, 38°1'W, 8 Apr. 2000, *E.A. Rocha et al.* (K, photos); l.c., road to Ibimirim, 30 June 1952, *Andrade-Lima* 52-1156 (R, IPA), 24 July 1994, *L.P. Felix et al.* 6702 (PEUFR); l.c., Serra de São Gonçalo, no alto da serra, 23 May 1971, *Heringer & Andrade-Lima* 856 (RB, IPA); Mun. Ibimirim, Lagoa de Areia, 23 July 1994, *L.P. Felix et al.* 6696 (PEUFR); l.c., 12 Sep. 1954, *I.I.A. Falcão et al.* 1056 (RB); Mun. Nova Petrolândia, 12 km NW of town, 12 Feb. 1991, *Taylor & Zappi* (obs.).

ALAGOAS: W Alagoas, Mun. Mata Grande, Serra Verde, 30 July 1981, *R. de Lyra & G.L. Esteves* 692 (MAC); Mun. Delmiro Gouveia, near bridge over Rio São Francisco at border with Bahia, 12 Feb. 1991, *Taylor & Zappi* 1618C (K, photo); Mun. Piranhas, Xingó, 9 Oct. 1993, *R. de Lyra-Lemos et al.* 2812 (MAC).

SERGIPE: Mun. Canindé de São Francisco, 11 Feb. 1991, *Taylor & Zappi* (obs.); Mun. Poço Redondo, 13 July 1983, *M. Fonseca* 570 (IPA 30960, ASE 2959), 1 Aug. 1986, *Fonseca* s.n. (ASE 4452); Mun. Porto da Folha, 9.5 km NW of Monte Alegre de Sergipe, 11 Feb. 1991, *Taylor & Zappi* (obs.); Mun. Nossa Senhora da Glória, 8–9 km NW of town, 11 Feb. 1991, *Taylor & Zappi* (obs.).

BAHIA: Mun. Rodelas, Salgado do Melão, 3 Aug. 1994, *M.C. Ferreira* 600 (HRB); Mun. [Nova] Glória, Brejo do Burgo, 30 Nov. 1992, *F.P. Bandeira* 112 (ALCB, HUEFS), 27 Aug. 1995, *F.P. Bandeira* 265 (HUEFS); Mun. Paulo Afonso, 20 Mar. 1959, *Castellanos* 22451 (R); Mun. Jaguarari, 64 km S of Juazeiro towards Senhor do Bonfim, between Maçaroca and Barrinha, 8 Jan. 1991, *Taylor et al.* 1379, 1380 (K, HRCB, ZSS, CEPEC); l.c., Barrinha, 7–8 June 1915, *Rose & Russell* 19787 (US); ibid., Flamengo, July 1988, *E. Esteves Pereira* 274 (UFG); l.c., Faz. Suçuarana, 9°58'S, 39°52'W, 8 Aug. 1983, *G.C.P. Pinto & S.B. Silva* 195/83 (HRB, CEPEC); Mun. Uauá, c. 51 km N of Monte Santo, 25 Aug. 1996, *L.P. de Queiroz* 4635 (HUEFS, K); Mun. Jeremoabo, Xuquê [Chuquê], May 1921, *Luetzelburg* 12673 (M); Mun. Monte Santo, c. 5 km N of town towards Uauá, 25 Aug. 1996, *L.P. de Queiroz* 4608 (HUEFS, K); Mun. Itiúba, road to Filadélfia, 10°39'S, 39°44'W, 11 May 2002, *Nascimento & Nunes* 101 (HUEFS); Mun. Queimadas, 9–11 June 1915, *Rose & Russell* 19865 (US); Mun. Tucano, Aug. 1961, *Pabst* s.n. (HB); Mun. São Gabriel & Mun. João Dourado, APA Gruta dos Brejões, 4 Aug. 2002, *Taylor & Machado* (obs.); Mun. Morro do Chapéu, NE of Gruta dos Brejões, 4 Aug. 2002, *Taylor* (K, photo); l.c., W of Icó, Brejão (Formosa), 'Lagedo Bordado', 5 Aug. 2002, *Taylor* (K, photos); l.c., 2.5 km E of América Dourada, 0.5 km from SE bank of Rio Jacaré, 17 July 1989, *Taylor & Zappi* (obs.); l.c., c. 18 km S of town, 'Buraco do Possidônio', 2002, *M. Machado & Taylor* (obs.); Mun. Araci, 18 Jan. 1994, *M.A.S. das Neves* 84 (HUEFS); l.c., 14.5 km N of town towards Tucano, 5 Jan. 1991, *Taylor et al.* 1357 (K, HRCB, ZSS, CEPEC); Mun. Jacobina, 27–28 km WNW of Jacobina towards Lages, 12 Jan. 1991, *Taylor et al.* 1397 (K, HRCB, ZSS, CEPEC); Mun. Miguel Calmon, 28 km S of Jacobina, 26 Apr. 1992, *Taylor & Zappi* (obs.); Mun. Olindina, Mata da Faz. Olhos D'agua, 13 July 1993, *O.B. Borges* 38 (HRB); Mun. Riachão do Jacuípe, 10 km SE of town on road BR 324, 10 July 1985, *Noblick & Lemos* 4011 (K, HUEFS); l.c., road BR 324, 26 km NW of Tanquinho, 25 Apr. 1992, *Taylor & Zappi* (K, photo); Mun. Ipirá, 12°22'S, 39°41'W, 4 Oct. 1986, *L.P. de Queiroz* 959 (HUEFS); l.c., near Rio do Peixe, 12°13'S, 39°48'W, 6 Aug. 1979, *J.E.M. Brazão* 115 (HRB); l.c., Faz. Caldeirão, 18 July 1984, *E.L.P.G. Oliveira* 756 (BAH 6115); Mun. Ibitiara, c. 52 km W of Seabra towards Ibotirama, Serra Malhada, 14 Jan. 1991, *Taylor et al.* 1418A (HRCB, K, fr.); Mun. Muritiba (?), Vale dos rios Paraguaçu e Jacuípe, Ilha do Umbuzeiro, 12°32'S, 39°5'W, Aug./Sep. 1980, *Iscardino et al.* in Grupo Pedra do Cavalo 595 (CEPEC, HRB, ALCB, BAH 3383); Mun. Itatim, Morro do Agenor, 12°42/43'S, 39°42'W, 14 Oct. 1995, *F. França* 1405 (HUEFS), 31 Mar. 1996, *E. de Melo* 1566 (HUEFS); Mun. Milagres, 18 July 1979, *Hatschbach* 42449 (CEPEC, MO), 16 July 1982, *Hatschbach & Guimarães* 45071 (MO); l.c., road BA 046 c. 4 km from junction with road BR 116 towards Amargosa, 12°51'S, 39°46'W, 2 June 1993, *L.P. Queiroz & T.S.N. Sena* 3196 (K, HUEFS); l.c., 13°56'S, 39°43'W, 26 Oct. 1978, *A. Araújo* 122 (RB); Mun. Botuporã, c. 60 km SE of Macaúbas towards Paramirim, 26 Aug. 1988, *Eggli* (obs.); Mun. Maracás, c. 20 km N of town towards Planaltino, Faz. Tanquinho, 29–30 June 1993, *L.P. Queiroz & V.L.F. Fraga* 3260 (K, HUEFS); l.c., 6–7 km E of Porto Alegre on road to Jequié, 4 Feb. 1991, *Taylor et al.* 1552 (K, HRCB, ZSS, CEPEC); Mun. Livramento do Brumado, 11 km S of town towards Brumado, 23 Nov. 1988, *Taylor & Zappi* in Harley 25543 (SPF, K); Mun. Ituaçu, 13°48'S, 41°16'W, 22 June 1987, *L.P. de Queiroz* 1656 (HUEFS); Mun. Jequié, 4 km from town on road BR 116, 13 Oct. 1977, *T.S. dos Santos* 3146 (CEPEC); Mun. Caitité, Brejinho das Ametistas, 1 km from village, 26 July 1989, *Taylor & Zappi* (K, photo); Mun. Urandi, near railway line, c. 3 km S of town, 22 July 1989, *Taylor & Zappi* (obs.); Mun. Vitória da Conquista, Gameleira, 14°50'25"S, 41°W, 17 April 2003, *Taylor & Zappi* (obs.).

CONSERVATION STATUS. Least Concern [LC] (1); PD=2, EI=1, GD=2. Short-list score (1×5) = 5. A widespread and frequent species.

The pollination biology of this species has been studied by Locatelli & Machado (1999a) and their observations of hummingbird visitation are confirmed by the present authors. It exhibits considerable regional and local variation.

5. Tacinga saxatilis *(F. Ritter) N. P. Taylor & Stuppy* in Succ. Pl. Res. 6: 115 (2002). Holotype: Brazil, Minas Gerais, Montes Claros, 1959, *Ritter* 1035 (U).

Platyopuntia saxatilis F. Ritter, Kakt. Südamer. 1: 32–33 (1979). *Opuntia saxatilis* (Ritter) P.J. Braun & E.

Esteves Pereira in Cact. Succ. J. (US) 63: 82 (1991).

O. *saxatilis* var. *pomosa* P. J. Braun & E. Esteves Pereira in ibid. 124–129 (1991). O. *saxatilis* subsp. *pomosa* (P. J. Braun & E. Esteves Pereira) P. J. Braun & E. Esteves Pereira in Schumannia 3: 188 (2002). Holotype: Brazil, Minas Gerais, border region of Goiás and Minas Gerais, 1979, *E. Esteves Pereira* 156 (UFG 12368).

O. *saxatilis* var. *occibahiensis* P. J. Braun & E. Esteves Pereira in ibid. 311–315 (1991). O. *saxatilis* subsp. *occibahiensis* (P. J. Braun & E. Esteves Pereira) P. J. Braun & E. Esteves Pereira in Schumannia 3: 188 (2002). Holotype: Brazil, W Bahia, 1982, *E. Esteves Pereira* 188 (UFG 12376).

O. *saxatilis* var. *minutispina* P. J. Braun & E. Esteves Pereira in Cact. Succ. J. (US) 67: 108 (1995). O. *saxatilis* subsp. *minutispina* (P. J. Braun & E. Esteves Pereira) P. J. Braun & E. Esteves Pereira in Schumannia 3: 188 (2002). Holotype: Brazil, Minas Gerais, ['próximo Manga/Januária'], W of Rio São Francisco, c. 550 m, ['VII.1978'], *E. Esteves Pereira* 228 (UFG 12382).

Shrubby, to 50 × 400 cm, without main stem. Branches erect, stem-segments 10–16 × 5–11 cm, 7–17 mm thick, orbicular to ovate, grey-green, sometimes glaucous. Areoles 1–14 mm apart, to 0.5 mm diam., sunken into raised podaria, with felt, glochids immersed in the stem; spines 0–6, delicate, generally bristly, off-white to dark brownish, erect, 1–19 mm, very thin. Flowers at the apex and margins of the terminal stem-segments, 3–4 × 2.6–4.2 cm; pericarpel 17–25 × 17 mm, green, with yellowish or dull red, acute, fleshy bract-scales subtending bristles; perianth-segments 12–22 × 9–14 mm, outer narrowly lanceolate, deep red to yellow-orange, spreading; innermost spathulate, yellow to orange, spreading; stamens 2–9 mm, exposed; style 7–20 mm; stigma-lobes exserted. Fruit to 38 × 28 mm diam., globose to depressed-globose, slightly beaked, the beak 2–6 mm, brownish green to wine-red or brownish red when ripe; funicular pulp translucent, greenish, placenta yellowish. Seeds many per fruit, to 4.8 × 3.6 × 2.8 mm.

CONSERVATION STATUS. Least Concern [LC]; see subspecies below.

The spiny sister-species of *T. inamoena* (see below), with which it is narrowly sympatric in western Bahia, replacing it in the Rio São Francisco valley further south on limestone outcrops. It links *T. inamoena* to the preceding species in its tendency to having somewhat beaked fruits. Its range extends westwards on limestone outcrops into north-western Minas Gerais, slightly beyond the limits of Eastern Brazil as defined here. Two subspecies are recognized:

1. Areoles 7–14 mm apart; perianth-segments spathulate, commonly yellowish (W & cent.-N Minas Gerais, and W of the Rio São Francisco in SW Bahia) . . . . . . **5a.** subsp. **saxatilis**
1. Areoles very densely disposed, 1–6 mm apart; perianth-segments lanceolate, deep orange (Mun. Iuiú, Bahia) . . . . **5b.** subsp. **estevesii**

5a. subsp. **saxatilis**

VERNACULAR NAME. Palma.

Stem-segments 10–12 × 5–6 cm.

Southern Rio São Francisco *caatinga* element: on ± forest-covered limestone (*Bambuí*) outcrops surrounded by *caatinga, mata seca semidecídua* and *cerradão*, c. 450–700 m, western Bahia (west of the Rio São Francisco) to north-western, northern and central Minas Gerais (to c. 17°55'S). Map 19.
BAHIA: Mun. Santana, 28 km S of town towards Santa Maria da Vitória, 15 Jan. 1991, *Taylor et al.* 1429 (K, HRCB, ZSS, CEPEC); Mun. Santa Maria da Vitória, E of town, Sep. 1982, *E. Esteves Pereira* 188 (UFG); Mun. Bom Jesus da Lapa, Faz. Serra Solta, 17 July 1975, *Andrade-Lima* 75-8189 (IPA 45216); Mun. Cocos, W of town, reported by Braun & Esteves Pereira (1999b).
MINAS GERAIS: Mun. Manga – Mun. Januária, June 1978, *E. Esteves Pereira* 228 (UFG); Mun. Januária, 'distrito de Fabião, antes do cerrado do Judas, Buraco dos Macacos', 15°7'23"S, 44°14'24"W, 16 Nov.

1998, *J.A. Lombardi & L.G. Temponi* 2224 (K, BHCB); [Mun. Unaí, N of the town, Apr. 1979, *E. Esteves Pereira* 156 (UFG), Jan. 1991, *Taylor et al.* (obs.) — record outside of study area]; Mun. Itacarambi/Manga, 15°26'S, 43°55'W, Sep. 1999, *I. Ribeiro* (Fund. Zoo-Bot. BH, photos); Mun. Capitão Enéas, limestone outcrop visible from road BR 122, 9 June 2002, *M. Machado & G. Charles* (K, photos); Mun. Montes Claros, c. 13 km NW of town on road to Januária, 6 Nov. 1988, *Taylor & Zappi* in *Harley* 25509 (SPF, K), 1959, *Ritter* 1035 (U); l.c. (?), 1980, *E. Esteves Pereira* 37 (UFG); Mun. Bocaiúva, dirt road to Engenheiro Dolabela, 500 m from highway to Montes Claros, 17°28'S, 44°1'W, 4 Nov. 1988, *Taylor & Zappi* in *Harley* 25504 (K, SPF); Mun. Buenópolis, 7 km S of town, 11 Oct. 1988, *Taylor & Zappi* in *Harley* 24839 (K, SPF).

CONSERVATION STATUS. Least Concern [LC] (1); PD=1, EI=1, GD=2. Short-list score (1×4) = 4. Least Concern at present, but found only on limestone and therefore potentially threatened by quarrying activities in the long term at some of its sites.

At the first Bahian locality cited above the hybrid *T. saxatilis* subsp. *saxatilis* × *T. inamoena* has been observed and collected (*Taylor et al.* 1430, CEPEC, HRCB, K, ZSS) together with both parental taxa. This subspecies shows significant regional variation (cf. synonymy above).

5b. subsp. **estevesii** *(P. J. Braun) N. P. Taylor & Stuppy* in Succ. Pl. Res. 6: 118 (2002). Holotype: Brazil, Bahia, [Serra de Iuiú], 1984, *E. Esteves Pereira* 191 (UFG; ZSS, iso.).

Opuntia estevesii P. J. Braun in Cact. Succ. J. (US) 62: 165–169 (1990).

Stem-segments to 16 × 11 cm.

Southern Rio São Francisco *caatinga* element: on exposed *Bambuí* limestone outcrops in *caatinga*, c. 500–550 m, east of the Rio São Francisco, Mun. Iuiú, southern Bahia. Endemic to the above locality. Map 19.
BAHIA: Mun. Iuiú, Serra de Iuiú [south of the town], July 1984, *E. Esteves Pereira* 191 (UFG, ZSS).

CONSERVATION STATUS. Vulnerable [VU D2] (2); area of occupancy < 20 km^2; PD=1, EI=1, GD=1. Short-list score (2×3) = 6. It is presently known only from the location cited above (number of individuals unknown) and could be affected by quarrying in the future.

6. Tacinga inamoena *(K. Schum.) N. P. Taylor & Stuppy* in Succ. Pl. Res. 6: 119 (2002). Type (Taylor *et al.* 2002): Brazil, Rio de Janeiro [*fide* Schumann]; 'Minas Gerais, Serra dos Ilheos, Sítio' [according to Glaziou 1909: 327, but see below; probably north-eastern Minas Gerais], '1883–84', [*anon.* in] *Glaziou* 14864 (B†; K, lecto.).

Opuntia inamoena K. Schum. in Martius, Fl. bras. 4(2): 306 (1890). *Platyopuntia inamoena* (K. Schum.) F. Ritter, Kakt. Südamer. 1: 32 (1979).
Opuntia inamoena var. *flaviflora* Backeberg, Descr. Cact. Nov. [1:] 10 (1956, publ. 1957) and Die Cact. 1: 613, illus. (1958). Type: a cultivated plant (not known to have been preserved).

VERNACULAR NAMES. Quipá, Guibá, Guipá, Palmatória, Palmatória-miúda, Iviro, Gogóia, Palma-de-ovelha.

Subshrub, 10–50 × 50–300 cm, without main stem; stem-segments erect or sprawling, very variable in shape, size and thickness, 5–15 × 4–12 cm, orbicular to ovate etc., light green, with ducts exuding sticky, whitish mucilage when cut; areoles 10–15 mm apart, sunken and expanded inwards. Spines absent except sometimes on very young stem-segments (but see *T.* ×*quipa* below), glochids numerous but mostly concealed inside the sunken areoles. Flowers from near the apex of terminal segments, c. 50 × 35–40 mm; pericarpel globose, to 10 mm, green, with orange-red, acute, fleshy bract-scales subtending bristles; outer perianth-segments 16 mm, narrowly lanceolate, deep red or pinkish orange, spreading, inner segments 18–20 mm, spathulate, orange, spreading to slightly reflexed at apex; stamens exposed, erect; stigma-lobes exserted. Fruit to c. 40 mm diam., globose to depressed globose, yellow or orange when ripe, the depressed umbilicus small in relation to fruit diameter, funicular pulp translucent, edible. Seeds many per fruit, to 3 mm diam.

Widespread Eastern Brazil element: usually on rocks (including inselbergs) or very stony ground, open *caatinga* and *campo rupestre*, c. 100–1550 m, from the middle drainage of the Rio Jequitinhonha (MG) northwards to northernmost Piauí, and westwards on sandstone outcrops in the *cerrado* of western Bahia. Endemic to Eastern Brazil. Amongst the commonest of cacti from the region. Only records marking its approximate eastern, western and southern limits are given below. Map 19.

PARAÍBA: Mun. Tacima, border with Mun. Caiçara, W margin of Rio Curimataú, 6°36'S, 35°28'W, 1 Apr. 2000, *E.A. Rocha et al.* (K, photo).

BAHIA: W Bahia, Mun. Barreiras, 34 km W of town, 2 Mar. 1972, *Anderson et al.* 36462 (MO); E Bahia, Mun. Muritiba, nr São José do Itaporã, Pau-Ferro, Faz. Oito de Dezembro, 6 Aug. 2002, *M. Machado* (HUEFS, K, photos).

MINAS GERAIS: NW Minas Gerais, Mun. Itacarambi/Januária, vale do Rio Peruaçu, illustrated in Costa *et al.* (1998: Fig. 29); cent.-N Minas Gerais, Mun. Porteirinha, 8 km S of town towards Riacho dos Machados, 7 Nov. 1988, *Taylor & Zappi* (obs.); Mun. Grão Mogol, subida da Trilha da Tropa, 27 May 1987, *Zappi et al.* in CFCR 11973 (SPF); Mun. Cristália, road to town from Grão Mogol, 14 Apr. 1981, *Pirani et al.* in CFCR 917 (SPF, K); NE Minas Gerais, Mun. Itinga, on road BR 367, 41°48'W, 15 Feb. 1988, *W.W. Thomas et al.* 5971 (SPF, NY); Mun. Itaobim, 1 km W of town, 0.5 km N of Rio Jequitinhonha, 18 Nov. 1988, *Taylor & Zappi* (K, photos).

CONSERVATION STATUS. Least Concern [LC] (1); PD=2, EI=2, GD=2. Short-list score (1×6) = 6. This species is locally a co-dominant element of the *caatinga* flora, where it is widespread in more open areas.

The type locality, as given by Glaziou (1909: 327), is assumed to be false and probably an invention to disguise the fact that he was not the real collector (cf. Wurdack 1970). It may be no coincidence that the Glaziou number for '*Opuntia rubescens*' [*sensu* Schumann (1890), *non* DC.], *Gl.* 14865, which is *Tacinga braunii* (q.v.), immediately follows that for *T. inamoena* (*Gl.* 14864) and both species grow together in the valley of the Rio Jequitinhonha, north-eastern Minas Gerais, where *T. braunii* is endemic. It seems probable, therefore, that the type of *T. inamoena* came from north-eastern Minas Gerais, where it reaches its south-eastern limit.

Braun & Esteves Pereira (1989: 272) remark that *Opuntia inamoena* occurs in the state of Espírito Santo, but have so far apparently failed to substantiate this claim with a definite locality.

Marlon Machado has pointed out a very distinct dwarf race of the species, found on gneissic outcrops towards the limits of the *caatinga* in eastern Bahia and north-eastern Minas Gerais. This plant forms a small erect shrub with miniature, almost shortly cylindrical stem-segments of regular size and with reduced numbers of areoles; likewise the pericarpel. It seems to be genetically determined, since it does not change habit when grown under divergent conditions and appears to come true from seed. Its easternmost locality in Mun. Muritiba is cited above, but identical populations occur in Mun. Itatim, Ipirá and Itaberaba (*Horst & Uebelmann* HU 896; M. Machado, pers. comm.). It was first collected in the early 1970s at Araçuaí, Minas Gerais, near the south-eastern limit of the species (HU 336). It probably deserves to be named as a subspecies, especially if this will ensure its conservation.

The true *T. inamoena* has spineless stem-segments as in *T. funalis* and *T. braunii*, but, as in those species, its abundant fine glochids demand that it be treated with appropriate respect and handled only with forceps. Forms with mature stems bearing occasional spines are usually to be referred to the following hybrid (but see below *T. werneri* also):–

Tacinga ×**quipa** *(F. A. C. Weber) N. P. Taylor & Stuppy* in Succ. Pl. Res. 6: 120 (2002). Type: Brazil, Pernambuco, without date or collector (P†).

Opuntia quipa F.A.C. Weber in Bois, Dict. hort.: 894 (1898). NB. Although previously treated as a synonym of *T. (Opuntia) inamoena*, Weber's diagnosis mentions the presence of occasional spines indicating hybridity.

[*T. inamoena* × *T. palmadora*]

VERNACULAR NAME. Quipá.

Like *T. inamoena*, but stem-segments with occasional scattered spines, mainly at apex. Flowers not seen. Fruit variable in shape and colour.

Caatinga, c. 200–1080 m, of sporadic occurrence throughout the range of *T. palmadora* where *T. inamoena* is also present. Endemic to North-eastern Brazil.
PERNAMBUCO: without locality (*locus classicus* of accepted name).
ALAGOAS: W Alagoas, Mun. Delmiro Gouveia, near bridge over Rio São Francisco at border with Bahia, 12 Feb. 1991, *Taylor & Zappi* 1618D (K, photo).
BAHIA: Mun. (?), Raso da Catarina, Sede da Reserva, 25 Oct. 1982, *L.P. de Queiroz* 428 (ALCB); Mun. Uauá, 1.7 km ESE of town towards Bendengó and Canudos, 6 Jan. 1991, *Taylor et al.* 1363 (K, HRCB, ZSS, CEPEC); Mun. Jacobina, 27–28 km W of town towards Lages, 12 Jan. 1991, *Taylor et al.* 1398 (K, HRCB, ZSS, CEPEC); Mun. Morro do Chapéu, W of Icó, Brejão (Formosa), 'Lagedo Bordado', 5 Aug. 2002, *Taylor* (K, photos).

The commonest and most widespread hybrid amongst the cacti of Eastern Brazil, but the two parental species certainly do not hybridize as often as they co-occur, perhaps because hummingbirds seem to prefer the flowers of *T. palmadora* to those of *T. inamoena*. Care is needed in determining plants as this hybrid, since another involving *T. inamoena* and *T. werneri* is very similar and it is likely that all 3 species can grow together (eg. nr Morro do Chapéu, BA).

Insufficiently known taxa
The following are of uncertain identity for lack of extant type specimens. Werdermann's species has previously been referred to either the above hybrid or as a form of *Tacinga (Opuntia) palmadora* (see below). It could also refer to *T. werneri*, for which it would be an older name, although the fruit shape as described probably rules this out. The hybrid between this species and *T. inamoena* is another plant known from the region of Werdermann's type to which his name could refer. Braun & Esteves Pereira's assignment of it may be correct, since there exists a rare aberrant form of *T. palmadora* in *caatinga* not far from the type locality with few spines and yellowish white fruit. The details are:

Opuntia catingicola Werderm. in Notizbl. Bot. Gart. Berlin–Dahlem 12: 223 (1934). *O. palmadora* subsp. *catingicola* (Werderm.) P. J. Braun & E. Esteves Pereira in Succulenta 74: 133 (1995). Type: Brazil, Bahia, between Mundo Novo and Ventura, 600 m, Apr. 1932, *Werdermann 3998* (B†). *O. catingicola* var. *fulviceps* Haage in Backeberg, Kakteenlex., ed. 3: 495 (1976) and Cactus Lex.: 356 (1977), nom. inval. (Art. 36.1 & 37.1). Based on a cultivated plant (not known to have been preserved).

4. BRASILIOPUNTIA (K. Schum.) A. Berger

Entwicklungslin. Kakt.: 94 (1926). Type: *Cactus brasiliensis* Willd.

A very distinct, highly specialized, monotypic, arborescent genus, which is allied to *Opuntia sens. str.* on the basis of seed-envelope morphology/anatomy (Stuppy 2002), but with a cylindrical, apical leader shoot of indeterminate (unjointed) growth, markedly different pollen (cf. Leuenberger 1976) and non-sensitive stamens. The adult tree is similar in habit to an old *Araucaria* and displays unique shoot-morphology, the erect, cylindric leader, giving rise to progressively more flattened, lateral stem-segments, the ultimate of which are very thin, hardly succulent, almost leaf-like and drought-deciduous. However, the first shoot (plumule) of the seedling, which arises between massive cotyledons, is thin, clearly flattened and early-determinate (cf. *Opuntia sens. str.*), soon giving rise to one or more equally flattened, subapical secondary segments and sometimes supplanted by stronger shoots of indeterminate growth arising from the axils of the cotyledons. The cylindrical, indeterminate leader shoot(s) develop from either of these sources and may be cylindric from the beginning or remain ± flattened for some time. Such leader stems show very early periderm formation, and the production of glochids from their areoles is very limited or absent. Plates 8.2–10.1.

This may be the tallest member of the Cactaceae, with specimens attaining heights of between 20 and 25 metres in the *mata de cipó*, a part of the *agreste*, the Atlantic Forest and *caatinga* ecotone.

1. Brasiliopuntia brasiliensis *(Willd.) A. Berger*, l.c. (1926). Type: probably a living plant in the Berlin botanical garden; no material extant at B-W. Lectotype (Taylor *et al.* 2002): W. Piso, Historia naturalis Brasiliae: illustration, p. 100, below (1648); cf. Willdenow, l.c. infra.

Cactus brasiliensis Willd., Enum. pl. suppl.: 33 (1814). *Opuntia brasiliensis* (Willd.) Haw., Suppl. pl. succ.: 79 (1819).
Cactus arboreus Vell., Fl. flumin.: 207 (1829). *Opuntia arborea* (Vell.) Steud., Nomencl. Bot. Ed. 2, 1: 220 (1841). Holotype: not extant. Lectotype (Taylor *et al.* 2002): Vellozo, tom. cit., Icones 5: tab. 28 (1831).
Cactus heterocladus A. St. Hil., Voy. Rio de Janeiro 2: 103 (1830). Type: Rio de Janeiro, not known to have been preserved.
Opuntia bahiensis Britton & Rose, Cact. 1: 210–211 (1919). *Brasiliopuntia bahiensis* (Britton & Rose) A. Berger, l.c. (1926). *Opuntia brasiliensis* subsp. *bahiensis* (Britton & Rose) P. J. Braun & E. Esteves

Pereira in Succulenta 74: 132 (1995). *Brasiliopuntia brasiliensis* subsp. *bahiensis* (Britton & Rose) P. J. Braun & E. Esteves Pereira in Schumannia 3: 188 (2002). Lectotype (Taylor *et al*. 2002): Brazil, Bahia, near Toca da Onça [Jaguaquara], 27–29 June 1915, *Rose & Russell* 20068 (US; NY, lectopara.). *Brasiliopuntia subacarpa* Rizzini & A. Mattos in Revista Brasil. Biol. 46(2): 323–328 (1986). *Opuntia brasiliensis* subsp. *subacarpa* (Rizzini & A. Mattos) P. J. Braun & E. Esteves Pereira in Succulenta 74: 132 (1995). *Brasiliopuntia brasiliensis* subsp. *subacarpa* (Rizzini & A. Mattos) P. J. Braun & E. Esteves Pereira in Schumannia 3: 188 (2002). Holotype: Brazil, Minas Gerais, Itinga, 17 Oct. 1984, *Rizzini & Mattos-Filho* 38 (RB; UFMT, iso.).

VERNACULAR NAMES. Urumbeba, Rumbeba, Cumbeba, Mumbeca, Mumbebo, Facho-de-renda, Palmatória-grande, Palmatória-do-diabo, Ambeba, Arumbeva, Gerumbeba, Jurubeba, Palmadora, Palmatória, Xiquexique-do-sertão.

Tree 4–20(–25) m, trunk to 15–20 × 35 cm diam., cylindric, with clusters of spines to 9 cm long; pith chambered. Branches dimorphic, patent; intermediate joints cylindric, 20–100 cm, ultimate joints (4–)6–15 × 3–6(–7) cm, rhomboid to obovate, irregular in outline, narrow at base, thin, bright to dark green, deciduous. Areoles on the lateral stem-segments 15–30 mm apart, with white felt, spineless or with 1 spine to 40 mm, those on the leader shoot with 1–3 or more spines, or those that have given rise to lateral stem-segments spineless and often lacking glochids also; leaves early deciduous, ovoid or more elongate on the leader shoot, fleshy, bright green. Flowers from near the apex of the leader shoot or terminal stem-segments, or by proliferation from the pericarpels of old flowers, c. 2.5–3.5 × 4.5 cm; pericarpel globose, obovoid, elongate or elongate-flattened, c. 8–28 × 9–12 mm, green, tuberculate, bearing scale leaves c. 1 mm long, subtending areoles with white wool; outer perianth-segments to 15 mm, ovate, greenish to yellowish, erect to spreading, innermost to 20 mm long and 10 mm broad, lanceolate to spathulate, yellow, spreading; stamens non-sensitive, to 7 mm, anthers c. 0.6 mm, whitish; style c. 9 × 1.5 mm, whitish, stigma-lobes 3–6, exserted, to 4.5 mm. Fruit 2–4 cm diam., solitary or clustered-proliferous, globose to obovoid or pear-shaped, purplish, red, orange-red or pale yellow, with areoles bearing conspicuous clusters of dark brown glochids; funicular pulp fibrous, white, yellowish or deep red (*fide* Britton & Rose), placenta greenish. Seeds 1–5, mostly 2, per fruit, to 8–10 mm. Chromosome number: 2n = 22 (Pinkava 2002: 61, Table 1).

Widespread southern neotropical element: *restinga,* drier phases of *Mata atlântica, agreste (mata de cipó), caatinga* (especially along temporary water courses), *mata de brejo, mata seca* (on limestone), *mata de galeria* and *mata do planalto,* especially on deep sandy substrates and as a lithophyte, near sea level to c. 1000 m, western Paraíba, eastern and central-southern Pernambuco, Alagoas, Sergipe, north-western, northern and eastern Bahia, north-eastern and central-southern Minas Gerais and Espírito Santo; semi-humid forests of extra-Amazonian Brazil; Atlantic drainage of Andes eastwards (Peru, Bolivia, northern Argentina, Paraguay). Map 18.

PARAÍBA: Mun. Cazajeiras, plant seen growing outside pharmacy in the town on road near eastern access to highway BR 230, *ex* Fazenda Capoeiras, east of town, where said to be frequent and increasing (property of owner of pharmacy), 3 Apr. 2000, *E.A. Rocha et al*. (K, photo).

PERNAMBUCO: Mun. Gravatá, Russinha, 27 Jan. 1933, *Pickel* 4480 (IPA), l.c., *Pickel* 3197 (IPA, *fide* Andrade-Lima 1966: 1454); Mun. Caruaru, 18 May 1992, *F.A.R. Santos* 28 (HUEFS, PEUFR), l.c., 9 km NE of town, 8°14'21"S, 35°54'52"W, Dec. 1999, *E.A. Rocha et al*. (K, photos); Mun. Inajá, Reserva Biol. Serra Negra, 17 Sep. 1995, *A. Laurênio et al*. 198 (K, PEUFR).

ALAGOAS: W Alagoas, Mun. Mata Grande, 20 Dec. 1974, *Andrade-Lima* 74-7773 (IPA); S Alagoas, Mun. Piaçabuçu, Ponta da Terra, 17 Nov. 1987, *G.L. Esteves et al*. 1954 (SPF).

SERGIPE: Mun. Nossa Senhora da Glória, Faz. Olhos D'agua, 30 Nov. 1981, *G. Viana* 261 (ASE 1778); Mun. Frei Paulo, 5 km após o povoado de Mocambo, 26 Mar. 1981, *M. Fonseca et al*. 452 (ASE 851); Mun. Simão Dias, 3 Feb. 1975, *A.C. Barreto* s.n. (ASE 288); l.c., Faz. Mercador, 30 Oct. 1981, *G.N. Silva* 37 (ASE 1651); Mun. Riachão do Dantas, Faz. São José, 17 June 1982, *E. Gomes* 77 (ASE 2566); l.c., Mata de Riachão, 27 Aug. 1982, *G. Viana* 632 (ASE 2742); l.c., Faz. Salobre, 2 Feb. 1983, *E. Gomes* 179 (ASE 3008), 19 Feb. 1987, *G. Viana* 1679 (ASE 4628).

BAHIA: NW Bahia, Mun. Formosa do Rio Preto, 4 km from Rio Riachão towards the town, white

sandy substrate, 13 Oct. 1989, *Walter et al.* 478 (K); N Bahia, Mun. Senhor do Bomfim, c. 12 km N of town, W of Estiva, Serra da Jacobina, below TV mast, 1 Mar. 1974, *Erskine* 132, cult. RBG Kew 1974–1138 (K); Mun. Morro do Chapéu, a few km S of the summit, Apr. 2002, *M. Machado* (obs.); E Bahia, Mun. Serrinha, 1 Dec. 1968, *A.L. Costa* s.n. (ALCB, in spirit); Mun. Lamarão, 48 km N of Feira de Santana on road BR 116, 2 Nov. 1972, *J. Ratter* (K, photos); Mun. (?), 'Pimenta', between Feira de Santana and Serrinha, 20 July 1959, *Gomes & Laboriau* 849 (RB); Mun. Ipirá, 12°22'S, 39°41'W, 1 Oct. 1985, *L.R. Noblick* 4400 (HUEFS); Mun. Serra Preta, Faz. Manoino, 12°9'S, 39°19'W, 7 Dec. 1992, *L.P. de Queiroz et al.* 2923 (K, MBM, HUEFS); Mun. Feira de Santana, 16 Feb. 1994, *F. França* 949 (HUEFS); l.c., 12°10'S, 39°11'W, 13 Nov. 1986, *L.P. de Queiroz* 1031 (HUEFS); l.c., Serra de São José, 12°38'S, 39°28'W, 18 June 1985, *H.P. Bautista & G.C.P. Pinto* 1024 (HRB, CEPEC, HUEFS); Mun. Rui Barbosa, road BR 407, Faz. Buriti, 13 Oct. 1978, *A.P. Araújo* 74 (HRB, RB); Mun. Cachoeira, forest in front of EMBASA, July 1980, *Iscardino et al.* in Grupo Pedra do Cavalo 462 (CEPEC, ALCB, HRB, BAH, HUEFS); Mun. Castro Alves, Km 22 on road BR 242, 11 May 1975, *A.L. Costa* s.n. (ALCB, in spirit); Mun. Itiruçu, 22 Jan. 1965, *R.P. Belém & J.M. Mendes* 220 (RB); Mun. Jaguaquara [Toca da Onça], 27–29 June 1915, *Rose & Russell* 20068 (US, NY); Mun. Uruçuca, Serra Grande village, in garden, probably from a wild plant in nearby *restinga* forest, 7 Feb. 1995, *Taylor* (obs.); Mun. Vitória da Conquista, 12 & 20 km S of town, 17 & 18 April 2003, *Taylor & Zappi* (K, photos); Mun. Itapetinga, 15°13'S, 40°10'W, 24 Nov. 1979, *A.P. Araújo* 174 (HRB, RB); Mun. Itaju do Colônia, Km 2 on road to Pau Brasil, 24 Jan. 1969, *T.S. dos Santos* 350 (K, CEPEC); l.c., Lagoa Bonita, 11 Feb. 1975, *A.L. Costa* s.n. (ALCB, in spirit); Mun. Potiraguá, 15°23'S, 39°58'W, 2001, *J. Jardim* 3144 (CEPEC); Mun. Itapebi, Faz. Ventania, Ventania–Itapebi road, 8 Nov. 1967, *R.S. Pinheiro* 378 & *T.S dos Santos* 41 (CEPEC); l.c., between Itapebi and Belmonte, 13 Apr. 1967, *Castellanos* 26354 (CEPEC); Mun. Itamaraju, c. 5 km NW of town, Faz. Pau-brasil, 3 July 1979, *A. Mattos Silva et al.* 554 (K, RB, CEPEC).
MINAS GERAIS: NE Minas Gerais, Mun. Itinga, 17 Oct. 1984, *Rizzini & Mattos-F.* 38 (RB, UFMT); cent. & S Minas Gerais, Mun. Santana do Riacho, margin of the Rio Cipó, growing as hedge, 19 Nov. 1989, *Zappi* 192 (HRCB, SPF); Mun. Matozinhos, Faz. Cauaia, 31 Oct. 1996, *Lombardi* 1446 (BHCB, K); Mun. Lagoa Santa, A.P.A. Carste de Lagoa Santa, Oct. 1995 to Feb. 1996, *Brina & Costa* in BHCB 32769 (BHCB, K); Mun. Entre Rios de Minas, Fazenda de Pedra, Nov. 1969, *L. Krieger* 7764 (CESJ); Mun. Arcos, *fide* Uebelmann (1996): HU 956.
ESPÍRITO SANTO: Mun. Colatina, 15 km S of São Domingos towards Colatina, 16 Dec. 1990, *Taylor & Zappi* 784 (K, HRCB, ZSS, MBML); Mun. Linhares, Reserva Florestal da CVRD, Estação Flamengo, 20 Nov. 1987, *D.A. Folli* 668 (Univ. Fed. ES); Mun. Santa Teresa, S. J. de Petrópolis, Escola Agrotécnica Federal, 8 Oct. 1985, *W. Boone* 825 (MBML, HRCB); l.c., Barra de Santa Júlia, 14 May 1986, *H.Q.B. Fernandes & W. Boone* 1975 (MBML, HRCB); Mun. Vitória, campus of Univ. Federal, May 1990, *Taylor & Zappi* (obs.); Mun. Guarapari, Km 32 on road ES 060, 12 Sep. 1987, *O.J. Pereira* 1048 (Univ. Fed. ES); Mun. Piúma, Monte Agá, south of the town, 20°52'23"S, 40°46'24"W, 26 Nov. 1999, *Zappi et al.* 461 (UEC, K).
RIO DE JANEIRO: Mun. São Francisco do Itabapoana, Faz. São Pedro, 21°24'S, 41°4'W, *H.C. de Lima et al.* (poster presentation at 1998 Congresso Nac. de Botânica, Salvador, BA); Mun. Santa Maria Madalena, access road to the town, 22°0'55"S, 42°7'30"W, 21 Nov. 1999, *Zappi et al.* 354 (K, UEC).

CONSERVATION STATUS. Least Concern [LC] (1); PD=4, EI=1, GD=2. Short-list score (1×7) = 7. Much of its habitat, especially the Atlantic Forest and *agreste*, has been destroyed in Brazil, but it does occur in some national parks and has a wide distribution in the southern Neotropics, although its status outside of Brazil is little known.

This widespread species, which is broadly circumscribed here, is rather variable in fruit shape and colour and in the number, shape and colour intensity of its perianth-segments. In the population sampled as *Erskine* 132 (see above) pericarpel shape varied from globose to very elongate or flattened and across the full range of the taxon fruit colour does not appear to present a consistent geographical pattern (there are disjunct occurrences of both yellow and reddish to purple colorations). At least some forms from the Brazilian Nordeste have reddish, ovoid fruit, and one such was distinguished as *Opuntia bahiensis* by Britton & Rose (1919), whom, it seems, assumed that the type of *Cactus brasiliensis* came from Rio de Janeiro, where the species has globose to depressed, yellow fruit. Willdenow, however, did not state the origin of the plant grown at Berlin, but made reference to Piso, ie. *Historia naturalis Brasiliae* (Piso 1648). One of Piso's illustrations has now been designated as lectotype and, since Piso was based at the Dutch colony at Recife, Pernambuco (Stafleu & Cowan 1983: 276) and a coloured copy of the same illustration in the contemporary work of Marcgraf (see Whitehead & Boeseman 1989: tab. 3a) shows red fruits, it is clear that the name *C. brasiliensis* should be applied in its strictest sense to a red-fruited form from the Nordeste (Plate 9). Therefore, Britton & Rose's *Opuntia bahiensis* would be a synonym, even if the species was interpreted in a narrow sense (likewise *Brasiliopuntia subacarpa* Rizzini & A. Mattos).

★5. NOPALEA Salm-Dyck

Cact. hort. dyck. 1849: 63–64 (1850). Lectotype: *Cactus cochenillifera* L.

Seed-morphology/anatomy indicates that this small genus is very closely related to *Opuntia sensu stricto* (Stuppy 2002), but it differs markedly in flower- and pollen-morphology (Leuenberger 1976). Only the following introduced species is found in Eastern Brazil:

★1. Nopalea cochenillifera *(L.) Salm-Dyck*, Cact. hort. dyck. 1849: 64 (1850). Lectotype (Howard & Touw 1982: 173): Dillenius, Hort. eltham.: tab. 297, fig. 383 (1732).

Cactus cochenillifera L., Sp. pl.: 468 (1753). *Opuntia cochenillifera* (L.) Mill., Gard. Dict. ed. 8: no. 6 (1768).

VERNACULAR NAMES. Palmatória, Palma, Palma-miuda, Palma-doce, Palma-de-engorda.

Shrubby, to 3 m tall, main stem with segments flattened at first, cylindric in age. Branches erect, sometimes pendent, stem-segments (10–)12–26 × (3–)8–10 cm, narrowly elliptic or obovate, bright green, sometimes glaucous. Areoles 25–30 mm apart, with white felt; spineless. Flowers from the apex of the terminal segments, 5–6 × 2–2.5 cm; pericarpel obovoid, to 30 mm, green, with areoles with white wool and few, dull pink bract-scales; outer perianth-segments to 15 mm, lanceolate, deep pink, erect, innermost 20 mm, lanceolate to spathulate, deep pink, erect, clasping the stamen filaments, these exserted, dark pink; stigma-lobes long-exserted, green. Fruit rarely formed in Brazil, to 4 cm, solitary, obovoid, red; funicular pulp translucent, purplish, placenta greenish. Seeds few per fruit, to 4 mm diam. Chromosome number: 2n = 22 (Pinkava 2002: 88).

Introduced: on cultivated land throughout North-eastern Brazil (native to Mexico and Central America). The following are only representative records of the species and do not adequately account for its extensive use in NE Brazil:
PERNAMBUCO: Mun. São Lourenço da Mata, São Bento, 4 Feb. 1933, *Pickel* 9216 (IPA); Mun. Poção, 21 Oct. 1991, *F.A.R. Santos* 1 (HUEFS); Mun. Pesqueira, 5 Nov. 1934, *Pickel* 3660 (IPA).
ALAGOAS: Mun. Delmiro Gouveia, Oct. 1993, *R. de Lyra-Lemos* s.n. (MAC).
BAHIA: W Bahia, Mun. Vanderlei, between the town and road BR 242, 19 July 1989, *Taylor & Zappi* (K, photos); N Bahia, Mun. Glória, povoado de Brejo do Burgo, 26 Aug. 1995, *Bandeira* 257 (K, HUEFS); E Bahia, Mun. Conceição da Feira, Porto Castro Alves, 12°32'S, 39°5'W, Dec. 1980, *Iscardino et al.* in Grupo

Pedra do Cavalo 1037 (ALCB, HUEFS); Mun. Iramaia, 27.5 km NW of Pé-de-Serra, 5 Feb. 1991, *Taylor et al.* (ZSS, photos); Mun. (?), between Gandu and Itaibó, 9 Oct. 1972, *R.S. Pinheiro* 2004 (CEPEC).

Like *Opuntia ficus-indica* (see below), this species is cultivated as cattle fodder for use during drought and is also suitable as a host for the cochineal insect. Plate 10.2.

6. OPUNTIA Mill.

Gard. Dict. Abr. ed. 4: [unpaged] (1754). Type (cf. Leuenberger 1993): *Cactus opuntia* L. (= *O. ficus-indica* (L.) Mill.).

Shrubs or occasionally treelike, 0.5–6.0 m; branches not dimorphic; stem-segments compressed, orbicular, obovate to elliptic or rhomboid in outline, sometimes becoming cylindric through prolonged secondary growth at base of plant; areoles in the axils of deciduous leaves, with glochids and felt, spines usually present, or absent in some forms of No. 3. Leaves minute, subulate, sessile, fleshy, early deciduous. Flowers solitary, mainly on the margin of the joint or born by proliferation from the flower receptacle; epigynous, pericarpel globose or turbinate, with areoles and leaf-like, fleshy, green or coloured bract-scales; perianth multiseriate, tube 0, outer perianth-segments coloured, fleshy, inner segments coloured, delicate, spreading; stamens numerous, spreading, sensitive and closing around the style when touched; pollen with reticulate exine (see Cact. Syst. Initiatives 14: 3 (2002)). Fruit solitary or clustered-proliferous; turbinate with a pedicillate base or globose, with a broad, not very deep umbilicus and deciduous flower remnants; funicular pulp translucent or opaque, fibrous; placenta white, greenish or coloured. Seeds few to numerous, lenticular reniform, to 5 mm, enclosed in a hard, bony funicular envelope.

A genus of at least 150 species, even when (as here) narrowly circumscribed, ranging from Canada to southern South America, but with only 1, marginally represented species native to sandy places in the *Mata atlântica* zone within the core area of Eastern Brazil (plus 2 spp. introduced from the Northern Hemisphere). Plates 10.3–11.2.

1. Areoles with clusters of numerous golden spines; segments ± orbicular (cultivated and naturalised) . . . . ***2. dillenii**
1. Areoles spineless or with few or dark brownish spines; segments obovate, elliptic or rhomboid . . 2

2. Segments dark green; fruits proliferating (cent.-E Minas Gerais and SE Espírito Santo; occasionally cultivated elsewhere) . . . . **1. monacantha**
2. Segments generally somewhat glaucous; fruits never proliferating (cultivated everywhere) . . . . ***3. ficus-indica**

1. Opuntia monacantha *Haw.*, Suppl. pl. succ.: 81 (1819). Type: 'Lesser Antilles, Barbados', *fide* Haworth, not extant. Neotype (Leuenberger 2002): Lindley, Bot. Reg. 20: tab. 1726 (1835). NB. Leuenberger's neotypification was published on 26 March 2002, c. 2 days before that proposed by Taylor *et al.* (2002).

Cactus urumbeba Vell., Fl. flumin.: 207 (1829). *Opuntia urumbeba* (Vell.) Steud., Nomencl. Bot. Ed. 2: 222 (1841) ('*O. umbrella*'). Type: not extant. Lectotype (Taylor *et al.* 2002): Vellozo, Fl. flumin., Icones 5: tab. 32 (1831).

O. arechavaletae Speg. in Anales Mus. Nac. Hist. Nat. Buenos Aires, III, 4: 520 (1905) ('*O. arechavaletai*'); Scheinvar, Fl. Ilustr. Catarin., pt I (Cactáceas), 61, 63–68, tt. 27 & 28 (1985) ('*O. arechevaletai*'). Type: Uruguay, near Montevideo, Pan de Azúcar, Nov. 1903, ex Herb. J. Arechavaleta (LPS 14705, holo., n.v.).

[*O. vulgaris sensu* Britton & Rose (1919) et auctt. p.p. (cf. Leuenberger 1993), non Mill. (1768).]

VERNACULAR NAMES. Urumbeba, Palmatória, Monducuru.

Shrubby, to 3 m tall, main stem flattened. Branches erect to decumbent; stem-segments 8–20 × 4–10 cm, rhomboid to obovate, narrow at base and irregular in outline, dark green. Areoles 20–40 mm apart, with white to pale brown felt; spines 0–1(–2), (5–)15–50(–60) mm, pungent, pale grey. Flowers arising at the apex of the terminal segments or born by proliferation from the flower receptacle, 7–10 × 3–6 cm; pericarpel narrowly turbinate, to 30–70 mm, bright green, with areoles with white wool and few, dull red bract-scales; outer perianth-segments to 20 mm, lanceolate, bright yellow with deep red shades outside, spreading; innermost 25 mm, spathulate, bright yellow, spreading; stamens exposed, sensitive; stigma-lobes exserted; pollen exine reticulate (illustrated in Cact. Syst. Initiatives 14: 3 (2002)). Fruit to 5–10 cm, solitary or clustered-proliferous, narrowly turbinate to obovoid, curved at base, greenish to reddish outside, eventually falling to the ground and turning yellow; funicular pulp translucent, whitish, placenta yellowish. Seeds many per fruit, to 4 mm diam. Chromosome number: 2n = 22, 32, 33 (Pinkava 2002: 103).

Southern humid/subhumid forest element: sand-dunes in open *carrasco*, c. 1000 m, central-eastern Minas Gerais, and open *restinga* near sea level, southern Espírito Santo (and presumably northern Rio de Janeiro); South-eastern and Southern Brazil; Paraguay, Uruguay and northern and eastern Argentina; frequently naturalized or planted elsewhere (including North-eastern Brazil). Map 18.
SERGIPE: Mun. Arauá, Faz. Thuy, [cultivated], 17 Sep. 1982, *E. Gomes* 125 (ASE 2781).
BAHIA: Mun. Salvador ('Bahia'), [cultivated], 24 Apr. 1918, *Curran* 23 (US, photo).
MINAS GERAIS: Mun. Diamantina, Mercês, Apr. 1959, *Ritter* 1037 (see Eggli *et al.* 1995: 454); Mun. Rio Vermelho, Pedra Menina, Faz. do Sr José Batista, 7 Mar. 1988, *Zappi & Prado* in CFCR 11823 (SPF, K).
ESPÍRITO SANTO: Mun. Itapemirim, Rodovia do Sol, S of Marataízes, 21°4'51"S, 40°50'24"W, 26 Nov. 1999, *Zappi et al.* 470 (UEC, K), 18 Jan. 1970, *Krieger* 7608 (CESJ, SPF).

CONSERVATION STATUS. Least Concern [LC] (1); PD=2, EI=1, GD=1. Short-list score (1×4) = 4.

This species has previously been known as either *Opuntia monacantha* Haw. or *O. vulgaris* Mill., but both of these names are beset with nomenclatural difficulties. The former, which is maintained here, was based on a collection from Barbados (Lesser Antilles), whence only *O. dillenii* (Ker Gawl.) Haw. is currently recorded as native (Howard 1989). In order to maintain its use for the plant now widely associated with Haworth's name, neotypification has been necessary. This assumes that the provenance data given by Haworth were erroneous, or that *O. monacantha* as now understood had been introduced to Barbados by the early years of the 19th Century. In any case, Haworth's brief and unsatisfactory diagnosis does not agree with *O. dillenii* and so neotypification or rejection are the only realistic options open. The above action seems marginally preferable to taking up the next available, well-typified name, *O. urumbeba* (Vell.) Steud., which unfortunately has never been used, even though its epithet repeats a distinctive, vernacular name for the plant. The name *Cactus monacanthos* Willd. 1814 was cited by Haworth with a question mark as a possible synonym of his *Opuntia monacantha*. This indication of doubt rules out any consideration of Haworth's name as a combination based on that of Willdenow, which cannot be typified.

Opuntia vulgaris Mill. has been used in two quite different senses (*O. humifusa* Raf. and *O. monacantha*), but is now considered to be a renaming of *Cactus opuntia* L., a taxonomic synonym of *O. ficus-indica* (L.) Mill. (Leuenberger 1993).

O. monacantha has been recorded only rarely as a native plant within the core area covered here, where it is at its north-eastern limit. The collections from central-eastern

Minas Gerais are rather disjunct on present knowledge, but at least that from Rio Vermelho was from a population which appeared to be native and in an edaphically appropriate sand-dune habitat. Similar disjunct populations are known from localities remote from the coast in the state of São Paulo and it is probably this species that is depicted growing near Lorena, São Paulo state, in Martius, *Flora brasiliensis* 1 (1, Tabulae Physiognomicae): tab. VII (1841).

*2. Opuntia dillenii *(Ker Gawl.) Haw.*, Suppl. pl. succ.: 79 (1819). Lectotype (Benson 1969: 126): l.c. infra, tab. 255 (1818).

Cactus dillenii Ker Gawl. in Edwards Bot. Reg. 3: tab. 255 (1818). *Opuntia stricta* (Haw.) Haw. var. *dillenii* (Ker Gawl.) L. D. Benson in Cact. Succ. J. (US) 41: 126 (1969).
O. dillenii var. *reitzii* Scheinvar in Feddes Repert. 95: 277 (1984) & Fl. Ilustr. Catarin., pt I (Cactáceas), 81–88, tt. 26 (as 'O. arechevaletai') & 38–39 (1985). Holotype: Brazil, Bahia, Mun. Salvador, Itapoã, 6 Jan. 1982, *I. & L. Scheinvar* 2892 (MEXU, n.v.). **Synon. nov.** [An escape from cultivation.]

VERNACULAR NAMES. Palmatória, Palma-de-espinho.

Shrubby, to 3 m tall, main stem flattened. Branches erect; stem-segments 12–23 × 8–20 cm, orbicular or obovate, grey-green, glaucous. Areoles 30–60 mm apart, with pale brown felt and a crown of glochids; spines (3–)4–12, the largest flattened, to 30 mm, pungent, golden yellow or reddish. Flowers from the apex of the terminal segments, 7–10 × 6–9 cm; pericarpel turbinate, 40–70 mm, grey-green, with areoles with white wool and few, dull red bract-scales; outer perianth-segments to 30 mm, lanceolate, bright yellow, spreading; innermost 35 mm, spathulate, bright yellow, spreading; stamens exposed; stigma-lobes exserted, green. Fruit to 7–9 cm, solitary, turbinate to obovoid, purplish red; funicular pulp translucent, purplish, placenta greenish. Seeds many per fruit, to 4 mm diam.

Introduced and sometimes escaping by the sea, becoming invasive; planted inland for hedging; native of Caribbean coasts and southwards to Ecuador; widely introduced elsewhere in warmer regions.
PARAÍBA: Mun. João Pessoa, commonly planted in and around the town and escaping in the vicinity of Cabo Branco, 9 Feb. 1995, *Taylor* (obs.).
PERNAMBUCO: Mun. Jaboatão dos Guararapes, Oct. 1964, *D. Andrade-Lima* 64-4267 (IPA).
ALAGOAS: Mun. Maceió, commonly planted near the sea shore, 13 Feb. 1995, *Taylor* (obs.).
SERGIPE: Mun. Aracaju, Feb. 1991, *Taylor & Zappi* (obs.).
BAHIA: Mun. Araci, 11.5 km N of town towards Tucano, planted as hedge, 5 Jan. 1991, *Taylor et al.* 1350 (K, HRCB, ZSS, CEPEC); Mun. Salvador, 1 Feb. 1989, *G.C.P. Pinto* 03/89 (HRB), 22 Apr. 1992, *F.A.R. Santos* 31 (HUEFS).

Benson (1982: 497–501) treats *O. dillenii* as a variety of the scarcely spiny, more narrowly segmented *O. stricta* (Haw.) Haw., and on such authority they were synonymized in the first edition of the CITES Cactaceae Checklist (Hunt 1992b), but subsequently retained as separate species (Hunt 1999a). Further studies are needed in the Caribbean region where these taxa are native (cf. Howard 1989). An opuntia agreeing closely with *O. stricta* is common about the region of Cabo Frio, RJ, just outside the area covered here, where it appears to have been introduced and has successfully naturalised.

*3. Opuntia ficus-indica *(L.) Mill.*, Gard. Dict. ed. 8: no. 2 (1768). Neotype (Leuenberger 1991: 625): 'Cactus articulato-prolifer, articulis ovatis-oblongis: spinis setaceis. Lin. Spec. plant. 468. 16' (S).

Cactus ficus-indica L., Sp. pl.: 468 (1753).

C. *opuntia* L., l.c. (1753). *Opuntia vulgaris* Mill., Gard. Dict. ed. 8 (1768). *O. opuntia* (L.) H. Karst., Deut. Fl.: 888 (1882), nom. inval. Lectotype (Leuenberger 1993): Burser Herbarium 24: 26 (UPS).

? O. fusicaulis Griffiths *fide* Scheinvar, Fl. Ilustr. Catarin., pt I (Cactáceas): 68–73, tt. 31 & 32 (1985).

VERNACULAR NAMES. Palmatória, Palma, Palma-de-gado, Palma-gigante, Figo-da-Índia, Figo-da-Espanha, Orelha-de-onça, Jamaracá, Jurumbeba.

Shrubby, to 6 m tall, main stem with segments flattened at first, cylindric and forming greyish bark in age. Branches erect; stem-segments 20–40 × 12–20 cm or more, elliptic or obovate, grey-green, sometimes glaucous. Areoles 20–50 mm apart, with white or brown felt; spineless or with occasional spines to 10 mm. Flowers from the apex of the terminal segments, 6–9.5 × 5–6 cm; pericarpel obovoid, to 50 mm, green, areoles with dull red bract-scales; outer perianth-segments to 20 mm, rounded, yellow to orange, sometimes reddish, spreading; innermost 30 mm, lanceolate to spathulate, yellow to orange, spreading; stamens exserted, filaments dark pink; stigma-lobes exserted. Fruit 5–10 × 4–8 cm, solitary, obovoid, yellow or deep orange (sometimes reddish); funicular pulp usually orange or yellowish, placenta greenish. Seeds many per fruit, to 5 mm diam. Chromosome number: 2n = 66, 88 (*fide* Kiesling 1999).

Introduced and increasingly planted about houses and on farms everywhere in North-eastern Brazil, often at the expense of other cactus vegetation. According to Kiesling (1999) originally domesticated in Mexico about 9000 years ago, having back-crossed with its putative wild ancestors, such as *O. streptacantha* and *O. megacantha*; subsequently introduced throughout the warmer parts of the world and sometimes becoming a serious pest. The following are representative records only, the plant being found in all states in Eastern Brazil.

PERNAMBUCO: Mun. Triunfo, 21 Nov. 1992, *F.A.R. Santos* 57 (HUEFS); Mun. Alagoinha, 12 Feb. . 1993, *F.A.R. Santos* 73 (HUEFS); Mun. Floresta, N of Serra Negra, 8°33'S, 38°1'W, 8 Apr. 2000, *E.A. Rocha et al.* (K, photos).

BAHIA: Mun. Morro do Chapéu, W of Icó, Brejão (Formosa), 5 Aug. 2002, *Taylor* (K, photos); Mun. Iramaia, 27.5 km NW of Pé-de-Serra, 5 Feb. 1991, *Taylor et al.* (ZSS, photos); Mun. Poções, road BR 116, 9 km S from the town, Fazenda Boa Esperança, 800 m, 5 Apr. 1988, *L.A. Mattos Silva et al.* 2334 (K, CEPEC).

MINAS GERAIS: Mun. Santana do Riacho, Serra do Cipó, Pensão Chapéu de Sol, cult., 22 Nov. 1989, *Zappi* 204 (HRCB, SPF); Mun. Entre Rios de Minas, Fazenda de Pedra, Nov. 1969, *L. Krieger* 7696 (CESJ).

An important source of cattle fodder during drought in the *sertão*. Also producing delicious fruits for human consumption, but sometimes replacing stands of native cacti, which is to be regretted.

5.5 CACTOIDEAE

Stems leafless or leaves replaced by minute fugitive scales; glochids lacking, spines never barbed; pollen mostly tricolpate; seeds with the testa exposed.

The deletion of an approximately 700 base-pair intron in the chloroplast-encoded gene *rpo*C1 supports a monophyletic origin for the subfamily Cactoideae of the Cactaceae (Wallace & Cota 1996). The tribes and their generic composition adopted here is that employed by Barthlott & Hunt (1993) modified on the basis of unpublished phylogenies derived from DNA gene sequence data, presented at IOS meetings or otherwise communicated by R. Wallace (Iowa State Univ., USA) since 1993.

Tribe **HYLOCEREEAE** Buxb.

DNA gene sequence data indicate that this tribe is one of the basal elements of Cactoideae. Its Brazilian representatives are robust climbers or large epiphytes, with flattened or trigonous stems and large or elongate flowers, the latter > 20 cm long.

7. HYLOCEREUS (A. Berger) Britton & Rose

in Contr. U.S. Natl. Herb. 12: 428 (1909). Type: *Hylocereus triangularis* (L.) Britton & Rose (*Cactus triangularis* L.).

Including *Selenicereus* sect. *Salmdyckia* D. R. Hunt (1989) (*Mediocactus* Britt. & Rose, excl. typ.).

Sprawling and clinging to rocks or climbing in trees; freely producing aerial roots; stems stout, trigonous or 3-winged, irregularly constricted, the juvenile bearing numerous, fine bristle-like spines, the adult with few, very short, stout, conical spines per areole. Flowers nocturnal, very large, to 32 × 25–34 cm, the pericarpel and tube bearing conspicuous podaria subtending large bract-scales and/or spiny areoles, the scales or spines sometimes becoming more pronounced on the reddish, ± globose fruit, spines ultimately deciduous. Seeds to 3 mm, testa smooth, black.

The circumscription of *Hylocereus* adopted here is influenced by an unpublished phylogeny, based on DNA gene sequence data, presented by R. Wallace (Iowa State Univ., USA) at the IOS Congress, Bologna, September, 1996. This indicates that when *Selenicereus* is circumscribed to include Sect. *Salmdyckia* D. R. Hunt (1989) it is paraphyletic in respect of *Hylocereus*. Hitherto *Hylocereus* has been distinguished from the very similar trigonous-stemmed members of *Selenicereus* sect. *Salmdyckia* on the basis of large scales versus spiny areoles on the pericarpel, flower-tube and fruit. However, such a separation was weakened from the start by the occurrence of occasional pericarpel spines in the otherwise typical *Hylocereus* species, *H. trigonus* (Haw.) Saff. (syn. *Cereus plumieri* Gosselin, see Hunt 1984: 41 & fig. 12), native of the SE Caribbean. The gene sequence data imply that the shared trigonous stems, whose similarity has been the cause of confusion between the two taxa treated below, are in fact a character uniting them generically and that their floral differences are perhaps only significant at subgeneric or sectional level. Plates 11.3–12.1.

1. Pericarpel, flower-tube and fruit bearing spines; stem edges green, never horny (native, widespread) . . . . **1. setaceus**
1   Pericarpel, flower-tube and fruit bearing large bract-scales only; stem edges often with a horny margin (introduced, common near habitations) . . . . ***2. undatus**

1. Hylocereus setaceus *(Salm-Dyck) R. Bauer* in Cact. Syst. Initiatives 17: 29 (2003). Type: assumed not to have been preserved. Neotype (Bauer, l.c.): Pfeiffer & Otto, Abbild. Beschr. Cact. 1: tab. 16 (1839).

Cereus setaceus Salm-Dyck in De Candolle, Prodr. 3: 469 (1828).
Selenicereus setaceus (Salm-Dyck) Werderm., Bras. Säulenkakt.: 87 (1933); Hunt in Bradleya 10: 31–32, tab. V (1992). *Mediocactus setaceus* (Salm-Dyck) Borg, Cacti, ed. 2: 213 (1951).
Cactus triangularis Vell., Fl. flumin.: 206 (1829), Icones 5: tab. 24 (1831), non L. (1753).

Selenicereus rizzinii Scheinvar in Revista Brasil. Biol. 34: 249–256 (1974). Holotype: Brazil, Rio de Janeiro, Araruama, Barra de Maricá, *Rizzini* 29L-1974 (RB).

[*Mediocactus coccineus sensu* (Salm-Dyck) Britton & Rose, Cact. 2: 211, fig. 290 (1920) non *Cereus coccineus* Salm-Dyck ex DC.; see Hunt (1989).]

[*C. triangularis sensu* K. Schum. in Martius, Fl. bras. 4(2): 208–209 (1890), pro parte, tab. excl., non (L.) Haw.]

[*C. undatus sensu* Luetzelburg, Estud. Bot. Nordéste 3: 68, 111 (1926), non Haw.]

VERNACULAR NAMES. Rainha-de-noite, Mandacaru-de-três-quinas, Espada-de-jacaré.

Scrambling, climbing or epiphytic on trees or on rocks, sometimes forming large clumps, with basitonic or mesotonic branching. Stems 3(–5)-angled, to 100 × 2–10 cm, sometimes constricted, very woody at base, margins straight or undulate, never horny (unless very old), podaria sometimes decurved for climbing; epidermis bright green, yellowish when exposed. Areoles 4–6 mm diam., 1.5–4.5 cm apart (in juvenile forms as close as 4 mm or less, with very slender, bristly spines and hairs). Central spines 3–6, conic, 1–6 mm, occasionally accompanied by bristly, more delicate radial spines. Flowers nocturnal, 20–32 × 22–25 cm; pericarpel greenish, densely covered in tubercles/podaria bearing areoles with 10–15 pungent spines; flower-tube infundibuliform, 10–13 × 1.5–2.5 cm, with scattered woolly areoles and acute, spreading bract-scales; perianth-segments to 12 cm, outer segments reflexed, lanceolate to linear, fleshy, dark red-green, inner segments to 43 mm wide, erect to spreading, lanceolate and fimbriate, delicate, white; stamens exserted in relation to the perianth-segments, curved, anthers linear; style 15–17 × 0.5 cm, stigma-lobes c. 16, exserted; ovary locule ovoid to oblong in longitudinal section. Fruit ovoid, to 9 × 7–8 cm, floral remnants deciduous; pericarp greenish then bright red when ripe, covered in spiny areoles at first, spines to 2 cm; funicular pulp white. Seeds c. 2.5–3.0 mm, cochleariform, black, shiny; testa-cells flat, smooth.

Widespread neotropical element: epiphyte, climber or lithophyte (on limestone or on gneiss/granite inselbergs) in *caatinga-agreste, cerradão, Mata atlântica, mata de brejo, mata de encosta, mata de planalto* and *restinga*, near sea level to c. 900 m, widespread in Eastern Brazil from northern Piauí southwards; Northern, Central-western and Southern Brazil (southern Pará southwards to Mato Grosso do Sul and Paraná); Central (?) and South America (southwards to E Bolivia, N Argentina and Paraguay). Map 20.

MARANHÃO: (?), near Peritoró, on the road connecting Teresina (PI) with Belém (PA), Feb. 1995, *Taylor* (obs.).

PIAUÍ: N Piauí, Mun. Cocal, *mata de encosta*, W of junction between roads PI 211 and BR 343, 19 Feb. 1995, *J.B. Fernandes da Silva* 365 (MG).

CEARÁ: between Pacajús and Fortaleza, 16 Feb. 1966, *Andrade-Lima* 66-4446 (IPA).

PERNAMBUCO: Mun. São Lourenço da Mata, São Bento and Veneza, Apr. 1928, *Pickel* 1686 (IPA); Mun. Moreno, Tapera, 29 Oct. 1932, *Pickel* 3120 (IPA); Mun. Bezerros, Serra Negra, 9 Nov. 1971, *Andrade-Lima* 71-6746 (IPA); Mun. Escada/Vitória de Santo Antão, 6 Apr. 1973, *Andrade-Lima* 73-7299 (IPA).

ALAGOAS: Mun. Marechal Deodoro, *restinga* S of Ponta do Cavalo Ruço, 14 Feb. 1995, *Taylor & Zappi* (K, photos); Mun. Piaçabuçu, *restinga*, 30 Sep. 1981, *R.F.A. Rocha* s.n. (MAC).

SERGIPE: Mun. Carmópolis, Faz. Sta Bárbara, 1 Feb. 1983, *E. Gomes* 169 (ASE 2998).

BAHIA: Mun. Jaguarari, Itumirim, 8 June 1915, *Rose & Russell* 19813 (US); Mun. Rio Real, 18 km N of Esplanada, 10 Feb. 1991, *Taylor & Zappi* (obs.); Mun. Morro do Chapéu, in forest at base of the Cachoeira do Ferro Doido, 2002, *fide* M. Machado; l.c., a few km S of the summit of Morro do Chapéu, Apr. 2002, *M. Machado* (obs.); Mun. Candeal, 18 Jan. 1994, *M.A.S. das Neves* 90 (HUEFS); Mun. Iaçu, 27 Apr. 1994, *L.P. de Queiroz* 3865 (HUEFS); Mun. Salvador, on trees in Praça Visconde de Cairu, beside the Mercado Modelo, 2 Aug. 1998, *Taylor* (obs.); Mun. Rio de Contas, 16 km E of town towards Jussiape, 25 Nov. 1988, *Taylor & Zappi* in *Harley* 25548 (K, SPF); Mun. Iramaia, 32 km N of Contendas towards Maracás, 4 Feb. 1991, *Taylor et al.* 1548 (K, HRCB, ZSS, CEPEC); Mun. Ituaçu, c. 2 km SW of town, limestone, 18 Aug. 1988, *Eggli* 1194 (ZSS); Mun. Ilhéus, Castelo Novo, Faz. Santa Rita, 12 Dec. 1968, *Castellanos* 26952 (HB); Mun. Vitória da Conquista, c. 7 km N of town, 3 km W of road BR 116, 17 Aug. 2002, *M. Machado* (obs.); l.c., 12 km from town towards Itambé, 16 Aug. 2002, *M. Machado* (obs.); Mun. Itaju do Colônia, Km 8 on road from Itaju to Pau Brasil, 13 Jan. 1969, *T.S. dos Santos* 359 (CEPEC).

MINAS GERAIS: Mun. Januária, Pandeiros, *fide* Uebelmann (1996): HU 719; Mun. Pedra Azul, 10 km E of town towards Almenara, 16°8'S, 41°12'W, 19 Oct. 1988, *Taylor & Zappi in Harley* 25189 (K, SPF); Mun. Montes Claros, 13 km N of town towards Januária, 16°38'S, 43°55'W, 6 Nov. 1988, *Taylor & Zappi in Harley* 25511 (K, SPF); Mun. Augusto de Lima, Santa Barbara, 6 Aug. 1988, *Eggli* (ZSS, photo); Mun. Santana do Riacho, calcareous rocks at foot of Serra do Cipó, nr Cardeal Mota, 28 Oct. 1988, *Taylor & Zappi* (obs.); Mun. Iapu, on road BR 458, 14 km W of crossing to the town, 13 Dec. 1990, *Taylor & Zappi* 752 (K, HRCB, ZSS, BHCB); Mun. Aimorés, Fazenda do Sr Hulbert Schumacher, 21 Oct. 1997, *M.F. de Vasconcelos* s.n. (K, BHCB); Mun. Matozinhos, Faz. Cauaia, 31 Oct. 1996, *Lombardi* 1466 (BHCB, K); Mun. Lagoa Santa, reported by Warming (1908: 146, 149, fig. 36; cf. Warming & Ferri 1973); Mun. Santa Bárbara, s.d., *C. Paganini*, cult. Santuário do Caraça, Catas Altas, Aug. 2001, Taylor (K, photo); Mun. São João del Rei, Oct. 1970, *Krieger* 7529 (CESJ); Mun. Laranjal, 8 km N of town towards Muriaé, 18 Dec. 1990, *Taylor & Zappi* 793 (K, HRCB, ZSS, BHCB); on road BR 040, between Juiz de Fora and Três Rios (RJ), 'Paraibuna', 25 July 1991, *Zappi* (obs.); Mun. Além Paraíba, road BR 116, 16 Oct. 1985, *Hatschbach & J.M. Silva* 49858 (MBM).

ESPÍRITO SANTO: Mun. São Mateus, road BR 101, Ponta do Ipiranga, 13 Oct. 1992, *Hatschbach et al.* 58057 (MBM); Mun. Mantenópolis, 5 km NE of town towards Mantena, 15 Dec. 1990, *Taylor & Zappi* 778 (K, HRCB, ZSS, MBML); Mun. Barra de São Francisco, 2 km W of town, 5.5 km E of the border ES/MG, 16 Dec. 1990, *Taylor & Zappi* 779 (K, HRCB, ZSS, MBML); Mun. Santa Teresa, *fide* Uebelmann (1996): HU 244; Mun. Afonso Cláudio, road to Itarana, 20°3'8"S, 41°3'3"W, 24 Nov. 1999, *Zappi et al.* 411 (K, UEC), l.c., 6 km NE of town towards Serra Pelada, 17 Dec. 1990, *Taylor & Zappi* 790 (K, HRCB, ZSS, MBML); Mun. Vila Velha, Jacaranema, 12 Oct. 1988, *O.J. Pereira* 1852 (Herb. Univ. Fed. ES); Mun. Guarapari, Setiba, near Laguna de Corais, *restinga*, 12 Sep. 1987, *O.J. Pereira* 1049 (Herb. Univ. Fed. ES); Mun. Itapemirim, Rodovia do Sol, S of Marataízes, 21°4'51"S, 40°50'24"W, 26 Nov. 1999, reported growing with *Opuntia monacantha* (*Zappi et al.* 470).

RIO DE JANEIRO: Mun. Santa Maria Madalena, access road to the town, 22°0'55"S, 42°7'30"W, 21 Nov. 1999, *Zappi et al.* 355 (K, UEC).

CONSERVATION STATUS. Least Concern [LC] (1); PD=2, EI=1, GD=1. Short-list score (1×4) = 4. Much of its presumed former habitat has been destroyed in North-eastern Brazil, but its range and frequency can be judged from the above-cited records.

In its vegetative state this widely distributed, native species is sometimes confused with the introduced *H. undatus* (see below), but the pericarpel and immature fruit is spiny and the stem-margins seldom horny. Its fruits are edible when red and mature. It may be related to a taxon cultivated in Colombia for the export of its yellow, egg-shaped, edible fruit, and to another known from Roraima in Northern Brazil (*A. R. Pontes* s.n., K, photo.), which is probably also the same as the plant from the Guianas illustrated and discussed by Leuenberger (1997: 48–51) as possibly identifiable with *H. extensus* (Salm-Dyck ex DC.) Britton & Rose (*Cereus extensus* DC., *tantum quoad typ.*; *Selenicereus extensus* (Salm-Dyck ex DC.) Leuenb.). The Colombian plant, known in the British supermarket trade as 'Pitaya', has been referred to *Selenicereus* (sect. *Salmdyckia*) *megalanthus* (K. Schum. ex Ule) Moran, based on a collection from the Amazonian drainage of Peru (Hunt 1992a). It is possible that, when better understood, *Hylocereus setaceus* will prove to be the oldest name for a widespread neotropical species comprising most of the above as regional subspecies, amongst which *Selenicereus tricae* D. R. Hunt (1989) might also figure, extending the range of this complex to Central America (including southern Mexico).

It is likely that *H. setaceus, sens. str.,* is significantly under-recorded in the northern half of its range within Eastern Brazil, as is suggested by the isolated collections from northern

Piauí and Ceará, which are more than 700 km distant from the next nearest records, in northern Bahia and eastern Pernambuco. Plants resembling the species have been seen in north-central Maranhão, near Peritoró, while travelling the road connecting Teresina (PI) with Belém (PA), and it may be expected to occur in Paraíba, perhaps in *brejo* forest. Its southern limit in Brazil is indicated as Paraná above, but Braun & Esteves Pereira (2002: 79) suggest it reaches Rio Grande do Sul.

*2. Hylocereus undatus *(Haw.) Britton & Rose* in Britton, Fl. Bermuda: 256 (1918); Cact. 2: 187–188 (1920). Type: a plant cultivated at the London Horticultural Society, originating from China, assumed not to have been preserved. Neotype (Taylor 1995: 119–120; superseding that designated by Scheinvar 1988): Bot. Mag. 44: tab. 1884 (1817), as 'Cactus triangularis'.

Cereus undatus Haw. in Philos. Mag. Ann. Chem. 7: 110 (1830).

Scrambling, climbing or epiphytic, forming large woody masses of branches, with basitonic or mesotonic branching. Stem-segments trigonous/3-winged, to 100 × 4–7(–8) cm, sometimes constricted, narrow and very woody at base, margins crenate, horny in age; epidermis bright green, yellowish when exposed to full sun. Areoles 4–6 mm diam., (2.5–)3–6(–8) cm apart, with 2–6 conic, c. 4 mm spines. Flower-bearing areoles in the axils of the stem's marginal crenations; flowers nocturnal, very showy, to 30 × 34 cm; pericarpel greenish, densely covered in rounded, broad-based, green or reddish bract-scales, but lacking spines; flower-tube infundibuliform, c. 12 × 1.5–2.5 cm, covered in broadly triangular, adpressed, to acute, spreading bract-scales; perianth-segments up to 14 cm, outer segments reflexed, lanceolate to linear, fleshy, dark red with green bases, inner segments erect to spreading, spathulate and apiculate, fimbriate, delicate, white; stamens disposed in a broad ring around the style, sometimes curved, anthers linear; style 14–20 × 0.8 cm, stout, stigma-lobes c. 20 or more, exserted; ovary locule oval in longitudinal section. Fruit ovoid to globose, to 12 cm diam., flower remnants deciduous; pericarpel covered in broadly triangular, fleshy, orange-yellow to bright red bract-scales; funicular pulp white. Seeds c. 3 mm, cochleariform, black, shiny; testa-cells flat, smooth (Barthlott & Hunt 2000: pl. 20.5–6).

Introduced as garden plant and sometimes escaping into roadside trees and maritime scrub; perhaps native in Mexico and Central America, commonly introduced elsewhere in the tropics and subtropics worldwide.
PERNAMBUCO: Mun. Nazaré da Mata, 21 Feb. 1955, *J.C. Moraes* (SPSF 4788); Mun. Caruaru, W of town, on roof of house beside road BR 232, 11 Feb. 1995, *Taylor* (obs.); Mun. Ibimirim, growing outside a house in the town, 8 Apr. 2000, *E.A. Rocha et al.* (obs.).
ALAGOAS: Mun. Maceió, Avenida Fernandes Lima, covering the top of a wall, 16 Feb. 1995, *Taylor* (obs.).
BAHIA: Mun. Morro do Chapéu, 3 Aug. 2002, *Taylor* (K, photos); Mun. Marcionílio Sousa, Machado Portella, 1915, *Russell* s.n. (US); Mun. Ilhéus, Pontal, near Praia do Sul, 20 m from beach, 20 Nov. 1981, *L.A. Mattos-Silva* 1386 (CEPEC).
MINAS GERAIS: Mun. Santana do Riacho, Serra do Cipó, 19 Nov. 1989, *Zappi et al.* 192C (HRCB, SPF).
ESPÍRITO SANTO: Mun. Colatina, 21 km S of Ângelo Frechiani towards Colatina, in a roadside tree, 16 Dec. 1990, *Taylor & Zappi* 787 (HRCB, K, ZSS).

Widely cultivated in South America, this species may be found at the sites of old houses, where it often scrambles to the tops of trees. However, it does not seem to be able to reproduce by means of seed in Eastern Brazil, perhaps because all or most individuals belong to the same clone.

Some published reports of this species actually refer to the native *H. setaceus* (see above).

*8. SELENICEREUS (A. Berger) Britton & Rose

in Contr. U.S. Natl. Herb. 12: 429 (1909).

Including *Cryptocereus* Alexander; *Selenicereus* sect. *Cryptocereus* (Alexander) D. R. Hunt (1989).

An ill-defined genus of c. 11 species, native of Mexico, the Caribbean and northern South America. Typically, the genus has cylindrical, scandent stems with 5 or more low ribs (Sect. *Selenicereus*), but the species encountered in Brazil have flattened stems. One such is native of Northern Brazil, *Selenicereus* (sect. *Strophocactus*) *wittii* (K. Schum.) G. D. Rowley (Amazônia, *igapó*), but some authors now refer this back to *Strophocactus* Britton & Rose. A detailed account of this species can be found in Barthlott *et al.* (1997).

Only the following Mexican plant is encountered in Eastern Brazil:

*Selenicereus anthonyanus *(Alexander) D. R. Hunt* in Bradleya 7: 93 (1989), native of southern Mexico, is the cactus most frequently cultivated as a house plant (in pots and hanging baskets) in Eastern Brazil. It is more rarely planted outdoors, where it has been observed climbing trees in a semi-naturalized state in South-eastern Brazil, outside the area treated here. It resembles the following genus in its vegetative state (especially the Mexican *Epiphyllum anguliger*), but has flowers with a stouter, much shorter tube and bristly pericarpel.

9. EPIPHYLLUM Haw.

Syn. pl. succ.: 197 (1812). Type and only species native of Eastern Brazil:

1. Epiphyllum phyllanthus *(L.) Haw.*, l.c. (1812). Lectotype (Leuenberger 1997): Dillenius, Hort. eltham.: tab. 64, fig. 74 (1732).

Cactus phyllanthus L., Sp. pl. 1: 469 (1753).

Epiphytic, forming woody clumps, with basitonic and mesotonic branching. Stems flat, alate, sometimes trigonous at base, 15–100(–150) × 3–6(–8) cm, to 1 cm thick, lanceolate, narrow at base, margins crenate, teeth 2.5–4.5 cm apart, apex obtuse, midrib rather woody; epidermis green, yellow-green to reddish or purplish when exposed, margins sometimes horny in age; areoles spineless, but sometimes bristly on old branches. Flower-bearing areoles in the axils of the teeth; flowers nocturnal, 15–25 × 4–6(–8) cm; pericarpel greenish, with areoles and minute triangular bract-scales, ridged; flower-tube more than 14 cm, cylindric, very narrow, 6–8 mm diam., with scattered linear bract-scales; perianth-segments 1.5–3.0 cm, reflexed, lanceolate to linear, pale pink to white; stamens disposed in a ring around the style, anthers linear; style 14–20 cm, stigma-lobes 6–8, exserted; ovary locule narrowly oblong in longitudinal section. Fruit narrowly ovoid to pear-shaped, apiculate, to 10 × 3.5 cm, dehiscent by lateral slit, flower remnants deciduous; pericarpel ridged, bright red or pink, with acute bract-scales; funicular pulp white, mucilaginous. Seeds 4–4.5 mm, cochleariform, black, shiny, with a broad white hilum; testa-cells flat, smooth (Barthlott & Hunt 2000: pl. 21.1–2).

Widespread neotropical element: epiphyte in *Mata atlântica, caatinga-agreste, mata do planalto, mata ciliar* and *cerrado*, near sea level to at least 1300 m, ± common throughout the more humid parts of Eastern Brazil;

Neotropics from Central America southwards. It is the most widespread cactus species in Eastern Brazil and only a few representative collections/observations are cited here. There are currently no records from Rio Grande do Norte.

CEARÁ: Mun. Maranguape, Serra de Maranguape, 24 Nov. 1955, *Andrade-Lima* 55–2404 (IPA 8631).

PERNAMBUCO: Mun. Nazaré da Mata, 7 Mar. 1955, *J.C. Moraes* (SPSF 4951); Mun. Moreno, Tapera, 17 Jan. 1930, *B. Pickel* 2309 (IPA); Mun. Recife, Jardim Botânico, 21 Mar. 1966, *E. Tenório* 66–86 (IPA), l.c., 30 Mar. 2000, *Taylor & Griffiths* (obs.).

ALAGOAS: Mun. Colônia Leopoldina, 1 km from border with Pernambuco, 21 Nov. 1966, *D. Andrade-Lima* 66–4739 (IPA); Mun. Murici, 15 Feb. 1995, *Taylor & Zappi* (obs.).

BAHIA: Mun. Rio Real, 18 km N of Esplanada, 10 Feb. 1991, *Taylor & Zappi* (obs.); Mun. Itatim, 1997, *F. França* 1494 (HUEFS; K, photo.); Mun. Conceição de Almeida, 8 Feb. 1981, *G.C.P. Pinto* 45/81 (HRB); Mun. (?), Ilha de Itaparica, 22 Dec. 1976, *Pereira de Souza* s.n. (ALCB); Mun. Ilhéus, área do CEPEC, Km 22 Ilhéus–Itabuna road, 7 Dec. 1988, *T.S. dos Santos* 4431 (CEPEC); Parque Nacional de Monte Pascoal, 21 Mar. 1968, *S.G. da Vinha & T. S. dos Santos* 83 (CEPEC).

MINAS GERAIS: Mun. Monte Azul/Espinosa, c. 12 km E of Monte Azul, Serra Geral, top of serra on path to village of 'Gerais', 28 Jan. 1991, *Taylor et al.* 1477 (BHCB, HRCB, K, ZSS); Mun. Mirabela, c. 10 km SSE of town on road BR 135, 10 Aug. 1988, *Eggli* 1132 (ZSS); Mun. Grão Mogol, Riacho Ribeirão, 16°33'S, 42°54'W, 28 May 1988 & 16 June 1990, *Zappi et al.* in CFCR 12058, 13124 (SPF); Mun. Rio Vermelho, Pedra Menina, Faz. Sr José Batista, 42°6'W, 17°54'S, 6 Mar. 1988, *Zappi & Prado* in CFCR 11820 (SPF); Mun. Conceição do Mato Dentro, Serra do Cipó, 26 Apr. 1988, *Zappi* in CFSC 11178 (SPF); Mun. Laranjal, 6 Feb. 1971, *Krieger* 9960 (CESJ).

ESPÍRITO SANTO: Mun. Santa Teresa, Fazenda Santa Lúcia, 19°57'24"S, 40°32'37"W, 25 Nov. 1999, *Zappi et al.* 438 (K, UEC).

RIO DE JANEIRO: Mun. Santa Maria Madalena, access road to the town, 22°0'55"S, 42°7'30"W, 21 Nov. 1999, *Zappi et al.* 356 (K, UEC).

CONSERVATION STATUS. Least Concern [LC] (1); PD=2, EI=1, GD=2. Short–list score (1×5) = 5. Frequency has been reduced due to destruction of forest formations, but nevertheless very widespread and records for Eastern Brazil too numerous to list.

Widespread in South America in various habitats, this is perhaps the only epiphytic cactus likely to be found in, or bordering on *cerrado* vegetation. Plate 12.2.

*Epiphyllum oxypetalum *(DC.) Haw.* from southern Mexico and Central America is occasionally cultivated and has been recorded planted or escaped outdoors in the states of Pernambuco: Mun. Moreno, 7 Mar. 1936, *B. Pickel* 4121 (IPA); Mun. Arcoverde, 14 May 1966, *D. Andrade-Lima* 66–18 (IPA); and Bahia: Mun. Vera Cruz, Ilha de Itaparica, 31 Mar. 1997, *Aline* 1 (ALCB). It is easily distinguished from *E. phyllanthus* by its very long, slender, cylindric to angled, basal extension shoots and flowers to 27 cm in diameter, the outer perianth–segments conspicuously reddish to deep pink.

Tribe **ECHINOCEREEAE** Buxb.

The tribal placement of the following genus is rather uncertain at present and awaits the results of analysis of DNA gene sequence data. It is placed in this tribe, including the former Leptocereeae, only because the genus from which it was separated, ie. *Acanthocereus*, is currently placed there.

10. PSEUDOACANTHOCEREUS F. Ritter

Kakt. Südamer. 1: 47 (1979). Type: *P. brasiliensis* (Britton & Rose) F. Ritter.

Literature: Taylor *et al.* (1992).

An isolated genus of only two species, comprising *P. sicariguensis* (Croizat & Tamayo) N. P. Taylor (NE Colombia & NW Venezuela, Map 11) and the following (Plates 12.3–13.1):

1. Pseudoacanthocereus brasiliensis *(Britton & Rose) F. Ritter*, l.c. Type: Brazil, Bahia, Mun. Marcionílio Sousa, Machado Portella, 1915, *Rose & Russell* 19903 (US, lecto. designated here; NY, lectopara.).

Acanthocereus brasiliensis Britton & Rose, Cact. 2: 125 (1920). *Cereus brasiliensis* (Britton & Rose) Luetzelb., Estud. bot. Nordéste 3: 111 (1926).

Pseudoacanthocereus boreominarum Rizzini & A. Mattos in Revista Brasil. Biol. 46(2): 327 (1986). *P. brasiliensis* f. *boreominarum* (Rizzini & A. Mattos) P.J. Braun & E. Esteves Pereira in Schumannia 3: 188 (2002). Holotype: Brazil, Minas Gerais, Mun. Itinga, *Rizzini & Mattos-Filho* 40 (RB 232057).

VERNACULAR NAME. Catana-de-jacaré.

Shrubby, scrambling or decumbent, sometimes forming large clumps, with basitonic or mesotonic branching. Roots tuberous. Stems erect to decumbent, 1.5–4.5 cm diam.; 2–6(–7)-ribbed, ribs high, acute, somewhat undulate; epidermis dark green to greyish. Areoles 3–6 mm diam., 2–7 cm apart, felt white, soon glabrescent, areolar growth ± indeterminate. Spines 8–14, central and radial difficult to distinguish, 5–50 mm. Flower-bearing region of stem not differentiated; flowers nocturnal, 15–17 × 11–12 cm; pericarpel green with dull pink tinge, bearing tubercles/podaria and areoles with bristles/spines to 5 mm; flower-tube infundibuliform, 9–10 × 1.4–2.0 cm, bearing scattered areoles with minute spines; perianth-segments to 60 mm, outer segments reflexed, lanceolate, fleshy, dull pink to reddish brown, inner segments spreading, lanceolate to spathulate and fimbriate, delicate, white; stamens exserted in relation to the perianth-segments, curved, anthers linear; style c. 100 × 3 mm, stigma-lobes 8–11, exserted; ovary locule rectangular to oblong in longitudinal section. Fruit globose, to 5.5–8.0 cm diam., floral remnants deciduous; pericarp greenish, becoming yellow when ripe and falling to the ground, smelling of pineapple, bearing podaria and deciduous areoles with white to pale brown felt; funicular pulp white. Seeds c. 4.5 mm, cochleariform, light brown, dull; testa-cells flat, elongate, cuticular folds coarse, dense (Taylor *et al.* 1992, Barthlott & Hunt 2000: pl. 9.1–2).

Eastern *caatinga-agreste* element: within and at the margins of *caatinga-agreste*, 40–700 m, east of the Chapada Diamantina crestline in northern and central-eastern Bahia, and in the drainage of the Rio Jequitinhonha (associated with gneiss/granite inselbergs and *lajedos*) of north-eastern Minas Gerais (apparently disjunct, but possibly under-recorded through destruction of habitat in intervening areas). Endemic to the core area of Eastern Brazil. Map 20.

BAHIA: Mun. Jaguarari, *Ritter* 1230 (SGO 125358, *fide* Eggli et al. 1995: 499); l.c., Catuni, 8 June 1915, *Rose & Russell* 19819 (US, NY); Mun. Senhor do Bomfim ['Villa Nova'], *Zehntner* in *Rose* 19735 (US, NY); Mun. Itiúba, road to Filadélfia, 10°39'S, 39°44'W, 11 May 2002, *Nascimento & Nunes* 102 (HUEFS); Mun. Morro do Chapéu, near Ventura, by the Rio Jacuípe, 24 Dec. 1988, *Taylor & Zappi* (obs.); Mun. Riachão do Jacuípe, 17 km NW of the town towards Nova Fátima, 7 Feb. 1970, *A.L. Costa* s.n. (ALCB, in spirit); Mun. Candeal, beside road BR 324, 20.5 km NW of Tanquinho, 25 Apr. 1992, *Taylor & Zappi* (obs.); Mun. Tanquinho, *W. Uebelmann* (mss.); Mun. Feira de Santana, Serra de São José, Nov. 2001, *Zappi* (obs.); l.c., 8 km SW of the town on road BR 116, margin of Rio Jacuípe, 19 Feb. 1981, *A.M. de Carvalho et al.* 571 (CEPEC); Mun. Rui Barbosa, 7–8 km SE of the town on road to village of Alagoas, 23 Aug. 1988, *Eggli* 1265 (ZSS); Mun. Itaberaba, 21 km SE of Rui Barbosa on road to Alagoas, 7 Feb. 1991, *Taylor et al.* 1579 (K, ZSS, HRCB, CEPEC); Mun. Conceição da Feira, NE of Barragem de Bananeiras, 12°32'S, 39°5'W, 16 Feb. 1981, *Iscardino et al.* in Grupo Pedra do Cavalo 1053 (ALCB); Mun. Iaçu, *fide* Uebelmann (1996: HU 138); Mun. Marcionílio Sousa, Machado Portella, Dec. 1912, *Zehntner* in IFOCS 5037 (M), 1915, *Rose & Russell* 19903 (NY, US); Mun. Jaguaquara [Toca da Onça], 27 June 1915, *Rose & Russell* 20107 (US).

MINAS GERAIS: Mun. Pedra Azul, 8 km W of the town towards road BR 116, 18 Oct. 1988, *Taylor & Zappi* in *Harley* 25185 (K, SPF); Mun. Itinga, 2 km E of the town, S of Rio Jequitinhonha, 14 Dec. 1990, *Taylor & Zappi* 768 (K, ZSS, HRCB, BHCB); Mun. Itaobim, 1 km W of Itaobim, 0.5 km N of Rio Jequitinhonha, 16°34'S, 41°31'W, *Taylor & Zappi* in *Harley* 25531 (K, SPF); Mun. Coronel Murta, 5 km SE of the town on road BR 342 to Araçuaí, 21 Feb. 1988, *Supthut* 8863 (ZSS).

CONSERVATION STATUS. Vulnerable [VU C1 + 2a(i)] (2); PD=3, EI=1, GD=2. Short-list score (2×6) = 12. Much of its preferred habitat has been destroyed and its original range may have been more extensive and less markedly discontinuous than it appears today.

This inconspicuous and, when out-of-flower, rather ugly plant appears to be of erratic occurrence, which may in part be the result of the widespread destruction of the *agreste/caatinga*, under whose shade it prefers to grow. While it is a variable species, there are no reliable differences between plants from Bahia and Minas Gerais and *P. boreominarum* Rizzini & A. Mattos cannot be justified even at infraspecific rank.

The report by Uebelmann (1996), under 'HU 1197', of a '*Pseudoacanthocereus sp.*' from Penedo, Alagoas ['Sergipe'] is considered improbable.

Tribe **RHIPSALIDEAE** DC.

In terms of numbers of species, this is the second largest tribe of Cactoideae in Eastern Brazil (after Cereeae). All the species treated here are epiphytes and/or lithophytes in the core area of Eastern Brazil, with flowers < 6 cm long and seeds with a mucilage sheath (either restricted to the hilum area or covering the entire surface).

11. **LEPISMIUM** Pfeiff.

in Allg. Gartenzeitung 3: 314–315 (1835). Lectotype: *L. commune* Pfeiff. (= *L. cruciforme* (Vell.) Miq.).

Literature: Barthlott (1987), Barthlott & Taylor (1995: 44) — now excluding subgenera *Pfeiffera, Acanthorhipsalis* & *Lymanbensonia;* see Nyffeler (2000c).

Epiphytic, *branching mesotonic or basitonic*, or branching subacrotonically only when damaged, freely producing aerial roots; stem-segments often angled or flat with the areoles subtended by crenate or serrate marginal teeth, *not deciduous as units when old*. Areoles sometimes densely hairy, scale leaves often clearly visible; composite terminal areoles lacking. Flower-buds always somewhat erumpent, the flowers either ± immersed in the sunken stem areoles or pendent, often campanulate; stamens generally much shorter than the perianth-segments, never exserted; pericarpel ridged or angled, rarely almost terete; tube inconspicuous. Fruit globose, ellipsoid or turbinate; *pericarp ridged when immature* (except in No. 3), white, magenta, red, dark wine-coloured or blackish.

In the narrower sense suggested by Nyffeler's recent molecular studies, a genus of only 6 species, from the eastern Andes (Bolivia & NW Argentina: 2 endemic, Map 14) and south-eastern South America (Brazil, Paraguay, Uruguay, NE Argentina: 3 endemic). Species of *Lepismium sens. str.* can be distinguished from other Brazilian Rhipsalideae by the combination of a basi- to mesotonic (not acrotonic) branching pattern and non-deciduous stem-segments. The genus is restricted to the *Mata atlântica* in Eastern Brazil. Plates 13.2–14.1.

1. Distal parts of stems flattened, with conspicuous marginal teeth to at least 4 mm long . . . .
. . . . **1. houlletianum**
1. Stems flattened or 3–6-winged/angled/ribbed, with strongly adpressed marginal teeth/ crenations to at most 2 mm long 2

2. Areoles woolly at anthesis and subsequently; flowers immersed in the areoles, 1 or more per areole, whitish to deep pink (Pernambuco southwards) . . . . **3. cruciforme**
2. Areoles not woolly; flowers 1 per areole, not sunken, whitish (SE Brazil) . . . . **2. warmingianum**

1. Lepismium houlletianum *(Lem.) Barthlott* in Bradleya 5: 99 (1987); Barthlott & N. P. Taylor, ibid. 13: 49, pl. 7 (1995). Type: Brazil, *Houllet*, s.n. (assumed not to have been preserved). Neotype (Barthlott & Taylor 1995: 46): Gürke, Blühende Kakt. 2: tab. 111 (1909).

Rhipsalis houlletiana Lem. in Ill. Hort. 5, misc.: 64 (1858). *R. houlletii* Hook. f. in Bot. Mag. 100: tab. 6089 (1874). Type: as above.
R. regnellii G. Lindb. in Gartenflora 39: 118 (1890). *Lepismium houlletianum* f. *regnellii* (G. Lindb.) Barthlott & N. P. Taylor in Bradleya 13: 47 (1995). *Rhipsalis houlletiana* var. *regnellii* (G. Lindb.) Kimnach in Cact. Succ. J. (US) 68: 156 (1996). Based on syntypes. Lectotype (Barthlott & Taylor, l.c.): Minas Gerais, Caldas, 1861, *Regnell* [III] 626 (S; S, NY, P, US, F, lectoparas.).

VERNACULAR NAME. Rabo-de-arara.

Plant 1–2 m, pendent; basal shoots erect at first; stem-segments 15–40 × 2–7.5 cm (towards apex), cylindric and woody at base, flattened distally and only to c. 2.5 mm thick (away from the vascular midrib), margins strongly serrate-dentate, teeth mostly acute, not adpressed but sometimes incurved, to 6 mm or more long, epidermis dark green to glaucous, sometimes purple tinged at the stem margin; seedlings with stems flattened at first. Areoles in the axils of the teeth, glabrous or with 1–3 minute bristles. Flower-buds somewhat erumpent; flowers scented, pendent, c. 20 × 15–26 mm, perianth-segments 10–15, white (yellowish post anthesis), c. 17 × 3 mm, not expanding fully; stamens in two weakly defined groups, the innermost clustered around the style, the remainder spreading out against the perianth and the largest slightly > half as long as the inner perianth-segments, white, but strikingly orange reddish at base (yellowish at base in the forma *regnellii*); style white, exceeding stamens, stigma-lobes 3–6, to c. 3 mm, outspread, white becoming yellow in age; pericarpel cylindric, c. 6–8 × 2.5–4 mm, longitudinally ridged. Fruit 5–8 mm diam., ovoid to globose, magenta to deep pink or black. Seed (Barthlott & Hunt 2000: pl. 1.7–8). Chromosome number: 2n = 22 (Barthlott 1976).

Southern humid forest element: epiphyte in *Mata atlântica*, including *mata de neblina*, 500–1900 m, central and south-eastern Minas Gerais; common in South-eastern and Southern Brazil; North-eastern Argentina (Misiones). Map 21.
MINAS GERAIS: Mun. Conceição do Mato Dentro, Rio Santo Antônio, 25 Apr. 1978, *Martinelli* 4298 (RB); Mun. Caparaó, Parque Nacional do Caparaó, 30 Apr. 1988, *Krieger et al.* in FPNC 93 (CESJ, SPF),

15 June 1991, *Hatschbach et al.* 55545 (MBM); Mun. Lima Duarte, Parque Estadual do Ibitipoca, 8 July 1990, *R.C. Oliveira* 25194 (CESJ).

CONSERVATION STATUS. Least Concern [LC] (1); PD=2, EI=1, GD=2. Short-list score (1×5) = 5. Least Concern taking its total range into account.

Luetzelburg (1925–26, 3: 111) reports this species from eastern Bahia, but it has not been encountered subsequently and his record may be an error or misidentification.

As in the case of *Schlumbergera kautskyi* (see below), the pointed, marginal, stem-segment teeth (podaria) in this species are assumed to function as, and represent the equivalent of leaf drip tips. It is rather variable in stem and floral characters.

2. Lepismium warmingianum *(K. Schum.) Barthlott* in Bradleya 5: 99 (1987). Type (syntypes): Brazil, Minas Gerais, Lagoa Santa, *Warming* s.n. (B†, lectotype designated by Britton & Rose 1923: 238); ibid., Caldas, *Lindberg* 511 (B†). Lectotype (Barthlott & Taylor 1995: 46): Minas Gerais, Caldas, 18 Oct. 1854, *Lindberg* 511 (S; BR & MO lectoparas. numbered '611').

Rhipsalis warmingiana K. Schum. in Martius, Fl. bras. 4(2): 291 (1890); G.A. Lindberg in Gartenflora 41: 8–12, figs 5–7 (1892).
R. linearis K. Schum. in ibid.: 296 (1890). *Lepismium lineare* (K. Schum.) Barthlott in Bradleya 9: 89 (1991). Lectotype (Barthlott & Taylor 1995): Paraguay, Nov. 1874, *Balansa* 2500 (K; P, lectopara.).
Rhipsalis gonocarpa F. A. C. Weber in Rev. Hort. 64: 427 (1892). Type: Brazil, São Paulo (assumed not to have been preserved).

Stems pendulous to 3 m; stem-segments 25–150 × 0.3–1.5(–2.5) cm, flat or 3–4-angled, flat segments only 1 mm thick away from the vascular midrib, margins serrate-crenate, teeth low, obtuse, to 2 mm proud of the stem and 1–2.5 cm apart on its margins, the pale, scarious scale leaves strongly adpressed, epidermis dark olive to yellow-green. Areoles in the axils of the scale-bearing teeth, felt 0 or very sparse prior to flowering, bristles 0. Flowers sweetly scented, pendent, to 27 × 23 mm; perianth-segments 8–12, largest to 15 × 5 mm, cream or white, not expanding fully; pericarpel green, 5–8 × 4 mm, distinctly angled; stamens white, yellow or reddish at base, arranged in two groups, the outermost longest, to c. 10 mm, the inner series clustered around the style; style c. 12 mm, stigma-lobes c. 5, 3 mm, cream. Fruit c. 12 × 10 mm, ridged and green to red when unripe, rounded, smooth, dark wine-coloured to black and opaque when ripe.

Southern humid forest element: epiphyte or lithophyte, *Mata atlântica*, c. 750 m (MG), central-southern Minas Gerais and southern Espírito Santo (Domingos Martins); common in South-eastern and southern Brazil; eastern Paraguay and north-eastern Argentina (Misiones). Map 21.
MINAS GERAIS: Lagoa Santa, reported by Warming (1908: 146 & 151; cf. Warming & Ferri 1973); Mun. Arcos, *fide* Uebelmann (1996): HU 957.
ESPÍRITO SANTO: Mun. Domingos Martins, May 1990, *R. Kautsky* in *Rauh* 70693, cult. Univ. Bonn, Germany, Accn. No. 05815 (K).

CONSERVATION STATUS. Least Concern [LC] (1); PD=2, EI=1, GD=2. Short-list score (1×5) = 5. Least Concern taking its total range into account, but apparently rare in the region treated here.

Also reported from Bahia and Paraíba by Luetzelburg (1925–26, 3: 111), but these are assumed to be misidentifications.

Lepismium warmingianum is the sister species of *L. lorentzianum* (Griseb.) Barthlott, from eastern Bolivia (Santa Cruz & Tarija) and north-western Argentina.

3. Lepismium cruciforme *(Vell.) Miq.* in Bull. Sci. Phys. Nat. Néerl.: 49 (1838). Type: not extant. *Typ. cons.* (Taylor 1994, Greuter *et al.* 2000: 385): Brazil, Rio de Janeiro, Vellozo, Fl. flumin., Icones 5: tab. 29 (1831).

Cactus cruciformis Vell., Fl. flumin., 207 (1829), *nom. cons.* (Brummitt 1996). *Cereus cruciformis* (Vell.) Steud., Nomencl. bot., ed. 2, 1: 333 (1840). *Hariota cruciformis* (Vell.) Kuntze, Revis. gen. pl. 1: 262 (1891). *Rhipsalis cruciformis* (Vell.) A. Cast. in Anales Mus. Nac. Hist. Nat. Buenos Aires 32: 496 (1925). *Cereus squamulosus* Salm-Dyck in DC., Prodr. 3: 469 (1828), *nom. rej.* Type: not known to have been preserved. Neotype (Taylor 1994): Vellozo, Fl. flumin., Icones 5: tab. 29 (1831).
C. tenuispinus Haw. in Philos. Mag. Ann. Chem. 1: 125 (1827), *nom. rej.* Type: not known to have been preserved. Neotype (Taylor 1994): Loddiges, Bot. Cab. 19: tab. 1887 (1832).
C. tenuis DC., Prodr. 3: 469 (1828), *nom. rej.*; *Lepismium tenue* (DC.) Pfeiff. in Allg. Gartenzeitung 3: 315 (1835). Type: a living plant seen by de Candolle. Neotype (Taylor 1994): 'Received in 1827 from Mr. Hitchin of Norwich. Cereus myosurus.', water colour illustration by Bond, 20 Jan. 1829 (K). *C. myosurus* Salm-Dyck in DC., Prodr. 3: 469 (1828), *nom. rej.* Type: not known to have been preserved. Neotype (Taylor 1994): as above. [Pfeiffer, l.c., treated *C. myosurus* as a synonym of *L. tenue.*]
Lepismium radicans Vöcht. in Jahrb. Wiss. Bot. 9: 399 (1874). *Rhipsalis radicans* (Vöcht.) F. A. C. Weber in Bois, Dict. hort.: 1047 (1898). Type: not known to have been preserved.
Note: for further synonymy, see Britton & Rose (1923: 215). All synonyms relating to this species are included in the Index to botanical names and epithets (Chapter 7.2).

Stems pendent, creeping or trailing, to 3 m, 15–40(–60) × 1–3(–7.5) cm, flat or 3–6-angled/ribbed, crenate, often irregular, epidermis green, red tinged around the edges and areoles, or entirely red when fully exposed. Areoles in the axils of the marginal crenations, subtended by conspicuous scales prior to flowering, sunken, becoming rather large in age, with abundant wool in tufts to 5 mm or more long and weak bristles to 6(–10) mm, to 20 mm or more apart on the stem margins/angles. Flowers immersed in the areoles, to c. 15 × 10 mm; perianth-segments 5–6 visible from within, white, cream or pale to deep pink, opening widely. Fruit broadly turbinate to depressed-globose, to 8.5 × 8.5 mm, magenta to deep red, shiny. Seed (Barthlott & Hunt 2000: 35, pl. 2.1–2) 1.4 × 0.7 mm, shiny brown. Chromosome number: 2n = 22 (Barthlott 1976).

Disjunct humid forest element: epiphyte or lithophyte, *Mata atlântica*, including *agreste* (*mata de cipó*) and *mata de brejo* (NE Brazil) and *restinga*, sea level to 1200 m, eastern Pernambuco to south-eastern Minas Gerais; common in South-eastern and Southern Brazil; south-eastern Paraguay and north-eastern Argentina. Map 21.
PERNAMBUCO: Mun. Taquaritinga do Norte, alto da serra, 12 Nov. 1983, *V.C. Lima et al.* in CFPE 819 (IPA), Aug. 2001, *Zappi* (obs.).
BAHIA: E & SE Bahia, Mun. Ubaíra, Três Braços (Ilha Formosa), Rio Risada, c. 5 km from the village, 13°30'S, 39°42'W, 16 Jan. 1994, *França et al.* 934 (K, HUEFS); Mun. Vitória da Conquista, 12 km from town towards Itambé, 16 Aug. 2002, *M. Machado* (K, photos); Mun. Camacã (Camacan), road to Canavieiras, 23 Jan. 1971, *T.S. dos Santos* 1408 (CEPEC), Feb. 2003, *J. Jardim* (obs.).
MINAS GERAIS: SE Minas Gerais, Mun. Marliéria, Parque Estadual do Rio Doce, 20 Oct. 1979, *Heringer* 13992 (UB); Mun. Viçosa, Rio Casca, 7 Sep. 1957, *A.G. Andrade* 27 (R); l.c., Distrito de São Miguel, 7 Sep. 1957, *E.C. Rente* 440 (R); Mun. Coronel Pacheco, Est. Experimental Cel. Pacheco, 10 Aug. 1944, *Heringer* 1507 (SP); Mun. Lima Duarte, Parque Estadual do Ibitipoca, 17 July 1987, *R.F. Novelino et al.* 21538 (CESJ); l.c., Serra do Ibitipoca, July 1987, *H. Magalhães* s.n. (R).
ESPÍRITO SANTO: Mun. Vitória, May 1990, *Zappi* (obs.); Mun. Venda Nova, W of town, c. 850 m, *J. Fowlie*, cult. HNT Accn. No. 28676, Apr. 1974 (HNT 2001).

CONSERVATION STATUS. Least Concern [LC] (1); PD=2, EI=1, GD=2. Short-list score (1×5) = 5. Least Concern taking its total range into account.

A common and highly variable plant in South-eastern Brazil, but apparently rare or seldom collected in the *brejo* and *agreste* forests of the Nordeste. Braun & Esteves Pereira

(2002: 79) indicate that *Lepismium* subg. *Lepismium* (ie. *L. cruciforme*) occurs in the states of Rio Grande do Norte, Paraíba, Alagoas, Sergipe and Goiás, but this may only be guesswork and should not be taken as fact in the absence of concrete records.

Its sister species is *L. incachacanum* (Cárdenas) Barthlott, from Bolivia (Cochabamba & La Paz).

12. RHIPSALIS Gaertn.

Fruct. sem. pl. 1: 137, tab. 28 (1788). Type: *R. cassutha* Gaertn. (= *R. baccifera* (J. S. Muell.) Stearn).

Including *Erythrorhipsalis* A. Berger (1920); *Rhipsalis* subg. *Erythrorhipsalis* A. Berger (1920).

Literature: Barthlott (1987), Barthlott & Taylor (1995: 48).

Habit pendent when epiphytic, or erect to decumbent when on rocks; stems segmented, *always ± acrotonically branched* when apex undamaged (basal extension shoots excepted) and lacking sharp spines, though sometimes bristly; stem-segments terete (unribbed), ribbed, angled, winged or flat (2-ribbed), the non-basal segments *deciduous when old*. Areoles small to almost absent (*R. pulchra*) or sunken and conspicuous and/or woolly only after bearing flowers, subtended by small to minute scale leaves; terminal composite areoles often present. Flowers 1–13 per areole, *whitish or tinged yellow or pink*, rarely red, buds erumpent from sunken areoles or developing at the stem surface; pericarp ± naked, tube almost 0, perianth-segments c. 5–18, reflexed or semi-patent (campanulate flowers); stamens strongly exserted or included. Fruit subspherical to ellipsoid, perfectly rounded (never angled), naked, white, orange, pink or purplish, pulp highly mucilaginous. Seeds 1–1.7 mm, black-brown; testa ± smooth.

As currently circumscribed a genus of 35 species (Barthlott & Taylor 1995, Taylor & Zappi 1997) with a centre of diversity in South-eastern Brazil (especially southern Espírito Santo, Rio de Janeiro and São Paulo). The genus is divided into 5 subgenera, each of which is represented in Eastern Brazil, mainly distributed in the *Mata atlântica* and in cloud forests associated with the *campos rupestres*. A minimum of 19 species is treated here (but see Nos 14 & 16, and subg. *Calamorhipsalis* below), of which only 4 taxa (2 spp., 2 subspecies) appear to be endemic, but the group remains poorly collected in North-eastern Brazil, where most of its habitat has been destroyed. Six species (Nos 2, 13–16 & 18) only just enter the area covered, having the major part of their ranges further south and west. Plates 14.2–22.1.

Besides the taxa treated below, Luetzelburg (1925–26, 3: 111) records *R. clavata* F. A. C. Weber (SE Brazil) and *R. robusta* Lem. (= *R. pachyptera* Pfeiff., SE to S Brazil) from eastern Bahia, which probably represent misidentifications (the former may be *R. cereuscula,* the latter *R. russellii* or *R. crispata*). A record for *R. trigona* Pfeiff. from Espírito Santo is likewise assumed to be an error (Barthlott & Taylor 1995: 55). Braun & Esteves Pereira (2002: 79) indicate many state records for *Rhipsalis* subgenera in the Nordeste that seem improbable and can in most cases be put down to guesswork. Luetzelburg also uses the names *R. cribrata* (Lem.) N. E. Br. and *R. platycarpa* Pfeiff., both being of uncertain application.

1. Flower red; flower-buds conspicuously erumpent (splitting the stem's epidermis); stems perfectly terete (S Espírito Santo) . . . . **19. hoelleri**
1. Flower not red and/or flower-buds not erumpent; stems various 2

2. Stems flat, angled, winged, strongly to weakly ribbed or terete with ± raised podaria subtending ± fleshy scale leaves (or cylindric and smooth but with conspicuously erumpent flower-buds exposing floccose areoles after the fall of the fruit); flower-buds etc. 1–many per lateral areole 3
2. Stems perfectly terete or only the shortest, ultimate segments somewhat angled; scale leaves minute, not fleshy or soon scarious; flower-buds etc. 1 per lateral areole or flowers terminal . . . 11

3. Flower-buds conspicuously erumpent from sunken, often very woolly areoles that were absent, hidden by scale-leaves or minute prior to flower development, solitary 4
3. Flower-buds not erumpent or only so on close inspection, the areoles not more obviously woolly after flowering, or flowers 2 or more per areole, at least on the older stem-segments 7

4. Stem-segments of indeterminate growth, branching often only subacrotonic, with numerous discontinuous ribs (S Espírito Santo) . . . . **7a. pacheco-leonis** subsp. **catenulata**
4. Stem-segments of determinate growth, branching acrotonic from terminal composite areoles, terete, angled or with low ± continuous ribs 5

5. Stem-segments terete, not ribbed, but sometimes with raised podaria subtending the scale-leaves (widespread) . . . . **5. floccosa**
5. Stem-segments ribbed or angled 6

6. Stem-segments weakly to more strongly (3–)5-ribbed, ribs continuous (S Espírito Santo) . . . .
. . . . **9. sulcata**
6. Stems 3–4-angled in cross-section, the angles discontinuous and not forming ribs (widespread) . . . . **6a. paradoxa** subsp. **septentrionalis**

7. Fruit white or faintly tinged pink in part; stems mostly flat 8
7. Fruit entirely pink to red or purplish; stems flat or 3–4(–5)-angled/winged 9

8. Ultimate stem-segments very thin, only c. 1 mm thick (excluding midrib) in living material, margins shallowly crenate, the areoles 2–4 mm from the outermost part of margin; flowers mostly 1 per areole, developing during the rainy season (SE Bahia, Espírito Santo) . . **3. oblonga**
8. Ultimate stem-segments stouter, to 2 mm or more thick (excluding midrib) in living material, margins strongly crenate-lobed, the areoles to 5–6 mm from the outermost part of margin; flowers 1–5 per areole, developing during the dry season (Pernambuco, E Bahia) . . . . **4. crispata**

9. Stems mostly pendulous or sprawling, > 25 mm wide, flat or 3–5-winged 10
9. Stems mostly erect, to 25(–30) mm wide, 3–5-angled (S Espírito Santo) . . . . **8. cereoides**

10. Flowers/fruits 1–5 per areole, flower 12–20 mm diam., yellowish, fruit globose to oblong . . . . . .
. . . . **2. elliptica**
10. Flowers/fruits (1–)3–9 per areole, flower to 9 mm diam., whitish, fruit globose . . . . . **1. russellii**

11. Flowers < 8 mm long, with 4–7, patent to reflexed perianth-segments visible from within 12
11. Flowers > 8 mm long, campanulate, with 8–15 concave perianth-segments visible from within . . 14

12. All stems of ± indeterminate growth, not forming composite areoles at apex; flowers and fruits always lateral (widespread) . . . . **10. lindbergiana**
12. Higher order stem-segments of determinate growth, forming composite areoles at apex; flowers/fruits both lateral and terminal 13

13. Pericarpel shorter than perianth; fruit globose to shortly barrel-shaped, to c. 5 × 4 mm; flowers/fruit mostly lateral (SE Brazil) . . . . **11. teres**
13. Pericarpel as long or longer than perianth both in bud and at anthesis; fruit ovoid, c. 7 × 6 mm; flowers/fruit mostly from the terminal composite areoles (Bahia & Pernambuco northwards) . . . . **12. baccifera**

14. Basal extension shoots of indeterminate growth apparently lacking or to only c. 12 cm long, all other stem-segments determinate and < 7 cm; stamens white at base . . . . . . **16. clavata**
14. Basal and higher order extension shoots of indeterminate growth present and > 12 cm long, other stem-segments usually decreasing in size towards distal parts of plant; stamens yellow, red or purplish at base 15

15. Stem-segments > 7 cm long, extension shoots and shorter higher order segments not markedly different, terminal composite areoles mostly lacking; flowers lateral and subterminal (S Minas Gerais, > 1500 m) . . . . **13. pulchra**
15. Stems clearly differentiated into long extension shoots and much shorter secondary segments, some < 7 cm long and all with terminal composite areoles; flowers terminal on the higher order stem-segments 16

16. Ultimate stem-segments swollen and sometimes angled/ribbed; fruit white, rarely with reddish scales (fruit sometimes red outside Brazil) . . . . **17. cereuscula**
16. All stems terete, never angled/ribbed; fruit purple, magenta, red, orange or greenish tinged maroon 17

17. Stems clothed in semi-adpressed, bristle-like spines; fruit bristly . . . . **18. pilocarpa**
17. Stems and fruit naked 18

18. Flower 20–25 × 20 mm; fruit globose-ovoid, c. 10 mm long, unscented . . . . **14. burchellii**
18. Flower c. 15 × 12 mm; fruit truncate, c. 6 mm long, smelling of blackcurrants . . . . **15. juengeri**

Subg. *Phyllarthrorhipsalis* Buxb. (Nos 1–4): seedlings, where known, flattened (2-ribbed) at first; adult stem-segments of determinate size and acrotonically branched (except secondary segments arising laterally from the somewhat indeterminate basal extension shoots); new stem-segments and flower-buds scarcely erumpent; lateral areoles visible before flowering; stem-segments mostly flattened, or with 3–5 continuous angles or wings, relatively thin; flowers lateral and terminal, remaining open day and night, one to many at a time per areole, pericarpel exposed, areoles flowering repeatedly, enlarging and bearing more flowers each time; fruit white, pink or purplish. Central and South America. Type: *R. pachyptera* Pfeiff.

1. Rhipsalis russellii *Britton & Rose*, Cact. 4: 242 (1923). Type (Barthlott & Taylor 1995: 60): Brazil, Bahia, Mun. Jaguaquara [Toca da Onça], 27–29 June 1915, *Rose & Russell* 20106 (NY, lecto.; US, lectopara.).

VERNACULAR NAME. Mandacaru-da-serra.

Sprawling lithophyte to 80 cm, or pendent epiphyte with mostly dichotomous branching, to 200 cm or more; stem-segments usually flattened in epiphytic plants, or mostly 3–4(–5)-angled in lithophytes, irregularly elliptic, often narrowly so, to c. 15 × 3–5.8 cm, 2–3 mm thick away from the vascular midrib; areoles exposed from the start, indented 3–11 mm from the crenate stem margin and to c. 40 mm apart, with conspicuous felt and 0–2 minute bristle-spines. Flower-buds whitish or pinkish, flowers (1–)3–9 per areole, inconspicuous, to c. 7 × 5–9 mm, sometimes scarcely opening, with only the stigma-lobes visible; perianth-segments whitish to greenish white, 4–5 visible from above, boat-shaped and rounded to acuminate at apex, patent to ± reflexed; stamens few, c. 12–14, but sometimes only 2, 2–3 mm, anthers minute, whitish; style c. 2.5 mm, stigma-lobes 3–6, c. 1–1.5 mm, whitish; pericarpel larger than the perianth until just before anthesis, green, c. 2 × 3 mm, sometimes with a broad bract-scale below apex. Fruit densely clustered, globose, to c. 7 mm diam., whitish to pale pink at first, later purplish/pinkish magenta or red.

Widespread humid forest/*campo rupestre* element: epilithic or epiphytic, *campo rupestre, mata de brejo* and *Mata atlântica* and their ecotones with *caatinga-agreste*, 50–1050 m, from the Chapada Diamantina and Serra do Espinhaço eastwards, Bahia, Minas Gerais and Espírito Santo; Central-western Brazil (Goiás and Mato Grosso) and Southern Brazil (Paraná: Guaíra [HU188]). Map 22.

BAHIA: cent. Bahia (Chapada Diamantina), Mun. Lençóis, Morro da Chapadinha, 12°27'24"S, 41°27'10"W, 23 Nov. 1994, *Melo et al.* in PCD 1283 (K, ALCB, HUEFS); Mun. Andaraí, 12°46'59"S, 41°20'58"W, 6 Aug. 2001, *F.R. Nonato et al.* 1015 (HUEFS 55760); l.c., 5 km S of town, near the Rio Paraguaçu, 21 Aug. 1988, *Eggli* 1240 (ZSS, CEPEC); l.c., Igatu, 28 Jan. 1988, *L.P. Bautista* s.n. (HRB); Mun. Mucugê, 0.5 km W of town, 22 Dec. 1988, *Taylor & Zappi* in *Harley* 25600 (CEPEC, SPF, K); Mun. Ituaçu, c. 2 km NE of town, 18 Aug. 1988, *Eggli* 1200 (ZSS); E & SE Bahia, Mun. Jaguaquara [Toca da Onça], 27–29 June 1915, *Rose & Russell* 20106 (US, NY); Mun. Ipiaú, just west of town on road to Jequié, Aug. 2001, *Zappi* (obs.); Mun. Ilhéus, Faz. Almada, 12 Dec. 1967, *S.G. Vinha & A. Castellanos* 4 (CEPEC); l.c., grounds of CEPLAC, 7 Dec. 1988, *T.S. dos Santos* 4433 (ALCB); Mun. Floresta Azul, 500 m, [before 1966], *Martins*, ex coll. Brieger, ESALQ, Piracicaba, SP, cult. R.B.G. Kew (K, spirit coll. 32129, 39004); Mun. Vitória da Conquista, 12 km from town towards Itambé, 16 Aug. 2002, *M. Machado* (K, photos); Mun. Camacã (Camacan), road to Pau Brasil, 19 Jan. 1971, *T.S. dos Santos* 1352 (CEPEC); Mun. Itabela, 16 km S of Eunápolis, 22 Sep. 1966, *P.R. Belém & R.S. Pinheiro* 2644 (CEPEC, IAN); Mun. Teixeira de Freitas, vale do Rio Alcobaça, 12 May 1971, *T.S. dos Santos* 1627 (RB).

MINAS GERAIS: cent. Minas Gerais, Mun. Conceição do Mato Dentro, 11 km S of town on road to Morro do Pilar, 20 Nov. 1989, *Zappi et al.* 195 (HRCB); Mun. Lima Duarte, Parque Estadual de Ibitipoca, 4 Aug. 2003, *Taylor* (obs.).

ESPÍRITO SANTO: reported from the vicinity of Domingos Martins by Rauh (*Rauh & Kautsky* 67568, det. W. Barthlott from living material grown at Univ. Bonn, Germany, n.v.).

CONSERVATION STATUS. Vulnerable [VU B2ab(iii, iv)] (2); area of occupancy estimated to be < 2000 km^2; PD=2, EI=1, GD=2. Short-list score (2×5) = 10. It is markedly disjunct and its habitat has contracted severely and especially in south-eastern Bahia. Elsewhere, it is found within some reasonably secure protected areas (eg. Parque Nacional Chapada Diamantina, BA), but its ecological niches are invariably limited.

A distinctive species with flowers minute in relation to the stem-segments, which somewhat resemble those of the larger-flowered *R. elliptica* (see below) and *R. pachyptera* (Rio de Janeiro to Rio Grande do Sul). Its clustered fruits are quite variable in colour. In cultivation, at least, the collection from Minas Gerais sometimes produces flowers almost devoid of fertile stamens.

A living collection, at the University of Bonn's botanic garden, said to be from Goiás (*P.J. Braun* s.n.), which is yet to flower, seems referable here, and unlocalised living material of this species from 'Mato Grosso' (*Uebelmann* s.n.) has recently flowered and fruited at Zürich, ZSS (Accn No. 82-1444; BONN, photo.). Clearly, its range westwards from Eastern Brazil is incompletely understood at present and could well include Mato Grosso do Sul and eastern Paraguay also, Guaíra (PR) being on the border with these (HU 188).

2. Rhipsalis elliptica *G. A. Lindb. ex K. Schum.* in Martius, Fl. bras. 4(2): 293 (1890). Type (syntypes): Brazil, *Mosén* 3630, *Glaziou* 14859 & *Schenck* 1218 (B†). Lectotype (Barthlott & Taylor 1995: 60): São Paulo, Sorocaba, *Mosén* 3630 (S).

R. chloroptera F. A. C. Weber in Bois, Dict. hort.: 1045 (1898). Type: Brazil, São Paulo, Santos, assumed not to have been preserved.

Pendent to 1 m or more; basal extension shoots to 35 cm; higher order stem-segments mostly flattened, rarely 3-winged, broadly to rather narrowly elliptic, rounded to narrowly truncate at apex, mostly 8–14(–18) × 3–6(–8.5) cm, c. 2–2.5 mm thick away from the vascular midrib, dark green, but often turning a striking shade of magenta or purple when drought-stressed or in strong light, usually at least the margins purplish tinged, margins undulate to broadly crenate; areoles ± inconspicuous or hidden prior to anthesis, glabrous or bearing minute bristles, indented up to c. 3 mm from the crenate stem margin and to 2 cm apart. Flowers 1–5 per areole, buds sometimes pinkish, at anthesis to 14 mm long (from tips of reflexed perianth-segments to stigma-lobes), 12–20 mm diam. before full anthesis, perianth-segments pale to golden yellow, the mostly 5 principal to 9 × 4.5 mm, strongly reflexed hiding the pericarpel, bluntly rounded at apex; stamens conspicuous to 8 mm, white; style white, scarcely exserted from the stamens, stigma-lobes 3–4, to 3.5 × 1 mm, white; pericarpel 3.5–4.0 × 3.5–4.0 mm, pale green. Fruit deep reddish pink, globose to elongate.

Southern humid forest element: epiphyte or lithophyte in *Mata atlântica*, including *mata de neblina*, near sea level to c. 1500 m, central, south-eastern and southern Minas Gerais; common elsewhere in South-eastern and Southern Brazil. Map 22.

MINAS GERAIS: Mun. Catas Altas, Reserva Particular de Patrimônio Natural (RPPN) do Caraça, between Bocaina and Gruta da Bocaina, 2000, *J.O. Rego et al.* (BHCB, Herb. Fund. Zoo-Bot. de Belo Horizonte), l.c., 1 Aug. 2001, *Taylor* (K, photos); Parque Nacional do Caparaó, Córrego do Inácio, 17 Dec. 1988, *Krieger et al.* in FPNC 706 (CESJ); Mun. Lima Duarte/Rio Preto, Chapadão da Serra Negra, 3 Oct. 1959, *Castellanos* 22541 (R).

CONSERVATION STATUS. Least Concern [LC] (1); PD=2, EI=1, GD=1. Short-list score (1×4) = 4. Least Concern taking its total range and abundance into account.

Luetzelburg (1925–26, 3: 111) reports the synonymous *R. chloroptera* F. A. C. Weber from Sergipe, but in the absence of documented material this record must remain rather doubtful and no species of Subg. *Phyllarthrorhipsalis* is currently known from that state.

3. Rhipsalis oblonga *Loefgr.* in Arch. Jard. Bot. Rio Janeiro 2: 36–37, tab. VIII (1918). Type: Brazil, Rio de Janeiro, Ilha Grande, 1915, *Löfgren & Rose* (holo. not found at SP, RB or R; material at US is doubtfully authentic and judged to be *R. elliptica/R. goebeliana*; material at NY is dated 1917). Lectotype (Barthlott & Taylor 1995: 61): Löfgren, l.c., tab. VIII (1918).

R. crispimarginata Loefgr., l.c. tab. IX (1918). Type: l.c., 1915, *Löfgren, Rose & Campos-Porto* (holo. not found at SP, RB or R). Lectotype (designated here): l.c., 11 July 1915, *Rose et al.* 20401 (US).

Pendent to 1 m or more; stem-segments dark to pale green or yellow-green, usually flat (rarely 3-angled), the margins sometimes tinged reddish when exposed to sun, frequently emitting aerial roots along the vascular midrib; basal or sub-basal extension shoots to 10–20 × 1–2 cm, gradually cuneate towards the almost cylindric base, ultimate segments to 12(–23) × 2–6(–7) cm, cuneate to truncate at base, ± truncate to rounded at apex, only c. 1 mm thick away from the vascular midrib, often contorted, margins crisped and tortuous, crenate beneath the areoles; areoles c. 1 mm diam., 14–35 mm apart, sunken 2–4(–10) mm from stem margin, bearing 1 or more fine, bristly, pale, inconspicuous spines. Flowers 1–2(–3) per areole, to 15 × 18 mm when fully expanded (maximum diameter of perianth-segments when patent, but before full expansion when the spreading stamens reach a maximum of 15 mm diam., at which stage perianth-segments are strongly reflexed back and hiding pericarpel); perianth-segments c. 8, the 5 largest to 8 × 4 mm, greenish-yellow or sometimes tipped reddish in bud, turning yellow post-anthesis; stamens to 6–8 mm, white, conspicuous; style to 7 mm, white, stigma-lobes (3–)4–5, to 2–3 × 1 mm, white; pericarpel c. 4 × 4 mm, ovoid (narrowed to apex), greenish yellow. Fruit c. 6–7 × 7–8 mm, entirely white or pale pink at apex.

Humid forest element: epiphyte in perhumid *Mata atlântica*, low elevations, south-eastern Bahia (and Espírito Santo?); to South-eastern Brazil (São Paulo, Serra do Mar). Map 22.

BAHIA: Mun. Una, 8.7 km E of São José [da Vitória?] on road to Una, 4 Mar. 1986, *T.S. dos Santos & E. Judziewicz* 4143 (US, CEPEC, ASE 4495); ibid., 9 km on the road from São José da Buararema, 28 Oct. 1983, *R. Callejas et al.* 1561 (CEPEC); ibid., ± 35–40 km S of Olivença on the Ilhéus–Una road, 15°16'S, 39°4'W, 2 Dec. 1981, *G.P. Lewis & A.M. de Carvalho* 729 (CEPEC, RB, K, NY, M); Mun. Camacã, road to Pau Brasil, 19 Jan. 1971, *T.S. dos Santos* 1352 (CEPEC, RB); Mun. Canavieiras, road BA 001, 4 km N of junction to Santa Luzia, 23 April 2003, *Taylor & Zappi* (obs.); Mun. Teixeira de Freitas, Vale do Rio Alcobaça, 12 May 1971, *T.S. dos Santos* 1627 (CEPEC).

ESPÍRITO SANTO: unlocalized, s.d., *Luetzelburg* 12 (M).

CONSERVATION STATUS. Near Threatened [NT] (1); PD=2, EI=1, GD=2. Short-list score (1×5) = 5. Taking its total range into account the taxon does not qualify for any of the threatened Red List categories, but much of its former habitat in southern Bahia has suffered ongoing destruction since the above cited collections were made in 1971.

R. oblonga is very similar to *R. goebeliana* Backeb. from Bolivia (Yungas) and to *R. occidentalis* Barthlott & Rauh from northern Peru, southern Ecuador and Suriname. They differ from *R. oblonga* in their stem-segments being consistently narrowly cuneate at base and in their flowers with much shorter and less conspicuous stamens.

4. Rhipsalis crispata *(Haw.) Pfeiff.*, Enum. cact.: 130 (1837); Britton & Rose, Cact. 4: 245, fig. 232, tab. XXXV fig. 3 (1923). Type: Brazil, not known to have been preserved. Neotype (Barthlott & Taylor 1995: 61): Brazil, São Paulo, Mun. Rio Claro, Fazenda São José, edge of lake, 1991, *A. Cardoso* in *Zappi* 249, cult. N. P. Taylor, 10 Jan. 1994 (K, in spirit).

Epiphyllum crispatum Haw. in Philos. Mag. Ann. Chem. 7: 111 (1830).
Rhipsalis rhombea sensu Loefgr. in Arch. Jard. Bot. Rio Janeiro 1: 89, tab. 16 (1915) non *Cereus rhombeus* Salm-Dyck.

Like the preceding, but flat or 3-angled, aerial roots produced mainly from the junctions of the stem-segments, extension shoots larger, to 40 cm or more, ± erect or ascending at first; ultimate segments to 7(–10) cm wide, yellow-green to dark green, much stouter, c. 2 mm or more thick, the areoles sunken 5–6 mm or more (to 20 mm exceptionally) into the marginal crenations; flowers 1–5 per areole, 12–20 mm diam., greenish white; stamens shorter than perianth-segments; stigma-lobes 2–4; fruit 6 × 5 mm, white or greenish, ripening slowly. Chromosome number: 2n = 22 (Barthlott 1976).

Humid/subhumid forest element: epiphyte in *mata de brejo* and *Mata atlântica*, 200–800 m, eastern Pernambuco (Mun. Caruaru) and eastern Bahia; South-eastern and Southern Brazil (coast E of Rio de Janeiro between Niterói and Cabo Frio, inner São Paulo, Santa Catarina). Map 22.

PERNAMBUCO: Mun. Caruaru, [Brejo dos Cavalos?], Faz. Caruaru, 28 Nov. 1970, *Andrade-Lima* 70-6200 (IPA 23024, K, photos).

BAHIA: Mun. Amélia Rodrigues, road BR 324, epiphyte on mango tree in front of roadside restaurant, 'Café da Manhã', 1999, *M. Machado*, cult. ibid., Cruz das Almas, 2002 (HUEFS); Mun. Santa Teresinha, Serra da Jibóia, Nov. 2001, *Zappi & Jardim* (K, photos), 7 Aug. 2002, *Taylor & Machado* (K, photos).

CONSERVATION STATUS. Vulnerable [VU B2ab(iii)] (2); area of occupancy estimated to be < 2000 km²; PD=2, EI=1, GD=1. Short-list score (2×4) = 8. Apparently widely distributed, but rarely observed and Vulnerable despite its range in view of the widespread destruction of the habitat it favours. Only the last of the above cited populations is known to be extant. Occurring at only 7 other localities, all subject to anthropogenic change.

The 3 records from NE Brazil are markedly disjunct from other known sites in Rio de Janeiro (Cabo Frio, Silva Jardim, Saquarema & Itacoatiara), São Paulo (Rio Claro & Altinópolis) and Santa Catarina (30 km E of Blumenau), but similar disjunctions are known in Araceae from Pernambuco (eg. *Philodendron eximium* Schott and *P. corcovadense* Kunth, *fide* S. Mayo, pers. comm.). *Rhipsalis crispata* is a species of more markedly seasonal or drier habitats, where it has been found as an epiphyte or lithophyte, both near the coast and far inland, but it can also grow in the sand of the *restinga* (eg. at Cabo Frio, RJ). It is closely related to *R. oblonga*, but has thicker and often broader stem-segments, which are deeply crenate-sinuate at the margin and, in cultivation, appears to flower following a dry or cool period, whereas *R. oblonga* flowers during the warm, humid growing period.

The use of the name *R. crispata* for the plant described here is clearly supported by early herbarium records from the 19th Century in Europe, where it was becoming widespread in cultivation under Haworth's epithet (eg. in the garden at Leuven (Louvain), Belgium, 1837, ex Herb. Martens [BR!], and from Munich, 3 Jan. 1850, ex Herb. Kummer [M!]).

Subg. *Epallagogonium* K. Schum. (including Subg. *Trigonorhipsalis* A. Berger and Subg. *Goniorhipsalis* K. Schum.) (Nos 5–9): stems ± angled to ribbed or terete with raised podaria, never flattened (except in the first shoot of seedlings of No. 5 and very rarely in shade forms of No. 8); all adult stem-segments determinate, except in *R. pacheco-leonis*; flower-buds 1 per areole in the species treated below (except *R. cereoides*), strongly erumpent and the pericarpel sunken into the stem (less so in Nos 8–9), areoles sometimes flowering > once. South America. Type: *R. paradoxa* (Pfeiff.) Salm-Dyck.

5. Rhipsalis floccosa *Salm-Dyck ex Pfeiff.*, Enum. cact.: 134 (1837). Type: a living plant, presumed not to have been preserved. Neotype (Barthlott & Taylor 1995: 55): Brazil, Bahia, Mun. Ilhéus, grounds of CEPLAC, 3 Mar. 1974, *C. Erskine* 164, cult. R.B.G. Kew, 11 Nov. 1992 (K, in spirit).

Lepismium floccosum (Salm-Dyck ex Pfeiff.) Backeb. in Backeberg & Knuth, Kaktus-ABC: 155 (1935, publ. 1936).

Pendent to 3 m or more; basal stems to 1.5 cm thick, new stem-segments arising singly or in groups of 2–5 from the apex of existing segments, at maturity to 30 cm long, pliable and rubbery in feel, but not flaccid; primary stem of seedling flattened, subsequent segments angled, mature growth terete but commonly with prominent podaria subtending the scale leaves, sometimes falsely ribbed in dried material, c. 6–10 mm thick, grey-green, matt, minutely roughened, podaria to c. 25 mm apart; areoles sunken and hidden until flowering, to 4 mm diam. and floccose after bearing fruits. Flowers to 12–25 mm diam. or slightly larger, perianth-segments pale green to greenish white, pericarpel immersed in the sunken, woolly areole. Unripe fruits light green, sometimes dark reddish above, ripe fruit turbinate, c. 7 × 6 mm, white, with a faint reddish ring around the perianth scar.

CONSERVATION STATUS. Least Concern [LC]; found in various countries in South America (see under subsp. *floccosa* below) and frequently seen in Eastern Brazil.

Together with *R. lindbergiana*, the most widespread and commonest *Rhipsalis* species in Eastern Brazil, from Pernambuco southwards, but rather variable and requiring further study in the field. The following, which includes the type of the species (subsp. *floccosa*), is a frequent plant of the *Mata atlântica* and *brejo* forest.

5a. subsp. **floccosa**

Ultimate stem-segments to 10 mm thick, podaria subtending the leaf-scales usually very prominent. Flowers small, to c. 12 mm diam. Fruit whitish, often with a reddish ring around the perianth remains, half-immersed in stem.

Widespread humid forest element: epiphytic or epilithic in *agreste* and *Mata atlântica*, including *mata de brejo,* near sea level to c. 1200 m, eastern Pernambuco, eastern foot of the Chapada Diamantina and coastal Bahia to Espírito Santo; re-appearing south-westwards in the interior of South-eastern Brazil (Minas Gerais and São Paulo), where frequent; replaced by subsp. *oreophila* in the East Brazilian Highlands, by subsp. *pulvinigera* in mountains and coastal zones of South-eastern and Southern Brazil, by subsp. *hohenauensis* in eastern Paraguay and north-eastern Argentina, by subsp. *tucumanensis* in the eastern Andes of Argentina, Bolivia and Peru (Junín) and by subsp. *pittieri* in northern Venezuela. Map 23.

PERNAMBUCO: Mun. Taquaritinga do Norte, epiphyte on fruit trees, Aug. 2001, *Zappi* (obs.); Mun. São Lourenço da Mata, Engenho São Bento, 26 Dec. 1963, *D. Andrade-Lima* 63-4190 (IPA); Mun. Moreno, Tapera, 12 Mar. 1931, *B. Pickel* 2560 (IPA), 23 Aug. 1934, *B. Pickel* 3630 (IPA, F); Mun. Brejo da Madre de Deus, 19/20 Aug. 1980, *A. Perrucci & M.A. Maia Filho* 19 (IPA), 5 Feb. 1965, *D. Andrade-Lima* 65-4300 (IPA); Mun. Caruaru, Brejo dos Cavalos, 28 Oct. 1971, *D. Andrade-Lima* 71-6585 (IPA), 26 Nov. 1991, *F.A.R. Santos* 14 (HUEFS, PEUFR); Mun. Bezerros, Serra Negra, 8°10'S, 35°46'W, 17 Oct. 1999, *Krause & Liebig* 250 (PEUFR); Mun. Cabo de Santo Agostinho, Gùrjaú, 8 Dec. 1964, *D. Andrade-Lima* 64-4268 (IPA); Mun. Águas Belas, Serra do Cumanati, 29 Dec. 1969, *D. Andrade-Lima* 69-5627 (IPA); Mun. Brejão, 7 km N of Teresinha, 1 km S of junction to Brejão, 13 Feb. 1991, *Taylor & Zappi* 1628 (UFP, HRCB, K, ZSS).
SERGIPE: Mun. Siriri, 1 Sep. 1983, *G. Viana* 750 (ASE 3409); Mun. Itabaiana, forest at edge of Estação Ecológica da Serra da Itabaiana, 5 km W of Areia Branca on road BR 235, 10°41'15"S, 37°23'20"W, 22 Jan. 1992, *W.W. Thomas et al.* 8894 (CEPEC, HNT, HUEFS); Mun. Estância, 0.5 km N of the Rio Piauí, 10 Feb. 1991, *Taylor & Zappi* 1615 (ASE, HRCB, K, ZSS).
BAHIA: Cent. Bahia, Mun. Mundo Novo, 6 km E of town on road BA 052, 26 Apr. 1992, *Taylor & Zappi* (obs.); E & SE Bahia: Mun. Alagoínhas, 24 km S of Inhambupe, 15 Feb. 1991, *Taylor & Zappi* 1634 (CEPEC, HRCB, K, ZSS); Mun. Araçás, 2 km S of town, 10 Feb. 1991, *Taylor & Zappi* 1612 (CEPEC, HRCB, K, ZSS); Mun. Feira de Santana, Serra de São José, Nov. 2001, *Zappi* (obs.); Mun. Conceição do Jacuípe, 87 km NW of Salvador on road to Feira de Santana, 12 Mar. 1989, reported in Barthlott & Taylor (1995: 53, pl. 17); Mun. Muritiba, Faz. 8 de Dezembro, epiphyte on *Syagrus coronata*, 6 Aug. 2002, *Taylor & Machado* (K, photos); Mun. Nazaré ['Nazareth'], 30 June 1915, *Rose & Russell* 20145 (US); Mun. (?), Ilha de Itaparica, Sep. 1974, *A. Leal Costa* s.n. (ALCB 022913); l.c., Mun. Vera Cruz, Mar Grande, 2 Jan. 1992, *C.S.N. Guimarães* 8 (ALCB); Mun. Ubaíra, 13°30'S, 39°42'W, 16 Jan. 1994, *F. França* 932 (HUEFS); Mun. Valença, exit to the town from road BR 101, Aug. 2001, *Zappi* (obs.); Mun. Jitaúna/Jequié, c. 30 km NW of Ipiaú, Aug. 2001, *Zappi* (obs.); Mun. Ipiaú, road to Ibirataia, 30 Oct. 1970, *T.S. dos Santos* 1244 (CEPEC); Mun. Ilhéus, grounds of CEPEC, Km 22 on the Ilhéus–Itabuna road (BR 415), 2 Sep. 1981, *J.L. Hage & H.S. Brito* 1279 (CEPEC), 7 Dec. 1988, *T.S. dos Santos* 4444 (ALCB, HUEFS); Mun. Vitória da Conquista, 12 km from town towards Itambé, 16 Aug. 2002, *M. Machado* (K, photos); Mun. Barra do Choça, Barra Nova, Morro do Criminoso, c. 15°S, 44°33.5'W, Sep. 2003, *M. Machado* (obs.); Mun. Belmonte, Estação Experimental de CEPLAC, 18 Sep. 1970, *T.S. dos Santos* 1109 (CEPEC).
ESPÍRITO SANTO: Mun. São Mateus, Reserva Biológica de Sooretama, Lagoa do Macuco, 15 May 1977, *Martinelli et al.* 2084 (RB, UEC); Mun. Santa Teresa, road Itarana – Santa Teresa, 19°55'50"S, 40°41'7"W, 24 Nov. 1999, *Zappi et al.* 424 (K, UEC); l.c., Fazenda Santa Lúcia, 19°57'24"S, 40°32'37"W, 25 Nov. 1999, *Zappi et al.* 439 (K, UEC); Mun. Santa Leopoldina, road to Santa Teresa, estate of Floriano Bremekamp, 20°3'24"S, 40°32'55"W, 25 Nov. 1999, *Zappi et al.* 450 (K, UEC).

CONSERVATION STATUS. Least Concern [LC] (1); PD=1, EI=1, GD=2. Short-list score (1×4) = 4.

Although much of its former habitat has been destroyed, it has managed to colonize cultivated trees such as mangoes, as well as other long-lived species planted as street trees and in village squares.

Variable in the length and especially thickness of its stem-segments.

5b. subsp. **oreophila** *N. P. Taylor & Zappi* in Cact. Consensus Initiatives 6: 7 (1998). Holotype: Brazil, Minas Gerais, mountains E of Monte Azul, 1964, *Ritter* 1247 (SGO 125604, lectotype of *R. monteazulensis* F. Ritter, l.c. infra).

Rhipsalis monteazulensis F. Ritter, Kakt. Südamer. 1: 42 (1979). Type (Taylor & Zappi, l.c.): as above.

Ultimate stem-segments to 6 mm thick, almost perfectly cylindrical, without obvious podaria subtending the scale-leaves. Flowers small, to c. 12 mm diam., greenish white. Fruit white, scarcely immersed in stem.

Northern *campo rupestre* element: epiphytic or epilithic in *mata de neblina (capão de mato)*, *campo rupestre*, c. 1200–1750 m, Chapada Diamantina and Serra do Espinhaço, Bahia and northern Minas Gerais. Endemic to the core area of Eastern Brazil. Map 23.

BAHIA: cent. Bahia (Chapada Diamantina), Mun. Lençóis, W towards Caeté-Açu, Serra Larga, 23 May 1987, *B. Stannard et al.* in CFCR 10931 (SPF, K, B, ZSS); Mun. Abaíra, below Tanquinho, 13°17'S, 41°53'W, 1480 m, on rocks, 14 Sep. 1992, *W. Ganev* 1094 (HUEFS, K); l.c., Distrito de Catolés, Mata do Tijuquinho, 19 Apr. 1998, *L.P. de Queiroz* 5001 (HUEFS); Mun. Rio de Contas, Pico das Almas, 1750 m, on rocks, 12 Nov. 1988, *Harley et al.* 26400 (CEPEC, SPF, K); l.c., NW of Campo do Queiroz, 1500 m, 13 Dec. 1988, *Taylor* in Harley 25566 (CEPEC, SPF, K).

MINAS GERAIS: N Minas Gerais (N Serra do Espinhaço), Mun. Monte Azul/Espinosa, c. 12 km E of Monte Azul, E of Vila Angical, Serra Geral, E side near top of serra on path to village of Gerais, epiphyte, 1220 m, 28 Jan. 1991, *Taylor et al.* 1476 (BHCB, HRCB, K, ZSS).

CONSERVATION STATUS. Least Concern [LC] (1); PD=1, EI=1, GD=1. Short-list score (1×3) = 3. Only the southernmost of the 4 sites listed above is undergoing habitat modification. The northernmost site is within a national park.

The forms of this species from above 1200 metres in the northern sector of the East Brazilian Highlands (Chapada Diamantina, BA, and northern Serra do Espinhaço, MG) have almost perfectly cylindrical stems devoid of podaria and seem sufficiently distinct to be treated as subspecifically different from those of the lowland forests and South-eastern *campos rupestres*. They have smaller flowers than plants from the latter area, which are referred to the following:

5c. subsp. **pulvinigera** *(G. A. Lindb.) Barthlott & N. P. Taylor* in Bradleya 13: 55 (1995). Lectotype (Barthlott & Taylor, l.c.): Gartenflora 38: 184, fig. 34 (1889).

R. pulvinigera G. A. Lindb. in Gartenflora 38: 182 (1889). *Lepismium pulvinigerum* (G. A. Lindb.) Backeb. in Backeb. & F. Knuth, Kaktus-ABC: 155 (1935, publ. 1936).
Rhipsalis gibberula F. A. C. Weber in Rev. Hort. 64: 426 (1892). *Lepismium gibberulum* (F. A. C. Weber) Backeb. in Backeb. & F. Knuth, Kaktus-ABC: 155 (1935, publ. 1936). Type: see Barthlott & Taylor, l.c. 57.

VERNACULAR NAME. Dedinho.

Like subsp. *floccosa*, but with generally smaller stem-segments and larger flowers, 18–20(–30) mm diam.; fruit deep magenta-pink or pure white.

Southern humid forest element: epiphytic and epilithic, to 1850 m, southern Serra do Espinhaço, Serra da Mantiqueira and Serra do Caparaó, central and southern Minas Gerais to southern Espírito Santo; abundant in South-eastern and Southern Brazil (to Rio Grande do Sul). Map 23.

MINAS GERAIS: cent.–S Minas Gerais, Mun. Santana do Riacho, Serra do Cipó, 20 Oct. 1973, *A.B. Joly et al.* in CFSC 4665 (UEC, SP), 20 Feb. 1968, *H.S. Irwin et al.* 20574 (UB), 20 June 1987, *Zappi & C. Kameyama* in CFSC 10189 (SPF); l.c., Serra das Bandeirinhas, 27 July 1991, *Giulietti et al.* in CFSC 12513 (SPF, K); l.c., caminho para Cachoeira das Flores, 9 Sep. 1987 *J. Prado et al.* in CFSC 10644 (SPF, K); Mun. Caeté, Serra da Piedade, 29 Oct. 1971, *L. Anne & Strang* 1920/1461 (GUA), 15 Jan. 1971, *Irwin et al.* 30425 (NY), Nov. 1915, *Hoehne* 6352 (SP, R), 24 Feb. 1987, *Zappi et al.* in CFCR 10343A (HRCB), 20 July 1987, *Zappi et al.* in CFCR 11167 (SPF), 3 Nov. 1988, *Taylor & Zappi* in Harley 25499 (K, SPF); Mun. Catas Altas, Serra do Caraça, Gruta do Padre Caio, 23 May 1987, *Zappi & Scatena* in CFCR 10931 (SPF), 11 Sep. 1990, *Zappi et al.* 241 (K, HRCB, BHCB, SPF); Mun. Lima Duarte, Parque Estadual do Ibitipoca, 15 May 1970, *Krieger* 8589 (CESJ), 27 July 1991, *Zappi* 261 (K, SPF, CESJ); l.c., Gruta do Fugitivo, 16 Oct. 1986, *H. C. Sousa et al.* 9829 (BHCB, K, SPF); SE Minas Gerais, Parque Nacional do Caparaó, 15 Oct. 1988, *M. Brugger et al.* in FPNC 356 (CESJ, HRCB).

ESPÍRITO SANTO: Mun. Domingos Martins, May 1986, *Rauh* 67567, cult. Univ. Bonn, Germany, Accn. No. 04504 (K), also observed in flower at Bonn, 18 March 2000.

CONSERVATION STATUS. Least Concern [LC] (1); PD=1, EI=1, GD=2. Short-list score (1×4) = 4. Widespread and common outside of the area treated here.

Typical *R. floccosa* subsp. *pulvinigera* is distinguished from its northern relatives by its generally smaller and more evenly sized, somewhat shiny, smoother, purple mottled stem-segments (usually to only 20 cm), larger, more expanded flowers (18–25 mm diam. or more), less woolly flower-bearing areoles and strongly exserted fruits, usually turning bright pinkish red or pure white when ripe. Beyond Eastern Brazil it ranges from Rio de Janeiro (Serra dos Órgãos) to Rio Grande do Sul (Mun. Guaíba) and replaces subsp. *floccosa* in the coastal regions of South-eastern and Southern Brazil. They can be reliably distinguished only on the basis of fertile material. The form of subsp. *pulvinigera* from the Reserva Particular de Patrimônio Natural (RPPN) do Caraça is unusual in having ± erect stems and very large flowers.

6. Rhipsalis paradoxa *(Salm-Dyck ex Pfeiff.) Salm-Dyck*, Cact. Hort. Dyck. 1849: 228 (1850). Type: Brazil, assumed not to have been preserved. Neotype (Barthlott & Taylor 1995: 57): Lem. in Hort. Universel 2: tab. 50 (1840).

Lepismium paradoxum Salm-Dyck ex Pfeiff., Enum. cact.: 140 (1837). *Hariota paradoxa* (Salm-Dyck ex Pfeiff.) Kuntze, Revis. gen. pl. 1: 263 (1891). *H. alternata* Lem. in Hort. Universel 2: 39, tab. 50 (1840), nom. illeg. *Rhipsalis alternata* (Lem.) Lem., Cactées: 80 (1868). Type: as above.

CONSERVATION STATUS. Least Concern [LC]; the widespread subsp. *paradoxa* is included in various protected areas.

Only the following heterotypic subspecies is found in the area covered here:

6a. subsp. **septentrionalis** *N. P. Taylor & Barthlott* in Bradleya 13: 57 (1995). Holotype: Brazil, SE Bahia, Mun. Juçari, 250 m, [before 1966], *Martins* in coll. Brieger, Piracicaba, SP, cult. Royal Botanic Gardens, Kew, Accn. No. 1966.48946, 18 Nov. 1991 (K in spirit; iso. dried).

Pendent to 3 m or more; branching strictly acrotonic from terminal composite areoles; stems of seedlings of indeterminate growth, with 4–5 continuous ribs bearing closely set areoles and fine bristly spines, new stem-segments arising singly or in groups of 2–5 from the apex of existing segments, to c. 30 × 0.7–1.1 cm, determinate, with prominent winglike angles subtending each scale-leaf/areole, these alternating giving the stem a ± triangular cross-section, epidermis yellow-green to pale grey-green, matt, markedly roughened to the touch, areoles sunken and hidden before flowering, subtended by minute, adpressed scale leaves positioned 1.5–3.5 cm apart, areole scar to 5 mm diam. after bearing fruit, but almost glabrous. Flower-buds strongly erumpent, flowers yellowish, 15 mm diam., perianth-segments to 7 × 2–3 mm; stigma-lobes 5, white. Fruit turbinate, to 7.5 × 7.5 mm, white.

Humid forest element: epiphyte in *Mata atlântica*, low elevations to c. 950 m, eastern Pernambuco, eastern Bahia, central-eastern Minas Gerais and Espírito Santo. Endemic to Eastern Brazil. Map 23.
PERNAMBUCO: Mun. Primavera, Apr. 1988, *P.J. Braun* s.n., cult. Univ. Bonn, Germany, 1995 (K).
BAHIA: Mun. Conceição do Jacuípe, 87 km NW of Salvador on road to Feira de Santana, 12 Mar. 1989, *Supthut* 8900 (ZSS); Mun. Barra do Choça, Barra Nova, Morro do Criminoso, c. 15°S, 40°33.5'W, Sep. 2003, *M. Machado* (obs.); Mun. Juçari, [before 1966], *Martins*, ex coll. Brieger, cult. R. B. G. Kew, 18 Nov. 1991 (K); Mun. Camacã, near the town, coll. before 1979, *G. Daniels*, cult. HNT Accn No. 41403, May 1982 (HNT 5191).
MINAS GERAIS: Mun. Rio Vermelho, between the town and Pedra Menina, [Mun. Coluna, *fide* Uebelmann (1996)], June 1993, *Uebelmann* 1507, cult. Univ. Bonn, Germany, 1996 (K).
ESPÍRITO SANTO: Mun. Linhares, 6–8 km do Ramal do Lado leste, próximo ao Vale do Rio Doce, 2 Nov. 1971, *T.S. dos Santos* 2047 (CEPEC); Mun. Domingos Martins, May 1986, *Rauh* 67565, cult. Univ. Bonn, Germany, 1995 (K).

CONSERVATION STATUS. Endangered [EN B2ab(iii)] (3); PD=1, EI=1, GD=1. Short-list score (3×3) = 9. Endangered in view of continuing forest destruction throughout the region of its occurrence, in which it is apparently a rarity. Only tiny fragments of its original habitat remain (< 500 km²) and at least 3 of the above-cited populations are likely to have been destroyed.

This taxon differs from subsp. *paradoxa* (Rio de Janeiro to Santa Catarina) in its consistently narrower vegetative parts and darker flowers. Its northernmost habitat, in Pernambuco, is illustrated in Braun & Esteves Pereira (2002: 20, Abb. 30).

7. Rhipsalis pacheco-leonis *Loefgr.* in Arch. Jard. Bot. Rio Janeiro 2: 38 (1918) ('pacheco-leoni'). Type: region of Cabo Frio, 1915, *Campos-Porto & Rose* (not found at RB or SP). Lectotype (Barthlott & Taylor 1995: 59): Rio de Janeiro, Iguaba Grande, 1915, *Rose [& Campos-Porto]* 20707 (US).

Lepismium pacheco-leonis (Loefgr.) Backeb. in Backeberg & F. Knuth, Kaktus-ABC: 154 (1935, publ. 12 Feb. 1936) ('pacheco-leonii').

CONSERVATION STATUS. Data Deficient [DD]; requiring survey work in its 6 known locations, but decline in these can be expected in view of their proximity to urban and touristic developments.

Only the following heterotypic subspecies is found in the area covered here:

7a. subsp. **catenulata** *(Kimnach) Barthlott & N. P. Taylor* in Bradleya 13: 59 (1995). Holotype: Rio de Janeiro, Mun. Nova Friburgo, '2000 ft below Sansão', 1976, *Fowlie* s.n. (HNT, n.v.; US, iso.).

Rhipsalis paradoxa var. *catenulata* Kimnach in Cact. Succ. J. (US) 64: 91 (1992).

Like *R. paradoxa* subsp. *septentrionalis*, but smaller in all its parts, stem-segments indeterminate, to only c. 0.7 cm diam., branching sub-acrotonic, terminal composite areoles only rarely produced, the winglike stem angles more crowded and compressed, giving the stem a ± quadrangular cross-section; areoles/ scale-leaves to only 1 cm apart, some bearing bristle-spines, or these developing post-anthesis, the scar left after fruit production being c. 3 mm diam.; flowers < 10 mm diam., brownish salmon without, pinkish salmon to orange within, anthesis associated with conspicuous nectar production from extra-floral nectaries on pericarpel (ant-pollination syndrome?); fruit to 5 × 4.5 mm, white with a pink circle marking the perianth scar.

Southern humid forest element: at c. 900 m, southern Espírito Santo (Domingos Martins); Rio de Janeiro (Mun. Nova Friburgo). Endemic to South-eastern Brazil. Map 23.
ESPÍRITO SANTO: Mun. Domingos Martins, May 1986, *Rauh & Kautsky* 67560, cult. Univ. Bonn, Germany, Accn. No. 04502 (K, BONN).

CONSERVATION STATUS. Data Deficient [DD]; known from only two localities that await detailed surveys.

Although not studied by the authors in habitat, living material from both localities has been observed in cultivation.

The homotypic subspecies, *R. pacheco-leonis* subsp. *pacheco-leonis,* is known from the regions of Macaé, Cabo Frio (Iguaba Grande & Búzios) and Pedra da Gávea, Rio de Janeiro at low elevations. It has often rather weakly developed stem angles, bears deep pinkish fruit and was confused with *R. dissimilis* by Britton & Rose (1923).

8. Rhipsalis cereoides *(Backeb. & Voll) Backeb.*, Kakteen Pflanzen Samen 1927–1937, 10 Jahre Kakteenforschung [cat.]: 39 (1937–1938). Type: Brazil, *Voll*, assumed not to have been preserved. Neotype (Barthlott & Taylor 1995: 61): Rio de Janeiro, Mun. Maricá, Itaipu-Açu, Apr. 1936, *Voll & Brade* s.n. (RB 10258).

Lepismium cereoides Backeb. & Voll in Backeberg & F. Knuth, Kaktus-ABC: 411 (1935, publ. 12 Feb. 1936).

Shrubby, forming dense clumps of ± upright axes (or pendulous when rarely epiphytic); stem-segments all of determinate size, to 14 × 3 cm, (2–)3–5-angled, new stem-segments arising singly or in groups of 2–3(–4) from the apex of existing segments. Flowers 1–4 per areole, 18–20 mm diam. or more, whitish, with 6 or more perianth-segments visible from above. Fruit ovoid-oblong, to 10 × 7 mm, but usually smaller, pink to whitish. Chromosome number: $2n = 22$ (Barthlott 1976).

Southern humid forest (inselberg) element: lithophyte on gneissic inselbergs (rarely epiphytic on nearby trees), southern Espírito Santo (Domingos Martins); Rio de Janeiro (both sides of the Baía de Guanabara). Map 23.
ESPÍRITO SANTO: Mun. Domingos Martins, May 1986, *Rauh & Kautsky* 67557, cult. Univ. Bonn, Germany, Accn. No. 04462 (K).

CONSERVATION STATUS. Vulnerable [VU B2ab(iii)] (2); area of occupancy estimated to be < 2000 km². PD=2, EI=1, GD=1. Short-list score (2×4) = 8. Altogether known from < 10 locations, some being close to the cities of Rio de Janeiro and Niterói and affected by urban expansion and coastal tourism, resulting in ongoing decline.

This species has not been seen at the site noted above, but the authors were able to study an extensive population in fruit at Itacoatiara, Rio de Janeiro, near the type locality, in 1996.

9. Rhipsalis sulcata *F. A. C. Weber* in Bois, Dict. hort.: 1046 (1898). Type: cultivated material of unknown wild origin, assumed not to have been preserved (P, not found). Neotype (Barthlott & Taylor 1995: 60): cult. New York Bot. Gard., Mar. 1912 (fls), received from Simon, ex Muséum [Nat. d'Histoire Naturelle] Paris [where Weber worked], in 1902 (NY).

Stems ± pendent, apparently all of determinate growth, the basal segments to c. 8–10 mm diam., new stem-segments arising singly or in groups of 2–5 from the apex of existing segments, vigorous segments to 250 × 6–7 mm, matt grey-green, weakly but recognizably (3–)5-angled/ribbed whether turgid or drought-stressed, shorter segments sometimes almost terete, the weakest ultimate segments tapering to c. 3–4 mm diam. near their apices, scale-leaves minute, upon raised podaria, often subtending inconspicuous, pale bristles to 1 mm, podaria/scales to 70 mm apart on the same angle/rib, 2–4 clustered around the small, terminal composite areoles at the stem apices. Flowers like those of the preceding species but c. 15 mm diam. Fruit pinkish.

Southern humid forest element: ecology and range poorly understood; only the following collection of documented provenance has been seen:
ESPÍRITO SANTO: Mun. Domingos Martins, 8 May 1986, *Rauh & Kautsky* 67562, cult. Univ. Bonn, Germany, Accn. No. 04490, 1996 (K).

CONSERVATION STATUS. Data Deficient [DD]; see below.

Until recently known only in cultivation, where it was misidentified as *R. micrantha* (Andes to Central America). It seems to be a member of the *R. pacheco-leonis* / *R. pentaptera* A. Dietr. complex, although it strongly resembles *R. floccosa*, differing most obviously in its clearly angled/ribbed stem-segments, less conspicuously erumpent flower-buds and scarcely woolly fertile areoles post-anthesis. It is possible that it was first collected in the vicinity of Rio de Janeiro and may not be an endemic of the core area of Eastern Brazil, although its range is presently documented only from the above-cited collection from southern Espírito Santo.

Scheinvar (1985) misapplies the name *R. sulcata* to specimens of the very different *R. trigona* Pfeiff. (São Paulo to Santa Catarina).

Subg. *Rhipsalis* (Nos 10–12): seedlings 3–6-ribbed/angled; adult stems usually terete, branching ± acrotonic, but producing indeterminate, greatly elongated, basal extension shoots, the ultimate stem-segments usually the shortest; new stem-segments and flower-buds inconspicuously erumpent; flowers lateral and sometimes terminal, one per areole, areoles flowering once only, pericarpel fully exposed; fruits as above. Neotropics and palaeotropics eastwards to Sri Lanka.

10. Rhipsalis lindbergiana *K. Schum.* in Martius, Fl. bras. 4(2): 296 (1890). Type: Rio de Janeiro, based on various syntypes (B†, LE, P, S, C etc.). Lectotype (here designated, replacing that by Barthlott & Taylor (1995), which is inadmissible under ICBN Art. 9.10): *Widgren s.n.* (S). Amongst the 6 collections cited in the protologue is that of the Swede, 'Widgren 84'. At Stockholm (S) Schumann has annotated *Reidel* 84 and *Widgren s.n.* as his new species and it is here assumed that his later reference to 'Widgren 84' is an error caused by the omission of the word 'Reidel' from the list in the protologue.

R. densiareolata Loefgr. in Arch. Jard. Bot. Rio Janeiro 2: 41, tab. XII (1918). Type: Rio de Janeiro, Tijuca, 1915, *Rose & Campos-Porto* (holo. not found RB, R, SP, US). Lectotype (designated here): Löfgren, l.c., tab. XII (1918).

VERNACULAR NAME. Enxerto.

Pendent to 4 metres or more, forming a dense mass of closely adjacent stems; branching sub-acrotonic, new stem-segments arising singly or in groups of 2–5 from near the apex of existing segments, of indeterminate growth, lacking terminal composite areoles, to 60–90 × 3–6(–12) mm, cylindric, but sometimes longitudinally ridged when drought-stressed or when dried for the herbarium, light grey-green to dark green; the soon scarious scale-leaves marking the position of the hidden areoles often arranged in whorls, these to c. 10 mm apart; areoles apparent only after flowering, developing sparse wool and usually a few bristle-spines, areole scar c. 1.5 mm diam. Flowers to 9 mm diam.; perianth-segments 5–9 visible from above the flower, greenish white, at full anthesis strongly reflexed to show off the stamens; stigma-lobes 3–4, white, c. 1 mm; pericarpel green to reddish. Fruit 3–5 × 2.5–4 mm, white or pink.

Disjunct/widespread humid/subhumid evergreen forest element: epiphyte or less commonly a lithophyte, in high *restinga* forest, *Mata atlântica, mata de brejo* and rarely at edges of *caatinga-agreste* and in *campo rupestre*, near sea level to c. 1000 m, southern Pernambuco southwards, mostly within 150 km of the coast but reaching the northen parts of the Chapada Diamantina and Serra do Espinhaço (MG), rarely on the latter's western flanks in the south; to South-eastern Brazil (to W Rio de Janeiro & SE São Paulo). Map 24.

PERNAMBUCO: Mun. Inajá, Reserva Biol. Serra Negra, topo da serra, c. 1000 m, 14 Sep. 1995, *A. Laurênio et al.* 159 (K, PEUFR), 9/10 Dec. 1995, *M. Tschá et al.* 369, 387 (K, PEUFR); Mun. Brejão, 7 km N of Terezinha, 1 km S of turning to Brejão, 13 Feb. 1991, *Taylor & Zappi* 1629 (UFP, HRCB, K, ZSS).

SERGIPE: Mun. Campo de Brito, Tapera da Serra, 11 Nov. 1981, *G. Viana* 202 (IPA, ASE 1658).

BAHIA: cent.-N Bahia, Mun. Saúde, 68.5 km S of Campo Formoso on road to Jacobina, 12 Jan. 1991, *Taylor et al.* 1396 (CEPEC, HRCB, K, ZSS); Mun. Jacobina, Barracão de Cima, 11°1'7"S, 40°32'43"W, 6 July 1996, *H.P. Bautista et al.* in PCD 3465 (K, ALCB, HUEFS); Mun. Morro do Chapéu, as epiphyte and lithophyte, both sides of road BA 052, nr Pousada Ecológica das Bromélias, 1000 m, 3 Aug. 2002, *Taylor & Machado* (K, photos); Mun. Mundo Novo, 2 km E of town on road BA 052, 26 Apr. 1992, *Taylor & Zappi* (obs.); NE & E Bahia, Mun. Candeal, 8 July 1994, *F.A.R. Santos* 92 (HUEFS); Mun. Alagoinhas, 24 km S of Inhambupe, 15 Feb. 1991, *Taylor & Zappi* 1635 (CEPEC, HRCB, K, ZSS); Mun. Entre Rios, 12 Feb. 1991, *Taylor & Zappi* 1614 (HRCB, K, ZSS); Mun. Mundo Novo, 2 km E on road BA 052, 26 Apr. 1992, *Taylor & Zappi* (obs.); Mun. Baixa Grande, 1.5 km E of junction of road between Rui Barbosa and Juazeiro towards Baixa Grande, 9 Feb. 1991, *Taylor et al.* 1606 (K, HRCB, ZSS); Mun. Ipirá, 24 km E of town towards Feira de Santana on road BA 052, 9 Feb. 1991, *Taylor et al.* 1611 (K, HRCB, ZSS); Mun. Rui Barbosa, 7 km N of town, *Ritter* 1245 (SGO 125541, *fide* Eggli et al. 1995: 503–504); Mun. Feira de Santana, 12°15'16"S, 39°2'46"W, 7 Dec. 1992, *L.P. Queiroz et al.* 2926 (K, CEPEC, HUEFS); l.c., 12°18'S, 39°W, 6 Feb. 1997, *F. França* 2079 (HUEFS); Mun. Conceição do Jacuípe, 87 km NW of Salvador on road to Feira de Santana, 12 Mar. 1989, reported in Barthlott & Taylor (1995: 53, pl. 17); Mun. Cachoeira, Vale dos Rios Paraguaçu e Jacuípe, 39°5'W, 12°32'S, Nov. 1980, Grupo Pedra do Cavalo 931 (ALCB, IPA, HRB); Mun. Cachoeira/São Felix, 13 Feb. 1920, *Zehntner* s.n. (US); Mun. Cruz das Almas, 18 June 1915, *Rose & Russell* 19896 (US, NY); Mun. Lajedo do Tabocal, 25.5 km W of road BR 116 towards Maracás, 5 Feb. 1991, *Taylor et al.* 1556 (CEPEC, HRCB, K, ZSS); Mun. Jaguaquara [Toca da Onça], 27 June 1915, *Rose* 21408 (US); Mun. Ipiaú, Aug. 2001, *Zappi* (obs.); Mun. Vitória da Conquista, 12 km from town towards Itambé, 16 Aug. 2002, *M. Machado* (obs.); Mun. Juçari (Jussari), 27 May 1966, *R.P. Belém & R.S. Pinheiro* 2328 (UB, CEPEC, F); Mun. Una, 5 km N of Comandatuba, 1977, *Harley & Storr* 69, cult. R.B.G. Kew Accn. No. 1977-634, 1995 (K).

MINAS GERAIS: N Minas Gerais, Mun. Montes Claros, 13 km N of town on road to Januária, 16°38'S, 43°55'W, 6 Nov. 1988, *Taylor & Zappi* in *Harley* 25510 (SPF, K); E, cent.-S & SE Minas Gerais, Mun. Conselheiro Pena, 26 km NE of the Rio Doce, road to Cuparaque via Penha do Norte, 15 Dec. 1990, *Taylor & Zappi* 774 (BHCB, HRCB, K, ZSS); Mun. Lagoa Santa, A.P.A. Carste de Lagoa Santa, Oct. 1995 to Feb. 1996, *Brina & Costa* in BHCB 32768 (BHCB, K); Mun. Marliéria, Parque Estadual do Rio Doce, 20 Sep. 1974, *E.P. Heringer* 13991 (UB); l.c., Revés de Belém, 1 Sep. 1975, *D. Sucre et al.* 10190 (RB); Mun. São João del Rei, Oct. 1895, *A. Silveira et al.* 93 (R); Mun. Ubá, s.d., *Saint-Hilaire* 159 p.p. (P); Mun. Muriaé/Laranjal, 2.5 km N of São João da Sapucaia and 8 km N of Laranjal on road BR 116, 18 Dec. 1990, *Taylor & Zappi* 792 (BHCB, HRCB, K, ZSS); Mun. Laranjal, 8 km em direção a Palma, 24 Nov. 1982, *J.R. Pirani et al.* 252 (SP, SPF); Mun. Leopoldina, 16 Oct. 1981, *P.L. Krieger & M. Brugger* 18710 (CESJ, SPF); Mun. Lima Duarte, May 1897, *F. Brandão* s.n. in Herb. Com. Geog. Geol. de Minas No. 2298 (R). **ESPÍRITO SANTO:** Mun. Linhares, Canal da Lagoa Juparanã, 13 Aug. 1965, *R.P. Belém* 1588 (CEPEC), 21 July 1976, *H.F. Martins* s.n. (MBML); Mun. Santa Teresa, Museu de Biologia, 4 Oct. 1988, *H.Q.B. Fernandes* 2587 (MBML); Mun. Afonso Cláudio, road to Itarana, 20°3'8"S, 41°3'3"W, 24 Nov. 1999, *Zappi et al.* 410 (K, UEC).

CONSERVATION STATUS. Least Concern [LC] (1); PD=2, EI=1, GD=2. Short-list score (1×5) = 5.

In Eastern Brazil this species can be readily distinguished from *R. baccifera* in the living state by its uniformly very long shoots lacking composite terminal areoles and giving rise to subacrotonic secondary segments. Its fruits are generally smaller than those of *R. baccifera* and sometimes pinkish. Unfortunately, these species are less easy to separate in the herbarium and have often been confused, although this can be easily avoided on the basis of provenance, since they have discrete ranges, being sympatric only in parts of the Hiléia Baiana of eastern Bahia, where they differ markedly. *R. lindbergiana* is somewhat variable in stem thickness, the stoutest forms including that described as *R. densiareolata*.

R. lindbergiana most closely resembles *R. baccifera* subsp. *shaferi* (Britton & Rose) Barthlott & N. P. Taylor, which ranges westwards and south-westwards from inner São Paulo (Campinas) to Paraguay, northern Argentina and southern Bolivia.

11. Rhipsalis teres *(Vell.) Steud.*, Nomencl. bot., ed. 2, 2: 449 (1841). Type: not extant. Lectotype (Barthlott & Taylor 1995: 64): Brazil, Rio de Janeiro, Ilha de Santa Cruz, Vellozo, Fl. flumin. Icones 5: tab. 30 (1831).

Cactus teres Vell., Fl. flumin.: 196 (1829); ibid., Icones 5: tab. 30 (1831).

Hariota prismatica Lem. in Ill. Hort. 10, misc.: 84–85 (1863). *Rhipsalis prismatica* (Lem.) Rümpler in C.F. Först., Handb. Cact. ed. 2: 884 (1886). *R. teres* f. *prismatica* (Lem.) Barthlott & N.P. Taylor in Bradleya 13: 65 (1995). Type: Brazil, ex Hort. Verschaffelt, Oct. 1863, not known to have been preserved. Neotype (Barthlott & Taylor, l.c.): Rio de Janeiro, Mun. Maricá, A.P.A., 1 Mar. 1991, *M. Botelho & L.O. Melande* 507 (GUA).

R. virgata F.A.C. Weber in Rev. Hort. 64: 425 (1892). Type: Brazil, c. 1883–1884, [collector?], not known to have been preserved.

R. tetragona F.A.C. Weber, l.c. 429 (1892), nom. inval. (Art. 34)? Type: cult. material, not known to have been preserved.

R. gracilis N.E. Br. in Gard. Chron. ser. 3, 33: 18 (1903). Holotype: Hort. J. Corderoy, cult. Kew (K).

R. heteroclada Britton & Rose, Cact. 4: 224 (1923). *R. teres* f. *heteroclada* (Britton & Rose) Barthlott & N.P. Taylor, l.c. 65 (1995). Type (Barthlott & Taylor, l.c.): Brazil, Rio de Janeiro, *Rose* 20371 (US, lecto.; NY).

R. alboareolata F. Ritter, Kakt. Südamer. 1: 41 (1979). Holotype: Brazil, Rio Grande do Sul, 1964, *Ritter* 1286 (U).

R. clavellina F. Ritter, l.c. 43 (1979). Holotype: l.c., 1964, *Ritter* 1487 (U).

R. maricaensis Scheinvar in Arch. Jard. Bot. Rio Janeiro 31: 71–78 (1992 publ. 1993). Holotype: Rio de Janeiro, Maricá, *Casari* 1143 (GUA).

[*R. penduliflora sensu* K. Schum. in Martius, Fl. bras. 4(2): 276 (1890), non N.E. Br.]

CONSERVATION STATUS. Least Concern [LC]; widespread and very common in SE & S Brazil, where at least some of its habitat remains intact or is included within protected areas.

A variable and complex taxon like *R. baccifera*, of which it is assumed to be the southern sister-species. Plants from the core area of Eastern Brazil are referred to the following form:

11a. f. **capilliformis** *(F. A. C. Weber) Barthlott & N. P. Taylor* in Bradleya 13: 65 (1995). Type: Brazil, cult. Hort. Chantin, not known to have been preserved. Neotype (Barthlott & Taylor, l.c.): Rio de Janeiro, Serra dos Órgãos, 1966, *Hunt* 6510 (K, neo. & isoneo.).

R. capilliformis F.A.C. Weber in Rev. Hort. 64: 425 (1892).

Like the following and especially *R. baccifera* subsp. *hileiabaiana*, but basal extension shoots to 100 cm, ultimate stem-segments only 1.3–2.0 mm thick; flowers apparently self-incompatible, pericarpel c. 2–2.5 × 2–2.5 mm, smaller than the perianth during most of the bud's development, flowers borne laterally, laterally and terminally, or rarely terminally only, c. 7 × 7 mm or wider before the perianth-segments become strongly reflexed, these c. 3.5 mm or larger, often turning yellow post-anthesis; stamens more numerous, 3–4 mm or more; stigma-lobes 3–5; fruit globose to shortly barrel-shaped, to 4–5 × 4 mm, white or pink to deep purplish red, leaving a more prominent scar on the stem. Chromosome number: 2n = 22 (Barthlott 1976).

Southern humid forest element: epiphyte or lithophyte in *Mata atlântica*; common in the Serra da Mantiqueira and Serra do Mar of South-eastern and Southern Brazil (to Rio Grande do Sul). Map 24.
MINAS GERAIS: Mun. Entre Rios [de Minas], Jan. 1898, *A. Jaguaribe* 2894 (R); Mun. Juiz de Fora, Museu Mariano Procópio, 30 Oct. 1986, *Krieger & Coelho* 77 (RB).
ESPÍRITO SANTO: Mun. Piúma, Monte Agá, south of the town, 20°52'23"S, 40°46'24"W, 26 Nov. 1999, reported growing with *Coleocephalocereus fluminensis* (*Zappi et al.* 469).

CONSERVATION STATUS. Least Concern [LC] (1); PD=2, EI=1, GD=2. Short-list score (1×5) = 5.

It is probable that *R. teres* will be found elsewhere in the southern part of the core area.

12. Rhipsalis baccifera *(J. S. Muell.) Stearn* in Cact. J. (Croydon) 7: 107 (1939). Lectotype (Aymonin 1983): J. S. Mueller, l.c. infra, tab. 29.

Cassyta baccifera J. S. Muell., Ill. syst. sex. Linnaei. class IX. ord. 1, tab. 29 (1770–77).
Rhipsalis cassutha Gaertn., Fruct. sem. pl. 1: 137 (1788). Lectotype (designated here): Gaertner, l.c., tab. 28 (1788).
R. cassuthopsis Backeb., Die Cact. 2: 660 (1959). *R. cassythoides* Loefgr. in Arch. Jard. Bot. Rio Janeiro 2: 40 (1918) ('cassythoydes') non G. Don (1834). Type: Brazil, Pará, Belém, *Simão da Costa*, cult. RB (holo. not found RB, R, SP etc.). Lectotype (designated here): Löfgren, l.c., tab. XI (1918).

VERNACULAR NAMES. Enxerto, Conambaia, Tripa-de-galinha, Irahuka'arã.

Pendent to 2 m or more, freely producing aerial roots, pale to dark green; all stems slender cylindric, naked except for minute whitish scale-leaves occasionally subtending very fine whitish bristles, basal extension shoots to 500 × 3–5 mm, eventually rather woody, subsequent branching acrotonic from felted terminal composite areoles, the secondary segments arising in clusters of 3 or more, subsequent orders of segments decreasing in size, the penultimate c. 90 × 2 mm, giving rise to clusters of 2 or more ultimate segments measuring 20–65 × 2 mm. Flowers self-compatible, terminal or subterminal from or around the composite areoles of the ultimate and lower order stem-segments, sometimes 2 or more clustered together, often also lateral, especially on the extension shoots; buds minutely erumpent, the pericarpel at least twice the size of the developing perianth until shortly before anthesis; flowers 6–7 × 4–5 mm, principal perianth-segments 4–6, to 3 mm, patent to reflexed, greenish white; stamens 1–2.5 mm, relatively few, whitish; style 3 mm, stigma-lobes 3, 1 mm, white, exserted; pericarpel barrel-shaped, 3 × 2 mm, pale green. Fruit translucent white, sometimes pinkish, or violet near apex, ovoid, to 7.5 × 6 mm, juice extremely mucilaginous and sticky. Seed (Barthlott & Hunt 2000: 35–36, pl. 5.7–8) 1.2 × 0.6 mm, glossy black-brown.

CONSERVATION STATUS. Least Concern [LC]; at the level of species it has the widest range of any member of the Cactaceae.

Note: the type of R. *baccifera* is assumed to have come from the Caribbean, whence it was introduced to England by Philip Miller in 1758 (Stearn, l.c.). The above description accounts only for what appear to be typical, slender-stemmed forms found in Northern and North-eastern Brazil and the Caribbean (eg. Jamaica), since it is clear that the species represents a complex entity requiring further detailed study. Thicker-stemmed forms, such as are known from the Guianas and elsewhere (as well as Old World plants referred to its various heterotypic subspecies) are not accounted for above, although some of these from the Americas were previously included with subsp. *baccifera* by Barthlott & Taylor (1995: 63).

In Eastern Brazil this species is divisible into the following subspecies:

1. Higher order stem-segments short, densely clustered, 6 or more axes arising from the apex of the longer lower order segments (coastal E Bahia and region of Catolés, Mun. Abaíra, Chapada Diamantina) . . . .**12b.** subsp. **hileiabaiana**
1. Stem-segments not as above (Amazônia; and Maranhão to E Pernambuco) . . **12a.** subsp. **baccifera**

12a. subsp. **baccifera**

Higher order stem-segments 40–90 mm, solitary or in clusters of 2–3 from the apex of previous segments. Flowers with perianth-segments mostly patent; stamens 1–2 mm.

Amazonian forest element: epiphyte in *mata de brejo* and *mata de tabuleiro*, near sea level to c. 600 m, North-eastern Brazil southwards as far as coastal Pernambuco; replaced by subsp. *hileiabaiana* in central & eastern Bahia; common throughout more humid parts of the neotropics, from Northern Brazil and the central Andes of Bolivia etc. northwards to eastern Mexico, the Caribbean islands and Florida (replaced by subspp. *erythrocarpa, mauritiana & horrida* in the palaeotropics). Map 24.

MARANHÃO: Mun. Santa Luzia do Paruá, Nova Esperança, Rio Alto Turiaçu, 2°55'S, 45°45'W, 30 Nov. 1978, *J. Jangoux & R.P. Bahia* 88 (K, MG); Mun. Monção, Ka'apor Indian Reserve, 7 km from 'Urutawy', 17 Feb. 1985, *Batée & Ribeiro* 840 (K, NY).

CEARÁ: N Ceará (Serra de Baturité), Mun. Pacatuba, 'serra na mata ao sítio Pitaguari', 27 Aug. 1979, *J.E. de Paula & R.C. de Mendonça* 1232 (MG); Mun. Guaramiranga, 4 km W of town, Sítio Uruguaiana, 23 Mar. 1945, *H.C. Cutler* 8319 (F).

PARAÍBA: Mun. Itaporororca, 27 Oct. 1998, *E.A. Rocha* 509 (IPA); Mun. Areia, Centro de Ciências Agrárias, 6°58'12"S, 35°42'15"W, 29 Dec. 1980, *V.P.B. Fevereiro et al.* M417 (K), l.c., Campus III da UFPB, 30 Apr. 1998, *E.A. Rocha* 404 (EAN, JPB, IPA, UFP); Mun. João Pessoa, on old tree near city centre, 10 Feb. 1995, *Taylor* (obs.).

PERNAMBUCO: Mun. São Lourenço da Mata, Usina Tiúma, 1930, *Pickel* 2469 (IPA 6874, K, photo); Mun. Olinda, 9 Apr. 2000, *E.A. Rocha et al.* (obs.); Mun. Recife, Dois Irmãos, 10 Dec. 1999, *E.A. Rocha* (PEUFR); l.c., on trees in the campus of Univ. Federal de Pernambuco (UFPE), 13 Dec. 1999, *Zappi* (K, photos), 30 Mar. 2000, *Taylor & Griffiths* (obs.); l.c., Jardim Botânico, 30 Mar. 2000, *Taylor & Griffiths* (K, photo).

CONSERVATION STATUS. Least Concern [LC] (1); PD=1, EI=1, GD=2. Short-list score (1×4) = 4.

Records of *R. baccifera* from South-eastern and Southern Brazil refer to *R. lindbergiana* and *R. teres*; see above. However, *R. baccifera* subsp. *shaferi* (Britton & Rose) Barthlott & N. P. Taylor is known from the state of São Paulo, where it has been collected in Mun. Campinas (*Ratter* s.n.).

12b. subsp. **hileiabaiana** *N. P. Taylor & Barthlott* in Bradleya 13: 63 (1995). Holotype: Brazil, Bahia, Mun. Ilhéus, 7 Aug. 1983, *J. L. Hage & H. S. Brito* 2113 (CEPEC; K, MBM, HRB isos.).

Higher order stem-segments to c. 30 mm, in dense clusters (to 10 or more from the apex of vigorous, lower order segments), strongly differentiated from the much longer extension shoots. Flowers with perianth-segments reflexed to expose the 2–2.5 mm stamens.

Disjunct Bahian humid forest element: epiphyte (rarely lithophyte) in *Mata atlântica (Hileia Baiana/mata de brejo)*, at low elevations to c. 800 m, and in *mata de neblina*, 1650–1800 m, region of Catolés (Mun. Abaíra), Chapada Diamantina, eastern and central Bahia. Endemic to Bahia. Map 24.
BAHIA: cent. Bahia (Chapada Diamantina), Mun. Abaíra, 13°15–16'S, 41°54–55'W, Catolés de Cima, Tijuquinho, 10 Jan. 1992, *Nic Lughadha et al.* in Harley 50699 (SPF, CEPEC, HUEFS, K), 16 Nov. 1992, *W. Ganev* 1468 (HUEFS, K), l.c., Mata do Cigano, 26–28 Feb. 1992, *P.T. Sano & T. Laessøe* in Harley 52352, 52395 (SPF, CEPEC, HUEFS, K), l.c., Riacho da Taquara, 3 Feb. 1992, *Stannard et al.* in Harley 51154 (SPF, HUEFS, CEPEC, K); E Bahia, Mun. Santo Amaro, Oliveira dos Campinhos, Mar. 1952, *A. Leal Costa* 1096 (ALCB); Mun. Santa Teresinha, Serra da Jibóia, 23 Nov. 2001, *Zappi* (K, photos); Mun. Maragogipe, Guapira, 16 June 1987, *G.C.P. Pinto* 24/87 (HRB); município unknown, basin of the Rio Gongoji, Oct.–Nov. 1915, *H.M. Curran* 135 (US), Apr. 1917, *Luetzelburg* 8 (M); Mun. Almadina/Coaraci, 9 Aug. 1977, *L.A. Mattos* 93 & *J.L. Hage* 179 (CEPEC, K); Mun. Ilhéus, CEPEC, Km 22 on the road Ilhéus–Itabuna, 12 Apr. 1968, *S.G. da Vinha* 180 (ALCB), 19 Nov. 1970, *J.L. Hage* 29 (CEPEC), 23 Feb. 1981, *J.L. Hage & E.B. dos Santos* 465 (CEPEC), 20 May 1981 & 7 Aug. 1983, *J.L. Hage & H.S. Brito* 695 (CEPEC, K), 2113 (CEPEC, MBM, K, HRB), 24 June 1965, *R.P. Belém & M. Magalhães* 524 (UB); Mun. Floresta Azul, [before 1966], *Martins* in coll. Brieger, Piracicaba, SP (No. 43), cult. R.B.G. Kew (K, Spirit Coll. No. 52524); Mun. Vitória da Conquista, 12 km from town towards Itambé, 16 Aug. 2002, *M. Machado* (K, photos); Mun. Una, 6 km SE of Una, 1977, *Harley & Storr* 86, cult. R.B.G. Kew Accn. No. 1977-652 (K); Mun. Canavieiras, road BA 001, 0.5 km N of junction with road to Santa Luzia, 23 April 2003, *Taylor & Zappi* (obs.); unlocalised, s.d., *Blanchet* 1518 (P).

CONSERVATION STATUS. Vulnerable (2) [VU B2ab(i, ii, iii, iv)]; area of occupancy estimated to be < 2000 km^2; PD=1, EI=1, GD=1. Short-list score (2 × 3) = 6. It is assumed that a significant proportion of the above-cited localities (c. 5) have suffered severe modification.

This endemic subspecies, which is restricted to the region of Bahia receiving most rainfall (within the 1750 mm annual isohyet along the coast and from very humid woodland in *brejo* forest on serras further inland and in the highest part of the Chapada Diamantina), strongly resembles some forms of *R. teres* in habit, but has flowers and fruits typical of *R. baccifera*.

Subg. *Erythrorhipsalis* A. Berger (Nos 13–18). Like Subg. *Rhipsalis*, but flowers campanulate (except sometimes in No. 18), pendent, scented, one or more at a time from or around the margins of the terminal collective areole of ultimate and sometimes lower order stem-segments (also occasionally lateral, especially in *R. pulchra*, but obliquely oriented on the segments); perianth-segments 8–18 or more; stamen filaments usually highly coloured at base giving the flower a coloured throat; fruit white, pink, purplish, red or orange. SE South America. Type: *R. pilocarpa* Loefgr.

13. Rhipsalis pulchra *Loefgr.* in Arch. Jard. Bot. Rio Janeiro 1: 75–76, tab. 5 (1915). Holotype: Brazil, Serra da Mantiqueira, *O. A. Derby* 4394 in Comm. Geogr. Geol. São Paulo 8834 (SP; US, iso.).

Rhipsalis macahensis Glaziou in Mém. Soc. Bot. France 1(3): 326 (1909), nom. nud. Intended type: Rio de Janeiro, Alto Macahé, *Glaziou* 18262 (P, K).

Pendulous to 6 metres or more, stems erect at first but eventually always pendulous, weak and pliable, almost as if flaccid, emitting aerial roots, to 5 mm diam., plain green, the scales sometimes with reddish marks below; branching acrotonic or subacrotonic, the stem-segments of indeterminate size but generally decreasing in length further away from the rootstock, their apices often not forming recognisable composite areoles, when present these very inconspicuous, actively growing segments gradually tapered to a fine point at apex; areoles almost 0 except on juvenile stems, represented by a few minute trichomes visible at the distal edges of the inconspicuous scales, these < 0.25 × 0.5 mm, truncate, apiculate, pale at maturity, triangular and reddish when young. Flower-buds inconspicuously erumpent, leaving a glabrous scar to 1.5 mm diam. on the stem, pinkish to dark reddish; flowers with a distinctive sweetish scent, downwardly directed, lateral and aligned with the stem and/or ± terminal and sometimes in groups of 3, in lateral flowers the pericarpel attached sublaterally; perianth not expanding fully, ± campanulate, 15 × 15–25 mm; perianth-segments c. 11, 8–9 visible from within the flower, outermost ovate, c. 6 × 5 mm, whitish tinged purplish-pink at apex, inner segments 10–13 × 4.5 mm, creamy white, concave (boat-shaped) from within; stamens in two groups, the outer series spreading towards the perianth, the inner adpressed to style and varying from very short to equalling the style, all except the outermost with dark orange filaments, anthers pale yellow; style 5 mm, stigma-lobes 3–6, to 3 × 0.7 mm, white, spreading; pericarpel 4–5 × 4.5 mm, pale green to pinkish, naked or with 1–2 fleshy bract-scales. Fruit slightly depressed-globose, to c. 7 × 7.5 mm, translucent whitish or purplish red.

Southern humid forest element: epiphyte in *capão de mato*, > 1500 m, Serra da Mantiqueira, southern Minas Gerais; South-eastern Brazil (Rio de Janeiro & São Paulo). Map 25.
MINAS GERAIS: Mun. Lima Duarte, Parque Estadual do Ibitipoca, 27 Sep. 1970, *Krieger* 9269 (CESJ, HRCB), 27 July 1991, *Zappi* 260 (K, SPF, CESJ), 3–5 Aug. 2003, *Taylor & Zappi* (K, photos).

CONSERVATION STATUS. Least Concern [LC] (1); PD=2, EI=1, GD=2. Short-list score (1×5) = 5. Possibly Conservation Dependent, since it is found within various protected areas.

This poorly known species was originally described with and commonly bears purplish magenta fruits and deep pink flowers, but the population cited here from southern Minas Gerais (*Zappi* 260) has white fruits and rather pale flowers. However, there can be no doubts about its identity.

14. Rhipsalis burchellii *Britton & Rose*, Cact. 4: 225 (1923). Lectotype (Barthlott & Taylor 1995: 69): São Paulo, Mun. São Paulo, Jabaquara, 15 Aug. 1915, *Rose & Russell* 20857 (US; NY, K, lectoparas.).

Plant to 2 m or more; stems freely emitting aerial roots; primary shoots to 50.0 × 0.5 cm, crimson-red, secondary to fifth order to 2.5–7.0 cm, terminal segments 2.5–3.5 cm × 1–2 mm, sometimes slightly thickened at the tips; flowers terminal, 1–3 together, or appearing at the joints of the subterminal stem-segments, campanulate, only half-expanded, to 20–25 × 20 mm, pale silvery pink; pericarpel c. 4 × 3.5 mm, stigma-lobes 4–5, to 5 mm, whitish. Fruit ovoid, c. 10 mm, magenta.

Southern humid forest element: epiphyte in *mata de neblina/galeria*, c. 900 m, southern Espírito Santo; South-eastern and Southern Brazil. Map 25.

ESPÍRITO SANTO: Mun. Domingos Martins, May 1986, *Rauh*, living material observed in a private collection in California, U.S.A., 1997, Taylor (obs.); l.c., *R. Whitman* s.n., cult. Univ. Bonn, Accn. No. 13488, 18 Mar. 2000, Taylor (obs.).

CONSERVATION STATUS. Not Evaluated [NE] in view of taxonomic uncertainties (see below).

This complex of species, amongst which *R. burchellii* has the oldest typifiable name, is difficult to resolve from herbarium materials alone, and it is possible that as many as 5 species of this relationship are present in the area. The following, recently described species is one of these, as is No. 16.

The oldest name within this complex is *R. cribrata* (Lem.) N. E. Br. (*Hariota cribrata* Lem. 1857), but this is too poorly typified to be applied with confidence and has been variously misapplied by previous authors (Barthlott & Taylor 1995).

15. Rhipsalis juengeri *Barthlott & N. P. Taylor* in Bradleya 13: 69, 72, pl. 29 & 30 (1995). Holotype: origin unknown, cult., Bot. Gard. Univ. Bonn, Germany, Accn. No. 01700, Mar. 1995 (BONN).

Like *R. burchellii* in habit and *R. clavata* in its flowers, but plant pendulous to at least 3 m and sometimes much longer; stems scarcely succulent, plain green or tinged dull red; primary long extension shoots to 200 × 0.3 cm, terminal segments 1.25–1.75 mm thick; flowers solitary or two together, sometimes appearing at the joints of the subterminal stem-segments, campanulate, only half-expanded, c. 12–15 × 12–15 mm, whitish, pericarpel c. 2 × 2.5 mm, stamens yellow at base; fruit globose-truncate, c. 6 mm or more, purplish to translucent greenish tinged maroon, smelling strongly of blackcurrants (*fide* Barthlott).

Southern humid forest element: epiphyte in *Mata atlântica*, c. 1500–1600 m, south-eastern/southern Minas Gerais; range and/or endemic status uncertain. Map 25.

MINAS GERAIS: Mun. Lima Duarte, Parque Estadual do Ibitipoca, 27 July 1991, *Zappi* 259 (K, SPF, CESJ); (?) l.c., 12 May 1970, *Krieger* 8594 (CESJ), 1 Nov. 1973, *Krieger* 13174 (CESJ, HRCB), 29 & 30 Sep. 1970, *D. Sucre et al.* 7193, 7228 (RB); Mun. Lima Duarte/Rio Preto, Serra Negra, sterile living plant shown to the authors at the July 1998 Congresso Nac. de Botânica, Salvador, Bahia.

CONSERVATION STATUS. Not Evaluated [NE] in view of taxonomic uncertainties.

Of the above, only the collection by Zappi and that from the Serra Negra have been seen as living plants and suggest the above identity, although neither has been seen in flowering condition.

16. Rhipsalis clavata *F. A. C. Weber* in Rev. Hort. 64: 429 (1892); N. P. Taylor in Bot. Mag. 19: 160–164, tab. 446 (2002). Type: Brazil, Rio de Janeiro, Petrópolis, 1886, *Binot*, probably a living plant (assumed not to have been preserved). Neotype (Barthlott & Taylor 1995: 72): l.c., Teresópolis, Granja Comari, 11 Feb. 1964, *Castellanos* 24569 (GUA).

Hariota clavata F.A.C. Weber, l.c. (1892). *Hatiora clavata* (F.A.C. Weber) Moran in Gentes Herb. 8: 343 (1953).

Pendent to 3 m or more, basal extension shoots of indeterminate growth rare, but when present to c. 12 cm; primary determinate segments 3–7 cm long, secondary to fifth order segments often swollen at tips, up to 2.5(–6.0) cm long, terminal segments 1–4 cm long, 1.0–1.5 mm thick, thickened at the tips to 1.5–3.0 mm, up to 10 or more arising at once from the collective areole at apex of the previous segment. Flowers terminal, 1–3 together, sometimes appearing at the joints of the subterminal segments, pendent, campanulate to fully expanded, 13–15 × 13–23 mm; perianth-segments oblanceolate, to 10 × 2.5 mm, pure white, (7–)8–11 visible from within; stamens differentiated into long and short, not coloured at base; style 8 mm long, stigma-lobes 3–5, 2–3 mm long, slender, spreading; pericarpel pale whitish green, c. 2–3 × 2.5 mm; fruit (excluding withered perianth remains) c. 4.5–5.0 × 4.5–5.0 mm, greenish white with a magenta ring at apex around the perianth remains/scar, rarely red. Chromosome number: 2n = 22 (Barthlott 1976).

Southern humid forest element: epiphyte in *Mata atlântica*, c. 800–1140 m, southern Espírito Santo and northern Rio de Janeiro (and perhaps south-eastern Minas Gerais); South-eastern Brazil, from sea level to high elevations (westwards to Ilha São Sebastião, São Paulo). Map 25.

MINAS GERAIS (all of the collections cited from this state are only provisionally identified as this species in the absence of studies based on living material): SE Minas Gerais, Parque Nacional do Caparaó, 15 Oct. 1988, *M. Brugger et al.* in FPNC 328, 330 (CESJ, HRCB), Oct. 1941, *Brade* 17115 (RB); l.c., Cachoeira Bonita, 18 Sep. 1988, *Krieger* in FPNC 233 (CESJ, HRCB, SPF).
ESPÍRITO SANTO: Mun. Domingos Martins, 1986, *Kautsky* in *Rauh* 68384 (BONN) — red-fruited form.
RIO DE JANEIRO: Mun. Santa Maria Madalena, Parque do Desengano, Santa Clara, 21 Jan. 1976, *Araujo* 950 (GUA, MEXU); l.c., 'subida da Pedra Dubois', 21°56'36–42"S, 41°59'29–33"W, 22 Nov. 1999, *Zappi et al.* 367 (K, UEC).

CONSERVATION STATUS. Not Evaluated [NE] in view of taxonomic uncertainties.

The collections cited above from Rio de Janeiro and Espírito Santo are definitely *R. clavata*, but those from Minas Gerais cannot be confidently identified at present from dried specimens and may include representatives of the preceding species as well as possibly the orange-fruited *R. campos-portoana* Loefgr., which appears to be wide-ranging in Southern and South-eastern Brazil. A specimen of *R. clavata* said to have been collected in the grounds of CEPLAC, between Itabuna and Ilhéus, Bahia (coll. 1975, *G. Daniels*, cult. Huntington Bot. Gard.) and conserved in HNT is suspected as having incorrect provenance data (probably due to switched labels during cultivation). This Bahian locality is amongst the most well-collected by local botanists, who presumably would have obtained the distinctive *R. clavata* by now, if it occurred there.

17. Rhipsalis cereuscula *Haw.* in Philos. Mag. Ann. Chem. 7: 112 (1830). Type: Brazil, a living plant not known to have been preserved or illustrated. Neotype (Barthlott & Taylor 1995: 69): Brazil, São Paulo, Mun. Piracicaba, campus of ESALQ, 3 Dec. 1993, *V. C. Souza* 4970 (ESA; K, isoneo.).

Hariota cereuscula (Haw.) Kuntze, Revis. gen. pl. 1: 262 (1891). *Erythrorhipsalis cereuscula* (Haw.) Volgin in Vestn. Moskovsk. Univ., Ser. 16, Biol. 36(3): 19 (1981).
Hariota saglionis Lem., Cact. aliq. nov.: 39 (1838). *Rhipsalis saglionis* (Lem.) Otto in Walp., Repert. bot. syst. 2: 936 (1843). Type: a living plant of unknown provenance in the collection of Monville; assumed not to have been preserved.
R. penduliflora N.E. Br. in Gard. Chron. ser. 2, 7: 716 (1877). Holotype: Hort. Pfersdorff, 1875, cult. Kew (K).
[*R. cribrata sensu* K. Schum. in Martius, Fl. bras. 4(2): 277–278 (1890), non (Lem.) N.E. Br.]

Stems pale green, freely emitting aerial roots, these attaching the basal extension shoots and terminal clusters of stem-segments to the trunk and branches of the host tree in new positions; extension shoots to c. 60 cm × 3 mm, erect then spreading and finally pendent, densely clothed in fine adpressed bristle-spines at first, later mostly caducous, these subtended by minute reddish scales, the bristles forming a small erect tuft at the stem-segment apices, second order segments arising in clusters of up to 10 or more at or near apex of extension shoots, c. 4–10 cm, or these lacking and extension shoots bearing specialized ultimate segments directly; highest order stem-segments densely clustered, sausage-like to ± globose, of irregular thickness and shape, 5–15 × 2–4 mm or thicker, sometimes slightly angled or ribbed, often partially shrunken or appearing swollen with sap, branching strictly acrotonic, with 1–5 new segments developing at apex from the bristly composite areole. Flowers in early spring, c. 17 × 16 mm, campanulate, whitish or very pale creamy yellow, 1–3 together from the terminal areole or its vicinity; pericarpel obovoid, c. 5.5 × 4.0 mm, pale yellow-brown to pale green, ± naked or bearing reddish bract-scales with trichomes and/or occasional bristles in their axils; perianth-segments c. 15, to 9 × 4 mm, partially differentiated into an outer somewhat spreading series and an inner series remaining erect and tending to enclose the stamens and style at the start of anthesis; stamens whitish, ± reddish at base, the innermost clustered around the whitish, exserted style, some of the outermost almost as long as the perianth-segments; stigma-lobes 3–5, c. 1 mm, whitish, exserted well beyond the perianth and stamens. Fruit c. 7 × 6 mm, white (occasionally red outside Brazil), sometimes marked with reddish bract-scales. Chromosome number: 2n = 22 (Barthlott 1976).

Disjunct humid forest element: epiphyte in *Mata atlântica*, including *mata de brejo* and *agreste* (NE Brazil) and *mata do planalto*, c. 500–1400 m, north-eastern Pernambuco, eastern Bahia and central-southern to southern Minas Gerais; South-eastern and Southern Brazil; Bolivia (La Paz), Argentina, Paraguay and Uruguay. Map 25.

PERNAMBUCO: Mun. Taquaritinga do Norte, Sítio Cafundó, 28 Dec. 1972, *Andrade-Lima* 72-7127 (IPA); l.c., road to Vertentes, 6 Mar. 1966, *Andrade-Lima* 66-4483 (IPA); Mun. Brejo da Madre de Deus, Fazenda Bituri, 25 May 1995, *D.C. Silva et al.* 60 (K, UFP).

BAHIA: Mun. Almadina, Faz. Beija Flor, Serra da Pancadinha, entrada a 1 km da cidade na estrada para Floresta Azul, 11 Aug. 1972, *R.S. Pinheiro* 1910 (CEPEC); Mun. Vitória da Conquista, 12 km from town towards Itambé, 16 Aug. 2002, *M. Machado* (K, photos).

MINAS GERAIS: Mun. Lagoa Santa, reported by Warming (1908: 146; cf. Warming & Ferri 1973) as *R. saglionis*; Mun. Catas Altas, Reserva Particular de Patrimônio Natural (RPPN) do Caraça, Capela do Sagrado Coração ("Capelinha"), 1997, *Padre C.M. Dell'Amore*, cult. Santuário do Caraça, Aug. 2001, Taylor (K, photos); Mun. São João del Rei, Sep. 1893, *A. Silveira* in Herb. Com. Geogr. Geol. Minas 1593 (R), Mar. 1994, *Rui Alves*, cult. Jard. Bot. Rio de Janeiro, Feb. 1995, Taylor (obs.).

CONSERVATION STATUS. Least Concern [LC] (1); PD=2, EI=1, GD=1. Short-list score (1×4) = 4. Least Concern in terms of its overall range, but frequency assumed to have been much reduced in North-eastern Brazil due to destruction of its habitat. Apparently rare, but nevertheless protected, in the RPPN do Caraça, Catas Altas, Minas Gerais.

This is another good example of a Rhipsalideae with a markedly disjunct distribution in the *brejos/agrestes* (*mata de cipó*) of North-eastern Brazil. The sometimes irregularly swollen ultimate stem-segments may function as a water store, permitting the development of flowers at their apices during the close of the dry winter season, when they become visibly shrunken through water loss.

18. Rhipsalis pilocarpa *Loefgr.* in Monatsschr. Kakt.-Kunde 13: 52 (Apr. 1903) and in Revista Centro Sci. Campinas 1903(4): 188 (July 1903); N. P. Taylor in Bot. Mag. 14: 125–129, tab. 320 (1997). Type: Brazil, São Paulo, Mun. Itu, forests of Ipanema and Itu, cult. Hort. Bot. São Paulo, Feb. 1903, *Löfgren* (not found at SP, RB or R). Lectotype (Barthlott & Taylor 1995: 69): Löfgren, l.c.: 55, illus. (Apr. 1903).

Erythrorhipsalis pilocarpa (Loefgr.) A. Berger in Monatsschr. Kakt.-Kunde 30: 4 (1920).

Like the preceding in overall habit, but all stem-segments cylindric, of ± uniform thickness and conspicuously white or yellowish bristly, seldom emitting aerial roots, often strongly tinged with dark red, otherwise dark green except the actively growing pale green apices, the lower parts of basal extension shoots c. 5 mm or more diam., the penultimate and ultimate flower-bearing segments c. 35 × 3.5–4.0 mm, lateral areoles bearing c. 8–15 strongly adpressed bristle-spines to 6 mm, these forming an erect cephalium-like tuft at the terminal collective areole and becoming displaced and patent at the joints of the earlier-formed segments (forming and performing like the terminal and ring cephalia of *Arrojadoa rhodantha*); flowers in early winter, mostly 1–4 together at or around the terminal composite areole, fragrant (moth-pollination syndrome?), buds densely bristly, the perianth strongly tinged pinkish red at first, flowers rather variable in size and form, c. 10–30 mm diam., pendent, all perianth-segments patent to reflexed, silvery cream, tinged pink to purple at apex, c. 18 or more in total, to 9 × 2.5 mm, linear-tapered, minutely apiculate, pericarpel ± obconic, clothed in areoles and bristles identical to the stem-segments, green with minute reddish scales, stamens numerous, to 7 mm or more, whitish, reddish at base, expanded as in a loose brush and very conspicuous when the perianth-segments are strongly reflexed (cf. Subg. *Rhipsalis* & Subg. *Phyllarthrorhipsalis* etc.), the innermost shorter, tightly clustered around the style and effectively protecting the nectar-chamber around its base, style and 6–8 spreading, whitish 2 mm stigma-lobes exceeding the brush of stamens by 1–2 mm; fruit spherical, to 8–12 mm diam., pericarp bright red to crimson, bearing numerous areoles and whitish bristles. Seed 1.6–1.7 mm (Barthlott & Hunt 2000: pl. 4.5–6). Chromosome number: 2n = 22 (Barthlott 1976).

Southern humid forest element: epiphyte (rarely lithophyte) in *Mata atlantica*, 500–900 m, southern Minas Gerais (Rio Preto) and southern Espírito Santo (Domingos Martins); South-eastern and Southern Brazil (to Paraná). Map 25.

MINAS GERAIS: Mun. Rio Preto, *F.S. Pires & M. Brugger* 21615 (CESJ, SPF, HRCB); Mun. Bocaina de Minas, 1994, *J.C.E. Correia*, cult. Jard. Bot. Rio de Janeiro, Feb. 1995, Taylor (obs.).

ESPÍRITO SANTO: Mun. Domingos Martins, May 1986, *Rauh & Kautsky* 67561, 67564, cult. Univ. Bonn, Germany, Accn. Nos 04512, 04508; l.c., July 1989, *Rauh* 68383 (BONN, photo).

CONSERVATION STATUS. Vulnerable (2) [VU B2ab(i, ii, iii, iv)]; area of occupancy estimated to be < 2000 km^2; PD=2, EI=1, GD=2. Short-list score (2×5) = 10. Known from only 8 localities throughout its entire range and apparently rare (Taylor, l.c.).

This, the rare and geographically more restricted and variable sister species of *R. cereuscula*, appears to have evolved the stamen-brush floral syndrome convergently with *Rhipsalis* subgenera *Rhipsalis, Epallagogonium, Trigonorhipsalis* & *Phyllarthrorhipsalis*, and contrasts strongly with other members of Subg. *Erythrorhipsalis* in lacking truly campanulate flowers. As in its sister species, the bristly stem-segments may assist in the collection of moisture from mists and night-time dews.

Subg. *Calamorhipsalis* K. Schum. (No. 19): seedlings 3–4-ribbed/angled; adult branching sub-acrotonic or acrotonic; flower-buds and new stem-segments conspicuously erumpent; trichome-bearing, composite terminal and normal lateral areoles apparently lacking or hidden at first, visible only after flowering, scale-leaves minute, not fleshy; stem-segments perfectly terete, of indeterminate growth in the species treated below; flower-buds strongly erumpent, leaving a prominent scar on the stem, lateral to subterminal, solitary, areoles flowering only once; fruit red, magenta or orange. Lectotype (Backeberg 1942): *R. neves-armondii* K. Schum.

Only the following species is currently accepted for Eastern Brazil. However, a recent collection by Jomar Jardim (Feb. 2003, photo), from near Prado, Bahia, strongly suggests that *R. neves-armondii* is native here.

19. Rhipsalis hoelleri *Barthlott & N. P. Taylor* in Bradleya 13: 50, pl. 9 (1995). Holotype: Brazil, Espírito Santo, (?) Mun. Domingos Martins, 1987, *Orssich* s.n., cult. Univ. Bonn, Germany, Accn. No. 04841 (BONN).

Pendent, to 150 cm or more; stems perfectly terete, limp and growing vertically downwards, only 3–4 mm diam., green with faint purplish marks at the minute and very inconspicuous scales, growth indeterminate. Flowers c. 10 mm diam., not opening widely; perianth-segments, style and stigma-lobes intense carmine red, stamens colourless, at least distally. Fruit subglobose, 8 mm diam., not translucent, dull olive-red when immature, ripening to intense tomato red.

Southern humid forest element: habitat details unknown; awaiting rediscovery in the region indicated below. Map 25.
ESPÍRITO SANTO: Mun. (?) Domingos Martins, 1987, *Orssich* s.n., cult. Univ. Bonn, Germany (BONN, holo.).

CONSERVATION STATUS. Data Deficient [DD], but to date collected only once.

This recently described species is closely related to *R. puniceodiscus* (Rio de Janeiro to Santa Catarina), which it strongly resembles in vegetative characters. Its unusual red flowers are presumed to be an adaptation for pollination by hummingbirds.

13. HATIORA Britton & Rose

in Bailey, Stand. cycl. hort. 3: 1433 (1915). Type: *Hatiora salicornioides* (Haw.) Britton & Rose.

Literature: Barthlott & Taylor (1995, 1996).

Like *Rhipsalis*, but habit ± erect, all stem-segments of strictly determinate growth and flowers always terminal from composite areoles, highly coloured.

A genus of 6 species endemic to the *Mata atlântica* of Brazil, between Bahia and Rio Grande do Sul. Two species from Subg. *Hatiora* are represented here, although *H. epiphylloides* (Porto & Werderm.) Buxb. (Subg. *Rhipsalidopsis*) may occur in Minas Gerais (Bocaina de Minas), just outside the southern limits of the core area of Eastern Brazil. The genus is weakly distinguished from *Rhipsalis*, especially in view of the floral similarities presented by *H. cylindrica*. Plate 23.

1. Stem-segments globose, ovoid, cylindric or inverted bottle-shaped; inner perianth-segments erect; fruit whitish, segments erect; often with a reddish ring at apex . . **1. salicornioides**
1. Stem-segments uniformly cylindric, or somewhat thicker at base; all perianth-segments spreading; fruit entirely red or purplish . . . . **2. cylindrica**

1. Hatiora salicornioides *(Haw.) Britton & Rose* in Bailey, Stand. cycl. hort. 3: 1433 (1915). Type: 'Ind. Occident.' (assumed not to have been preserved). Neotype (Barthlott & Taylor 1995: 73, pl. 27): 'Brazil [Rio de Janeiro], Messrs Bowie & Cunningham, Duncanson del.', water colour illustration (K).

Plate 14. 14.1 *Lepismium cruciforme*, in fruit. Cult. Univ. Bonn (WB). **14.2** *Rhipsalis russellii*. Bahia, SE of Vitória da Conquista, 2002 (MM). **14.3** *Ibid*., in fruit. Cult. Univ. Bonn (WB). **14.4** *Rhipsalis elliptica*. Minas Gerais, RPPN do Caraça, 2001 (NT).

Plate 15. 15.1 *Rhipsalis elliptica*, in fruit. São Paulo, *Zappi* 246, cult. Kew (K). **15.2** *R. oblonga*. São Paulo, Ilhabela, *Taylor & Zappi* 1645A, cult. Kew (K). **15.3** *R. crispata*, in young fruit. Bahia, Serra da Jibóia, 2001 (DZ). **15.4** *R. floccosa* subsp. *floccosa*. Bahia, Muritiba, 2002 (NT).

Plate 16. 16.1 *Rhipsalis floccosa* subsp. *oreophila*. Bahia, *Harley et al.* 25566 (NT). **16.2** *R. floccosa* subsp. *pulvinigera*. Minas Gerais, *Zappi et al.* 241, cult. Kew (K). **16.3** *R. paradoxa* subsp. *septentrionalis* (Pernambuco, *Braun* s.n.), with subsp. *paradoxa* (right). Cult. Univ. Bonn (WB). **16.4** *Ibid* (subsp. *paradoxa*, left). Ibid. (WB).

Plate 17. **17.1** *Rhipsalis pacheco-leonis* subsp. *catenulata*. Type collection, cult. Kew (K). **17.2** *Ibid*. Ibid. (K). **17.3** *Ibid*., in fruit. Ibid. (K). **17.4** *R. cereoides*. Cult. Univ. Bonn (WB).

Plate 18. **18.1** *Rhipsalis cereoides*, in fruit. Rio de Janeiro, Itacoatiara, 1996 (AJ). **18.2** *R. sulcata*. Cult. Univ. Bonn (WB). **18.3** *Ibid*. Ibid. (WB). **18.4** *R. lindbergiana*. Minas Gerais, *Taylor & Zappi* 792 (NT).

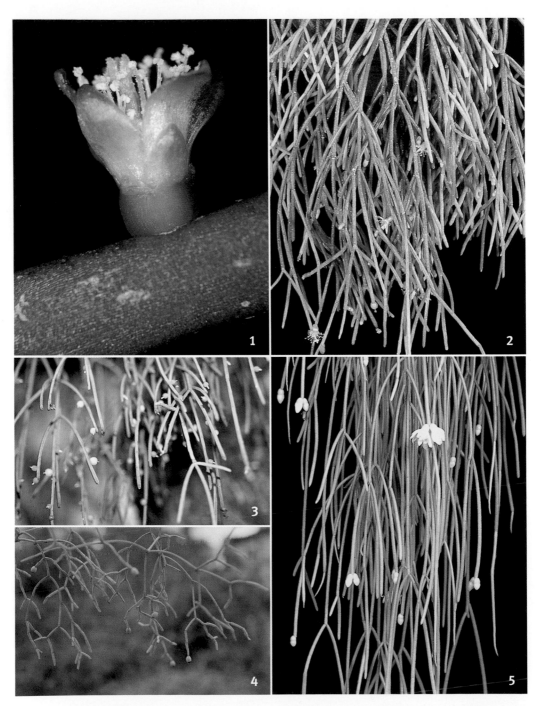

Plate 19. 19.1 *Rhipsalis lindbergiana*. Bahia, *Harley & Storr* 69, cult. Kew (K). **19.2** *R. teres* f. *capilliformis*. Cult. Univ. Bonn (WB). **19.3** *R. baccifera* subsp. *baccifera*. Pernambuco, Recife, 1999 (DZ). **19.4** *R. baccifera* subsp. *hileiabaiana*. Bahia, Serra da Jibóia, 2001 (DZ). **19.5** *R. pulchra*. S Minas Gerais, *Zappi* 260, cult. T. Hewitt (K).

Plate 20. **20.1** *Rhipsalis pulchra*. S Minas Gerais, *Zappi* 260, cult. T. Hewitt (K). **20.2** *R. burchellii*. Cult. Univ. Bonn (WB). **20.3** *R. juengeri*. Type collection, cult. Univ. Bonn (WB). **20.4** *Ibid.*, in fruit. Ibid. (WB). **20.5** *R. cereuscula*. SE Bahia, S of Vitória da Conquista, 2002 (MM).

Plate 21. 21.1 *Rhipsalis clavata*. Cult. Univ. Bonn (WB). **21.2** *Ibid*., in fruit. Ibid. (WB). **21.3** *R. pilocarpa*. Cult. Univ. Bonn (WB). **21.4** *Ibid*. Espírito Santo, Domingos Martins, *Rauh* 68383 (WR).

Plate 22. *Rhipsalis hoelleri*. Type collection, cult. Univ. Bonn (WB).

Plate 23. 23.1 *Hatiora salicornioides*. Cult. Univ. Bonn (WB). **23.2** *H. cylindrica*. Bahia, Itapoã, cult. R. Augusto, 2002 (MM). **23.3** *Schlumbergera kautskyi*. Type collection, cult. Univ. Bonn (WB). **23.4** *S. opuntioides*. Cult. Univ. Bonn (WB).

Plate 24. 24.1 *Schlumbergera microsphaerica*, in fruit. Minas Gerais, Pico da Bandeira, *Orssich* (WB). **24.2** *Ibid.*, flower. Cult. Ralf Bauer (RB). **24.3** *Brasilicereus phaeacanthus*. Bahia, *Harley et al.* 25541 (NT). **24.4** *Ibid.*, in bud. Minas Gerais, *Taylor & Zappi* 762 (NT).

Plate 25. 25.1 *Brasilicereus phaeacanthus*, in fruit. Bahia, Jequié, 2002 (MM). **25.2** *B. markgrafii*. Minas Gerais, Serra da Bocaina, 2002 (GC). **25.3** *Ibid*. Minas Gerais, Grão Mogol, *Harley et al*. 25069 (NT). **25.4** *Ibid*. Loc. cit., *Preston-Mafham* 111, cult. A. Hofacker (AH).

Plate 26. **26.1** *Cereus mirabella*. Minas Gerais, Mirabela, *Horst & Uebelmann* HU 540, cult. A. Hofacker (AH).
26.2 *C. albicaulis*, in fruit. Bahia, *Taylor et al.* 1392 (NT). **26.3** *Ibid*. Pernambuco, *Taylor & Zappi* 1622 (NT).

Plate 27. 27.1 *Cereus fernambucensis* subsp. *fernambucensis*, in fruit. Bahia, *Harley et al.* 27874 (NT). **27.2** *Ibid.* Neotype collection (NT). **27.3** *C. fernambucensis* subsp. *sericifer*. Espírito Santo, 1990 (NT). **27.4** *C. insularis.* Cult. A. Hofacker (AH).

Plate 28. 28.1 *Cereus jamacaru* subsp. *jamacaru,* with Marlon Machado. Bahia, APA Gruta dos Brejões, 2002 (NT). **28.2** *C. jamacaru* subsp. *calcirupicola,* juvenile spination. Minas Gerais, Engenheiro Dolabela, 2002 (GC). **28.3** *C. sp. nov.* (5c), seedling. SE Bahia, S of Vitória da Conquista, cult. M. Moreira, 2002 (MM). **28.4** *Ibid.*, immature plant, c. 7 metres tall. Loc. cit., 2003 (NT).

Plate 29. 29.1 *Cereus sp. nov.* (5c). SE Bahia, S of Vitória da Conquista, 2003 (NT). **29.2** *C. hildmannianus*, with ripe fruit. Cult. Kew (K). **29.3** *Cipocereus laniflorus*. Type collection, cult. Kew (K). **29.4** *Ibid.*, in fruit. Minas Gerais, RPPN do Caraça, 2001 (NT).

Plate 30. 30.1 *Cipocereus crassisepalus*. Minas Gerais, Serra do Ambrósio, *Zappi & Prado* in CFCR 11822 (DZ). **30.2** *C. pusilliflorus,* in fruit. Minas Gerais, *Horst & Uebelmann* HU 400, cult. Werner van Heek (WvH). **30.3** *C. bradei.* Minas Gerais, *Eggli* 1117 (UE). **30.4** *Ibid.,* in fruit. Minas Gerais, *Harley et al.* 24508 (NT).

Plate 31. 31.1 *Cipocereus minensis* subsp. *leiocarpus,* in bud. Minas Gerais, SW of Datas (GC). **31.2** *Ibid.* Minas Gerais, *Harley et al.* 25136 (NT). **31.3** *Ibid.*, with young fruit. Ibid. (NT). **31.4** *C. minensis* subsp. *minensis.* Minas Gerais, Serra do Cipó (WR).

Plate 32. *Stephanocereus leucostele.* Bahia, APA Gruta dos Brejões, 2002 (NT).

Plate 33. 33.1 *Stephanocereus leucostele*. Bahia, *Taylor et al.* 1521 (NT). **33.2** *S. luetzelburgii*. Bahia, *Harley et al.* 25551 (NT). **33.3** *Ibid.*, cephalium apex. Bahia, *Harley et al.* 25559 (NT).

Plate 34. **34.1** *Arrojadoa bahiensis*. Bahia, *Taylor et al.* 1557A (NT). **34.2** *Ibid.* Ibid., cult. Kew (K). **34.3** *Ibid.* Ibid. (K).

Plate 35. 35.1 *Arrojadoa dinae* subsp. *dinae*. Minas Gerais, *N. Roque et al*. in CFCR 15407 (RH). **35.2** *A. dinae* subsp. *eriocaulis*. Minas Gerais, *Harley et al*. 25518 (NT). **35.3** *Ibid*. Bahia, cult. G. Charles (GC). **35.4** *A.* ×*albiflora* (*A. dinae* × *A. rhodantha*). Ibid. (GC).

Plate 36. **36.1** *Arrojadoa penicillata*. Bahia, *Zappi* 124 (K). **36.2** *Ibid.*, in fruit. Bahia, *Taylor et al.* 1542 (NT). **36.3** *A. rhodantha,* plant 3.5 metres tall! Bahia, *Taylor et al.* 1370 (NT). **36.4** *Ibid.,* in bud. Minas Gerais, *Taylor et al.* 1459 (NT).

Plate 37. 37.1 *Arrojadoa rhodantha*. Pernambuco, *Zappi* 226 (NT). **37.2** *Ibid.*, in fruit. Bahia, Morro do Chapéu, June 2002 (MM). **37.3** *A. sp. nov.* (4b). Bahia, nr Sussuarana, 2003 (NT). **37.4** *Ibid.,* in flower. Loc. cit., cult. A. Oliveira Soares Filho, 2003 (NT).

Plate 38. 38.1 *Pilosocereus tuberculatus*. Pernambuco, Mun. Floresta, 2000 (NT). **38.2** *P. gounellei* subsp. *gounellei*. Bahia, Araci, 1991 (NT). **38.3** *P. tuberculatus*. Pernambuco, *Taylor & Zappi* 1623 (NT). **38.4** *P. gounellei* subsp. *gounellei*. Bahia, *Harley et al.* 25556 (NT). **38.5** *Ibid.*, in fruit. N Paraíba, Tacima, 2000 (NT).

Plate 39. 39.1 *Pilosocereus gounellei* subsp. *zehntneri*. N Bahia, APA Gruta dos Brejões, 2002 (NT). **39.2** *Ibid.* SW Bahia, *Taylor et al.* 1425 (NT). **39.3** *P. catingicola* subsp. *catingicola*. Bahia, *Zappi 117* (NT). **39.4** *P. catingicola* subsp. *salvadorensis*, with DZ. NW Sergipe, *Taylor & Zappi* 1618A (NT).

Plate 40. 40.1 *Pilosocereus azulensis,* young plant. Type locality, 2002 (MM). **40.2** *P. floccosus* subsp. *quadricostatus* × *P. multicostatus.* Minas Gerais, Pedra Azul, 2002 (MM). **40.3** *P. arrabidae.* Espírito Santo, *O.J. Pereira et al.* 2119 (NT). **40.4** *Ibid.,* in fruit. Rio de Janeiro, Arraial do Cabo, *Zappi* 476 (DZ).

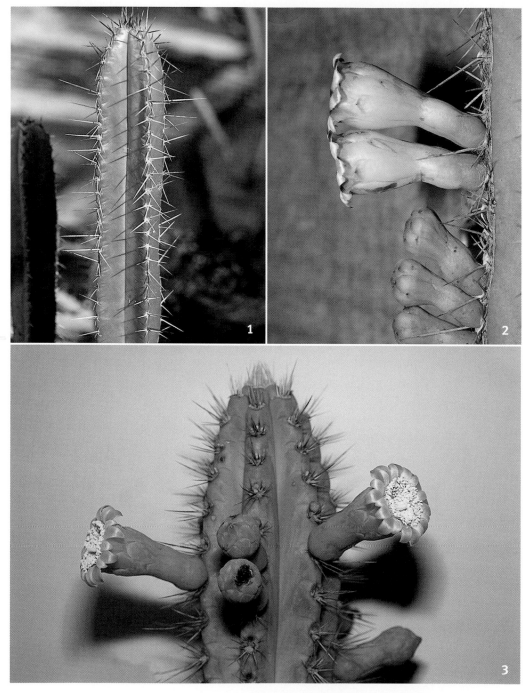

Plate 41. 41.1 *Pilosocereus brasiliensis* subsp. *brasiliensis*. Espírito Santo, *O.J. Pereira et al.* 2120, cult. UNESP Rio Claro (NT). **41.2** *P. brasiliensis* subsp. *ruschianus*. E Minas Gerais, *Taylor & Zappi* 770 (NT). **41.3** *P. flavipulvinatus*. Maranhão, *Zappi* 213 (NT).

Plate 42. 42.1 *Pilosocereus pentaedrophorus* subsp. *pentaedrophorus*. Bahia, *Zappi* 131A (NT). **42.2** *Ibid.,* in bud and fruit. Bahia, Serra de São José, 2001 (DZ). **42.3** *P. pentaedrophorus* subsp. *robustus*. Bahia, *Taylor et al.* 1515C (NT). **42.4** *P. glaucochrous*. Bahia, Morro do Chapéu, 2002 (NT). **42.5** *Ibid.,* in bud. Ibid. (NT).

Plate 43. 43.1 *Pilosocereus floccosus* subsp. *floccosus*. Minas Gerais, near Rodeador, 2002 (GC). **43.2** *P. floccosus* cf. subsp. *quadricostatus*. NE Minas Gerais, *Harley et al.* 25146 (NT). **43.3** *P. floccosus* subsp. *quadricostatus*. NE Minas Gerais, *Taylor & Zappi* 765 (NT). **43.4** *P.* ×*subsimilis*. NE Minas Gerais, *Harley et al.* 25535 (NT).

Plate 44. 44.1 *Pilosocereus fulvilanatus* subsp. *fulvilanatus*. Minas Gerais, Grão Mogol, 2002 (GC). **44.2** *P. fulvilanatus* subsp. *rosae*. Type locality (KH). **44.3** *Ibid*. Loc. cit., cult. A. Hofacker (AH). **44.4** *P. pachycladus* subsp. *pachycladus*. N Minas Gerais, *Harley et al.* 25520 (NT).

Plate 45. 45.1 *Pilosocereus pachycladus* subsp. *pachycladus*. Bahia, W of Morro do Chapéu, 2002 (NT). **45.2** *P. pachycladus* subsp. *pernambucoensis*. Piauí, Mun. Padre Marcos, 2000 (NT). **45.3** *Ibid*. (NE form). Pernambuco, Belém de São Francisco, 2000 (NT). **45.4** *P. magnificus*. NE Minas Gerais, *Taylor & Zappi* 755 (NT).

Rhipsalis salicornioides Haw., Suppl. pl. succ.: 83 (1819) ('salicornoides'). *Hariota salicornioides* (Haw.) DC.,
Mém. Cact.: 23 (1834).

H. *villigera* K. Schum. in Martius, Fl. bras. 4(2): 265 (1890). *Rhipsalis salicornioides* var. *villigera* (K. Schum.)
Loefgr. in Arch. Jard. Bot. Rio de Janeiro 1: 85 (1915). *Hatiora salicornioides* f. *villigera* (K. Schum.)
Supplie, Rhipsalidinae: [66] (1990). Type: Brazil, *Sello* 5239 (B†; MO, No. '523').

Epiphyte or lithophyte, rarely terrestrial on sand, to 2 m, ± erect. Stems articulate, segments globose,
ovoid, cylindric or inverse bottle-shaped and then very narrow at base, often with an irregular outline and
presenting bristly spines at apex, 10–25 × 0.8–3.0(–5.0) mm, basal segments thick, with brownish bark,
apical segments 1–3.5 cm, epidermis bright green, with red shades or spots and minute areoles at the sides,
sometimes bearing up to 6 bristly spines, terminated by a large composite areole with felt and minute
bristles. Flowers terminal from the composite areole, solitary or 2–3 together, orange or bright yellow, c.
15 × 10–15 mm; pericarpel pale greenish, obconic, smooth, c. 3.5 × 4 mm; tube inconspicuous, almost 0;
inner perianth-segments erect, to 11 × 3 mm, others sometimes more expanded, ovate, broader, to 4 mm
wide; stamens all at the same level, anthers rounded; style 5 mm, stigma-lobes 4–7, ± exserted; ovary locule
obtriangular in longitudinal section. Fruit globose to turbinate, c. 10 × 5–8 mm, translucent, whitish,
funicular pulp transparent, very mucilaginous. Seeds ovoid, 1.2 × 0.9 mm, brown-black, shiny; testa-cells
flat, smooth (Barthlott & Hunt 2000: pl. 3.5–6). Chromosome number: 2n = 22 (Barthlott 1976).

Disjunct humid forest element: usually epiphytic or lithophytic, *Mata atlântica, mata de grotão* and *mata de
neblina,* c. 800–1750 m, Bahia (E Chapada Diamantina), central-southern Minas Gerais (S Serra do
Espinhaço and Serra da Mantiqueira), Espírito Santo and northern Rio de Janeiro, but northern records
markedly disjunct; South-eastern and Southern Brazil (to Paraná). Map 26.

BAHIA: cent. Bahia, Mun. Lencois, Serra da Chapadinha, 8 July 1996, *H.P. Bautista et al.* in PCD 3486
(ALCB, K), 27 Feb. 1997, *Gasson et al.* in PCD 5882 (ALCB, K); Mun. (?) Barra da Estiva, Serra do
Sincorá [above Sincorá Velho?], reported by Ule (1908).

MINAS GERAIS: Mun. Santana do Riacho, Serra do Cipó, 20 Oct. 1973, *Joly et al.* in CFSC 4606 (SP);
l.c., Alto do Palácio, 28 Dec. 1948, *Balegno & Cuezzo* 3992 (R); Mun. Jaboticatubas, Serra das
Bandeirinhas, 9 Sep. 1987, *Zappi et al.* in CFCR 10849 (SPF); Mun. Caeté, Serra da Piedade, Sep. 1915,
F.C. Hoehne 6349 (SP), 6348 (R), 16 Jan. 1971, *Irwin et al.* 30460 (MO), 24 Feb. 1987, *Zappi* in CFCR
10343 (SPF), s.d., *V.C. Souza & C.M. Sakuragui* 2047 (ESA, K); Mun. Santa Bárbara/Catas Altas, Serra
do Caraça, 1936, *Brade* s.n. (RB); l.c., 11 Sep. 1990, *Zappi* 239 (HRCB, BHCB); l.c., Mun. Catas Altas,
between Bocaina and Gruta da Bocaina, 1 Aug. 2001, *Taylor* (K, photo); Mun. Itabirito, c. 50 km SE of
Belo Horizonte, Pico do Itabirito, 17 Feb. 1968, *Irwin et al.* 19845 (MO); Mun. Ouro Preto, Mar. 1892,
Ule s.n. (R 91021); Mun. Caparaó, Parque Nacional da Serra do Caparaó, 30 Apr. 1988, *Krieger et al.* in
FPNC 92 (CESJ, SPF), 17 Dec. 1988, *Krieger et al.* in FPNC 625 (CESJ), 19 Nov. 1988, *Krieger et al.* in
FPNC 848 (CESJ), 24 Oct. 1989, *Mello-Silva et al.* 103 (SPF); Mun. Viçosa, Fazenda de Aguada, 27 Nov.
1930, *Y. Mexia* 5373A (US, F, MO); Mun. São João del Rei/Tiradentes, June 1896, *A. Silveira* 1279 (R);
l.c., Serra de São José, 1300–1400 m, 3 Oct. 1987, *M. Perón* 291 & 315 (RB); Mun. Lima Duarte, Serra
da Capoeira Grande, 10 July 1987, *F.R.S. Pires et al.* 21541 (SPF); l.c., Parque Estadual do Ibitipoca, 16
Dec. 1986, *H.C. Sousa et al.* 9829 (BHCB); Mun. São Tomé das Letras, 27 Sep. 1968, *M.C. Vianna* 357
(GUA); Mun. Lima Duarte/Rio Preto, Chapadão da Serra Negra, 3 Oct. 1959, *Castellanos* 22537 (R).

ESPÍRITO SANTO: Mun. Santa Teresa, Reserva Biológica de Nova Lombardia, 14 July 1976, *A.G. Pedrine*
s.n. (MBML); l.c., morro da estação de TV, 14 Nov. 1985, *W. Pizziolo* 200 (MBML); l.c., Fazenda Santa
Lúcia, 19°57'24"S, 40°32'37"W, 25 Nov. 1999, *Zappi et al.* 440 (K, UEC).

RIO DE JANEIRO: Mun. Santa Maria Madalena, 'subida da Pedra Dubois', 21°56'36–42"S,
41°59'29–33"W, 22 Nov. 1999, *Zappi et al.* 368 (K, UEC).

CONSERVATION STATUS. Least Concern [LC] (1); PD=2, EI=1, GD=2. Short-list score (1×5) = 5.

This species presents rather diverse morphology, partly related to the conditions under
which it grows (Zappi 1991). Its flowers are also unusually variable, especially in the degree
to which the outer perianth-segments expand, some forms scarcely opening except to reveal
the anthers and stigma-lobes. Most of these variations appear to be determined genetically.

The above synonymy is almost certainly incomplete, since many of the names of potential variants are inadequately typified and lack details as to the key floral characters that distinguish the two species treated here.

2. Hatiora cylindrica Britton & Rose, Cact. 4: 219 (1923); Backeberg, Die Cact. 2: 709, Abb. 652 (1959). Holotype: Brazil, Rio de Janeiro, Ilha Grande, 22–24 July 1915, *Rose & Löfgren* (US).

Rhipsalis cylindrica (Britton & Rose) Vaupel, Die Kakteen 1: 39 (1925), nom. illeg., non (Vell.) Steud. (1841) nec Kuntze (1891). *Hatiora salicornioides* f. *cylindrica* (Britton & Rose) Supplie, Rhipsalidinae: unpaged [p. 65] (1990). *Rhipsalis salicornioides* var. *cylindrica* (Britton & Rose) Kimnach in Cact. Succ. J. (US) 68: 156 (1996).

Small shrub to c. 50 cm with erect, strictly acrotonic branching, up to 4 new stem-segments arising from the apex of existing segments, cylindrical, 10–40 × 2–3 mm, dark green to reddish-purple when exposed to intense sunlight, with a conspicuous terminal composite areole, bearing a mass of very short trichomes and few acicular spines to c. 1 mm, and sometimes with a few lateral areoles, but these inconspicuous and inactive. Flower-buds reddish yellow, flowers one per terminal areole, orange-yellow, fully expanded, funnelform, c. 10 × 10 mm, perianth-segments lanceolate, acute, 6 × 2 mm, the outer series c. 7, reddish orange, inner segments c. 8, orange-yellow; stamens c. 20, to 5 mm, yellowish, anthers small, yellow; pericarpel 4 mm, red; style creamy white, 5 mm, stigma-lobes 4, slender, curled. Fruit dark red-magenta, turbinate to globular, 6 mm in diam.

Humid forest element: epiphytic and terrestrial (in sand-dunes and probably as a lithophyte), *restinga, mata de brejo* and *Mata atlântica*, from near sea level to c. 1200 m, eastern Bahia and (?) southern Espírito Santo; South-eastern Brazil (but rarely collected). Map 26.
BAHIA: E Bahia, Mun. Salvador, Itapuã, in sand-dunes away from the beach, Mar. 1989, *R. Augusto* s.n., cult. ibid., Cruz das Almas, 7 Aug. 2002 (HUEFS); Mun. Venceslau Guimarães, c. 40 km W of Teolândia, near Nova Esperança, 1 Apr. 1993, *S.C. Sant'Ana & L.A.M. Silva* 319 (CEPEC, K), l.c., 8 km above Rio Vermelho, 2 km above junction of road to Taquara, 13°36'S, 39°47'W, 15 May 1992, *W.W. Thomas et al.* 9359 (CEPEC); Mun. Barra do Choça, Sep. 2003, *M. Machado* (photos).
ESPÍRITO SANTO: (?) Mun. Castelo, Forno Grande, June 1949, *Brade* 19978 (RB) — identification provisional (could be referable to the preceding species).

CONSERVATION STATUS. Vulnerable [VU B2ab(iii)] (2); area of occupancy estimated to be < 2000 km^2; PD=2, EI=1, GD=1. Short-list score (2×4) = 8. Bahian and doubtless other localities are declining due to growth in tourism and deforestation, but it may be included in protected areas.

The above description, kindly prepared by Marlon Machado, is based on living material seen by NT from Itapuã, Bahia, where the plant was discovered by Sra Rita Augusto (Cruz das Almas) growing directly in the sand of the coastal dunes. The fully expanded flowers of this species readily distinguish it from other, more or less cylindrical forms referable to *H. salicornioides,* which have erect inner perianth-segments adpressed to the stamens. This may indicate divergence in their pollination syndromes.

The erect stem-segments of ± constant size and absence of indeterminate extension shoots are the only significant characters that separate this taxon from superficially similar *Rhipsalis* species.

Excluded taxon

A single collection of the magenta-flowered sister-species, *Hatiora herminiae* (Porto & A. Cast.) Backeb. ex Barthlott, made in October 1942, from the Estação Experimental Coronel Pacheco, Minas Gerais (*Heringer* 911, SP), is assumed to represent a plant that was in cultivation.

14. SCHLUMBERGERA Lem.

in Rev. Hort., ser. 4, 7: 253 (1858). Type: *S. epiphylloides* Lem. (= *S. russelliana* (Hook.) Britton & Rose).

Including *Epiphyllanthus* A. Berger (1905).

Literature: Hunt (1969), Barthlott & Rauh (1975), Barthlott & Taylor (1995: 74), Taylor & Zappi (1995).

Epiphytic or epilithic; stems terete, few-ribbed, 2–3-winged or flattened, segmented, all stem-segments of determinate growth, oblong to obovate, sometimes truncate. Areoles scattered over the surface of the joints or confined to the ribs or to the crenate or serrate margins, crowded into composite areoles at the apex of segments and there originating new joints and flowers (except in *S. opuntioides*). Spines bristly, short or absent. Flowers scarcely to strongly zygomorphic, red, pink or purplish (rarely white); pericarpel terete or ribbed, greenish to brownish, naked, flower-tube elongate, with bract-scales only at its very base, straight to oblique at apex; perianth-segments spreading or recurved, in various series, giving the impression of a flower inserted within a flower; stamens numerous, comprising an outer and a distinct inner series united at their bases to form a short tube around the style and partially enclosing the nectar-chamber; style and stamens exserted; stigma-lobes erect, connivent. Fruit globose to obconic, ribbed or terete, the perianth deciduous. Seeds subreniform to ovate in outline, 1–1.7 mm, testa dark brown-black, shiny, with intercellular depressions.

An endemic Brazilian genus of 6 species, ranging from southern Espírito Santo (Domingos Martins) and adjacent Minas Gerais (Serra do Caparaó) to Rio de Janeiro, southernmost Minas Gerais and south-eastern São Paulo, in the mountains of the Serra do Mar and Serra da Mantiqueira (*Mata atlântica,* to 2700 metres altitude). Three species are native to the area covered here, the first being endemic. Plates 23.3–24.2.

1. Terminal stem-segments globose, short-cylindric or linear, rounded in cross-section . . . .
. . . . **2. microsphaerica**
1. Terminal stem-segments orbicular to obovate or truncate in outline, laterally compressed 2
2. Stem-segments almost unarmed, but margins toothed; flowers scarcely zygomorphic . . **1. kautskyi**
2. Stem-segments covered in areoles with pungent, very thin spines; flowers strongly
zygomorphic . . . .**3. opuntioides**

1. Schlumbergera kautskyi *(Horobin & McMillan) N. P. Taylor* in Bradleya 9: 90 (1991); McMillan & Horobin in Succ. Pl. Res. 4: 23–26, pl. 3.1–3.3 (1995). Holotype: Brazil, Espírito Santo, Mun. Domingos Martins, 900 m, May 1986, *R. Kautsky* in *Rauh* 67558, cult. Univ. Bonn, Germany (BONN).

S. truncata subsp. *kautskyi* Horobin & McMillan in Epiphytes 15(57): 8 (1991). *S. truncata* var. *kautskyi* Horobin & McMillan in ibid. 14: 111–115 (1990), nom. inval. (Art. 37.1).

Epilithic or less often epiphytic, pendent; stem-segments 2.2–3.5(–4.0) × 1.4–1.8(–2.5) cm; compressed, truncate, with 1–2 sharp teeth on each side, or at least at apex on either side of the composite areole; epidermis dark green, often with reddish marks around the margins; primary stem of seedling flattened. Areoles in the axils of the marginal teeth and at apex; spines bristly or absent. Flowers self-compatible, scarcely zygomorphic, to c. 5 × 2.7 cm; pericarpel 12 × 5–6 mm, narrow, cylindric, 4-angled, green to dark reddish; flower-tube pink, with magenta outer perianth-segments, inner segments spreading or recurved at anthesis, lanceolate, magenta-pink. Fruits 25 × 19 mm, turbinate, 4-angled, pericarp yellow-green when ripe, with a reddish tinge. Seed 1 mm.

Southern humid forest (inselberg) element: lithophytic on inselbergs, rarely epiphytic, *Mata atlântica*, 900–1300 m, central-southern Espírito Santo. Endemic. Map 26.
ESPÍRITO SANTO: Mun. Domingos Martins, Pico da Pedra Azul, *Kautsky* s.n. (K, in spirit), May 1986, *R. Kautsky* in *Rauh* 67558, cult. Univ. Bonn, Germany (BONN, holo.); Mun. Alfredo Chaves, São Bento de Urânia, *R. Kautsky*, reported by Horobin & McMillan, l.c. (1990).

CONSERVATION STATUS. Endangered [EN B2ab(iii)] (3); area of occupancy estimated to be < 500 km²; PD=2, EI=1, GD=1. Short-list score (3×4) = 12. Area being developed as a residential centre.

Disjunct from its nearest relatives in the Serra dos Órgãos (RJ) by over 250 kilometres.

2. Schlumbergera microsphaerica *(K. Schum.) Hövel* in Kakt. and. Sukk. 21: 186 (1970); Heath in Calyx 2: 64 (1992); McMillan & Horobin in Succ. Pl. Res. 4: 27, pl. 6.1–6.4 (1995). Type: Brazil, Rio de Janeiro, *Glaziou* s.n. (B†). Neotype (Taylor 1991b; Heath 1992b): Arch. Jard. Bot. Rio de Janeiro 2: tab. 5, fig. B (1918).

Cereus microsphaericus K. Schum. in Martius, Fl. bras. 4(2): 197 (1890). *Epiphyllanthus microsphaericus* (K. Schum.) Britton & Rose, Cact. 4: 181 (1923). *Arthrocereus microsphaericus* (K. Schum.) A. Berger, Kakteen: 337 (1929), tantum quoad typ. *Cereus damazioi* Weing. in Monatsschr. Kakt.-Kunde 21: 91–94 (1911), nom. illeg. (Art. 52.1). *Trichocereus damazioi* (Weing.) Werderm., Bras. Säulenkakt.: 94 (1933), nom. illeg. Type: as above.
Cereus obtusangulus K. Schum., l.c. 198 (1890). *Epiphyllanthus obtusangulus* (K. Schum.) A. Berger in Rep. (Annual) Missouri Bot. Gard. 16: 84 (1905). *Zygocactus obtusangulus* (K. Schum.) Loefgr. in Arch. Jard. Bot. Rio de Janeiro 2: 28 (1918). *Epiphyllum obtusangulum* (K. Schum.) Vaupel, Die Kakteen, Lfg 2: 88 (1926). *Schlumbergera obtusangula* (K. Schum.) D.R. Hunt in Kew Bull. 23: 260 (1969). *S. microsphaerica* f. *obtusangula* (K. Schum.) P. V. Heath in Calyx 2: 64 (1992). Type: Brazil, Rio de Janeiro, *Glaziou* s.n. (C, missing). Neotype (Taylor 1991b): Schumann, Gesamtb. Kakt.: fig. 30 (1897).
? *Cereus parvulus* K. Schum., l.c. 197 (1890). *Schlumbergera microsphaerica* f. *parvula* (K. Schum.) P. V. Heath in Calyx 2: 64 (1992). Type: Brazil, Rio de Janeiro, *Glaziou* s.n. (B†).
Zygocactus candidus Loefgr. in Arch. Jard. Bot. Rio de Janeiro 2: 30 (1918). *Epiphyllanthus candidus* (Loefgr.) Britton & Rose, Cact. 4: 182 (1923). *Schlumbergera microsphaerica* subsp. *candida* (Loefgr.) D.R. Hunt in Succ. Pl. Res. 4: 79 (1995). Type: Rio de Janeiro, Itatiaia, Várzea de Ayuruóca (not extant).

Epiphytic or epilithic, ± erect, to 40 cm, densely branched; stem-segments 5–40 × 2–5 mm, globose, short-cylindric, angled or not, or linear, sometimes the terminal segments compressed, epidermis bright green, often with reddish markings or entirely reddened. Areoles scattered over the surface, or mainly at apex of the terminal segments; spines bristly, to 5 mm, brown or yellowish. Flowers self-compatible, zygomorphic, to 4–4.5 cm; pericarpel rounded, smooth or ridged; flower-tube pale pink, with pink bract-scales; perianth-segments spreading at anthesis, lanceolate, pale or deep pink; pericarpel reddish brown. Fruits 5–10 mm diam., turbinate to rounded, indistinctly 5-ribbed, green or white. Seed c. 1 mm.

Southern humid forest element: lithophytic or epiphytic, *mata de neblina*, at > 2000 m, Serra do Caparaó (Pico de Bandeira), Minas Gerais/Espírito Santo; Rio de Janeiro (Serra de Itatiaia). Map 26.
MINAS GERAIS: Serra do Caparaó, 2200 m, Sep. 1941, *Brade* s.n. & 17024 (RB); Parque Nacional do Caparaó, 2300–2700 m, 29 Sep. 1977, *L. Krieger* 15161 (CESJ, SPF), 4 Oct. 1966, *Strang* 715 (HB), 29 June 1955, *N. Santos & Z. Campos* s.n. (R 52203); ibid., Terreirão, Cachoeira, 30 Apr. 1989, *L. Krieger et al.* in FPNC 956 (CESJ).

CONSERVATION STATUS. Data Deficient [DD], but much of its range is within protected areas.

The name *S. microsphaerica* must be used in preference to the more familiar *S. obtusangula*, since the former, as *Epiphyllanthus microsphaericus*, was the name accepted by Britton & Rose (1923: 181), the first authors to treat these two equally priorable names as synonymous.

Scheinvar (1985) reports *S. obtusangula* from northern Santa Catarina in Southern Brazil, but this is assumed to be an error and it has not been possible to locate the collection she cited. It is probably no coincidence that the locality she cites is a site for *Hatiora rosea* (Lagerh.) Barthlott, which sometimes develops cylindric, ribbed stem-segments like those of *S. microsphaerica*. This may also explain why Braun & Esteves Pereira (2002: 79) indicate that *Schlumbergera* occurs in Santa Catarina.

The distribution of *S. microsphaerica* is markedly disjunct between the localities cited above and its *locus classicus* on Itatiaia, by some 350 kilometres. Living plants from these two areas should be compared in view of the distances involved, but photographs of flowering plants in habitat at the eastern locality, shown to the authors by Inês Ribeiro de Andrade have confirmed the identity accepted here (see also Plate 24.1).

3. Schlumbergera opuntioides *(Loefgr. & Dusén)* D. R. *Hunt* in Kew Bull. 23: 260 (1969); McMillan & Horobin in Succ. Pl. Res. 4: 26–27, pl. 3.4, 3.5, 5.1 & 13.6 (1995). Holotype: Brazil, Rio de Janeiro, Itatiaia, 2400 m, 11 June 1902, *Dusén* 1530 (R).

Epiphyllum opuntioides Loefgr. & Dusén in Arq. Mus. Nac. Rio de Janeiro 13: 49 (1905). *Zygocactus opuntioides* (Loefgr. & Dusén) Loefgr. in Arch. Jard. Bot. Rio de Janeiro 2: 26, tab. 4 (1918). *Epiphyllanthus opuntioides* (Loefgr. & Dusén) Moran in Gentes Herb. 8: 338 (1953). *E. obovatus* Britton & Rose, Cact. 4: 180 (1923), nom. illeg. (Art. 52.1).
Epiphyllum obovatum Engelm. ex K. Schum., Gesamtbeschr. Kakt.: 224 (1897), nom. inval. (provisional name based on *Sellow* 884, B†).

Epiphyte or lithophyte, pendent, to 50 cm; stem-segments 15–70 × 5–30 mm, to 9 mm thick, orbicular to obovate, compressed, epidermis dark green to purplish. Areoles woolly, scattered over the surface of the segments, composite terminal areoles lacking; spines to 5 mm, pungent but delicate, golden, more densely developed on the older segments where up to 80 per areole. Flowers self-incompatible, 1–3 per stem-segment, strongly zygomorphic, to 6 × 4.5 cm; pericarpel obconic or 4-angled, smooth, red or white; flower-tube c. 3–4 cm, pale pink, with magenta outer perianth-segments, inner segments spreading at anthesis, those on the lower side of the flower reflexed, lanceolate, magenta-pink; stamens red, anthers bearing red-brown pollen; stigma-lobes 4–7, white; pericarpel slightly 5–7-angled, red. Fruits c. 10 mm diam., spherical to 4–5-angled, green. Seed 1.6–1.7 × 1.0 mm, glossy brown (Barthlott & Hunt 2000: pl. 2.5–6). Chromosome number: 2n = 22 (Barthlott 1976).

Southern humid forest element: lithophytic/epiphytic, *capão de mato*, c. 1700 m, Serra da Mantiqueira, southern Minas Gerais; to north-western Rio de Janeiro (Itatiaia) and eastern São Paulo (Campos do Jordão). Map 26.
MINAS GERAIS: Mun. Lima Duarte, Parque Estadual do Ibitipoca, 27 July 1991, *Zappi* 258 (SPF, CESJ).

CONSERVATION STATUS. Near Threatened [NT] (1); area of occupancy estimated to be < 500 km²; PD=2, EI=1, GD=1. Short-list score (1×4) = 4. Included within some protected areas and dependent on their maintenance.

Tribe **CEREEAE** Salm-Dyck

Columnar, treelike to shrubby or semi-erect, to low-growing and ± globose; stems ribbed; pericarpel (and tube) of flower and fruit with minute bract-scales or naked, glabrous and lacking bristles/spines (except in 2 *Cipocereus* spp. and *Cereus* subg. *Mirabella*), but sometimes immersed in a woolly/bristly, lateral or terminal cephalium. Type: *Cereus* Mill.

The most important tribe in Eastern Brazil. Its circumscription follows that employed by Taylor & Zappi (1989) and Barthlott & Hunt (1993), but this is not supported by recent DNA gene sequence studies (Soffiatti unpubl.) and will very likely have to be revised. Analysis of the chloroplast segment *rpl*16 suggests that *Brasilicereus*, *Praecereus* and *Cereus* are not closely related to the remainder of Brazilian taxa here included in the tribe, of which the Trichocereeae could be the sister group. However, an earlier survey of the family, utilizing surface waxes (*n*-alkanes), lends support to the distinctiveness of tribe Cereeae as employed here (Maffei *et al.* 1997). The following genus, displaying a conspicuously scaly pericarpel, may in any case be closer to *Browningia* and its relatives (Browningieae):

15. **BRASILICEREUS** Backeb.

in Blätt. Kakteenf. 1938(6): [20] (1938). *Cereus* subg. *Brasilicereus* (Backeb.) P. J. Braun (1988). Type: *Brasilicereus phaeacanthus* (Gürke) Backeb. (*Cereus phaeacanthus* Gürke).

Erect plants, usually branched above the ground, 0.5–5.0 m, roots fibrous. Stems erect or inclined and partly supported by surrounding vegetation, with seasonal growth increments separated by slight constrictions, giving a jointed appearance; ribs 7–14, low, sinuses straight; central cylinder very woody, cortex ± lacking mucilage. Areoles felted, longer hairs absent, developing golden to brownish, adpressed to spreading spines, central spine(s) porrect, much longer than radials. Flower-bearing region of stems not differentiated, apparently quite random; flowers nocturnal, shortly tubular, to 7 cm; pericarpel and tube ridged, usually covered in acute or broad, rounded bract-scales; pericarpel and ovary locule depressed; stamens inserted in two series, the uppermost forming a 'throat ring' (as in *Echinopsis*, Trichocereeae), filaments slender, mobile; pollen 16-colpate (in *B. phaeacanthus*, Leuenberger 1976). Fruit ovoid, indehiscent, floral remnants persistent, erect, *pale brown* when not decayed (not strongly blackening as in some other Cereeae), deeply sunken into fruit apex, pericarp somewhat ridged and bearing large, rounded, greenish, red, brownish or purplish bract-scales; funicular pulp white. Seeds black, cochleariform, applanate, shiny; testa-cells flat.

An endemic genus of 2 species, possibly related to *Praecereus* Buxb. (with 2 South American species, *P. euchlorus* being the most widespread, ranging into Central-western & South-eastern Brazil) and replacing it in Eastern Brazil; Map 7 (Taylor 1992a: 25). The genus is restricted to the South-eastern *campos rupestres* and Southern *caatingas-agrestes*. Plates 24.3–25.4.

1. Shrubby to tree-like, branched above ground; flowers with ovate, truncate bract-scales; pericarpel c. 2–3× wider than long; stems 2–6 cm diam.; hilum–micropylar region forming an angle of 20–30° with the long-axis of seed (*caatinga/agreste* etc., E & S Bahia & N Minas Gerais) . . . . **1. phaeacanthus**
1. Stem solitary or poorly branched at base; flowers with acute to acuminate bract-scales; pericarpel < 2× wider than long; stems 1.5–2.0 cm diam.; hilum–micropylar region forming an angle of 60° with the long-axis of seed (*campo rupestre/carrasco*, Mun. Grão Mogol etc., Minas Gerais) . . . . **2. markgrafii**

1. Brasilicereus phaeacanthus *(Gürke) Backeb.* in Jahrb. Deutsch. Kakteen-Ges. 1941(2): 50 (1942). Type: Bahia, nr Maracás, Sep. 1906, *Ule* 7022 (HBG, lecto. designated here [K, photo.]).

Cereus phaeacanthus Gürke in Monatsschr. Kakt.-Kunde 18: 57 (1908). *Cephalocereus phaeacanthus* (Gürke) Britton & Rose, Cact. 2: 57 (1920). *Pilocereus phaeacanthus* (Gürke) Backeb. in Backeberg & F. M. Knuth, Kaktus-ABC: 333 (1935, publ. 1936).

Brasilicereus breviflorus F. Ritter, Kakt. Südamer. 1: 228 (1979). *Cereus phaeacanthus* var. *breviflorus* (F. Ritter) P. J. Braun in Bradleya 6: 87 (1988). *Brasilicereus phaeacanthus* subsp. *breviflorus* (F. Ritter) P. J. Braun & E. Esteves Pereira in Succulenta 74: 83 (1995). Holotype: Brazil, Minas Gerais, Itaobim, 1965, *Ritter* 1337 (U).

Shrubby to tree-like, mostly somewhat branched above ground level, to 5 m. Stems erect, inclined on surrounding vegetation or sometimes semi-decumbent on rocks, arching over and rooting at apex, 2–6 cm diam.; ribs 7–14, to 5 × 6 mm; epidermis dull or dark to grey-green. Areoles 2–3 mm diam., 4–8 cm apart on the ribs. Central spines 1–3, to 3 cm; radials 10–12, to 1.5 cm. Flowers 4–7 × 4–5.2 cm; pericarpel very strongly depressed, 5–7 × 15–18 mm, pale green; flower-tube infundibuliform, c. 3.2 × 1.7–2.0 cm, pericarpel and tube bearing broad, ovate, truncate, reddish to purplish bract-scales, these rarely absent from the pericarpel; perianth-segments to 2.7 cm, outer segments spreading, lanceolate, fleshy, dark red at tips, inner segments spreading to erect, spathulate, delicate, white to very pale green; nectar-chamber 11 × 13 mm; style 40–48 × 2 mm, stigma-lobes c. 10, exserted in relation to the anthers; ovary locule strongly depressed in longitudinal section. Fruit globose to ovoid, to 3.5 cm diam.; pericarp greenish to dull red or slightly purplish. Seeds c. 2–3 × 1.4 mm, cochleariform, hilum–micropylar region forming an angle of 20–30° with long-axis.

Southern *caatinga* element: in *caatinga-agreste,* often associated with granite/gneiss inselbergs, 40–920 m, central-eastern to central-southern Bahia and central-northern and north-eastern Minas Gerais. Endemic to the core area of Eastern Brazil. Map 27.

BAHIA: Mun. Macajuba, 26 km N of Rui Barbosa on road to Baixa Grande, 9 Feb. 1991, *Taylor et al.* (obs.); Mun. Rui Barbosa, 1.5 km SW of town towards main road to Itaberaba, 7 Feb. 1991, *Taylor et al.* 1570 (HRCB, CEPEC, K, ZSS); Mun. Ibipitanga, 5–10 km N of Ibiajara towards Novo Horizonte, 12°56'S, 42°10'W, Aug. 2003, *M. Machado* (obs.); Mun. Cachoeira (?), Vale dos Rios Paraguaçu e Jacuípe, Morro Belo, 12°32'S, 39°5'W, Dec. 1980, *Iscardino et al.* in Grupo Pedra do Cavalo 976 (CEPEC); Mun. Elísio Medrado/Castro Alves, S of Monte Cruzeiro, 25 June 1915, *Rose & Russell* 20053 (US, NY); Mun. Marcionílio Sousa, Machado Portella, 19–23 June 1915, *Rose & Russell* 19912 (US, NY); Mun. Iramaia, 40 km NW of 'Pé-de-Serra' on road from Maracás, 5 Feb. 1991, *Taylor et al.* (obs.); Mun. Santa Inês, nr Caldeirão, 26 June 1915, *Rose & Russell* 20058 (US); Mun. Jaguaquara [Toca da Onça], 27–29 June 1915, *Rose & Russell* 20095 (US, NY); Mun. Maracás, Sep. 1906, *Ule* 7022 (K, photo ex HBG); Mun. Livramento do Brumado, 7 km from Rio de Contas, 13°38'S, 41°50'W, 23 Nov. 1988, *Taylor & Zappi* in *Harley* 25541 (SPF, K); Mun. Jequié, reported as *Horst & Uebelmann* 746 by Hofacker & Braun (1998); l.c., 29.5 km E of Porto Alegre towards Jequié, 4 Feb. 1991, *Taylor et al.* (obs.); Mun. Tanhaçu, 21 km N of Sussuarana towards Contendas do Sincorá, 3 Feb. 1991, *Taylor et al.* 1547 (HRCB, CEPEC, K, ZSS); município unknown, between Vitória da Conquista and Jequié, 22 Sep. 1946, *Duarte* 9310, *Pereira* 10023 & *Barroso* (RB, HB); Mun. Caitité, c. 9 km ESE from town towards Brumado, then c. 2 km S,

27 Aug. 1988, *Eggli* (ZSS, photo); Mun. Aracatu, c. 20 km N of Anajé towards Sussuarana, 17 Aug. 1988, *Eggli* (ZSS, photo); Mun. Urandi, Jan. 1964, *Ritter* 1239 (SGO 124989, *fide* Eggli *et al.* 1995: 501–502); Mun. Vitória da Conquista, c. 32 km SE of town towards Itambé, 16 Aug. 1988, *Eggli* 1176 (ZSS). **MINAS GERAIS:** cent.-N Minas Gerais, Mun. Espinosa, 4 km S of BA/MG border, 25 July 1989, *Taylor & Zappi* (obs.); Mun. Monte Azul, 7 km E of town, beyond Vila Angical, Serra Geral, 920 m, 28 Jan. 1991, *Taylor et al.* 1468 (HRCB, BHCB, K, ZSS); NE Minas Gerais, Mun. Taiobeiras / Águas Vermelhas, c. 5 km NW of Curral de Dentro towards junction with road BR 116, 13 Aug. 1988, *Eggli* 1162 (ZSS); Mun. Pedra Azul, 21 Sep. 1965, *Duarte* 9308 (HB, RB), *Duarte* 9289 & *Pereira* 10002 (RB); l.c., 8 km W of town, 15°57'S, 41°22'W, 18 Oct. 1988, *Taylor & Zappi in Harley* 25184 (SPF, K); Mun. Itinga, 5 km E of town, road BR 367, 16°36'S, 41°43'W, 19 Nov. 1988, *Taylor & Zappi in Harley* 25534 (SPF, K); Mun. Itaobim, 8 km W of town on road BR 367, 14 Dec. 1990, *Taylor & Zappi* 762 (HRCB, BHCB, K, ZSS); Mun. Padre Paraíso, c. 10 km N of town, 13 Dec. 1990, *Taylor & Zappi* (obs.).

CONSERVATION STATUS. Least Concern [LC] (1); PD=3, EI=1, GD=2. Short-list score (1×6) = 6.

This species is rather variable in stem thickness and rib-number, but this variation seems to lack any kind of geographical pattern. A form with unusually short flowers has been described from near the Rio Jequitinhonha, north-eastern Minas Gerais (*B. breviflorus* F. Ritter), but it is otherwise unremarkable and does not merit recognition at any rank when the overall variation of the species is taken into consideration. A rather different impression is given in a recent article by Hofacker & Braun (1998), in which they distinguish two subspecies in a key and illustrate each by a single collection. Their key implies that these two entities are geographically separated, the heterotypic subsp. *breviflorus* (F. Ritter) P. J. Braun & E. Esteves Pereira representing the species in Minas Gerais, while the homotypic subspecies is restricted to southern Bahia. However, this is not so, since the short-flowered population named by Ritter is only one amongst a number of variants found in Minas Gerais, none of which differs significantly from contiguous Bahian populations. The Bahian form they illustrate from Jequié under the number 'HU 746' is very far from typical of the species, having a rather peculiar naked pericarpel.

2. Brasilicereus markgrafii *Backeb. & Voll* in Arch. Jard. Bot. Rio de Janeiro 9: 155 (1949, publ. 1950); Hofacker & P. J. Braun in Kakt. and. Sukk. 49: 267–268, illus. (1998). Holotype: Brazil, Minas Gerais, Grão Mogol, Nov. 1938, *Markgraf et al.* s.n. (RB 65043).

Cereus markgrafii (Backeb. & Voll) P. J. Braun in Bradleya 6: 87 (1988).

Shrubby, solitary or poorly branched, to 2.5 m. Stems ± erect, 1.5–2.0 cm diam.; ribs 8–14, 2 × 3 mm; epidermis bright to grey-green. Areoles 2 mm diam., 4–8 cm apart on the ribs. Central spines 1–2(–3), to 4(–5) cm, radials 10–12, to 1 cm. Flowers 6 × 4 cm; pericarpel greenish, 11–12 × 8–9 mm; flower-tube infundibuliform, 4 × 1.5–1.8 cm, pericarpel and tube bearing broad, triangular, acute to acuminate bract-scales, green to strongly reddish at tips; perianth-segments to 2.8 cm, outer segments spreading, lanceolate, fleshy, dark red at tips, inner segments spreading to erect, spathulate, delicate, white to very pale green; stamens with greenish filaments; nectar-chamber 7–8 × 10–12 mm; style 60–65 × 3 mm, stigma-lobes c. 8, at the same level as the anthers; ovary locule hemiglobose to depressed in longitudinal section. Fruit globose to ovoid, to 3 cm diam.; pericarp greenish to dull red or slightly purplish. Seeds c. 2.4 × 1.8 mm, black, shiny; hilum-micropylar region forming an angle of 60° with long-axis (Barthlott & Hunt 2000: pl. 47.3).

South-eastern *campo rupestre* (Grão Mogol) element: *carrasco*, 850–1000 m, region of Grão Mogol, northern Minas Gerais. Endemic to the core area within Minas Gerais. Map 27.

MINAS GERAIS: Mun. Grão Mogol, Serra da Bocaina, 48 km W of bridge over Rio Vacaria, 28 km E of Caveira, 31 Jan. 1991, *Taylor et al.* 1505 (HRCB, BHCB, K, ZSS); Grão Mogol, [near town], Nov. 1938, *Markgraf, Mello-Barreto & Brade* s.n. (RB), 23 July, *Zappi et al.* in CFCR 9829 (SPF), 4 Sep. 1986, *Mello-Silva & Cordeiro* in CFCR 10118 (SPF); l.c., c. 18 km W of town, 21 Feb. 1969, *Irwin et al.* 23684 (K, NY, MO); 7 km S of town, 16°37'S, 42°56'W, 15 Oct. 1988, *Taylor & Zappi* in *Harley* 25069 (SPF, K); Mun. Cristália, May 1988, *Zappi* (obs.).

CONSERVATION STATUS. Endangered [EN B1ab(iii) + 2ab(iii)] (3); extent of occurrence = 434 km^2; PD=3, EI=1, GD=1. Short-list score (3×5) = 15. Ongoing deterioration of habitats in this region of Minas Gerais is occurring. Part of its range is included in the Parque Estadual Serra do Barão, Grão Mogol.

This species is clearly differentiated from the widespread and variable *B. phaeacanthus* by its acute pericarpel bract-scales and curved seeds. Its pollen-morphology merits investigation.

16. CEREUS Mill.

Gard. Dict. Abr. ed. 4 [unpaged] (1754).

Including *Mirabella* F. Ritter (1979); *Cereus* subg. *Mirabella* (F. Ritter) N. P. Taylor (1992a).

Literature: Taylor (1992a: 17, 25).

Treelike, shrubby or semi-scandent, with (2–)3–10-ribbed, triangular to cylindric or constricted-cylindric stems; roots tuberous or fibrous; vascular tissues becoming very woody; epidermis and cuticle thick and tough, often covered in whitish or bluish wax; areoles well-spaced along the ribs, usually rather spiny at least in young plants. Flowers large, elongate funnelform to salverform, nocturnal, ± sweetly scented, pericarpel and tube terete or somewhat angled, naked except for few minute scales, glabrous or with small tufts of trichomes in their axils; perianth white inside; floral remnants strongly blackening post-anthesis. Fruit ovoid to barrel-shaped, splitting laterally or opening at apex, pericarp deep pink to reddish or yellow, sometimes strongly glaucous until ripe, floral remnant persistent or early-deciduous leaving a depressed scar. Seeds c. 2–3 mm, black, testa smooth to ruminate.

A genus of some 20 poorly understood, South American species, divided between 4 subgenera, of which two are represented in Eastern Brazil (Subg. *Cereus* & Subg. *Mirabella*) by a total of 6 or 7 species (one probably undescribed, see 5c). A third subgenus (Subg. *Ebneria* (Backeb.) D. R. Hunt, with *Praecereus* excluded) is represented in Central-western Brazil by 4 described species, *C. spegazzinii* F. A. C. Weber, *C. kroenleinii* N. P. Taylor (1995b), *C. adelmarii* (Rizzini & A. Mattos) P. J. Braun and *C. saddianus* (Rizzini & A. Mattos) P. J. Braun, although Braun & Esteves Pereira (2002: 79) indicate that is also found in Bahia, which seems unlikely. Subgenera *Ebneria* and *Mirabella* are considered to be vicariant groups, the latter replacing the former in Eastern Brazil (Taylor 1992a): Map 8.

In Eastern Brazil the genus has one or more representatives in all of the major vegetation types (see subgenera, below). Plates 26–29.2.

1. Floral remnant early-deciduous, leaving a well-defined scar at the apex of the ± terete (not strongly angled) developing fruit; rootstock fibrous, not rhizomatous-tuberous (in Brazilian spp.); semi-decumbent shrubs with branches to > 5 cm diam. or erect and tree-like to > 4 m (Subg. *Cereus*) 2

1. Floral remnant not deciduous from fruit or breaking off above its base to leave a blackened appendage on the strongly angular unripe fruit; rootstock rhizomatous-tuberous; semi-scandent shrubs to < 3.5 m, with non-erect branches to 5 cm diam. (Subg. *Mirabella*) 5

2. Fruit (when undamaged) opening from apex into c. 3 segments (S Minas Gerais, N Rio de Janeiro) . . . . **6. hildmannianus**
2. Fruit opening by a single lateral split, often from near base 3

3. Tree-like to > 4 m, with a ± well-defined, much thickened trunk; fruit pinkish- to purplish-red (NE Brazil to cent. Minas Gerais) . . . . **5. jamacaru**
3. Semi-decumbent or low shrub, rarely treelike, 0.5–4.0 m; fruit pinkish red (coastal sand & rocks) or yellow (W Espírito Santo, S edge of Minas Gerais & N Rio de Janeiro) 4

4. Ribs 3–5; flowers 14–25 cm (mainland Brazil) . . . . **3. fernambucensis**
4. Ribs 5–9; flowers c. 13 cm (Fernando de Noronha) . . . . **4. insularis**

5. Ribs (2–)3–4(–6), acute; wood yellow beneath the bark (*caatinga*, Piauí, Ceará, Pernambuco, N & S Bahia to (?) cent.-N Minas Gerais) . . . . **2. albicaulis**
5. Ribs (3–)4–6, rounded; wood whitish beneath the bark (*cerrado & cerrado-caatinga* ecotone, SW Maranhão, W Bahia, N & cent.-E Minas Gerais) . . . . **1. mirabella**

Subg. *Mirabella* (F. Ritter) N. P. Taylor (Nos 1 & 2): roots tuberous; stems semi-scandent, slender; bract-scales of pericarpel and tube with conspicuous trichomes and sometimes fine spines in their axils; fruit bearing persistent floral remnant at apex. More or less restricted to sandy substrates in *cerrado* and *caatinga*.

Kiesling (1994) has transferred the two species treated here to *Monvillea* Britton & Rose, which is typified by *Cereus cavendishii* Monv. ex Lem. Hunt (1988) drew attention to the view already expressed by others that this name and its better-known synonym, *C. paxtonianus* Monv. ex Salm-Dyck, had been misapplied by J. D. Hooker, and later by Britton & Rose, to plants now correctly known as either *Praecereus euchlorus* (F. A. C. Weber) N. P. Taylor or *P. saxicola* (Morong) N. P. Taylor (both described from central-eastern South America). As to type, Hunt suggested that *C. cavendishii* and, therefore also *Monvillea*, were referable to *Acanthocereus* Britton & Rose. Heath (1992a) characteristically disagreed with Hunt's view and neotypified *C. cavendishii* with an illustration published by J. D. Hooker (1899), to maintain the usage established by Britton & Rose. Heath argued that the type locality given for *C. cavendishii*, namely Cartagena, Colombia, was an error. However, photographs of *Acanthocereus tetragonus* (L.) Hummelinck from northern Colombia preserved at NY, showing juvenile growth stages, strongly suggest that *Cereus cavendishii* could have been based on a juvenile *Acanthocereus,* or possibly *Pseudoacanthocereus,* from the region of Cartagena, rather than a *Praecereus* from central South America, and that Heath's neotypification should be superseded. It also should be noted that the genus *Praecereus*, represented by *P. euchlorus* subsp. *smithianus* (Britton & Rose) N. P. Taylor, does occur in northern South America (Colombia & Venezuela), but this subspecies has stems with 8–15 ribs (not 4–6 as required by the original description of *C. cavendishii*): see Map 7.

1. **Cereus mirabella** *N. P. Taylor* in Bradleya 9: 85 (1991) et ibid. 10: 21, fig. F (1992). Holotype: Brazil, Minas Gerais, c. 15 km W of Água Boa, *Ritter* 1238 (U).

Mirabella minensis F. Ritter, Kakt. Südamer. 1: 111 (1979), non *Cereus minensis* Werderm. (= *Cipocereus minensis* (Werderm.) F. Ritter). *Monvillea minensis* (F. Ritter) R. Kiesling in Cact. Succ. J. (US) 66: 165 (1994).

Rootstock tuberous, with a large yam-like taproot; plant shrubby, usually ± decumbent upon or scrambling over other vegetation, stems ascending to 1 m above the ground, to 2 m long and 3 cm diam., strongly glaucous when new, wood whitish beneath the bark; ribs (3–)4–6, rounded (5–6 in seedlings); areoles 3–5 mm diam., with blackish glandular trichomes at first, later whitish, 2–4 cm apart on the ribs; spines 3–6 (more in seedlings and young basal shoots), needle-like, brownish, paler tipped, the longest ± central, to 25 mm. Flower (Taylor 1992a: 21, Fig. F) narrow funnelform, c. 18.5 × 10 cm; tube angular, c. 12 × 1.3–1.5 cm, bearing occasional bract-scales with trichomes in their axils, stamens inserted in uppermost 2–2.5 cm, to 5 cm; style c. 14 cm, stigma-lobes c. 10, to 9 mm, whitish; perianth-segments c. 4–5 cm, narrow, white; pericarpel oblong, c. 22 × 15 mm, angular, bearing minute bract-scales with trichomes and occasional minute spines in their axils. Fruit 3.5–4.0 × 2–3 cm, tapered at apex, with a few minute scales bearing trichomes and sometimes minute spines near base, pinkish, bearing the blackened perianth remains. Seed c. 2.4 × 1.6 mm, testa ± smooth, shiny black.

Cerrado element: mostly in sandy phases of the *cerrado* and more open places of the *cerrado-caatinga* ecotone, c. 150–750 m, south-western Maranhão and western Bahia to cent.-N and W Minas Gerais (Rio São Francisco drainage), and disjunctly in E-cent. Minas Gerais (Rio Doce/Rio Jequitinhonha watershed); Tocantins (?) and Goiás. Map 28.

MARANHÃO: Mun. Carolina, 16 km E of town, road BR 230, Serra da Madeira, 7°21'S, 47°25'W, 5 July 1993, *T.M. Sanaiotti* (K, photo).

BAHIA: W Bahia, Mun. Formosa do Rio Preto, Faz. Estrondo, 11°7'7"S, 45°29'36"W, 12 Nov. 1997, *D. Alvarenga et al.* 1072 (K); Mun. Barreiras, 31.5 km W of town on road BR 020, 19 Jan. 1991, *Taylor et al.* 1442 (HRCB, CEPEC, K, ZSS); (?) Mun. Angical, 20 km W of Cristópolis on road BR 242, *Horst & Uebelmann* 206 [Uebelmann, mss.]; Mun. Carinhanha, 14°14'S, 43°52'W, 2001, *J. Jardim* 3574 (CEPEC).

MINAS GERAIS: W Minas Gerais, Mun. Januária, 'distrito de Fabião, 2 km na estrada partindo do Abrigo do Malhador', 15°7'16"–15°8'57"S, 44°15'20"–44°14'13"W, 26 Oct. 1998, *J.A. Lombardi* 2123 (K, BHCB); Mun. São Francisco, near Urucuia, 1987, illustrated by Braun & Esteves Pereira (2002: 143, Abb. 156); Mun. João Pinheiro, *fide* Uebelmann (1996): HU 584 [record outside E Brazil area]; cent.-N Minas Gerais, Mun. Varzelândia, 36 km ENE of town on dirt road to Cachoerinha (near where the road MG 401 crosses the Rio Verde Grande), 23 July 1989, *Zappi* 167 (K, fr & seed only); Mun. São João da Ponte, 20 km from Lontra on dirt road to Varzelândia, 11 Aug. 1988, *Eggli* 1145 (ZSS); l.c., 17 km SSW of Varzelândia, 23 July 1989, *Zappi* 166 (SPF, HRCB); l.c., Km 265–263 on road BR 135 from Montes Claros, 10 Aug. 1988, *Eggli* 1131 (ZSS); Mun. Brasília de Minas, 27 km N of Mirabela on road BR 135, 16°2'S, 44°18'W, 6 Nov. 1988, *Taylor & Zappi in Harley* 25513 (SPF, K); Mun. Mirabela, c. Km 310 on road BR 135 from Montes Claros, 10 Aug. 1988, *Eggli* 1129 (ZSS); Mun. Francisco Dumont, *fide* Uebelmann (1996): HU 710; cent.-E Minas Gerais, Mun. São Sebastião do Maranhão (?), c. 15 km W of Água Boa, *Ritter* 1238 (U).

CONSERVATION STATUS. Vulnerable [VU B2ab(iii)] (2); area of occupancy estimated to be < 2000 km²; PD=2, EI=1, GD=1. Short-list score (2×4) = 8. *Cerrado* habitats in Eastern Brazil are suffering considerable modification by agricultural development and the populations of this species, although numbering > 10, are severely fragmented.

Together with *Arrojadoa dinae, Cipocereus crassisepalus* and 3 *Discocactus* spp., this is one amongst few terrestrial cacti that inhabits *cerrado*, although it cannot be said to be a common component of this vegetation, being of erratic occurrence in sandy places. The record of *C. mirabella* from south-western Maranhão is separated from those from western Bahia by c. 500 km, but this probably reflects the lack of collections from this relatively inaccessible region and it is possible that its distribution is ± continuous in the *cerrados* in between. Of greater interest is the apparent disjunction in its distribution across the Serra do Espinhaço, between central-eastern (W of Água Boa, the type locality) and central-northern Minas Gerais — c. 250 km. Nevertheless, it is possible that geographically intermediate populations await discovery. Braun & Esteves Pereira (2002: 79) indicate that '*Mirabella*' occurs in Tocantins, which is well possible given the distribution of this species recorded above.

2. Cereus albicaulis *(Britton & Rose) Luetzelb.*, Estud. bot. Nordéste 3: 111 (1923, publ. 1926); N. P. Taylor in Bradleya 10: 24, pl. III (1992). Type: Brazil, Bahia, Barrinha, 1915, *Rose & Russell* 19808 (US, lecto. designated here; NY, lectopara.).

Acanthocereus albicaulis Britton & Rose, Cact. 2: 125–126 (1920). *Mirabella albicaulis* (Britton & Rose) F. Ritter, Kakt. Südamer. 1: 110 (1979). *Monvillea albicaulis* (Britton & Rose) R. Kiesling in Cact. Succ. J. (US) 66: 165 (1994).

Very similar to the preceding, but scrambling to at least 3 m high; tuberous taproot to 40 cm; stem to 5 cm diam., cortex yellow beneath the bark; ribs (2–)3–4(–6), usually sharply triangular-acute; central spine better defined. Flower (Taylor, l.c.) nocturnal, 13.5–18.3 cm (more slender in all its parts than *C. mirabella* in the few examples seen), stamens c. 20 mm. Fruit to 7.5 × 4 cm, funicular pulp white. Seed c. 2 × 1.4 mm (Barthlott & Hunt 2000: pl. 46.5–6).

Widespread *caatinga* element: in *caatinga de altitude*, *caatinga* and *carrasco*, often on sandy substrates of the Cipó soil series, c. 470–1020 m, northern and south-eastern Piauí, north-western Ceará, western and central-southern Pernambuco to western, northern and north-eastern Bahia and southwards through the Chapada Diamantina to the Serra do Espinhaço, southern Bahia and (?) northernmost Minas Gerais. Endemic to Eastern Brazil. Map 28.

PIAUÍ: N Piauí, Mun. Piracuruca, Alto Alegre, *fide* Uebelmann (1996): HU 941; SE Piauí, Mun. Dom Expedito Lopes, Gaturiano, Dec. 1963, *Ritter* 1236 (SGO 121755, *fide* Eggli *et al.* 1995: 501); Mun. Padre Marcos, Serra Velha, 27 Dec. 1993, *M.E. Alencar* 37 (PEUFR), without date, *M.E. Alencar* 375 (PEUFR); Mun. Simões, 7°36'S, 40°45'W, 5 Apr. 2000, *E.A. Rocha et al.* (K, photos); Mun. Canto do Buriti, *fide* Uebelmann (1996): HU 837; Mun. São Raimundo Nonato, near foot of Serra da Capivara, 6 Feb. 1990, *Zappi* 218 (SPF, HRCB); l.c., Zabelê, 8°44'1"S, 42°29'21"W, 20 Dec. 1998, *J.R. Lemos* 74 (PEUFR).

CEARÁ: NW Ceará, Mun. Ubajara, Planalto da Ibiapaba, 21 July 1994, *F.S. Araújo* s.n. (UEC 96574).

PERNAMBUCO: Mun. Araripina, *fide* Uebelmann (1996): HU 932; Mun. (?), between Jutaí and Cruz de Malta (Santa Cruz), 26 Nov. 1959, *Andrade-Lima* 59-3411 (IPA); Mun. Petrolina, Rajada, *fide* Uebelmann (1996): HU 1755; Mun. Floresta, N of Serra Negra, 8°33'S, 38°1'W, 8 Apr. 2000, *E.A. Rocha et al.* (K, photos), l.c., 38 km from town on road PE 360 to Floresta, 18 Jan. 1992, *F.A.R. Santos* 45 (HUEFS, PEUFR); Mun. Nova Petrolândia, 18 km W of town towards Floresta, 12 Feb. 1991, *Taylor & Zappi* 1622 (UFP, HRCB, K, ZSS).

ALAGOAS: reported by Luetzelburg (1925–26, 3: 111), without locality.

BAHIA: W Bahia, Mun. Barra, Ibiraba, dunes by the Rio São Francisco, s.d., *P. Rocha* 47 (SPF); Mun. Santa Maria da Vitória, near airstrip, 4 Oct. 1985, *Horst & Uebelmann* 833, cult. 1986 (K, photo); N & cent. Bahia, Mun. Remanso, 9°29'S, 42°16'W, 27 Feb. 1978, *Miranda* 315 (IPA); l.c., 11 km E of Remanso, 9°33'S, 42°W, 7 Apr. 2000, *E.A. Rocha et al.* (K, photos); l.c., 47 km E, 9°22'S, 41°48'W, 7 Apr. 2000, *E.A. Rocha et al.* (obs.); Mun. Juazeiro, 16 July 1959, *Castellanos* 22444 (R); Mun. Sento Sé, c. 0.5 km N of Cabeluda, 11 Jan. 1991, *Taylor et al.* 1392 (CEPEC, HRCB, K, ZSS); Mun. Jaguarari, 64 km S of Juazeiro on road to Senhor do Bomfim, 8 Jan. 1991, *Taylor et al.* 1376 (CEPEC, HRCB, K, ZSS); l.c., Barrinha, 1915, *Rose & Russell* 19808 (US, NY); l.c., Flamengo, *Ritter* 1237 (SGO 121756, *fide* Eggli *et al.* 1995: 501); Mun. Uauá ['Curaçá'], Lagoa da Fulô, 9°51'S, 39°41'W, 16 Aug. 1983, *G.C.P. Pinto & S.B. Silva* 187/83 (HRB); l.c., c. 10.5 km NW of town towards Poço de Fora, 7 Jan. 1991, *Taylor et al.* 1367 (CEPEC, HRCB, K, ZSS); Mun. Jacobina, 27–28 km WNW of town towards Lajes, 12 Jan. 1991, *Taylor et al.* 1399 (CEPEC, HRCB, K, ZSS); Mun. Euclides da Cunha, *fide* Uebelmann (1996): HU 731; Mun. Morro do Chapéu, 19 km W of town on road BA 052, 11°28'S, 41°19'W, 25 Dec. 1988, *Taylor & Zappi in Harley* 27390 (CEPEC, SPF, K); Mun. Cafarnaum, 9 km N of town towards road BA 052, 24 Aug. 1988, *Eggli* 1278 (ZSS); Mun. Seabra/Iraquara, 29 km N of town on road to Água de Rega, 27 Feb. 1971, *Irwin et al.* 31216 (K, UB); Mun. Seabra, 8 km E of Seabra, 20 Mar. 1989, *Supthut* 8951 (ZSS); Mun. Rio do Pires, Ibiajara, *fide* Uebelmann (1996): HU 129; NE Bahia, Mun. Biritinga, 18 Jan. 1994, *M.A.S. das Neves* 81 (HUEFS); S Bahia, Mun. Iaçu, Morro da Garrafa, 12°45'S, 39°51'W, 22 Feb. 1997, *E. de Melo* 2076 (HUEFS); Mun. Guanambi, 19 km from the town on road to Caitité, 25 July 1989, *Zappi* 170 (SPF, HRCB); Mun. Caitité, Brejinho das Ametistas, 26 July 1989, *Zappi* 173 (SPF); Mun. Jacaraci, 2 km NW of town, 1982, *Horst & Uebelmann* 577 (K, ZSS); Mun. Condeúba, 'Serra de Condeúba', reported by Werdermann (1933: 99); Mun. Cordeiros, 15°4'S, 41°53.5'W, Sep. 2003, *M. Machado* (obs.).

MINAS GERAIS: (?) cent.-N Minas Gerais, reported by Braun & Esteves Pereira (1991b: 190; 2003: 32A, illus.).

CONSERVATION STATUS. Least Concern [LC] (1); PD=2, EI=1, GD=2. Short-list score (1×5) = 5. Least Concern in view of its extensive range, but the plant is not common and its habitat is declining.

This species has a patchy geographical distribution, being widely spread in the *caatinga* towards its northern limits, but more or less restricted to the East Brazilian Highlands in the southern parts of its range. This distribution pattern may in part correspond with that of the 'Cipó' soil series, upon which a distinct type of *caatinga* vegetation is found (type No. 5 of Andrade-Lima 1981: 159), but it has also been seen growing upon inselbergs and occasionally on other substrates. It is variable in the robustness of its stems.

It was originally described by Britton & Rose as an *Acanthocereus*, whose species it vaguely resembles in vegetative characters, although not in flower and fruit-morphology, which Britton & Rose unfortunately did not know.

Subg. *Cereus* (Nos 3–6): rootstock fibrous (in Brazilian taxa); treelike, semi-decumbent or creeping, stems stout; bract-scales of pericarpel and tube glabrous in their axils (rarely with inconspicuous trichomes in *C. fernambucensis*); floral remnant early-deciduous from fruit apex.

This is the least understood subgenus, species delimitation being hampered by lack of data on fruit and seedling morphology. Found in various phases of the *Mata atlântica, caatinga-agreste, cerrado* (but only on limestone outcrops) and South-eastern *campos rupestres* (rare).

3. Cereus fernambucensis *Lem.*, Cact. gen. nov. sp.: 58 (1839). Type: 'patria est Fernambuco' [Pernambuco = Recife]; not known to have been preserved. Neotype (designated here): Brazil, Pernambuco, Mun. Jaboatão dos Guararapes, Candeias, by the sea, 20 Feb. 1990, *Zappi* 228 (HRCB) — nearest extant population (in 1990), since already extinct in Mun. Recife.

? *C. obtusus* Haw., Rev. pl. succ.: 70 (1821). *Piptanthocereus obtusus* (Haw.) F. Ritter, Kakt. Südamer. 1: 231 (1979). Type: not known to have been preserved.

? '*C. neotetragonus*' Backeb., Die Cact. 4: 2363 (1960), nom. inval. (Art. 37). Supposedly based on a plant from Brazil (Rio de Janeiro?), but not typified. The plant in the plate 77 that Backeberg cites from Werdermann, Blühende Kakt. und andere Sukk. (1934), does not match his description and probably represents the Andean *C. hankeanus* F. A. C. Weber.

[*C. variabilis* auctt. non Pfeiff., Enum. cact.: 105 (1837), nom. illeg. (Art. 52.1).]

[*C. pitajaya sensu* Hook. in Bot. Mag. 70: tab. 4084 (1844), non (Jacq.) DC. (= *Acanthocereus tetragonus* (L.) Hummelinck).]

Shrubby or caespitose, erect or sprawling, always branched at or very near base, 0.5–4.0 m; stems with numerous constrictions, pale green to glaucous, non-mucilaginous; ribs 3–5, to 65 × 10–30 mm; areoles to 10 mm diam., 15–35 mm apart on mature stems, rather woolly, felt often blackish at first, hairs whitish and conspicuous on young growth. Spination variable, pale yellow-brown to brown when young, later grey, sometimes almost lacking; central spines 1–4, to c. 50 mm; radials to 8 or more, to c. 25 mm.

Flowers slender funnel-form, 14–25 × 9–15 cm, white, sometimes pinkish without; pericarpel bract-scales sometimes bearing trichomes in their axils, tube pale green. Fruit 5–7.2 × 3–6 cm, pinkish red or yellow, laterally dehiscent by a single split in the pericarp, the base of the floral nectar-chamber forming a deep depression (to 8 mm) at its apex. Seed c. 2–3 × 1.5 mm, black, shiny, cuticular folds almost absent.

CONSERVATION STATUS. Least Concern [LC].

The specific epithet is correctly spelled with 'f' and should not be corrected to 'pernambucensis' as advocated by various authors, since Pernambuco was written Fernambuco by some europeans during the 19th Century (see Brummitt & Taylor 1990: 302–303; Werdermann 1933: 89–90). The illegitimate name *C. variabilis* Pfeiff. as to type belongs in the synonymy of *Acanthocereus tetragonus* (L.) Hummelinck, a Caribbean taxon, and cannot be used even though Pfeiffer's concept included elements referable to the *Cereus* treated here (cf. ICBN Arts 7.5 & 52.1/2). Haworth's *C. obtusus* has also been used for this species, but its original description and typification are unsatisfactory and its provenance is uncertain. Prior to conservation of the name *C. jamacaru* DC. with a new type (see below), De Candolle's name was in fact based on an illustration of *C. fernambucensis*.

Two subspecies are recognized and are quite separate geographically, except in the lowlands of the northern half of Rio de Janeiro:

1. Fruit pinkish red; flower to c. 17 cm (sand & rocks by the sea) . . . . . **3a.** subsp. **fernambucensis**
1. Fruit yellow; flower to c. 25 cm (rocks inland, W Espírito Santo, S Minas Gerais &
   N Rio de Janeiro) . . . . **3b.** subsp. **sericifer**

3a. subsp. **fernambucensis**

VERNACULAR NAMES. Cardo-ananá, Cardo-de-praia, Cardo-vinagre, Figueira-do-inferno.

Low-growing, caespitose, mostly < 80 cm tall, occasionally semi-pendent or scrambling through vegetation and reaching 2.5 m high; stems very variable in size, 4–11 cm diam., pale to blue-green but not strongly glaucous; areoles spiny. Flowers c. 14–17 × 9–11 cm; pericarpel 13–18 × 7.5–12.0 mm, tube narrowed to 7–8.5 mm, then flared to 13–30 mm diam. at apex; perianth-segments to 4.5–6 × 1.4–2.0 cm; style 11–16.5 cm; stigma-lobes 8–13, 8–15 mm, green. Fruit pinkish red.

Humid/subhumid forest (*restinga*) element: on sand-dunes, rocks and growing through shrubs of the *restinga*, sometimes within reach of the sea spray, to c. 100 m, throughout the coast of Eastern Brazil from Rio Grande do Norte southwards (to São Paulo: Ilha do Cardoso). Endemic to North-eastern and South-eastern Brazil. Map 29.

RIO GRANDE DO NORTE: Mun. Natal, 5 km S of Ponta Negra, 2 Sep. 1946, *Wurdack* B-136 (US); Mun. Búzios, *E.A. Rocha* (K, photo).
PARAÍBA: Mun. Mataraca, reported by Oliveira-Filho (1993: 224); Mun. Cabedelo, 6°58'40"S, 34°50'18"W, reported and illustrated by Locatelli & Machado (1999b); Mun. João Pessoa, Cabo Branco, edge of *restinga* forest, 9 Feb. 1995, *Taylor* (K, photos).
PERNAMBUCO: Mun. Jaboatão dos Guararapes, Prazeres [Estação], 21 Feb. 1933, *B. Pickel* 3233 (IPA); l.c., Candeias, by the sea, 20 Feb. 1990, *Zappi* 228 (HRCB); Mun. Ipojuca, Praia de Porto de Galinhas, 4 Sep. 1992, *H.B. Martins* 2 (PEUFR).
ALAGOAS: Mun. Maragogi, Peroba, 7 June 1993, *I.A. Bayma* 68 (MAC); Mun. Marechal Deodoro, Pontal da Barra, 5 Dec. 1977, *G.L. Esteves* 162 (MAC); l.c., A.P.A. de Santa Rita, Saco da Pedra, 20 Feb. 1991, *C.S.S. Barros et al.* 108 (MAC), 23 Feb. 1989, *Esteves & Lyra-Lemos* 2190 (MAC); Mun. Penedo,

7 km S of turnoff to town on road to Piaçabuçu, 14 Feb. 1991, *Taylor & Zappi* (obs.); Mun. Piaçabucu 30 Sep. 1981, *R.F.A. Rocha* s.n. (MAC).

SERGIPE: Mun. Barra dos Coqueiros, 4 Aug. 1998, *Taylor* (obs.); Mun. Aracaju, Praia de Atalaia, 11 July 1980, *M. Fonseca* s.n. (ASE 766).

BAHIA: Mun. Conde, Praia de Seribinha, 28 Apr. 1996, *E.C. Medeiros Neto* 36 (HUEFS); Mun. Mata de São João, Sauípe, 27 Aug. 1996, *R. Soeiro* 39/95 (HRB); Mun. Camaçari, estrada do Coco, próx. a Guarajuba, 19 Nov. 1981, *G.C.P. Pinto* 419/81 (HRB); Mun. Salvador, Itapuã, 12°56'S, 38°21'W, 2 Mar. 1980, *L.R. Noblick et al.* 1695 (ALCB, HUEFS); ibid., Jardim de Alá, 1959, *A.L. Costa* s.n. (ALCB); ibid., Lagoa de Abaeté, 4 Jan. 1991, *Taylor et al.* 1345 (CEPEC, HRCB, K, ZSS); Mun. Itacaré, between the praias do Farol and da Ribeira, 14 Dec. 1992, *A.M. Amorin et al.* 963 (CEPEC, K); Mun. Ilhéus, Sambaituba, 17 Mar. 1993, *W.W. Thomas et al.* 9554 (CEPEC); l.c., 6 km S of town, W of coast road, 3 Jan. 1989, *Taylor & Zappi in Harley* 27874 (CEPEC, K, SPF); Mun. Una, 3 km N of Comandatuba, 15°20'S, 39°2'W, 3 Jan. 1989, *Taylor & Zappi in Harley* 27875 (CEPEC, SPF, K); Mun. Canavieiras, 32 km from town on road to Camacã, 8 Sep. 1965, *R.P. Belém* 1729 (CEPEC); Mun. Porto Seguro, 7 km N of town towards Santa Cruz Cabrália, Taperapuã, 20 Apr. 1982, *A.M. de Carvalho et al.* 1225 (CEPEC); Mun. Alcobaça, road BA 001, 5 km S of Alcobaça, 17 Mar. 1978, *S.A. Mori et al.* 9619 (CEPEC, NY); Mun. Nova Viçosa, 18 June 1985, *Hatschbach* 49455 (CEPEC, K).

ESPÍRITO SANTO: Mun. Itaúnas, 29 Mar. 1995, *J. Araújo* (K, photo); Mun. Linhares, Vale do Rio Doce, Km 30–40 from Linhares, 5 Oct. 1971, *T.S. dos Santos* 2065 (CEPEC); ibid., Res. Biológica dos Comboios, IBDF, Regência, próx. à reserva da FUNAI, 24 Jan. 1990, *D.A. Folli* 1075 (Herb. Univ. Fed. ES); Mun. Guarapari, Setiba, 14 Oct. 1988, *O.J. Pereira* 1844 (Herb. Univ. Fed. ES); Mun. Anchieta, Rodovia do Sol, between Meaípe (Mun. Guarapari) and Anchieta, 20°45'13"S, 40°32'37"W, 25 Nov. 1999, reported growing with *Pilosocereus brasiliensis* (*Zappi et al.* 451); Mun. Itapemirim, Rodovia do Sol, S of Marataízes, 21°4'51"S, 40°50'24"W, 26 Nov. 1999, reported growing with *Opuntia monacantha* (*Zappi et al.* 470), Jan. 1970, *L. Krieger* 7607 (CESJ).

RIO DE JANEIRO: Mun. Quiçamã (Quissamã), road to Barra do Furado, 22°6'5"S, 41°26'W, 23 Nov. 1999, *Zappi et al.* 396 (K, UEC).

CONSERVATION STATUS. Least Concern [LC] (1); PD=1, EI=1, GD=2. Short-list score (1×4) = 4. Least Concern because of its frequency, although the majority of littoral habitats in Eastern Brazil are undergoing significant modification and substantial reductions in numbers of adult individuals can be projected.

The above key and description of this variable subspecies does not account for southern forms from western Rio de Janeiro and São Paulo, which are considerably larger in their stems and flowers. Its pollination biology has been studied by Locatelli & Machado (1999b).

3b. subsp. **sericifer** *(F. Ritter) N. P. Taylor & Zappi* in Cact. Consensus Initiatives 3: 7 (1997). Holotype: Brazil, Rio de Janeiro, Três Rios, 1965, *Ritter* 1410 (U).

Piptanthocereus sericifer F. Ritter, Kakt. Südamer. 1: 232–233, Abb. 185–187 (1979). *Cereus sericifer* (F. Ritter) P. J. Braun in Bradleya 6: 86 (1988).

Erect or semi-decumbent shrub to 4 m tall, branching mainly near base or from decumbent parts of the branches; stem 10 cm diam. or more, green to strongly glaucous; areoles almost naked to very spiny. Flowers to c. 25 × 15 cm; pericarpel c. 25 × 20 mm, tube narrowed to not < 10 mm, then flared to 35 mm or more at apex; perianth-segments to 11 × 4 cm; style c. 20 cm; stigma-lobes c. 15, 15–20 mm, pale yellow. Fruit yellow.

Southern humid/subhumid forest (inselberg) element: on ± naked rock outcrops (especially gneiss/granite inselbergs) inland in the *Mata atlântica* zone, 50–400 m, western and central Espírito Santo and southernmost Minas Gerais, to adjacent Rio de Janeiro (Rio Paraíba drainage). Endemic to South-eastern Brazil. Map 29.

MINAS GERAIS: on road BR 040, between Juiz de Fora and Três Rios (RJ), 'Paraibuna', 25 July 1991, *Zappi* (obs.).

ESPÍRITO SANTO: Mun. Águia Branca, 29.5 km from Barra de São Francisco on road ES 080, 16 Dec. 1990, *Taylor & Zappi* 781 (HRCB, MBML, K, ZSS); l.c., c. 5 km S of Águia Branca, 16 Dec. 1990, *Taylor & Zappi* 782 (HRCB, MBML, K, ZSS); Mun. São Gabriel da Palha, photo of plant in habitat with ripe yellow fruit, supplied by A. Hofacker, 28 Sep. 2003 (by e-mail); Mun. Colatina, 15 km S of São Domingos on road to Colatina, 16 Dec. 1990, *Taylor & Zappi* (obs.); município unknown, Rio Doce, Feb. 1917, *Luetzelburg* 7227 in part (M); Mun. Itarana, road to Praça Oito, 19°54'35"S, 40°49'46"W, 24 Nov. 1999, *Zappi et al.* 416 (K, UEC); Mun. Santa Teresa, 8 km S of Santo Antônio coming from Colatina, 17 Dec. 1990, *Taylor & Zappi* 788 (HRCB, K); l.c., Rio Cinco de Novembro, *H.Q.B. Fernandes* 2259, cult. HRCB, May 1990 (HRCB); Mun. Jerônimo Monteiro, illustrated by Uebelmann (1996) together with *Coleocephalocereus pluricostatus*; Mun. Bom Jesus do Norte, 7 km N of town coming from São José do Calçado, 18 Dec. 1990, *Taylor & Zappi* 791 (HRCB, MBML, K, ZSS).

RIO DE JANEIRO: Mun. Itaocara, 21 Aug. 1963, *Castellanos* 24010 (GUA); Mun. Santa Maria Madalena, access road to the town, 22°0'55"S, 42°7'30"W, 21 Nov. 1999, *Zappi et al.* 353 (K, UEC); Mun. Tres Rios road BR 040, at crossing of Rio Preto, rocks W of road, 3 Aug. 2003, *Taylor* (obs.).

CONSERVATION STATUS. Least Concern [LC] (1); PD=1, EI=1, GD=1. Short-list score (1×3) = 3. Habitats not significantly declining at present.

This subspecies represents a distinct inland race of the otherwise littoral *C. fernambucensis*. Apart from its yellow fruit it differs from subsp. *fernambucensis*, as seen in Eastern Brazil, by being larger in all its parts, although forms of the latter from further south approach it in their stems and flowers.

4. Cereus insularis *Hemsl.*, Rep. Challenger, Bot. 1(2): 16 (1884). Type: Brazil, Fernando de Noronha (St Michael's Mount), *Moseley* s.n. (K, BM).

? C. ridleii Andrade-Lima ex Backeb., Die Cact. 4: 2352, Abb. 2247 (1960). Holotype: Brazil, Fernando de Noronha, Oct. 1955, *Andrade-Lima* 55-2221 (IPA).

Like *C. fernambucensis* subsp. *fernambucensis,* but stems more evenly cylindrical (less obviously constricted/jointed), to 2(–5) m tall; ribs 5–9, 1–2 cm high, spines usually more numerous, to 15 or more, partially hiding the stem; flowers very similar, but to only c. 13 cm. Seed: Barthlott & Hunt (2000: pl. 45.3–4).

Humid/subhumid forest (*restinga*) element: rocky habitats and cliffs, Fernando de Noronha, North-eastern Brazil. Endemic. Map 1.

PERNAMBUCO: Archipelago of Fernando de Noronha, 1887, *Ridley et al.* 23 (K, BM; IPA, photos), Oct. 1955, *Andrade-Lima* 55-2221 (IPA), 1987, *Duranton* s.n. (K, SPF 48093, 48094, 48095); l.c., próximo a salina, 20 Oct. 1955, *Andrade-Lima* 55-2222 (IPA); Praia do Bode, 3°50'S, 32°25'W, 3 June 1993, *A.M. Miranda et al.* 999 (PEUFR); l.c., Praia Ataleia, 4 June 1993, *A.M. Miranda* 1042, 1043 (PEUFR, HUEFS).

CONSERVATION STATUS. Least Concern [LC] (1); PD=2, EI=1, GD=2. Short-list score (1×5) = 5. The Archipelago of Fernando de Noronha is an effectively managed protected area.

This taxon is clearly very similar to *C. fernambucensis*, from which it is separated by c. 350 km of Atlantic Ocean, but further studies are required to determine how close they are and also the status of the much larger *C. ridleii*. *Cereus insularis* is apparently very variable in habit and spination, some forms being completely covered in fierce spines, the stems 5–9–ribbed (cf. *Duranton* s.n.). Its flowers, which are known from carefully preserved material, scarcely differ from those borne by smaller forms of *C. fernambucensis*, such as are found along the coast of Pernambuco. The poorly known *C. ridleii*, which was discovered in the archipelago by the respected Dárdano Andrade-Lima, remains something of an enigma (Braun 1990). It is known for certain only from his original habitat photograph, depicting a distinctive tree-like plant (c. 3–5 m), and from its holotype preserved at Recife (IPA). Contrary to Backeberg's statements (l.c.), the type specimen, collected in October 1955 (*Andrade-Lima* 55-2221), is 6-ribbed and bears flowers. (It is possible that wild material was cultivated at Recife until it flowered, but the label does not give any clue about this.) Were it not for the photograph and the 6-ribbed stem, the holotype could easily be identified as *C. fernambucensis*. The photograph published by Backeberg (1960: Abb. 2247) shows the tree-like *C. ridleii* surrounded by plants of the low-growing *C. insularis*. It is therefore tempting to suppose that *C. insularis* could be some kind of stabilized juvenile, yet reproductive (neotenic) form, which occasionally develops into the erect adult stage represented by *C. ridleii*, such behaviour being known, for example, in the Madagascan cactus relative, *Didierea trollii* (Rowley 1992). An alternative explanation would be that an arborescent *Cereus*, presumably *C. jamacaru*, was introduced to the islands in historic times and what Andrade-Lima observed was a rare product of hybridization of this with *C. insularis*. This is not such an unlikely scenario, because 'mandacaru' is regarded as a useful plant in the Nordeste and is frequently transported and planted around habitations as a living hedge.

Recently, Brazilian botanist, Emerson Rocha (pers. comm.), has visited the archipelago and has observed distinctively tall, somewhat branched forms of *C. insularis*, although these are much more spiny than Andrade-Lima's *C. ridleii*. In various respects, apart from habit and size, *C. fernambucensis* and *C. jamacaru* are morphologically comparable, and *C. ridleii*, if it is not the product of hybridization, would appear to potentially link them even further, combining the small flowers of the former with the treelike habit of the latter.

5. Cereus jamacaru DC., Prodr. 3: 467 (1828), *nom. cons.* Holotype (Taylor & Zappi 1992c): Brazil, Bahia, Mun. Curaçá, 7 Jan. 1991, *Taylor et al.* 1369 (CEPEC; HRCB, K, ZSS isos.), *typ. cons.* (Greuter *et al.* 2000: 387).

Piptanthocereus goiasensis Ritter, Kakt. Südamer. 1: 234 (1979). *Cereus goiasensis* (Ritter) P. J. Braun in Bradleya 6: 86 (1988). *C. jamacaru* subsp. *goiasensis* (Ritter) P. J. Braun & E. Esteves Pereira in Succulenta 74: 84 (1995). Holotype: Brazil, Maranhão, Carolina, 1963, *Ritter* 1282 (U).

VERNACULAR NAMES. Mandacaru, Mandacaru-de-boi, Mandacaru-facheiro, Mandacaru-de-faixo, Cardeiro, Jamacaru, Jamaracurú, Jumucurú, Jumarucú, Cumbeba, Urumbeba.

Tree-like, to at least 12 m (rarely to 18 m), with a spread of up to 10 m or more, often with a short but well-defined, bark-covered, greyish trunk 50–200 × 30–80(–100) cm, regrowing from base to form a shrub when cut down, densely branched; stems 7–20 cm diam., often strongly glaucous when young, later dark blue-green, epidermis very tough, vascular cylinder extremely woody when old; ribs 4–10, sometimes only 3 in juvenile plants < 1.5 m tall, to 60 × 18–45 mm, high and narrow to broader and more rounded, with or without transverse fold lines alongside the areoles, margin weakly to more strongly crenate; areoles ± circular, 4–8 mm diam. at first, 15–40 mm apart, dirty white to grey woolly felted and hairy at first, often showing marked indeterminate growth in age, to 15 mm diam. Spination very variable, often ± absent on mature branches high above the ground, but sometimes spiny throughout and/or very spiny near base and then with many more spines per areole than described below, spines pale yellow-brown to reddish brown at first, later grey to blackish; central spines 1–4 or more, very variable in size, occasionally to 150 × 3 mm; radials 7–12, to c. 35 mm. Flowers developing from areoles almost anywhere on the branches, funnel-form, 21–30 × c. 10–20 cm; pericarpel 20–28 × 12–18 mm, like the tube bearing green or reddish bract-scales glabrous in their axils, apex of tube to 50 mm diam.; nectar-chamber to 65 mm; perianth-segments 50–100 × 20–27 mm, outer greenish sometimes tinged crimson near apex, inner pure white; style white to pale green, stigma-lobes 12–16, 11–19 × 1 mm, greenish. Fruit 6–10 × 4–8 cm, crimson to pinkish red, dehiscent by a longitudinal split to reveal white funicular pulp. Seed black, testa with cuticular folds.

CONSERVATION STATUS. Least Concern [LC].

This well-known taxon, whose name has already been conserved for purely nomenclatural reasons so that it can continue to be used in its familiar sense (Taylor & Zappi 1992c), may in future be liable to taxonomic union with *C. hexagonus* (Linnaeus) Miller. The Linnaean species is traditionally known from Venezuela southwards to the Guianas and northernmost Brazil (Roraima), but there exist populations further to the south-east, in the states of Pará and Maranhão, whose identity has not been determined. Since, on present knowledge, there do not appear to be any clear characters to distinguish between this pair of taxa, the absence of any significant geographical disjunction makes it tempting to suppose that there may be only a single widespread and variable species involved. Nevertheless, the name *C. jamacaru* is maintained for the time being, at least until the situation in the field has been thoroughly researched.

Even in the restricted sense adopted here, this is a very wide-ranging and rather variable species divisible into the following subspecies:

1. Juvenile plants (between 10 cm and 1 m high) passing through a stage with only 3–7 ribs and yellow to orange-brown spines of variable length; mature stem-segments variously shaped; flower 15–20 cm diam. or more; pericarpel and tube to c. 16 cm, bract-scales red, conspicuous; largest perianth-segments 8–10 cm (NE Minas Gerais northwards, from c. 18°25'S in the drainage of the Rio Jequitinhonha) . . . . **5a.** subsp. **jamacaru**
1. Juvenile plants (between 10 cm and 1 m high) passing through a stage with 5–8 ribs and uniformly short, dark red-brown spines; mature stem-segments broadest near base; flower 10–15 cm diam.; pericarpel and tube to 21 cm, bract-scales green or brownish, inconspicuous; largest perianth-segments 5–7 cm (W Bahia to central Minas Gerais, to 20°S, Rio São Francisco drainage, often on limestone) . . . .**5b.** subsp. **calcirupicola**

5a. subsp. **jamacaru**

Widespread *caatinga* element: in stony to sandy soil and on rocks of various kinds, *caatinga-agreste*, entering into *Mata atlântica* (and *restinga*) in NE Brazil, c. 50–1000(–1200) m, widespread in Eastern Brazil, but less frequent west of the Rio São Francisco in south-western Bahia (where replaced by subsp. *calcirupicola* on limestone outcrops) and uncommon within the southern part of the Chapada Diamantina, southwards to central-northern and north-eastern Minas Gerais (extending as a frequent plant in the Rio Jequitinhonha valley to at least 17°S and beyond in the region of Diamantina, at c. 18°25'S), sometimes cultivated further south, as well as within its natural range, ranging north-westwards to Maranhão, northernmost Piauí and Ceará; Northern Brazil (Tocantins & Pará)? Only a few records marking the approximate western, eastern and southern limits of natural range of the subspecies are cited (Map 29):
MARANHÃO: Mun. Carolina, 1963, *Ritter* 1282 (U); Mun. Loreto, Rio Balsas, 26 May 1962, *G. & L.T. Eiten* 4722 (SP).
PERNAMBUCO: Mun. Moreno, Tapera, 29 Feb. 1930, *Pickel* 2260 (IPA 4469).

ALAGOAS: S Alagoas, road BR 101, between Miaí de Cima and Piaçabuçu, *restinga* forest, Aug. 2001, *Zappi* (obs.).

SERGIPE: Mun. Japoatã, Aug. 2001, *Zappi* (obs.).

SERGIPE/BAHIA: Mun. Cristinápolis to Mun. Jandaíra, surviving in fields derived from cut-over Atlantic Forest along road BR 101, 2 & 4 Aug. 1998, *Taylor* (obs.).

BAHIA: NW Bahia, Mun. Santa Rita de Cássia, Monte Alegre, reported by Andrade-Lima (1975: 229); E Bahia, Mun. Conceição da Feira, Porto Castro Alves, 12°32'S, 39°5'W, Dec. 1980, *Iscardino et al.* in Grupo Pedra do Cavalo 1038 (ALCB, CEPEC); SE Bahia, Mun. Itambé, road BR 415 15 km E of town, 16 April 2003, *Taylor & Zappi* (obs.).

MINAS GERAIS: cent.-N Minas Gerais, Mun. Januária, [Pedras de] Maria da Cruz, Faz. São Francisco, 11 Aug. 1997, *I. Pimenta* (K, photos); Mun. Brasília de Minas, 11 km N of Lontra on road to Januária, 15°50'S, 44°20'W, 6 Nov. 1988, *Taylor & Zappi* in *Harley* 25512 (K, SPF); Mun. Porteirinha, 5 km N of the town, 30 Jan. 1991, *Taylor et al.* 1500 (K, ZSS, HRCB, BHCB); ibid., 8 km S of town, 8 Nov. 1988, *Taylor & Zappi* (obs.); NE Minas Gerais, Mun. Coronel Murta, 5 km SE of town on road BR 342, 21 Feb. 1988, *Supthut* 8862 (ZSS); Mun. Araçuaí, reported by Rizzini & Mattos-Filho in Rodriguésia 31(54): 64 (1980); Mun. Itinga, near road BR 116, c. 36 km N of Padre Paraíso, 13 Dec. 1990, *Taylor & Zappi* (obs.); Mun. Diamantina, 25 km NE of town on road to Mendanha, by the Rio Jequitinhonha, 18°6'S, 43°28'W, 31 Oct. 1988, *Taylor & Zappi* in *Harley* 25485 (SPF, K), ibid., 25 km E, 17 Mar. 1970, *Irwin et al.* 27786 (NY, K).

CONSERVATION STATUS. Least Concern [LC] (1); PD=1, EI=1, GD=2. Short-list score (1×4) = 4. One of the commonest cacti in Eastern Brazil and with various uses (see below). Sometimes regenerating from stumps and recovering, or intentionally spared, when the forest it inhabits is cut down.

Very important as a source of cattle fodder during times of drought and sometimes planted for hedging purposes. Formerly used as a source of timber, but most larger specimens have gone. A widespread regional variant from western Rio Grande do Norte and adjacent Paraíba differs from 'typical' plants in having more ribs, both as a seedling and subsequently, as well as the stem being less obviously constricted, with ± parallel-sided false seasonal growth segments. Local variations are common and include specimens with up to 10 ribs and marked differences in number and length of spines. One of the finest specimens seen stands in a garden behind the filling station ('posto') south of road BA 052 on the west side of Morro do Chapéu, Bahia. Its height has been estimated at c. 18 metres, while its trunk is 2 × 0.8 m and bears a massive crown composed of 100s of branches. It is obviously of considerable age and cannot have been planted, almost certainly being a remnant of the *caatinga de altitude* vegetation surrounding the town.

Apparently very similar to subsp. *jamacaru* is a plant from Depto Florida (W Santa Cruz), Bolivia, which Ritter (1980: 553) named *Piptanthocereus colosseus*. Braun & Esteves Pereira (1995) refer this taxon to *C. lamprospermus* K. Schum., but its relationship deserves further investigation.

5b. subsp. **calcirupicola** *(F. Ritter) N. P. Taylor & Zappi* in Cact. Consensus Initiatives 3: 7 (1997). Holotype: Brazil, Montes Claros, 1959, *Ritter* 1011 (U).

Piptanthocereus calcirupicola F. Ritter, Kakt. Südamer. 1: 234–235, Abb. 188 & 189 (1979). *Cereus calcirupicola* (F. Ritter) Rizzini in Revista Brasil. Biol. 46: 782, figs 1–4 (1986); Zappi in Bol. Bot. Univ. São Paulo 12: 46–48, figs 12–27 (1990, publ. 1991).

Piptanthocereus cipoensis F. Ritter, op. cit. 236, Abb. 191 (1979). *Cereus calcirupicola* var. *cipoensis* (F. Ritter) P. J. Braun in Bradleya 6: 86 (1988). *C. calcirupicola* subsp. *cipoensis* (F. Ritter) P. J. Braun & E. Esteves Pereira in Succulenta 74: 83 (1995). Holotype: Brazil, Minas Gerais, Serra do Cipó, 1964, *Ritter* 1411 (U).

Piptanthocereus cabralensis F. Ritter, op. cit. 235, Abb. 190 (1979). *Cereus calcirupicola* var. *cabralensis* (F. Ritter) P. J. Braun in Bradleya 6: 86 (1988). *Cereus calcirupicola* subsp. *cabralensis* (F. Ritter) P. J. Braun & E. Esteves Pereira in Succulenta 74: 83 (1995). Holotype: Brazil, Minas Gerais, Serra do Cabral, 1964, *Ritter* 1412 (U).

Piptanthocereus calcirupicola var. *pluricostatus* F. Ritter, l.c. (1979). *Cereus calcirupicola* var. *pluricostatus* (F. Ritter) Rizzini in ibid. 784 (1986). Holotype: Brazil, Minas Gerais, Granjas Reunidas, 1964, *Ritter* 1011a (U).

C. calcirupicola var. *albicans* Rizzini, l.c. (1986). Type: Brazil, Minas Gerais, Cordisburgo, 15 July 1983, *Rizzini & Mattos* s.n. (RB, not found Feb. 1989). Lectotype (designated here): Rizzini in ibid. 783, fig. 2 (1986).

[*C. coerulescens sensu* Warm., Lagoa Santa: 150–151, figs 36 & 37 (1908), non Salm-Dyck (1834).]

Rio São Francisco (Rio das Velhas, MG) *caatinga/mata seca* element: on ±forest-covered limestone (Bambuí) outcrops, where locally co-dominant with other arborescent cacti, more rarely on arenitic rock or sand (at higher elevations only), amongst *caatinga, cerradão, cerrado* and rarely *campo rupestre*, c. 450–1200 m, western Bahia to central Minas Gerais (to c. 19°40'S). Endemic to the core area of Eastern Brazil. Map 29.

BAHIA: W Bahia, Mun. Santana, 28 km S of town towards Santa Maria da Vitória, Faz. São Geraldo, path to Gruta do Padre, 15 Jan. 1991, *Taylor et al.* (ZSS, K, photos); 7 km from Porto Novo coming from Santana on dirt road, 16 Jan. 1991, *Eggli* (ZSS, photos); Mun. Cocos, W of town, reported (as '*Cereus* sp.') and illustrated by Braun & Esteves Pereira (1999b).

MINAS GERAIS: Mun. Montalvânia, S of the ferry crossing of the Rio Carinhanha, 1999, *Klaassen et al.* (photo); Mun. Januária, 'distrito de Fabião, 2 km na estrada partindo do Abrigo do Malhador', 15°7'16"–15°8'57"S, 44°15'20"–44°14'13"W, 26 Oct. 1998, *J.A. Lombardi* 2122 (K, BHCB); Mun. Itacarambi/Manga, 15°26'S, 43°55'W, Sep. 1999, *I. Ribeiro* (photos); Mun. Varzelândia, edge of town, 23 July 1989, *Taylor & Zappi* (obs.), 12 Aug. 1988, *Eggli* s.n. (ZSS, photos); Mun. Capitão Enéas, limestone outcrop visible from road BR 122, June 2002, *M. Machado* (obs.); Mun. Montes Claros, 12 km N of town on road to Januária, 16°38'S, 43°55'W, *Taylor & Zappi in Harley* 25507 (SPF, K); Mun. Bocaiúva, Engenheiro Dolabela, 0.5 km E of road BR 135, 17°28'S, 43°58'W, 4 Nov. 1988, *Taylor & Zappi in Harley* 25502 (SPF, K); Mun. Buenópolis, Serra do Cabral, 7 km from town on dirt road to Barreira da Lapa, 17°53'S, 44°15'W, 12 & 13 Oct. 1988, *Taylor & Zappi in Harley* 24910, 24992 (SPF, K); Mun. Cordisburgo, near Gruta de Maquiné, see Eiten (1983: fig. 24); Mun. Paraopeba & Mun. Sete Lagoas, reported by Rizzini (1986); Mun. Santana do Riacho, Serra do Cipó, Cardeal Mota, 8 May 1987, *Zappi et al. in CFSC* 10124 (SPF), 27 Oct. 1988, *Taylor & Zappi in Harley* 25424 (SPF, K); Mun. Matozinhos, Faz. Cauaia, 31 Oct. 1996, *Lombardi* 1463 (BHCB, K); Mun. Lagoa Santa, Lapa Vermelha, reported by Warming (1908: fig. 37; cf. Warming & Ferri 1973).

CONSERVATION STATUS. Least Concern [LC] (1); PD=1, EI=2, GD=2. Short-list score (1×5) = 5. Its habitat is declining in some places due to quarrying of limestone.

This subspecies is distinguished from typical subsp. *jamacaru* by relatively minor yet recognizable differences in juvenile stem-morphology and flower shape etc. It ranges to the south-west of the region occupied by subsp. *jamacaru* and is at its eastern limit in the upper drainage of the Rio São Francisco (Rio das Velhas). Some populations on isolated limestone outcrops have developed into distinctive variants, such as that described as var. *pluricostatus* by Ritter (see synonymy above), but these are matched by variations seen in subsp. *jamacaru* and do not seem worthy of being named.

5c. Cereus sp. nov. (?)

Possibly keying out here to *C. jamacaru* (but mode of dehiscence of ripe fruits at present unknown) is a recently discovered Bahian taxon requiring further study (Plates 28.3–29.1).

It is tall and treelike, with stems constantly 3-ribbed for the first 3–8 metres, the massive trunk sometimes attaining 10 metres! The ribs are extremely thin, very broad and strongly crenate and it has very stout flowers. It inhabits a high *agreste* forest (*mata de cipó*), including characteristic *Mata atlântica* cacti (*Pereskia aculeata, Lepismium cruciforme,* 4 *Rhipsalis* spp.), as follows: Mun. Vitória da Conquista, 20 km from town towards Itambé, 18 April 2003, *Taylor et al.* (K, photos), l.c., campus of UESB, 900m, 18 April 2003, *Taylor et al.* (obs.); Mun. Cândido Sales, 2002, *M. Moreira* (obs.) This remarkable plant, which is probably related to *C. trigonodendron* K. Schum. ex Vaupel (syn. *C. vargasianus* Cárdenas; eastern Peru), will be reported on further as soon as ripe fruits have been studied. These are reported to be yellow by Avaldo de Oliviera Soares Filho.

6. Cereus hildmannianus *K. Schum.* in Martius, Fl. bras. 4(2): 202 (1890). Type: Brazil, Rio de Janeiro (Schumann, l.c.), or Minas Gerais, Queluz [= Conselheiro Lafaiete] (*fide* Glaziou 1909: 325), *Glaziou* s.n. (B†). Lectotype (designated here): Schumann, l.c., tab. 41, fig. I (1890) [depicting a 6-ribbed, spineless stem apex bearing a flower].

? *C. alacriportanus* Pfeiff., Enum. cact.: 87 (1837). *Piptanthocereus alacriportanus* (Pfeiff.) F. Ritter, Kakt. Südamer. 1: 236–237 (1979). Type: Brazil, Rio Grande do Sul, Porto Alegre, assumed not to have been preserved or lost.

Cereus milesimus E. C. Rost in Desert Pl. Life 4: 43 (1932). Type: Brazil, São Paulo, Ribeirão Preto, living plant assumed not to have been preserved. Lectotype (designated here): Rost, l.c. 42, photo.

Piptanthocereus neonesioticus F. Ritter var. *interior* F. Ritter, Kakt. Südamer. 1: 237–238 (1979). *Cereus neonesioticus* (F. Ritter) P. J. Braun var. *interior* (F. Ritter) P. J. Braun in Bradleya 6: 86 (1988). Holotype: Brazil, São Paulo, Piedade, *Ritter* 1421 (U).

[*C. jamacaru sensu* Scheinvar, Fl. Ilus. Catarinense, I, Cact.: 96–103, tab. 45–47 (1985), non DC. (1828).]

[*C. peruvianus* auctt. pro parte, non (L.) Mill.]

CONSERVATION STATUS. Not Evaluated [NE]; not studied in sufficient detail.

Only the following subspecies is found in the area covered here:

6a. subsp. **hildmannianus**

Very similar to *C. jamacaru,* but tree-like or shrubby, to 15 m or more tall, sometimes with a very well developed trunk to > 2 m; stem tissues highly mucilaginous; ribs 5–12, very variable in number, to 50–70 × 10–30 mm, sometimes higher and thinner, juvenile plants 3–5-ribbed; flowers 10–14 cm diam.; fruit 5–12 × 7–12 cm, yellow, orange or reddish depending on exposure, usually splitting open from the apex along c. 3 lines to fully expose the white funicular pulp (Plate 29.2); seeds c. 3 × 2.8 mm, testa with cuticular folds (Barthlott & Hunt 2000: pl. 45.1–2).

Southern humid/subhumid forest element: in rocky places and on dry shallow soils in *mata de planalto,* c. 800–1000 m, southern and western Minas Gerais, from (?) Conselheiro Lafaiete southwards and westwards, but frequently cultivated for ornament outside its natural range in Eastern Brazil; South-eastern and Southern Brazil; Mato Grosso do Sul (?); central and south-eastern South America (E Paraguay, Uruguay, NE & E Argentina etc., where replaced by the shorter-flowered subsp. *uruguayanus* (R. Kiesling) N. P. Taylor). Map 29.

MINAS GERAIS: Mun. Conselheiro Lafaiete, *fide* Glaziou (1909: 325); S of Juiz de Fora near road BR 040, 25 July 1991, *Zappi* (obs.).

RIO DE JANEIRO: Mun. Tres Rios, W of road BR 040, c. 10 km S of border with Minais Gerais, 3 Aug. 2003, *Taylor & Zappi* (obs.); l.c., road to Petrópolis, 21 July 1996, *Taylor* (obs.); Mun. Niterói/Maricá,

Itacoatiara, on rocks above the sea, 28 July 1996, *Taylor* (obs.) — at both these sites sympatric with *C. fernambucensis* subspp. [records outside of E Brazil area].

CONSERVATION STATUS. Least Concern [LC] (1); PD=1, EI=1, GD=2. Short-list score (1×4) = 4. Least Concern, given its very extensive range in eastern South America.

This species has not been studied in the field within the core area treated here. However, plants from Rio de Janeiro, south-western Minas Gerais, São Paulo and Paraná have been examined in habitat, though seldom with ripe fruit. The fruits that have been observed in the field and in cultivation generally display the characteristic mode of dehiscence described above and, together with the highly mucilaginous stem tissues, provide an effective means for distinguishing *C. hildmannianus* from *C. fernambucensis* subspp. *fernambucensis* and *sericifer*, with whose ranges it slightly overlaps, and from the related but allopatric *C. jamacaru*. All three species exist as spineless forms and when these are cultivated outside their natural range their identification can be very difficult unless ripe fruits are present. Unnecessary confusion between *C. hildmannianus* (*C. peruvianus* auctt.) and *C. jamacaru* has been created by Leia Scheinvar (see synonymy above) and propagated by her elsewhere, eg. Gutman *et al.* (2001: 709), but there can be little doubt that these are two quite distinct and geographically separate, naturally occurring species.

C. hildmannianus may be widespread in drier phases of the semi-humid and humid, subtropical and tropical planalto forests east of the Chaco in south-eastern South America, and has a potentially much more extensive synonymy than that tentatively given above (Hunt 1992b). However, this will remain uncertain until the *Cereus* species of Paraguay and Bolivia are better understood.

Schumann (l.c.) stated that *C. hildmannianus* came from Rio de Janeiro, whereas Glaziou (1909: 325), its collector, later gave a precise locality in southern Minas Gerais as its provenance (see type citation above). There are only the above records of the plant growing wild in the state of Rio de Janeiro, where it is much more commonly cultivated (as elsewhere), but it is certainly native in Minas Gerais and is not uncommon in the adjacent states of São Paulo and Paraná. The commonly cultivated form has ± spineless stems, as does that depicted in Schumann's plate, which is here selected as lectotype (in the absence of any extant original herbarium material). This is in contrast to some wild forms, which can have very spiny stems, and it is possible, therefore, that what Schumann received from Glaziou and described was actually a cultivar from Rio de Janeiro and not a wild plant. Braun & Esteves Pereira (2003) report it from Espírito Santo, but if it occurs there at all, it is most likely only to be as a cultivated plant.

17. CIPOCEREUS F. Ritter

Kakt. Südamer. 1: 54 (1979); N. P. Taylor & Zappi in Bradleya 7: 16–17 (1989). Type: *C. pleurocarpus* F. Ritter (= *C. minensis* (Werderm.) F. Ritter).

Including *Floribunda* F. Ritter (1979); *Pilosocereus* subg. *Floribunda* (F. Ritter) P. J. Braun (1988); *tantum quoad typ.*

Shrubby or tree-like, 0.4–3.5 m, branched or not above the ground; vascular cylinder strongly to weakly woody; tissues sometimes stiff, scarcely to extremely mucilaginous; branches showing slight constrictions or ± constant growth, epidermis green, grey or pale waxy blue, smooth; ribs 4–18, acute to rounded, sinuses straight. Areoles with felt and variable spination; spines straight, opaque to translucent. Fertile part of stem not differentiated, random (lateral) and/or subapical. Flowers 1.5–9.0 × 1–7 cm, anthesis commencing at evening or diurnal; flower exterior bluish, covered in wax, sometimes reddish, especially on the distal half, but still waxy, smooth to ridged, glabrous or with areoles, and then hairs and minute spines as well as areolar felt also present; pericarpel oblong, nearly cylindric; flower-tube straight, terete, with some scattered acute bract-scales, more frequent towards its apex; perianth-segments reflexed, outer segments fleshy, inner segments thin, white or cream-coloured; nectar-chamber straight, not protected by the stamens (except in *C. pusilliflorus*); stamens many, anthers 1–3 mm, verrucose, forming a compact mass; style 9–65 mm, stigma-lobes 8–10, exserted or included in relation to the anthers. Fruit ovoid to globose, indehiscent; floral remnant persistent, blackening, erect, with rounded, deep insertion; pericarp covered in blue wax, dark purplish, brownish, pinkish or green beneath; funicular pulp aquose, translucent, greenish or yellowish. Seeds 1.2–1.8 mm, brown or dark brown-black, cochleariform, hilum-micropylar region differentiated, intercellular depressions present, mainly forming craterae, sometimes only slightly pronounced, testa-cells convex, conic or flat, cuticular folds coarse, very dense.

A genus of 5 very distinct species endemic to the Serra do Espinhaço and Serra do Cabral of Minas Gerais (*campo rupestre* and sandy phases of the *cerrado*); see Maps 5 & 30. The waxy and mostly light bluish, indehiscent, ovoid to globose fruits with translucent pulp are characteristic. Species Nos 1 & 2 may be related on the basis of seed- and stem-morphology (including seedlings); likewise Nos 3–5. However, natural hybridization between Nos 2 and 4 has been observed (Taylor & Zappi 1989). Recent preliminary gene sequence studies employing the chloroplast gene segment *rpl*16 (Soffiatti unpubl.) suggest that *Cipocereus* is monophyletic and basal to the remainder of the Brazilian 'Cereeae' treated below (as well as to *Uebelmannia*). Plates 29.3–31.4.

1. Ribs 4–7, triangular in section; areoles with dense felt and long hairs; central spines larger than radials 2
1. Ribs > 8, rounded in section; areoles with short, brown or white felt, long hairs absent; central and radial spines alike or upper areoles unarmed 3

2. Roots not tuberous; flowers and fruits bearing areoles with spines and long hairs (Serra do Caraça, MG) . . . . **1. laniflorus**
2. Roots tuberous (*Dahlia*-like); flowers and fruits naked (E & N of Diamantina, MG)
. . . . **2. crassisepalus**

3. Stems 5–8 cm diam., unarmed or bearing very few black spines, transverse folds between adjacent areoles on the same rib well marked, epidermis sky blue, waxy (Serra do Cabral & adjacent W slope of Serra do Espinhaço, MG) . . . . **3. bradei**
3. Stems 2.5–5.0 cm diam., densely spiny; transverse folds absent; epidermis grey-green or bright green 4

4. Plants to c. 50 cm tall; flowers small, < 2 cm, diurnal; fruits to 1.3 cm diam., pinkish, with a translucent waxy bloom (W slope of Serra Geral, northern Minas Gerais) . . . . . **5. pusilliflorus**
4. Plants > 50 cm tall; flowers > 4 cm, nocturnal; fruits > 2 cm diam., dark blue-black covered in a pale blue waxy bloom, or whitish (MG: Serra da Bocaina, Grão Mogol southwards) . . . . **4. minensis**

1. Cipocereus laniflorus *N. P. Taylor & Zappi* in Cact. Consensus Initiatives 3: 7 (1997). Holotype: Brazil, Minas Gerais, Serra do Caraça, 11 Sep. 1990, *Zappi et al.* 240 (SPF; HRCB, BHCB, isos.).

Pilosocereus laniflorus (N.P. Taylor & Zappi) P. J. Braun & E. Esteves Pereira in Schumannia 3: 188 (2002).

Erect, 0.5–1.5(–2) m, tapered to base, unbranched or shrublike, rarely with 25 or more actively growing stem apices; vascular cylinder strongly woody; inner tissues mucilaginous; branches 6–9 cm diam., epidermis olive-green, with strikingly pruinose, bluish wax on the young growth; ribs 5–7, 2 × 2 cm, triangular in section. Areoles to 6 mm diam., ± contiguous, felt at first pale yellow to light-brown, becoming grey to blackish, long hairs whitish to yellowish, very dense at apex. Spines opaque, dark brown to reddish; central spines 7–9, 1.5–2.0(–3.0) cm, porrect; radials 10–14, to 1 cm, adpressed. Fertile part of stem lateral to subapical, not differentiated, but fertile areoles producing dense hairs to 1 cm in some individuals. Flower-buds woolly, acute 1–2 days before anthesis; flower (May–August) c. 7 × 3.5 cm, exterior dark blue, ridged, with areoles, these bearing 3–5 minute reddish spines and long white hairs; pericarpel 18–20 × 12–15 mm; flower-tube 45 × 24 mm at its widest point, ± parallel-sided, straight, with acute bract-scales at apex; outer perianth-segments 12–14 × 7 mm, dark blue to purplish cream, inner segments 18 × 8 mm, white or pale creamy yellow, spathulate, apiculate; nectar-chamber c. 20 × 6–7 mm; anthers 2–2.3 mm; style 40 × 1.6 mm, tapering; stigma-lobes c. 8, 6 mm, acute at apex; ovary locule c. 7 × 6 mm, quadrangular in longitudinal section. Fruit (July–Sep.) to 3.5–4.0 × 3.5–4.0 cm, ovoid, up to 20 per plant; pericarp purplish-blue, strongly 7–8-ribbed, with woolly areoles and spines, the latter growing somewhat as the fruit develops. Seeds (1.5–)1.6–1.7 × 1–1.1 mm, blackish, hilum-micropylar region 0.6(–0.7) mm, forming an angle of 30° with long-axis; testa-cells domed, with intercellular depressions and coarse, dense cuticular folds.

South-eastern *campo rupestre* element: deeply fissured quartzitic outcrops on steep hill slopes and cliffs in *campo rupestre,* in constant association with *Philodendron cipoense*, 1000–1500 m, Serra do Caraça, south-central Minas Gerais. Endemic to the above-cited area within Minas Gerais. Map 30.
MINAS GERAIS: Mun. Santa Bárbara, Reserva Particular de Patrimônio Natural (RPPN) do Caraça, steep hill slopes below the Oratório and above the track to Buraco da Boiada, side valley west of the Ribeirão Caraça and the Cascatona, 1000–1200 m, 3 Aug. 2001, *Taylor* (K, photos); l.c., Campo de Fora, rocks and cliffs above the Córrego do Eugênio/Capivari, 2 Aug. 2001, *Taylor* (K, photos); Mun. Catas Altas, RPPN do Caraça, Gruta do Padre Caio, 24 May 1987, *Zappi & Scatena* in CFCR 10987 (SPF, ZSS), 11 Sep. 1990, *Zappi et al.* 240 (SPF, HRCB, HBCB), ibid., cult. R.B.G. Kew, Accn. No. 1991-1422, 17 & 25 Feb. 1997 (K, fls in spirit, photos), l.c. Aug. 1996, *Soffiatti et al.* (K, photos), 1 Aug. 2001, *Taylor* (K, photos); l.c., 17–20 minutes' walk on the path from the Caraça Santuário towards the Grande Funil and Cascatona, steep hill slope east of path, 3 Aug. 2001, *Taylor* (K, photos); l.c., between Bocaina and Gruta da Bocaina, 1 Aug. 2001, *Taylor* (K, photos).

CONSERVATION STATUS. Endangered [EN D] (3); extent of occurrence < 10 km² and area of occupancy probably not more than 1% of this (0.1 km²); PD=3, EI=1, GD=1. Short-list score (3×5) = 15. Presently known from 8 small areas situated sufficiently close together to be considered a single population, which is entirely within the Reserva Particular de Patrimônio Natural do Caraça. Field surveys during 2001 (by Taylor and staff from the Fundação Zoo-Botânica de Belo Horizonte) and throughout 2002 (by Juliana Rego, FZ-BBH, supported by the RBG Kew Threatened Plants Appeal), have so far documented a total of 321 individuals, less than 250 of these being adult plants. Knowledge of the whole area suggests it is unlikely that further populations exist and the Caraça massif itself is isolated from other parts of the Serra do Espinhaço range by areas of lower elevation, formerly covered in high forest. Two of the small areas have 80 or more individuals, while the remainder have smaller numbers (2–38). Only a handful of plants is actually visible from the footpaths that traverse the Caraça RPPN and entry/exit to and from the reserve is controlled through a single manned gate on the only paved access road. Collecting on the reserve is strictly by permission from its director, which is not freely given, and there are signs advising the public that such activity is prohibited. Thus, the existing population, though small, appears to be safe from human interference at present.

In its flowers and fruits bearing well-developed, woolly and spiny areoles this species seems to be a relict, which has the most plesiomorphic floral characters within tribe Cereeae, only

Cipocereus minensis subsp. *minensis* and *Cereus* subg. *Mirabella* having comparably primitive floral features. The strongly glaucous young growth and approximate, darkly spined areoles of this species are strongly convergent with those of *Pilosocereus fulvilanatus*, q.v.

The plant is most abundant at lower elevations, below 1400 metres. Above this elevation the plants are few and appear to be limited by winter frosts. Plants at these elevations occur only under the protection of rocks or bushes and exposed stems are invariably damaged and re-sprout from below.

2. Cipocereus crassisepalus *(Buining & Brederoo) Zappi & N. P. Taylor* in Bradleya 9: 86 (1991). Holotype: Brazil, Minas Gerais, Diamantina, 500–1000 m, *Horst & Uebelmann* 169 (U).

Cereus crassisepalus Buining & Brederoo in Krainz, Die Kakteen, Lfg 53 (1973). *Piptanthocereus crassisepalus* (Buining & Brederoo) F. Ritter, Kakt. Südamer. 1: 238 (1979).

Erect shrub or treelet, 1–2 m, branched above the ground; root system well developed, roots tuberous as in a *Dahlia*; cauline vascular cylinder rather woody; succulent tissues mucilaginous. Young plants slender, stems with 5–8 low ribs, the areoles with dense, white or greyish, long hairs and reddish spines. Mature stems 5–6 cm diam., epidermis olive-green, rather glaucous when young; ribs 4–6, 1.5–2.0 × 2.5–3.0 cm, triangular in section. Areoles to 15 mm diam., 10–30 mm apart, sometimes almost contiguous, with dense felt and long hairs, at first golden red, later blackish. Spines opaque, yellowish, becoming dark brown or black; central spines 1–2(–3), to 4 cm, porrect; radials 3–8, to 2 mm, adpressed. Fertile part of stem not differentiated, subapical. Flower-buds obtuse before anthesis; flowers c. 7.5–9 × 4–5 cm, dull, dark blue, ridged, glabrous except for scattered apiculate bract-scales; pericarpel 15–20 × 16 mm; tube 50 × 30–40 mm at its widest point, ± parallel-sided, straight, with ovate, acute bract-scales at apex; outer perianth-segments 14 × 9–11 mm, dark blue to purplish-cream, inner segments 18 × 10 mm, white or cream, spathulate, apiculate; nectar-chamber c. 20 × 6–7 mm; anthers 2–3 mm; style 45–50 mm, tapering; stigma-lobes c. 9, c. 10 mm; ovary locule c. 13 × 8–9 mm, rectangular in longitudinal section. Fruit to 4.5 × 3.0 cm, narrowly oblong to ovoid; pericarp dark blue, smooth, naked. Seeds 1.6–1.7 × 1.0–1.2 mm, blackish, hilum-micropylar region c. 0.5 mm, forming an angle of 35–40° with long-axis; testa-cells domed, with intercellular depressions and coarse, dense cuticular folds.

South-eastern *campo rupestre (cerrado)* element: in sandy *cerrado/carrasco* associated with crystalline rock outcrops, 500–1200 m, north of Diamantina, Serra Negra and east side of Serra do Espinhaço, Minas Gerais. Endemic to the core area within Minas Gerais. Map 30.

MINAS GERAIS: Mun. Itamarandiba, c. 20 km SW of airport, towards Penha de França, 3 Aug. 1988, *Eggli* 1061 (ZSS); Mun. Diamantina, *Horst & Uebelmann* 169 (U); l.c., 25 km along road to Mendanha, E of road towards Rio Jequitinhonha, 31 Oct. 1988, *Taylor & Zappi in Harley* 25487 (SPF, K) — population including hybrids with sympatric *Cipocereus minensis*; l.c., between Mendanha and Inhaí, 17°59'44"S, 43°36'4"W, *M. Machado* (obs.); Mun. Rio Vermelho, Serra do Ambrósio, 1959, *Ritter* s.n. [1012] (U), 14 July 1984, *Varanda et al.* in CFCR 4885 (SP, SPF), 6 Mar. 1988, *Zappi & Prado* in CFCR 11822 (SPF). Some of the location data used in Map 30 were supplied by M. Machado (pers. comm.), for localities where this species is sympatric with *Uebelmannia*, but the coordinates are not reproduced here for the protection of the latter.

CONSERVATION STATUS. Vulnerable [VU B1ab(iii) + 2ab(iii)] (2); extent of occurrence = 1407 km²; PD=3, EI=1, GD=1. Short-list score (2×5) = 10. Habitat destruction by charcoal producers continues to be a problem for this species comprising < 10 populations.

As noted above, this species hybridizes with *C. minensis* when they come into contact.

3. Cipocereus bradei *(Backeb. & Voll) Zappi & N. P. Taylor* in Bradleya 9: 86 (1991). Type: Brazil, Minas Gerais, Diamantina, *Brade* s.n. (apparently not preserved). Lectotype (designated here): Backeberg & Voll, l.c. infra, [p.3], photograph (1935).

Pilocereus bradei Backeb. & Voll in Jahrb. Deutsch. Kakteen-Ges. 1941: 78 (1942); Backeb. & Voll in Blätt. Kakteenf. 1935 (1:) [p. 3] (1935), *nom. inval.* (Art. 36). *Pilosocereus bradei* (Backeb. & Voll) Byles & G. D. Rowley in Cact. Succ. J. Gr. Brit. 19(3): 66 (1957). *Cephalocereus bradei* (Backeb. & Voll) Borg, Cacti, ed. 2: 147 (1951); E. Y. Dawson in Los Angeles County Mus. Contr. Sci. 10: 6 (1957). *Pseudopilocereus bradei* (Backeb. & Voll) Buxb. in Beitr. Biol. Pflanzen 44: 252 (1968).

VERNACULAR NAME. Quiabo-da-lapa.

Erect shrub or treelike, 1–3.5 m, branched above the ground; vascular cylinder weakly woody; succulent tissues mucilaginous. Seedlings and young plants with 10 low ribs and areoles with dense reddish spination. Mature stems 5–8 cm diam., sometimes slightly constricted, epidermis covered in thick, striking, light blue wax; ribs 8–11, 0.5 × 1.2–1.7 cm, rounded in section, produced below the areoles. Areoles to 3 mm diam., 20–23 mm apart, felt short, at first white, the areoles frequently unarmed, becoming blackish in age. Spines on young areoles 0 or few, if present, black or dark reddish brown, to 15 mm; old areoles with 5–7 spines, to 2.5 cm, porrect. Fertile part of stem not differentiated, flowers numerous, often from one side of stem and/or subapical region. Flower-buds obtuse before anthesis; flowers c. 6.5–8.0 × 3.5–4.0 cm, exterior dark blue, ridged, glabrous except for scattered bract-scales; pericarpel 15 × 15 mm; tube 50 × 20–25 mm at its widest point, ± parallel-sided, straight, with ovoid to spathulate, apiculate bract-scales at apex; outer perianth-segments 9–12 × 6–8 mm, dark blue to purplish cream, inner segments 18 × 8 mm, white or cream, spathulate, fimbriate and apiculate; nectar-chamber c. 20–23 × 8–10 mm; anthers 2.2–3.0 mm; style 45–48 mm, tapering; stigma-lobes c. 10, c. 9 mm; ovary locule c. 9 × 7 mm, rectangular in longitudinal section. Fruit to 4.5 × 3.5 cm, ovoid; pericarp dark-blue, smooth, naked, but covered in dense, sky blue wax. Seeds 1.6–1.8 × 1.1–1.2 mm, brownish, hilum-micropylar region 0.7 mm, forming an angle of 35–40° with long-axis; testa-cells conic, with intercellular depressions and coarse, dense cuticular folds.

South-eastern *campo rupestre* element: crystalline rocks in *campo rupestre*, *carrasco* or *cerrado*, 500–1200 m, Serra do Cabral and west slope of Serra do Espinhaço, Minas Gerais. Endemic. Map 30.
MINAS GERAIS: Mun. Francisco Dumont, 2 km from the town, 7 Aug. 1988, *Eggli* 1117 (ZSS, SPF); Mun. Joaquim Felício, road to Várzea da Palma, 7 May 1990, *Zappi et al.* in CFCR 13235 (SPF, HRCB); Mun. Buenópolis, above and west of the town, 12 Oct. 1988, *Taylor & Zappi* in Harley 24508, 24990 (SPF, K); Mun. Augusto de Lima, Santa Barbara, 6 Aug. 1988, *Eggli* s.n. (ZSS, photo); Mun. Diamantina, N of Sopa, c. 18°10'S, 43°44'W, *M. Machado* (obs.); Mun. Conselheiro Mata, 18°17'S, 43°58'W, 8 June 2002, *G. Charles & M. Machado* (photos).

CONSERVATION STATUS. Endangered [EN B1ab(iii) + 2ab(iii)] (3); extent of occurrence = 1887 km²; area of occupancy < 500 km²; PD=3, EI=1, GD=1. Short-list score (3×5) = 15. Of very limited range, comprising 6 somewhat fragmented localities, none included within any protected area. Subject to commercial seed harvest for the international trade and likely to suffer a decline in habitat quality in the future due to local extractive activities.

At the second locality cited above *C. bradei* was found growing with *C. minensis* and there was evidence of introgression.

4. Cipocereus minensis *(Werderm.) F. Ritter*, Kakt. Südamer. 1: 57 (1979). Type: Minas Gerais, June 1932, *Werdermann* 3992 (B†). Lectotype (designated here): Werdermann, l.c. infra, 112, photograph captioned '*Cereus minensis* Werderm.' (1933).

Cereus minensis Werderm., Bras. Säulenkakt.: 113 (1933). *Pilocereus minensis* (Werderm.) Backeb. in Backeberg & Knuth, Kaktus-ABC: 328 (1935, publ. 1936). *Pilosocereus minensis* (Werderm.) Byles & G. D. Rowley in Cact. Succ. J. Gr. Brit. 19(3): 67 (1957). *Cephalocereus minensis* (Werderm.) E. Y. Dawson in Los Angeles County Mus. Contr. Sci. 10: 6 (1957). *Coleocephalocereus minensis* (Werderm.) F. H. Brandt in Succulenta 60: 117–118 (1981).

Cipocereus pleurocarpus F. Ritter, Kakt. Südamer. 1: 54–55 (1979). *Coleocephalocereus pleurocarpus* (F. Ritter) F. H. Brandt in Succulenta 60: 117 (1981). *Pilosocereus pleurocarpus* (Ritter) P. J. Braun in Bradleya 6: 88 (1988). *Cipocereus minensis* subsp. *pleurocarpus* (F. Ritter) N.P. Taylor & Zappi in Bol. Bot. Univ. São Paulo 12: 48 (1990, publ. 1991). Holotype: Brazil, Minas Gerais, Serra do Cipó, Nov. 1964, *Ritter* 1327 (U).

VERNACULAR NAMES. Rabo-de-raposa, Quiabo-da-lapa, Quiabo-do-inferno.

Erect shrub to 2 m, branched above the ground, branches sometimes decumbent; vascular cylinder woody; inner tissues stiff, not mucilaginous. Stems 3–5 cm diam., sometimes constricted, epidermis olive-green; ribs 12–16, 4–5 × 5 mm, rounded in section. Areoles to 2–3 mm diam., 7–8 mm apart, felt short, at first brown, becoming greyish and glabrescent. Spines golden brown to dark reddish brown; central spines 4–6, 5–40 mm, porrect; radials 7–10, 5–10 mm. Fertile part of stem not differentiated, flowers numerous, randomly from lateral and subapical regions. Flower-buds obtuse before anthesis; flowers c. 4.5–7.5 × 2.5–6.5 cm, ridged or quite terete, glabrous or with spiny areoles; pericarpel 10–18 × 15–18 mm; flower-tube to 50 × 15–30 mm at its widest point, ± parallel-sided, straight, with ovoid, apiculate bract-scales at apex; outer perianth-segments 8–10 × 4–5 mm, inner segments 9 × 8 mm, white or cream, spathulate, fimbriate and apiculate; nectar-chamber c. 15–20 × 8 mm; anthers c. 2.5 mm; style 45–50 mm, tapering; stigma-lobes c. 8, c. 7 mm; ovary locule c. 6–8 × 7–8 mm, rectangular or quadrangular in longitudinal section. Fruit ovoid. Seeds 1.5–2.0 × 1.1–1.2 mm, brownish, hilum-micropylar region 0.6–0.8 mm, forming an angle of 30–40° with long-axis; testa-cells domed to conic, with intercellular depressions and coarse, dense cuticular folds.

CONSERVATION STATUS. Least Concern [LC]; see subspecies below.

The precise application of the name *Cereus minensis* Werderm. (1933, 1942) is here changed from that established by Ritter (1979) when describing *Cipocereus pleurocarpus*. Unfortunately, Werdermann did not see the plant in fruit, nor did he indicate a type locality, and the holotype itself is no longer extant. In these circumstances it seems Ritter may have assumed that the plants Werdermann described and illustrated (sterile) were referable to the more widespread smooth-fruited race of the species, since Werdermann indicates that the plant was 'scattered throughout Minas' (Werdermann 1942: 62). However, the only localities actually mentioned were the Serra do Cipó and Serra do Caraça, the former subsequently being the origin of Ritter's new species, the latter apparently the source of the 'partly dried floral remnants' described in Werdermann's protologue (compare l.c.: 61, 93). Curiously, the reference to Caraça seems to be inaccurate, because this well preserved mountain habitat has been extensively botanized in recent years and does not appear to hold a population of *C. minensis* (rather it is the home of the endemic and very different *C. laniflorus* — certainly not that illustrated by Werdermann, l.c.: 62). However, in 2001, a population of *C. minensis* came to light only some 40 km north of the Serra do Caraça, near Cocais, which is close to or on the historical routes connecting Diamantina, Belo Horizonte and Caraça, key places visited and mentioned by Werdermann. Furthermore, this plant, which was recently encountered by Taylor and staff from the Fundação Zoo-Botânica de Belo Horizonte, occurs on north-facing cliffs, as described by Werdermann (l.c.: 61), and closely resembles that illustrated by him (l.c.: 62). Besides this, his description of the floral remnants refers to the flower-tube as 'bearing together with the ovary sparse minute scales, *these with a tufted lanose pubescence in the axils*' (l.c.: 93). This feature is only rarely present in the more widespread race that Ritter identified with *C. minensis*, but is characteristic of and constant for the plant found at the Serra do Cipó and Cocais. Therefore, reluctantly, we have concluded that the names *C. minensis* and *C. pleurocarpus* refer to one and the same race of the species, leaving the more wide-ranging northern subspecies without a name. This is rectified here.

The two subspecies recognized are distinguished as follows:

1. Flower 7.5 × 6.5 cm or larger; ripe fruit not ribbed, smooth, blue (Grão Mogol to E edge
   of Serra do Cabral and region of Diamantina) . . . . **4a.** subsp. **leiocarpus**
1. Flower 4.5–6.0 × 2.5–4.5 cm; fruit ribbed, with a few spine-bearing areoles, brownish,
   pale green, whitish or bluish (Serra do Cipó southwards) . . . . **4b.** subsp. **minensis**

4a. subsp. **leiocarpus** *N. P. Taylor & Zappi* **subsp. nov.** *differt a subsp.* minensis *floribus maioribus, pericarpio fructus laevi.* Holotype: Minas Gerais, Mun. Grão Mogol, Serra da Bocaina, 48 km W of bridge over Rio Vacaria, 28.5 km E of Caveira on road BR 251, 31 Jan. 1991, *Taylor et al.* 1506 (HRCB; K, ZSS, BHCB, isotypes).

[*C. minensis sensu* F. Ritter, Kakt. Südamer. 1: 56 (1979). *C. minensis* subsp. *minensis sensu* Taylor & Zappi in D. R. Hunt, CITES Cactaceae Checklist, ed. 2: 170 (1999).]

Spines generally golden brown (at least in northern part of range), central spines not exceeding 15 mm. Flowers to 65 mm diam., blue without; tube smooth, glabrous; outer perianth-segments dark blue or purplish, patent at anthesis. Fruit not ridged, smooth, bright to deep blue-waxy, only rarely bearing areoles.

Widespread South-eastern *campo rupestre* element: mostly amongst crystalline rocks in *campo rupestre*, 500–2000 m, Serra do Espinhaço and eastern part of Serra do Cabral, Minas Gerais. Endemic. Map 30.
MINAS GERAIS: Mun. Grão Mogol, Serra da Bocaina, 48 km W of bridge over Rio Vacaria, 28.5 km E of Caveira on road BR 251, 31 Jan. 1991, *Taylor et al.* 1506 (K, HRCB, ZSS, BHCB); ibid., near the town, 4 Sep. 1985, *Zappi et al.* in CFCR 8356 (SPF), 5 Nov. 1986, *Cordeiro & Mello-Silva* in CFCR 10140 (SPF), 27 May 1988, *Zappi et al.* in CFCR 11971 (SPF), 15 Oct. 1988, *Taylor & Zappi* in *Harley* 25070, 25136 (SPF, K); Mun. Bocaiúva, E of Engenheiro Dolabela, 7 Aug. 1988, *Eggli* 1115 (ZSS, SPF); Mun. Joaquim Felício, road to Várzea da Palma, Serra do Cabral, 7 May 1990, *Zappi et al.* in CFCR 13236, fl. (SPF, HRCB); Mun. Couto de Magalhães de Minas, 3 Aug. 1988, *Eggli* 1055 (ZSS); Mun. São Gonçalo do Rio Preto, Parque Estadual do Rio Preto, Ribeirão das Éguas, 18°9'S, 43°22'W, 18 Nov. 1999, *Lombardi* 3566 (BHCB, K); Mun. Diamantina, 12 Jan. 1963, *Duarte* 7902 (RB), 29 May 1979, *V.F. Ferreira* 875 (RB); l.c., 5 km from Biri-Biri, 5 June 1985, *H.F. Leitão Filho et al.* 17603 (UEC); road to Curvelo 4.5 km from Diamantina, 4 Aug. 1990, *Sakuragui & Souza* 183 (ESA, K); road to Conselheiro Mata, June 1934, *Brade* s.n. (RB 35757), 3 July 1985, *Pirani et al.* in CFCR 7958 (SPF), 29 Jan. 1986, *Zappi et al.* in CFCR 9425 (SPF), 10 Sep. 1986, *T.B. Cavalcanti et al.* in CFCR 10305 (SPF), 25 Feb. 1987, *Zappi et al.* in CFCR 10394 (SPF), 18 July 1987, *Zappi et al.* in CFCR 11129 (SPF), 19 July 1987, *Zappi et al.* in CFCR 11293 (SPF); road to Mendanha, 31 Oct. 1988, *Taylor & Zappi* in *Harley* 25483, 25489 (SPF, K); Mun. Rio Vermelho, c. 9 km SSE from Pedra Menina, (?) Rio Mundo Velho, 5 Aug. 1988, *Eggli* 1087 (ZSS); Mun. Gouvea, *Ritter* 1013 (SGO 125024, *fide* Eggli et al. 1995: 445); Mun. Datas, Córrego Água Limpa, 23 May 1989, *Hatschbach et al.* 53108 (MBM); Mun. Serro, Milho Verde, 2 May 1995, *T. Grandi* (K, photo); Mun. Santo Antônio do Itambé, Pico de Itambé, c. 2000 m, 16 Aug. 1987, *Zappi et al.* in CFCR 11258 (SPF); Mun. Presidente Kubitschek, Trinta Réis, 19 Apr. 1987, *Zappi et al.* in CFCR 10667 (SPF).

CONSERVATION STATUS. Least Concern [LC] (1); PD=3, EI=1, GD=1. Short-list score (1×5) = 5. Modification of its habitat is relatively slow at present, but needs monitoring.

Hybridizes with *C. crassisepalus* and *C. bradei,* q.v., where they come into contact.

4b. subsp. **minensis**

Spines generally reddish brown to blackish, central spines to 40 mm. Flowers with distal half brownish, reddish or yellowish without; tube with ridges and areoles with minute spines; perianth-segments brownish or bright yellow, erect at anthesis. Fruit ribbed, with a few spiny areoles, brownish, pale green, whitish or bluish. Seed: Barthlott & Hunt (2000: pl. 47.5–6).

South-eastern *campo rupestre* element: on cliffs and between rocks at 900–1300 m, Serra do Cipó and Lapinha, southwards to Cocais and Itabirito, central-southern Minas Gerais. Endemic to the above area. Map 30.

MINAS GERAIS: Mun. Santana do Riacho, Serra do Cipó, 7 Dec. 1949, *Duarte* 2398, and 16 Sep. 1950, *Duarte* 3261 (RB), Nov. 1964, *Ritter* 1327 (U), 5 July 1972, *Hatschbach* 29886 (MBM), 23 July 1980, *Menezes et al.* in CFSC 6390 (SPF), 30 June 1981, *Giulietti et al.* in CFSC 7358 (SPF), 17 Feb. 1982, *Mantovani et al.* in CFSC 7786 (SPF), 25 Jan. 1988, *Martinelli et al.* 11348 (RB, SPF), 2 Nov. 1986, *Zappi & Kameyama* in CFSC 9867 (SPF), 30 Aug. 1988, *Kameyama et al.* in CFSC 11206 (SPF), 27 Oct. 1988, *Taylor & Zappi* in *Harley* 25409, 25413 (K, SPF), 18 Nov. 1989, *Zappi et al.* 190, 191 (HRCB, SPF); l.c., Juquinha, May 1997, *Soffiatti et al.* (K, photos), l.c., near Cachoeira Capivara, *Soffiatti et al.* (K, photo.); Mun. Jaboticatubas, Lapinha [on road to Santana do Riacho], 9 Sep. 1990, *M.H.L. Arndt & A.M. Silva* s.n. (K, BHCB); Mun. Cocais, W of town, on the road near to and at Pedra Pintada, 4 Aug. 2001, *J.O. Rego et al.* 850 (BHCB, Herb. Fund. Zoo-Bot. de Belo Horizonte 647 & fls/frs in alcohol [Carpoteca 761], K, photos), l.c. (?), 'N side on the way up Mount Caraça', 1932, cf. Werdermann (1942: 61, photo p. 62); Mun. Itabirito, Pico de Itabirito, 9 Aug. 1993, *W. Antunes Teixeira* s.n. (K, BHCB).

CONSERVATION STATUS. Endangered [EN B2ab(iii)] (3); area of occupancy < 500 km^2; PD=3, EI=1, GD=2. Short-list score (3×6) = 18. Endangered from excessive burning of its habitat (2 populations including one within the Parque Nacional da Serra do Cipó) and by mining operations (1 population); an extensive part of the habitat at Cocais is private land and inaccessible sheer cliff.

Flower and fruit colour is quite variable in this taxon, the more colourful flower forms suggesting that the plants are evolving towards hummingbird pollination.

5. **Cipocereus pusilliflorus** *(F. Ritter) Zappi & N. P. Taylor* in Bradleya 9: 86 (1991). Holotype: Brazil, Minas Gerais, Monte Azul, Jan. 1964, *Ritter* 1232 (U).

Floribunda pusilliflora F. Ritter, Kakt. Südamer. 1: 58–60 (1979). *Pilosocereus pusilliflorus* (F. Ritter) P. J. Braun in Bradleya 6: 90 (1988) and in Kakt. and. Sukk. 42(10): Karteiblatt 29 (1991).

Erect shrub to 50 cm, branched at ground level, branches semi-erect to decumbent; vascular cylinder weakly woody; inner tissues mucilaginous, turning orange when cut. Stems 4–5 cm diam., not constricted, epidermis bright green; ribs 13–18, 2–3 × 3–9.5 mm, rounded in section, sinuses straight. Areoles to 2.5 mm diam., to 7 mm apart, felt short, brown, becoming glabrescent. Spines golden brown to dark reddish brown; central spines 2–4, 1–2 cm, porrect; radials 10–12, 3–6 mm. Fertile part of stem not differentiated, flowers near or below stem apex at one side, buds sometimes developing on the underside of horizontal stems. Flowers c. 1.6 × 1.0 cm, glabrous, ridged; pericarpel 4 × 10 mm, purplish outside; tube parallel-sided, straight, reddish pink, with small bract-scales at apex; perianth-segments erect, whitish, inner segments acute; nectar-chamber c. 2 × 4 mm, closed at apex by the innermost layer of filaments; style 9 mm, tapering. Fruit globose, 6–13 mm diam., reddish, with a blue-waxy bloom. Seeds (1.3–)1.4–1.5 × 1.1 mm, brownish, hilum-micropylar region 0.6 mm, forming an angle of 20–30° with long-axis of seed; testa-cells flat, with slight intercellular depressions and coarse, dense, marginal cuticular folds (Barthlott & Hunt 2000: pl. 47.7–8).

Northern *campo rupestre* element: on cliffs and ledges of crystalline rocks, 800–1000 m, west slope of Serra Geral (northern Serra do Espinhaço), east of Monte Azul, central-northern Minas Gerais. Endemic. Map 30.

MINAS GERAIS: Mun. Monte Azul, Jan. 1964, *Ritter* 1232 (U), *Horst & Uebelmann* HU 400 (U), 29 Jan. 1991, *Taylor et al.* 1486 (K, ZSS, HRCB, BHCB). A further report by Uebelmann from Santa Barbara (Mun. Augusto de Lima), under the number HU 840, cannot be confirmed with certainty and is probably to be discounted (U. Eggli, pers. comm.).

CONSERVATION STATUS. Critically Endangered [CR D] (4); PD=3, EI=1, GD=1. Short-list score (4×5) = 20. Critically Endangered on present knowledge (< 5 individuals seen at the only certain locality) and urgently in need of further field studies to determine if its range is more extensive. Maintained in a few *ex situ* collections in Europe, including that at the Royal Botanic Gardens, Kew.

In its stem areoles lacking long trichomes and in its seed-micromorphology this species is clearly allied with *C. minensis* and *C. bradei* and certainly not directly related to *Arrojadoa bahiensis* (syn. *Floribunda bahiensis*), q.v., with which it is convergent in its floral hummingbird syndrome and habitat.

18. STEPHANOCEREUS A. Berger

Entwicklungslin. Kakt.: 59, 97 (1926). Type: *Stephanocereus leucostele* (Gürke) A. Berger (*Cereus leucostele* Gürke).

Including (?) *Coleocephalocereus* subg. *Lagenopsis* Buxb. (1972); *Stephanocereus* subg. *Lagenopsis* (Buxb.) N. P. Taylor & Eggli (1991).

Columnar-segmented and ± branched, or bottle-shaped and solitary, 0.5–3.5(–6.0) m high. Juvenile plants globose and without long-hairy stem areoles at first. Stems erect, 3–15(–20) cm diam., constricted or not, vascular cylinder very woody; tissues highly mucilaginous; epidermis dull to grey-green; ribs 12–20, sinuses straight. Areoles with felt and abundant, white, long hairs; spines stiff, golden to greyish, central spines porrect to deflexed, longer than the radials; indeterminate growth of basal areoles present. Flower-bearing region of stem differentiated into a slender, elongate chlorophyllous cephalium or a terminal cephalium tuft, this becoming transformed (through continued vegetative growth) into a lateral, ring-like flowering region (cf. *Arrojadoa* spp.) with long bristles, spines and hairs. Flowers nocturnal, white within, 3.5–10.0 × 2.5–5.5 cm; pericarpel and tube ± naked, pericarpel smooth, subglobose to obpyramidal; tube almost terete, with broad, fleshy scales near apex; perianth-segments triangular to lanceolate or spathulate, spreading to reflexed, innermost white to slightly pinkish; innermost filaments curved towards the style. Fruit subglobose to ovoid-ellipsoid, impressed or not at apex, (at least in subg. *Stephanocereus*) opening by a basal pore upon its detachment from the plant, naked, flower remnant blackened, pendent to erect, deeply or shallowly inserted in fruit apex; pericarp smooth, greenish or dark purple to blackish, with a blue waxy coating (as in *Cipocereus*), funicular pulp solid. Seeds 1.2–2.2 mm, cochleariform; testa-cells convex; cuticular folds present.

A genus endemic to Bahia, comprising two monotypic subgenera, *Stephanocereus* and *Lagenopsis* (Buxb.) N. P. Taylor & Eggli, the first characteristic of the Bahian *caatinga*, the second of the *campo rupestre* of the Chapada Diamantina. Recent preliminary chloroplast gene sequence studies (Soffiatti unpubl.) have cast doubt on the circumscription adopted here, each of the species having more in common with different elements within *Arrojadoa*. As currently circumscribed *Stephanocereus* differs from the latter in having a globose juvenile phase and larger, strongly smelling flowers adapted for nocturnal pollination by bats, rather than diurnal/crepuscular for hummingbirds (Taylor & Zappi 1996). This said, Aloísio Cardoso has photographed an opportunistic hummingbird visiting the expanding flowers of *S. luetzelburgii* at dusk. Plates 32 & 33.

1. Columnar, segmented, at maturity at least 1.75 m, with flowering region terminal and in rings at the articulations of the stem (*caatinga* surrounding the East Brazilian Highlands of N to S Bahia) . . . . **1. leucostele**
1. Bottle-shaped, to 1.5 m, rarely more, with subapical flowers on a continuous, elongate, terminal, chlorophyllous cephalium (*campo rupestre,* Chapada Diamantina, central Bahia) . . . . **2. luetzelburgii**

1. S. leucostele *(Gürke) A. Berger*, Entwicklungslin. Kakt.: 59, 97 (1926). Holotype: Brazil, Bahia, Mun. Maracás, 'Calderão', *Ule* 2 (B).

Cereus leucostele Gürke in Monatsschr. Kakt.-Kunde 18: 53 (1908). *Cephalocereus leucostele* (Gürke) Britton & Rose, Cact. 2: 36 (1920). *Pilocereus leucostele* (Gürke) Werderm., Bras. Säulenkakt.: 111–112 (1933).

Single-stemmed or tree-like, 1.5–6.0 m, ± branched, branches arising at the stem-segment joints or more rarely between these. Stems erect, 5–8 cm diam., slightly thickened below the cephalia; vascular cylinder very woody; ribs 13–18, 7–8 × 5 mm; epidermis grey-green; areoles 3–4 mm, to 10 mm apart, with abundant, long white trichomes partly obscuring the stem. Spines golden or whitish, stout; central spines 3–5, 12–20(–30) mm, spreading to deflexed, radials 15–20, to 10 mm. Fertile part of stem apical at first, later forming up to 5 or more rings on the same branch as the vegetative growth overtops successive cephalia, flower-bearing areoles with abundant, loose white wool and long, golden or whitish bristles to 40–90 mm. Flowers nocturnal, 9–10 × 5–5.3 cm, smelling strongly of cabbage; pericarpel and tube almost naked, with only a few small, triangular scales, these sometimes slightly woolly in their axils, pericarpel 13–16 × 22 mm, green; tube 50–60 × 22–35 mm, cylindric, somewhat curved, greenish cream, with fleshy, brownish scales at apex; perianth-segments 15–18 × 6–8 mm, spathulate to lanceolate, outer segments reflexed, inner segments spreading to reflexed, pure white; stamens included in relation to the perianth-segments; style to 60 mm, stigma-lobes 7–8, exserted; ovary locule 10 × 10 mm, obtriangular in longitudinal section. Fruit globose to pyramidal, 4.4–4.7 × 3.5–4.5 cm; pericarp impressed at apex, greenish to dark blue, wall to 8 mm thick, funicular pulp white or red. Seeds 2–2.2 mm long, cochleariform, black, dull; testa-cells convex, with cuticular folds (Barthlott & Hunt 2000: pl. 49.1–4).

Central-southern (Bahian) *caatinga* element: in *caatinga* surrounding the Chapada Diamantina, northern Serra do Espinhaço and Serra Geral, Bahia, 300–1100 m, east of the Rio São Francisco. Endemic. Map 31.
BAHIA: Mun. Sento Sé, 10°7'S, 41°25'W, 11 July 2000, *G. Charles* (photos); Mun. Campo Formoso, 20 km from Maçaroca towards Fazenda Lagoa do Angico, 8 Jan. 1991, *Taylor et al.* 1385 (K, HRCB, ZSS, CEPEC); Mun. Itaguaçu da Bahia, 31 km SE of town towards Irecê, 13 Jan. 1991, *Taylor et al.* 1411 (K, HRCB, ZSS, CEPEC); Mun. (?), 10°53'S, 41°35'W, 13 July 2000, *G. Charles* (photos); Mun. Ourolândia, E of the town and also NE of Olho D'Água do Facundo, 4 Aug. 2002, *Taylor* (obs.); Mun. João Dourado & Mun. Morro do Chapéu, APA Gruta dos Brejões, 4 Aug. 2002, *Taylor & Machado* (obs. & K, photos); Mun. Jacobina, Catinga do Moura, 10 km from Lages, 16 July 1989, *Zappi* 133 (SPF); Mun. América Dourada, c. 17 km S of road BA 052, 2 km E of Belo Campo, 26 Dec. 1988, *Taylor &*

Zappi (obs.); Mun. Morro do Chapéu, Icó, Sep. 2001, *M. Machado* (obs.); l.c., between Cafarnaum and road BA 052, 26 Dec. 1988, *Taylor & Zappi* (obs.); Mun. Canarana, 6 km E towards Cafarnaum, 24 Aug. 1988, *Eggli* (obs.); Mun. Ipupiara, Vanique, 11°47'S, 42°32'W, 23 June 1978, *C.A. Miranda 292* (RB); Mun. Ibitiara, 49 km W of Seabra on road to Ibotirama, 1100 m, 14 Jan. 1991, *Taylor et al.* (obs.); Mun. Oliveira dos Brejinhos, 131 km W of Seabra on road BR 242, 14 Jan. 1991, *Taylor et al.* (obs.); Mun. Marcionílio Sousa, Machado Portela, 19 June 1915, *Rose & Russell 19902* (US); Mun. Iaçu, Jan. 1964, *Ritter 1227* (SGO 121290, *fide* Eggli *et al.* 1995: 498); Mun. Botuporã, c. 50 km S of Macaúbas on road to Paramirim, 26 Aug. 1988, *Eggli* (obs.); Mun. Rio de Contas, 15 May 1983, *Hatschbach 46404* (MBM); Mun. Paramirim, near Caraíbas, 30 Nov. 1988, *Taylor* (K, photos); Mun. Dom Basílio, 13°59'S, 41°45'W, 26 June 1978, *A.P. Araújo 43* (RB); Mun. Barra de Estiva, 17 km NE of Contendas on road to Maracás, 4 Feb. 1991, *Taylor et al.* (obs.); Mun. Ituaçu, c. 2 km NE of town, 18 Aug. 1988, *Eggli 1202, 1203* (ZSS, photos, seeds); Mun. Contendas do Sincorá, 27 km N of Sussuarana, 3 Feb. 1991, *Taylor et al.* (obs.); Mun. Maracás, 6–7 km E of Porto Alegre, 300 m, 4 Feb. 1991, *Taylor et al.* (obs.); Mun. Rio do Antônio, 43 km on road from Brumado to Caitité, 14 Mar. 1983, *Leuenberger et al. 3067* (CEPEC, B, K); Mun. Guanambi, 19 km NE of town, N of road to Caitité, 25 July 1989, *Taylor & Zappi* (obs.); Mun. Brumado, 7 km N of town on road to Livramento, 14°8'S, 41°40'W, 22 Nov. 1988, *Taylor & Zappi in Harley 25539* (K, SPF, CEPEC); Mun. Manuel Vitorino, road BR 116, 27 July 1989, *Taylor & Zappi* (obs.); Mun. Palmas de Monte Alto, near Telebahia tower above the town, 21 July 1989, *Taylor & Zappi* (obs.); Mun. Aracatu, 95 km W of Vitória da Conquista towards Brumado, 41 km SE of Brumado, on road BA 262, 2 Feb. 1991, *Taylor et al. 1521* (K, HRCB, ZSS, CEPEC); Mun. Anajê, c. 10 km N towards Sussuarana, 17 Aug. 1988, *Eggli* (ZSS, photos); Mun. Poções, c. 15–20 km NE of Planalto on road BR 116, 27 July 1989, *Taylor & Zappi* (obs.).

CONSERVATION STATUS. Least Concern [LC] (1); PD=3, EI=1, GD=1. Short-list score (1×5) = 5. Least Concern at present, but its habitat is continuing to decline and its status needs to be monitored.

One of the most characteristic cacti of the Bahian *caatinga*, but absent from north-eastern Bahia, being closely associated with the East Brazilian Highlands, where the following species replaces it in the *campos rupestres*. This suggests that their ancestor was a plant of montane origin (cf. *Cipocereus*).

2. S. luetzelburgii *(Vaupel)* N. P. Taylor & Eggli in Bradleya 8: 91 (1991). Type: Brazil, Bahia, Serra das Almas [Pico das Almas], July 1913, *Luetzelburg 22* (B, holo., n.v.).

Cereus luetzelburgii Vaupel in Z. Sukkulentenk. 1: 57 (1923). *Pilocereus luetzelburgii* (Vaupel) Werderm., Bras. Säulenkakt.: 111 (1933). *Cephalocereus luetzelburgii* (Vaupel) Borg, Cacti: 111 (1937). *Pilosocereus luetzelburgii* (Vaupel) Byles & G. D. Rowley in Cact. Succ. J. Gr. Brit. 19: 67 (1957). *Pseudopilocereus luetzelburgii* (Vaupel) Buxb. in Beitr. Biol. Pflanzen 44: 252 (1968). *Coleocephalocereus luetzelburgii* (Vaupel) Buxb. in Krainz, Kakteen, Lfg. 48–49 (1972).

Solitary, 0.5–1.0(–1.5) m, often not erect but inclined ± westwards, bottle-shaped, sometimes branched at base, normally unbranched above unless damaged, except for populations found around Morro do Chapéu, which are frequently branched; central cylinder very woody at base. Stems erect to decumbent at apex, 12–15(–20) cm diam. at base, 2.5–4 cm diam. at apex, in the flower-bearing region; ribs 12–20, 20 × 18 mm on juvenile stem, gibbose beneath the areoles; epidermis grey-green to dark green. Areoles 5–6 mm, to 1 cm apart, long hairs abundant on the fertile part, white to pale cream. Spines golden or whitish, blackish when old; central spines 4–6, 1.2–2(–3) cm, ascendent to spreading, radials 10–16, to 1.2 cm. Fertile part of stem elongate, with lateral to subapical flower-bearing areoles with abundant, loose white to pale brown hairs and long, golden, whitish or greyish bristles, these 2–4 cm. Flowers 2–5 or more borne together, 4–6 × 2.5–3.0 cm, smelling strongly of cabbage; pericarpel 10 × 5–6 mm, greenish; tube 2.5 × 1.2–2.0 cm, narrowed above nectar-chamber, slightly curved, greenish cream, with fleshy, brownish bract-scales only at apex; perianth-segments 8–10 × 3–5 mm, spathulate to lanceolate, outer segments reflexed, inner segments spreading to reflexed, white; stamens included in relation to the perianth-segments; style to 28 mm, stigma-lobes 6–8, exserted; ovary locule 4 × 4 mm, obtriangular in longitudinal section. Fruit obovoid, 2.2–2.5 × 1.5–2 cm; pericarpel smooth, dark blue to blackish, but covered in lighter blue wax, scarcely impressed at apex (not observed in fully ripe state, but claimed by some observers to be dehiscent and reddish). Seeds 1.2–1.5 mm, cochleariform, black, dull; testa-cells convex, with cuticular folds.

Northern *campo rupestre* (Chapada Diamantina) element: on and between crystalline and sandstone rocks and gravels, *campo rupestre*, 380–1550 m, Bahia. Endemic. Map 31.

BAHIA: Mun. Jacobina, 2.5 km E of bridge over the river E of town, 16 July 1989, *Zappi* 129 (SPF); Mun. Morro do Chapéu, W of Icó, N of Brejão (Formosa), nr 'Lagedo Bordado', 5 Aug. 2002, *Taylor & Machado* (obs.); l.c., c. 3.5 km SE of Morro do Chapéu, 25 July 1993, *L.P. de Queiroz & N.S. Nascimento* 3441 (HUEFS, K); l.c., c. 18 km E of town, margins of Rio Ferro Doido, 17 Feb. 1971, *Irwin et al.* 32438 (K, NY, WAG); l.c., 20 km W of town, 2002, *M. Machado* (obs.); Mun. Seabra, c. 28 km W of town S of road BR 242, 18 July 1989, *Taylor & Zappi* (K, photos); Mun. Palmeiras, Morro do Pai Inácio, 12°27'S, 41°28'W, 13 Feb. 1994, *V.C. Souza et al.* 5252 (K, SPF); Mun. Lençóis, 24 Sep. 1965, *A.P. Duarte* 9363 & *E. Pereira* 10076 p.p. (RB); l.c., Serra da Chapadinha, 12°27'S, 41°25'W, 24 Nov. 1994, *Melo et al.* in PCD 1315 (K, ALCB); Mun. Mucugê, c. 15 km NW of town, 12°58'S, 41°28'W, 26 Mar. 1980, *Harley et al.* 20976a (K, CEPEC); l.c., c. 7 km SW of town towards Barra da Estiva, 21 Aug. 1988, *Eggli* 1251 (ZSS); Mun. Piatã, 13°8'S, 41°50'W, 27 Aug. 1992, *W. Ganev* 976 (HUEFS); Mun. Érico Cardoso [Água Quente], 26 km E of town, between Santa Rosa and the Pico das Almas, 13°32'S, 42°W, 1 Dec. 1988, *Taylor* in *Harley* 25559 (K, SPF, CEPEC); Mun. Rio de Contas, 16 km NW of town, near the Pico das Almas, 13°32'S, 41°54'W, 27 Nov. 1988, *Taylor* in *Harley* 25551 (K, SPF, CEPEC); l.c., 'near Bom Jesus [do Rio de Contas]', 1915, *Luetzelburg* s.n. (US); Mun. Jussiape, Cachoeira da Fraga, próx. da cidade, 17 Feb. 1987, *Harley et al.* 24372 (SPF); Mun. Barra da Estiva, 3–13 km W of town towards Jussiape, 23 Mar. 1980, *Harley et al.* 20837 (K, CEPEC); Mun. Ituaçu, c. 2 km NE of town, 19 Aug. 1988, *Eggli* 1204 (ZSS); unlocalized, Nov. 1913, *Luetzelburg* 742 (M).

CONSERVATION STATUS. Least Concern [LC] (1); PD=3, EI=1, GD=2. Short-list score (1×6) = 6. Habitat modification affecting this species is currently limited.

Unmistakable for its bottle-shaped stem, this species and *Micranthocereus purpureus* are the most characteristic elements of the Chapada Diamantina's cactus flora. Variable in stem shape and rib number between populations.

19. ARROJADOA Britton & Rose

Cact. 2: 170 (1920). Type: *Arrojadoa rhodantha* (Gürke) Britton & Rose (*Cereus rhodanthus* Gürke).

Including (?) *Pierrebraunia* E. Esteves Pereira (1997), quoad typ.

Literature: Taylor & Zappi (1996).

Columnar, shrubby, 0.2–1.5(–3.5) m, branched below, at or above ground, but not forming a definite trunk (except rarely in *A. rhodantha*); vascular cylinder rather woody; tissues very mucilaginous; subterranean stem-tuber system sometimes present. Stems erect or rarely decumbent, 1–7(–12) cm diam., constricted, epidermis dull to bright green; ribs 6–14(–17), low, crenate, sinuses straight. Areoles with felt and long hairs at first; spines stiff, very pungent, central spines porrect to deflexed, longer than the radials; indeterminate growth of areoles near stem base present. Flower-bearing region of stem apical, more or less dilated and forming a cephalium with bristles and wool, or cephalium lacking (*A. bahiensis*) or flowers also arising partly in a lateral position (*A. dinae* subsp. *eriocaulis*). Flower-buds coloured bright pink from the earliest stages of their development; flowers diurnal, opening at morning, or late afternoon to evening, and sometimes remaining open through part or all of the night, brightly coloured, 15–42 × 5–20 mm; pericarpel smooth, naked, subglobose and clearly delimited from the tube; tube broadened below, cylindric above, reddish, pink or magenta, with broad, fleshy bract-scales only near the apex; innermost filaments curved towards the style, protecting the nectar-chamber; perianth-segments triangular to lanceolate or spathulate, erect to spreading, opening very little, innermost white, yellow, pinkish or purplish. Fruit obovate to turbinate, indehiscent, naked, flower remnant drying black, normally erect, broad at base, forming a deep and often broad point of insertion at apex of fruit; pericarp smooth, variously coloured, funicular pulp white, scarce, ± solid or aqueous. Seeds to 1.5 mm, cochleariform; testa-cells convex; cuticular folds present or nearly absent.

A genus of 4 or 5 very distinct, but rather variable, hummingbird-pollinated species, characteristic of the *caatinga-agreste* and *campos rupestres* (and included *cerrados*). The potential fifth species is a plant discovered in August 2000, in north-eastern Goiás, by botanist, Rafaela Forzza (RB). It appears to be related to *A. dinae*. Prior to this the genus was considered to be endemic to Eastern Brazil. Another novelty of uncertain status is mentioned under *A. rhodantha* (see 4b).

P. J. Braun & E. Esteves Pereira (1995b: 81) have established the Subgenus *Albertbuiningia* based on the second species treated here, *A. dinae* Buining & Brederoo. The following key does not attempt to distinguish the hybrids that occasionally occur between species Nos 2–4 (2×3, 2×4 & 3×4, see below). Plates 34–37.

1. Stem 4–10 cm diam., neither strongly constricted nor terminated or interrupted by cephalia, flowers from ± undifferentiated areoles at stem apex (cent. Bahia, *campo rupestre*, 1000–2000 m) . . . . **1. bahiensis**
1. Stems constricted, thickened at apex or < 4 cm diam., the flowers developed in cephalia composed of wool and long bristle-spines 2

2. Stems 10–50 × 2 cm, sometimes arising from a tuberous, rhizomatous rootstock; flowers bicoloured (anthesis p.m.), inner perianth-segments contrasting with flower-tube (Serra do Espinhaço: Caitité BA to Bocaiúva MG, *campo rupestre, cerrado* & ecotones with *caatinga*) . . . **2. dinae**
2. Stems usually > 50 cm or > 2 cm diam., rootstock fibrous; flowers concolorous 3

3. Stems to 1.8 cm diam., but expanded and much broader below the cephalia; anthesis p.m., outer perianth-segments expanding (N, E & S Bahia & NE Minas Gerais, *caatinga-agreste*) . . . . **3. penicillata**
3. Stems 2–6 cm or more in diameter, not as above; anthesis a.m., perianth-segments hardly expanding, erect (widespread, *caatinga* and ecotones with *campo rupestre*) . . . . **4. rhodantha**

1. Arrojadoa bahiensis

1. Arrojadoa bahiensis *(P. J. Braun & E. Esteves Pereira) N. P. Taylor & Eggli* in Kew Bull. 49: 98 (1994); N. P. Taylor & Zappi in Bot. Mag. 13: 72–75, tab. 291 (1996). Type: Bahia, 'Chapada Diamantina', *E. Esteves Pereira* 337 (UFG 13007, holo. n.v.).

Floribunda bahiensis P. J. Braun & E. Esteves Pereira in Pabstia 4: 11–16, illus. (1993); A. Hofacker in Kakt. and. Sukk. 45: 120–123, illus. (1994). *Pierrebraunia bahiensis* (P. J. Braun & E. Esteves Pereira) E. Esteves Pereira in Cact. Succ. J. (US) 69: 296 (1997).
Arrojadoa cremnophila Taylor et al. in sched., nom. nud.

Subglobose to elongate plants (depending on exposure), normally to 30 cm, but occasionally to 60 cm or more, unbranched or branched below ground or at base to form clumps of stems; central cylinder not woody. Stems erect or inclined, 4–10 cm diam.; ribs (9–)10–14(–17), 5 × 15(–25) mm, with or without lateral folds; epidermis bright yellow-green to dark green. Areoles 2–3 mm, approximate, with felt and long, white to grey trichomes. Spines brittle, pale pinkish brown or golden, black when old; central spines 1–4, to 30 mm, ascendent, radials 6–8(–10), to 5 mm, spreading; weak indeterminate growth of spines occurring at basal stem areoles. Fertile part of stem apical, flower-bearing areoles scarcely differentiated. Flowers (reported in February and August, but produced in all months in cultivation) few or up to 25 developing together, opening first at night and closing around noon the following day (timing and degree of perianth expansion apparently linked to light and low temperatures), c. 29–39(–42) × to 11 mm at apex of tube, the inner perianth-segments expanding to give an opening 8–13 mm in diam., pericarpel and tube together to c. 30 mm, naked or the latter with a few minute bract-scales in distal half, magenta-pink outside, pericarpel c. 5 × 7 mm, nectar-chamber externally 6 × 8.5 mm, tube constricted above and below; outer perianth-segments short and fleshy, magenta-pink, inner thin and delicate, white, 3.9–4.7 × 2.3–2.8 mm; stamens 3–14 mm, the lowermost with thickened filament bases protecting the nectar-chamber, anthers c. 0.8–1.1 mm, pale yellow, with abundant pollen; style to 28 × 0.8–0.4 mm, white; stigma-lobes 5–7, white, pointed, partly exserted in relation to the anthers or hidden amongst them, c. 4 × <1 mm. Fruit c. 17 × 15 × 12 mm, broadly ovoid, somewhat compressed, ridged and contracted or beaked at apex, deep shocking pink, with a paler waxy bloom, funicular pulp liquid, translucent, floral remnants ± blackened. Seeds to 1.3 mm, cochleariform, black, dull; testa-cells convex, cuticular folds present.

Northern *campo rupestre* (Chapada Diamantina) element: on cliffs and rock ledges in sun or deep shade, *campo rupestre*, c. 1000–2000 m, central Bahia. Endemic. Map 32.

BAHIA: Mun. Mucugê, 27 km NW of the town on road to Guiné, then 1 km E, 12°51'S, 41°31'W, 23 Dec. 1988, *Taylor & Zappi in Harley* 27382 (K, SPF, CEPEC), 7 Sep. 1981, *A. Furlan et al. in CFCR* 2020 (SPF); W of town, above cemitério, 15 Mar. 1989, *Supthut* 8937 (ZSS), 6 Feb. 1991, *Taylor et al.* 1557A & cult. Zappi HRCB, then cult. RBG Kew, 1995 (K, ZSS, photos), coll. unknown, cult. A. Hofacker, 1993 (ZSS – fr., K, photos); Mun. Abaíra, '9 km N de Catolés, caminho de Ribeirão de Baixo a Piatã, subida da Serra do Atalho', 13°14'S, 41°55'W, 10 July 1995, *L.P. de Queiroz* 4384 (HUEFS); Mun. Rio de Contas, Pico do Itobira, 2 Jan. 2003, *Harley* (K, photos); l.c., top of Pico das Almas, 1958 m, 20 Feb. 1987, *Harley* 24522 (SPF, K, CEPEC), 15 Dec 1988, *Taylor in Harley* 25565 (K, SPF, CEPEC).

CONSERVATION STATUS. Vulnerable [VU D2] (2); extent of occurrence = 1614 km²; area of occupancy estimated to be < 20 km²; PD=3, EI=1, GD=2. Short-list score (2×6) = 12. Known from only c. 5 populations (one within the Parque Nacional da Chapada Diamantina) and at risk from collection of plants and seeds, even though many individuals are protected by the plant's preference for steep slopes and cliffs inaccessible to the collector.

This unique and remarkable Bahian endemic was first collected by scientists only in 1981, but could prove to be quite widespread in the Chapada Diamantina when the inaccessible cliff habitats it occupies have been further investigated. Its flowers, fruit and seed would at first sight seem to ally it with the following allopatric-vicariant species. However, small, sterile individuals strongly resemble juvenile plants of the sympatric *Stephanocereus luetzelburgii* and in morphology (but not size and colour) their flowers are also very similar. Furthermore, recent preliminary chloroplast gene sequence studies (Soffiatti unpubl.) suggest that these two may indeed be sister species and that the absence of a proper cephalium distinguishes them from the remainder of the *Arrojadoa-Stephanocereus* alliance.

   A. bahiensis is superficially similar and convergent with *Cipocereus (Floribunda) pusilliflorus*, but has rather different, woolly adult areoles (long hairs lacking in *C. pusilliflorus*), mature fruits and seeds, and its perianth is clearly and abruptly differentiated into coloured fleshy outer segments and white thinner inner segments, as in *A. dinae*.

2. **Arrojadoa dinae** Buining & Brederoo in Kakt. and. Sukk. 24: 99–101 (1973). Holotype: Bahia, Brazil, Mun. Urandi, 900 m, *Horst & Uebelmann* 399 (U).

A. *beateae* P. J. Braun & E. Esteves Pereira in Kakt. and. Sukk. 40: 250–256 (1989). Holotype: Brazil, Minas Gerais, Mun. Itacambira, July 1987, *Esteves Pereira* 261 (UFG); *Braun* 830 (ZSS, B, paratypi).

A. *dinae* var. *nana* P. J. Braun & E. Esteves Pereira in ibid. 42: 190–195 (1991). A. *dinae* subsp. *nana* (P. J. Braun & E. Esteves Pereira) P. J. Braun & E. Esteves Pereira in Succulenta 74: 82 (1995). Holotype: Minas Gerais, *Braun* 415 (ZSS).

A. *heimenii* van Heek & W. Strecker in Kakt. and. Sukk. 50: 137–138 (1999); W. van Heek in Piante Grasse, Speciale 1997: 29, 31, 'Foto' 70 & 71 (1998), nom. inval. Holotype: Minas Gerais, Mun. Engenheiro Dolabela, *P. Braun* 20528 (UFG).

VERNACULAR NAME. Rabo-de-raposa.

Subshrub, sometimes creeping/suckering or semidecumbent, 10–60 cm, poorly branched above ground; vascular cylinder moderately woody; subterranean stems not or slightly thickened, or strongly tuberous, sprouting when above-ground parts are damaged. Aerial stems erect, 1–2 cm diam., slightly to strongly thickened below the flower-bearing apical cephalium and sometimes to 3.5 cm diam. near rootstock; ribs 6–11, 3–4 × 2–4 mm; epidermis olive-green. Areoles 2 mm, to 5 mm apart, with few to many, long, white to grey trichomes, conspicuously hairy in some forms. Spines reddish at first, paler to off-white in age, delicate, needle-like; central spines 1–2, to 10 mm, radials 8–10, 3–5 mm. Cephalium with much loose wool and long reddish bristle-spines to 20 mm, up to 6 cephalia per stem. Flowers arising from the cephalia or sometimes partly from the stem beneath, 14–29 × 4–9 mm, opening late in the afternoon; pericarpel 4 × 3.5 mm; tube to 10 × 4–5 mm, cylindric, coral-red, orange-red or magenta, with fleshy bract-scales only at apex; perianth-segments 2–3 × 2 mm, triangular to lanceolate, outer segments erect, coral-red to reddish pink, inner segments opening very slightly, rather variable in colour, pale creamy yellow, white or lilac-pink to purplish; stamens included in relation to the perianth; style 8–9 mm, stigma-lobes 7–8, included; ovary locule 1.8–2.0 × 2 mm, hemicircular in longitudinal section. Fruit obovoid, broadly beaked to truncate at apex, to 7–21 × 5–18 mm, pericarp smooth, greenish, pink, reddish or purplish brown at apex. Seeds 0.9–1.8 × 0.8–1.4 mm, cochleariform, black, dull; testa-cells convex, with cuticular folds (Barthlott & Hunt 2000: pl. 49.5–7).

CONSERVATION STATUS. Near Threatened [NT]; extent of occurrence > 20000 km² and at least 17 locations known, but most of these are projected to decline in quality and numbers of individuals held (the subspecies, below, are threatened).

The following subspecies are recognized — both are rather variable in flower and fruit colour:

1. Well-developed subterranean stem-tubers present; stem areoles very woolly, giving the stem a felted appearance .... **b.** subsp. **eriocaulis**
1. Subterranean part of stem sometimes thickened, but well-developed tubers lacking; above-ground vegetative part of stem not as above .... **a.** subsp. **dinae**

2a. subsp. **dinae**

? A. *multiflora* F. Ritter, Kakt. Südamer. 1: 89–90 (1979). Holotype: Brazil, Bahia, Mun. Caitité, Brejinho das Ametistas, Jan. 1964, *Ritter* 1243 (U).

Northern *campo rupestre* element: sandy *cerrado, campo rupestre* (sometimes on rocks), *gerais* and in the *caatinga/campo rupestre* ecotone, 550–1400 m, central-southern Bahia (from c. 13°55'S southwards) and northern Minas Gerais (south to Bocaiúva) in the Serra do Espinhaço (and Serra Geral). Endemic to the core area within Eastern Brazil. Map 32.

BAHIA: Mun. Caitité, 14°10'S, 42°32'W, 27 Oct. 1993, *L.P. de Queiroz* 3605 (HUEFS); (?) l.c., 15 km NW towards Riacho de Santana, *Ritter* 1243 (SGO 121896, *fide* Eggli *et al.* 1995: 503); l.c., hill above the town with Telebahia tower, 27 Aug. 1988, *Eggli* 1305 (ZSS); ibid., 17 Mar. 1989, *Supthut* 8944 (ZSS) — showing some influence (?) of introgression with *A. rhodantha*; (?) l.c., Brejinho das Ametistas, 1964, *Ritter* 1243 (U); Mun. Urandi, *Horst & Uebelmann* 399 (U); l.c., SE of town, 1983, *Braun* 411 (ZSS); Mun. Jacaraci, 5.5 km from town on road to Caetité, *H. Britsch* in I.S.I. 1691, cult. R. Mottram, 1988 (K); Mun. Cordeiros, 15°4'S, 40°53.5'W, Sep. 2003, *M. Machado* (photos).
MINAS GERAIS: Mun. Rio Pardo de Minas/São João do Paraíso (?), 'SW of Condeúba, Bahia', 1983, *Braun* 415 (ZSS); Mun. Monte Azul, Serra Geral, c. 12–15 km E of town, 24 July 1989, *Zappi* 169 (SPF, K); Mun. Rio Pardo de Minas, Montezuma, *Horst & Uebelmann* 578 (ZSS); l.c., divisa entre Mun. Espinosa e Montezuma, Serra do Pau d'Arco, 15°4'55"S, 42°38'27"W, 15 Mar. 1994, *N. Roque et al.* in CFCR 15407 (K, SPF); Mun. Itacambira, Alto da Serra, c. 28 km de Pau D'oleo (42 km de Juramento), 16 June 1991, *R. Mello-Silva et al.* 524 (SPF, K); l.c. (?), *Braun* 830 (ZSS), July 1987, *E. Esteves Pereira* 261 (UFG); Mun. Bocaiúva, NE of Engenheiro Dolabela, c. 15 & 40 km towards Sítio, 7–8 Aug. 1988, *Eggli* 1116, 1123 (ZSS).

CONSERVATION STATUS. Vulnerable [VU B1ab(iii)] (2); extent of occurrence = c. 20,000 km², PD=3, EI=1, GD=2. Short-list score (2×6) = 12. Continuing decline in habitat quality caused by charcoal producers and forestry activities.

Very variable in stem morphology and flower colour. On present knowledge, the typical northern populations and the southern forms (described as *A. beateae* and *A. heimenii*) appear to be disjunct. Ritter's *A. multiflora* is clearly a synonym of the species treated here, but its referal to subsp. *dinae* is currently uncertain. The plant recently discovered in north-eastern Goiás by Rafaela Forzza (RB) is, from photographs at least, vegetatively hard to distinguish from subsp. *dinae*.

2b. subsp. **eriocaulis** *(Buining & Brederoo)* N. P. Taylor & Zappi in Cactaceae Consensus Initiatives 3: 7 (1997). Holotype: Brazil, Minas Gerais, Mun. Mato Verde, 900 m, *Horst & Uebelmann* 349 (U).

A. eriocaulis Buining & Brederoo in Kakt. and. Sukk. 24: 241–244 (1973).
A. eriocaulis var. *albicoronata* van Heek *et al.* in Kakt. and. Sukk. 33: 224–227 (1982). *A. eriocaulis* subsp. *albicoronata* (van Heek *et al.*) P. J. Braun & E. Esteves Pereira in Succulenta 74: 82 (1995). Type: Brazil, Minas Gerais, Grão Mogol, 15 Aug. 1981, *van Heek et al.* 81/113 (KOELN, holo., n.v.).
A. eriocaulis var. *rosenbergeriana* van Heek & W. Strecker in Kakt. and. Sukk. 44: 258–262 (1993). Type: Brazil, Minas Gerais, SE of Mato Verde, *W. van Heek & W. Strecker* 85/216 (KOELN, holo., n.v.).
A. multiflora subsp. *hofackeriana* P. J. Braun & E. Esteves Pereira in Kakt. and. Sukk. 53: 75 (2002). Holotype: Bahia, nr Piatã, *I. Horst* HU1394 (UFG 22429, n.v.; ZSS, iso., n.v.). **Synon. nov.**

Northern *campo rupestre* element: in sandy *cerrado*, 700–950 m, SE edge of the Chapada Diamantina and eastern drainage of Serra do Espinhaço, central-southern Bahia and northern Minas Gerais. Endemic to the core area within Eastern Brazil. Map 32.
BAHIA: Mun. (?) Piatã, c. 95 km S of road BR 242, *Horst & Uebelmann* HU 1394, see Bohle (2000: 48), l.c., 13°40'30"S, 47°16'12"W, 15 July 2000, cult. G. Charles, June 2001 (photos).
MINAS GERAIS: Mun. Mato Verde/Rio Pardo de Minas, 12 km E of Mato Verde towards Santo Antônio do Retiro, 15°23'S, 42°45'W, 9 Nov. 1988, *Taylor & Zappi* in *Harley* 25518 (K, SPF), 13 Aug. 1988, *Eggli* 1154 (ZSS); Mun. Grão Mogol, on road to Francisco Sá, 5 Jan. 1986, *Kameyama et al.* in CFCR 8839 (SPF, MBM); l.c., Córrego Escurinha, 23 July 1986, *Zappi et al.* in CFCR 9828 (SPF), 28 May 1988, *Zappi et al.* in CFCR 12019 (SPF).

CONSERVATION STATUS. Endangered [EN B2ab(iii)] (3); area of occupancy assumed to be < 500 km²; PD=3, EI=1, GD=2. Short-list score (3×6) = 18. Endangered from habitat destruction by charcoal producers; needs regular monitoring in view of this and its apparent rarity.

Variable in flower colour. The plants described and illustrated as *A. multiflora* by Ritter, l.c. and *A. eriocaulis* var. *rosenbergeriana* by van Heek & Strecker, l.c., appear to be somewhat intermediate between the two subspecies recognized. The form from Bahia requires further study, but does not seem to be specifically distinct as some commentators have suggested.

Arrojadoa dinae is unique amongst the cacti of Eastern Brazil in its possession, at least in subsp. *eriocaulis*, of well-developed true stem-tubers (despite claims that these also occur in the root-tuberous *Cereus* subg. *Mirabella* by Braun & Esteves Pereira 2002: 42).

3. Arrojadoa penicillata *(Gürke) Britton & Rose*, Cact. 2: 171 (1920); N. P. Taylor & Zappi in Bot. Mag. 13(2): 75–78, tab. 292 (1996). Type: Brazil, Bahia, Mun. Maracás, 'Calderão', Oct. 1906, *Ule* 7052 (B, holo. in spirit, HBG, iso. [K, photo.]).

Cereus penicillatus Gürke in Monatsschr. Kakt.-Kunde 18: 70–71 (1908). *Cephalocereus penicillatus* (Gürke) Werderm., Bras. Säulenkakt.: 116 (1933).

Arrojadoa penicillata var. *decumbens* Backeb. & Voll in Arch. Jard. Bot. Rio de Janeiro 9: 164–166, illus. p. 159 (1949, publ. 1950). Lectotype (Taylor & Zappi, l.c.): illus., l.c. 159.

A. penicillata var. *spinosior* Brederoo & S. Theun. in Succulenta 59: 20–27 (1980). Holotype: Brazil, [Minas Gerais], Rio Jequitinhonha, *Horst* 113 (U).

VERNACULAR NAME. Rabo-de-raposa.

Shrubby or semi-scandent to decumbent (especially when growing in sand-dunes), to 4 m long, ± branched; rootstock fibrous. Stems erect or more rarely decumbent, 1–1.8 cm diam., not very fleshy, expanded to 2.5(–3.0) cm diam. below the cephalia, vascular cylinder very woody; ribs 7–12, 2–4 × 3 mm; epidermis olive-green. Areoles 2 mm, to 7 mm apart, long hairs present but soon falling. Spines reddish brown, delicate; central spines 2–4, 4–30 mm, ± deflexed, radials 6–10, 3–5 mm. Cephalia sometimes forming 3 or 4 rings on the same axis, with abundant, loose wool and long red to brownish bristle-spines to 20–30 mm. Flowers to 30 × 20 mm, opening shortly before dusk, pinkish magenta or reddish pink; pericarpel 7–8 × 5 mm; tube to 12 × 4–7 mm, cylindric, with fleshy bract-scales only at apex; perianth-segments bright pink with a very fine darker line around the margin, outer segments to 7 × 2.8 mm, triangular to lanceolate, patent (well-expanded), innermost segments remaining erect and forming a tube which encloses the stamens etc., its opening 3–4 mm in diam.; stamens included in relation to the perianth-segments; style 10 mm, stigma-lobes 6–8, included; ovary locule 2–2.5 × 2 mm, obtriangular in longitudinal section. Fruit obovate to globose, sometimes laterally compressed, 12–25 × 12 mm; pericarp smooth, greenish to dull red, olive-brownish or whitish. Seeds c. 1.2 mm, cochleariform, black, dull; testa-cells convex, with or without cuticular folds.

Central-southern *caatinga* element: on granite/gneiss inselbergs or *lajedos,* sand-dunes and stony ground (rarely on limestone), growing under or through shrubs in the *caatinga-agreste* and *caatinga/campo rupestre* ecotone, c. 200–850 m, north-western Bahia (Pilão Arcado & Barra), northern to southern Bahia east of the Chapada Diamantina, and north-eastern Minas Gerais (Rio Jequitinhonha valley). Endemic to the core area of Eastern Brazil. Map 33.

BAHIA: NW Bahia, Mun. Pilão Arcado, 42°54'W, 10°2'S, 21 Nov. 1978, *Miranda* 303 (IPA, HRB, F); Mun. Barra, dunes by the Rio São Francisco, 10°47'16"S, 42°46'22"W, 23 June 1996, *Hind* s.n. (K, photos); N, NE, E & S Bahia, Mun. Jaguarari, vicinity of Pilar, 9°54'S, 39°57'W, 18 Aug. 1983, *G.C.P. Pinto & S.B. Silva* 191/83 (HRB); l.c., 9.5 km N of Jaguarari, 15 July 1989, *Zappi* 124, cult. Taylor &

Zappi, 1995 (K, K, photos); l.c., 20 km N of Senhor do Bomfim towards Juazeiro, 25 Dec. 1984, *Harley in CFCR 7577* (SPF, K); Mun. Monte Santo, c. 5 km N, na estrada para Uauá, 25 Aug. 1996, *L.P. de Queiroz & N.S. Nascimento 4609* (HUEFS, K); Mun. Jacobina, 10 km from Jacobina towards Morro do Chapéu, 14 Mar. 1990, *A.M. de Carvalho & J. Saunders 2787* (CEPEC); ibid., 'margem esquerda do Rio Jacuípe', 21 Jan. 1985, *Passos 26* (MO, BAH 6021); Mun. Santa Luz, 10 km from town on road BA 120 to Queimadas, 16 Nov. 1986, *L.P. de Queiroz 1135* (NY, HUEFS); Mun. Valente, c. 5 km SW of town towards São Domingos, 11°25'S, 39°30'W, 29 Dec. 1992, *L.P. de Queiroz et al. 3041* (MBM, HUEFS); Mun. Morro do Chapéu, Ventura, Sep. 2001, *M. Machado* (obs.); Mun. Riachão do Jacuípe, 9 km NW of town, road BR 324, 25 Apr. 1992, *Taylor & Zappi* (obs.); l.c., 13 km SE of town on road BR 324, 10 July 1985, *L.R. Noblick & Lemos 4106* (CEPEC, HRB, ALCB, HUEFS); l.c., 24.5 km NW of Tanquinho, road BR 324, 25 Apr. 1992, *Taylor & Zappi* (obs.); Mun. Santa Barbara/Feira de Santana, 21 km 'SW' [SE] of Tanquinho along BR 116, 31 Mar. 1976, *Davidse et al. 11672* (MO); Mun. Serra Preta, 42.5 km E of Baixa Grande on road BA 052, 9 Feb. 1991, *Taylor et al.* (obs.); Mun. Feira de Santana, 10 km N of town on road BR 101, 26 Jan. 1980, *Andrade-Lima 80-8812* (IPA); Mun. Iraquara, Faz. Torrinha, entrando 13 km após a entrada para Iraquara, 18 Apr. 1991, *G.L. Esteves & R. de Lyra-Lemos 2534* (SPF, K, MAC); Mun. Rui Barbosa, 1.5 km SE of town near road to Itaberaba, 7 Feb. 1991, *Taylor et al. 1569* (K, HRCB, ZSS, CEPEC); Mun. Seabra, 6 km E of town, then 1 km behind posto Esso, 25 Aug. 1988, *Eggli 1282* (ZSS); Mun. Itaberaba, 12°30'S, 39°59'W, 9 Oct. 1987, *L.P. de Queiroz 1773* (HUEFS); l.c., lowlands along river below town, 22 Aug. 1988, *Eggli 1252* (ZSS); l.c., 63 km from Argoim on road BR 242, 16 Apr. 1991, *G.L. Esteves & R. de Lyra-Lemos 2529* (MAC); Mun. Iaçu, Faz. Lapa, 12°42'S, 39°56'W, 26 Feb. 1983, *G.C.P. Pinto 171/83* (HRB); Mun. Itatim, Morro do Agenor, 12°42'S, 39°46'W, 26 Nov. 1995, 21 Apr. 1996 & 1 Sep. 1996, *F. França 1490, 1617, 1817* (HUEFS; K, photo); l.c., Morro das Tocas, 25 Nov. 1995 & 28 Sep. 1996, *F. França 1453, 1853* (HUEFS); Mun. Milagres, road BA 046 c. 4 km from junction with road BR 116 towards Amargosa, 12°51'S, 39°46'W, 2 June 1993, *L.P. de Queiroz & T.S.N. Sena 3197* (K, HUEFS); Mun. Castro Alves, Oct. 1972, *G.C.P. Pinto s.n.* (ALCB); l.c., 11 km from road BR 116 (Argoim), 12°32'S, 39°55'W, 13 Mar. 1984, *L.C.O. Filho 86* (HRB); Mun. Milagres, Km 5 towards Iaçu from road BR 116, 13 Oct. 1981, *A.M. de Carvalho & G.P. Lewis 962* (CEPEC, K); Mun. Milagres/Amargosa, 19 Oct. 1967, *A.P. Duarte 10584* (HB, RB); Mun. Marcionílio Sousa, Machado Portela, 19 June 1915, *Rose & Russell 19913 p.p.* (US); Mun. Santa Inês, Km 7–9 from town towards road BR 116, 17 Oct. 1972, *R.S. Pinheiro* (CEPEC); Mun. Maracás, Faz. Tanquinho, 2 Jan. 1989, *L. Paganucci 158 & M.L. Guedes 124* (ALCB); l.c., 6–7 km E of Porto Alegre, 4 Feb. 1991, *Taylor et al. 1554* (K, ZSS, CEPEC, HRCB) — showing influence from *A. rhodantha*; Mun. Ituaçu, c. 2 km NE, 18 Aug. 1988, *Eggli 1199* (ZSS); l.c., c. 10 km SE of town towards Tanhaçu, 18 Aug. 1988, *Eggli 1196* (ZSS); Mun. Jequié (?), between the town and Porto Alegre, 1974, *Gottsberger s.n.*, cult. (K, ZSS); between Jequié and Manuel Vitorino, 18 Jan. 1975, *Read et al. 3443* (US); Mun. Jequié, E of the town, Aug. 2001, *Zappi* (obs.); Mun. Aracatu, close to the border with Mun. Brumado, 25 km on road BR 030 towards Sussuarana, 3 Feb. 1991, *Taylor et al. 1542* (K, HRCB, ZSS, CEPEC); l.c., c. 20 km N of Anagê (Anajé) towards Sussuarana, 17 Aug. 1988, *Eggli 1188* (ZSS); Mun. Poções, 5 km towards Boa Nova, 25 Jan. 1973, *I. & G. Gottsberger 16-25173a* (K); cent.-S Bahia (W of Serra do Espinhaço), (?) see hybrid, Nos 2 × 3, below.
MINAS GERAIS: NE Minas Gerais, Mun. Itaobim, 1 km W of town, 0.5 km N of the Rio Jequitinhonha, 16°34'S, 41°31'W, 18 Nov. 1988, *Taylor & Zappi in Harley 25529* (K, SPF); l.c., 8 km W of town on road BR 357 towards Itinga, 14 Dec. 1990, *Taylor & Zappi 761* (K, HRCB, ZSS, BHCB), 9 Apr. 1983, *Martinelli & Leuenberger 9224* (K); Mun. Itinga, 2 km E of town on road BR 357, 14 Dec. 1990, *Taylor & Zappi* (obs.).

CONSERVATION STATUS. Least Concern [LC] (1); PD=3, EI=1, GD=2. Short-list score (1×6) = 6. Least Concern, but its extensive habitat continues to decline.

This species has a more restricted range than the following, but is not nearly as common, nor as variable. The record by Uebelmann (1996) from western Pernambuco seems to be a misidentification of the following species, which is abundant in the region. Their different habitat preferences deserve further analysis — they are rarely sympatric despite the considerable overlap in distribution.

4. Arrojadoa rhodantha *(Gürke) Britton & Rose*, Cact. 2: 171–172 (1920). Type: Brazil, Piauí, Caatinga de São Raimundo [Nonato], 1907, *Ule* 11 (B, holo.; K [photo ex B]).

Cereus rhodanthus Gürke in Monatsschr. Kakt.-Kunde 18: 69 (1908). *Cephalocereus rhodanthus* (Gürke) Werderm., Bras. Säulenkakt.: 116 (1933).

Arrojadoa canudosensis Buining & Brederoo in Cact. Succ. J. (US) 44: 111–113 (1972). *A. rhodantha* subsp. *canudosensis* (Buining & Brederoo) P. J. Braun in Bradleya 6: 90 (1988). Holotype: Brazil, Bahia, Canudos, *Horst & Uebelmann* 251 (U).

A. aureispina Buining & Brederoo in Kakt. and. Sukk. 23: 95–98 (1972). *A. rhodantha* subsp. *aureispina* (Buining & Brederoo) P. J. Braun & E. Esteves Pereira in Succulenta 74: 81 (1995). Holotype: Brazil, Bahia, Mun. Caitité, 1966, *Horst & Uebelmann* 154 (U).

A. aureispina var. *anguinea* P. J. Braun & E. Esteves Pereira in Cact. Succ. J. (US) 60: 174–180 (1988). *A. rhodantha* subsp. *anguinea* (P. J. Braun & E. Esteves Pereira) P. J. Braun & E. Esteves Pereira in Succulenta 74: 81 (1995). Holotype: Brazil, Bahia, *Braun* 80a (ZSS).

A. aureispina var. *guanambensis* P. J. Braun & Heimen in Kakt. and. Sukk. 31: 334–337 (1980). *A. rhodantha* subsp. *guanambensis* (P. J. Braun & Heimen) P. J. Braun & E. Esteves Pereira in Succulenta 74: 81 (1995). Holotype: Brazil, Bahia, Mun. Guanambi, June 1979, *Braun & Heimen* 80 (KOELN, n.v.; ZSS, iso.).

A. theunisseniana Buining & Brederoo in Krainz, Die Kakt., Lfg 52 (1973). *A. rhodantha* var. *theunisseniana* (Buining & Brederoo) P. J. Braun in Bradleya 6: 90 (1988). Holotype: Brazil, Bahia, Botuporã, *Buining* 1002 (U).

A. horstiana P. J. Braun & Heimen in Kakt. and. Sukk. 32: 186–190 (1981). Holotype: Brazil, Minas Gerais, *Braun & Heimen* 83 (KOELN, n.v.; ZSS, iso.).

A. rhodantha subsp. *reflexa* P. J. Braun in Kakt. and. Sukk. 35: 34–38 (1984). Holotype: Brazil, Bahia, 1000 m, *Braun* 68 (KOELN, n.v.).

A. rhodantha var. *occibahiensis* P. J. Braun in Kakt. and. Sukk. 36: 114–115 (1985). Holotype: Brazil, Bahia, W of the Rio São Francisco, *Horst & Uebelmann* HU 616 (ZSS).

VERNACULAR NAMES. Rabo-de-raposa, Rabo-de-onça.

Shrubby, to 2 m, rarely reaching 3.5 m, ± branched. Stems erect or semi-decumbent, 1.7–6.0 cm diam., not or scarcely expanded below the cephalia, vascular cylinder very woody; ribs 8–14, 4–5 × 5 mm, epidermis olive-green. Areoles 2–3 mm, to 10 mm apart, long hairs soon lost from young growth. Spines reddish, golden, greyish or brown, stiff, very sharp; central spines 4–5, 1.5–4.0(–5.0) cm, ascendent to spreading, radials 6–10, to 1 cm. Fertile part of stem apical, later forming up to 4 and sometimes many more rings on the same axis, cephalium areoles with abundant, loose white hairs and long red, golden or brownish bristles, these 2–3 cm. Flowers 20–35 × 8–10 mm (at apex), opening at morning only, pinkish magenta or reddish pink; pericarpel 6–7 × 4–5 mm; tube to 20 × 10–14 mm, cylindric or constricted above the nectar-chamber, with fleshy bract-scales at apex; perianth-segments 5–6 × 5 mm, obovate, outer segments erect, inner segments expanding partially to give an aperture of only a few mm; stamens included in relation to the perianth-segments; style 20 mm, stigma-lobes 6–8, included; ovary locule 2–3 × 2 mm, obtriangular in longitudinal section. Fruit globose, 20–25 × 10–18 mm, often somewhat laterally compressed; pericarp smooth, dark red. Seeds c. 1.2 mm, cochleariform, black, dull; testa-cells convex, with or without cuticular folds.

Central-southern *caatinga* element: found on various substrates (including in the shade of dense *caatinga* forest and in the open on inselbergs) and entering the *caatinga/campo rupestre* ecotone, 220–1200 m, southwestern Piauí and western Pernambuco to central-northern Minas Gerais. Endemic to the core area of Eastern Brazil. Map 33.

PIAUÍ: Mun. Picos, road to Jaicós, 7°13'S, 41°17'W, 6 Apr. 2000, *E.A. Rocha et al.* (K, photos); Mun. Padre Marcos, Serra Velha, 14 May 1993, *M.E. Alencar* 33 (PEUFR, IPA), s.d., *M.E. Alencar* 138 (PEUFR, IPA); Mun. Simões, 7°36'S, 40°45'W, 5 Apr. 2000, *E.A. Rocha et al.* (K, photos); Mun. Paulistana, 11 km N of town, 6 Apr. 2000, *E.A. Rocha et al.* (K, photos), l.c., 7 km N of border with Pernambuco, 6 Apr. 2000, *E.A. Rocha et al.* (obs.); Mun. São Raimundo Nonato, 1907, *Ule* 11 (B, K, photo), *Horst & Uebelmann* 422 (ZSS); Parque Nac. Serra das Confusões, cult. Vitória da Conquista,

Bahia, April 2003, *Avaldo de Oliveira Soares Filho* s.n., Taylor (obs.) — form with very long central spines.
PERNAMBUCO: Mun. Araripina, Sítio Jatobá, 20 Nov. 1992, *F.A.R. Santos* 54 (HUEFS, PEUFR); l.c., behind Posada do Araripe, 15 Feb. 1990, *Zappi* 226 (SPF, K, photos); Mun. Serrita, 7°48'S, 39°14'W, 4 Apr. 2000, *E.A. Rocha et al.* (K, photos); Mun. Ouricuri, 28 Feb. 1984, *G. Costa-Lima* 17 (IPA, HRB); Mun. São José do Belmonte, 2 Mar. 1981, *A. Peruci* 1 (IPA); Mun. Parnamirim, 23 km along road to Petrolina, 25 May 1984, *F. Araújo* 100 (IPA, PEUFR); l.c., 18 km from the town, Sítio Favela, 11 Feb. 1993, *F.A.R. Santos* 72 (ALCB, HUEFS); Mun. (?) Ibimirim, [Açude] Poço da Cruz [Açude Engenheiro Francisco Saboia], 12 Apr. 1968, *Andrade-Lima* 68–5353 (IPA); Mun. Dormentes, road to Lagoa, 33 km from Dormentes, 19 Nov. 1992, *F.A.R. Santos* 52 (PEUFR); Mun. Afrânio, 5 km N of Rajada, 8°47'S, 40°52'W, 6 Apr. 2000, *E.A. Rocha et al.* (K, photos); Mun. Santa Maria da Boa Vista, 2 May 1971, *E.P. Heringer* 451 (IPA); Mun. Petrolina, 17 km S of Rajada, W of road BR 407, 8°55'S, 40°44'W, 6 Apr. 2000, *E.A. Rocha et al.* (K, photos); l.c., Distrito de Nilo Coelho, 9°15–30'S, 40°30–45'W, 4 Apr. 1991, *P.E. Nogueira et al.* 285 (SPF).
BAHIA: W Bahia, Mun. Vanderlei, 12 km N of road BR 242 towards the town, 19 July 1989, *Taylor & Zappi* (obs.); Mun. Santana, 30 km from the town towards Porto Novo, 16 Jan. 1991, *Taylor et al.* 1435 (K, HRCB, ZSS, CEPEC); Mun. Sítio do Mato, *Horst & Uebelmann* 208, cult. Braun (ZSS); Mun. Serra do Ramalho, Serra do Ramalho, 13°31'S, 43°45'W, 2001, *J. Jardim* (CEPEC); Mun. Carinhanha, 14°1'S, 43°43'W, 2001, *J. Jardim* 3500 (CEPEC); N & NE Bahia, Mun. Casa Nova, c. 50 km from Remanso towards Petrolina, 10 Feb. 1990, *Zappi* 221 (SPF); Mun. Curaçá, 43 km N of Barro Vermelho towards Curaçá, 7 Jan. 1991, *Taylor et al.* 1370 (K, HRCB, ZSS, CEPEC) — giant form to 3.5 m high (!); Mun. Juazeiro, Km 12 from town towards Sento Sé, 8 Dec. 1971, *Andrade-Lima et al.* 1209 (IPA, MAC); Mun. Sento Sé, 158 km W of Juazeiro, then 15.5 km S on road to Cabeluda, 11 Jan. 1991, *Taylor et al.* 1393 (K, HRCB, ZSS, CEPEC); Mun. Uauá, 12.5 km N of town towards Patamuté, 6 Jan. 1991, *Taylor et al.* 1366 (K, HRCB, ZSS, CEPEC); Mun. Canudos, 1983, *Braun* 464 (ZSS); Mun. Jaguarari, S of the border of Mun. Juazeiro, 64 km S of Juazeiro, 8 Jan. 1991, *Taylor et al.* 1374 (K, HRCB, ZSS, CEPEC); l.c., between Barrinha and Flamengo, Serra dos Icós, 27 Nov. 1968, *I. Pontual* 68–781 (IPA); Mun. Monte Santo, c. 5 km N of town towards Uauá, 25 Aug. 1996, *L.P. de Queiroz* 4609 (HUEFS); Mun. (?), 10°53'S, 41°35'W, 13 July 2000, *G. Charles* (photos); Mun. Queimadas, 10°55'S, 39°36'W, 17 Nov. 1986, *L.P. de Queiroz* 1159 (HUEFS); Mun. Araci, 18 Jan. 1994, *M.A.S. das Neves* 87 (HUEFS); l.c., 14.5 km N of town towards Tucano, 5 Jan. 1991, *Taylor et al.* 1355 (K, HRCB, ZSS, CEPEC); Mun. Morro do Chapéu, near Buracão, 19 km from the town, 11°38'57"S, 41°16'16"W, 16 Mar. 1996, *Giulietti et al.* 2440 (ALCB, UEC); l.c., 15 km W of town beside road BA 052, 5 Aug. 2002, *Taylor* (obs.); Mun. Cafarnaum/Mun. Morro do Chapéu, 26 Sep. 1965, *A.P. Duarte* 9551 (HB, RB); Mun. Canarana, 6 km E of town towards Cafarnaum, 24 Aug. 1988, *Eggli* 1277 (ZSS); cent. & S Bahia, Mun. Ibitiara, Serra Malhada, 38 km W of Seabra towards Ibotirama, 1050 m, 25 Aug. 1988, *Eggli* 1292 (ZSS); l.c., 52 km W, 1020 m, 14 Jan. 1991, *Taylor et al.* 1419 (K, HRCB, ZSS, CEPEC); Mun. Botuporã, c. 50 km S of Macaúbas on road to Paramirim, 26 Aug. 1988, *Eggli* 1296 (ZSS); Mun. Bom Jesus da Lapa, Km 8 on road to Ibotirama, then 3 km E, 17 Apr. 1983, *Leuenberger et al.* 3080 (CEPEC, B); 28 km SE of Bom Jesus da Lapa on road to Caitité, 43°13'W, 13°23'S, 16 Apr. 1980, *Harley et al.* 21424 (SPF, K); Mun. Abaíra, Morro do Zabumba, Engenho de Baixo, 13°18'S, 41°48'W, 13 Mar. 1992, *Stannard et al. in Harley* 51947 (SPF, CEPEC, HUEFS, K); Mun. Marcionílio Sousa, Machado Portela, see collection cited for hybrid with *A. penicillata*, below; Mun. Riacho de Santana, 22 km from town towards Bom Jesus da Lapa, 16 Apr. 1983, *Leuenberger et al.* 3073 (CEPEC, B); Mun. Livramento do Brumado, 10 km S, 14 Mar. 1991, *G.P. Lewis* s.n. (K, photos); Mun. Caitité, estrada para Igaporã, 14°1'S, 42°33'W, 31 Mar. 1984, *G.B.A. Bohrer* 29 (HRB); l.c., hill above town with Telebahia tower, 27 Aug. 1988, *Eggli* 1303 (ZSS); Mun. Dom Basílio, 13°59'S, 41°42'W, 1999, *M. Machado et al.* (obs.); Mun. Tanhaçu, 44 km N of Anagé (Anajé) towards Sussuarana, 27 Jan. 1965, *G. Pabst* 8658 & *E. Pereira* 9769 (HB, RB); Mun. Guanambi, 5–10 km E on road BR 030, 15 Jan. 1995, *Hatschbach et al.* 61900 (MBM, K); l.c., 19 km NE on road to Caitité, 25 July 1989, *Zappi* 169A (SPF); Mun. Palmas de Monte Alto, hill with Telebahia tower, 28 Aug. 1988, *Eggli* 1313 (ZSS); Mun. Aracatu, 33 km along road BR 030 from Brumado towards Sussuarana, 3 Feb. 1991, *Taylor et al.* 1544 (CEPEC, K, HRCB, ZSS); Mun. Anagé, 10 km N towards Sussuarana, 17 Aug. 1988, *Eggli* 1181 (ZSS); Mun. Malhada, Iuiú, c. 2 km S of town, 28 Aug. 1988, *Eggli* 1317 (ZSS); Mun. Urandi, c. 3 km S of town centre, near railway line, 22 July 1989, *Taylor & Zappi* (obs.).
MINAS GERAIS: Cent.-N Minas Gerais, Mun. Januária/Mun. Manga, reported by Braun & Esteves Pereira (1995a); Mun. Monte Azul/Espinosa, 5.5 km E of Monte Azul, beyond Vila Angical, 28 Jan.

1991, *Taylor et al.* 1459 (K, HRCB, ZSS, BHCB); l.c., 'Serra do Espinhaço', 14 Mar. 1995, *Hatschbach et al.* 61889 (MBM, K); Mun. Rio Pardo de Minas (?), 'W of Rio Pardo', 1979, *Braun & Heimen* 83 (ZSS); Mun. Varzelândia, c. 42 km E of town towards the Rio Verde Grande, 23 July 1989, *Taylor & Zappi* (obs.); Mun. Porteirinha, 8 km S of town on road to Riacho dos Machados, 15°49'S, 43°4'W, 7 Nov. 1988, *Taylor & Zappi in Harley* 25516 (K, SPF).

CONSERVATION STATUS. Least Concern [LC] (1); PD=3, EI=1, GD=2. Short-list score (1×6) = 6. Least Concern, but its extensive habitat continues to decline.

This extremely variable species comprises many locally distinct forms but cannot be conveniently divided into a manageable and meaningful number of infraspecific taxa. A part of this variation can be attributed to gene exchange with species Nos 2 and 3 (see below), but much of it is probably inherent and independent of such influence, occurring in areas remote from the known ranges of its congeners. *A. rhodantha* is one of the most characteristic cactus species of the central and southern *caatingas*, but is absent from north-eastern Minas Gerais (Rio Jequitinhonha drainage), where *A. penicillata* occurs.

4b. Arrojadoa (?) sp. nov. ('A. marylanae', nom. prov.)

Possibly keying out here is a remarkable plant recently discovered by Marylan Coelho from the Univ. Estadual do Sudoeste da Bahia and communicated by Marlon Machado. It has a massive erect, 27–28-ribbed stem (cf. *Coleocephalocereus goebelianus*), that rarely branches, and dense, porrect, needle-like spination. The ring cephalia are numerous and closely spaced, but with wool and bristles similar to *C. goebelianus* and the flowers are quite different to those of other *Arrojadoa* species, the perianth strongly resembling that of *Coleocephalocereus* or *Melocactus*. At present it is known only from southern Bahia, where it grows in large numbers on quartzitic ridges together with *Espostoopsis dybowskii*, as follows: Mun. Tanhaçu, nr Sussuarana, Serra Escura, Sep. 2001, *Avaldo de Oliveira Soares Filho* s.n. (photos), 19 April 2003, *Taylor et al.* (K, photos). Its generic disposition is at present unclear, but there can be little doubt that it is a new taxon, which could be some kind of stabilised intergeneric hybrid (between *Arrojadoa* and *Coleocephalocereus* ?). Plate 37.3–4.

2 × 3: *Arrojadoa dinae* subsp. *dinae* × (?) *Arrojadoa penicillata*

Vegetatively intermediate between the presumed parents, the ends of some stems strongly expanded beneath the cephalium. Flowers described as pink, green within.

BAHIA: Mun. Palmas de Monte Alto / Mun. Guanambi, road BR 030, 31 Mar. 1984, 14°14'S, 43°1'W, *J.E.M. Brazão* 303 (HRB, K photo).

Although *A. penicillata* is not otherwise recorded west of the Serra do Espinhaço in southern Bahia, the above-cited collection may confirm its presence. The characters displayed by this presumed hybrid seem best explained as having come from *A. penicillata* and *A. dinae*.

2 × 4: *Arrojadoa dinae* subsp. *dinae* × *Arrojadoa rhodantha*

A. albiflora Buining & Brederoo in Succulenta 54: 21–27 (1975). Holotype: Brazil, Bahia, Mun. Urandi, 1070 m, *Horst* 401 (U).

Intermediate between the parents, but variable in flower colour. See Ritter (1979: Abb. 61) for a good illustration.

BAHIA: (?) Mun. Caetité, São Francisco, caminho para Lagoa Real, 14°S, 42°12'W, 8 Feb. 1997, *B. Stannard et al.* in PCD 5224 (ALCB, K, CEPEC, UEFS, SPF); Mun. Urandi, mountains above town, between sandstone rocks, 1964, *Ritter* 1244 (U); l.c., 1070 m, *Horst* HU 401 (U), 6 Oct. 1985, cult. ZSS, 10 May 1989, *Horst & Uebelmann* 401 (ZSS).

Probably to be found at other sites where the species are almost or quite sympatric, eg. east of Monte Azul (MG).

3 × 4: *Arrojadoa penicillata* × *Arrojadoa rhodantha*

Sometimes forming a complete range of intermediates between the parents in places within the areas of sympatry cited below.

Caatinga in the drainage of the Rio Paraguaçu and Rio Brumado/Rio de Contas, 300–450 m, southern Bahia.
BAHIA: Mun. Marcionílio Sousa, Machado Portela, 19 June 1915, *Rose & Russell* 19913 p.p. (NY); Mun. Maracás, c. 18 km S of the Contendas–Maracás road towards Porto Alegre, 4 Feb. 1991, *Taylor et al.* (obs.); Mun. Aracatu, 33 km on road BR 030 from Sussuarana, 3 Feb. 1991, *Taylor et al.* 1544 (K, HRCB, ZSS, CEPEC); (?) Mun. Brumado, 7 km N of town on road to Livramento, 14°8'S, 41°40'W, 22 Nov. 1988, *Taylor & Zappi* in *Harley* 25538 (K, SPF, CEPEC) — could also be merely a very slender form of *A. rhodantha*.

20. PILOSOCEREUS Byles & G. D. Rowley

in Cact. Succ. J. Gr. Brit. 19: 66–69 (1957). Type: *Pilosocereus leucocephalus* (Poselg.) Byles & G. D. Rowley [NE Mexico].

Including *Pseudopilocereus* Buxb. (1968).

Literature: Zappi (1994).

Shrubby or tree-like, rarely epiphytic (on carnaúba palm trunks), 0.3–10.0 m; branched or not above ground; vascular cylinder strongly to weakly woody; tissues mostly extremely mucilaginous; branches showing more or less constant growth, rarely constricted; ribs triangular, (3–)4–28. Areoles with felt and generally with long hairs, spination diverse; spines straight, rarely curved at base, opaque to translucent. Fertile part of stem not or various differentiated, when strongly so flower-bearing areoles with long hairs and/or bristly spines, lateral cephalium superficial, rarely sunken into the branch. Flowers 2.5–9.0 × 2–7 cm, anthesis starting at evening; flowers dull or sometimes bright pink-red without, smooth to slightly striate, glabrous; pericarpel with few and minute bract-scales, symmetry of flowers dependent on shape of flower-tube, which can be straight or curved, terete or infundibuliform, distal half or third with broad bract-scales; outer perianth-segments thick, inner segments thin, white or rarely pinkish; nectar-chamber broad, ± protected by the innermost filaments; stamens many, anthers 1.2–2.5 mm, forming a compact mass; style 25–65 mm, stigma-lobes 8–12, exserted or included in relation to the anthers. Fruit depressed-globose, very rarely globose, always dehiscent by lateral, abaxial, adaxial or apical slit, floral remnants persistent, blackening, pendent or rarely erect; pericarp green, purplish, brownish, magenta, deep red or wine-coloured, smooth, striate or rugose; funicular pulp solid, white, red or magenta. Seeds 1.2–2.5 mm, dark-brown or black, cochleariform, rarely hat-shaped, hilum–micropylar region differentiated, intercellular depressions present to absent, testa-cells flat, convex or conic, cuticular folds fine, coarse or absent (for seed SEM images of all species treated below, except *P. azulensis*, see Zappi 1994).

A genus of some 37 species in Mexico, the Caribbean (incl. Florida Keys), Venezuela, Suriname, Guyana, Peru, Ecuador, Brazil and Paraguay. Its range in Brazil includes Roraima, Pará, Maranhão, Piauí, Ceará, Rio Grande do Norte, Paraíba, Pernambuco, Alagoas, Sergipe, Bahia, Minas Gerais, Espírito Santo, coastal Rio de Janeiro, northern São Paulo, Tocantins, Goiás, Mato Grosso and Mato Grosso do Sul. This is the largest and most important genus of Cactaceae in Eastern Brazil, represented by at least 20 species, plus various heterotypic subspecies, and occurring in a wide variety of vegetation types, though sometimes restricted to rock outcrops within habitats such as *cerrado* and *Mata atlântica*. On occasions certain species may even dominate the vegetation in which they occur, e.g. *P. gounellei, P. catingicola* and *P. pachycladus*. The Brazilian species are classified into two subgenera and various species-groups. Plates 38–49.3.

1. Branching candelabriform, the new branches arising near the apex of the subtending stems; floral remnants erect to pendent, not deeply immersed in apex of fruit, forming a circular insertion point (Subg. *Gounellea*) 2
1. Branching erect and/or plants branched only at base, the new branches first developing well below the apex of the stems subtending them; floral remnants pendent, deeply immersed in apex of fruit, forming a linear insertion point (Subg. *Pilosocereus*) 3

2. Treelike, with a well-defined trunk, lacking branches near base; ribs 4–7; flower-bearing areoles without long hairs . . . . **1. tuberculatus**
2. Shrublike, usually without a well-defined trunk, branching near base; ribs (8–)10–15; flower-bearing areoles with abundant, silky, long hairs enveloping the flowers . . . . **2. gounellei**

3. Treelike or shrubby, branching above the base; vascular cylinder strongly to moderately woody . . . 4
3. Shrubby, branching only at base when undamaged; vascular cylinder weakly woody 18

4. Apical, subapical and flower-bearing areoles without conspicuous long hairs (except when young) 5
4. Apical, subapical and flower-bearing areoles with abundant long hairs 7

5. Epidermis covered in blue wax . . . . **8. pentaedrophorus**
5. Epidermis greyish, pale or olive-green, not strongly waxy 6

6. Spines brown to greyish, opaque; flower-buds acute; flower opening wide, 6–7 × 4–5 cm, tube straight (*restinga*, S Bahia & SE Brazil) . . . . **5. arrabidae**
6. Spines golden to translucent yellow; flower-buds obtuse; flower narrow, 5.8–6.0 × 3.5–3.7 cm, tube curved (*caatinga* and littoral zone, Ceará, Piauí and S Maranhão) . . . . . . **7. flavipulvinatus**

7. Epidermis rough, green to grey-green, not covered in wax . . . . **10. floccosus**
7. Epidermis smooth, frequently bluish or obviously covered in wax 8

8. Areoles contiguous, difficult to isolate from one another 9
8. Areoles > 2 mm apart, distinct 10

9. Flower-bearing areoles with scarce, white long hairs; young spines golden, flexible . . . . **13. magnificus**
9. Flower-bearing areoles with abundant dark yellow to reddish brown long hairs; spines dark brown to blackish . . . . **11. fulvilanatus**

10. Spines golden, translucent 11
10. Spines brownish, reddish or blackish, opaque 15

11. Long hairs equally abundant on both flower-bearing areoles and vegetative areoles; flower-tube narrow, curved, pinkish or reddish without . . . . **9. glaucochrous**
11. Long hairs more abundant on the flower-bearing areoles; tube ± straight, broad at apex, greenish, brownish or dark purplish brown 12

12. Ribs 5–12; central spines thicker and longer than the radials; stem blue
. . . . **12a. pachycladus** subsp. **pachycladus**
12. Ribs > 12; central spines delicate, nearly indistinguishable from the radials, stem green
or bluish 13

13. Fertile part of stem with golden bristles to 3–4 cm and few long hairs; ribs (15–)18–26
. . . . **17. multicostatus**
13. Fertile part of stem without or with few bristles, long hairs white, silky; ribs 13–19 14

14. Spines golden; ribs wider than high; fertile part of the stem subapical, not strongly modified
(*caatinga*, N Bahia northwards) . . . . **12b. pachycladus** subsp. **pernambucoensis**
14. Spines pinkish yellow to brownish; ribs higher than wide; fertile part of stem lateral,
a true, sunken cephalium sometimes present (*Bambuí* limestone outcrops, cent.-N Minas
Gerais and SW Bahia) . . . . **20. densiareolatus**

15. All stems with 7 or more ribs (seedlings excepted), and/or flowers > 4 cm in diameter at
full anthesis 16
15. Ribs 4–6, or some stems with up to 7 ribs 17

16. Areoles 8–16 mm apart on the ribs; seeds 2–2.3 mm (NE Brazil) . . . . **3. catingicola**
16. Areoles 5–7 mm apart on the ribs, seeds 1.5–1.6 mm (NE Minas Gerais) . . . . **4. azulensis**

17. Treelike or shrubby, primary branches 4–6-verticillate; flower-buds acute, with triangular
bract-scales, flowers ± solitary, 4.7–7.0 cm in diameter, tube straight, wide (NE Brazil)
. . . . **3. catingicola**
17. Shrubby, sparsely branched, or if treelike branching not as above; flower-buds obtuse, with
obovate to truncate bract-scales, flowers aggregated, 2.5–3.0 cm in diameter, tube curved,
narrow (SE Bahia & SE Brazil) . . . . **6. brasiliensis**
18. Ribs (5–)6–8, transverse folds visible; flower-bearing areoles not differentiated; spines
opaque, brown to grey, central spines well differentiated from the radials (*restinga*) . . . **5. arrabidae**
18. Ribs 8–23, transverse folds not visible; flower-bearing areoles ± differentiated; spines
red, dark brown or golden, translucent, central spines not very different from the radials
(*cerrado*, *caatinga* and *campo rupestre*) 19

19. Stem blue-green or strongly glaucous; flower-tube infundibuliform; ribs 8–17 20
19. Stem pale to dark green; flower-tube cylindric, straight to curved; ribs 12–28 21

20. Fruit dehiscent by an apical transverse slit; pericarp rugose, red to wine-coloured when ripe;
seeds with flat testa-cells (Serra do Espinhaço, Minas Gerais) . . . . **15. aurisetus**
20. Fruit dehiscent by lateral slit, pericarp smooth, green or bluish; seeds mostly with domed
to highly conical testa-cells (Bahia, S Piauí) . . . . **14. machrisii**

21. Flower-bearing areoles strongly differentiated, forming a lateral cephalium sunken into the
branch, with abundant wool and bristles to 3–6 cm; stem epidermis pale green . . . **19. chrysostele**
21. Flower-bearing areoles not strongly differentiated, appearing randomly or at apex of branches,
usually with bristles and some hairs 22

22. Vegetative areoles with long hairs; fruit pulp white; seeds dull, testa-cells domed to conic
(SEM: cuticular folds coarse, dense); flowers c. 2.5 cm wide, pinkish without (Bahia, slopes
E of the Rio São Francisco) . . . . **16. aureispinus**
22. Vegetative areoles without long hairs; fruit pulp magenta; seeds shiny, testa-cells slightly
domed to flat (SEM: cuticular folds scarce to absent); flowers c. 3 cm wide or more,
greenish-white without 23

23. Flowers curved, > 5 cm, flower-buds obtuse before anthesis (SE Piauí northwards)
. . . . **18. piauhyensis**
23. Flowers straight, < 5 cm, flower-buds acute before anthesis (NE Minas Gerais)
. . . . **17. multicostatus**

Subg. *Gounellea* Zappi (Nos 1 & 2). Endemic to the *caatinga* and *mata calcária* of NE Brazil and northern Minas Gerais.

1. Pilosocereus tuberculatus *(Werderm.) Byles & G. D. Rowley* in Cact. Succ. J. Gr. Brit. 19(3): 69 (1957). Type: Brazil, Pernambuco, 1932, *Werdermann* (B†). Lectotype (Zappi 1994): l.c., Serra Negra, '900 m', Mar. 1932, Werdermann, l.c. infra, photograph, p. 21 (iconotype).

Pilocereus tuberculatus Werderm., Bras. Säulenkakt.: 101 (1933). *Pseudopilocereus tuberculatus* (Werderm.) Buxb. in Beitr. Biol. Pflanzen 44: 253 (1968).

VERNACULAR NAMES. Caxacubri, Mandacaru-de-laço.

Tree, 2–5(–6) m, with a spread of up to 6 m diam., trunk to c. 20 cm diam.; vascular cylinder strongly woody up to the apex of mature axes; branches 3–6 cm diam., apex erect to inclined, older parts ± horizontal, new branches arising subapically, epidermis olive-green, with pruinose wax on the tips when young; ribs 4–7(–8–12, in seedlings), 1–2 × 0.5–2.0 cm, with sinuate sinuses and pronounced, oblique, transverse folds above the areoles. Areoles 8–10 × 6–7 mm, 2–3.3 cm apart, situated upon prismatic stem-projections (podaria), felt at first light brown, becoming black. Spines opaque, at first light brown, sharp, later greyish, brittle, porrect, centrals 3–5, 30–65 mm, lowermost generally longest, often decurved near base, radials 10–12, 4–7(–10) mm, adpressed, decurved. Fertile part of stem not differentiated, subapical. Flower-buds acute 1–2 days before anthesis, with active extra-floral nectaries; flowers 6–6.7 × c. 3 cm; pericarpel obconic, tube 45 × 20–23 mm at its widest point, ± parallel-sided, straight, nearly terete, olive-green, striate, distal third with a few acute bract-scales; style 40–42 × 2–3 mm, tapering; stigma-lobes c. 8, 6 × 0.7 mm; ovary locule obtriangular in longitudinal section. Fruit to 4 cm diam., globose to depressed, dehiscent by a lateral slit, floral remnant erect, with rounded, shallow insertion; pericarp olive-green to purplish, rugose, with active extra-floral nectaries; funicular pulp magenta. Seeds (1.2–)1.3–1.4(–1.5) × 0.9–1.0 mm, testa-cells domed, with intercellular depressions and fine cuticular folds.

Central-southern *caatinga* element: in dense or sparse *caatinga* vegetation, on fine, white or reddish, sandy soil (especially of the Cipó series), c. 200–790 m, northern Bahia, western Sergipe and Pernambuco. Endemic to the core area within North-eastern Brazil. Map 34.
PERNAMBUCO: Mun. Sertânia, just N of border with Mun. Ibimirim, road PE 360, 8°27'S, 37°28'W, 8 Apr. 2000, *E.A. Rocha et al.* (obs.); Mun. Floresta, 20 km E of Airi on road PE 360 towards Ibimirim, N side of Serra Negra, 12 Feb. 1991, *Taylor & Zappi* 1623 (HRCB, K, ZSS, UFP); l.c., 8°33'S, 38°1'W, 8 Apr. 2000, *E.A. Rocha et al.* (K, photos); l.c., fazenda ± 35 km após Ibimirim, 4 Apr. 1989, *Rodal 74* (IPA); Mun. Ibimirim, triângulo irrigado de Moxotó, Nov. 1992, *F.A.R. Santos 39, 44* (PEUFR, K, photo); Mun. Buíque, Brejo de São José, 8°37'S, 37°10'W, 15 Nov. 1994, *Rodal 365* (PEUFR, K); Mun. Nova Petrolândia, 12 km NE. of turnoff to the old town, 12 Feb. 1991, *Taylor & Zappi* 1621 (HRCB, K, ZSS, UFP); Mun. Petrolina, 5 km from the city towards Recife, 1966, *Hunt* (obs.).
SERGIPE: Mun. Canindé de São Francisco, povoado Curituba, 23 Apr. 1981 & 1992, *M. Fonseca et al.* 480 (ASE 895, K, photo).
BAHIA: Mun. Remanso, 20 km E of town towards Casa Nova, 10 Feb. 1990, *Zappi* 219 (HRCB); l.c., 47 km E, 9°22'S, 41°48'W, 7 Apr. 2000, *E.A. Rocha et al.* (obs.); Mun. Glória, povoado de Brejo do Burgo, 28 Nov. 1992, & 26 Aug. 1995, *Bandeira 74* (ALCB), *Bandeira 263* (ALCB, K, HUEFS); Mun. Santa Brígida, c. 10 km W of junction for the town with road BR 110, Raso da Catarina, c. 9°29'S, 38°12'W, 19 Dec. 1993, *L.P. de Queiroz & Nascimento 3736* (K, HUEFS); Mun. Canudos, Reserva Biológica de Canudos, 10 Feb. 2001, *T. Andrade* (K, photos); Mun. Barra, dunes by the Rio São Francisco, 10°47'16"S, 42°46'22"W, 23 June 1996, *Giulietti et al.* in PCD 2965 (K, ALCB, UEC); l.c., Ibiraba, road to Brejos, 10°48'S, 42°50'W, 23 Feb. 1997, *L.P. de Queiroz 4813* (K, HUEFS); l.c., dunes by the Rio São Francisco, 5 Oct. 1987, *M.T. Rodrigues s.n.* (SPF 48476), 1988, *P.L. Bernardes 189* (SPF); Mun. Morro do Chapéu, between Gruta dos Brejões and Olho D'Água do Facundo, 4 Aug. 2002, *M. Machado* (photos); Mun. Ourolândia, NE of Olho D'Água do Facundo, 4 Aug. 2002, *Taylor*

(obs.); Mun. Araci, 14.5 km N of town towards Tucano, 5 Jan. 1991, *Taylor et al.* 1353 (K, HRCB, ZSS, CEPEC).

CONSERVATION STATUS. Least Concern [LC] (1); PD=2, EI=1, GD=1. Short-list score (1×4) = 4. Least Concern, but some habitat is declining due to agricultural development, eg. for plantations of *Opuntia ficus-indica*.

This species shares unique apomorphic characters with *P. gounellei*, such as the sub-apical branching-pattern and the morphology of the fruits, lacking a deeply sunken floral remnant. This pair of species are the only representatives of Subg. *Gounellea* and are endemic to Eastern Brazil.

P. tuberculatus is characteristic of the sand-dunes and sandy soils of the Rio São Francisco, but is also known from NE Bahia, between Araci and Glória, including the Raso da Catarina. A collection by Luetzelburg (No. 19, M!), bearing a label indicating southern Paraíba (Monteiro), is clearly a labelling error, since the vernacular name given on the label, 'Rabo de Raposa', was applied to *Harrisia adscendens* by Luetzelburg, which is common about Monteiro, where *P. tuberculatus* is absent. Andrade-Lima (1981: 159) lists this species as one of those characteristic of his *caatinga* type No. 5, which is found on the 'Cipó' soil series.

The biology of *P. tuberculatus* is interesting in that it is able to secrete nectar from the outer bract-scales of the flower-buds, tube and fruit, which attracts ants. In large specimens, these insects inhabit the hollow pith of old, dead branches, suggesting a symbiotic relationship of attraction/defence. The slight damage or sudden movement of a branch of this plant is immediately followed by a quick defence reaction by the ants, which run out from inside the dead branches to attack the supposed agressor.

2. Pilosocereus gounellei *(F. A. C. Weber) Byles & G. D. Rowley* in Cact. Succ. J. Gr. Brit. 19: 67 (1957). Type: Brazil, Pernambuco, *Gounelle* s.n. (P†). Neotype (Zappi 1994): Brazil, Paraíba, São Gonçalo [Várzea de Souza], Jan. 1936, *P. Luetzelburg 26921* (M; K, photo.; IPA).

> *Pilocereus gounellei* F.A.C. Weber in K. Schum., Gesamtbeschr. Kakt.: 188 (1897). *Cephalocereus gounellei* (F.A.C. Weber) Britton & Rose, Cact. 2: 34 (1920). *Pseudopilocereus gounellei* (F.A.C. Weber) Buxb. in Beitr. Biol. Pflanzen 44: 252 (1968).
> *Cereus setosus* Gürke in Monatsschr. Kakt.-Kunde 18: 19 (1908), nom. illeg. (Art. 53), non Loddiges (1832). *Pilocereus setosus* Gürke in ibid. 18: 52–53 (1908). Type: Brazil, Bahia, Mun. Maracás, 'Calderão', [1906], *Ule 1* (B [photo K, SPF ex B]).

Shrub, more rarely treelike, 0.5–4.0 m, with many branches arising from the main stem; vascular cylinder moderately woody; branches with apex erect to inclined, older parts horizontal, new branches arising sub-apically, epidermis olive-green, or strongly glaucous, especially when young; ribs 8–15, with sinuate sinuses and transverse folds above the areoles. Areoles ovate, felt cream, grey or brownish, with white long hairs to 35 mm, growth indeterminate near base of plant. [Spination: see subspecies.] Flower-buds completely encircled by silky long hairs, obtuse 2 days before anthesis; flower 4–9 × 2.5–6.0 cm; pericarpel obconic; tube 32–60 × 16.5–17.0 mm at base, widening towards apex to 20–25 mm, curved, greenish brown to pinkish, slightly striated, with decurrent triangular bract-scales, these more frequent and larger towards tube apex; style 2.6–3.5 × 2.5–1.2 mm, tapering, stigma-lobes c. 10, 4 mm long; ovary locule obtriangular in longitudinal section. Fruit 3.5–4.5 × 4.5–6.0 cm, globose to depressed, dehiscent by adaxial or abaxial slit, floral remnant erect or pendent, with rounded, shallow insertion; pericarp pinkish red, smooth or striate. Seeds (1.6–)1.7–1.8 × 1.4–1.6 mm, hat-shaped, symmetric, hilum-micropylar region expanded, forming an angle of 70–80° with long axis of seed, testa-cells flat, without intercellular depressions or cuticular folds.

CONSERVATION STATUS. Least Concern [LC]; see subspecies below.

The neotype chosen (Zappi 1994) was collected by Luetzelburg in the state of Paraíba (*Luetzelburg* 26921), and is a form which agrees with Weber's description, where the spines are given as relatively short (1 cm long), this being somewhat atypical for the species as a whole.

P. *gounellei* is the type of Subg. *Gounellea,* including its sister species, *P. tuberculatus,* from which it differs in its mostly shrubby, not treelike habit, mature branches with a higher number of ribs (8–)9–15, only moderately woody vascular cylinder, flowering areoles with white silky hairs and very distinct seeds, which may be adapted for dispersal by water (Zappi 1994). In the Brazilian Nordeste it is commonly known by the vernacular name of xique-xique, and represents one of the most characteristic plants of the *caatinga*, where, together with *Cereus jamacaru* DC. (vernacular: mandacaru) and *Tacinga inamoena* (K. Schum.) N. P. Taylor & Stuppy (quipá), it is one of the most common and widespread cacti of Eastern Brazil.

This species is divided in two subspecies:

1. Spines stout, strong, brownish to greyish, opaque, (0.9–)1.0–1.9 mm diam. near base, centrals distinctly longer than the radials (NE Brazil) . . . . **2a.** subsp. **gounellei**
1. Spines slender, fragile, golden to reddish, translucent, 0.25–0.6(–0.8) mm diam. near base, centrals and radials ± equal (N Minas Gerais to W & N Bahia) . . . . **2b.** subsp. **zehntneri**

2a. subsp. **gounellei**

VERNACULAR NAMES. Xique-xique, Alastrado.

Shrub; branches 5–9 cm diam.; ribs (8–)9–13, 7–12 × 12–20 mm. Spines (0.9–)1.0–1.9 mm diam. at base, brown to greyish, opaque; centrals 1–5(–6), 1–13(–15) cm, ascending to porrect, the lowermost porrect to reflexed, radials 12–15, 6–30 mm, adpressed. Flower-bearing areoles occupying the subapical region of 1–2 ribs, with silky long hairs covering the flower-buds only. Fruit with magenta (rarely white) funicular pulp.

Caatinga element: widely distributed, common and locally dominant in low, sparse *caatinga* and along road sides, on shallow, rocky or sandy soils and gneissic/granitic outcrops/inselbergs, including those surrounded by more humid vegetation, rarely as epiphyte in seasonally flooded carnaúba (*Copernicia prunifera*) forest, near sea level to c. 1200 m, E Maranhão, Piauí, Ceará, Rio Grande do Norte, Paraíba, Pernambuco, Alagoas, Sergipe and Bahia (N of 15°S). Endemic to North-eastern Brazil. Map 34.
MARANHÃO: Mun. Barão de Grajaú, Morro do Cruziero, 3 Feb. 1990, *Zappi* 213A (HRCB).
PIAUÍ: N Piauí, Mun. Luís Correia, near Camurupim, gneissic inselberg, 19 Feb. 1995, *Taylor* (K, photos); Mun. Cocal, W of junction between road PI 211 and BR 343, 19 Feb. 1995, *Taylor* (obs.); Mun. Campo Maior, S of Rio Jenipapo, beside road BR 343, epiphytic in flooded carnaúba forest, 18 Feb. 1995, *Taylor* (K, photos); Mun. Dom Expedito Lopez, 1 Feb. 1990, *Zappi* 211A (SPF); Mun. Jaicós, 7°22'S, 41°9'W, 6 Apr. 2000, *E.A. Rocha et al.* (obs.); l.c., 74 km S of Picos, 6 Apr. 2000, *E.A. Rocha et al.* (K, photos); Mun. Canto do Buriti and São Raimundo Nonato, Feb. 1990, *Zappi & Taylor* (obs.);

Mun. Paulistana, 54 km N of town beside road BR 407, 6 Apr. 2000, *E.A. Rocha et al.* (K, photos).
CEARÁ: Mun. Camocim, between Chaval and Camocim, gneissic inselbergs, 19 Feb. 1995, *Taylor* (K, photos); Serra de Meruóca, between Coreaú and Alcântaras, 20 Feb. 1995, *Taylor* (K, photos); Mun. Quixadá, 27 Jan. 1990, *Zappi* 206 (HRCB), l.c., 12 km SW of town, 5°2'S, 39°2'W, 16 Feb. 1985, *Gentry et al.* 50209 (MO); Mun. Tauá, 30 Jan. 1990, *Zappi* 209A (HRCB); Mun. Milagres, 2 km E of town on road BR 116, 4 Apr. 2000, *E.A. Rocha et al.* (obs.).
RIO GRANDE DO NORTE: Chapada do Apodi, Pedra das Abelhas, 11 July 1960, *Castellanos* 22895 (RB); Mun. Natal/Parnamirim, 5 km S of Ponta Negra, 2 Sep. 1946, *Wurdack* B-136 (US); Mun. Pau dos Ferros, 2 Apr. 2000, *E.A. Rocha et al.* (obs.); Mun. Tangará, W of town on road BR 226 to Currais Novos, 6°14'S, 35°51'W, 1 Apr. 2000, *E.A. Rocha et al.* (photo); Mun. Jardim do Seridó, 1 Apr. 2000, *E.A. Rocha et al.* (obs.).
PARAÍBA: Mun. Brejo do Cruz, serra W of town, 6°22'S, 37°28'W, 2 Apr. 2000, *E.A. Rocha et al.* (obs.); Mun. Tacima, 6°36'S, 35°28'W, 1 Apr. 2000, *E.A. Rocha et al.* (K, photos); Mun. Guarabira, Serra da Jurema, 6°50'S, 35°29'W, 1 Apr. 2000, *E.A. Rocha et al.* (K, photos); Mun. Patos, *Vasconcellos Sobrinho* s.n. (IPA); Mun. Pocinhos, 8 July 1994, *L.P. Felix & Miranda* 6581 (PEUFR); Mun. Piancó, 5 km from Sta Terezinha, 29 Nov. 1971, *Andrade-Lima et al.* 1085 (PA, RB); Mun. Itaporanga, Serra Água Branca, 7–10 Apr. 1994, *M.F. Agra et al.* 2823 (JPB); Mun. Maturéia, Serra de Teixeira, Pico do Jabre, 7°11'10"S, 37°8'22"–25'53"W, 1994, *M.F. Agra et al.* (JPB); Mun. Sumé, 29–30 June 1994, *M.F. Agra et al.* 2833, 2834, 2835 (JPB).
PERNAMBUCO: Mun. Bodocó, 22 km NE of Ouricuri, 7 Mar. 1970, *Eiten & Eiten* 10859 (SP); Mun. Ouricuri, July 1984, *Costa-Lima* 023 (IPA); Parnamirim–Ouricuri, 4 Jan. 1961, *Andrade-Lima* 61-3587 (IPA); Mun. Serra Talhada, rock outcrop N of the town, 13 Feb. 1990, *Zappi* 222A (HRCB); Mun. Salgueiro, 17 Feb. 1990, *Zappi* 227B (SPF, K); Mun. Taquaritinga do Norte, Gravatá do Ibiapina, 8 Aug. 2001, *Zappi* (obs.); Mun. Moreno, Tapera, Dec. 1928, *Pickel* 1814 (IPA); Mun. Caruaru, Sítio Gaibeira, 22 Oct. 1991, *F.A.R. Santos* 7, 8 (HUEFS, PEUFR); Mun. Afrânio, towards Piauí, 23 Apr. 1971, *Heringer et al.* 309 (IPA, RB); Mun. Belém de São Francisco, 8°40'S, 38°44'W, 8 Apr. 2000, *E.A. Rocha et al.* (obs.); Mun. Ibimirim, Poço do Ferro, 6 km from Ibimirim towards Floresta, 4 May 1989, *M.J. Rodal* 38 (PEUFR); Mun. Petrolina, Distrito de Nilo Coelho, 9°15–30'S, 40°30–45'W, 4 Apr. 1991, *P.E. Nogueira et al.* 284 (SPF).
ALAGOAS: Mun. Santana do Ipanema, reported by Rizzini (1982: 67); Mun. Traipu, June 1990, *R. de Lyra-Lemos* (MAC, photo).
SERGIPE: Mun. Canindé de São Francisco, Mun. Poço Redondo, Mun. Monte Alegre and Mun. Nossa Senhora da Glória, Feb. 1991, *Taylor & Zappi* (obs.).
BAHIA: W Bahia, Mun. Barra, Ibiraba, road to Brejos, 10°48'S, 42°50'W, 26 Feb. 1997, *L.P. de Queiroz* 4873 (K, HUEFS); N, E & S Bahia, Mun. [Nova] Glória, Brejo do Burgo, 27 Aug. 1995, *Bandeira* 264 (HUEFS, K); l.c., Raso da Catarina, 31 Jan. 1982, *E. de S.F. da Rocha et al.* 799 (GUA); Mun. Remanso, 47 km E of town on road to Casa Nova, 9°22'S, 41°48'W, 7 Apr. 2000, *E.A. Rocha et al.* (obs.); Mun. Juazeiro, 2–6 June 1915, *Rose & Russell* 19725 (US, NY); l.c., 14.5 km E of the town, 7 Jan. 1991, *Taylor et al.* 1372A (K, ZSS, photos); Mun. Jaguarari, Barrinha, 7–8 June 1915, *Rose & Russell* 19789 (US, NY); l.c, 71 km S of Juazeiro on road BR 407 towards Senhor do Bonfim, 15 July 1989, *Zappi* 126 (SPF, HRCB); Mun. Araci, 18 Jan. 1994, *M.A.S. das Neves* 83 (HUEFS); Mun. Queimadas, 9–11 June 1915, *Rose & Russell* 19846 (US, NY); Mun. Boa Vista do Tupim, 27 Apr. 1994, *L.P. de Queiroz* 3879 (HUEFS); Mun. Itatim, Morro do Agenor, 12°42'S, 39°46'W, 26 Nov. 1995, *F. França* 1489 (HUEFS, K, photo.); Mun. Milagres, July 1989, *Taylor & Zappi* (obs.); Mun. Marcionílio Souza, Machado Portela, 19–23 June 1915, *Rose & Russell* 19945 (US, NY); Mun. Jaguaquara [Toca da Onça], 27–29 June 1915, *Rose & Russell* 20081 p.p. (US, NY); Mun. Paramirim, 5 km N of town on road to Macaúbas, 17 Jan. 1997, *Hatschbach et al.* 65897 (MBM, K); l.c., 43 km E of Livramento do Brumado, road to Paramirim, 13°38'S, 42°10'W, 28 Nov. 1988, *Taylor* in *Harley* 25556 (K, SPF, CEPEC); Mun. Maracás, 29 June 1993, *L.P. de Queiroz* 3261 (HUEFS); l.c., 'Calderão', [1906], *Ule* 01 (K, SPF photos ex B); Mun. Dom Basílio, 12 Mar. 1991, *Lewis* s.n. (K, photos); Mun. Tanhaçu, 47 km N of Anajé, road to Sussuarana, 17 Aug. 1988, *Eggli* 1191 (ZSS); Mun. Palmas de Monte Alto, July 1989, *Zappi* s.n. (K, photos).

CONSERVATION STATUS. Least Concern [LC] (1); PD=1, EI=2, GD=2. Short-list score (1×5) = 5. Found throughout most of North-eastern Brazil and common.

Subspecies *gounellei* is characterized by stout and sometimes very long spines, to nearly 2 mm diam. and to 15 cm long. It received the nickname of 'tyre-killer' from Werdermann (1933, 1942) and is typical of low, very dry *caatinga* on sandy or stony soil and common on gneiss/granite outcrops. It can also be seen as an epiphyte on *Copernicia* palms in the northern *caatinga* vegetation type defined in Andrade-Lima (1981: 160).

2b. subsp. **zehntneri** *(Britton & Rose) Zappi* in Succ. Pl. Res. 3: 43 (1994). Type (Zappi, l.c.): Brazil, Bahia, district of Chique-Chique [Xique-Xique], Serra de Tiririca, Nov. 1917, *Zehntner* s.n. (US, lecto.; NY, lectopara.).

Cephalocereus zehntneri Britton & Rose, Cact. 2: 35 (1920). *Pilocereus gounellei* var. *zehntneri* (Britton & Rose) Backeb. in Backeberg & Knuth, Kaktus-ABC: 331 (1935, publ. 1936). *Pilosocereus gounellei* var. *zehntneri* (Britton & Rose) Byles & G. D. Rowley in Cact. Succ. J. Gr. Brit. 19: 67 (1957); Backeberg, Die Cactaceae 4: Abb. 2283 (1960). *P. zehntneri* (Britton & Rose) F. Ritter, Kakt. Südamer. 1: 74–75 (1979). *Pseudopilocereus zehntneri* (Britton & Rose) P. V. Heath in Calyx 4: 141 (1994).

P. superfloccosus Buining & Brederoo in Cact. Succ. J. (US) 46: 60–63, excl. figs 1 & 2 (1974). *Pilosocereus superfloccosus* (Buining & Brederoo) F. Ritter, Kakt. Südamer. 1: 84 (1979). Holotype: Brazil, Bahia, Mun. Santana, 8 km NW of Porto Novo, 460 m, 1975, *Horst* 394 (U).

P. braunii E. Esteves Pereira in Kakt. and. Sukk. 38: 132 (1987) and ibid. 40: 6–13 (1989). Holotype: Brazil, Bahia, 1979, *P. J. Braun* 70 (ZSS, K, iso.).

VERNACULAR NAMES. Xique-xique-das-pedras, Chique-chique, Cheque-cheque.

Shrub or treelike, with a definite trunk to 1 m or more, branches 3.7–7.0 cm diam.; ribs 9–15, 3–8 × 6–13 mm. Spines 0.25–0.6(–0.8) mm diam. at base, whitish or golden to reddish, translucent, centrals 4–10, 10–20 mm, porrect to reflexed, radials 10–20, 7–10(–20) mm, adpressed. Flower-bearing areoles slightly to strongly modified, apical and subapical, on 1–3 ribs, often forming a true lateral cephalium sunken into the branch, with greyish white long hairs to 4 cm. Fruit with white or magenta funicular pulp.

Rio São Francisco *caatinga* element: rupicolous, on outcrops of *Bambuí* limestone (and sandstone in northern part of range), c. 450–1000 m, northern Minas Gerais and central-western and northern Bahia. Endemic to the core area of Eastern Brazil. Map 34.
BAHIA: W Bahia, Mun. Santa Maria da Vitória, Santana and Porto Novo, on limestone, 1979, *Braun* 70 (ZSS, K, UFG); Mun. Santana, 16 Jan. 1991, *Taylor et al.* 1433 (K, HRCB, ZSS, CEPEC); l.c., 28 km S of Santana on road to Santa Maria da Vitória, 15 Jan. 1991, *Taylor et al.* 1425 (K, HRCB, ZSS, CEPEC); Mun. Santa Maria da Vitória, 1983, *Braun* 371 (ZSS T-06122); Mun. Cocos, W of town, reported by Braun & Esteves Pereira (1999b). N Bahia, Mun. Sento Sé, Limoeiro, 'Serra da Mimosa' [Minas do Mimoso], 1985, *Horst & Uebelmann* 458 (ZSS); Mun. Umburanas, region of Delfino, NW of Lagoinha, road to Minas do Mimoso, 41°20'W, 10°22'S, 950–1000 m, 4 Mar. 1974, *Harley* 16729 (K); Mun. Xique-Xique/Itaguaçu da Bahia (?), Serra da Tiririca, Nov. 1917, *Zehntner* s.n. (US, NY); Mun. Morro do Chapéu, APA Gruta dos Brejões, 4 Aug. 2002, *Taylor* (K, photo); l.c., limestone escarpment forming E bank of Rio Salitre north of Brejão (Formosa), 5 Aug. 2002, *Taylor* (obs.); l.c., 21–22 km W of the town towards América Dourada, arenitic rocks, 900 m, 13 Jan. 1991, *Taylor et al.* 1405 (K, HRCB, ZSS, CEPEC); l.c., 2 km E of América Dourada, SE bank of Rio Jacaré (Rio Vereda do Romão Gramacho), *Bambuí* limestone, 13 Jan. 1991, *Taylor et al.* 1408A (K, ZSS, photos); Mun. Ibotirama, road BA 242, 12°45'S, 43°10'W, 2001, *L. Aona* 738 (UEC, photo); cent.-S Bahia, Bom Jesus da Lapa (?), 1975, *Andrade-Lima* 75-8165 (IPA); Bom Jesus da Lapa, Morro da Lapa, 17 Apr. 1983, *Leuenberger et al.* 3077 (CEPEC); Mun. Iuiú, 2 km S of town, 21 July 1989, *Zappi* 157 (SPF, HRCB).
MINAS GERAIS: Mun. Manga (?), Mocambinho, Serra Azul, 15°6'S, 43°59'W, 18 Mar. 1998, *I. Pimenta* (K, photos); Mun. Januária, 5 km W from the town, 5 Sep. 1985, *Horst & Uebelmann* 716 (ZSS); l.c., 18 km E of Rio São João, 5 Sep. 1985, *Horst & Uebelmann* 715 (ZSS); l.c., 3 km W of the town on road

MG 479 towards Pandeiros, João Menegal, 11 Aug. 1988, *Eggli* 1136 (ZSS, HRCB); ibid., [Pedras de] Maria da Cruz, Faz. São Francisco, 11 Aug. 1997, *I. Pimenta* (K, photos); Mun. Montes Claros, 13 km N of town towards Januária, 16°38'S, 43°55'W, 27 Jan. 1991, *Taylor et al.* 1455 (K, HRCB, ZSS, BHCB).

CONSERVATION STATUS. Least Concern [LC] (1); PD=1, EI=1, GD=2. Short-list score (1×4) = 4. Least Concern, but since many populations are found on limestone it may decline due to future quarrying activities.

Ritter (1979), having visited Montes Claros (MG) and Bom Jesus da Lapa (BA), recognized *P. zehntneri* as a good species and combined it under *Pilosocereus*. Not taking account of this, in 1987, E. Esteves Pereira (l.c.) described the populations from Santana and Bom Jesus da Lapa (BA) as a new species, *P. braunii*, based on the presence of a true cephalium and glaucous epidermis. Study of diverse populations of *P. zehntneri* and *P. braunii* suggests that there is clinal variation in these characters, which become less obvious towards the eastern limits of its distribution, the floriferous areoles being much less hairy and modified in populations from Montes Claros (MG) and Morro do Chapéu and América Dourada (BA). It is clear that *P. braunii* is represented by populations of extreme plants that belong to a more wide-ranging and variable taxon, recognized by Ritter (1979) as *P. zehntneri*, and treated by Zappi (1994) as a subspecies of *P. gounellei*.

The recognition of two subspecies for *P. gounellei* is based on the absence of absolute discontinuities between the taxa concerned. The incomplete geographical isolation of these taxa in regions such as west of Morro do Chapéu (BA) explains the difficulties of delimitation between them.

In localities such as those near Montes Claros, Varzelândia, Cocos, Santana and Bom Jesus da Lapa, *P. gounellei* subsp. *zehntneri* is sympatric with *P. densiareolatus* (see notes under this species), and has been confused with it. *P. superfloccosus* was based on a mixture of material (including photographs) of *P. gounellei* subsp. *zehntneri* and *P. densiareolatus*, and this name has been used by some authors for the latter species.

Subg. *Pilosocereus* (Nos 3–20):

PILOSOCEREUS ARRABIDAE Group (Nos 3–5)
This group ranges from the coastal *restinga* into the *caatinga* via the *agreste*.

3. Pilosocereus catingicola *(Gürke) Byles & G. D. Rowley* in Cact. Succ. J. Gr. Brit. 19: 66 (1957). Lectotype (Zappi 1994): Gürke, l.c., photograph p. 55 'Cereus catingicola Gürke in den sandigen Gebieten der Catinga bei Bahia [de São Salvador] nach einer von Herrn E. Ule aufgenommenen Photographie'.

Cereus catingicola Gürke in Monatsschr. Kakt.-Kunde 18: 84 (1908). *Cephalocereus catingicola* (Gürke) Britton & Rose, Cact. 2: 56 (1920). *Pilocereus catingicola* (Gürke) Werderm., Bras. Säulenkakt.: 104–105 (1933). *Pseudopilocereus catingicola* (Gürke) Buxb. in Beitr. Biol. Pflanzen 44: 252 (1968).
Pilocereus arenicola Werderm., ibid.: 109 (1933). *Pilosocereus arenicola* (Werderm.) Byles & G. D. Rowley in Cact. Succ. J. Gr. Brit. 19: 66 (1957). *P. catingicola* subsp. *arenicola* (Werderm.) P. J. Braun & E. Esteves Pereira in Succulenta 74: 134 (1995). Type: Brazil, Northern Bahia, between Sauré and Aracy, c. 350 m, Apr. 1932, *Werdermann* 3143 (B†).

P. robustus F. Ritter, Kakt. Südamer. 1: 72, Abb. 42 & 224 (1979). *Pseudopilocereus robustus* (F. Ritter)
    P. V. Heath in Calyx 4: 141 (1994). *Pilosocereus catingicola* subsp. *robustus* (F. Ritter) P. J. Braun & E.
    Esteves Pereira in Succulenta 74: 134 (1995). Holotype: Brazil, Bahia, Ourives [Mun. Tanhaçu],
    1964, *Ritter* s.n. (U).
[*P. superbus* F. Ritter, Kakt. Südamer. 1: 67 (1979), tantum quoad Abb. 35.]

Treelike or shrubby, branched above base, branching pattern 3–6-verticillate; central vascular cylinder
somewhat to strongly woody; cortex only moderately mucilaginous; epidermis olive-green to glaucous,
smooth; ribs (3–)4–12, sinuses straight, sometimes with pronounced transverse folds above the areoles.
Areoles brownish to greyish, with long hairs, presenting weak indeterminate growth. Spines yellowish
brown when young, opaque. Fertile region of stem not or slightly modified, on the apical and subapical
areoles of 1–2(–3) ribs; flower-bearing areoles with white to grey long hairs to 30 mm. Flower-buds acute
3 days before anthesis; flowers 5.5–6.7 × 4.7–7.0 cm, opening widely, somewhat compressed; pericarpel
hemiglobose, tube 38–40 × 15–22 mm at base, to 30–45 mm wide at apex, pale green, glaucous, smooth,
straight, distal half with some broad bract-scales; style 48–53 × 2–5 mm tapering to 1.8–3.5 mm, stigma-
lobes 9–12, 5 mm, exserted; ovary locule quadrangular, triangular or depressed in longitudinal section.
Fruit to 4.2–6.0 cm diam., depressed-globose, dehiscent by lateral slits, floral remnant erect or pendent,
deeply inserted; pericarp purple; funicular pulp magenta. Seeds 2–2.3 mm, with intercellular depressions.

CONSERVATION STATUS. Least Concern [LC]; very wide-ranging, but significant reductions in its
habitats have already occurred and are likely to continue.

This species is divided in two subspecies:

1. Branches (6–)8–12 cm diam., ribs 4–6, spines stout, 10–40 mm (Bahia) . . . . **3a.** subsp. **catingicola**
1. Branches 3.5–6.0(–8.0) cm diam., ribs (5–)6–12, spines slender, 2–10 mm (Bahia northwards from
    Salvador to Rio Grande do Norte and inland as far as cent.-S Pernambuco) . . . **3b.** subsp. **salvadorensis**

3a. subsp. **catingicola**

VERNACULAR NAMES. Facheiro, Mandacaru-babão, Mandacaru-de-facho.

Tree to 6–7 m, with a well-defined trunk; vascular cylinder strongly woody; branches 8–12 cm diam.,
erect; ribs (3–)4–6, 30–35 × 15–30 mm. Areoles 7–10 mm, 14–16 mm apart, long hairs brown at base,
white at apex, to 10–15 mm. Spines (0.4–)0.6–1.6 mm diam. at base, centrals 5–6, 10–40 mm, ascending
to porrect, the largest porrect, radials 10–12, 4–13 mm, adpressed, sometimes decurved. Flower 6–6.7 ×
to 7.0 cm; inner perianth-segments sometimes with pink veins; style to 53 × 3.5–5.0 mm, stigmas c. 12,
exserted. Fruit to 6 cm diam. Seeds 2–2.3 × 1.4–1.7 mm, testa-cells domed to flat, with intercellular
depressions and fine to coarse, dense cuticular folds.

Eastern *caatinga* element: locally co-dominant with other arborescent plants in *caatinga* and '*caatinga de
altitude*', c. 100–1100 m, eastwards from the Chapada Diamantina, northern, north-eastern, central-
eastern and southern Bahia. Endemic. Map 35.
BAHIA: Mun. Canudos, c. 30 km N of Euclides da Cunha on road to Canudos, 5 Jan. 1990, *Taylor et al.*
1358 (K, ZSS, HRCB); Mun. Jeremoabo, Serra do Xuquê [Chuquê], s.d., *Luetzelburg* s.n. (M, photo
filed with sheet of *Melocactus zehntneri*, *Luetzelburg* 52); Mun. Jacobina, 27–28 km W of town, towards
Lajes, 12 Jan. 1991, *Taylor et al.* 1398B (K, ZSS, photos); l.c., 28 km E on road to Capim Grosso, 26 Apr.
1992, *Taylor & Zappi* (obs.); Mun. Queimadas, 9–11 June 1915, *Rose & Russell* 19847 (US, NY); Mun.
Araci, 18 Jan. 1994, *M.A.S. das Neves* 86, 86A (HUEFS); l.c., 14.5 km N of town, 5 Jan. 1991, *Taylor
et al.* 1351 (K, HRCB, ZSS, CEPEC); Mun. Valente, c. 5 km SW of town towards São Domingos,
11°25'S, 39°30'W, 29 Dec. 1992, *L.P. de Queiroz et al.* 3038 (K, HUEFS); Mun. Conceição de Coité, 7
km NW of town on road BA 420 towards Valente, 14 July 1989, *Zappi* 117A (SPF, HRCB); Mun.
Morro do Chapéu, 15 km NE of town towards Várzea Nova, 12 Jan. 1991, *Taylor et al.* 1401 (K, HRCB,

ZSS, CEPEC); l.c., 15–20 km W of town, 5 Aug. 2002, *Taylor & Machado* (obs.); Mun. Biritinga, 18 Jan. 1994, *M.A.S. das Neves* 82 (HUEFS); Mun. Riachão do Jacuípe, 3 km N on road to Conceição do Coité, 8 July 1994, *F.A.R. Santos* 95 (HUEFS, K, photo); Mun. Santa Bárbara, 37 km N of town on road BR 116 towards Serrinha, 13 July 1989, *Zappi* 117 (SPF, HRCB); l.c., at the road junction to Juazeiro/Jacobina, 4 Jan. 1991, *Taylor et al.* 1348 (K, HRCB, ZSS, CEPEC); Mun. Ipirá, 19 km E of town on road BA 052 towards Feira de Santana, 9 Feb. 1991, *Taylor et al.* 1610 (K, HRCB, ZSS, CEPEC); Mun. Rafael Jambeiro, road BR 242, 4 km W of Argoim, 21 Mar. 1989, *Supthut* 8971 (ZSS); Mun. Castro Alves, 13 Apr. 1995, *L.P. de Queiroz* 4326 (HUEFS); Mun. Milagres, 10 km towards Itaberaba, then 2 km towards the Fazendas Morros and Antônio Romeu, 27 Jan. 1973, *I. & G. Gottsberger* 12-27173 (K); Mun. Marcionílio Souza, Machado Portela, 19–23 June 1915, *Rose & Russell* 19915 (US, NY); Mun. Jaguaquara [Toca da Onça], 27–29 June 1915, *Rose & Russell* 20085 (US, NY); Mun. Tanhaçu, Ourives, 1964, *Ritter* s.n. (U); l.c., 54 km on road BR 030 from Brumado towards Sussuarana, 3 Feb. 1991, *Taylor et al.* 1545 (K, HRCB, ZSS, CEPEC); Mun. Anagé, 1985, *Horst & Uebelmann* 472 (ZSS); Mun. Poções, 22 Sep. 1965, *A.P. Duarte* 9308 *& E. Pereira* 10021 (HB, RB); without precise locality, road between Caitité and Jequié, Dec. 1912, *Zehntner* 645 (US).

Besides the records cited above, this taxon has also been observed in the following municípios of Bahia: Itapicuru, Olindina, Aporá, Inhambupe, Euclides da Cunha, Filadélfia, Itiúba, Ponto Novo, Várzea Nova, Capim Grosso, Miguel Calmon, São José do Jacuípe, Gavião, Baixa Grande, Candeal, Serra Preta, Santa Teresinha, Iaçu, Maracás, Barra da Estiva, Contendas do Sincorá, Jequié, Planalto.

CONSERVATION STATUS. Least Concern [LC] (1); PD=1, EI=2, GD=2. Short-list score (1×5) = 5. Least Concern, but its habitat continues to decline and its status needs to be monitored.

Subspecies *catingicola* is represented by populations of arborescent plants that occur inland in the *caatinga* zone of Bahia and ascend into the Chapada Diamantina, presenting branches with 4–6 ribs and strong spination. In the highland areas it is occasionally sympatric with *P. pachycladus* (e.g. in various sites within Mun. Morro do Chapéu), but hybrids have not been observed. Southern forms are often quite glaucous and the habit of montane forms differs somewhat from those found in the lowland *caatinga*, but it is clear that only one taxon is involved. This subspecies provisionally includes *P. arenicola*, which was probably based on marginal populations intermediate with subsp. *salvadorensis*, and thus could not be safely neotypified.

3b. subsp. **salvadorensis** *(Werderm.) Zappi* in Succ. Pl. Res. 3: 55 (1994). Holotype: not extant (B†). Lectotype (Zappi, l.c.): Brazil, Bahia, between Bolandeiras and São Salvador, Apr. 1932, Werdermann, l.c. infra, 37, photograph (iconotype).

Pilocereus salvadorensis Werderm., Bras. Säulenkakt.: 110 (1933). *Cephalocereus salvadorensis* (Werderm.) Borg, Cacti: 107 (1937). *Pilosocereus salvadorensis* (Werderm.) Byles & G. D. Rowley in Cact. Succ. J. Gr. Brit. 19: 67 (1957). *Austrocephalocereus salvadorensis* (Werderm.) Buxb. in Krainz, Die Kakteen, Lfg 33 (1966). *Pseudopilocereus salvadorensis* (Werderm.) Buxb. in Beitr. Biol. Pflanzen 44: 253 (1968).
Pilocereus hapalacanthus Werderm., Bras. Säulenkakt.: 110–111 (1933). *Cephalocereus hapalacanthus* (Werderm.) Borg, Cacti: 107 (1937); E. Y. Dawson in Los Angeles County Mus. Contr. Sci. 10: 6 (1957). *Pilosocereus hapalacanthus* (Werderm.) Byles & G. D. Rowley in Cact. Succ. J. Gr. Brit. 19: 67 (1957). *Pseudopilocereus hapalacanthus* (Werderm.) Buxb. in Beitr. Biol. Pflanzen 44: 252 (1968). *Pilosocereus catingicola* subsp. *hapalacanthus* (Werderm.) P. J. Braun & E. Esteves Pereira in Succulenta 74: 134 (1995). Holotype: not extant (B†). Lectotype (Zappi 1994): Brazil, Pernambuco, sand-dunes N of Recife, Feb. 1932, Werdermann, l.c., 104, photograph captioned '*Pilocer. hapalacanthus* Werd.'
Pilocereus sergipensis Werderm., Bras. Säulenkakt.: 106 (1933). *Cephalocereus sergipensis* (Werderm.) Borg, Cacti: 108 (1937). *Pilosocereus sergipensis* (Werderm.) Byles & G. D. Rowley in Cact. Succ. J. Gr. Brit.

19: 67 (1957). *Pseudopilocereus sergipensis* (Werderm.) Buxb. in Beitr. Biol. Pflanzen 44: 253 (1968). Type: Sergipe, Jaboatão, 1932, *Werdermann*, not extant (B†). Neotype (Zappi 1994): Brazil, Sergipe, Mun. Nossa Senhora das Dores, 500 m S of the town, 110 m, 11 Feb. 1991, *Taylor & Zappi 1616* (HRCB; K, isoneo.).

Pilocereus rupicola Werderm., Bras. Säulenkakt.: 109 (1933). *Cephalocereus rupicola* (Werderm.) Borg, Cacti: 111 (1937). *Pilosocereus rupicola* (Werdermann) Byles & G. D. Rowley in Cact. Succ. J. Gr. Brit. 19: 67 (1957). *Pseudopilocereus rupicola* (Werderm.) Buxb. in Beitr. Biol. Pflanzen 44: 252 (1968). Holotype: Brazil, Sergipe, Serra de Itabaiana, 550 m, Mar. 1932, *Werdermann 3092* (B†).

VERNACULAR NAMES. Facheiro, Facheiro-da-praia.

Shrubs or trees 2–10 m, with or without a definite trunk, branching at ground level and upwards; vascular cylinder slightly woody; branches 3.5–6.0(–8.0) cm diam., erect; ribs (5–)6–12, 10–20 × 16–20 mm; areoles 4 mm diam., 8–10 mm apart, rounded, reddish brown to grey, slightly hairy when young. Spines 0.3–0.7(–1.0) mm diam. at base, central(s) 1–11, 5–12 mm, ascending to porrect, radials 8–12, 3–4(–5) mm, adpressed. Flowers 5.5 × 4.7 cm, inner perianth-segments white; style 48 × 1.8–2.0 mm, tapering, stigma-lobes c. 9, exserted. Fruit to 4.2 cm diam. Seeds black, shiny, 2–2.1 × 1.1–1.3 mm, hilum-micropylar region 6–8 mm, forming an angle of 40° with long axis of seed, testa-cells domed to flat, with intercellular depressions and coarse, sparse cuticular folds.

Widespread humid forest (*restinga*) / *caatinga* element: in dense or sparse *restinga* on sand-dunes north of Salvador (BA) to Rio Grande do Norte and (?) NE Ceará, extending up the São Francisco River valley and westwards to the *caatinga* region around the borders of Bahia (Raso da Catarina), Alagoas, Sergipe and Pernambuco, where locally co-dominant with other arborescent vegetation, near sea level to 550 m. Endemic to North-eastern Brazil. Map 35.

RIO GRANDE DO NORTE: Mun. Natal, Mãe Luiza, dunes, 9 Mar. 1962, *Tavares* 905 (US), l.c., Praia do Meio, 1988, illustrated by Braun & Esteves Pereira (2002: 21, Abb. 34); Mun. Nísia Floresta, Praia de Búzios, 1997, *E.A. Rocha* (obs.).

PARAÍBA: Mun. Mataraca, reported by Oliveira-Filho (1993: 224); Mun. Mamanguape, 1997, *E.A. Rocha* 191 (JPB); Mun. Cabedelo, Praia do Poço, 21 Feb. 1962, *Castellanos* 23245 (GUA), 9 Feb. 1995, *Taylor & Zappi* (K, photos), Mar.–Apr. 1996, see Locatelli *et al.* (1997: figs 1–5); Mun. Pedras de Fogo, Santa Emília, 24 Sep. 1962, *Tavares* 1053 (US).

PERNAMBUCO: Mun. Itamaracá & Mun. Abreu e Lima (Rio Doce), *fide* Andrade-Lima (1966: 1457); Mun. Recife, sand-dunes, Feb. 1932, reported by Werdermann (1933: 104, photo); Mun. Vitória de Santo Antão, 25 Apr. 1935, *Pickel* 3785 (IPA 4473); Mun. Inajá, Reserva Biol. Serra Negra, 17 Sep. 1995, *A. Laurênio et al.* 196 (K, PEUFR); Mun. Nova Petrolândia, 3 km N of the Barragem de Itaparica, 250 m, 12 Feb. 1991, *Taylor & Zappi* 1619 (HRCB, K, ZSS, UFP).

ALAGOAS: Mun. Marechal Deodoro, *restinga* S of Ponta do Cavalo Ruço, 14 Feb. 1995, *Taylor & Zappi* (K, photos); Mun. Traipu, June 1990, *R. de Lyra-Lemos* (MAC, photo); Mun. (?), between Juquia and Lagoa do Pau beach, Aug. 2001, *Zappi* (obs.); Mun. Igreja Nova, 6 km SE of the town on road AL 225, 14 Feb. 1991, *Taylor & Zappi* 1632 (HRCB, K, ZSS, MAC); Mun. (?), between Miaí de Cima and Piaçabuçu, Aug. 2001, *Zappi* (obs.); Mun. Piaçabuçu, Tapera, 4 Feb. 1988, *I.S. Moreira et al.* 113 (MAC, SPF); Mun. Penedo, 7 km S of town on road to Piaçabuçu, 14 Feb. 1991, *Taylor & Zappi* 1631 (HRCB, K, ZSS, MAC).

SERGIPE: Mun. Canindé de São Francisco, border of SE/BA, to Xingozinho, 11 Feb. 1991, *Taylor & Zappi* 1618A (HRCB, K); Mun. Nossa Senhora das Dores, 0.5 km S of town, 11 Feb. 1991, *Taylor & Zappi* 1616 (HRCB, K, ZSS, ASE); Mun. Jaboatão, type locality of synonymous *Pilocereus sergipensis* Werderm. (see above); Mun. Feira Nova, 2 km NW of town, 11 Feb. 1991, *Taylor & Zappi* 1617 (HRCB, K, ZSS, ASE); Mun. Nossa Senhora da Glória, 7 km NW of town, 11 Feb. 1991, *Taylor & Zappi* 1617A (HRCB, K, ZSS); Mun. Itabaiana, just below western edge of highest part of Serra de Itabaiana, c. 550 m, 3 Aug. 1998, *Taylor* (K, ASE, photos); Mun. Barra dos Coqueiros and Mun. Aracaju (Praia José Sarny), *restinga*, 3/4 Aug. 1998, *Taylor* (obs.).

BAHIA: Mun. Glória, Brejo do Burgo, Faz. Torrão, 28 Nov. 1992, *F.P. Bandeira* 84 (ALCB); Mun. Paulo Afonso, 2 km from crossing towards town, coming from Canindé de São Francisco, 11 Feb. 1991, *Taylor*

& *Zappi* 1618 (HRCB, K, ZSS, CEPEC); Mun. Santa Brígida, c. 10 km W of junction for the town with road BR 110, Raso da Catarina, c. 9°29'S, 38°12'W, 19 Dec. 1993, *L.P. de Queiroz & Nascimento* 3733 (K, HUEFS); Mun. Mata de São João, Aug. 2001, *Zappi* (obs.); Mun. Camaçari, 42 km N of Salvador airport, towards Arembepe, *restinga*, 23 Mar. 1989, *Supthut* 8975 (ZSS); Mun. Salvador, Jardim de Alá, 1960, *A.L. Costa* s.n. (ALCB 02912); l.c., Lagoa de Abaeté, 1 km N of Praia de Itapoã, 13 July 1989, *Zappi* 116 (SPF, HRCB), 4 Jan. 1991, *Taylor et al.* 1342 (K, HRCB, ZSS, CEPEC); l.c., near Salvador, 29 May 1915, *Rose & Russell* 19676 (US, NY).

CONSERVATION STATUS. Near Threatened [NT] (1); PD=1, EI=2, GD=2. Short-list score (1×5) = 5. Although its range is extensive, its habitat continues to decline at a significant rate, especially near the coast, and so its status needs to be monitored. For example, the once extensive population on the peninsula of Cabedelo, Paraíba has all but disappeared since 1995, due to building developments, and only a few plants are protected inside the local IBAMA forest reserve. The same can be said of habitats along the road 'Linha Verde' on the coast of NE Bahia and of the NE coast of Alagoas.

The pollination of this subspecies by the phyllostomid bat, *Glossophaga soricina* Pallas, and by hawkmoths, has been documented and photographed by Locatelli *et al.* (1997).

Maria Luiza Santos (Aracaju) has shown the first author photographs of what is assumed to be dense stands of this taxon growing on the rocky banks of the lower reaches of the Rio São Francisco, where it apparently develops into exceptionally tall specimens with many erect branches.

4. Pilosocereus azulensis N. P. *Taylor & Zappi* in Cact. Consensus Initiatives 3: 8 (1997). Holotype: Brazil, Minas Gerais, Mun. Pedra Azul, 16°3'S, 41°14'W, 20 Oct. 1988, *Taylor & Zappi* in *Harley* 25220 (SPF; K, iso.).

Treelike, c. 4 m, branched above the ground, with a well-defined trunk; vascular cylinder weakly woody; branches 8–9.5 cm diam., erect, epidermis dark green, slightly rough; ribs 8–10, 10–15 × 15 mm, sinuses straight, and without visible transverse folds. Areoles 3–4 mm diam., 5–7 mm apart, felt greyish with brown long hairs to 3–5 mm. Spines 0.3–0.5 mm diam. at base, opaque, brown to blackish, centrals 1–4, 10–15 mm, ascending, radials 10–14, 2–8 mm, adpressed. Fertile part of stem slightly differentiated, with few apical or sub-apical flower-bearing areoles with bristly, black spines to 20 mm and white tufts of hair to 10 mm. Flowers not seen. Fruits (dried) depressed-globose, floral remnant pendent, deeply inserted. Seeds 1.5–1.6 × 1.0 mm, black, shiny, cochleariform, hilum-micropylar region 0.6–0.7 mm, forming an angle of 25–35° with long axis of seed, testa-cells domed, with inconspicuous intercellular depressions and hardly any cuticular folds.

South-eastern *caatinga* (inselberg) element: associated with gneissic inselbergs in *caatinga-agreste*, 650 m; known only from the region of Pedra Azul, Minas Gerais (and from perhaps the same area recorded as a vaguely localized collection from south of Vitória da Conquista, Bahia). Endemic to the core area within Eastern Brazil. Map 35.
MINAS GERAIS: Mun. Pedra Azul, '6 km ao sul da cidade na estrada para Jequitinhonha', 16°3'S, 41°14'W, 20 Oct. 1988, *Taylor & Zappi* in *Harley* 25220 (K, SPF); l.c., between Pedra Azul and Salinas, 10 Sep. 1985, *Horst & Uebelmann* 272 (ZSS) — NB. this HU number has also often been applied to living collections of *P. pachycladus*. Unlocalized: Bahia/Minas Gerais (?), S of Vitória da Conquista, 10 Apr. 1968, *Castellanos* 27073 (HB).

CONSERVATION STATUS. Critically Endangered [CR D] (4); PD=2, EI=1, GD=1. Short-list score (4×4) = 16. This species is currently known from only a small remnant of dry forest vegetation, which is being cleared for agriculture and charcoal production. Only the above 3 collections have been recorded to date and two of these are poorly localized and all may refer to the same small area of the type. Further study of this plant and its habitat is urgently required. In August 2002, Marlon Machado and staff from the Univ. Estadual do Sudoeste da Bahia located c. 30 juvenile plants near the type locality (Plate 40.1), but the adjacent forest in which the type was collected has gone.

In 1988 a single mature specimen of this taxon was observed in semi-shade of dry forest (*agreste*), sympatric with *Pilosocereus floccosus* subsp. *quadricostatus*. The shape of the ribs and spination, as well as the only slightly differentiated flowering region are reminiscent of the P. ARRABIDAE Group, especially of some populations of the geographically distant *P. catingicola* subsp. *salvadorensis.*

Although flowering material has yet to be examined, vegetative morphology and seeds suggest that this species belongs to the P. ARRABIDAE Group, which is otherwise unrepresented in the region of the Rio Jequitinhonha drainage, where all other Species Groups in Subg. *Pilosocereus* are present. The alternative explanation, that it is a hybrid involving *P. floccosus* subsp. *quadricostatus* and *P. multicostatus*, does not seem plausible given the characters it displays. Furthermore, plants that are almost certainly representative of such a hybrid have recently been seen around Pedra Azul by Marlon Machado (in Oct. 1999) and Graham Charles (K, photos, June 2002) and these are clearly not comparable with *P. azulensis* (Plate 40.2).

5. Pilosocereus arrabidae *(Lem.) Byles & G. D. Rowley* in Cact. Succ. J. Gr. Brit. 19: 66 (1957). Lectotype (Zappi 1994): Brazil, Rio de Janeiro, Vellozo, Fl. flumin., Icones 5: tab. 18 (1831).

Pilocereus arrabidae Lem., Rev. Hort. 34: 429 (1862). *Cephalocereus arrabidae* (Lem.) Britton & Rose, Cact. 2: 42 (1920). *Pseudopilocereus arrabidae* (Lem.) Buxb., Beitr. Biol. Pflanzen 44: 252 (1968).
Cereus warmingii K. Schum. in Martius, Fl. bras. 4(2): 204 (1890), *quoad typ.* Type: Brazil, Rio de Janeiro, restingas de Copacabana, *Warming* s.n. (B†).
[*Pilocereus exerens* K. Schum. in Engler & Prantl, Pflanzenfamilien 3(6): 181 (1894), typ. excl., nom. illeg. Art. 52.1.]
[*Cephalocereus exerens* (K. Schum.) Rose in Bailey, Stand. Cycl. Hort. 2: 715 (1914), typ. excl., nom. illeg. Art. 52.1.]
[*Cactus hexagonus sensu* Vell., Fl. flumin.: 205 (1825, publ. 1829), non Linnaeus (1753).]
[*Cactus heptagonus sensu* Vell., Fl. flumin.: 205 (1825, publ. 1829), non Linnaeus (1753).]
[*Cereus macrogonus sensu* K. Schum. in Martius, Fl. bras. 4(2): 202, tab. 40 (1890), non Salm-Dyck (1850).]
[*Pilocereus virens sensu* Ule, Monatsschr. Kakt.-Kunde 13: 28 (1903), non *Cereus virens* DC.]

VERNACULAR NAME. Facheiro-da-praia.

Shrub, 0.1–3.0(–4.0) m, or rarely with a short trunk, branching near base or above, sometimes 3–5-verticillate; vascular cylinder weakly woody; stems 4.5–9.5 cm diam., straight, slightly upcurved, epidermis yellow-green to dark green, sometimes slightly greyish; ribs (5–)6–8, 13–15 × 12–20 mm, sinuses straight, with oblique transverse folds above the areoles. Areoles (5–)6–8 mm diam., (7–)15–32 mm apart, situated upon more or less conspicuous stem-projections (podaria), felt whitish to brownish or greyish. Spines 0.6–1.1 mm diam. at base, translucent only when young, brown to greyish, centrals (0–)2–4(–8), 15–40 mm, sometimes decurved near base, often 1 ascending and two porrect, radials 7–10, 2–20 mm, adpressed. Fertile part of stem not differentiated, flowers distributed along the branches. Flower-buds acute 2–3 days before anthesis; flowers 6–7 × 4–5 cm; pericarpel subglobose, tube flared from 11 to 30–32 mm wide, infundibuliform, straight, creamy green, smooth, distal half with wide, broadly triangular bract-scales; style 15–18 × 2–3 mm, tapering, stigma-lobes c. 9, 6–7 mm, exserted; ovary locule obovate, compressed in longitudinal section. Fruit 3–5 × 5–5.8 cm, depressed-globose, dehiscent by a lateral or abaxial slit, floral remnant pendent, deeply inserted; pericarp green to deep reddish pink, slightly rugose; funicular pulp magenta. Seeds (1.4–)1.5–1.7(–1.8) × 1–1.2 mm, testa-cells domed, with intercellular depressions and dense, coarse cuticular folds.

Southern humid forest (*restinga*) element: in dense or sparse, sandy *restinga*, near sea level, southern Bahia and Espírito Santo to Rio de Janeiro. Map 35.
BAHIA: Mun. Santa Cruz Cabrália, 20–24 km N of Porto Seguro, 31 July 1989, *Zappi* 184 (SPF, HRCB); Mun. Prado, 13 Dec. 1998, *Fonseca & Guedes* 1156 (ALCB, UEC).
ESPÍRITO SANTO: Mun. Conceição da Barra, Ilha de Guriri, June 1996, *M. Canal* s.n. (Herb. Univ. Fed. ES); Mun. Guarapari, on sand, [1973], *Horst & Uebelmann* 240 (ZSS); l.c., road ES 060, *restinga* close to Lagoa Vermelha, 21 May 1990, *O.J. Pereira et al.* 2119 (Herb. Univ. Fed. ES, HRCB); Mun. Itapemirim, Rodovia do Sol, S of Marataízes, 21°4'51"S, 40°50'24"W, 26 Nov. 1999, reported growing with *Opuntia monacantha* (*Zappi et al.* 470).
RIO DE JANEIRO: Mun. São João da Barra, c. 10 km from town by the crossing towards Gruçaí, 23 Mar. 1982, *Rocha et al.* 893 (RB); Mun. Quiçamã (Quissamã), road to Barra do Furado, 22°6'5"S, 41°26'W, 23 Nov. 1999, *Zappi et al.* 397 (K, UEC).

CONSERVATION STATUS. Near Threatened [NT] (1); PD=2, EI=1, GD=1. Short-list score (1×4) = 4. Although its range is extensive so is the ongoing and accelerating modification of its coastal habitat.

Inhabiting a long stretch of *restinga* vegetation, from between Santa Cruz Cabrália and Porto Seguro, Bahia, to west of the city of Rio de Janeiro, *P. arrabidae* has been rather frequently confused with *P. brasiliensis* subsp. *brasiliensis*, with which it is sympatric, at least along the coast of Espírito Santo (cf. *Pereira* 2119). There are some superficial similarities, such as the undifferentiated flower-bearing areoles and green epidermis, but *P. arrabidae* presents (5–)6–8 ribs, thicker branches, acute, straight flower-buds, large flowers, 4–5 cm diam. at anthesis, whereas *P. brasiliensis* subsp. *brasiliensis* has 4–5 ribs, thinner branches, 4.5–5.5 cm diam. and obtuse, curved flower-buds, with narrow flowers up to 3 cm diam. at anthesis. Their seeds also differ considerably. Contrary to the impression given in Braun & Esteves Pereira (2003: 17), *P. arrabidae* is restricted to coastal rocks and dunes in Espírito Santo, being replaced by *P. brasiliensis* subsp. *ruschianus* inland.

PILOSOCEREUS PENTAEDROPHORUS Group (Nos 6–10). Found in *Mata atlântica* through to *caatinga/cerrado de altitude* and in the *caatinga-cerrado* ecotone, but avoiding the driest areas and completely absent from the *caatingas* of the Rio São Francisco valley.

6. Pilosocereus brasiliensis *(Britton & Rose) Backeb.*, Die Cact. 4: 2423 (1960). Type (Zappi 1994): Brazil, Rio de Janeiro, Mun. Rio de Janeiro, Corcovado, 10 July 1915, *Rose & Russell* 20190 (US, lecto.; NY, lectopara.).

Cephalocereus brasiliensis Britton & Rose, Cact. 2: 57, fig. 84 (1920). *Pilocereus brasiliensis* (Britton & Rose) Werderm., Bras. Säulenkakt.: 105 (1933).
? *Cereus sublanatus* Salm-Dyck, Hort. Dyck. 1834: 337 (1834). *Pilocereus sublanatus* (Salm-Dyck) Backeb. & F. M. Knuth, Kaktus-ABC: 333 (1935, publ. 1936). *Pilosocereus sublanatus* (Salm-Dyck) Byles & G. D. Rowley in Cact. Succ. J. Gr. Brit. 19: 69 (1957). Type: a living plant of unknown provenance, assumed not to have been preserved or lost.

Erect or semi-scandent shrub, 2–7 m, usually without a well-defined trunk, main stem little branched; vascular cylinder moderately woody, branches 4.5–7.5 cm diam., semi-decumbent to prostrate when shaded, erect when in full light; ribs 4–6(–7), 15–17 × 8–18 mm, sinuses straight, transverse folds above the areoles oblique. Areoles 4–7 mm diam., 6–8 mm apart, situated upon stem-projections (podaria), felt white, brown to blackish, with white, greyish or brown long hairs to 15 mm, indeterminate growth not observed. Spines opaque, at first reddish or yellowish brown, later greyish. Fertile part of stem not or only slightly differentiated, flowers appearing randomly over the stem. Flower-buds obtuse 1–2 days before anthesis; flowers 4.5 × 2.5–3.0 cm; pericarpel subglobose, tube 20–40 × 10–12 mm at base, flared to 23 mm at apex, constricted beneath and above nectar-chamber, curved, nearly cylindric, pale green, smooth, apex with few, thick, rounded bract-scales; style 26–32 × 1–1.5 mm, tapering; stigma-lobes 7–8, 8 mm; ovary locule hemicircular in longitudinal section. Fruit 2 × 3–3.5 cm, depressed-globose, dehiscent by a lateral slit, floral remnant pendent, deeply inserted; pericarp wine-coloured, smooth; funicular pulp magenta. Seeds (1.4–)1.6–1.7(–1.8) × 1.1–1.2 mm, testa-cells domed, with intercellular depressions and dense, coarse cuticular folds.

CONSERVATION STATUS. Least Concern [LC]; see subspecies below.

Two subspecies are recognized as follows:

1. Branches dark green; ribs 4–5 (coastal) . . . . **6a.** subsp. **brasiliensis**
1. Branches greyish green to glaucous; ribs (4–)5–7 (inland) . . . . **6b.** subsp. **ruschianus**

6a. subsp. **brasiliensis**

Shrub. Branches semi-decumbent to prostrate when shaded, erect in exposed habitats, epidermis shiny dark green, sometimes slightly greyish; ribs 4–5; areoles usually hairy only at the stem apex (except in cultivation under glass), long hairs pale brown, to 10 mm, later glabrescent, brown to black. Spines 0.8–1.0 mm diam. at base, centrals (0–)1–3, 10–30 mm, ascending to porrect; radials 4–8, 3–16 mm, adpressed. Fertile part of stem not or only slightly differentiated, flowers solitary or in groups of 2–3; anthers 1.5–1.8 mm; stigma-lobes c. 7, not exserted much beyond anthers.

Southern humid forest (*restinga*) element: in *restinga* and gneissic inselbergs of the coast, to 100 m, Espírito Santo; Rio de Janeiro. Endemic to South-eastern Brazil. Map 36.

ESPÍRITO SANTO: Mun. Vitória, Campus of the Univ. Fed. do Espírito Santo, May 1990, *O.J. Pereira* 2120B (HRCB); Mun. Vila Velha, *restinga* de Jacaranema, 21 May 1990, *O.J. Pereira* 2120 (HRCB, Herb. Univ. Fed. ES); Mun. Guarapari, dense *restinga,* 21 May 1990, *O.J. Pereira* 2120A (HRCB); Mun. Anchieta, Rodovia do Sol, between Meaípe (Mun. Guarapari) and Anchieta, 20°45'13"S, 40°32'37"W, 25 Nov. 1999, *Zappi et al.* 451 (K, UEC); Mun. Piúma, Monte Agá, south of the town, 20°52'23"S, 40°46'24"W, 26 Nov. 1999, reported growing with *Coleocephalocereus fluminensis* (*Zappi et al.* 469, q.v.).

CONSERVATION STATUS. Vulnerable [VU B2ab(iii)] (2); area of occupancy estimated to be < 2000 km²; PD=1, EI=1, GD=2. Short-list score (2×4) = 8. Has apparently declined in the southern part of its range (around the city of Rio de Janeiro) and will be affected by ongoing touristic developments along much of the length of its coastal habitat.

Variable in the degree of areolar wool and long hairs developed in some populations.

6b. subsp. **ruschianus** *(Buining & Brederoo) Zappi* in Succ. Pl. Res. 3: 64 (1994). Holotype: not extant. Lectotype (Zappi, l.c.): Brazil, Espírito Santo, Mun. Colatina, Buining & Brederoo, l.c., 33, photograph, above right (iconotype).

Pseudopilocereus ruschianus Buining & Brederoo in Kakt. and. Sukk. 31: 33–35 (1980). *Pilosocereus ruschianus* (Buining & Brederoo) P. J. Braun in Bradleya 6: 88 (1988).

Sometimes treelike, to 7 metres. Branches erect or sometimes procumbent on inselbergs, epidermis grey-green, slightly glaucous at apex; ribs (4–)5–7, variable in number between populations; areoles with white to greyish long hairs, mainly at the apex of stems. Spines 0.5–0.8 mm diam. at base, centrals 1–3, 10–38 mm, ascending to porrect, radials 8–15, 5–12 mm, adpressed. Fertile part of stem slightly differentiated, occupying many consecutive flowering areoles, with greyish or brownish long hairs; anthers 2.8–3.0 mm, slightly verrucose; stigma-lobes c. 8, exserted in relation to the anthers.

Southern humid/subhumid forest (inselberg) element: on gneissic inselbergs associated with *agreste* and *mata seca*, c. 80–700 m, southern Bahia, Espírito Santo and eastern Minas Gerais. Endemic to the core area within Eastern Brazil. Map 36.

BAHIA: Mun. Vitória da Conquista, 32 km SE of town on road to Itambé, 1 km from Rio Periquito, granitic outcrop, 16 Aug. 1988, *Eggli* 1175 (ZZS, HRCB); Mun. Floresta Azul, 2 km W on road BR 415, 16 April 2003, *Taylor & Zappi* (K, photo); Mun. Itaju do Colônia, 2 km on road between Itaju and Pau Brasil, 24 Jan. 1969, *T.S. dos Santos* 349 (CEPEC); between Santa Cruz da Vitória and Itaju do Colônia, 19 Dec. 1967, *Castellanos* 27058 (CEPEC).

MINAS GERAIS: Mun. Carlos Chagas, c. 25 km W of Nanuque, 1994, *S. Porembski et al.* (K, photo); Mun. Galiléia, towards the border with Espírito Santo, 1982, *Horst & Uebelmann* 512, cult. (ZSS); l.c., road BR 259, 23 km from São Vítor towards Galiléia, 15 Dec. 1990, *Taylor & Zappi* 770 (HRCB, K, ZSS); Mun. Conselheiro Pena, 7.5 km E of Goiabeira on road to Cuparaque, 15 Dec. 1990, *Taylor & Zappi* 777 (K, HRCB, ZSS, BHCB); Mun. Aimorés, Terreno do Varejão, 7 July 1997, *M.F. de Vasconcelos* s.n. (K, BHCB).

ESPÍRITO SANTO: E of Mantena, 1983, *Braun* 488 (ZSS); Mun. Nova Venézia, Pedra do Elefante, Apr. 1996, *S. Porembski* (K, photo); Mun. São Domingos, 15 km S of town on road to Colatina, 16 Dec. 1990, *Taylor & Zappi* 783 (K, HRCB, ZSS, MBML); Mun. Pancas, Bridge over Rio Pancas, 16 Dec. 1990, *Taylor & Zappi* s.n. (K, photos); Mun. (?), 'Rio Doce', Feb. 1917, *Luetzelburg* 7227 (M); Mun. Itarana, road to Praça Oito, 19°55'34"S, 40°48'44"W, 24 Nov. 1999, *Zappi et al.* 417 (K, UEC); Mun. Santa Teresa, S. João de Petrópolis, Escola Agrotécnica Federal, 8 Oct. 1985, *H.Q.B. Fernandes* 1554 (MBML); l.c., Rio Cinco de Novembro, Dec. 1987, *H.Q.B. Fernandes* 2259 (MBML, HRCB).

CONSERVATION STATUS. Least Concern [LC] (1); PD=1, EI=1, GD=2. Short-list score (1×4) = 4. Habitat is not significantly declining at present.

Variable in rib number between populations. At the northern limits of its range this subspecies occasionally becomes a tree in semi-humid forest, such as between Itororó and Itapetinga (BA).

7. Pilosocereus flavipulvinatus *(Buining & Brederoo) F. Ritter*, Kakt. Südamer. 2: 707 (1980). Holotype: not extant. Lectotype (Zappi 1994): Brazil, Piauí, Simplício Mendes, c. 350 m, Buining & Brederoo, l.c., 138, photograph (iconotype).

Pseudopilocereus flavipulvinatus Buining & Brederoo in Succulenta 58: 137–143 (1979).
Pilosocereus carolinensis F. Ritter, Kakt. Südamer. 1: 80 (1979). *P. flavipulvinatus* var. *carolinensis* (F. Ritter) F. Ritter, op. cit. 2: 707 (1980). *Pseudopilocereus carolinensis* (F. Ritter) P. V. Heath in Calyx 4: 140 (1994). *Pilosocereus flavipulvinatus* subsp. *carolinensis* (F. Ritter) P. J. Braun & E. Esteves Pereira in Succulenta 74: 134 (1995). Holotype: Brazil, Maranhão, Carolina, 1963, *Ritter* 1217 (U).
P. carolinensis var. *robustispinus* F. Ritter, Kakt. Südamer. 1: 81 (1979). *Pseudopilocereus carolinensis* var. *robustipinus* (F. Ritter) P. V. Heath in Calyx 4: 140 (1994). *Pilosocereus flavipulvinatus* var. *robustispinus* (F. Ritter) P. J. Braun & E. Esteves Pereira in Succulenta 74: 134 (1995). Holotype: Brazil, Maranhão, Barão de Grajaú, 1963, *Ritter* 1217A (U).

Slender, sparsely branched tree, 2–5(–8) m, with a well-defined trunk; vascular cylinder strongly woody; branches 3–12(–15) cm diam., closely grouped and erect or sometimes semi-decumbent, epidermis olive-green, glaucous at apex when young; ribs 6–12, 8–17 × 9–15 cm, sinuses straight, with oblique transverse folds above the areoles. Areoles 5–10 mm, 10–15 mm apart, situated upon rounded stem projections, felt yellowish white, long hairs pale brown to grey, apical areoles sometimes unarmed. Spines 0.5–0.7 mm diam. at base, translucent, pale yellow and ascending when young, later greyish; centrals 3–5(–7), 15–70 mm, porrect, lowermost largest; radials 12–15, 5–14 mm, adpressed. Fertile part of stem not differentiated, subapical, on 2–4 ribs. Flower-buds acute 2 days before anthesis; flowers 5.8–6.0 × 3.5–3.7 cm, pericarpel subglobose, tube 40 × 13 mm at base to 25 mm at apex, constricted above and below the nectar-chamber, strongly curved, slightly compressed, pale green, glaucous, smooth, distal half with acute bract-scales; style 42–46 × 2–2.5 mm, tapering, stigma-lobes c. 9, 5 mm, exserted in relation to the stamens; ovary locule hemicircular in longitudinal section. Fruit to 2–2.5 × 3–3.5 cm, depressed-globose, dehiscent by a lateral slit, floral remnant pendent, deeply inserted; pericarp pale green, slightly rugose; funicular pulp magenta. Seeds (1.6–)1.7–1.8(–1.9) × 1.1–1.2 mm, testa-cells domed to flat, with intercellular depressions and coarse, sparse cuticular folds.

Northern *caatinga* element: in the *caatinga/cerrado* ecotone, dense, high and low shrubby *caatinga*, open, seasonally flooded carnaúba (*Copernicia prunifera*) forest (often as epiphyte) and *caatinga*-mangrove ecotone, in northern draining river valleys and at the coast, sea level to c. 350 m, northern and eastern-central Ceará, northern to central-south-eastern Piauí and along the border regions between Piauí and Maranhão; (?) to northern Tocantins. Endemic to North-eastern Brazil? Map 36.
MARANHÃO: Mun. Carolina, 1963, *Ritter* 1217 (U); Mun. Barão de Grajaú, 1963, *Ritter* 1217A (U); l.c., just W of Rio Parnaíba near road BR 230, 3 Feb. 1990, *Zappi* 213 (HRCB, ZSS).
PIAUÍ: N Piauí, Mun. Luís Correia, W of Chaval [CE], 19 Feb. 1995, *Taylor* (K, photos); Mun. Buriti dos Lopes, 5 km N of town, rocks by the Rio Pirangi, 19 Feb. 1995, *Taylor* (K, photos); Mun. Cocal, c. 17 km E of town towards border with Ceará, 19 Feb. 1995, *Taylor* (K, photos); Mun. Piracuruca, c. 12 & 29 km W of border with Ceará, 20 Feb. 1995, *Taylor* (K, photos); Cent.-SE Piauí, Mun. Picos, 23 km from town along road BR 407 towards Jaicós, 7°13'S, 41°17'W, 6 Apr. 2000, *E.A. Rocha et al.* (K, photos); Mun. Jaicós, 6 km S of town, 6 Apr. 2000, *E.A. Rocha et al.* (obs.); Mun. Canto do Buriti, 15 km S of town, 5 Feb. 1990, *Zappi* 216 (HRCB, ZSS); Mun. Simplício Mendes, see type citation above; Mun. Paulistana, 51 km N of town on road BR 407, 6 Apr. 2000, *E.A. Rocha et al.* (obs.).
CEARÁ: N Ceará, between Chaval and Camocim, various sites including mangrove, 19 Feb. 1995, *Taylor* (K, photos); Mun. Granja, c. 1 km S, 20 Feb. 1995, *Taylor* (K, photo); Mun. Viçosa de Ceará, E of Cocal (PI), 19 Feb. 1995, *Taylor* (K, photos); Mun. Coreaú, 20 Feb. 1995, *Taylor* (K, photos); Mun. Maranguape, 'Kagado', 31 Oct. 1935, *F. Drouet* 2652 (F, US); E-cent. Ceará, Banabuiú ['Banabuiá'], see Eiten (1983: fig. 62).

CONSERVATION STATUS. Least Concern [LC] (1); PD=2, EI=1, GD=2. Short-list score (1×5) = 5. Least Concern at present, but needs to be monitored in view of increasing habitat destruction in the areas of higher rainfall that it inhabits.

This species ranges through the northern part of the *caatinga,* reaching that vegetation's north-western limits, and occupying the ecotones with the *cerrado* and forests transitional to those of Amazônia. It is the only *Pilosocereus* that enters the coastal mangrove-*caatinga* ecotone (NW Ceará) and is also frequently epiphytic on the trunks of carnaúba palms in seasonally flooded palm forest at its northern limit (*caatinga* type No. 12 of Andrade–Lima 1981: 160). Its stems vary considerably in thickness, those of plants from the drier vegetation of central-southern Piauí and adjacent parts of Maranhão being much more slender than those from further north.

8. Pilosocereus pentaedrophorus *(J. F. Cels) Byles & G. D. Rowley* in Cact. Succ. J. Gr. Brit. 19: 67 (1957). Type: not extant. Neotype (Zappi 1994): Brazil, Bahia, Mun. Serrinha, 16 km N of the town on road BA 409 towards Conceição do Coité, 14 July 1989, *Zappi* 120 (SPF; HRCB, isoneo.).

Cereus pentaedrophorus J. F. Cels, Cat. Cact. Aloées: 9 (1858); *Cereastreae* (subg. *Cereus*) *pentaedrophorus*
Labour., Monogr. Cact.: 365 (1853), nom. inval. (Art. 43.1). *Pilocereus pentaedrophorus* (J. F. Cels)
Console ex K. Schum., Gesamtbeschr. Kakt.: 174 (1897). *Cephalocereus pentaedrophorus* (J. F. Cels)
Britton & Rose, Cact. 2: 31 (1920). *Pseudopilocereus pentaedrophorus* (J. F. Cels) Buxb. in Beitr. Biol.
Pflanzen 44: 253 (1968). *Pilocereus polyedrophorus* Lem. in Rev. Hort. 1862: 428 (1862), nom. illeg.
(Art. 52.1). *Pseudopilocereus polyedrophorus* (Lem.) P. V. Heath in Calyx 2: 54 (1992), nom. illeg.

VERNACULAR NAMES. Facheiro, Facheiro-fino, Mandacaru-de-veado.

Shrub or treelike to 6(–15) m; vascular cylinder woody; branches blue-green, glaucous; ribs 4–8(–10),
sinuses sinuate and pronounced, with horizontal transverse folds above the areoles. Areoles 3–7(–8) mm,
12–22 mm apart, felt white to greyish black, apical young areoles frequently unarmed, growth
indeterminate, strong near base of plant. Spines 0.5–1.0 mm diam. at base, translucent, yellowish brown.
Fertile part of stem not differentiated, subapical, occupying several consecutive areoles of more than one
rib. Flower-buds curved and obtuse, constricted in the region of the nectar-chamber, varying markedly
in colour in different populations between green, reddish or purplish blue; flower 3.5–5.5 × 2.8 cm; tube
35–45 × 11 mm at base to 20–21 mm diam. at apex, constricted below and above the nectar-chamber
region, strongly curved, smooth, terete or nearly so, apex with few ovoid to spathulate bract-scales; style
35 × 1.5–2.0 mm, tapering, stigma-lobes c. 10; ovary locule hemicircular to depressed in longitudinal
section. Fruit 1.5–2.0 × 2–3 cm, depressed-globose, dehiscent by a lateral slit, with pentagonal to
hexagonal outline, floral remnant pendent, deeply inserted; pericarp pale green to wine-coloured;
funicular pulp purplish or magenta. Seeds (1.6)–1.7–1.9(–2.0) × (1–)1.1–1.3(–1.4) mm, testa-cells flat,
with intercellular depressions nearly absent and fine, sparse cuticular folds.

CONSERVATION STATUS. Least Concern [LC]; see subspecies below.

Two subspecies are recognized within this taxon, the typical one inhabiting forest
vegetation east of the Chapada Diamantina (Bahia) and northwards, reaching
Pernambuco, and subsp. *robustus*, distributed towards the southern limit of the species, in
southern Bahia and north-eastern Minas Gerais. These are distinguished as follows:

1. Branches slender, long and leaning, to 4.5(–6.0) cm diam. below apex, ribs 4–6(–7), obtuse
   (Bahia to Pernambuco) . . . . **8a.** subsp. **pentaedrophorus**
1. Branches stout, never leaning, to 7.5 cm diam. below apex, ribs (5–)6–10, acute (S Bahia
   & NE Minas Gerais) . . . . **8b.** subsp. **robustus**

8a. subsp. **pentaedrophorus**

Treelike, sparsely branched, to 6 m or more in high forest; branches sometimes leaning on other
vegetation for support, 3–4.5(–6.0) cm diam.; ribs 4–6(–7), 9–18 × 8–20 mm, obtuse. Central spines 1–3,
10–25 mm, porrect to deflexed, radials 3–12, 4–12 mm, adpressed, all deflexed.

Eastern *caatinga* element: in *agreste* and dense *caatinga*, on rocky substrates, rarely reaching into *restinga*
sand-dunes (N of Salvador, BA), c. 5–1000 m, north-eastern Pernambuco, western Sergipe, and north-
eastern and eastern Bahia. Endemic to the core area within North-eastern Brazil. Map 36.
PERNAMBUCO: Mun. Brejo da Madre de Deus, 5 Feb. 1965, *Andrade-Lima* 65-4293 (IPA), l.c., Faz. Bituri,
26 May 1995, *D.C. Silva* 72 (UFP); Mun. Caruaru, 9 km NE of town, 8°14'21"S, 35°54'52"W, Dec.
1991, *E.A. Rocha et al.* (K, photos); Mun. Garanhuns, 'em capoeiras', *fide* Andrade-Lima (1966: 1457).
SERGIPE: Mun. Nossa Senhora da Glória, Faz. Olhos D'agua, 13 May 1982, *M.N. Almeida* 80 (ASE
2384); Mun. Frei Paulo, 26 June 1981, *M. Fonseca* 518 (ASE 944); l.c., Faz. Serras Pretas, 16 Apr. 1985,
G. Viana 1139 (ASE 3524); Mun. Lagarto, 0.5 km S of Rio Vaza Barris, coming from São Domingos,
14 Feb. 1991, *Taylor & Zappi* 1633 (HRCB, K, ZSS).
BAHIA: Mun. Jeremoabo, *Uebelmann* (mss, HU 253); Mun. Jaguarari, *Ritter* 1221 (SGO 125489, *fide* Eggli

et al. 1995: 497); Mun. Senhor do Bonfim, 8–9 June 1915, *Rose & Russell* 19921 (US, NY); Mun. Itiúba, 16 km E of Filadélfia, 25 Apr. 1992, *Taylor & Zappi* (obs.); Mun. Ponto Novo, 41 km N of Capim Grosso, 25 Apr. 1992, *Taylor & Zappi* (obs.); Mun. Jacobina, 21 km NW of town on road BR 324 towards Lages, 16 July 1989, *Zappi* 131A (SPF, HRCB); l.c., 2 km S of town towards Miguel Calmon, 26 Apr. 1992, *Taylor & Zappi* (obs.); Mun. Capim Grosso, 16 km N of town, 25 Apr. 1992, *Taylor & Zappi* (obs.); Mun. São José do Jacuípe, 10 km SE of Capim Grosso, 16 km NW of bridge over the Rio Jacuípe, 25 Apr. 1992, *Taylor & Zappi* (obs.); Mun. Santaluz, c. 18 km N of Valente, 29 Dec. 1992, *L.P. de Queiroz et al.* 3015 (K, CEPEC, HUEFS); Mun. Araci, 18 Jan. 1994, *M.A.S. das Neves* 89 (HUEFS); Mun. Itapicuru, 11°8'S, 38°13'W, 20 Feb. 1978, *R.P. Orlandi* 166 (RB); Mun. Morro do Chapéu, W of Ventura, 24 Dec. 1988, *Taylor & Zappi* s.n. (K, photos); Mun. Serrinha, 16 km W of town on road BA 409 towards Conceição do Coité, 14 July 1989, *Zappi* 120 (SPF, HRCB); Mun. Piritiba, 2 km S of França, 26 Apr. 1992, *Taylor & Zappi* (obs.); l.c., 13 km S, 26 Apr. 1992, *Taylor & Zappi* (obs.); Mun. Ichu (Ixu), 2 km S of town, 8 July 1994, *F.A.R. Santos* 91 (HUEFS); Mun. Biritinga, 18 Jan. 1994, *M.A.S. das Neves* 76, 79, 80 (HUEFS); Mun. Mundo Novo, 7 & 15 km E of town on road BA 052, 26 Apr. 1992, *Taylor & Zappi* (obs.); Mun. Lamarão, on road between Feira de Santana and Serrinha, 4 Jan. 1991, *Zappi* in *Taylor et al.* 1348B (K, ZSS, photos); Mun. Santa Bárbara, on road between Feira de Santana and Serrinha, 4 Jan. 1991, *Taylor et al.* 1348A (HRCB); Mun. Baixa Grande, 13 km SE of town, road BA 052, 26 Apr. 1992, *Taylor & Zappi* (obs.); Mun. Ipirá, 26 Apr. 1992, *Taylor & Zappi* (obs.); Mun. Feira de Santana, 12°15'S, 38°58'W, 5 Nov. 1983, *L.R. Noblick* 2751 (HUEFS); Mun. Castro Alves, 26 Jan. 1956, *Andrade-Lima* 56-2508 (IPA, PEUFR), 13 Apr. 1995, *L.P. de Queiroz* 4325 (HUEFS); Mun. Muritiba, Faz. 8 de Dezembro, nr Itaporã, 6 Aug. 2002, *Taylor* (obs.); Mun. Lauro de Freitas, bairro de Stella Maris, dunas do aeroporto, Jan. 2001, *E.A. Rocha* (K, photo); Mun. Boninal/Piatã, 12 km S of Boninal on road towards Piatã, 12°48'S, 41°48'W, 22 Dec. 1988, *Taylor & Zappi* in *Harley* 25597 (K, SPF, CEPEC); Mun. Itatim, Morro das Tocas, 12°42'S, 39°46'W, 24 Feb. 1996, *E. de Melo* 1444 (HUEFS); Mun. Mucugê, 0.5 km from town, near the cemetery, 13°S, 41°28'W, 22 Dec. 1988, *Taylor & Zappi* in *Harley* 27380 (K, SPF, CEPEC); Mun. Milagres, 26 Oct. 1978, *A. Araújo* 121 (RB); Mun. Marcionílio Souza, Machado Portela, 19–23 June 1915, *Rose & Russell* 19921 (US, NY); Mun. Jaguaquara [Toca da Onça], 27–29 June 1915, *Rose & Russell* 20081 p.p. (US, NY).

Besides the examined specimens cited above, this taxon has been observed in the following municípios in Bahia: Tapiramutá, Andaraí, Olindina, Rui Barbosa, Itaberaba, Jequié, Vitória da Conquista.

CONSERVATION STATUS. Least Concern [LC] (1); PD=1, EI=1, GD=1. Short-list score (1×3) = 3. Least Concern at present, but its extensive habitat continues to decline. It is rare in the north-eastern part of its range.

8b. subsp. **robustus** *Zappi* in Succ. Pl. Res. 3: 74 (1994). Holotype: Brazil, Bahia, Mun. Livramento do Brumado, 11 km S of town on road to Brumado, 450 m, 13°45'S, 41°49'W, 23 Nov. 1988, *Taylor & Zappi* in *Harley* 25544 (SPF; CEPEC, K, isos.).

Shrubby or treelike, densely branched, rarely more than 4 m tall; branches erect, to 7.5 cm diam.; ribs (5–)6–10, 15–21 × 8–19 mm, acute. Central spines 0–2(–3), 20–26 mm, 1–2 ascending and sometimes 1 porrect; radials 3–8, 4–20 mm, well-developed, adpressed.

Eastern *caatinga* element: in dense *caatinga-agreste* of the Rio de Contas (Rio Gavião) and Rio Pardo drainage systems, c. 400–900 m, southern and south-eastern Bahia and north-eastern Minas Gerais. Endemic to the core area of Eastern Brazil. Map 36.

BAHIA: Mun. Ituaçu, 2 km NE of town, 18 Aug. 1988, *Eggli* 1197 (ZSS); Mun. Rio do Antônio, Km 43 on Brumado–Caitité road, 14 Apr. 1983, *Leuenberger et al.* 3068 (CEPEC, K), l.c., 2 Feb. 1991, *Taylor et al.* 1523 (K, ZSS, HRCB, CEPEC); Mun. Maracás, Km 16 of Maracás–Tamburi road, 20 Apr. 1983, *Leuenberger et al.* 3085 (CEPEC, K); Mun. Caitité, 8 km from Caitité towards Brumado, track to São João, 27 Aug. 1988, *Eggli* 1319 (ZSS); Mun. Livramento do Brumado, 11 km S on road to Brumado, 13°45'S, 41°49'W, 23 Nov. 1988, *Taylor & Zappi* in *Harley* 25544 (K, SPF, CEPEC); Brumado–Caitité, *Horst & Uebelmann* HU 121 (ZSS); Mun. Vitória da Conquista, 7 km NE of town on road BR 116, then

3.5 km W, 16 April 2003, *Taylor & Zappi* (K, photos); l.c., c. 20 km S of town, 15°1'21"S, 40°49'51"W, 18 April 2003, *Taylor et al.* (obs.).
MINAS GERAIS: Mun. Taiobeiras, 35 km E of town towards road BR 116, E of bridge over Rio Atoleiro, 15°53'S, 41°57'W, 17 Oct. 1988, *Taylor & Zappi* in *Harley* 25149 (K, SPF); Mun. Águas Vermelhas, 1 km from Curral de Dentro towards road BR 116, 31 Jan. 1991, *Taylor et al.* 1515D (K, HRCB).

Besides the examined specimens cited above, this taxon has been observed in the following municípios of Bahia: Iramaia, Itaetê, Barra da Estiva, Contendas do Sincorá.

CONSERVATION STATUS. Near Threatened [NT] (1); PD=1, EI=1, GD=2. Short-list score (1×4) = 4. At present with an extent of occurrence of > 20000 km², but its habitat is discontinuous and continues to decline and regular monitoring is desirable.

Variable in rib number and stem thickness.

Intergeneric hybrids between *P. pentaedrophorus* and *Micranthocereus purpureus*, cited by Ritter (1979), have been observed in the region of Andaraí and Ituaçu, Bahia. At the southern limit of distribution of *P. pentaedrophorus*, in the drainage of the Rio Pardo (*Taylor et al.* 1515D), *P. p.* subsp. *robustus* can be found sympatric with *P. floccosus* subsp. *quadricostatus*.

9. Pilosocereus glaucochrous *(Werderm.) Byles & G. D. Rowley* in Cact. Succ. J. Gr. Brit. 19: 67 (1957). Type: Brazil, Bahia, near Morro do Chapéu, Serra do Espinhaço, c. 1000 m, Apr. 1932, *Werdermann* (B†). Lectotype (Zappi 1994): Werdermann, l.c., 102, photograph (iconotype).

Pilocereus glaucochrous Werderm., Bras. Säulenkakt.: 106–107 (1933). *Cephalocereus glaucochrous* (Werderm.) Borg, Cacti: 108 (1937). *Pseudopilocereus glaucochrous* (Werderm.) Buxb. in Beitr. Biol. Pflanzen 44: 252 (1968).

Treelike, to 3–5 m, sparsely branched, sometimes with a well-defined trunk; vascular cylinder woody; branches 3–6.5(–7.0) cm diam., inclined, epidermis bluish or greyish green, glaucous; ribs (3–4[in seedlings/juveniles]–)5–11, 7–12 × 7–18 mm, sinuses straight, transverse folds above the areoles straight. Areoles 4–6 mm, 6–10 mm apart, situated upon rounded, low projections (podaria), with blackish felt and white long hairs to 20 mm, mainly on the apical region of branches. Spines 0.5–1.0 mm diam. at base, translucent when young, golden yellow to greyish, centrals 2–4(–5), ascending to porrect, 20–35(–40) mm, radials 7–14, adpressed, mainly horizontal, 8–15 mm. Fertile part of stem not or slightly differentiated, apical to subapical, on many consecutive areoles of 2–5 ribs. Flower-buds obtuse 2 days before anthesis; flowers 4–5.2 × 2.5 cm; pericarpel subglobose, tube 34–38 × 10–12 mm at base, flared to 15–19 mm diam. at apex, curved, cylindric, constricted above and below the nectar-chamber, pink to orange-red, smooth, glaucous, distal third with ovate, wine-coloured bract-scales; style 24–29 × 1.2–1.7 mm, tapering, stigma-lobes 7–8, included in relation to the stamens; ovary locule hemicircular to depressed in longitudinal section. Fruit depressed-globose, dehiscent by a lateral slit, floral remnant pendent, deeply inserted; pericarp olive-green to purplish, smooth; funicular pulp purplish. Seeds 1.6–1.9 × 1.1–1.2 mm, testa-cells flat with inconspicuous intercellular depressions and fine, sparse cuticular folds.

Caatinga / Northern *campos rupestres* (Chapada Diamantina) element: in *'caatinga/cerrado de altitude'*, c. 740–1150 m, on various substrates including sand and calcareous materials, Chapada Diamantina, central Bahia. Endemic. Map 36.
BAHIA: Mun. Sento Sé, Serra São Francicsco (S of Minas do Mimoso *fide* G. Charles), Herm *et al.* (2001: 118, fig. 3); Mun. Jacobina, 28 km W of town on road BR 324 towards Lages (Lajes), 12 Jan. 1991, *Taylor et al.* 1398A (K, HRCB, ZSS); Mun. Morro do Chapéu, 18 June 1994, *L.P. de Queiroz* 4014 (HUEFS); l.c., 15 km NW of town towards Várzea Nova, 12 Jan. 1991, *Taylor et al.* 1402 (K, HRCB, ZSS, CEPEC); l.c., 5 km E of town, 3 Mar. 1973, *I. & G. Gottsberger* 16-3273 (K); l.c., road to Morrão, 11°35'3"S, 41°11'31"W, 5 Aug. 2001, *F.R. Nonato et al.* 974 (HUEFS 55719); Mun. América Dourada,

Nova América, *Horst & Uebelmann* 218, cult. ZSS (ZSS); Mun. Bonito, 46 km S of Morro do Chapéu on road from Utinga, 23 Dec. 1988, *Taylor & Zappi in Harley* 27384 (K, SPF, CEPEC); l.c., 10 km N of Bonito, 23 Aug. 1988, *Eggli* 1268 (ZSS); Mun. Iraquara, 8 km S of Souto Soares towards Seabra, 13 Jan. 1991, *Taylor et al.* 1412 (K, HRCB, ZSS, CEPEC); Mun. Seabra, 16 km W of Seabra on road BR 242 to Ibotirama, 25 Aug. 1988, *Eggli* 1290 (ZSS); l.c., 27 km W, July 1989, *Taylor & Zappi* (K, photos).

CONSERVATION STATUS. Near Threatened [NT] (1); PD=2, EI=1, GD=2. Short-list score (1×5) = 5. Near Threatened, since its habitat continues to decline and the number of populations is small (c. 10), although the plant is abundant over a wide area around the type locality.

The brightly coloured tube of the flower of this species is unusual in the genus, otherwise occurring only in *P. machrisii* and *P. aurisetus*. The nocturnal flowers of *P. glaucochrous* remain open for part of the following morning and, therefore, may be adapted for pollination by both bats and hummingbirds. Variable in rib number between populations.

10. Pilosocereus floccosus *Byles & G. D. Rowley* in Cact. Succ. J. Gr. Brit. 19: 67 (1957). Holotype: Brazil, Minas Gerais, Diamantina, June 1934, *A. C. Brade* s.n. (RB).

Pilocereus floccosus Backeb. & Voll in Arch. Jard. Bot. Rio de Janeiro 9: 150 (1949), nom. illeg. (Art. 53) non Lemaire in Ill. Hort. 13: tab. 470 (1866). *Pseudopilocereus floccosus* (Byles & G. D. Rowley) Buxb. in Beitr. Biol. Pflanzen 44: 252 (1968).
[*Cereus macrogonus sensu* Warming (1892), cf. Warming (1908: 151, footnote, 254), non Salm-Dyck (1850).]

Shrub, not branched above base, or treelike, 1–4(–5) m, with or without a well-defined trunk, sparsely or much branched; vascular cylinder weakly woody; branches erect, epidermis verrucose, greyish green; ribs with straight sinuses and without visible transverse folds. Areoles 3–10 mm, (2–)4–10 mm apart, with white to brownish grey felt and white or brownish long hairs to 10 mm, growth indeterminate near base of plant. Spines 0.3–1.1 mm diam. at base, opaque, yellowish brown or reddish, becoming greyish. Fertile part of stem strongly differentiated, flower-bearing areoles with pale brown or grey, woolly, long hairs to 15–20 mm and dark bristly spines to 35 mm, forming a ring around the subapical region of the stem. Flower-buds obtuse 2–3 days before anthesis; flowers 4–5 × 3 cm; pericarpel subglobose, tube to 40 × 12 mm at base, flared to 25 mm diam. at apex, curved, constricted above and below nectar-chamber, green, glaucous, smooth, distal third with broad ovate bract-scales; style 30 × 1.5–1.7 mm, tapering, stigma-lobes 9–10, exserted or included in relation to the stamens; ovary locule obtriangular in longitudinal section. Fruit dehiscent by a lateral, abaxial or adaxial slit; pericarp wine-coloured, smooth to somewhat rugose; funicular pulp bright red. [Seeds: see subspecies.]

CONSERVATION STATUS. Near Threatened [NT]; see subspecies below.

When first described by Backeberg & Voll, l.c., this species was placed in the genus *Pilocereus*, as *P. floccosus* Backeb. However, this name is illegitimate because of the prior-published *Pilocereus floccosus* Lem., which relates to a Caribbean taxon (nowadays a synonym of *Pilosocereus royenii* (L.) Byles & G. D. Rowley). The legitimate name is, therefore, to be attributed to Byles & Rowley (1957), who published *Pilosocereus floccosus* Byles & G. D. Rowley based on the type of Backeberg's illegitimate name.

Easy to distinguish from all the other species of the genus by its rough, verrucose epidermis, *P. floccosus* is also notable for its long-hairy flower-bearing areoles, forming a crown or zone at the apical or subapical region of the branches. A potentially closely related taxon occurs in north-western Minas Gerais — *P. albisummus* P. J. Braun & E. Esteves Pereira in Kakt. and. Sukk. 38: 126–131 (1987). It has not been possible to study this plant at close quarters in the field, but living and herbarium material at Zürich (ZSS), photographs of the plant, flower-buds and seeds (SEM), and its known ecological preference for *Bambuí* limestone outcrops are strongly indicative of an affinity to *P. floccosus* subsp. *floccosus*.

The typical subspecies occurs only on limestone rock outcrops of the *Bambuí* formation, in central Minas Gerais, where it does not normally become very tall, unless growing in very dense forest. North-east of the distribution of subsp. *floccosus*, *P. f.* subsp. *quadricostatus* lives in the *caatinga* associated with gneissic outcrops of the semiarid region of the Rio Jequitinhonha and Rio Pardo drainage basins (Taylor & Zappi 1992a). This species is subdivided as follows:

1. Branches 5–9 cm diam.; ribs 5–8; seeds shiny, testa-cells without cuticular folds (SEM)
.... **10a.** subsp. **floccosus**
1. Branches 8–11 cm diam.; ribs 4–5; seeds dull, testa-cells with dense cuticular folds
(SEM) .... **10b.** subsp. **quadricostatus**

10a. subsp. **floccosus**

Shrub, branching near base or above ground, rarely treelike, with or without a well-defined trunk; branches 5–8 cm diam., ribs 5–8, 12–30 × 20–30 mm. Central spines 3–6, 8–25 mm, ascending to porrect, radials 5–16, 2–25 mm, ascending or adpressed. Fruit 2.5–3.0 × 3–3.5 cm, depressed-globose. Seeds (1.8–)1.9–2.2 × (1.2–)1.3–1.4(–1.5) mm, hilum-micropylar region forming an angle of 50° with long-axis of seed, testa-cells flat with inconspicuous intercellular depressions and lacking cuticular folds.

Rio São Francisco (Rio das Velhas) *caatinga/mata seca* element: mostly on *Bambuí* limestone outcrops west of the Serra do Espinhaço, c. 600–800 m, Minas Gerais. Endemic to the core area within Minas Gerais. Map 36.
MINAS GERAIS: Mun. Buenópolis, 7 km S of town, on road BR 135 towards Curvelo, 17°56'S, 44°9'S, 11 Oct. 1988, *Taylor & Zappi in Harley* 24838 (K, SPF); Diamantina, June 1934, *Brade* s.n. (RB); Mun. Monjolos, *Horst & Uebelmann* 583 (ZSS); l.c., near Rodeador, 1983, *Braun* 436 (ZSS T-06119); l.c. (?), W of Conselheiro Mata, *Braun* 436 *bis* (ZSS TP-58-180); Mun. Santana do Riacho, Serra do Cipó, limestone outcrop near Cardeal Mota, 30 Mar. 1988, *Zappi in CFSC* 10923 (SPF); l.c., 27 Oct. 1988, *Taylor & Zappi in Harley* 25425 (K, SPF); Mun. Matozinhos, Faz. Cauaia, 31 Oct. 1996, *Lombardi* 1462 (BHCB, K); Mun. Lagoa Santa, Lapa Vermelha, reported by Warming (1908), cf. Warming & Ferri (1973).

CONSERVATION STATUS. Near Threatened [NT] (1); PD=1, EI=1, GD=1. Short-list score (1×3) = 3. A small number of populations is currently known (c. 8) and its habitat on limestone outcrops will in future be at risk from quarrying operations. Forest surrounding outcrops may also be cleared affecting local climate.

10b. subsp. **quadricostatus** *(F. Ritter)* Zappi in Succ. Pl. Res. 3: 86 (1994). Holotype: Brazil, Minas Gerais, Padre Paraíso (formerly Água Vermelha), 1965, *Ritter* 1342 (U).

Pilosocereus quadricostatus F. Ritter, Kakt. Südamer. 1: 78, Abb. 51 (1979). *Pseudopilocereus quadricostatus* (F. Ritter) P. V. Heath in Calyx 4: 140 (1994).

Shrubby or treelike, branched above ground, with a well-defined trunk; branches 8–11 cm diam., ribs 4–5(–6), 3–7 in young plants/seedlings (*fide* Ritter), 26–30 × 15–40 mm. Areoles either with very short spines or central-spines 1–3, 2–50 mm, ascending to porrect, and radials 6–8, c. 2–10 mm, ascending to adpressed. Fruit to 6 × 5 cm, depressed but often compressed due to the presence of many fruits close together on the same rib. Seeds 1.7–1.8 × 1.1–1.3 mm, hilum-micropylar region forming an angle of 40° with the long-axis of seed, testa-cells domed to flat with inconspicuous intercellular depressions and dense, coarse cuticular folds.

South-eastern *caatinga* element: in *caatinga* and *mata seca decídua* on associated gneissic inselbergs within the drainage of the Rio Jequitinhonha and Rio Pardo, c. 250–800 m, north-eastern Minas Gerais. Endemic to the core area within Minas Gerais. Map 36.

MINAS GERAIS: Mun. Águas Vermelhas, 1 km E of Curral de Dentro, 31 Jan. 1991, *Taylor & Zappi* s.n. (K, photos); Mun. Grão Mogol, 14 km NE of town on road to Salinas, 16 Oct. 1988, *Taylor & Zappi in Harley* 25146 (K, SPF) — aberrant form perhaps intermediate with subsp. *floccosus*; Mun. Coronel Murta, 5 km SE of town on road to Araçuaí, 21 Feb. 1988, *Supthut* 8861 (ZSS); Mun. Pedra Azul, 6 km S of town on road to Jequitinhonha, 16°3'S, 41°14'W, 20 Oct. 1988, *Taylor & Zappi in Harley* 25221 (K, SPF); Mun. Itaobim, road BR 367, 12–21 km W of road BR 116, 14 Dec. 1990, *Taylor & Zappi* 765 (K, HRCB, ZSS, BHCB); Mun. Itinga, *caatinga* of Faz. Alta Viagem, 17 Oct. 1983, *Rizzini & Mattos-F.* s.n. (RB 232050); Mun. Padre Paraíso [Água Vermelha], *Ritter* 1342 (U); l.c., 'Km 788' of road BR 116, *Horst & Uebelmann* 172, cult. (ZSS).

CONSERVATION STATUS. Vulnerable [VU B2ab(iii)] (2); remaining area of occupancy < 2000 km²; PD=1, EI=1, GD=1. Short-list score (2×3) = 6. Vulnerable from the ongoing destruction of its *caatinga* habitat and low number of known populations.

10b × 13: *Pilosocereus floccosus* subsp. *quadricostatus* × *Pilosocereus magnificus*

P. subsimilis Rizzini & A. Mattos in Revista Brasil. Biol. 46: 327 (1986). Holotype: Brazil, Minas Gerais, Itinga, 18 Dec. 1984, *Rizzini & Mattos* 41 (RB).

Treelike or shrubby, 2–5 m, usually with a ± well-defined trunk, sparsely branched above ground; vascular cylinder weakly woody; branches 7–10 cm diam., erect, epidermis dark green to olive-green, more or less glaucous, slightly rough; ribs 4–7, 17–30 × 20–35 mm, sinuses straight, transverse folds not visible. Areoles 4–6 mm diam., 2–5 mm apart, felt greyish, with white to brownish, long hairs 6–10 mm. Spines 0.3–0.5 mm diam. at base, translucent when young, yellowish, pale brown or reddish, centrals 4–5, 7–15 mm, ascending to porrect, radials 10–14, 6–10 mm, ascending to porrect or deflexed. Fertile part of stem slightly to strongly differentiated, flower-bearing areoles situated on all ribs of the apex of branches forming a ring, developing black, bristly spines to 30 mm and white long hairs to 10–20 mm; flowers 5 × 2.5 cm; tube 27 × 12 mm at base, to 20 mm diam. at apex, curved, constricted above and below the nectar-chamber, olive-green, smooth, distal third with broad bract-scales. Fruit depressed-globose; funicular pulp magenta. Seeds (1.5–)1.6(–1.7) × 1–1.1(–1.2) mm, hilum-micropylar region forming an angle of 40° with the long-axis of seed, testa-cells domed to flat, with intercellular depressions nearly inconspicuous, and coarse, sparse cuticular folds.

On gneissic inselbergs and in associated *caatinga-agreste*, c. 250–600 m, north-eastern Minas Gerais.

MINAS GERAIS: Mun. Padre Paraíso (formerly Água Vermelha), on road BR 116, 12 km N of town, 13 Dec. 1990, *Taylor & Zappi* 756 (K, HRCB, ZSS, BHCB); Mun. Itinga, 18 Oct. 1984, *Rizzini & Mattos-F.* 41 (RB); l.c., 5 km E of town on road BR 367, 16°36'S, 41°43'W, 19 Nov. 1988, *Taylor & Zappi in Harley* 25535 (K, SPF).

P. floccosus subsp. *quadricostatus* is also suspected of hybridizing with *P. multicostatus*, q.v.

PILOSOCEREUS ULEI Group (Nos 11–13).

Found in *caatinga-agreste, caatinga* and *campo rupestre*.

11. **Pilosocereus fulvilanatus** *(Buining & Brederoo) F. Ritter*, Kakt. Südamer. 1: 84 (1979). Holotype: Brazil, Minas Gerais, N Serra do Espinhaço, [Grão Mogol], 800–1000 m, 1968, *Horst* 277 (U).

Pseudopilocereus fulvilanatus Buining & Brederoo in Kakt. and. Sukk. 24: 145–147 (1973).

VERNACULAR NAME. Quiabo-da-lapa.

Shrubby to tree-like, 2–3(–4) m, branching above the ground; vascular cylinder weakly woody; branches 5–12 cm diam., erect, epidermis greyish green, coated with intense sky blue wax near apex, grey-green below; ribs 4–8, sinuses straight, transverse folds absent. Areoles almost contiguous, felt dark brown or black with dark long hairs, paler to glabrescent with age. Spines opaque, yellowish brown to blackish, thickened at base. Fertile part of stem strongly differentiated, flower-bearing areoles subapical, on one or several ribs, with abundant reddish brown and golden brown long hairs to 20–40 mm. Flowers 3.6–6.0 × 3–4 cm; pericarpel subglobose, greenish; tube straight, infundibuliform, distal third with thick acute bract-scales; nectar-chamber swollen; ovary locule hemicircular to triangular in longitudinal section. Fruit depressed-globose, frequently compressed, dehiscent by lateral slit, floral remnant pendent, deeply inserted; pericarp greenish, pink or dark purplish; funicular pulp magenta. Seeds with flat testa-cells, intercellular depressions nearly absent, cuticular folds absent.

CONSERVATION STATUS. Vulnerable [VU B1ab(iii)]; extent of occurrence estimated to be < 20000 km² and presently known from c. 6 locations, some of which are projected to decline in the near future.

This distinctive species can be easily differentiated from the rest of the genus by the unusual combination of intensely blue epidermis and dark reddish brown to golden brown areolar hairs, which are more abundant when the plants are flowering. The areoles with short dark spines are so closely arranged that it is sometimes difficult to isolate one from the next. It is the sister species of *P. ulei,* from the region of Cabo Frio, Rio de Janeiro.

This taxon is divided into two subspecies, found on crystalline rock outcrops associated with the *campos rupestres*, on both sides of the Serra do Espinhaço, Minas Gerais, where the species is endemic. The subspecies are differentiated as follows:

1. Branches 8–12 cm diam.; ribs 4–7; fruit dark pink to dark purple . . . . . **11a.** subsp. **fulvilanatus**
1. Branches to 5.5 cm diam.; ribs (5–)6–8; fruit green to brownish red . . . . **11b.** subsp. **rosae**

11a. subsp. fulvilanatus

Treelike or shrubby, 2–3(–4) m, branching above the ground; branches 8–12 cm diam.; ribs 4–7, 30–40 × 30–40(–45) mm. Areoles 3–7 × 2–5 mm, 1–3 mm apart, very congested. Spines 0.6–0.9 mm diam. at base, centrals 1–7, 15–30(–45) mm, ascending to porrect, radials 8–10, 8–15 mm, adpressed. Flower-bearing areoles extremely hairy. Flowers 3.6–5.2 × 3 cm; tube 21 × 11 mm at base, flared to 20 mm diam. at apex, greenish; style 20 mm, stigma-lobes c. 12. Fruit 3 × 3.5–4.0 cm; pericarp dark pink to dark purple, rugose. Seeds 1.5–1.6 × 1–1.1 mm.

South-eastern *campo rupestre* (Grão Mogol region) element: locally co-dominant with other woody vegetation on quartzitic rock outcrops, *campo rupestre*, Serra do Espinhaço, in the drainage of the Rio Jequitinhonha, c. 720–1000 m, northern Minas Gerais. Endemic to the core area within Minas Gerais. Map 37.
MINAS GERAIS: Mun. Grão Mogol, Serra da Bocaina, 28 km E of Caveira and 48.5 km W of Rio Vacaria on road BR 251, 31 Jan. 1991, *Taylor et al.* 1513 (K, HRCB, ZSS, BHCB); Serra de Grão Mogol, 1968, *Horst* 277 (U); l.c., 18 km W of town, 21 Feb. 1969, *Irwin et al.* 23676 & 23684 p.p. (NY); l.c., *campo rupestre* NE of the town, 22 May 1982, *M.C.H. Mamede et al.* in CFCR 3475 (SPF); Mun. Cristália, slopes of the Serra das Cabras, 28 May 1988, *Zappi et al.* in CFCR 12046 (SPF); l.c., margins of the Rio Itacambiruçu, 16°37'S, 42°56'W, 15 Oct. 1988, *Taylor & Zappi* in *Harley* 25071 (K, SPF); Mun. Botumirim, Ribeirão Gigante, 1991, *M.G. Carvalho* s.n. (K, photos).

CONSERVATION STATUS. Vulnerable [VU B1ab(iii)] (2); extent of occurrence = 724 km²; PD=1, EI=2, GD=1. Short-list score (2×4) = 8. Future inundation by a dam-lake and destruction of its habitat for charcoal production make this taxon Vulnerable. It is more secure in the higher parts of its range.

Known from only three populations between Grão Mogol and Botumirim (MG), subsp. *fulvilanatus* presents a rather restricted distribution, but it has to be said that this region is still rather under-explored.

11b. subsp. **rosae** *(P. J. Braun) Zappi* in Succ. Pl. Res. 3: 100 (1994). Holotype: Brazil, Minas Gerais, Mun. Augusto de Lima, nr Santa Bárbara, 6 km from road BR 135, west slopes of the Serra do Espinhaço, 800 m, 1982, *Horst & Uebelmann* 546 (ZSS; K, iso.).

Pilosocereus rosae P. J. Braun in Kakt. and. Sukk. 35(8): 178–181 (1984).

Shrubby, to 3 m, sparsely branched above ground, without a well-defined trunk; branches 5.5 cm diam.; ribs (5–)6–8, 15–20 × 15–25 mm. Areoles 3–4 × 3–4 mm, 2–4 mm apart. Spines 0.4–0.6 mm diam. at base; centrals 3–4, 10–20 mm; radials 8–10, to 12 mm, adpressed. Fertile part of stem differentiated, with long hairs to 3–4 cm, and black bristly spines to 2–3 cm. Flowers 3.8–6.0 × 3–4 cm; tube 30–45 × 11 mm at base, flared to 25 mm diam. at apex, straight, greenish brown to pinkish; style 25–30 mm, stigma-lobes c. 10. Fruit to 6.5 cm diam.; pericarp green to brownish red, striate. Seeds 1.6–1.8 × 1–1.2 mm. [Description based on dried material and protologue.]

South-eastern *campo rupestre* (Rio São Francisco drainage) element: on quartzitic rock outcrops, *campo rupestre*, c. 800 m, Serra do Espinhaço, in the drainage of the Rio das Velhas, central-northern Minas Gerais. Endemic to the core area within Minas Gerais. Map 37.
MINAS GERAIS: Mun. Augusto de Lima, near Santa Bárbara, 6 km from road BR 135, west slopes of the Serra do Espinhaço, 1982, *Horst & Uebelmann* 546 (ZSS, K); l.c., 6 Aug. 1988, *Eggli* 1104 (ZSS).

CONSERVATION STATUS. Critically Endangered [CR B1ab(iii) + 2ab(iii)] (4); area of occupancy < 10 km²; PD=1, EI=1, GD=1. Short-list score (4×3) = 12. The single site known is now a private park of difficult access, but needs to be visited to determine current threats.

Although it has not been possible to observe the population of subsp. *rosae* in the field, its striking morphological and ecological similarities with *P. fulvilanatus, sens. str.*, have led to its present position (Zappi 1994). It has a less robust habit, with branches less than 6 cm in diameter, and a tendency to develop a higher number of ribs (6–8) than in mature stems of subsp. *fulvilanatus* (4–6). Otherwise, the remaining characteristics of both taxa are very similar, and only the geographical separation on either side of the Serra do Espinhaço — the two taxa being found c. 170 km apart — justifies the acceptance of *P. rosae* as a very restricted western subspecies, which is so far known from only a single population, found near the village of Santa Bárbara, in the Município of Augusto de Lima, Minas Gerais. Its extent of occurrence remains uncertain.

12. Pilosocereus pachycladus *F. Ritter*, Kakt. Südamer. 1: 69–70, Abb. 5, 40–41 (1979). Holotype: Brazil, Bahia, Urandi, 1964, *Ritter* 1223 (U).

Pseudopilocereus azureus Buining & Brederoo in Kakt. and. Sukk. 26: 241–243 (1975), non *P. azureus* (F. Ritter) P. V. Heath (1994) nec *Pilosocereus azureus* F. Ritter (1979), see below. Holotype: Brazil, Bahia, Mun. Brumado, Umburanas, near the Rio Brumado, 450 m, s.d., *Horst* HU 237 (U). *Pilosocereus cyaneus* F. Ritter, Kakt. Südamer. 4: 1516 (1981). Type: as above. [Nom. nov. pro *Pseudopilocereus azureus* Buining & Brederoo.]
P. pachycladus (F. Ritter) P. V. Heath in Calyx 4: 140 (1994).
Pilosocereus azureus F. Ritter, Kakt. Südamer. 1: 73 (1979). *Pseudopilocereus azureus* (F. Ritter) P. V. Heath,

l.c. (1994), nom. illeg. (Art. 53.1). Holotype: Brazil, Minas Gerais, Januária, 1959, *Ritter* 958 (U).

Pilosocereus atroflavispinus F. Ritter, ibid.: 68 (1979). *Pseudopilocereus atroflavispinus* (F. Ritter) P. V. Heath, l.c. (1994). Holotype: Brazil, Bahia, Ituaçu, on rocks, 1964, *Ritter* 1349 (U).

Pilosocereus oreus F. Ritter, ibid.: 69–70 (1979). *Pseudopilocereus oreus* (F. Ritter) P. V. Heath, l.c. (1994). Holotype: Brazil, Minas Gerais, Monte Azul, 1964, *Ritter* 1226 (U).

Pilosocereus pernambucoensis var. *montealtoi* F. Ritter, ibid.: 66 (1979). *Pseudopilocereus pernambucoensis* var. *montealtoi* (F. Ritter) P. V. Heath, l.c. (1994). Holotype: Brazil, Bahia, Palmas de Monte Alto, 1964, *Ritter* 1225 (U).

? *Pilosocereus splendidus* F. Ritter, ibid.: 69, Abb. 36 (1979). *Pseudopilocereus splendidus* (F. Ritter) P. V. Heath, l.c., 141 (1994). Holotype: Brazil, Bahia, Urandi, 1964, *Ritter* 1224 (U).

Pilosocereus superbus F. Ritter, ibid.: 67 (1979). *Pseudopilocereus superbus* (F. Ritter) P. V. Heath, l.c. (1994). Holotype: Brazil, Bahia, Anajé [Anagé], 1964, *Ritter* 1347 (U).

Pilosocereus superbus var. *regius* F. Ritter, ibid.: 67, Abb. 34 (1979). *Pseudopilocereus superbus* var. *regius* (F. Ritter) P. V. Heath, l.c. (1994). Holotype: Brazil, Bahia, Brumado, Rio Brumado, 1964, *Ritter* 1343 (U).

Pilosocereus superbus var. *gacapaensis* F. Ritter, ibid.: 67, Abb. 36 (1979). *Pseudopilocereus superbus* var. *gacapaensis* (F. Ritter) P. V. Heath, l.c. (1994). Holotype: Brazil, Bahia, Mun. Riacho de Santana/Igaporã, Serra da Garapa ['Gacapa'], *Ritter* 1343A (U).

Pilosocereus superbus var. *lanosior* F. Ritter, ibid. 67 (1979). *Pseudopilocereus superbus* var. *lanosior* (F. Ritter) P. V. Heath, l.c. (1994). Holotype: Brazil, Bahia, Caitité, 15 km W of town, *Ritter* 1343B (U).

Pilosocereus cenepequei Rizzini & A. Mattos in Revista Brasil. Biol. 46: 324–327 (1986). Holotype: Brazil, Minas Gerais, Pedra Azul, 15 Oct. 1984, *Rizzini & A. Mattos-Filho* 39 (RB).

P. schoebelii P. J. Braun in Cact. Succ. J. (US) 58: 150–156 (1987). Holotype: Brazil, N Minas Gerais [probably near Janaúba], 600 m, *Braun* 426 (ZSS).

[*Pilocereus glaucescens sensu* Werderm., Bras. Säulenkakt.: 107, photo (1933).]

[*Cereus ulei sensu* Luetzelburg, Estud. bot. Nordéste 3: 69, 111 (1926) non (Gürke) A. Berger.]

VERNACULAR NAMES. Facheiro, Facheiro-azul, Mandacaru-de-facho.

Treelike or shrubby, 2–10 m or more; vascular cylinder weakly to moderately woody; branches 5.5–10.0(–15.0) cm diam., erect, epidermis glaucous, grey-green to blue, smooth; ribs 5–19, sinuses straight, transverse folds above the areoles visible at apex of young branches. Areoles 3–10 mm diam., 2–10 mm apart, felt white to greyish, with white long hairs. Spines translucent, golden-yellow, greyish with age. Fertile part of stem slightly to strongly differentiated. Flower-buds acute or obtuse 1–2 days before anthesis; flowers very variable, 4–7 × 2.2–4.5 cm, sometimes compressed; pericarpel subglobose; tube 25–50 × (8–)12–25 mm at base, flared to 24–35 mm diam. at apex, constricted below nectar-chamber, slightly to strongly curved, pale green to dark brown, distal third with broad bract-scales; inner perianth-segments white to pink; style 20–58 × 1.2–3.0 mm, tapering; stigma-lobes 9–13, exserted in relation to the anthers; ovary locule broadly hemicircular to triangular in longitudinal section. Fruit depressed-globose, dehiscent by a lateral slit, floral remnant pendent, deeply inserted; pericarp purplish; funicular pulp magenta. [Seed morphology: see subspecies.]

Although *P. pachycladus* is one of the most conspicuous species in Eastern Brazil, being both common and widely distributed, it has an involved history of taxonomic confusion and was first unequivocably named less than 30 years ago by Buining & Brederoo (1975), whose chosen epithet is blocked within *Pilosocereus*. One of the earlier names associated with this taxon, *Pilocereus glaucescens* Linke 1858, is of doubtful application, having been based on sterile living material collected in Brazil, without a precise locality (see 'Insufficiently known taxa', following *P. densiareolatus*). Ritter (1979) argues that the description is very ambiguous and is impossible to attribute to a single taxon with certainty. From this description, in fact, one cannot exclude *P. glaucochrous* (Werderm.) Byles & G. D. Rowley, some of the forms of *P. pachycladus*, or even the possibility of the original plant being a specimen from another country. Werdermann (1933, 1942) used the name *Pilocereus glaucescens* for populations here included within *Pilosocereus pachycladus* subsp. *pachycladus*. A second name incorrectly used for this species in the broad sense was *Cereus ulei* by Luetzelburg (1925–26, 3: 69), while a third once applied thus is *P. piauhyensis* (see *P. pachycladus* subsp. *pernambucoensis* and *P. piauhyensis*).

This very variable species comprises a number of heterotypic synonyms described by Ritter, Buining & Brederoo, and Braun (see above). The species concepts utilized by these authors have proved to be too narrow and, if applied to all the forms now known, would lead to a new species name for each slightly different population of this complex.

P. pachycladus presents a broad range of forms, including tree-like populations, widely distributed in the dense or sparse *caatinga* forests of the Nordeste, and more shrubby forms in rupicolous populations near its southern limit, in northern Minas Gerais. To the south of Pedra Azul, is the endemic *P. magnificus*, which has probably arisen by a process of allopatric speciation following isolation from a population of common ancestry with *P. pachycladus*, of which it is assumed to be the sister species.

CONSERVATION STATUS. Least Concern [LC]; a very widespread species.

Two subspecies are recognized:

1. Ribs 5–12, high and broad; central spines long, well differentiated from radials; flower-bearing areoles densely hairy (S of 10°S) . . . . **a.** subsp. **pachycladus**
1. Ribs (10–)13–19, low and close together; central spines poorly differentiated from radials equalling them or slightly longer; flower-bearing areoles scarcely hairy (N of 10°S)
. . . . **b.** subsp. **pernambucoensis**

12a. subsp. **pachycladus**

Treelike or shrubby, 1–5(–8) m, branched only at base or forming a well-defined trunk; branches 5.5–10.0(–11.0) cm diam., epidermis pale blue; ribs 5–12, 15–35 × 12–24 mm. Areoles 3–10 mm diam., 2–12 mm apart, woolly. Spines 0.4–1.2 mm diam. at base; golden yellow to brownish or greyish, centrals 1–8, (5–)15–30(–45) mm, ascending to porrect, radials 8–18, (3–)5–15 mm, adpressed. Fertile part of stem differentiated, with 1–4 subapical flower-bearing areoles on more than one rib, with white long hairs and sometimes bristly spines to 30 mm. Flower-tube pale green to very dark brown, slightly to strongly curved; inner perianth-segments white. Seeds 1.5–2.0 × 1–1.3 mm, testa-cells domed to flat, with intercellular depressions and coarse, dense to sparse cuticular folds.

Central-southern *caatinga* / Northern *campo rupestre* element: on quartzitic outcrops and in scrub associated with the *campos rupestres*, and locally co-dominant with other woody vegetation in *caatinga*, on stony ground within and on either side of the Chapada Diamantina, central Bahia, south of 10°S, on limestone outcrops in the Rio São Francisco valley (N Minas Gerais and W Bahia), in the northern part of the Serra do Espinhaço, Minas Gerais, eastwards on gneissic inselbergs, and disjunctly in *campo rupestre*, northern part of the Serra do Cabral, Minas Gerais, c. 400–1550 m. Endemic to the core area within Eastern Brazil. Map 37.

BAHIA: W Bahia, Mun. Vanderlei, 3 Oct. 1985, *Horst & Uebelmann* 570, cult. (ZSS); l.c., 19 July 1989, *Zappi* 148 (SPF); Mun. Santana, 28 km S of Santana, 15 Jan. 1991, *Taylor et al.* 1426 (K, HRCB, ZSS, CEPEC); N to S Bahia, Mun. Jaguarari, Barrinha, 7–8 June 1915, *Rose & Russell* 19788 (US); l.c., border with Mun. Juazeiro towards Senhor do Bonfim, between Maçaroca and Barrinha, 8 Jan. 1991, *Taylor et al.* 1375 (K, HRCB, ZSS, CEPEC); l.c., 91 km S of Juazeiro on road BR 407 towards Senhor do Bonfim, 15 July 1989, *Zappi* 128 (SPF, HRCB); Mun. Gentio do Ouro, Serra do Piquizeiro, 11°27'S, 42°25'W, 30 Nov. 1987, *Miranda* 47 (RB); l.c., Serra de Santo Inácio, Feb. 1907, reported by Ule (1908:

photo II); Mun. João Dourado, APA Gruta dos Brejões, 4 Aug. 2002, *Taylor* (obs.); Mun. Jacobina, 28 km W of town on road BR 324 to Lages, 16 July 1989, *Zappi* 132 (SPF, HRCB); l.c., 27 km W of town, 12 Jan. 1991, *Taylor et al.* 1400 (K, HRCB, ZSS, CEPEC); Mun. Morro do Chapéu, NE of Gruta dos Brejões, 4 Aug. 2002, *Taylor* (K, photos); l.c., N of Brejão (Formosa), 'Lagedo Bordado', 5 Aug. 2002, *Taylor* (K, photo); l.c., 19–21 km W of town on road BA 052 to América Dourada, 25 Dec. 1988, *Taylor & Zappi* in *Harley* 27391 (K, SPF, CEPEC), 13 Jan. 1991, *Taylor et al.* 1406 (K, HRCB, ZSS, CEPEC), 3 Aug. 2002, *Taylor* (K, photos); Mun. Seabra, E of town, 25 Aug. 1988, *Eggli* 1284 (ZSS, HRCB); Mun. Lençois, Morro do Pai Inácio, Jan. & Mar. 1997, *A. Conceição* s.n. (K, photos); Mun. Mucugê, 7 km S from the town on road towards Barra da Estiva, 21 Aug. 1988, *Eggli* 1248 (ZSS); l.c., 500 m from the town, near cemetery, 22 Dec. 1988, *Taylor & Zappi* in *Harley* 27379 (K, SPF); Mun. Riacho de Santana, 55 km from Bom Jesus da Lapa on road BA 430 towards Riacho de Santana, 20 July 1989, *Zappi* 151 (SPF, HRCB); Mun. Rio de Contas, road to Faz. Brumadinho, 1 km from the Fazenda, 13°32'S, 41°54'W, 9 Nov. 1988, *Taylor* in *Harley* 25562 (K, SPF); Mun. Livramento do Brumado, 5 km SW of Rio de Contas, 23 Aug. 1988, *Eggli* 1302 (ZSS, HRCB); l.c., 12 km on Livramento–Brumado road, 13°45'S, 41°49'W, 23 Nov. 1988, *Taylor & Zappi* in *Harley* 25546 (K, SPF); Mun. Riacho de Santana/Igaporã, Serra da Garapa ['Gacapa'], *Ritter* 1343A (U); Mun. Dom Basílio, 21 km S of Livramento do Brumado, 12 Mar. 1991, *Lewis* s.n. (K, photo); Mun. Ituaçu, 1964, *Ritter* 1349 (U); l.c., 2 km W of the town, 18 Aug. 1988, *Eggli* 1193 (ZSS); Mun. Brumado, 19 km from the town towards Sussuarana, 2 Feb. 1991, *Eggli* (ZSS, photos); l.c., Umburanas, near the Rio Brumado, 1964, *Ritter* 1343 (U); Mun. Aracatu, 25 km from Brumado on road BR 030 towards Sussuarana, 2 Feb. 1991, *Eggli* (ZSS, photo); Mun. Anagé, 1964, *Ritter* 1347 (U), 1985, *Horst & Uebelmann* 472 (ZSS); Mun. Palmas de Monte Alto, 1964, *Ritter* 1225 (U); l.c., hill with TV tower, 28 Aug. 1988, *Eggli* 1314 (ZSS), 21 July 1989, *Zappi* 153 (SPF, HRCB); Mun. Malhada, 9 km on road BR 030 towards Iuiú, 28 Aug. 1988, *Eggli* 1318 (ZSS); Mun. Caitité, 15 km W from the town, 1964, *Ritter* 1343B (U); l.c., 9 km SE from the town towards Brumado, 27 Aug. 1988, *Eggli* 1310 (ZSS); l.c., Brejinho das Ametistas, *Horst & Uebelmann* 421 (ZSS), 26 July 1989, *Zappi* 177 (SPF, HRCB); Mun. Urandi, 3 km S of the town, 22 July 1989, *Zappi* 159 (SPF, HRCB); Mun. Vitória da Conquista, Gameleira, 14°50'25"S, 41°0'23"W, 17 April 2003, *Taylor et al.* (obs.); Mun. Licínio de Almeida, Sep. 2003, *M. Machado* (photos).

MINAS GERAIS: Mun. Montalvânia, S of the ferry crossing of the Rio Carinhanha, 1999, *Klaassen et al.* (photo); Mun. Espinosa, *Horst & Uebelmann* 402 (ZSS); l.c., 2 km N from the town on road BR 122 to Urandi, 22 July 1989, *Zappi* 162 (SPF, HRCB); l.c., 8 km S from the town on road BR 122 to Monte Azul, 22 July 1989, *Zappi* 162A (SPF, HRCB); Mun. Monte Azul, 1964, *Ritter* 1226 (U); l.c., 12 km E from the town, E of Vila Angical, 28 Jan. 1991, *Eggli* (ZSS, photos), *Taylor* (K, photos); Mun. Januária, 1959, *Ritter* 958 (U); l.c., 8 km E of Rio São João, 5 Sep. 1985, *Horst & Uebelmann* 714 (ZSS); l.c., 15 km on road BR 135, S of the Rio São Francisco, Serra do Bom Sucesso, 11 Aug. 1988, *Eggli* 1141 (ZSS); l.c., [Pedras de] Maria da Cruz, 11 Aug. 1997, *I. Pimenta* (K, photos); Mun. Mato Verde, 12 km E from the town on road to Santo Antônio do Retiro, 13 Aug. 1988, *Eggli* 1156 (ZSS); l.c., 15°23'S, 42°45'W, 9 Nov. 1988, *Taylor & Zappi* in *Harley* 25520 (K, SPF), 22 July 1989, *Zappi* 164 (SPF, HRCB); Mun. São Francisco, Faz. Daniela, 11 Aug. 1997, *I. Pimenta* (K, photo); Mun. Porteirinha, 28 km N from the town towards Mato Verde on road BR 122, 27 Jan. 1991, *Taylor et al.* 1456 (K, HRCB, ZSS, BHCB); Mun. Janaúba, *Horst & Uebelmann* 581 (ZSS); Mun. Taiobeiras, c. 35 km E from the town towards road BR 116, after bridge over Rio Atoleiro, 15°53'S, 41°57'W, 17 Oct. 1988, *Taylor & Zappi* in *Harley* 25150 (K, SPF), 13 Aug. 1988, *Eggli* 1158 (ZSS, HRCB); Mun. Salinas (?), road to Pedra Azul, 1 km W from 'Sítio', 22 Feb. 1988, *Supthut* 8876 (ZSS); Mun. Pedra Azul, 15 Oct. 1984, *Rizzini & Mattos-Filho* 39 (RB); l.c., 10 km from the town towards road BR 116, 21 Oct. 1988, *Taylor & Zappi* in *Harley* 25401 (K, SPF); Mun. Francisco Sá/Grão Mogol, vicinity of Barrocão, 950 m, reported by Buining (1975: 26, as *Pseudopilocereus* sp.); Mun. Francisco Dumont, E of the Serra do Cabral, 2 km from the town towards Jequitaí, 7 Aug. 1988, *Eggli* 1118 (ZSS); N of Serra do Cabral (Francisco Dumont?), 9 Sep. 1985, *Horst & Uebelmann* 147, cult. (ZSS).

Besides the examined specimens cited above, this taxon has also been observed in the following municípios of Bahia: Caém, Jacobina, Ibitiara, Brejolândia, Igaporã/Caitité, Tanhaçu.

CONSERVATION STATUS. Least Concern [LC] (1); PD=1, EI=2, GD=2. Short-list score (1×5) = 5. Least Concern, but the *caatinga* habitats continue to decline (*campo rupestre* habitats are less threatened).

Further field study of this complex, geographically variable taxon may justify its division into additional subspecies, especially for the distinctive regional forms from the Rio São Francisco valley (BA/MG), Rio de Contas drainage (BA) and north-eastern Minas Gerais.

Marlon Machado has detected a population of typical subsp. *pachycladus* sympatric with a similar but greener-stemmed plant, which appeared to be a different species, perhaps Ritter's *P. splendidus*, near Licínio de Almeida, Bahia.

12b. subsp. **pernambucoensis** *(F. Ritter) Zappi* in Succ. Pl. Res. 3: 109 (1994). Holotype: Brazil, Pernambuco, Araripina, 1963, *Ritter* 1219 (U).

Pilosocereus pernambucoensis F. Ritter, Kakt. Südamer. 1: 65 (1979). *Pseudopilocereus pernambucoensis* (F. Ritter) P. V. Heath in Calyx 4: 140 (1994) (*'pernambucensis'*).
Pilosocereus pernambucoensis var. *caesius* F. Ritter, tom. cit.: 66 (1979). *Pseudopilocereus pernambucoensis* var. *caesius* (F. Ritter) P. V. Heath, l.c. (1994). Holotype: Brazil, Pernambuco, Petrolina, *Ritter* 1220 (U).
[*Cephalocereus piauhyensis sensu* (Gürke) Britton & Rose, Cact. 2: 49 (1920), fig. 72 tantum, typ. excl.]
[*Pilocereus piauhyensis sensu* (Gürke) Werderm., Bras. Säulenkakt.: 111 (1933), fig. p. 107 tantum, typ. excl.]
[*Pilosocereus piauhyensis* subsp. *piauhyensis sensu* Braun & Esteves Pereira in Kaktusy 39 (special): 12, A (2003).]

VERNACULAR NAME. Calumbi.

Treelike, 2–10 m, with a well-defined trunk; branches 7–15 cm diam., epidermis grey-green to glaucous, smooth; ribs (10–)13–19, 8–12 × 7–10 mm. Areoles 3–10 mm diam., 5–8 mm apart. Spines 0.3–0.8 mm diam. at base, golden yellow, centrals 8–12, 12–18 mm, ascending to porrect, radials 16–18, 5–8 mm, adpressed. Fertile part of stem slightly differentiated, with 1–3 subapical flower-bearing areoles on 1 or more ribs, with few white long hairs and sometimes also 15–20 mm, bristly spines. Tube pale green to brownish, curved; inner perianth-segments white to pink. Seeds (1.3–)1.4–1.6(–1.8) × (0.9–)1.0–1.1 mm, testa-cells flat, with intercellular depressions, cuticular folds coarse, few or absent.

Northern *caatinga* element: locally co-dominant with other woody vegetation in dense or sparse *caatinga* and *agreste,* on sandy or rocky substrates (including sandstones and gneissic inselbergs), c. 50–750 m, northern Bahia (north of 10°S), Alagoas, Pernambuco, Paraíba, Rio Grande do Norte, south-easternmost Piauí and southern and north-western Ceará (E, W & S sides of Chapada do Araripe, Chapada da Borborema & plateau of the Serra da Ibiapaba). Endemic to North-eastern Brazil. Map 37.
PIAUÍ: SE Piauí, Mun. Jaicós, near the town, 7°13'S, 41°17'W, 6 Apr. 2000, *E.A. Rocha et al.* (K, photo); Mun. Padre Marcos, 2 km along road to Simões from Marcolândia, 7°28'S, 40°41'W, 5 Apr. 2000, *E.A. Rocha et al.* (K, photos); Mun. Simões, road to town from Marcolândia, W escarpment of Chapada do Araripe, 7°36'S, 40°45'W, 5 Apr. 2000, *E.A. Rocha et al.* (K, photos); Mun. Paulistana, 54 km N and 24 km S of town on road BR 407, 6 Apr. 2000, *E.A. Rocha et al.* (obs.).
CEARÁ: NW Ceará, Mun. Tianguá, road BR 222, 10.5 km W of access road to Tianguá, 20 Feb. 1995, *J.B. Fernandes da Silva* 366 (MG), *Taylor* (K, photos); S Ceará, Mun. Jardim, SE escarpment of Chapada do Araripe, 7°34'S, 39°19'W, 4 Apr. 2000, *E.A. Rocha et al.* (K, photos) — glaucous form with only 10–12 ribs.
RIO GRANDE DO NORTE: 'Sebastianópolis' – Chapada do Apodi, 11 July 1960, *Castellanos* 22894 (R); Mun. Mossoró, 11 July 1960, *Castellanos* 22877 (R); Mun. Tangará, W of town on road BR 226 to Currais Novos, 6°14'S, 35°51'W, 1 Apr. 2000, *E.A. Rocha et al.* (obs.); Mun. São José do Campestre, 7 km S of town on road to Guarabira (PB), 1 Apr. 2000, *E.A. Rocha et al.* (obs.); Mun. Currais Novos, E of town, c. 36°30'W, 1 Apr. 2000, *E.A. Rocha et al.* (obs.).
PARAÍBA: Mun. Tacima, near border with Mun. Caiçara (near Logradouro), inselberg 'Pão de Açúcar', 6°36'S, 35°28'W, 1 Apr. 2000, *E.A. Rocha et al.* (K, photo); Mun. Guarabira, Serra da Jurema, 6°50'S, 35°29'W, 1 Apr. 2000, *E.A. Rocha et al.* (K, photos); Mun. Junco do Seridó, 22 Apr. 1978, *Andrade-Lima* 78-8628 (IPA); Mun. Soledade, see Eiten (1983: figs 67 & 68); Mun. Pocinhos, July 1920, *Luetzelburg*

12594 (M); Mun. Cruz de Espírito Santo, road BR 230, near bridge over Rio Paraíba, 10 Feb. 1995, *Taylor & Zappi* (obs.); Mun. São João do Cariri, road BR 412, 29–30 Oct. 1993, *M.F. Agra et al.* 2393 (JPB), 27–29 Apr. 1994, *M.F. Agra et al.* 3119, 3120, 3121 (JPB); Mun. Sumé, road BR 412, 29–30 June 1994, *M.F. Agra et al.* 2832 (JPB).

PERNAMBUCO: Mun. Exu, 11 km N of town, S escarpment of Chapada do Araripe, 7°25'S, 39°45'W, 5 Apr. 2000, *E.A. Rocha et al.* (K, photo); Mun. Araripina, 1963, *Ritter* 1219 (U); l.c., near the hotel 'Pousada do Araripe', 15 Feb. 1990, *Zappi* 225 (HRCB, ZSS); Mun. Taquaritinga do Norte, Gravatá do Ibiapina, 8 Aug. 2001, *Zappi* (obs.); Mun. Bom Jardim, road PE 090, 8 Feb. 1993, *F.A.R. Santos* 62 (ALCB, HUEFS); Mun. Brejo da Madre de Deus, 9 Feb. 1993, *F.A.R. Santos* 66 (HUEFS, PEUFR); Mun. Salgueiro, Morro do Cruzeiro, 17 Feb. 1990, *Zappi* 227A (HRCB); Mun. Gravatá, Russinha, *caatinga*, 4 Feb. 1933 & 26 Apr. 1934, *Pickel* 1915 (F, on 2 sheets); Mun. Caruaru, 9 km NE of town, 8°14'21"S, 35°54'52"W, Dec. 1991, *E.A. Rocha et al.* (K, photos); Mun. Poção, 21 Oct. 1991, *F.A.R. Santos* 3 (HUEFS); Mun. Pesqueira, 4 km SW from the town towards Garanhuns, 13 Feb. 1991, *Taylor & Zappi* 1625 (K, ZSS, HRCB, UFP); Mun. Alagoinha, 13 Feb. 1992, *F.A.R. Santos* 23 (HUEFS, PEUFR); Mun. Afrânio, 8 km S of border with PI on road BR 407, 6 Apr. 2000, *E.A. Rocha et al.* (obs.); l.c., 5 km N of Rajada, 8°47'S, 40°52'W, 6 Apr. 2000, *E.A. Rocha et al.* (K, photos) — form with stems 15 cm diam.; Mun. Belém de São Francisco, road BR 316 near border with Mun. Floresta, 8°40'S, 38°44'W, 8 Apr. 2000, *E.A. Rocha et al.* (K, photos); Mun. Ibimirim, 17 Nov. 1992, *F.A.R. Santos* 40 (HUEFS, PEUFR); Mun. (?), Petrolina–Cabrobó, 26 Apr. 1971, *Heringer et al.* 366B (RB); Mun. Nova Petrolândia, 12 km NE of crossing to old town, 12 Feb. 1991, *Taylor & Zappi* 1620 (K, ZSS, HRCB, UFP); Mun. Petrolina, road BR 428, 9°11'S, 40°23'W, 8 Apr. 2000, *E.A. Rocha et al.* (obs.); l.c., *Ritter* 1220 (U).

ALAGOAS: Mun. Palmeira dos Índios, 6 km S of border with Pernambuco on road to Arapiraca, 13 Feb. 1991, *Taylor & Zappi* 1630 (K, ZSS, HRCB, MAC).

BAHIA: Mun. Casa Nova, 10 Feb. 1990, 50 km E of Remanso, *Zappi* 220 (HRCB, ZSS); Mun. Curaçá, 39 km NW of Uauá, towards Poço de Fora, 7 Jan. 1991, *Taylor et al.* 1368 (K, HRCB, ZSS, CEPEC); Mun. Juazeiro, 2–6 June 1915, *Rose & Russell* 19744, 19758 (US, NY); Mun. Uauá, 5 km NW from the town towards Poço de Fora, 7 Jan. 1991, *Taylor et al.* 1366A (K, HRCB, ZSS); l.c., 13 km SE from the town towards Bendengó/Canudos, 5 Jan. 1991, *Taylor et al.* 1360 (K, HRCB, ZSS, CEPEC); Mun. Sento Sé, near Cabeluda, 11 Jan. 1991, *Taylor et al.* 1394 (K, HRCB, ZSS, CEPEC).

Besides the examined specimens cited above, this taxon has also been observed in the following municípios of Pernambuco: Lagoa Grande, Santa Maria da Boa Vista, Orocó, Floresta, Airi, Cruzeiro do Nordeste, Arcoverde, Sertânia, Garanhuns, Caetés, Terezinha, Bom Conselho; and Bahia: Remanso.

CONSERVATION STATUS. Least Concern [LC] (1); PD=1, EI=2, GD=2. Short-list score (1×5) = 5. Least Concern, but habitat continues to decline and has already declined very considerably.

This subspecies includes populations found from the northernmost curve of the Rio São Francisco valley northwards (Bahia, N of 10°S), that comprise distinctly treelike, sometimes massive plants, with usually high numbers of ribs (13–19) and fine, golden spination, which were described by Ritter (1979) as *Pilosocereus pernambucoensis*. The treatment of this taxon at subspecific level is justified by the existence of morphologically intermediate populations in the region of Juazeiro and Sento Sé (*Taylor* 1375, 1394), presenting 10–15 ribs and relatively fine spination. A 10-ribbed form has also been observed in southern Ceará (Mun. Jardim).

P. pachycladus subsp. *pernambucoensis* may present either coerulescent blue (eg. *Zappi* 225) or greyish green epidermis (*Taylor & Zappi* 1630), the greenish plants predominating in the eastern part of its range, glaucous forms being the norm to the west (eg. Plate 45.2). There is no reason to maintain Ritter's variety *caesius* (Ritter 1979), proposed on the basis of its glaucous epidermis. However, stem thickness does differ and is somewhat correlated with stem colour, the eastern populations, from central-eastern Pernambuco northwards to Rio Grande do Norte, being uniformly narrow-stemmed and greenish, suggesting that these could be recognized as a further subspecies, for which there is currently no name at any rank (Plate 45.3). This slender, greenish form obeys the distribution pattern of other taxa characteristic of the Eastern *caatingas-agrestes*, eg. *Tacinga palmadora*.

The collection cited above for north-western Ceará seems rather disjunct on present knowledge and underlines the need for field studies in the intervening areas of south-western Ceará.

Britton & Rose and Werdermann confused populations of *P. pachycladus* subsp. *pernambucoensis* with *Cereus piauhyensis* Gürke (1907), which they combined as *Cephalocereus piauhyensis* (Gürke) Britton & Rose (1920) and *Pilocereus piauhyensis* (Gürke) Werdermann (1933). The type material of *Cereus piauhyensis* Gürke was collected by Ule in the Serra Branca, north of São Raimundo Nonato, Piauí. It is clear that this name belongs to another species (see discussion under *P. piauhyensis*) and, to date, *P. pachycladus* has not been encountered further west than Mun. Paulistana in the south-easternmost corner of Piauí. This confused nomenclature also appears in Andrade-Lima (1989), where he follows the concept of Britton & Rose (1920) and Werdermann (1933), and illustrates and describes *P.* *pachycladus* subsp. *pernambucoensis* as *P. piauhyensis* (likewise Braun & Esteves Pereira 2003: 12, A). In the same work Andrade-Lima places the southern forms of *P. pachycladus* (ie. subsp. *pachycladus*) under *Pilosocereus glaucescens* (Linke) Byles & G. D. Rowley, a name of uncertain application (see notes at end of genus).

Egler (1951: 587, fig. 6) confuses this widespread northern subspecies with *Facheiroa squamosa*, which in Pernambuco is restricted to the south-westernmost part of the state.

13. Pilosocereus magnificus *(Buining & Brederoo) F. Ritter*, Kakt. Südamer. 1: 72–73, Abb. 1 (1979). Holotype: Brazil, Minas Gerais, Rio Jequitinhonha, 370 m, s.d. [1968], *Horst & Uebelmann* 224 (U).

Pseudopilocereus magnificus Buining & Brederoo in Cact. Succ. J. (US) 44: 66–70 (1972).
[*Pilosocereus glaucescens sensu* Rizzini & A. Mattos, Contrib. Conhecim. Fl. NE. Minas Gerais Bahia Med.: 35, Figs 8 & 35 (1992).]

VERNACULAR NAME. Facheiro.

Shrubby to treelike, 1.5–5.0 m, with or without a well-defined trunk; vascular cylinder weakly woody; branches 4–6(–7.5) cm diam., erect, epidermis pale blue, waxy, smooth; ribs (4–)5–12(–15), 15 × 18 mm, sinuses straight, transverse folds above the areoles only visible at apex of branches. Areoles 4 mm diam., 2 mm apart, felt white or blackish, with white long hairs. Spines 0.2–0.4 mm diam. at base, translucent, bristly, golden-yellow to brownish; centrals 8, to 15 mm, porrect; radials c. 16, to 10 mm, ascending. Fertile part of stem slightly modified, flower-bearing areoles in groups of 3–6, randomly distributed along the whole length of the branches, especially at the mid to basal part of the stems, bearing white long hairs. Flower-buds obtuse 2–3 days before anthesis; flowers to 6 × 2.3 cm; pericarpel subglobose, pale green; tube 35 × 10 mm at base, flared to 16 mm diam. at apex, constricted above and below the nectar-chamber, straight to slightly curved, greenish with purplish patches, smooth, apical region with acute bract-scales; style 41 × 1.3–2.0 mm, tapering; stigma-lobes c. 10, exserted in relation to the anthers; ovary locule hemicircular to depressed in longitudinal section. Fruit 1.5–2.5 × 2.5–3.0 cm, depressed-globose, dehiscent by a lateral, abaxial or adaxial slit, floral remnant erect to pendent, deeply inserted; pericarp thick, wine-coloured to purple, smooth; funicular pulp magenta. Seeds 1.3–1.5 × 0.8–1.2 mm, testa-cells domed, with intercellular depressions, cuticular folds lacking.

South-eastern *caatinga* (inselberg) element: locally co-dominant with other arborescent cacti on gneissic inselbergs in *caatinga-agreste*, c. 240–700 m, in the drainage of the Rio Jequitinhonha, north-eastern Minas Gerais. Endemic to the core area within Minas Gerais. Map 37.

MINAS GERAIS: Mun. Itinga, Jenipapo, Faz. Lajeadão, 16°40'S, 41°50'W, June 2003, *I. Andrade* (K, photo); l.c., 5 km E from the town on road BR 367 towards Itaobim, 16°36'S, 41°43'W, 19 Nov. 1988, *Taylor & Zappi in Harley* 25533 (K, SPF); Mun. Itaobim, near the Rio Jequitinhonha, *Horst & Uebelmann* 224, cult. (U, ZSS); l.c., 8 km W of town on road BR 367 towards Itinga, 14 Dec. 1990, *Taylor & Zappi* 763 (HRCB, K, ZSS, BHCB); 'Rio Jequitinhonha', s.d., *Horst & Uebelmann* 224 (U); Mun. Padre Paraíso, 12 km N of town, 13 Dec. 1990, recorded with *P.* ×*subsimilis* (Zappi 1994: 87); Mun. Caraí, on road BR 116, c. 9.5 km S of Padre Paraíso ['Água Vermelha'], 18 km N of Catuji, 13 Dec. 1990, *Taylor & Zappi* 755 (HRCB, K, ZSS, BHCB).

CONSERVATION STATUS. Near Threatened [NT] (1); PD=2, EI=2, GD=2. Short-list score (1×6) = 6. Not known to be decreasing at present, but range limited (5 localities) and requiring monitoring for habitat change.

Distinct within the genus for its approximate areoles bearing golden, bristly spines, which contrast with its strikingly pale blue, wax-covered epidermis. *P. magnificus* can also be distinguished by its small, narrow flowers, that appear randomly along the branches. Rather variable in rib number between populations.

PILOSOCEREUS AURISETUS Group (Nos 14–16)

Stems branching only at or near base. Found on or amongst rocks in *cerrado* and *campo rupestre*.

14. Pilosocereus machrisii *(E. Y. Dawson) Backeb.*, Die Cactaceae 4: 2419 (1960). Holotype: Brazil, Goiás, E from Ceres, road S from Uruaçu, 3 km from the town, 26 May 1956, *Dawson* 15110 (R; RSA, iso.).

'Pilocereus cuyabensis' Backeb. in Blätt. Kakteenforsch. 1935(1): [98] (1935), nom. inval. (Art. 36.1). *'Cephalocereus cuyabensis'* Borg, Cacti, ed. 2: 148 (1951), nom. inval. *'Pilosocereus cuyabensis'* Byles & G. D. Rowley in Cact. Succ. J. Gr. Brit. 19: 66 (1957), nom. inval. *'Pseudopilocereus cuyabensis'* Buxb. in Beitr. Biol. Pflanzen 44: 252 (1968), nom. inval.

Cephalocereus machrisii E. Y. Dawson in Los Angeles County Mus. Contr. Sci. 10: 1–8 (1957). *Pseudopilocereus machrisii* (E. Y. Dawson) Buxb. in Beitr. Biol. Pflanzen 44: 252 (1968); Buining & Brederoo in Succulenta 56: 90–94 (1977).

P. jauruensis Buining & Brederoo in Kakt. and. Sukk. 29: 153–155 (1978) (*'juaruensis'*). Holotype: Brazil, Mato Grosso do Sul, Rio Jauru ['Juaru'], Riacho Claro, 1974, *Horst & Uebelmann* 454 (U). *Pilosocereus jauruensis* (Buining & Brederoo) P. J. Braun in ibid. 35: 181 (1984) (*'juaruensis'*).

P. saudadensis F. Ritter, Kakt. Südamer. 1: 82 (1979). Holotype: Brazil, Mato Grosso, Serra da Saudade, 1963, *Ritter* 1216 (U).

P. pusillibaccatus P. J. Braun & E. Esteves Pereira, Cact. Succ. J. (US) 58: 240–247 (1986). Holotype: Brazil, Piauí, [Alto Parnaíba – Gilbués], southern border of Piauí, 500 m, Apr. 1982, *E. Esteves Pereira* 202 (ZSS; UFG, K, isos.).

P. cristalinensis P. J. Braun & E. Esteves Pereira in Kakt. and. Sukk. 38: 132 (1987). *P. machrisii* subsp. *cristalinensis* (P. J. Braun & E. Esteves Pereira) P. J. Braun & E. Esteves Pereira in Schumannia 3: 188 (2002). Holotype: Brazil, Goiás, Mun. Cristalina, near the town, 1050 m, Mar. 1973, *E. Esteves Pereira* 73 (ZSS; UFG, iso.).

P. lindanus P. J. Braun & E. Esteves Pereira in Kakt. and. Sukk. 38: 132 (1987) (*'lindaianus'*). Holotype: Brazil, Goiás, [Mun. Alto Paraíso de Goiás, Chapada dos Veadeiros], Oct. 1973, *E. Esteves Pereira* 60 (ZSS; UFG, iso.).

P. densivillosus P. J. Braun & E. Esteves Pereira in Kakt. and. Sukk. 45: 108–114 (1994). Holotype: Brazil,

W Goiás, Rio Araguáia drainage, near border with Mato Grosso do Sul, 1989, *E. Esteves Pereira* 49 (UFG 12350, n.v.).

'*P. circinnuspetalus*' P. J. Braun & E. Esteves Pereira in Innes & Glass, The Illustrated Encyclopedia of Cacti: 246, photo. top left (1991), nom. inval. (Art. 36.1). Based on material from Goiás, Mun. Mineiros.

? *P. estevesii* P. J. Braun in Cact. Succ. J. (US) 71: 74–77, with illus. (1999). Holotype: Bahia, W Bahia, 'Serra do Muquém', 550–600 m, 1982, *E. Esteves Pereira* 142 (UFG 20529, n.v.).

P. goianus P. J. Braun & E. Esteves Pereira in Brit. Cact. Succ. J. 20: 93–103 (2002). Holotype: NE Goiás, s.d., *E. Esteves Pereira* 89 (UFG 14873, n.v.). **Synon. nov.**

Shrubby, 0.4–3.2(–3.5) m, not normally branched above base; vascular cylinder weakly woody; branches 3.2–9.0 cm diam., erect, epidermis dark to bluish green, slightly greyish to glaucous, smooth; ribs 8–15, 7–15 × 13–20 mm, sinuses straight, transverse folds not visible; areoles 2–5 mm diam., 3–5 mm apart, felt white to greyish with white to brownish long hairs to 5–10 mm. Spines 0.25–0.60 mm diam. at base, translucent when young, reddish, pale brown or golden, centrals 3–8, 15–30(–40) mm, porrect or ascending, the largest deflexed on old areoles, radials 9–24, 3–25 mm, adpressed, the lowermost longer. Fertile part of stem strongly differentiated on 3 or more ribs of each branch, with yellowish to greyish long hairs to 20–30 mm, without bristles or these to 47 mm. Flower-buds acute 2 days before anthesis; flowers (measurements mainly based on *Taylor et al.* 1440 in cultivation) 3–4.5 × 3–4 cm; pericarpel subglobose to ovoid, 10 × 14 mm; tube c. 25 × 12 mm at base, flared to 18 mm diam. at apex, infundibuliform, straight, pinkish to brownish red, slightly striate, distal half with triangular, pinkish bract-scales; style 32–40 × 3–1 mm, tapering, stigma-lobes 8–12, exserted or equal to anthers, to 9 mm; ovary locule circular to compressed in longitudinal section. Fruit 1.5–2.5 × 2–3.5 cm, depressed-globose, dehiscent by a lateral slit, floral remnant suberect to pendent, deeply inserted; pericarp exterior yellowish, red or purplish, interior yellow, smooth; funicular pulp white or pinkish magenta. Seeds (1.3–)1.6–1.7(–1.8) × 1–1.3 mm, testa-cells domed to flat, with intercellular depressions, cuticular folds coarse, dense, sparse or absent.

Western *cerrado* element: on quartzitic, arenitic or limestone rock outcrops associated with *cerrado, cerrado de altitude* or *campo rupestre*, c. 500–800 m, southern Piauí and western Bahia; southern Pará (Araguatins), Goiás, Mato Grosso, Mato Grosso do Sul, western Minas Gerais (Serra da Canastra) and São Paulo (Altinópolis & Brotas); north-eastern Paraguay. Map 38.

PIAUÍ: Alto Parnaíba–Gilbués, south-western border of Piauí, Apr. 1982, *E. Esteves Pereira* 202 (ZSS, UFG, K).

BAHIA: W Bahia, Mun. Barreiras, Serra da Bandeira, hill with TV tower, 18 Jan. 1991, *Taylor et al.* 1436A (K, HRCB); l.c., road to the new airport, 2.5 km from junction of road to Brasília, 18 Jan. 1991, *Taylor et al.* 1440 (K, HRCB, ZSS, CEPEC), ibid., cult. RBG Kew, July 2002 (K, photos).

CONSERVATION STATUS. Least Concern [LC] (1); PD=2, EI=1, GD=2. Short-list score (1×5) = 5. Some populations may be affected by loss of habitat, but the species is widespread on rock outcrops.

The recently described *P. estevesii* P. J. Braun is tentatively referred here. Its supposed distinctive characteristics seem no more unusual than those of other synonyms listed above and, given the extensive range of *P. machrisii* and its pronounced variability throughout this range, it would be unwise to assume that this western Bahian plant (now said to be extinct at its only known locality) is worthy of specific status. As Braun, l.c., comments, it is somewhat reminiscent of the geographically proximal *P. flexibilispinus* P. J. Braun & E. Esteves Pereira, from the adjacent state of Tocantins, but the flowers of the latter confirm its proposed relationship with the *P.* PENTAEDROPHORUS Group (Zappi 1994), whereas those of *P. estevesii* offer no differences with those of *P. machrisii* and its allies. *P. estevesii* has fruit with magenta funicular pulp, whereas previously known populations of *P. machrisii* are reported to have white fruit pulp. However, variation in this character is common in the genus and is particularly so in the closely related *P. aurisetus*.

Other plants that may 'key out' to *P. machrisii* are the dwarf *P. parvus* (L. Diers & E. Esteves Pereira) P. J. Braun and the recently described *P. bohlei* Hofacker, both of which may well prove to be a 'good' species. These are discussed under 'Insufficiently known taxa', after *P. densiareolatus*. Unfortunately, the authors have not been able to study them in the field.

15. Pilosocereus aurisetus *(Werderm.) Byles & G. D. Rowley* in Cact. Succ. J. Gr. Brit. 19: 66 (1957). Type: Brazil, Minas Gerais, Serra do Cipó, *Werdermann* 3993 (B†). Lectotype (Zappi 1994): Werdermann, l.c., 104, photograph '*Pilocer. aurisetus* Werd.' (iconotype).

Pilocereus aurisetus Werderm., Bras. Säulenkakt.: 103 (1933). *Pseudopilocereus aurisetus* (Werderm.) Buxb. in Beitr. Biol. Pflanzen 44: 252 (1968).

P. werdermannianus Buining & Brederoo in Kakt. and. Sukk. 26: 74–77 (1975). *Pilosocereus werdermannianus* (Buining & Brederoo) F. Ritter, Kakt. Südamer. 1: 76–77, Abb. 46–49 (1979). *P. aurisetus* subsp. *werdermannianus* (Buining & Brederoo) P. J. Braun & E. Esteves Pereira in Succulenta 74: 134 (1995). Holotype: Brazil, Minas Gerais, Mun. Conceição do Mato Dentro, 660 m, s.d., *Horst & Uebelmann* 227 (U).

Pilosocereus saxatilis F. Ritter, Kakt. Südamer. 1: 77 (1979), nom. inval. (Art. 34(c)). Based on: Brazil, Minas Gerais, Conceição [do Mato Dentro], 1964, *Ritter* 960A (U).

Pilosocereus werdermannianus (Buining & Brederoo) F. Ritter var. *diamantinensis* F. Ritter, l.c. (1979). *Pseudopilocereus werdermannianus* var. *diamantinensis* (F. Ritter) P. V. Heath in Calyx 4: 141 (1994). Holotype: Brazil, Minas Gerais, Diamantina, 1959, *Ritter* 959 (U).

Pilosocereus werdermannianus (Buining & Brederoo) F. Ritter var. *densilanatus* F. Ritter, l.c. (1979). *P. saxatilis* var. *densilanatus* F. Ritter, l.c. (1979), nom. inval. *Pseudopilocereus werdermannianus* var. *densilanatus* (F. Ritter) P. V. Heath, l.c. (1994). *Pilosocereus aurisetus* subsp. *densilanatus* (F. Ritter) P. J. Braun & E. Esteves Pereira in Succulenta 74: 134 (1995). Holotype: Brazil, Minas Gerais, Penha de França, 1959, *Ritter* 960 (U).

Pilosocereus supthutianus P. J. Braun in Kakt. and. Sukk. 36: 100–103 (1985). *P. aurisetus* subsp. *supthutianus* (P. J. Braun) P. J. Braun & E. Esteves Pereira in Succulenta 74: 134 (1995). Holotype: Brazil, Minas Gerais, W of the Serra do Espinhaço [Mun. Engenheiro Dolabela, towards Sítio], 800 m, 1982, *Horst & Uebelmann* 547 (ZSS).

[*P. coerulescens sensu* (Lem.) F. Ritter, Kakt. Südamer. 1:75 (1979); Zappi in Bol. Bot. Univ. São Paulo 12: 51 (1991); non *Pilocereus coerulescens* Lem. (1862), nom. illeg. (Art. 52.1).]

Shrubby, 1–3 m, branching only at base; vascular cylinder weakly woody; branches erect; ribs 10–17, sinuses straight, transverse folds not visible. Areoles 2–4 mm diam., 4–6 mm apart, felt brownish to blackish, slightly hairy to tomentose, with white long hairs to 5–10 mm. Spines 17–25, translucent, white to golden-yellow. Fertile part of stem strongly differentiated, lateral or subapical, flower-bearing areoles with long hairs and golden bristles. Flower-buds acute 2–3 days before anthesis; flowers 3.2–5.0 × 3–4 cm; pericarpel subglobose, greenish to brownish; tube 25–30 × 10–12 mm at base, flared to 20–28 mm diam. at apex, infundibuliform, straight, pinkish to brownish red, slightly striate, distal half with pinkish bract-scales; style 22–25 × 0.8–1.5 mm, tapering, stigma exserted in relation to the anthers; ovary locule hemicircular to compressed in longitudinal section. Fruit depressed-globose, dehiscent by subapical slit beneath the floral remnant, which is pendent, deeply inserted; pericarp thick, folded and rugose; funicular pulp white, pink or magenta. Seeds with flat testa-cells, intercellular depressions visible to inconspicuous, cuticular folds lacking.

CONSERVATION STATUS. Least Concern [LC]; widespread and known from many localities that are not known to be significantly declining.

One of the specific features of *P. aurisetus* is found in its fruit, which characteristically splits across the apex, breaking the floral remnant. It also has very smooth seeds.

Two subspecies are distinguished as follows:

1. Plants to 2 m; branches 2.8–5.5 cm diam.; ribs 11–13; flower-bearing areoles with white hairs
   (Serra do Espinhaço) . . . . **15a.** subsp. **aurisetus**
1. Plants to 3 m; branches 4.5–7.0 cm diam., ribs 10–17; flower-bearing areoles with golden hairs
   (Serra do Cabral) . . . . **15b.** subsp. **aurilanatus**

15a. subsp. **aurisetus**

VERNACULAR NAMES. Rabo-de-raposa, Quiabo-da-lapa, Quiabo-do-inferno, Cabeça-de-velho.

Shrub, 1–2 m; branches 2.8–5.5 cm diam., epidermis glaucous or green; ribs 11–13, 5–6 × 9–11 mm. Areoles 2–4 mm diam., 4–6 mm apart, brown to blackish, slightly to densely hairy, hairs white. Spines 0.1–0.3(–0.35) mm diam. at base, whitish to golden yellow, centrals 5–7, 8–25(–30) mm, ascending to porrect, radials 12–16, 7–11 mm, adpressed. Fertile part of stem lateral to subapical, flower-bearing areoles with white long hairs to 20 mm and golden bristles to 30 mm. Stigma-lobes c. 10, 5–7 mm. Fruit 2–3 × 3.5–4.0 cm; pericarp brownish to wine-coloured; funicular pulp white, pink or magenta. Seeds 1.7–1.8 × 1.2–1.3 mm.

Widespread South-eastern *campo rupestre* element: quartzitic rock outcrops associated with *campo rupestre*, c. 650–1300 m, Serra do Espinhaço, central Minas Gerais. Endemic to the core area within Minas Gerais. Map 38.

MINAS GERAIS: Mun. Bocaiúva, 40 km NE of Engenheiro Dolabela, W of the Serra do Espinhaço, Fazenda Tabual, 8 Aug. 1988, *Eggli* 1126 (ZSS, HRCB); l.c., Engenheiro Dolabela, towards Sítio, 1982 & 7 Sep. 1985, *Horst & Uebelmann* 547 (ZSS); l.c., Fazenda Laginha, near Sítio, s.d., *Horst & Uebelmann* 722 (ZSS); Mun. Itamarandiba, 23 km SW of town towards Penha de França, Serra do Mato Virgem (Serra Negra), 4 Aug. 1988, *Eggli* 1077 (ZSS, HRCB); Mun. Rio Vermelho, Pedra Menina, Serra do Ambrósio, s.d., *Horst & Uebelmann* 109, cult. (ZSS); l.c., Faz. Sr J. Batista, 8 Mar. 1988, *Zappi & Prado* in CFCR 11825 (SPF); Mun. Diamantina, 1959, *Ritter* 959 (U); l.c., Mendanha, 18°6'S, 43°28'W, 31 Oct. 1988, *Taylor & Zappi* in *Harley* 25486 (K, SPF); l.c., road to Conselheiro Mata, 25 Feb. 1987, *Zappi et al.* in CFCR 10395 (SPF); Mun. Datas, 5.5 km N of Datas on road to Diamantina, 17 Feb. 1988, *Supthut* 8834 (ZSS); Mun. Gouveia, 14 Apr. 1987, *Zappi et al.* in CFCR 10945 (SPF); Mun. Presidente Kubitschek, Trinta Réis, 19 July 1987, *Zappi et al.* in CFCR 10666 (SPF); Mun. Serro, 10 Aug. 1972, *Hatschbach* 30133 (MBM); Mun. Conceição do Mato Dentro, 1964, *Ritter* 960A (U), *Horst & Uebelmann* 227 (U, ZSS); l.c., road to Morro do Pilar, bridge over the Rio Santo Antônio, 20 Nov. 1989, *Zappi* 195 (HRCB); Mun. Santana do Riacho, Serra do Cipó, 10 Dec. 1949, *A.P. Duarte* 2397 (RB), 22 July 1987, *Zappi et al.* in CFSC 10266 (SPF), l.c., 19°17'S, 43°34'W, 27 Oct. 1988, *Taylor & Zappi* in *Harley* 25410 (K, SPF); Mun. Cocais, W of town on road to Pedra Pintada, 4 Aug. 2001, *J.O. Rego et al.* 851 (BHCB, Herb. Fund. Zoo-Bot. de Belo Horizonte 650 & fls in alcohol [Carpoteca 762], K, photos).

CONSERVATION STATUS. Least Concern [LC] (1); PD=2, EI=1, GD=2. Short-list score (1×5) = 5. Some populations may be affected by excessive burning of their habitats, but threats are otherwise quite limited.

The easternmost populations from Itamarandiba and Rio Vermelho (subsp. *densilanatus* (F. Ritter) P. J. Braun & E. Esteves Pereira) have densely woolly stems and at first seem rather distinctive, but on close examination do not differ sufficiently to merit recognition as an additional subspecies. If it were to be recognized, the logical extension of such a move would be to give similar status to the northern- and southernmost populations cited above, which seem to have equally divergent characteristics in terms of size, white-woolly stems, glaucousness etc., and then other variants would also come into view.

15b. subsp. **aurilanatus** *(Ritter) Zappi* in Succ. Pl. Res. 3: 123 (1994). Holotype: Brazil, Minas Gerais, Joaquim Felício, 1964, *Ritter* 1325 (U).

Pilosocereus aurilanatus F. Ritter, Kakt. Südamer. 1: 77–78, Abb. 50 (1979). *Pseudopilocereus aurilanatus* (F. Ritter) P. V. Heath in Calyx 4: 140 (1994).

VERNACULAR NAME. Rabo-de-raposa.

Shrub, 1.5–3.0 m; branches 4–6 cm diam., epidermis olive-green, greyish, light blue waxy only at apex of branches; ribs 10–17, 4.5–7.0 × 9–10 mm. Areoles 3–6 mm diam., 4–6 mm apart, felt brownish, with white or yellowish long hairs to c. 5 mm, glabrescent when old. Spines 0.3–0.4 mm diam. at base, golden yellow, centrals 8–9, 10–25 mm, ascending to porrect, radials 14–16, adpressed, somewhat pectinate. Fertile part of stem lateral, with yellow to ferrugineous long hairs to 30 mm and golden bristles to 50 mm. Stigma-lobes c. 12, 5–12 mm. Fruit 3.5 × 3–4.3 cm; pericarp olive-green to wine-coloured; funicular pulp magenta. Seeds 1.8–2.0 × (1.3–)1.4–1.5 mm.

South-eastern *campo rupestre* (Serra do Cabral) element: locally co-dominant with other cacti on quartzitic rock outcrops, *campo rupestre*, 800–900 m, Serra do Cabral, Minas Gerais. Endemic to the core area within Minas Gerais. Map 38.

MINAS GERAIS: Mun. Joaquim Felício, 1964, *Ritter* 1325 (U); l.c., road to Várzea da Palma, 17 May 1990, *Zappi et al.* in CFCR 13234 (SPF, HRCB); Mun. Buenópolis, 17°53'S, 44°15'W, 12 Oct. 1988, *Taylor & Zappi* in Harley 24909 & 24991 (K, SPF).

CONSERVATION STATUS. Endangered [EN B1ab(iii) + 2ab(iii)] (3); extent of occurrence < 5000 km² and area of occupancy estimated to be < 500 km²; PD=1, EI=2, GD=1. Short-list score (3×4) = 12. Known only from a small region without protected areas, the surrounding habitats declining from charcoal production activities.

Although it has a very restricted distribution and somewhat different habit, this taxon described by Ritter (1979) from the Serra do Cabral, a disjunct mountain range west of the Serra do Espinhaço, is not considered worthy of more than subspecific rank (Zappi 1994). It is linked to subsp. *aurisetus* via the form of the latter described under the synonym, *P. supthutianus*. The most striking feature of subsp. *aurilanatus* is its stouter, taller stems with dense, golden hairs on the flower-bearing areoles.

16. Pilosocereus aureispinus *(Buining & Brederoo) F. Ritter*, Kakt. Südamer. 1: 83–84 (1979). Holotype: Brazil, Bahia, E of the Rio São Francisco [Ibotirama], Serra da Barriguda, 450 m, s.d., *Horst & Uebelmann* 391 (U).

Coleocephalocereus aureispinus Buining & Brederoo in Kakt. and. Sukk. 25: 73–75 (1974).

Shrubby, to 2 m, branched only at base; vascular cylinder weakly woody; branches 5–9 cm diam., erect, epidermis dark green, smooth; ribs (18–)20–24, 3–5 × 5–7 mm, sinuses straight, transverse folds not visible. Areoles 5 mm diam., 4–5 mm apart, felt white, with white long hairs to 10 mm, showing indeterminate growth near base of plant. Spines 0.3–0.4 mm diam. at base, translucent, golden to ferrugineous, sometimes bristly, centrals 8–16, 4–12(–18) mm, ascending to porrect, radials 14–16, 4–7(–12) mm, adpressed. Fertile part of stem weakly differentiated, flower-bearing areoles either lateral or forming rings around the branches, with sparse white long hairs and golden bristles to 50 mm. Flower-buds acute 5 days before anthesis; flowers 5 × 2–2.2 cm; pericarpel hemiglobose; tube 35 × 13 mm at base, flared to 15 mm diam. at apex, straight, terete, narrow, pink to brownish, smooth; perianth-segments pink to white; style 35 mm; stigma-lobes c. 10, exserted in relation to the anthers; ovary locule hemicircular in longitudinal section. Fruit 20–26 × 20–28 mm, globose to depressed-globose, dehiscent by a lateral slit, floral remnant erect to pendent, deeply inserted; pericarp reddish to deep purple, smooth, shiny, with folds around base of floral remnant; funicular pulp white. Seeds 1.4–1.5 × 1–1.2 mm, hilum-micropylar region narrow, forming an angle of 60° with long axis of seed, testa-cells conic, periclinal walls much elongated, with intercellular depressions and coarse, dense cuticular folds.

Western *cerrado* element: amongst arenitic rocks in *cerrado*, c. 450–550 m, central Bahia, east bank of the Rio São Francisco. Endemic to Bahia. Map 38.

BAHIA: Mun. Ibotirama, 14 km E from the town on road BR 242, 18 July 1989, *Zappi* 147 (SPF, HRCB); l.c., 13 km E from the town, 14 Jan. 1991, *Taylor et al.* 1422 (K, HRCB, ZSS, CEPEC); l.c. (?), 12°13'18"S, 43°8'48"W, 2000, *F.M. Sene* J78 'sp. A' (K, photos); E of Rio São Francisco [Ibotirama?], 'Serra da Barriguda', s.d., *Horst & Uebelmann* 391 (U); Mun. Oliveira dos Brejinhos, 30 km from Ibotirama, 'Rio São João', s.d., *Horst & Uebelmann* 391 bis (ZSS).

CONSERVATION STATUS. Data Deficient [DD]. Range poorly understood, but assumed to be more extensive than the currently known habitats; if not, this taxon may merit a conservation category of Endangered in view of its small population size.

Characteristic of this species is its small, narrow, dark-brown to pinkish flower exterior and unusual seeds, with remarkably conic testa-cells and narrow hilum-micropylar region. The only other species of this group whose seeds show any similarity to those of *P. aureispinus* is *Pilosocereus vilaboensis* from Goiás. Furthermore, the seeds of a probable synonym of the latter, *P. rizzoanus* P. J. Braun & E. Esteves-Pereira (1992), seem to present intermediate characteristics. The population described as *P. rizzoanus* is also geographically intermediate, occurring half way between those of *P. vilaboensis* and *P. aureispinus*.

The peculiar testa of the seeds of *P. aureispinus* may be related to dispersal by ants, that are especially abundant in the *cerrado* where this plant occurs, the conic testa-cells perhaps representing an adaptation related to the transport of the seed. Indeed, some of the plants were actually seen growing on top of anthills.

Known only from east of the Rio São Francisco, near Ibotirama, this species inhabits arenitic rock outcrops in a phase of the *cerrado* and is sympatric with *Facheiroa squamosa* (Gürke) P. J. Braun & E. Esteves Pereira. However, it may be expected to occur elsewhere to the north and south in this little-botanized region.

PILOSOCEREUS PIAUHYENSIS Group (Nos 17–20)
Stems branched at base and above. Found on or amongst rocks in *caatinga*.

17. Pilosocereus multicostatus *F. Ritter*, Kakt. Südamer. 1: 79–80, Abb. 52 (1979). Holotype: Brazil, Minas Gerais, Mun. Itaobim, 1965, *Ritter* 1346 (U).

Pseudopilocereus multicostatus (F. Ritter) P. V. Heath in Calyx 4: 140 (1994).

Shrubby, 1.5–5.0 m, branched at base or above, with or without a well-defined trunk; vascular cylinder weakly woody; branches 3.8–7.5 cm diam., erect, epidermis dark green, shiny, smooth; ribs (15–)18–25(–26), 3–6 × 6–9 mm, sinuses straight, transverse folds not visible. Areoles 2–3 mm diam., 4–7 mm apart, felt white to greyish, without long hairs. Spines 0.15–0.40 mm diam. at base, translucent, golden yellow to brownish, centrals 3–7, 10–20 mm, ascending to porrect, radials 15–18, 5–10 mm, adpressed. Fertile part of stem not to slightly differentiated, flower-bearing areoles randomly distributed along the branches, mostly subapical, with long, flexible, golden, to 40 mm bristles, and sparse white to grey, long hairs. Flower-buds acute 1–3 days before anthesis; flowers 4.7 × 2.9–3.0 cm; pericarpel nearly globose; tube 25 × 8.5 mm at base, flared to 18 mm diam. at apex, straight, terete, pale green, smooth, apex with triangular bract-scales; style 24–26 × 1–1.5 mm, tapering; stigma-lobes c. 10, not extending beyond the anthers; ovary locule circular to hemicircular in longitudinal section. Fruit 2.2–2.5 × 2.5–3.0 cm, depressed-globose, dehiscent by an adaxial or abaxial slit, floral remnant pendent, deeply inserted; pericarp bright green tinged purplish, opaque, smooth; funicular pulp magenta. Seeds 1.5 × 1.2 mm, testa-cells domed, with intercellular depressions and coarse, sparse cuticular folds.

South-eastern *caatinga* (inselberg) element: on gneissic inselbergs amongst *caatinga-agreste* in the drainage of the Rio Jequitinhonha, c. 670–900 m, north-eastern Minas Gerais. Endemic to the core area within Minas Gerais. Map 39.

MINAS GERAIS: Mun. Águas Vermelhas, 27.5 km E of Curral de Dentro towards road BR 116, 31 Jan. 1991, *Taylor et al.* 1517 (K, HRCB, ZSS, BHCB); Mun. Salinas, near Baixa Grande, road BR 251, 42°3'W, 31 May 2002, *M. Machado* (photos); Mun. Pedra Azul, 10 km SE from the town, 16°8'S, 41°12'W, 19 Oct. 1988, *Taylor & Zappi in Harley* 25188 (K, SPF); l.c., 10 km W from the town towards road BR 116, 1 Feb. 1991, *Taylor et al.* 1519A (K, HRCB, ZSS, BHCB); Mun. Itaobim, 1965, *Ritter* 1346 (U).

CONSERVATION STATUS. Near Threatened [NT] (1); extent of occurrence = < 5000 km²; PD=2, EI=1, GD=1. Short-list score (1×4) = 4. Needs to be monitored in view of its restricted distribution and the destruction of natural vegetation surrounding its rocky habitat.

Pilosocereus multicostatus is characterized by the high number of ribs and golden, flexible spines, together with the slender and delicate flowers. Besides the following obviously allied species, in its the bristly flower-bearing areoles, habit and ecology on inselbergs in *caatinga*, *P. multicostatus* recalls *P. chrysostele*, which inhabits similar rock outcrops in Pernambuco, Paraíba, Rio Grande do Norte and Ceará.

Marlon Machado has recently (June, 2002) photographed what is almost certainly a hybrid between *P. multicostatus* and *P. floccosus* subsp. *quadricostatus* in the surroundings of Pedra Azul, MG (Plate 40.2).

18. Pilosocereus piauhyensis *(Gürke) Byles & G. D. Rowley* in Cact. Succ. J. Gr. Brit. 19(3): 67 (1957). Holotype: Brazil, Piauí, 'Serra Branca' [Mun. São Raimundo Nonato], Jan. 1907, *Ule* 9 (B; K, SPF [photos ex B]).

Cereus piauhyensis Gürke in Monatsschr. Kakt.-Kunde. 18: 84–85 (1908). *Cephalocereus piauhyensis* (Gürke) Britton & Rose, Cact. 2: 49 (1920), tantum quoad typ., fig. excl. *Pilocereus piauhyensis* (Gürke) Werderm., Bras. Säulenkakt.: 111 (1933), tantum quoad typ. *Pseudopilocereus piauhyensis* (Gürke) Buxb. in Beitr. Biol. Pflanzen 44: 253 (1968).
P. *mucosiflorus* Buining & Brederoo in Kakt. and. Sukk. 28: 201–203 (1977). *Pilosocereus mucosiflorus* (Buining & Brederoo) F. Ritter, Kakt. Südamer. 1: 84 (1979). P. *piauhyensis* subsp. *mucosiflorus* (Buining & Brederoo) P. J. Braun & E. Esteves Pereira in Succulenta 74: 134 (1995). Holotype: Brazil, Piauí, São Raimundo Nonato, Serra da Capivara, 430–500 m, s.d., *Horst* 443 (U).
P. *gaturianensis* F. Ritter, Kakt. Südamer. 1: 81–82 (1979). *Pseudopilocereus gaturianensis* (F. Ritter) P. V. Heath in Calyx 4: 140 (1994). *Pilosocereus piauhyensis* subsp. *gaturianensis* (F. Ritter) P. J. Braun & E. Esteves Pereira in Succulenta 74: 134 (1995). Holotype: Brazil, Piauí, Gaturiano, 1963, *Ritter* 1218 (U).
? P. *chrysostele* subsp. *cearensis* P. J. Braun & E. Esteves Pereira in British Cact. Succ. J. 17: 21–27 (1999). Holotype: Brazil, Cent./S Ceará, [apparently some 150 km or more west of the eastern border of the state], 6 July 1981, *E. Esteves Pereira* 163 (UFG 13005, n.v.).

VERNACULAR NAMES. Facheiro, Rabo-de-raposa.

Shrubby, 1.5–2.5 m, branched only at base or also above; central cylinder weakly woody; branches 5–7.5 cm diam., erect, epidermis dark green, shiny, smooth; ribs 14–24, 5 × 7 mm, sinuses straight, transverse folds not visible. Areoles 2.5–5.0 mm diam., 5–7 mm apart, felt white to greyish, without long hairs, presenting indeterminate growth at base of stem. Spines 0.2–0.3 mm diam. at base, translucent, golden at first, later brownish to grey, centrals 5–9, 5–15 mm, ascending to porrect, radials 11–16, 3–8 mm, adpressed. Fertile part of stem not or only slightly differentiated, mainly on subapical areoles, with flexible, golden bristles and some white to grey long hairs. Flower-buds obtuse 1–3 days before anthesis; flowers 5.2–7.5 × 2.7–4.0 cm; pericarpel hemiglobose; tube 40–45 × 11–16 mm at base, flared to 25–29 mm diam. at apex, curved, terete, pale green, glaucous, smooth, with apical, rounded, impressed bract-scales; style 46–63 × 1.5–3.0 mm, tapering; stigma-lobes 8–11, exserted c. 1 cm in relation to the anthers; ovary locule hemicircular in longitudinal section. Fruit 2.1 × 3.8 cm, depressed-globose, dehiscent by a lateral slit, floral remnant pendent, deeply inserted; pericarp 4–6 mm thick, wine coloured, smooth; funicular pulp magenta. Seeds (1.3–)1.4–1.8 × (0.9–)1.0–1.3 mm, testa-cells flat, with intercellular depressions and coarse, sparse to absent cuticular folds.

Northern *caatinga* element: on arenitic/granitic rock outcrops associated with *caatinga*, c. 200–850 m, south-eastern Piauí, Ceará and Rio Grande do Norte. Endemic to North-eastern Brazil. Map 39.
PIAUÍ: Mun. Dom Expedito Lopez, Gaturiano, 1963, *Ritter* 1218 (U), *Horst & Uebelmann* 520, cult. (ZSS), 1983, *Braun* 465 (ZSS T-05418); l.c., Km 276 of the Petrolina–Teresina road, 5 km N of town, 1 Feb. 1990, *Zappi* 211 (HRCB, ZSS); Mun. São Raimundo Nonato, 'Serra Branca', Jan. 1907, *Ule* 09 (K, SPF photos ex B); l.c., Parque Nacional da Serra da Capivara, 6 Feb. 1990, *Zappi* 217 (HRCB), s.d., *Horst* 443, cult. (U, ZSS).
CEARÁ: N Ceará, Serra de Meruóca, between Coreaú and Alcântaras, 20 Feb. 1995, *Taylor* (K, photos); Cent. & S Ceará, unlocalized, see synonym cited above.
RIO GRANDE DO NORTE: Mun. Açu (Assu), road to Arcoverde, Jan. 1978, *Horst & Uebelmann* 479, cult. ZSS (ZSS).

CONSERVATION STATUS. Near Threatened [NT] (1); PD=2, EI=1, GD=2. Short-list score (1×5) = 5. Currently known from only c. 6 populations, which are not known to be declining and most plants are ± inaccessible on steep rocky slopes. The southernmost population is protected within a national park (see above).

Described by Gürke (1908) as *Cereus piauhyensis*, on the basis of material collected by Ule (*Ule* 09) at the Serra Branca, Piauí, its specific name was long misapplied to *P. pachycladus*, first by Britton & Rose (1920), who published *Cephalocereus piauhyensis* (Gürke) Britton & Rose, and later by Werdermann (1933), as *Pilocereus piauhyensis* (Gürke) Werderm. Both authors had not seen true material of *P. piauhyensis* and confused it with what is now known as *P. pachycladus* F. Ritter subsp. *pernambucoensis* (F. Ritter) Zappi, a very widespread taxon from North-eastern Brazil. The same mistake is made in the illustrated work of Andrade-Lima (1989).

The recently described, but poorly localized, *P. chrysostele* subsp. *cearensis,* is probably a northern smaller-flowered form of this species rather than that to which it is referred by its authors. The critical characters, which place it within *P. piauhyensis* rather than *P. chrysostele,* are the lack of a cephalium, the slender flower-tube and the morphology of the dehiscent fruit.

19. Pilosocereus chrysostele *(Vaupel) Byles & G. D. Rowley* in Cact. Succ. J. Gr. Brit. 19: 66 (1957). Type: not extant. Neotype (Zappi 1994): Brazil, Ceará, 'Serra do Cantim', June 1933, *Luetzelburg* 23755 (M; K, photo.; IPA, isoneo.).

Cereus chrysostele Vaupel in Zeitschr. Sukk.-Kunde 1: 58 (1923). *Pilocereus chrysostele* (Vaupel) Werderm., Bras. Säulenkakt.: 102–103 (1933). *Cephalocereus chrysostele* (Vaupel) Borg, Cacti: 104 (1937). *Pseudopilocereus chrysostele* (Vaupel) Buxb. in Beitr. Biol. Pflanzen 44: 252 (1968).

VERNACULAR NAMES. Facheiro, Facheiro-de-serra, Rabo-de-raposa.

Shrubby, branching from the base upwards or only above and treelike with a trunk to 1.5 m or more, 1.5–6.0 m; vascular cylinder weakly woody; branches 4–7 cm diam., erect, epidermis pale green, smooth; ribs 19–28, 5 × 5 mm, sinuses straight, transverse folds not visible. Areoles 2–3 mm diam., 4–5 mm apart, felt white to blackish, with white long hairs to 8 mm, growth indeterminate near base of branches. Spines 0.15–0.20 mm diam. at base, translucent, pale yellow, centrals 6–8, 5–12 mm, ascending to porrect, radials 9–12, 4–10 mm, adpressed. Fertile part of stem weakly to strongly differentiated, the flower-bearing areoles occupying the lateral and subapical region of the branches, with white hairs to 20 mm and usually conspicuous golden, 40–60 mm bristles. Flower-buds obtuse 4 days before anthesis; flowers 4.5–6.0 × 3.7 cm; pericarpel hemiglobose; tube 33–35 × 17 mm at base, flared to 22 mm diam. at apex, constricted above and below the nectar-chamber, straight, slightly compressed, brownish green to pinkish, distal third with thick, broadly ovate bract-scales; style 27–28 × 1.8–2.0 mm, tapering, stigma-lobes 12–13, included in relation to the anthers; ovary locule hemicircular to compressed in longitudinal section. Fruit 2–2.5 × 3–3.5 cm, depressed-globose, dehiscent by a lateral or adaxial slit, floral remnant pendent, deeply inserted; pericarp purplish, smooth; funicular pulp magenta. Seeds (1.2–)1.4–1.7 × (0.9–)1.0–1.2 mm, testa-cells flat, with intercellular depressions or these sometimes nearly absent, cuticular folds not visible.

Northern *caatinga* element: on grey or whitish, granitic inselbergs with surface broken into blocks of stone, associated with very dry, highland *caatinga*, 430–1190 m, central-northern Pernambuco, Paraíba, Ceará and Rio Grande do Norte. Endemic to North-eastern Brazil. Map 39.
CEARÁ: N Ceará, Mun. Irauçuba, 10 Jan. 1995, *D. Macêdo* s.n. (EAC, K, photos); W Ceará, Mun. Tamboril, *fide* Uebelmann (1996): HU 939; E Ceará, Mun. Jaguaribe, Fazenda Mulungu, 11 July 1994, *D. Macêdo et al.* s.n. (EAC, K, photos); unlocalized, 'Serra do Cantim', June 1933, *Luetzelburg* 23755 (M).
RIO GRANDE DO NORTE: 'Sebastianópolis'–Apodi, 11 July 1960, *Castellanos* 22633 (GUA); Mun. Angicos, between Lajes and Fernando Pedrosa, reported by Braun & Esteves Pereira (1999a: 21); Mun. Acari, beside road BR 427, S of Currais Novos, 6°23'S, 36°38'W, 1 Apr. 2000, *E.A. Rocha et al.* (K, photo).
PARAÍBA: Mun. Brejo da Cruz, serra W of town, 6°22'S, 37°28'W, 2 Apr. 2000, *E.A. Rocha et al.* (K, photos); Mun. Sousa, near border with Mun. Pombal, *fide* Uebelmann (1996 & mss): HU 460; Mun. Pedra Lavrada, reported by Luetzelburg (1925–26, 3: fig. 29); Mun. Santa Luzia, 1998, *E.A. Rocha* 517 (JPB); Mun. Monte Horebe, 7°11'S, 38°33'W, 3 Apr. 2000, *E.A. Rocha et al.* (obs.); Mun. Maturéia, Serra de Teixeira, Pico do Jabre, 1994, *M.F. Agra et al.* (JPB), 8 July 1994, *L.P. Felix & Miranda* 6583 (PEUFR); Mun. Desterro, 1999, *E.A. Rocha* (JPB); Mun. Serra Branca, Coxixola, 9 km from Serra Branca, 28 Jan. 1971, *Andrade-Lima et al.* 1065 (IPA); Mun. Camalaú, road to Monteiro, 9 Feb. 1993, *F.A.R. Santos* 69 (ALCB, HUEFS, PEUFR); Mun. Monteiro, border with PE, Serra da Jabitaca, *Luetzelburg*, original type locality cited in Vaupel (1923).
PERNAMBUCO: Mun. Triunfo, 8 Dec. 1991, *F.A.R. Santos* 16 (HUEFS); Mun. (?), Serra Talhada, towards Carnaubeira, 22 May 1971, *Heringer et al.* 851 (IPA, RB, UB); Mun. Serra Talhada, 21 Nov. 1992, *F.A.R. Santos* 56 (HUEFS, PEUFR 13128, K, photo); l.c., rock outcrop N of the town, 13 Feb. 1990, *Zappi* 222 (HRCB, ZSS); Mun. Floresta, Carqueja, Serra de São Gonçalo, 23 May 1971, *Heringer et al.* 858 (IPA); without precise locality, Serra da Borborema, *Löfgren* s.n. (US, NY, photos).

CONSERVATION STATUS. Least Concern [LC] (1); PD=2, EI=1, GD=2. Short-list score (1×5) = 5. Least Concern, since often occuring on steep, dry rocky hillsides where destruction of vegetation is less likely.

Variable in stem thickness and rib number.

20. Pilosocereus densiareolatus *F. Ritter*, Kakt. Südamer. 1: 73–74, Abb. 43 (1979). Holotype: Brazil, Minas Gerais, Montes Claros, 1959, *Ritter* 957 (U).

Pseudopilocereus densiareolatus (F. Ritter) P. V. Heath in Calyx 4: 140 (1994).
P. densiareolatus var. *brunneolanatus* P. J. Braun & E. Esteves Pereira in Kakt. and. Sukk. 50: 287 (1999). Holotype: Brazil, Bahia, limestone outcrops W of Cocos, Sep. 1984, *E. Esteves Pereira* 224 (UFG 12381, n.v., K, ZSS, isos. — labelled '*P. superfloccosus* var. *brunneolanatus*').
'*Facheiroa* Br. & R. Sp. nova' Innes & Glass, The Illustrated Encyclopaedia of Cacti: 112 (1991).
[*P. superfloccosus* Buining & Brederoo in Cact. Succ. J. (US) 46: 60–63 (1974) pro parte quoad figs 1 & 2, excl. typ. *Pilosocereus superfloccosus* auctt. pro parte, non (Buining & Brederoo) F. Ritter.]

VERNACULAR NAMES. Cabeça-de-velho, Facheiro-da-lapa.

Treelike, 2–5(–6.5) m, with a well-defined trunk; vascular cylinder weakly woody, pith wide; branches 3–7 cm diam., erect, epidermis olive-green to bluish, glaucous, smooth; ribs 13–22, 5–10 × 4–10 mm, sinuses straight, transverse folds not visible. Areoles 2.5–5.0(–6.0) mm diam., 2–6 mm apart, felt white to greyish with white long hairs to 20 mm at apex of branches, growth indeterminate near base of plant. Spines to 30, 0.2–0.4 mm diam. at base, translucent, brown to golden, with pinkish or reddish shades, greyish with age, centrals 6–15, 9–30(–35) mm, ascending to porrect, radials 14–16, 4–7 mm, adpressed. Fertile part of stem slightly to strongly differentiated, flower-bearing areoles consecutive, occupying 1–4 ribs, sometimes somewhat immersed in the stem forming a lateral cephalium, with white, greyish or brownish hairs to 3–8 cm and golden bristles. Flower-buds obtuse 5 days before anthesis; flowers 5–6 × c. 3–4 cm; pericarpel depressed-globose, olive-green; tube 3.8 × 10–17 mm at base, flared to 25 mm diam. at apex, slightly curved, olive-green, smooth, distal half with obovate to acuminate bract-scales; style 45 mm, tapering, stigma-lobes c. 12; ovary locule hemicircular to strongly depressed in longitudinal section. Fruit c. 2–3 × 3 cm, depressed-globose, pinkish at base, brownish olive-green above, dehiscent at apex and by lateral splits in the pericarp, floral remnant pendent, deeply inserted; funicular pulp white. Seeds brown-black, shiny, (1.7–)1.8–2.5 × 1.3–1.9 × 1.4–1.6 mm, testa-cells flat, intercellular depressions and cuticular folds absent.

Southern Rio São Francisco *caatinga* element: sometimes locally co-dominant with other arborescent cacti and other woody plants on *Bambuí* limestone outcrops in *caatinga*, c. 450–800 m, central-northern Minas Gerais and western Bahia. Endemic to the core area within Eastern Brazil. Map 39.
BAHIA: Mun. Santana, 8 km from Porto Novo, 1975, *Horst* 394 *bis*, cult. p.p., veg. (U, ZSS); l.c., 8 km N of Porto Novo, 2 Oct. 1985, *Horst & Uebelmann* 394 *ter*, cult. (ZSS), 16 Jan. 1991, *Taylor et al.* 1432 (K, HRCB, ZSS, CEPEC); l.c., 28 km S from Santana towards Santa Maria da Vitória, 15 Jan. 1991, *Taylor et al.* 1424 (K, HRCB, ZSS, CEPEC); Mun. Bom Jesus da Lapa, Morro da Lapa, 17 Apr. 1983, *Leuenberger et al.* 3077A (CEPEC); l.c., W from the town, 1972, *Horst & Uebelmann* 394, *pro parte* (ZSS T-06002); Mun. Cocos, limestone outcrops W from the town, Sep. 1984, *E. Esteves Pereira* 224 (ZSS T-05836, UFG, K).
MINAS GERAIS: Mun. Manga, region of Jaíba, Mar. 1998, *I. Pimenta* (K, photo); Mun. Varzelândia, 2 km E from the town, 23 July 1989, *Zappi* 168 (SPF, HRCB); l.c., 10 km N of Varzelândia on road to Jaíba, 12 Aug. 1988, *Eggli* 1149 (ZSS); Mun. Capitão Enéas, limestone outcrop visible from road BR 122, 9 June 2002, *G. Charles* (K, photo); Mun. Montes Claros, 13 km N from the town on road BR 135 towards Januária, 16°38'S, 43°55'W, 27 Jan. 1991, *Taylor et al.* 1454 (K, HRCB, ZSS, BHCB); 30 km NW of Engenheiro Dolabela towards Sítio, 8 Aug. 1988, *Eggli* 1127 (ZSS); Mun. Bocaiúva, Granjas Reunidas/Eng. Dolabela, 3 May 1936, *A.P. Duarte* 7756 (RB); l.c., access road to Engenheiro Dolabela, 17°28'S, 44°1'W, 4 Nov. 1988, *Taylor & Zappi* in *Harley* 25503 (K, SPF), 17 June 1990, *Zappi* 236 (HRCB); l.c., E of Engenheiro Dolabela, 7 Aug. 1988, *Eggli* 1109 (ZSS).

CONSERVATION STATUS. Least Concern [LC] (1); PD=2, EI=2, GD=2. Short-list score (1×6) = 6. Least Concern, but habitat liable to decline through limestone quarrying, so monitoring is desirable.

Variable in stature and extent of cephalium wool developed.

Probable hybrids with *P. pachycladus* subsp. *pachycladus:*

? P. occultiflorus P. J. Braun & E. Esteves Pereira in Cact. Succ. J. (US) 71: 310–315, figs 1–5, 7–9, tab. 1 & 2 (1999). Type: Minas Gerais, W of Rio São Francisco, '400 m', *E. Esteves Pereira* 223 (UFG, holo. n.v., BONN, ZSS, iso.).

Shrubby to treelike; branches to 8 cm diam., epidermis bluish green; ribs 8–11, 10 × 8 mm. Areoles c. 5 mm diam., 3 mm apart. Spines 0.3–0.7 mm diam. at base, centrals 10(–35) mm. Fertile part of stem with brownish hairs to 8 cm.

BAHIA: Mun. Santana, 8 km from Porto Novo, 460 m, 16 Jan. 1991, *Taylor et al.* 1434 (K, HRCB, ZSS, CEPEC); 28 km S from Santana towards Santa Maria da Vitória, 530 m, 15 Jan. 1991, *Taylor et al.* 1427 (K, HRCB, ZSS, CEPEC).
MINAS GERAIS: unlocalized (W of Rio São Francisco), *E. Esteves Pereira* 223 (see synonym tentatively cited above); Mun. Varzelândia, 10 km N from the town, on road to Jaíba, 12 Aug. 1988, *Eggli* 1150 (ZSS).

Braun & Esteves Pereira's recently described species is referred here as a putative hybrid with only little hesitation, since its characters are strongly suggestive of hybrid origin (N.B. stem morphology, cephalium, flower form, fruit, seed size etc.), involving *P. pachycladus* and *P. densiareolatus*, the latter being reported as sympatric, l.c. *Pilosocereus pachycladus* certainly also occurs in this region (see above) and obvious hybrids involving these two species have been studied by the present authors in the other habitats cited here.

Pilosocereus densiareolatus F. Ritter (1979) was described on the basis of plants inhabiting *Bambuí* limestone outcrops in central-northern Minas Gerais, with flower-bearing areoles only moderately differentiated.

Northwards from Minas Gerais, the same species presents an increasingly well-developed lateral cephalium, with flower-bearing areoles immersed in the branches, and those populations have become known as *Pilosocereus superfloccosus* (Buining & Brederoo) F. Ritter, described from western Bahia. This is, in fact, a confused and misapplied name, since its protologue (cf. *Pseudopilocereus superfloccosus* Buining & Brederoo 1974a) clearly illustrates reproductive parts of *Pilosocereus gounellei* subsp. *zehntneri*, and the positively identifiable elements amongst its type material (U, holo.) consist of fragments of flower and fruit of *Pilosocereus gounellei* subsp. *zehntneri*, and none of '*P. superfloccosus*' as interpreted by most authors.

The problem of confusion between *P. densiareolatus* and *P. gounellei* subsp. *zehntneri* is not an unusual one. Most material examined from these species was found to be mixed (Zappi 1994). Field study of such sympatric populations indicates that the arborescent *P. densiareolatus* only flowers when the branches are far away from the ground, ie. 2.5–4.0 metres high, and, furthermore, the flowers are hidden in a hairy lateral cephalium. *P. gounellei* subsp. *zehntneri* has a shrubby to treelike habit, but produces flowers when less than 1 metre tall, its reproductive parts being much more accessible and obvious to collectors. Young plants of *P. densiareolatus* frequently look like *P. gounellei* subsp. *zehntneri*, but subsequently the branching pattern is completely different, being candelabriform for *P. gounellei* subsp. *zehntneri* and erect for *P. densiareolatus*, whose spination is also denser and finer. The examination of the apex of the fruits has proved to be the best character to differentiate them: *P. densiareolatus* has a deeply sunken, pendent floral remnant, typical of

Subgenus *Pilosocereus*, while *P. gounellei* subsp. *zehntneri* has fruits with rounded and superficially inserted floral remnants, characteristic of Subgenus *Gounellea*.

Insufficiently known taxa

1) The following names refer to a taxon treated as *incertae sedis* here (cf. Zappi 1994: 103; Ritter 1979: 64–65): *Pilocereus glaucescens* Linke (1858); *Cephalocereus glaucescens* (Linke) Borg; *Pilosocereus glaucescens* (Linke) Byles & G. D. Rowley; *Pseudopilocereus glaucescens* (Linke) Buxb.; '*Cereastreae glaucescens*' Labour. (1853), nom. inval. (Art. 43.1); *Pilocereus coerulescens* Lem. (1862), nom. illeg. (type as for above); *Pilosocereus coerulescens* (Lem.) F. Ritter, nom. illeg. (The plant *described* by Lemaire under this illegitimate name originated from the Serra do Cipó, MG and is identifiable as *P. aurisetus*, q.v.) Being guided only by Labouret's original description it is clear that the small plant he had before him could have represented any one of at least 3 Brazilian species, even assuming that his statement that it came from Brazil was correct. This plant was not preserved and it is futile to speculate further on its identity in the absence of a definite locality.

2) The recently described *Pierrebraunia brauniorum* E. Esteves Pereira in Kakt. and. Sukk. 50: 311–314 (1999) will likely remain a botanical mystery until the secrecy about its geographical origin is overcome. Esteves Pereira gives away no more than to say that it emanates from high mountains in the Serra do Espinhaço of Minas Gerais (a mountain range more than 1000 km in extent) and inhabits an area where a decidedly improbable list of other cactus genera are said to grow (improbable in the sense that more than 15 years' study of the cacti of Eastern Brazil by the present authors has so far failed to reveal any instance where the genera *Facheiroa* and *Cipocereus* occur in proximity, especially since in Minas Gerais the former genus is restricted to limestone outcrops close to the Rio São Francisco, very far from the Serra do Espinhaço, where *Cipocereus* is found). The plant's extraordinary combination of characters is unequalled and its stated rarity lends support to the idea, ventured in Taylor (2000), that this is some sort of bizarre intergeneric hybrid, the likes of which are not unknown elsewhere in the family, eg. *Bergerocactus* with *Pachycereus* (×*Pachgerocactus*) and, separately, with *Myrtillocactus* (×*Myrtgerocactus*), from northern Baja California, Mexico. In the case of the vaingloriously named *Pierrebraunia brauniorum*, its combination of few-ribbed stems covered by a visibly roughened epidermis and small, deep pink, hummingbird syndrome flowers tempts the suggestion that this is a hybrid between two genera, the most plausible being a *Pilosocereus* (eg. *P. floccosus* subsp. *quadricostatus*) and either an *Arrojadoa* or a *Micranthocereus*, both of which are said to occur in the vicinity. There is a good illustration of the plant in Braun & Esteves Pereira (2002: 6, Abb. 1).

3) The following may be a distinct species of dwarf habit, but has not been studied *in vivo* by the authors, either in habitat or in cultivation. It was tentatively treated as a possible synonym of *P. machrisii* by Zappi (1994), whose distribution it overlaps in western Bahia (eg. it is reported from Mun. São Desidério, Sítio Grande, road to Barreiras, s.d., *Horst & Uebelmann* HU 203, cult. [ZSS], with *P. machrisii* being known from just north of Barreiras; Map 38). While it should 'key out' under *P. machrisii*, a photograph of the above collection in cultivation has been kindly supplied by Graham Charles as an additional aid to identifying this little-known taxon (see Plate 49.2): *Pilosocereus parvus* (L.

Diers & E. Esteves Pereira) P. J. Braun in Bradleya 6: 88 (1988); *Pseudopilocereus parvus* L. Diers & E. Esteves Pereira in Kakt. and. Sukk. 33: 100–104 (1982). Holotype: Brazil, Goiás, [Mun. Posse], Serra Geral de Goiás, border with Bahia, 900–1000 m, Apr. 1978, *E. Esteves Pereira* 94 (KOELN ['Succulentarium'], n.v.; UFG, iso.).

4) As the manuscript of the present work was being finalized, Andreas Hofacker published a new taxon, *Pilosocereus bohlei* Hofacker in Kakt. and. Sukk. 52: 253–257 (2001). This plant is known from a small population in the Serra São Francisco, northern Bahia, on quartzite (Map 38), where it is sympatric with *P. pachycladus*, *Micranthocereus flaviflorus* and *Melocactus zehntneri* (the holotype is cited, l.c. 255, as *A. Hofacker* 442, UFG 24356, n.v.). It will probably prove to be a 'good' species, but its relationship seems rather unclear at present for lack of adequate data on the morphology and dehiscence of its fruit. In the protologue, Hofacker compares it with *Stephanocereus luetzelburgii* and *Pilosocereus gounellei* — the swollen stem bases are reminiscent of the former, while the narrower apical fertile part of the stem strongly resembles flower-bearing growth of the latter. However, in response to a request for more information Hofacker (*in litt.*, 15.10.01) communicated that the fruit of his new taxon had deeply inserted perianth remains, which tends to rule out any affinity with *P. gounellei*. Besides, the flower of *P. bohlei* is different from either of the above, being short and S-shaped. In its basally branching habit and general stem morphology it could belong to the P. AURISETUS Group and may be 'keyed out' here under *P. machrisii* (q.v.). A brief summary of its morphology follows, based on the protologue: to 1.8 m, branched at base only, stems blue-green to strongly glaucous, to 12 cm diam. in the basal 35–40 cm, but narrowing to only 5 cm diam. in the fertile region above; ribs 9–12, tuberculate, to 15 × 15 mm; areoles oval, to 7 × 4 mm, long hairs (vegetative part) to 30 mm, white; spines red-brown to light brown or more golden in the fertile region, central 1, to 25 mm, radials 30–40, to 20 mm; areoles in fertile region bearing dense tufts of long white hairs to 70 mm, developing on up to 7 of the ribs; flower to 55 × 35 mm, white, tube curved; fruit 35 × 40 mm, greenish to bluish green, funicular pulp white; seed 1.0 × 0.7 × 0.4 mm, matt black, with strongly developed cuticula sculpturing (seed SEMs kindly supplied by A. Hofacker). See Plate 49.3.

21. MICRANTHOCEREUS Backeb.

in Blätt. Kakteenf. 1938(6): [22] (1938). Type: *M. polyanthus* (Werderm.) Backeb.

Including *Austrocephalocereus* Backeb. (1938) and *Siccobaccatus* P. J. Braun & E. Esteves Pereira (1990); *Micranthocereus* subg. *Austrocephalocereus* (Backeb.) P. J. Braun & E. Esteves Pereira (1991a); *M.* subg. *Siccobaccatus* (P. J. Braun & E. Esteves Pereira) N. P. Taylor (1991b).

Columnar, erect, unbranched or more commonly shrublike but branched only at base; vascular cylinder rather to scarcely woody, cortical tissues mostly very mucilaginous; seedlings frequently densely hairy and juvenile to adult stages also with long, flexible, bristly hypertrophic spines at stem base (except species nos. 1–3 and in populations of other species growing in sand). Branches erect or decumbent, 3–10 cm diam. or more, not constricted, epidermis dull to bright green or glaucous; ribs 14–33, low, crenate, sinuses straight. Areoles with felt and long hairs; spines weak, flexible to brittle, not very pungent, central spines porrect to deflexed, longer than the numerous radials; indeterminate growth at basal areoles present, sometimes forming long curled bristles. Flower-bearing region of stem lateral or subapical, differentiated and sometimes forming a cephalium modifying the stem's morphology. Flowers nocturnal or diurnal, when nocturnal strongly odoriferous, 15–42 × 5–32 mm; pericarpel smooth, naked, hemiglobose and often very clearly delimited from the tube; tube broadened below, then cylindric, white, greenish, yellowish, orange-red, pink or magenta, with broad bract-scales only near the apex; innermost filaments curved towards the style; perianth-segments triangular to lanceolate or spathulate, innermost white, yellowish, pinkish or purplish. Fruit obovate to turbinate, naked, indehiscent except in *M. albicephalus* and perhaps *M. auriazureus*, flower remnant drying brownish or black, erect, sometimes tardily deciduous, forming a broad scar at the fruit apex, but ± superficial; pericarp smooth, funicular pulp white or pinkish, solid. Seeds 1–1.7 mm, cochleariform or elongate; testa-cells flat or convex; cuticular folds present or absent.

The relationships of this endemic Brazilian genus within tribe Cereeae are unclear, but fruits with non-impressed apex, ± superficial floral remnants and stems bearing lateral cephalia are found in *Pilosocereus* subg. *Gounellea* (cf. *P. gounellei* subsp. *zehntneri*) and *Coleocephalocereus*. Some species in the latter genus display hypertrophic spine growth at the base of stems as in the majority of *Micranthocereus* species, but their seeds differ in shape and in the position of the hilum region. A hybrid between *Pilosocereus* (subg. *Pilosocereus*) *pentaedrophorus* and *M. purpureus* has been recorded from two distant sites at the eastern margins of the Chapada Diamantina, Bahia and implies that these two genera may be closely related.

The 8 species treated here seem distinct, though Nos 2 & 3, 4 & 5, and 6 & 7 may represent vicariant species-pairs, the last being difficult to distinguish, yet rather disjunct. Nos 1–7 are *campo rupestre* taxa (see Map 5), while No. 8 is found amidst *caatinga* forest on limestone outcrops. A ninth species closely related to No. 8 occurs on *Bambuí* limestone outcrops around the Serra Geral de Goiás, from southern Tocantins and Goiás to north-western Minas Gerais (*M. estevesii* (Buining & Brederoo) F. Ritter). Two of the species are single-site endemics (Nos 4 & 5) and two more are known from very small areas (Nos 1 & 6) and thus their conservation status needs to be carefully monitored. Plates 49.4–52.4.

1. Plants 0.3–2.5(–3.0) m, suffrutescent or with solitary inclined stems, vascular cylinder not woody, or, if woody, ribs 14–17; seeds cochleariform (crystalline rocks and sandstones, *campos rupestres*, Serra do Espinhaço and Chapada Diamantina) 2
1. Plants to > 3 m, maturing when > 1.2 m, erect with very woody vascular cylinder; ribs 21–30 or more; fruits drying inside the cephalium; seeds elongate, ± twisted (*Bambuí* limestone, SW Bahia and adjacent Minas Gerais) .... **8. dolichospermaticus**

2. Flowers > 30 × 20 mm, anthesis predominantly nocturnal 3
2. Flowers slender, < 25 × 11 mm, anthesis diurnal 4

3. Ribs 23–29(–32); cephalium wool white to yellowish; epidermis bright green; flowers greenish or pinkish white outside (Serra do Espinhaço: N Minas Gerais and southernmost Bahia) .... **2. albicephalus**
3. Ribs 10–26; cephalium wool pale brown, with pinkish or grey shades; epidermis grey-green or glaucous; flowers deep magenta outside (Chapada Diamantina, Bahia) .... **3. purpureus**

4. Stem solitary; floral remnants strongly blackened; central spines and bristles of flower-bearing areoles dark red to brown; ripe fruits green; seeds black (N Minas Gerais: Serra da Bocaina & Serranópolis) .... **1. violaciflorus**
4. Stems branched at base; floral remnants pale brown, not blackened; spines and bristles of flower-bearing areoles mostly golden or pale yellow; ripe fruits red or pinkish; seeds brownish ... 5

5. Flowers 15–18 mm, outer and inner perianth-segments of contrasting colours; stems 3–5.5 cm diam., erect to ± inclined or decumbent 6

5. Flowers 20–25 mm, perianth-segments ± concolorous or innermost paler; stems 5.5–7.0 cm diam., erect 7

6. Flowers orange-red/magenta-pink, with white to yellowish inner perianth-segments (N Bahia: Mun. Morro do Chapéu to Mun. Sento Sé) . . . . **7. flaviflorus**

6. Flowers pale purplish with pale cream or white inner perianth-segments (S Bahia: Mun. Caitité) . . . . **6. polyanthus**

7. Fertile part of stem not sunken, bristles and wool white or greyish (nr Grão Mogol, Minas Gerais) . . . . **4. auriazureus**

7. Cephalium sunken, bristles and wool golden or brownish (W of Seabra, Bahia) . . . . . **5. streckeri**

Subg. *Austrocephalocereus* (Backeb.) P. J. Braun & E. Esteves Pereira (Nos 1–3): Stems woody or lacking well-developed wood, erect and branched at base or solitary and then often inclined, 0.7–2.0(–3.0) m high, with a ± sunken, and sometimes discontinuous cephalium; lacking hypertrophic spine development at stem base; flowers diurnal and/or nocturnal; fruit with *persistent, blackened perianth remains* (inconsistently blackened in *M. purpureus*), pericarp fleshy; seeds black, testa-cells with convex periclinal walls and cuticular sculpturing. Endemic to the core area within Eastern Brazil and characteristic of the Northern *campos rupestres*.

1. Micranthocereus violaciflorus *Buining* in Kakt. and. Sukk. 20: 129–130 (1969). Holotype: Brazil, Minas Gerais, 'Chapada Diamantina' [Mun. Grão Mogol, Serra da Bocaina], 24 June 1968, *Buining* in *Horst* 275 (U).

Solitary, rarely branched at base, to 1 m high; vascular cylinder very woody. Stems erect to inclined, 3.5–4.0 cm diam.; ribs 14–17, 2 × 3–5 mm; epidermis green. Areoles 2 mm, to 3 mm apart, long hairs white to grey. Spines flexible, pale golden to golden brown, reddish or dark brown, central spines 6–8, longest one to 25 mm, radials 25–30, 5–15 mm. Fertile part of stem lateral, sunken, flower-bearing areoles with abundant, loose white hairs and long reddish or dark brown bristles to 3 cm. Flowers poorly known, 23 × 7 mm, diurnal, tube violet or pinkish red, perianth-segments deep pink, spreading to erect. Fruit obovate, truncate at apex, 11 × 9 mm, flower remnants black, pericarp smooth, dull, green. Seeds c. 1 mm, cochleariform, black; testa-cells convex, with cuticular folds (Barthlott & Hunt 2000: pl. 50.4).

Northern *campo rupestre* (N Serra do Espinhaço) element: amongst rocks at c. 900–1100 m, Serra da Bocaina and Serranópolis, northern Minas Gerais. Endemic to the core area within Minas Gerais. Map 40. **MINAS GERAIS:** Mun. Porteirinha, Serranópolis, 15°50'S, 42°49'W, Mar. 2000, *fide* M. Machado (*in litt.*, 11.04.00 and photo on Machado's website); Mun. Grão Mogol, 48 km W of bridge over Rio Vacaria, 28.5 km E of Caveira (road junction to Riacho dos Machados and Porteirinha) on road BR 251, 31 Jan. 1991, *Taylor et al.* 1514 (K, HRCB, ZSS, BHCB).

CONSERVATION STATUS. Vulnerable [VU D2] (2); area of occupancy estimated to be < 20 km²; PD=2, EI=1, GD=1. Short-list score (2×4) = 8. Will be under threat if charcoal producers move into the two very restricted areas where it occurs. Needs a population survey and regular monitoring.

This rare species is characterized by a suite of presumably plesiomorphic character states (a rather woody vascular axis, only moderately mucilaginous stem tissue, absence of hypertrophic spines at stem base, green pericarpel, strongly blackening, persistent perianth remnants and black seeds with convex testa-cells), which suggest that its phylogenetic position within *Micranthocereus* is basal. However, it has most characters in common with

Subg. *Austrocephalocereus*. It seems to be a relictual species, occupying a restricted habitat in a small area in the northern half of the Serra do Espinhaço (MG).

2. Micranthocereus albicephalus *(Buining & Brederoo) F. Ritter*, Kakt. Südamer. 1: 108 (1979). Holotype: Brazil, Minas Gerais, Mato Verde, 950 m, Aug. 1972, *Buining* in *Horst* 348 (U).

Austrocephalocereus albicephalus Buining & Brederoo in Kakt. and. Sukk. 24: 73–75 (1973).
  Coleocephalocereus albicephalus (Buining & Brederoo) F. H. Brandt in Kakteen Orchid. Rundschau 6: 124 (1981).
Micranthocereus monteazulensis F. Ritter, Kakt. Südamer. 1: 105–106 (1979). Holotype: Brazil, Minas Gerais, Serra de Monte Azul, 1964, *Ritter* 1214 (U).
M. aureispinus F. Ritter, tom. cit.: 107–108 (1979). Holotype: Brazil, Minas Gerais, SW of Monte Azul, 1971, *Ritter* 1482 (U).

Shrubby, to 0.5–2.5(–3.0) m, branched mainly at base; vascular cylinder not woody. Stems erect, 6–8.5 cm diam.; ribs 23–29(–32), 5 × 6.5 mm; epidermis bright green. Areoles 3–4 mm apart, long hairs white to grey. Spines flexible to brittle, golden, showing only weak indeterminate growth at basal areoles, central spines to 12, to 10 mm, radials 15–18, 5–15 mm, the lowermost slightly longer. Fertile part of stem lateral, sunken into the stem, flower-bearing areoles with abundant, white, compact hairs and long golden bristles to 6.5 cm. Flowers 45–50 × 26 mm, nocturnal; pericarpel 4–5 × 9 mm, whitish to cream; tube 27–30 × 10–25 mm, broadened at base, constricted at the middle, white or pale green with pink shades, with broad bract-scales only at apex; perianth-segments 10 × 3.5–5.5 mm, triangular to lanceolate, outer segments spreading to recurved, white, brown to magenta-pink at tips, inner segments spreading, white, pinkish to reddish brown at tips; stamens included in relation to the perianth-segments; nectar-chamber 7 × 8 mm; style 32 mm, stigma-lobes c. 8, exserted; ovary locule 3–4 × 4–5 mm, obtriangular in longitudinal section. Fruit broadly turbinate, 3.5 cm diam., flower remnants strongly blackened; pericarp smooth, shiny, pale to dark green or wine-coloured, apically dehiscent. Seeds 1.5–1.7 mm, cochleariform, black; testa-cells convex, with cuticular folds (Taylor 1991a: pl. 1.1).

Northern *campo rupestre* element: between crystalline rocks and on cliffs in *campo rupestre*, c. 800–1000 m, Serra do Espinhaço (Serra Geral) northern Minas Gerais and adjacent southernmost Bahia. Endemic to the core area of Eastern Brazil. Map 40.
BAHIA: Mun. Licínio de Almeida, *Ritter* 1415 (U); ibid. (?), 12 km from town towards Brejinho das Ametistas, 14°32'4"S, 42°31'51"W, 12 Mar. 1994, *N. Roque et al.* in CFRC 15039 (K, SPF), 10 June 2002, *G. Charles* (photos).
MINAS GERAIS: Mun. Monte Azul, mountains E of town (Serra Geral), 1964, *Ritter* 1214 (U); l.c., mountains SW of town, 1971, *Ritter* 1482 (U); Mun. Mato Verde/Rio Pardo de Minas, 12 km E of Mato Verde on road to Santo Antônio do Retiro, 15°23'S, 42°45'W, 9 Nov. 1988, *Taylor & Zappi* in *Harley* 25519 (K, SPF), 13 Aug. 1988, *Eggli* 1155 (ZSS).

CONSERVATION STATUS. Near Threatened [NT] (1); area of occupancy estimated to be < 2000 km²; PD=2, EI=1, GD=1. Short-list score (1×4) = 4. Potentially threatened due to its limited range and generally small size of its 4 known populations.

This is the southern sister species of the following.

3. Micranthocereus purpureus *(Gürke) F. Ritter* in Kakt. and. Sukk. 19: 157 (1968); Buining in Ashingtonia 2: 28–29 (1975). Holotype: Brazil, Bahia, Serra do Sincorá, 1906, *Ule* 4 (B, n.v.).

Cephalocereus purpureus Gürke in Monatsschr. Kakt.-Kunde 18: 86 (1908). *Austrocephalocereus purpureus* (Gürke) Backeb. in Jahrb. Deutsch. Kakteen-Ges. 1941(2): 53 (1942).

Cephalocereus lehmannianus Werderm. in Repert. Spec. Nov. Regni Veg. Sonderbeih. C: tab. 46 (1932); Bras. Säulenkakt.: 118–119 (1933). *Austrocephalocereus lehmannianus* (Werderm.) Backeb. in Cact. Succ. J. (US) 23: 149 (1951). *Micranthocereus lehmannianus* (Werderm.) F. Ritter in Kakt. and. Sukk. 19: 157 (1968), nom. inval. (Art. 33.2). *Coleocephalocereus lehmannianus* (Werderm.) F. H. Brandt in Kakteen Orchid. Rundschau 6: 124 (1981), nom. inval. (Art. 33.2). Holotype: Brazil, Bahia, Serra do Espinhaço, 1100 m, Apr. 1932, *Werdermann* 3295 (B†). Lectotype (designated here): Werdermann, l.c., tab. 46 (1932).

M. haematocarpus F. Ritter, Kakt. Südamer. 1: 105 (1979). Holotype: Brazil, Bahia, Ituaçu, 1964, *Ritter* 1328 (U).

M. ruficeps F. Ritter, Kakt. Südamer. 1: 106–107 (1979). Holotype: Brazil, Bahia, Jacobina, 1963, *Ritter* 1212 (U).

Shrubby, 0.6–2.5(–3.0) m, branched at base; vascular cylinder not woody. Stems erect, 4.5–10.0 cm diam.; ribs 12–26, rarely only 10 when immature, 4–5 × 8–9 mm; epidermis blue-green to strongly glaucous. Areoles 3–4 mm, to 6–8 mm apart, long hairs pale brown to grey. Spines flexible to brittle, golden or reddish, brown when old, showing only weak indeterminate growth at basal areoles, central spines 4–7, to 20(–25) mm, radials 20–25, to 10 mm. Fertile part of stem lateral, sunken into the stem, flower-bearing areoles with pale brown, pinkish or greyish, abundant, loose or compact hairs and few long golden or brownish bristles to 4 cm. Flowers 42 × 30–32 mm, partially opening at late afternoon, fully expanded at night; pericarpel 8 × 12 mm, whitish to pale pink; tube 20–25 × 20–25 mm, broadened at base, slightly constricted at the middle, deep pink-magenta, with broad bract-scales only at apex; perianth-segments 9–11 × 6–9 mm, triangular to lanceolate, outer segments recurved, deep pink to reddish, inner segments spreading to recurved, white; stamens included in relation to the perianth-segments; nectar-chamber 7 × 8 mm; style 25–28 mm, stigma-lobes c. 10, exserted; ovary locule 4–6 × 5–6 mm, obtriangular in longitudinal section. Fruit broadly turbinate, 1.7–2.0 cm diam., flower remnants brown, usually not blackened, pericarp smooth, shiny, purplish pink. Seeds 1.5–1.7 mm, cochleariform, black; testa-cells convex, with cuticular folds (Barthlott & Hunt 2000: pl. 50.2–3).

Northern *campo rupestre* (Chapada Diamantina) element: on crystalline rocks in *campo rupestre* and its ecotones with *caatinga* and *cerrado*, c. 350–1900 m, eastern flanks and highest peaks of the Chapada Diamantina. Endemic to Bahia. Map 40.

BAHIA: Mun. Jacobina, 2.5 km E of bridge to E of town, 16 July 1989, *Zappi* 129A (SPF); Mun. Morro do Chapéu, 19 June 1994, *L.P. de Queiroz* 4038 (HUEFS); l.c., 9 km NE of town towards Jacobina, 23 Aug. 1988, *Eggli* 1271 (ZSS); l.c., 19.5 km SE of town on road BA 052, Rio do Ferro Doido, 31 May 1980, *Harley et al.* 22900 (K, CEPEC); l.c., 11°49'S, 41°12'W, 11 Mar. 1996, *A. Conceição* 2248 (HUEFS); Mun. Seabra, c. 28 km W of town on road BA 242, 14 Jan. 1991, *Taylor et al.* (obs.); ibid., 16 km E of town, 18 July 1989, *Taylor & Zappi* (K, photos); Mun. Lençóis, 24 Sep. 1965, *A.P. Duarte* 9363 & *E. Pereira* 10076 (RB, HB); l.c., 14 km NW of town, Serra do Brejão, 12°27'S, 41°21'W, *Harley et al.* 22378 (K, CEPEC); Mun. Lençois, Morro do Pai Inácio, Mar. 1997, *A. Conceição* s.n. (K, photos); Mun. Andaraí, Rio Paraguaçu, 12°48'S, 41°20'W, 27 Dec. 1988, *Taylor & Zappi in Harley* 27429 (K, SPF, CEPEC), 20 Aug. 1988, *Eggli* 1223 (ZSS); Mun. Mucugê, 4 km WNW of town on the road to Boninal, 13°1'S, 41°26'W, 22 Dec. 1988, *Taylor & Zappi in Harley* 25599 (K, SPF, CEPEC); l.c., surroundings of town, 22 Dec. 1988, *Taylor & Zappi in Harley* 27381 (K, SPF, CEPEC); l.c., 3 km towards Igatu, 41°21'W, 22 Aug. 1986, *J.D.C. Arouck Ferreira et al.* 368 (HRB, RB), 10 Oct. 1987, *M.L. Guedes et al.* 1539 (ALCB); Mun. Piatã, road Piatã–Inúbia, c. 25 km NW of Piatã, 13°4'19"S, 41°55'24"W, 24 Feb. 1994, *Sano et al. in* CFCR 14539 (K, SPF); Mun. Abaíra, Serra do Rei, 13°17'S, 41°54'W, 12 Jan. 1994, *W. Ganev* 2791 (K, HUEFS), Feb. 1992, *Sano & Laessøe in Harley* 50897 (SPF, CEPEC); Mun. Rio de Contas, Pico das Almas, 'Campo do Queiroz', 13°32'S, 41°57'W, 11 Dec. 1988, *Taylor in Harley* 25564 (K, SPF, CEPEC); l.c., Cachoeira do Fraga, 13°35'S, 41°50'W, 24 Nov. 1988, *Taylor & Zappi in Harley* 25547 (K, SPF, CEPEC); Mun. Ituaçu, c. 2 km NE and 10 km N of town, 19 Aug. 1988, *Eggli* 1203, 1214 (ZSS).

CONSERVATION STATUS. Least Concern [LC] (1); PD=2, EI=1, GD=2. Short-list score (1×5) = 5. Its numerous habitats are not significantly affected by change at present.

The name *Cephalocereus purpureus* was misapplied to what is now known as *Coleocephalocereus goebelianus* by Britton & Rose (1920) and Werdermann (1933), the latter redescribing the true *Cephalocereus purpureus* as *C. lehmannianus*. Luetzelburg (1925–26, 1: fig. 38) illustrated the true *Micranthocereus purpureus* as 'Pilocereus na caatinga' and commented in the caption that it was always accompanied by *Cereus leucostele* (= *Stephanocereus leucostele*), which is not the case. He was evidently confusing it with *C. goebelianus* at this stage (however, cf. Luetzelburg 1925–26, 3: 69). The true identity of *M. (Austrocephalocereus) purpureus* was recognized by Ritter (1968) and Buining (1975b).

M. purpureus is the most wide-ranging and variable species in the genus and a characteristic element of the *campo rupestre* flora in the higher and eastern parts of the Chapada Diamantina, where it is constantly associated with *Stephanocereus luetzelburgii*. It hybridizes with *Pilosocereus pentaedrophorus* near Andaraí and Ituaçu.

Subg. *Micranthocereus* (Nos 4–7): Stems only rarely woody (*M.* cf. *flaviflorus*), branched at base, erect or semi-sprawling, 0.3–1.2 m high, with a superficial to sunken, and sometimes discontinuous cephalium and greater or lesser development of hypertrophic spines at stem base; seedlings densely woolly; flowers diurnal; *fruit with tardily deciduous, non-blackening perianth remains*, pericarp fleshy; seeds brown to brown-black, testa-cells with nearly flat periclinal walls lacking cuticular sculpturing. Endemic to the core area within Eastern Brazil and characteristic of the Northern and South-eastern *campos rupestres*.

4. Micranthocereus auriazureus *Buining & Brederoo* in Cact. & Succ. J. (US) 45: 120–123 (1973) ('*auri-azureus*'). Holotype: Brazil, Minas Gerais, Grão Mogol, 900–1000 m, 17–18 Aug. 1972, *Buining* in *Horst & Uebelmann* 346 ['348'] (U).

VERNACULAR NAME. Rabo-de-raposa.

Shrubby, to 1.2 m, much branched at base; vascular cylinder not woody. Stems erect, 6–7 cm diam.; ribs 15–19, 5 × 8–9 mm; epidermis blue-green, glaucous. Areoles 3–7 mm, to 4 mm apart, long hairs white to grey. Spines flexible to brittle, golden, showing indeterminate hypertrophic growth at basal areoles, forming long curled bristles, central spines to 6, 2–3(–8) cm, radials 25–30, 5–15 mm. Fertile part of stem lateral, superficial, flower-bearing areoles with white hairs and long bristles to 4(–8) cm. Flowers 20–25 × 10–11 mm, diurnal; pericarpel 2.5 × 5 mm, brownish to pale pink; tube to 16 × 7–9 mm, broadened at base, constricted at the middle, deep magenta pink, with broad bract-scales only at apex; perianth-segments 4–7 × 2.5–3.5 mm, triangular to lanceolate, outer segments spreading, magenta-pink, inner segments erect to spreading, lilac-pink; stamens included in relation to the perianth-segments; nectar-chamber 8 × 5–6 mm; style 15–19 mm, stigma-lobes c. 8, exserted; ovary locule 3–4 × 2–3 mm, obtriangular in longitudinal section. Fruit obovate, truncate at apex, 1.8 × 1.4–1.6 cm, flower remnants pale brown, deciduous when fruit ripe, pericarp smooth, dull, pale pink to reddish, sometimes splitting open transversely below apex. Seeds c. 1.5 mm, cochleariform, brown, shiny; testa-cells flat to convex, without cuticular folds.

South-eastern *campo rupestre* (Grão Mogol) element: between crystalline rocks and in quartz sand, c. 750–1000 m, Serra do Barão and vicinity, northern Minas Gerais. Endemic to the core area within Minas Gerais. Map 40.

MINAS GERAIS: Mun. Grão Mogol, 21 May 1982, *Giulietti et al.* in CFCR 3416 (SPF, K), 24 Aug. 1986, *Zappi et al.* in CFCR 9925 (SPF, MBM), 21 May 1987, *Pirani & Mello-Silva* in CFCR 10773 (SPF), 26 May 1988, *Zappi et al.* in CFCR 11944 (SPF), 15 Oct. 1988, *Taylor & Zappi* in Harley 25072 (SPF, K).

CONSERVATION STATUS. Endangered [EN B1ab(iii) + 2ab(iii)] (3); extent of occurrence < 5000 km²; area of occupancy estimated to be < 100 km²; PD=2, EI=1, GD=1. Short-list score (3×4) = 12. Endangered due to its very restricted distribution, part of which may be subject to inundation by a dam-lake in future. However, another part of its area is inside a recently created reserve (Parque Estadual da Serra do Barão).

According to the list in Uebelmann & Braun (1984) and the label on the type specimen at Utrecht, the type collection of *M. auriazureus* is HU 346, and the name should be considered legitimate, since it is clear that HU 348 was cited erroneously as its holotype in the protologue (HU 348 is the type number for the prior-published *M. albicephalus*, q.v.).

5. Micranthocereus streckeri *Van Heek & Van Criek.* in Kakt. and. Sukk. 37: 102–105, with illus. (1986). Holotype: Brazil, Bahia, W of Seabra, *Van Heek & Van Criekinge* 85/250 (KOELN ['Succulentarium'], n.v.).

Shrubby, to 0.8 m, densely branched at base. Stems erect, 3–5(–5.5) cm diam., vascular cylinder not woody; ribs 17–24(–25), 2–3 × 6–8 mm; epidermis grey-green, slightly glaucous. Areoles 2–3 mm, to 5 mm apart, with white to grey trichomes. Spines flexible, pale golden, hypertrophic growth of spines at basal areoles present, central spines 6–8, spreading, lowermost longest, to 25 mm, radials 20–30, 5–10 mm. Fertile part of stem lateral, superficial to sunken, flower-bearing areoles with compact white to brownish hairs and long pale golden to reddish brown bristles to 30 mm. Flowers 40–50 × 22 mm, diurnal; pericarpel 3.5 × 3.5 mm, pale pink; tube to 18 × 3–5 mm, broadened at base, cylindric above, deep pink, with bract-scales only at apex; perianth-segments 4–5 × 2.5 mm, triangular to lanceolate, outer segments spreading, deep pink, inner segments erect, purplish pink; stamens included in relation to the perianth; style 13–15 mm, stigma-lobes 5–6, included; ovary locule 2 × 2.5 mm, obtriangular in longitudinal section. Fruit obovoid, 11 × 10 mm, floral remnants pale brown, deciduous; pericarp smooth, dull, purplish. Seeds c. 1.2 mm, cochleariform, blackish, shiny; testa-cells flat to convex, without cuticular folds.

Northern *campo rupestre* (Chapada Diamantina) element: in the *campo rupestre/cerrado de altitude* ecotone, c. 1100 m, west of Seabra, central Bahia. Endemic. Map 40.
BAHIA: Mun. Seabra, 27–28.5 km W of town towards Ibotirama, 25 Aug. 1988, *Eggli* 1289 (ZSS), 17 July 1989, *Zappi* 145 (SPF), 14 Jan. 1991, *Taylor et al.* 1415 (K, HRCB, ZSS, CEPEC).

CONSERVATION STATUS. Critically Endangered [CR B1ab(iii) + 2ab(iii); C2a(ii); D] (4); PD=2, EI=1, GD=1. Short-list score (4×4) = 16. Critical due to its very restricted distribution, small population size (< 50 individuals seen) and potential for alteration of its habitat, which is above a main highway.

This species is poorly understood, being known from only the locality cited above where, as recently illustrated by Heek & Strecker (2002: 116), it appears to be undergoing introgression with the sympatric *M. purpureus*, making assessment of its typical morphological state rather difficult. Specimens which, in 1991, the authors interpreted to be least influenced by this introgression showed a certain resemblance to the geographically distant *M. auriazureus* (Grão Mogol, MG), but have ± sunken cephalia and darker seeds.

6. Micranthocereus polyanthus *(Werderm.) Backeb.* in Jahrb. Deutsch. Kakteen-Ges. 1941(2): 51 (1942). Holotype: Bahia, near Caitité, May 1932, *Werdermann* 3457 (B†). Lectotype (designated here): Werdermann, l.c. infra, tab. 43 (1932).

Cephalocereus polyanthus Werderm. in Repert. Spec. Nov. Regni Veg. Sonderbeih. C, Lfg 11, tab. 43 (1932); Bras. Säulenkakt.: 114–116 (1933). *Arrojadoa polyantha* (Werderm.) D. R. Hunt in Bradleya 5: 92 (1987).

Shrubby, to 1.2 m, much branched at base; vascular cylinder not woody. Seedlings with abundant white hairs and long curled, hypertrophic bristles at base. Stems erect, 3.5–5.0 cm diam.; ribs 15–20, 4–5 × 3–6 mm; epidermis grey-green, glaucous. Areoles 2–4 mm, to 10 mm apart, long hairs white to grey. Spines flexible, pale golden, showing indeterminate growth at basal areoles, forming long curled bristles, central spines 4–8, to 15 mm, radials 20–28, 5–10 mm. Fertile part of stem lateral, superficial to slightly sunken, flower-bearing areoles with abundant, loose white hairs and long, pale golden bristles to 20 mm. Flowers 18 × 7–8 mm, diurnal; pericarpel 2 × 3 mm, pale pink; tube to 16 × 5–6 mm, broadened at base, constricted at the middle, coral red to pale pink, with broad bract-scales only at apex; perianth-segments 4–5 × 1–2 mm, triangular to lanceolate, outer segments spreading to erect, coral red to reddish pink, inner segments opening very slightly, pale cream to white; stamens included in relation to the perianth-segments; style 13–15 mm, stigma-lobes c. 8, included; ovary locule 1–2 × 1.5–2.0 mm, obtriangular in longitudinal section. Fruit obovate, truncate at apex, 5–7 × 5 mm, flower remnants pale brown; pericarp smooth, dull, pale pink. Seeds c. 1 mm, cochleariform, dark brown, shiny; testa-cells flat to convex, without cuticular folds.

Northern *campo rupestre* element: in quartz sand amongst crystalline rocks, c. 900–1000 m, Mun. Caitité, southern Bahia. Endemic. Map 40.
BAHIA: Mun. Caitité, c. 9 km ESE from town on road to Brumado, then c. 2 km S on footpath towards 'São João', 27 Aug. 1988, *Eggli* 1309 (ZSS); l.c., 25 km S of town, 1 km SW of Brejinho das Ametistas, 26 July 1989, *Zappi* 176 (SPF), 2 Feb. 1991, *Taylor et al.* 1537 (K, HRCB, ZSS, CEPEC).

CONSERVATION STATUS. Endangered [EN B2ab(iii)] (3); area of occupancy estimated to be < 100 km²; PD=2, EI=1, GD=1. Short-list score (3×4) = 12. Endangered due to its restricted distribution and projected decline due to mining activities.

In its overall appearance, but especially in habit, soft stems, superficial cephalium and bicoloured flowers, it is assumed to be the southern sister species of the following.

7. **Micranthocereus flaviflorus** Buining & Brederoo in Kakt. and. Sukk. 25: 25–27 (1974). Holotype: Brazil, Bahia, Mun. Sento Sé / Umburanas, Serra do Curral Feio, 850 m, 22 July 1972, *Buining* in *Horst* 389 (U).

M. densiflorus Buining & Brederoo in Cact. & Succ. J. (US) 46: 113–116 (1974). *M. flaviflorus* subsp. *densiflorus* (Buining & Brederoo) P. J. Braun & E. Esteves Pereira in Succulenta 74: 132 (1995). Holotype: Brazil, Morro do Chapéu, 900 m, *Horst* 221 (U).
M. uilianus Brederoo & C. A. L. Bercht in Succulenta 63: 178–183 (1984). *M. flaviflorus* var. *uilianus* (Brederoo & C. A. L. Bercht) P. J. Braun & E. Esteves Pereira in Succulenta 74: 132 (1995), nom. inval. (Art. 33.2?). Holotype: Brazil, Bahia, Mun. Sento Sé, 'Serra da Mimosa, near Limoeiro, 1130 m', 4 July 1974, *Buining* in *Horst & Uebelmann* 439 (U; ZSS, iso.).

Shrubby, 30–100 cm, much branched at base; vascular cylinder not woody (except in the sand-dwelling populations from E side of Mun. Morro do Chapéu). Stems partly decumbent, 4–5 cm diam.; ribs 12–16, 3–5 × 5–7 mm; epidermis grey-green, glaucous. Areoles 2–3.5 mm, to 3 mm apart, long hairs white or cream to grey. Spines flexible, golden, sometimes reddish, showing indeterminate hypertrophic growth at basal areoles in some populations, forming long curled bristles, central spines 4–8, 10–25 mm, radials 20–25, 10–12 mm. Fertile part of stem lateral to apical, superficial, flower-bearing areoles with loose white to yellowish hairs and long pale to brownish golden bristles to 2.5 cm. Flowers 14–18 × 5–8 mm, diurnal; pericarpel 2 × 2.5–3.0 mm, pale pink; tube to 12 × 4–6 mm, broadened at base, constricted at the middle, orange-red to pink, with bract-scales only at apex; perianth-segments 2–4 × 1–3 mm, triangular to lanceolate, outer segments spreading to erect, orange-red to reddish or pink/magenta; inner segments opening very slightly, yellow to pale cream; stamens included in relation to the perianth-segments; style 10–12 mm, stigma-lobes 4–6, included; ovary locule 1.5 × 1.8–2.0 mm, obtriangular in longitudinal section. Fruit obovoid to globose, truncate at apex, 5–8 × 5–8 mm, flower remnants pale brown, deciduous; pericarp smooth, dull, red to reddish pink. Seeds c. 1–1.5 mm, cochleariform, blackish, shiny; testa-cells flat to convex, without cuticular folds (Barthlott & Hunt 2000: pl. 50.5–8).

Northern *campo rupestre* (Chapada Diamantina) element: in the *campo rupestre/caatinga* ecotone on quartz sand or sandstone between low shrubs, c. 700–1200 m, northern and western flanks of the Chapada Diamantina (northwards from the region of Morro do Chapéu), northern Bahia. Endemic. Map 40.
BAHIA: Mun. Sento Sé, 'Serra do Tinga', Aug. 1912, *Zehntner* 274 (US), ibid., s.n. (M); Serra do Mimoso, near Limoeiro, 4 July 1974, *Buining* in *Horst & Uebelmann* 439 (U, ZSS); l.c., 10°19'S, 41°24'W, 11 July 2000, *G. Charles* (photos); Mun. Sento Sé / Umburanas, Serra do Curral Feio [S of Minas do Mimoso], s.d., *Horst* 389 (U); l.c., 10°22'S, 41°19'W, 9 Apr. 1999, *L.P. de Queiroz et al.* 5120 (HUEFS 36691); Mun. Umburanas, 'estrada velha Delfino – Mimoso de Minas, Serra do Curral Frio', near Delfino, 10°24'7"S, 41°18'43"W, 9 Mar. 1997, *Harley et al.* (ALCB, UEC, K, photos), ibid., 10°24'14"S, 41°18'41"W, 3 Apr. 2002, *Oliveira et al.* 118 (HUEFS 59063); Mun. Várzea Nova, 11°16'S, 40°56'W, *Hofacker* (fide Machado, pers. comm.); Mun. Morro do Chapéu, north of town, W bank of the Rio Salitre, c. 8 km from Icó, N of Brejão (Formosa), 'lagedo bordado', 5 Aug. 2002, *Taylor & Machado* (K, photos); l.c., W of town, S of road BA 052 towards 'Pedra Duas Irmãs', c. 11°33'40"S, 41°17'50"W, June 2002, *M. Machado* (photos); l.c., SE of the town, 2 Aug. 2002, *Taylor* (K, photos); l.c., S of town, terreno de M. Machado, 3 Aug. 2002, *Taylor* (K, photos); l.c., Boqueirão dos Lages, 17 May 1975, *A.L. Costa & G.M. Barroso* s.n. (HBR, ALCB); l.c., 19–20 km W of town on road BA 052, 11°28'S, 41°19'W, 25 Dec. 1988, *Taylor & Zappi* in *Harley* 27392 (K, SPF, CEPEC), 14 Jan. 1977, *Hatschbach* 39565 (MBM); l.c., 22 km W, 17 July 1989, *Zappi* 135 (SPF, K), l.c., c. 26–27 km W, then 1 km S on road to Cafarnaum, 26 Dec. 1988, *Taylor & Zappi* (obs.).

CONSERVATION STATUS. Least Concern [LC] (1); extent of occurrence = 1024 km^2; PD=2, EI=1, GD=2. Short-list score (1×5) = 5. Area of occupancy unknown, but assumed to be < 10% of extent of occurrence.

The distribution of this species seems somewhat disjunct, but the region between its northern and southern sites has been little explored. However, this includes extensive areas of limestone and derived calcareous *caatinga* soils, which would not suit it and thus it is almost certain that its distribution is highly fragmented. However, populations observed contain abundant plants and numerous seedlings and some are rather extensive. It is highly variable between populations, and much more so than the above synonymy indicates. More importantly, its eastern and south-eastern populations (on more sandy terrain, Plate 52.1) may prove to be indistinguishable from the preceding species, when both are better understood. At present it is largely the remarkable geographical disjunction between these two taxa that causes the writers to hesitate in proposing their merger or re-shaping.

In August 2002, Taylor & Machado observed hummingbirds and a yellow butterfly actively visiting the flowers in two populations of the species in Mun. Morro do Chapéu.

Subg. *Siccobaccatus* (P. J. Braun & E. Esteves Pereira) N. P. Taylor (No. 8): Stems normally unbranched above ground unless damaged, forming erect, very woody columns to 5 m or more, with hypertrophic spine development at base when young; cephalium deeply sunken, continuous; flowers nocturnal; fruit drying out to become paper-thin, perianth remains deciduous; seeds slender elongate or with many, small testa-cells. Limestone west of the Rio São Francisco.

8. Micranthocereus dolichospermaticus *(Buining & Brederoo) F. Ritter*, Kakt. Südamer. 1: 108 (1979). Holotype: Brazil, Bahia, W of Bom Jesus da Lapa, 460 m, *Horst & Uebelmann* 395 (U)

Austrocephalocereus dolichospermaticus Buining & Brederoo in Kakt. and. Sukk. 25: 76–79 (1974). *Siccobaccatus dolichospermaticus* (Buining & Brederoo) P. J. Braun & E. Esteves Pereira in Succulenta 69: 7 (1990).

Erect, fertile from c. 1.2 m, reaching more than 3 m, unbranched above ground, but often sprouting at the base and forming a compact group of erect stems; vascular cylinder very woody. Seedlings with spines curved upwards and with long, curled, hypertrophic bristles at base. Stems 8–12 cm diam.; ribs 21–30(–32), 10–20 × 10–22 mm, in cephalium region 12 × 13 mm; epidermis blue-green, strongly glaucous. Areoles 3–5 mm, to 2–6 mm apart, long hairs white to grey, to 3 cm. Spines hard, brittle, golden to brown and black when old, central spines 4–7, to 15 mm, radials 8–12, to 5 mm. Fertile part of stem lateral, sunken into the stem, flower-bearing areoles with white to pale yellow, abundant, compact hairs and long brownish or reddish bristles to 4 cm. Flowers 4 × 2.5 cm, nocturnal, all white or cream-coloured; pericarpel 5 × 11–12 mm; tube 35–38 × 12–18 mm, slightly constricted at the middle, with broad bract-scales only at apex; perianth-segments 9–11 × 6–9 mm, triangular to lanceolate, outer segments recurved, inner segments spreading; stamens included in relation to the perianth-segments; nectar-chamber 10 × 9–10 mm; style 32–36 mm, stigma-lobes 8–9, exserted; ovary locule 4–5 × 2 mm, depressed in longitudinal section. Fruit broadly turbinate, 9–10 mm diam., flower remnants pale brown, deciduous; pericarp dry, brownish, wrinkled. Seeds 2 mm, elongate, brownish; testa-cells flat, hilum depressed (Barthlott & Hunt 2000: pl. 51.1–2).

Southern Rio São Francisco *caatinga* element: on *Bambuí* limestone outcrops surrounded by *caatinga* or forest with *caatinga* elements, c. 450–650 m, west of the Rio São Francisco, south-western Bahia and adjacent Minas Gerais. Endemic to Eastern Brazil. Map 40.

BAHIA: SW Bahia, Mun. Santana, 30 km from town towards Porto Novo, 16 Jan. 1991, *Taylor et al.* 1431 (K, HRCB, ZSS, CEPEC); 'Mun. Bom Jesus da Lapa, Faz. Serra Solta' (unlocalised), 17 July 1975, *Andrade-Lima* 75-8193 (IPA); l.c., W of Rio São Francisco, reported by Andrade-Lima (1977: figs 2, 6 & 9, as '*Austrocephalocereus purpureus*'); l.c., Mun. Serra do Ramalho, Serra do Ramalho, 13°31'S, 43°45'W, 2001, photographed by *J. Jardim* growing with *Facheiroa cephaliomelana* (*J. Jardim* 3486, CEPEC); near the Rio Carinhanha, 1986, illustrated by Braun & Esteves Pereira (2002: 178, Abb. 185).
MINAS GERAIS: Braun & Esteves Pereira, l.c., report this species from adjacent to the Bahian border.

CONSERVATION STATUS. Least Concern [LC]; PD=2, EI=1, GD=1. Short-list score (1×4) = 4. At one time thought to be threatened from the destructive collection of seeds and also potentially from the quarrying of limestone, but range and abundance now known to be much greater.

The peculiar, slender-elongate seeds of this species are assumed to be an adaptation for wind dispersal (W. Barthlott, pers. comm.). It is closely related to *M. estevesii* (Buining & Brederoo) F. Ritter, the only member of this genus occurring outside the geographical area treated here (NW Minas Gerais to S Tocantins).

22. COLEOCEPHALOCEREUS Backeb.

Blätt. Kakteenf. 1938(6): unpaged [22] (1938). Type: *C. fluminensis* (Miq.) Backeb.

Including *Buiningia* Buxb. (1971); *Coleocephalocereus* subg. *Buiningia* (Buxb.) P. J. Braun (1988).

Literature: Taylor (1991a: 17–19).

Tall columnar, sometimes unbranched (unless damaged), or low-growing and sometimes caespitose, with most branches arising near the rootstock or from the decumbent stem bases; stems depressed-globose to cylindric, often asymmetric at apex once cephalium has developed, vascular cylinder rather to scarcely woody, cortical tissues with or without mucilage, pith sometimes chlorophyllous; ribs 9–30 or more, low and rounded to higher and triangular; areoles large to rather small, close-set. Spination various, sometimes poorly developed; hypertrophic spines developed at stem base in some species. Cephalium lateral, always deeply sunken, composed of bristles and wool in variable proportions. Flowers rather small, c. 2–7.5 cm, diurnal (hummingbird-pollination syndrome) or nocturnal and odoriferous (bat syndrome), white, greenish or magenta, tubular, pericarpel naked, narrower than tube at anthesis, tube with few minute, naked bract-scales; perianth-segments spreading or the innermost remaining ± erect in the diurnal species. Fruit expelled from cephalium, obovoid-clavate, red or deep pink, opening by a small basal pore. Seeds small, 0.8–1.8 mm (pyriform in *C. goebelianus*), black, testa-cells flat and smooth to strongly convex and with considerable cuticular sculpturing; hilum ± perpendicular to long-axis, large (small and sunken in *C. goebelianus*).

A genus of 6 well-defined species (plus 2 heterotypic subspecies), all native to Eastern Brazil (5 endemic) and ranging between the *caatinga* (3 taxa) and *Mata atlântica* (5 taxa) regions, almost exclusively on or closely associated with gneiss/granite inselbergs (*C. goebelianus* also on other substrates). Recent chloroplast gene sequence studies (Cowan & Soffiatti unpubl.) have confirmed its monophyletic status and the distinctness of the subgenera *Simplex* and *Buiningia*, the former being basal, the latter the most derived. The 3 subgenera are allopatric and can be recognized on the basis of seed-morphology, presence/absence of stem mucilage, spination and floral pollination syndrome. Although there is currently little disagreement over the circumscription of the genus, *Coleocephalocereus* names have also been published for species here referred to *Cipocereus, Stephanocereus, Pilosocereus, Micranthocereus* and *Espostoopsis*. Plates 53.1–55.4.

1. Flowers nocturnal, expanding fully, white within; seeds verrucose, testa-cells domed; ribs with (rarely without) transverse folds above the areoles 2
1. Flowers diurnal (often a.m.), inner perianth-segments scarcely expanding, yellow-green or magenta; seeds smooth, testa-cells flat; ribs lacking transverse folds (Rio Jequitinhonha drainage, NE Minas Gerais) 5

2. Spines > 16 per areole, some strongly hooked in seedlings, lacking hypertrophic spination at ground level; stem tissues non-mucilaginous (cent.-N Minas Gerais & S Bahia) . . . **4. goebelianus**
2. Spines < 17 in areoles remote from the cephalium and stem base, not or scarcely hooked in seedlings, or hypertrophic spination developed at ground level; stem tissues mucilaginous (NE Minas Gerais & Espírito Santo southwards) 3

3. Stems with long, hypertrophic spines near base . . . . **1. buxbaumianus**
3. Stems lacking hypertrophic spines near base 4

4. Flowers c. 30–60 mm; cephalium bristles mostly yellowish or porrect, intermixed with abundant whitish wool; stem 6–19-ribbed . . . . **2. fluminensis**
4. Flowers c. 19–35 mm; cephalium bristles dark brown and adpressed, not intermixed with wool; stem 12–34-ribbed . . . . **3. pluricostatus**

5. Flowers yellow-green; seeds c. 1.35 mm; spines yellowish . . . . **5. aureus**
5. Flowers magenta; seeds c. 1.75 mm; spines reddish brown . . . . **6. purpureus**

Subg. *Coleocephalocereus* (Nos 1–3): stems short to tall (0.5–5.0 m), branched at or above base or solitary, tissues mucilaginous; ribs often with transverse epidermal folds above the areoles; spines finely needle-like; flowers nocturnal, whitish at least within; seeds with strongly convex testa-cells, these ornamented with characteristic, crown-like, cuticular folds encircling each convex periclinal wall (SEM), hilum ± broad. NE to S Minas Gerais and adjacent Espírito Santo (drainage of Rios Mucuri, Doce and others to the south); Rio de Janeiro and off-shore islands of São Paulo.

1. Coleocephalocereus buxbaumianus *Buining* in Succulenta 53: 28–33 (1974); Taylor in Bradleya 9: 17, *adnot.* (1991). Holotype: Minas Gerais, Mun. (?) Itambacuri, 1972, *Horst* 379 (U).

C. braunii L. Diers & E. Esteves Pereira in Kakt. and. Sukk. 36: 28–35 (1985). Holotype: Espírito Santo, Mun. (?) Afonso Cláudio, 500–600 m, 1983, *Horst, Braun* (470) & *E. Esteves Pereira* (172) HBE 4 (KOELN, n.v.); *Braun* 470 (ZSS, iso.), *E. Esteves Pereira* 172 (UFG, iso.).

VERNACULAR NAMES. Mandacaru-de-topete, Facheiro-das-pedras.

Stems solitary or branched at or near base, erect from the rootstock or the basal parts decumbent on the rock surface and the apical parts slightly inclined, to 220 × 5–13 cm. Ribs 14–36, 5–10 × 10–15 mm, ± tuberculate, with prominent folds above the areoles, epidermis dark green, shiny. Areoles 1.5–7.0 × 1.5–5.0 mm, 6–13 mm apart, with pale yellowish to brownish, later greyish felt, displaying very strong indeterminate growth on basal half of stem. Spines finely needle-like, yellow, golden or partly brownish at first, later pale greyish to black, c. 10–25 per areole above the stem base, central and radial spines weakly differentiated, centrals (0–)3–6(–9), 8–40 mm, radials 7–18, to 28 mm, basal areoles with abundant, sinuate, hypertrophic spines to 400 mm or more. Cephalium rather broad, sometimes as wide as the stem, with dense, usually golden bristle-spines to 60 mm, wool inconspicuous except near stem apex, bristle-spines blackish when old. Flowers (25–)30–75 × 23–40 mm, brownish to greenish white when in bud; pericarpel 3–6 × 5–9 mm, with or without small bract-scales, tube 20–50 × 10–13 mm at apex, bearing few, 2–11 × 0.2–2.0 mm bract-scales; nectar-chamber c. 7–12 × 6–7 mm; outer perianth-segments 7–17 × 3.5–6.0 mm, whitish with reddish, brownish or greenish apex and mid-stripe, inner segments 8–20 × 3.5–6.0 mm, acuminate, white; stamens 8–20 mm, white; style 18–60 mm, stigma-lobes c. 8–10, to 5 mm, pale yellow to whitish; ovary locule 1–5 × 2–6 mm. Fruit 18–35 × 10–22 mm, often somewhat laterally compressed or channelled, blood-red to shocking pink at apex, deep or paler pink below, whitish near the basal pore, floral remnants whitish near base, darker to blackish above. Seed c. 1.3–1.8 × 1–1.4 mm.

CONSERVATION STATUS. Least Concern [LC]; > 10 locations exist, the majority declining only rather slowly (except in the west of its range).

Two subspecies are recognized, subsp. *buxbaumianus* replacing subsp. *flavisetus* at the eastern edge of the species' range, which is poorly recorded due to the inaccessibility of many of its inselberg habitats:

1. Stems 5–8 cm diam., branching freely from the decumbent bases, forming loose, sprawling clumps; flowers c. 25–42 × 25 mm (NE/E Minas Gerais, E of 42°W, and Espírito Santo) . . . . **1a.** subsp. **buxbaumianus**
1. Stems 7–13 cm diam., solitary and erect or forming small compact clusters; flowers c. 50–75 × 30–40 mm (SE & SW Minas Gerais, W from 42°W) . . . . **1b.** subsp. **flavisetus**

1a. subsp. **buxbaumianus**

Ribs 14–22. Cephalium 4–7(–8) cm wide. Fruit quite naked, without bract-scales.

Southern humid/subhumid forest (inselberg) element: locally co-dominant with other cacti on gneissic inselbergs/*lajedos*, 100–700 m, eastern Minas Gerais and western Espírito Santo (Rio Doce drainage). Endemic to the core area within South-eastern Brazil. Map 41.

MINAS GERAIS: NE & E Minas Gerais, (?) Mun. Santa Maria do Salto, Sep. 2003, *J.R. Stehmann* (photos); Mun. Carlos Chagas, c. 25 km W of Nanuque on road to Teófilo Otoni, 1994, *S. Porembski et al.* (K, photos); Mun. Itambacuri, 20 km S of Teófilo Otoni on road BR 116, 14 Dec. 1990, *Taylor & Zappi* (obs.); l.c. (?), July 1974, *Buining & Horst* in *Horst* 379 (U); Mun. Galiléia, 23 km along road BR 259 from São Vítor towards Galiléia, 15 Dec. 1990, *Taylor & Zappi* 771 (K, ZSS, HRCB, BHCB).
ESPÍRITO SANTO: Mun. (?) Afonso Cláudio, (?) Pontões ['Mun. Três Pontões'], July 1983, *E. Esteves Pereira* 172 (UFG), *Braun* 470 (ZSS).

CONSERVATION STATUS. Least Concern [LC] (1); PD=1, EI=2, GD=2. Short-list score (1×5) = 5. Five populations are recorded above, but this is almost certainly less than the total that could be found if more intensive field studies were undertaken. Although the vegetation around the inselbergs where this taxon occurs has largely been destroyed, the actual habitat of the plant is less affected. However, disturbance is likely to increase and so monitoring of its status is essential.

The presence of abundant, long, basal hypertrophic spines in this taxon may have a moderating influence on the plant's temperature near to the sun-baked rock surface (Porembski *et al.* 1998: 115, fig. 6). It is quite variable in stature and spination, the first collection cited above having a distinctive habit and unusual dark cephalium bristles. It remains to be seen whether it belongs here.

1b. subsp. **flavisetus** *(F. Ritter) N. P. Taylor & Zappi* in Cactaceae Consensus Initiatives 3: 7 (1997). Holotype: Minas Gerais, Mun. Engenheiro Caldas, 1965, *Ritter* 1339 (U).

C. flavisetus F. Ritter, Kakt. Südamer. 1: 127–128 (1979).
C. estevesii L. Diers in Kakt. and. Sukk. 29: 201–205 (1978). Holotype: Minas Gerais, Mun. Pedra do Indaiá, 1974, *E. Esteves Pereira* 66 (KOELN, n.v.; UFG, iso.).

Ribs 18–36. Cephalium to 13 cm wide. Fruit bearing minute bract-scales.

Southern humid/subhumid forest (inselberg) element: on gneissic inselbergs/*lajedos*, 100–1000 m, south-eastern and south-western Minas Gerais (disjunct). Endemic to the core area within Minas Gerais. Map 41.

MINAS GERAIS: SE Minas Gerais, Mun. Engenheiro Caldas, 10 km S of the town, 1965, *Ritter* 1339 (U); Mun. São Joao do Oriente, 6 km E of the turning to Iapú on road BR 458, 13 Dec. 1990, *Taylor & Zappi* 754 (K, ZSS, HRCB, BHCB); Mun. (?) Caratinga, SW of the town, 1982, *Horst & Uebelmann* 511 (ZSS); SW Minas Gerais, Mun. Pedra do Indaiá ['Mun. Itapecerica, granito ao norte da cidade'], road MG 050, June 1974, *E. Esteves Pereira* 66 (UFG); Mun. Carmo da Mata, *F. Campos*, plant cultivated at Jardim Bot., Fund. Zoo-Bot. de Belo Horizonte, 31 July 2001, *Taylor* (obs.); Mun. (?), between São Joao del Rei and Morro do Ferro, *Horst & Uebelmann* 593 (Uebelmann, mss.).

CONSERVATION STATUS. Vulnerable [VU B2ab(iii)] (2); area of occupancy estimated to be < 2000 km^2; PD=1, EI=1, GD=2. Short-list score (2×4) = 8. More secure in the eastern sector of its range, especially where it occurs on larger, steep-sided inselbergs, but probably threatened in the west, where mining operations have destroyed much of its habitat. By no means a common plant.

The eastern part of the range of this species is mostly within the drainage of the Rio Doce in a region of relatively low rainfall (< 1000–1250 mm/yr, cf. map 'isoietas anuais 1914–1938' in Azevedo 1972; Nimer 1973: 40, fig. 18), where there is a mixture of

Mata atlântica and *caatinga*-like vegetation (Luetzelburg 1925–26, 2: 112–115). The following two species are also found in this region, but range further south into wetter areas. They have not yet been found truly sympatric with *C. buxbaumianus*, but occur in very close proximity.

2. Coleocephalocereus fluminensis *(Miq.) Backeb.* in Jahrb. Deutsch. Kakteen Ges. 1941(2): 53 (1942); Taylor in Bradleya 9: 17, *adnot.* (1991). Type: not extant. Lectotype (designated here): Brazil, Rio de Janeiro, Vellozo, l.c. infra, tab. 20 (1831).

Cactus melocactus Vell., Fl. flumin.: 205 (1829), Icones 5: tab. 20 (1831) non L. (1753). *Cereus fluminensis* Miq. in Bull. Sci. Phys. Nat. Neérl. 1838: 48 (1838). *C. vellozoi* Lem. in Rev. Hort. 1862: 427 (1862), nom. illeg. (Art. 52.1). *Cephalocereus melocactus* K. Schum. in Martius, Fl. bras. 4(2): 215 (1890), nom. illeg. (Art. 52.1). *Pilocereus melocactus* K. Schum. in Monatsschr. Kakt.-Kunde 3: 20 (1893), nom. illeg. (Art. 52.1). *Cereus melocactus* A. Berger in Annual Rep. Missouri Bot. Gard. 16: 62 (1905), nom. illeg. (Art. 52.1). *Cephalocereus fluminensis* (Miq.) Britton & Rose, Cact. 2: 29 (1920). *Austrocephalocereus fluminensis* (Miq.) Buxb. in Beitr. Biol. Pflanzen 44: 240 (1968).
? *Leocereus paulensis* Speg. in Anales Soc. Ci. Argent. 99: 116 (1925), pro parte. Type: see Zappi & Taylor (1992b).
Coleocephalocereus paulensis F. Ritter, Kakt. Südamer. 1: 161 (1979). *C. fluminensis* subsp. *paulensis* (F. Ritter) P. J. Braun & E. Esteves Pereira in Succulenta 74: 84 (1995). Holotype: São Paulo, Ilhabela, *Ritter* 1352 (U).
C. fluminensis var. *braamhaarii* P. J. Braun in Kakt. and. Sukk. 33: 118–121, with illus. (1982). *C. fluminensis* subsp. *braamhaarii* (P. J. Braun) P. J. Braun & E. Esteves Pereira in Succulenta 74: 84 (1995). Holotype: Brazil, Espírito Santo, coast near Vitória, *Braamhaar* 1/1980 (KOELN ['Succulentarium'], in spirit, n.v.).
C. diersianus P. J. Braun & E. Esteves Pereira in Kakt. and. Sukk. 39: 48–53 (1988). Holotype: Brazil, Minas Gerais, Mun. (?) Lajinha, c. 500 m, 1983, *Horst, Braun & E. Esteves Pereira* HBE 3 (KOELN, n.v.), *E. Esteves Pereira* 171 (UFG, iso.).

Stems solitary or more commonly branched from the decumbent basal parts, erect above or inclined towards the rock surface, to 2 m high and, including decumbent parts, > 300 × (3.5–)4.0–12.0 cm; ribs (5–)6–16(–19), ± triangular in cross-section, to 16 × 10–30 mm, with ± well-defined folds above the areoles, epidermis light to dark, shiny green or greyish green. Areoles 1–4 mm, to c. 8 mm apart, with white to greyish felt at first, indeterminate growth weak. Spines mostly few, 1–16 per areole, to 40 mm, finely needle-like, rarely stouter, whitish, yellowish or darker, central and radial spines poorly differentiated, the former 0–3, the latter 0–13. Cephalium mostly no more than half the stem's diameter, composed of abundant white wool interspersed with yellow or more rarely brownish bristle-spines, these straight or curved, to 30 mm. Flowers 30–60 × 20–40 mm; pericarpel c. 4 × 5–7 mm, tube 25–40 × 8–15 mm, with few narrow bract-scales, the uppermost to 7 × 2.5 mm; nectar-chamber 8–17 × 4–7.5 mm; outer perianth-segments c. 12–20 × 3.5–6.0 mm, greenish brown, purplish pink or creamy yellow, inner segments to 15 × 2.5–6.0 mm, white; stamens to 25 mm; style to 53 mm; stigma-lobes 8–15, to 6 mm, pale yellow. Fruit 15–26 × 13–20 mm, red, paler towards base. Seed 1.25–1.50 × 1.0 mm (Taylor 1991a: pl. 1.2).

CONSERVATION STATUS. Least Concern [LC]; widespread on usually inaccessible inselbergs.

The following subspecies are distinguished:

1. Stem erect except near base, to 12 cm diam. (southwards from border region between E Minas Gerais and SE Bahia) . . . . **2a.** subsp. **fluminensis**
1. Stem decumbent except at the inclined apex, to 6 cm diam. (NE Minas Gerais) . . . . **2b.** subsp. **decumbens**

2a. subsp. **fluminensis**

Ribs to 16(–19), epidermis light to dark shiny green. Spines all finely needle-like, mostly pale yellow to greyish white, rarely partly brownish. Cephalium bristle-spines slender hairlike, mostly pale.

Southern humid/subhumid forest (inselberg) element: locally dominant on gneissic inselbergs or *lajedos* within the *Mata atlântica* and *restinga* zones, near sea level to c. 900 m, north-eastern (MG/BA border region) to south-eastern Minas Gerais (border region between MG and ES/RJ) and Espírito Santo; Rio de Janeiro and off-shore islands of São Paulo (to Ilha Queimada Grande). Map 41.

MINAS GERAIS: Mun. Nanuque, 1994, *S. Porembski et al.* (K, photos); Mun. Serra dos Aimorés, c. 8 km E of Nanuque on road to state border (BA), 1994, *S. Porembski et al.* (K, photo); (?) Mun. Mantena, near frontier with ES, *Horst & Uebelmann* 335 (Uebelmann, mss.); Mun. Conselheiro Pena, 7.5 km E of Goiabeira on road to Cuparaque, with *C. pluricostatus*, 15 Dec. 1990, *Taylor & Zappi* 775 (K, ZSS, HRCB) — aberrant dwarf form; Mun. Aimorés, Pedra Lorena, 5 July 1997, *M.F. de Vasconcelos* s.n. (K, BHCB); Mun. Lajinha ['ES, Lagedinho'], arredores do cemitério, July 1983, *E. Esteves Pereira* 171 (UFG); Mun. Juiz de Fora, 'perto de Pedreira', 1999, *L. Aona* (UEC, K, photo); Mun. Além Paraíba, road BR 116, 5 km N of Rio Aventureiro, 18 Dec. 1990, *Taylor & Zappi* 793B (K, photo).

ESPÍRITO SANTO: Mun. Nova Venécia, 1989, *H.B.Q. Fernandes* s.n., cult. at MBML, May 1990 (HRCB) — somewhat intermediate with subsp. *decumbens*; ibid., Pedra do Elefante, Apr. 1996, *S. Porembski* (K, photo.); Mun. São Gabriel da Palha, 29.5 km SE of Barra de São Francisco near road ES 080, 16 Dec. 1990, *Taylor & Zappi* 780 (K, ZSS, HRCB, MBML); Mun. Pancas, 11 km NW of Ângelo Frechiani on road ES 341, 0.7 km W of bridge over Rio Pancas, 16 Dec. 1990, *Taylor & Zappi* 786 (K, ZSS, HRCB, MBML) — somewhat intermediate with subsp. *decumbens*; Mun. Colatina, 4 km W of Ângelo Frechiani on road ES 341, 16 Dec. 1990, *Taylor & Zappi* 785A (K, photo); Mun. Santa Teresa, Rio Cinco de Novembro, *H.B.Q. Fernandes* 2259, cult. MBML, May 1990 (HRCB); Mun. Vitória, May 1990, *Taylor & Zappi* (obs.); Mun. Conceição do Castelo, near road BR 262, 5 km E of Venda Nova, 850–900 m, 17 Dec. 1990, *Taylor & Zappi* (obs.); Mun. Guarapari, road ES 060, Km 30–35, Setiba–Guaraparí, near Lagoa Corais, 28 Aug. 1987, *O.J. Pereira* 1013 & *J.M.L. Gomes* 236 (SPF); Mun. Piúma, Monte Agá, south of the town, 20°52'23"S, 40°46'24"W, 26 Nov. 1999, *Zappi et al.* 469 (UEC, K).

RIO DE JANEIRO: Mun. (?), between Muriaé (MG) and Itaperuna, *Uebelmann* (mss.).

CONSERVATION STATUS. Least Concern [LC] (1); PD=1, EI=2, GD=2. Short-list score (1×5) = 5.

This subspecies is very variable in size, habit, rib number, cephalium colour etc., and especially so near the northern limits of its range, but apart from this the variation does not seem to show any obvious geographical pattern that would allow its division into additional subspecies, nor does there seem to be any point in naming every local form. It is sometimes sympatric with *C. pluricostatus*.

2b. subsp. **decumbens** *(F. Ritter) N. P. Taylor & Zappi* in Cact. Consensus Initiatives 3: 7 (1997). Holotype: Minas Gerais, Padre Paraíso ['Água Vermelha'], *Ritter* 1340 (U).

Coleocephalocereus decumbens F. Ritter in Kakt. and. Sukk. 19: 160–161 (1968).

Stem to c. 6 cm diam.; ribs to 13, epidermis dark or grey-green. Spines stout, stiff, dark brownish at first. Cephalium bristle-spines stiff, dark brown.

Southern humid/subhumid forest (inselberg) element: locally dominant on gneissic inselbergs and *lajedos*, in the *agreste*/*Mata atlântica* transition, c. 650 m, north-eastern Minas Gerais. Endemic to the core area within Minas Gerais. Map 41.

MINAS GERAIS: Mun. Padre Paraíso [Água Vermelha], 1965, *Ritter* 1340 (U); ibid., 2 km N of town on road BR 116, 23 Feb. 1988, *Supthut* 8890 (ZSS); Mun. Caraí, N of Catuji, 21 Oct. 1988, *Taylor & Zappi* in *Harley* 25403 (K, SPF); Mun. Teófilo Otoni, reported by W. Uebelmann (mss.); l.c., 1999, *L. Aona*, cult. UNICAMP, Campinas, São Paulo.

CONSERVATION STATUS. Endangered [EN B1ab(iii) + 2ab(iii)] (3); extent of occurrence = 91 km^2 and area of occupancy < 10 km^2; PD=1, EI=2, GD=1. Short-list score (3×4) = 12. Endangered from urban expansion at its extensive type locality, but range probably under-recorded, although the 3 known populations are fragmented.

The range of this taxon in north-eastern Minas Gerais is not well understood and more field studies are needed in the area between Padre Paraíso (MG) and the Rio Doce in Espírito Santo (eg. municípios Nova Venécia and Pancas), where plants intermediate with subsp. *fluminensis* have been observed and collected. At the type locality it is represented by a relatively uniform population occupying an area some kilometres in extent. It appears sufficiently distinct to warrant subspecific status at present, but may well prove to be of lesser significance once the variation of the species as a whole is better known.

3. Coleocephalocereus pluricostatus *Buining & Brederoo* in Krainz, Die Kakteen, Lfg. 46–47 (1971). Holotype: Minas Gerais, *Horst & Uebelmann* 245 (ZSS).

C. pluricostatus subsp. *uebelmanniorum* P. J. Braun & E. Esteves Pereira in Kakt. and. Sukk. 44: 150–154, with illus. (1993). Holotype: Brazil, S Espírito Santo, c. 250 km S of the type locality of the species, [Jerônimo Monteiro *fide* Uebelmann (1996)], 1991, *W. & R. Uebelmann* HU 1502 (UFG, n.v.).

Stem an unbranched, erect, solitary column or branched from the decumbent base, 100–500 × 7.5–11.0 cm; ribs (in seedlings 9–)12–34, 8–18 × 12–26 mm, epidermis dark green. Areoles to 2.5 mm, to c. 9 mm apart, at first with whitish to pale grey felt, soon almost naked, indeterminate growth almost lacking. Spines (0–)4–8, to 11 mm, brownish, not very conspicuous, sometimes lacking on older stems, central spine 0–1, radials to 8, slender. Cephalium becoming nearly as broad as the stem, composed of densely packed, dark, shiny brown, curved-adpressed bristle-spines, wool sparse, inconspicuous. Flowers c. 19–35 × 10–22 mm; pericarpel to 2.5–3.0 × 4.5–6.0 mm; tube to c. 24 × 12 mm, bearing few, greenish, to 1.5 mm bract-scales; nectar-chamber to 9 × 12 mm; outer perianth-segments 6–7 × 2.5 mm, pale green or yellowish, inner segments c. 7 × 2–2.5 mm, white; stamens to c. 12 mm, bearing pale yellow anthers; style 15–19 mm, stigma-lobes 6–10, 2–3 mm, greenish white; ovary locule 1–1.5 × 3 mm. Fruit 14–25 × 8–13 mm, red, glossy. Seed 1–1.2 × 0.8–1.1 mm, testa-cells strongly convex.

Southern humid/subhumid forest (inselberg) element: on gneissic inselbergs within the *Mata atlântica* zone, 100–300 m, near the eastern border of Minas Gerais and in adjacent Espírito Santo, from the region of Barra de São Francisco southwards for some 250 km. Endemic to the core area within South-eastern Brazil. Map 41.
MINAS GERAIS: Mun. (?) Mantena, 'im Osten des Staates Minas Gerais' [10 km from frontier with Espírito Santo (Uebelmann, mss.)], June 1968, *Horst & Uebelmann* 245 (ZSS); Mun. Conselheiro Pena, 7.5 km E of Goiabeira on road to Cuparaque, with dwarf form of *C. fluminensis*, 15 Dec. 1990, *Taylor & Zappi* 776 (K, ZSS, HRCB, BHCB).
UNLOCALIZED (MG/ES): [probably near the Rio Doce, c. 1917], *Luetzelburg* '32' (M).
ESPÍRITO SANTO: Mun. Barra de São Francisco, 1 km from border of Minas Gerais on road ES 080, 16 Dec. 1990, *Taylor & Zappi* 778A (K, photo); Mun. Itarana, Jatibocas, May 1946, *Brade* 18537 (RB); S Espírito Santo, Mun. Jerônimo Monteiro, 27 Oct. 1991, *W. & R. Uebelmann* 1502, cult. A. Hofacker, June 2003, Taylor (obs.).

CONSERVATION STATUS. Least Concern [LC] (1); PD=2, EI=1, GD=2. Short-list score (1×5) = 5. Least Concern, since its habitats are mostly inaccessible, steep slopes. Range and frequency of occurrence probably under-recorded.

The caespitose, southern form described by Braun & Esteves Pereira as subsp. *uebelmanniorum*, and published without precise locality information, is neither so geographically remote as its authors claim, nor the first collection from Espírito Santo, since Brade collected the species in the intervening area of the state in May 1946 (see above). Their use of subspecific status, which may be justified, needs to be evaluated in the light of studies of such intervening populations, but Herm *et al.* (2001: 100) illustrate a plant with slender yellowish flowers, which seem different from those observed further north by the present authors and might indicate specialization towards hummingbird pollination. The first collection of the species is even earlier than Brade's, dating from around 1917, by Luetzelburg, who is known to have visited the border region between Minas Gerais and Espírito Santo while travelling up the Rio Doce (Luetzelburg 1925–26). It should also be noted that the phrase 'Holotypus: Brasilia, Espirito Santo, *in regione septentrionali* ...' following the Latin diagnosis published by Braun & Esteves Pereira (l.c.: 152) is evidently an error (the locality is south, not north of the *locus classicus*).

Subg. *Simplex* N. P. Taylor (No. 4): stem tall (to 6.5 m), normally solitary unless damaged, tissues non-mucilaginous; ribs with transverse epidermal folds above the areoles; spines stout, hooked in seedlings; flowers nocturnal, white; seeds with strongly convex testa-cells, hilum narrow, sunken. S Bahia & cent.-N Minas Gerais. Type and only species:

4. Coleocephalocereus goebelianus *(Vaupel) Buining* in Kakt. and. Sukk. 21: 202–206 (1970); F. Ritter ex Backeb., Kakteenlex.: 92 (1966), nom. inval. (Art. 33.2). Type: Bahia, Serra das Almas, *Luetzelburg 32* (B†). Neotype (designated here): Bahia, Mun. Ituaçu, c. 10 km S of town towards Tanhaçu, 18 Aug. 1988, *Eggli* 1195 (ZSS).

Cereus goebelianus Vaupel in Z. Sukkulentenk. 1: 58 (1923).
Coleocephalocereus pachystele F. Ritter in Kakt. and. Sukk. 19: 157 (1968). Holotype: Bahia, Urandi, 1964, *Ritter* 1234 (U, not found Oct. 1991; SGO 122062, lecto. designated here, cf. Eggli *et al.* 1995: 500).
[*Cephalocereus purpureus* Britton & Rose, Cact. 2: figs 25, 27 & 28 (1920); Werderm., Bras. Säulenkakt.: 58, 117 (1933), non Gürke (1908).]
[*Austrocephalocereus purpureus sensu* (Gürke) Backeb. in Blätt. Kakteenf. 1938(6): [22] (1938) non *Cephalocereus purpureus* Gürke (1908).]
[*C. lehmannianus sensu* Werderm., quoad sem., ibid., 120, fig. 6 (1933), non Werderm. (1932).]

Stem solitary, normally branching only if damaged, but occasionally forming short branches at the point where the stem develops the cephalium, columnar-erect, 150–500(–650) × 7–12 cm, tissues non-mucilaginous; ribs c. 10 in seedlings, 14–30 or more in mature individuals, to 10 × 15 mm, somewhat tuberculate, with transverse folds above the areoles, epidermis mid- to dark green. Areoles to 5 mm diam., 6–12 mm apart, with sparse whitish felt. Spination variable depending on age and maturity, spines mostly curved, pale with dark tips and bases, mostly rather stiff, not flexible: seedlings with strongly recurved to hooked central spines 10–30 mm, juvenile plants with straight central spines 60–100 mm, adult stems with shorter and finer spination of c. 4–8 central spines to 60 mm and 12–18 radial spines to 15 mm. Cephalium developed when immature stem overtops surrounding vegetation (ie. when the plant is between 0.4 and 6.0 m high), becoming as broad as the stem, comprising creamy white wool interspersed with abundant dark, curved bristle-spines to 25–50 mm. Flower c. 40–55 × 25–28 mm; pericarpel white, c. 3–6 × 6–7 mm; tube c. 35–45, with few, 1 mm, pinkish bract-scales; nectar-chamber to 18 × 10 mm; perianth-segments c. 10–12 × 4–5 mm, the outermost red-brown to greenish, the remainder white; stamens to 20 mm, white; style to c. 45 mm, stigma-lobes c. 10, to 4 mm, whitish.

Fruit broadly clavate, c. 20–30 × 17–22 mm, magenta-red at apex, paler below, to white at base. Seeds 1.5 × 1.2 mm, rather narrow at the hilum; testa-cells strongly convex (Taylor 1991a: pl. 1.3).

Southern *caatinga* element: on quartz, gneiss/granite inselbergs (rarely on limestone outcrops) and stony soil of the *caatinga* and *caatinga/campo rupestre* ecotone, c. 300–1000 m, southern margins of the Chapada Diamantina and planalto de Maracás and western flank of the Serra do Espinhaço, central-southern Bahia to central-northern Minas Gerais. Endemic to the core area of Eastern Brazil. Map 41.

BAHIA: Mun. Ibipitanga, 5–10 km N of Ibiajara towards Novo Horizonte, 12°56'S, 42°10'W, Aug. 2003, *M. Machado* (obs.); Mun. Iramaia, 40 km W of Pé-de-Serra on road to Iramaia from Maracás via Placas, 5 Feb. 1991, *Eggli* (obs.); Mun. Riacho de Santana, *fide* Uebelmann (1996): HU 1496; Mun. Rio de Contas, by road at start of descent to Livramento, Serra das Almas, c. 1000 m, 23 Nov. 1988, *Taylor* (obs.); Mun. Livramento do Brumado, 26 Oct. 1978, *A. Araújo* 044 (RB); Mun. Dom Basílio, 21 km S of Livramento do Brumado, 21 Mar. 1991, *G.P. Lewis* (K, photos); Mun. Maracás, 5 km E of Porto Alegre above N bank of the Rio de Contas, 300 m, 4 Feb. 1991, *Taylor et al.* (obs.); Mun. Ituaçu, c. 10 km S of town towards Tanhaçu, 18 Aug. 1988, *Eggli* 1195 (ZSS); Mun. Tanhaçu, Ourives, reported by Ritter (1979: 123, Abb. 71–73); Mun. Brumado, 7 km N of town towards Livramento, 22 Nov. 1988, *Taylor* (K, photos); Mun. Aracatu, 25 km E of Brumado on road BR 030 towards Sussuarana, 3 Feb. 1991, *Taylor* (K, photos); Mun. Guanambi, 9.5 km from town on road to Caitité, 25 July 1989, *Zappi* 171A (K, fr. & seeds); Mun. Palmas de Monte Alto, *Ritter* 1234 (SGO 122060, *fide* Eggli *et al.* 1995: 500); Mun. Urandi, c. 3 km S of town above railway line, 22 July 1989, *Taylor & Zappi* (obs.); Mun. Vitória da Conquista, Pradoso, Gameleira, 17 April 2003, *Taylor et al.* (obs.); Mun. Cordeiros, 15°4'32"S, 41°54'W, Sep. 2003, *M. Machado* (obs.).

MINAS GERAIS: Mun. Espinosa, 2 km N towards MG/BA border, 22 July 1989, *Taylor & Zappi* (obs.); Mun. Monte Azul, Serra Geral, 'Serra Gillette', c. 10 km E of town, 28 Jan. 1991, *Taylor et al.* 1458 (HRCB, BHCB, K, ZSS); Mun. Porteirinha, 8 km S of town on road to Riacho dos Machados, 7 Nov. 1988, *Taylor & Zappi* in *Harley* 25515 (SPF, K).

CONSERVATION STATUS. Least Concern [LC] (1); PD=3, EI=1, GD=1. Short-list score (1×5) = 5. Least Concern at present, but its habitat continues to be lost and monitoring is desirable.

A specimen labelled as *Luetzelburg* 32 at Munich (M) and annotated as being the type number of *Cereus goebelianus* Vaupel by Werdermann cannot be considered a true duplicate of the collection described by Vaupel, since it represents *Coleocephalocereus pluricostatus* (E Minas Gerais and Espírito Santo), which is unknown from Bahia, where the type of the former was collected, and has much smaller flowers than those described by Vaupel, who does not appear to have studied it. As an examination of his materials has clearly shown, Luetzelburg's numbering and labelling of his collections was rather chaotic and this Munich specimen should not be allowed to further destabilize the nomenclature of the Bahian species (see below).

Until the late 1960s, this unmistakable plant was known by the misapplied name *Cephalocereus (Austrocephalocereus) purpureus* (= *Micranthocereus purpureus* (Gürke) F. Ritter). Vaupel's description of the stem, ribs and spination are not representative and probably referred to a juvenile plant or juvenile base of the stem, but the original details of cephalium and flower clearly refer to this species. The confusion with *M. purpureus* began with Britton & Rose (1920: fig. 25) and was compounded by Werdermann (1933, 1942), who believed *C. goebelianus* to be synonymous, having redescribed the true *M. purpureus* as *C. lehmannianus* in 1932. Ritter (1968) recognized the problem, but not being sure of the precise identity of Vaupel's name, redescribed the plant treated here as *Coleocephalocereus pachystele*.

Subg. *Buiningia* (Buxb.) P. J. Braun (Nos 5 & 6): plants low (< 1.2 m), caespitose; stems non-mucilaginous; ribs lacking transverse epidermal folds above the areoles; spines very long, finely to stoutly needle-like; flowers diurnal, coloured; seeds with flat testa-cells, hilum broad. NE Minas Gerais (drainage of Rio Jequitinhonha).

5. Coleocephalocereus aureus *F. Ritter* in Kakt. and. Sukk. 19: 158–160 (1968). Holotype: Minas Gerais, Itaobim, 1964, *Ritter* 1341 (U).

Buiningia aurea (F. Ritter) Buxb. in Krainz, Die Kakteen, Lfg. 46–47 (1971).
B. brevicylindrica Buining, l.c. (1971). *Coleocephalocereus brevicylindricus* (Buining) F. Ritter, Kakt. Südamer. 1: 122 (1979). Holotype: Minas Gerais, *Horst & Uebelmann* HU 167 (ZSS; ZSS, iso.).
Buiningia brevicylindrica var. *elongata* Buining, l.c. (1971). *Coleocephalocereus brevicylindricus* var. *elongatus* (Buining) F. Ritter, tom. cit. 122 (1979). *C. elongatus* (Buining) P. J. Braun in Bradleya 6: 92 (1988). Holotype: Minas Gerais, *Horst & Uebelmann* HU 271 (ZSS?).
Buiningia brevicylindrica var. *longispina* Buining, l.c. (1971). *Coleocephalocereus brevicylindricus* var. *longispinus* (Buining) F. Ritter, tom. cit. 122 (1979). Holotype: Minas Gerais, *Horst & Uebelmann* HU 167a (ZSS?).

Stem solitary or caespitose, mostly erect, 15–115 × 6–22 cm, ovoid to cylindric, rather variable in form; ribs 10–20, low and blunt-edged, epidermis yellow-green to slightly glaucous. Areoles 3–5 mm, white felted, 4–11 mm apart, showing indeterminate growth near stem base. Spines golden yellow (blackish in age and darker in seedlings), flexible, central spine(s) 1–4, to 50 × 1 mm at base, radials c. 10–15, to 15 mm, strong hypertrophic spine growth near stem base in some populations. Cephalium making the stem strongly assymetric in low-growing populations, reaching c. half the width of the stem, composed of abundant white wool mixed with curved, golden bristle-spines to 30 mm. Flowers 30–42 × 17–20 mm, greenish yellow to green; pericarpel to 5 × 5 mm; tube 20–25 mm; nectar-chamber c. 5 mm; outer perianth-segments to 10 × 3 mm, inner segments c. 8 × 2 mm, erect, expanding to give an opening of only 2–4 mm; stamens c. 15 mm; stigma-lobes c. 5–7, whitish. Fruit red to pinkish red, 16–24 × 13–17 mm. Seed c. 1.35 mm, smooth (Barthlott & Hunt 2000: pl. 51. 3–4).

South-eastern *caatinga* (inselberg) element: locally dominant on gneissic inselbergs/*lajedos*, c. 280–910 m, north-eastern Minas Gerais (Rio Jequitinhonha drainage and watershed with Rio Pardo). Endemic to the core area within Minas Gerais. Map 41.
MINAS GERAIS: Mun. Águas Vermelhas, c. 15 km E of Curral de Dentro on road BR 251, 31 Jan. 1991, *Taylor et al*. 1516A (K, ZSS, photos); Mun. Pedra Azul, 16 Jan. 1965, *G. Pabst* 8336 & *E. Pereira* 9447 (HB); ibid., 5 km do aeroporto, 21 Sep. 1965, *A.P. Duarte* 9287 (HB, RB); ibid., 3 km from town, 11 Dec. 1984, *Harley et al*. in CFCR 6692 (K, SPF); ibid., 8 km towards road BR 116, 18 Oct. 1988, *Taylor & Zappi* in *Harley* 25186 (K, SPF); ibid., 10 km along road to Almenara, 19 Oct. 1988, *Taylor* (K, photos); Mun. Salinas, near Baixa Grande, road BR 251, 1 km W of junction to [Amparo do] Sítio, 22 Feb. 1988, *Supthut* 8874 (ZSS); l.c., a few km further E, 42°3'W, 31 May 2002, *M. Machado & G. Charles* (K, photos); Mun. Medina, road BR 116, Km 87, 21 Oct. 1988, *Taylor & Zappi* in *Harley* 25402 (K, SPF); Mun. Itaobim, 500 m W of town on N bank of Rio Jequitinhonha, 18 Nov. 1988, *Taylor* (K, photos); ibid., S bank, 8 km W of road BR 116 on road BR 357 towards Itinga, 14 Dec. 1990, *Taylor & Zappi* 764 (K, HRCB, BHCB).

CONSERVATION STATUS. Near Threatened [NT] (1); PD=2, EI=2, GD=2. Short-list score (1×6) = 6. Some habitats significantly declining due to quarrying of gneiss/granite.

Very variable in habit and stem morphology. According to Marlon Machado the first and second localities along road BR 251 cited above have already all but disappeared through quarrying activities.

6. Coleocephalocereus purpureus *(Buining & Brederoo) F. Ritter,* Kakt. Südamer. 1: 128 (1979). Holotype: Brazil, Minas Gerais, 1972, *Horst & Uebelmann* 359 (U).

Buiningia purpurea Buining & Brederoo in Kakt. and. Sukk. 24: 121–123, with illus. (1973).

Like *C. aureus*, but stems to c. 80(–87) × 11.5 cm; ribs 12–18, to c. 15 × 23 mm, epidermis mid- to dark green; areoles to 5 mm, to 15 mm apart; spines yellow and reddish brown to red, the largest central directed downwards, to 70 × 1 mm at base; cephalium with wool and bristle-spines coloured like those of the stem; flowers magenta, 37–48 × 20–27 mm; style reddish above, stigma-lobes 4–5, 2.5 mm, reddish; fruit red, 16–25 × 11–17 mm; seed c. 1.75 mm (Taylor 1991a: pl. 1.4).

South-eastern *caatinga* (inselberg) element: on gneissic inselbergs and *lajedos* in *caatinga*, c. 240–300 m, near the Rio Jequitinhonha, north-eastern Minas Gerais. Endemic to the core area within Minas Gerais. Map 41.

MINAS GERAIS: Mun. Itinga, 4 km E [corrected from 'W' by W. Wayt Thomas, *in litt.*, 29.02.2000] of town on road BR 367, 15 Feb. 1988, *W.W. Thomas et al.* 5972 (SPF, MBM, NY), 19 Nov. 1988, *Taylor & Zappi* in *Harley* 25532 (K, SPF).

CONSERVATION STATUS. Critically Endangered [CR C2b] (4); area of occupancy < 10 km²; PD=2, EI=1, GD=1. Short-list score (4×4) = 16. Critically Endangered, since the principal habitat is close to a road and may be visited by collectors for plants and seed. It is on privately owned agricultural land whose future must be regarded as uncertain and it is reasonable to project decline in habitat quality and numbers of individuals. The owner of the property has recently begun cutting the forest at the top of the inselberg (M. Machado, pers. comm.).

The range of this species deserves further investigation, but is clearly much more restricted than that of its sister species, *C. aureus,* and is currently known only from the original locality cited above and from 3 very small, recently discovered subpopulations within a few kilometres of it (M. Machado and I. Ribeiro de Andrade, pers. comms.). Uebelmann (1996) states it is from Itaobim, which is here assumed to refer to the Itinga locality, since the former is the closest town of any size and is the likely starting point for the short journey to Itinga. However, it should not be difficult to determine the extent of the species' range more precisely, working eastwards from the known localities along the Rio Jequitinhonha valley.

23. MELOCACTUS (L.) Link & Otto

in Verh. Vereins Beförd. Gartenbaues Königl. Preuss. Staaten 3: 417 (1827), *nom. cons.* Type: *Cactus melocactus* L., *typ. cons. Cactus* [unranked] *Melocactus* L., Gen. Pl., ed. 5: 210 (1754).

VERNACULAR NAMES. Cabeça-de-frade, Coroa-de-frade.

Literature: Taylor (1991a).

Stem unbranched (unless damaged), highly succulent, scarcely woody except at extreme base, depressed-globose to cylindric, vegetative part rarely to 50 cm, with 7–21 vertical ribs, mucilage absent to abundant; areoles small to large, indeterminate growth absent. Spines (3–)4–21 or more per areole, usually only weakly differentiated into central and radial series, sometimes strongly curved to hooked at apex in seedlings. At reproductive maturity stem apex converted into a terminal cephalium of bristles and dense trichomes, vegetative growth ceasing; cephalium remaining small or sometimes attaining 30 cm in length, to 12 cm diam. Flowers short lived, expanding mid- to late afternoon, small (15–29 × 4–18 mm), tubular, naked, only the pink, magenta or red perianth and sometimes uppermost part of tube exserted from the cephalium (rarely cleistogamous); pericarpel very small, conspicuously narrower than the swollen nectar-chamber at base of tube; perianth-segments tapered to linear, arranged in 2–4 series; lowermost stamens with filaments enlarged at base to protect nectar-chamber, anthers minute, < 1 mm; style and 4–8 stigma-lobes very slender, equalling stamens or long-exserted and conspicuous. Fruit short- to elongate-clavate (c. 10–45 × 5–12 mm), naked except for the pale floral remains attached at apex, white, pink, magenta or reddish, expressed from the cephalium and sometimes falling down from the plant, funicular pulp often becoming liquid, translucent. Seeds small (0.85–1.75 mm), blackish, globose to ovoid; testa-cells few, large, flat to strongly convex or almost pointed (for SEM micrographs of all species treated below, except *M. lanssensianus*, see Taylor 1991a).

One of the most widely distributed genera of Cactaceae, comprising at least 34 species, ranging from South-eastern and North-eastern Brazil (16–17 spp.) and the Llanos-Amazonian region (4 spp.), northwards to the Caribbean (to N Cuba) (10 spp.), and W to the Andes and Central America (S Peru to W Mexico) (5 spp.). Braun & Esteves Pereira (2002: 79) indicate that *Melocactus* also occurs in the Northern Brazilian state of Tocantins, but it is unclear which species this could be. The greatest concentration of taxa and centre of diversity is in Eastern Brazil (especially Bahia), and 18 out of the total of 22 species and heterotypic subspecies recognized here are endemic to the core area. The genus is characteristic of the *caatingas-agrestes* and Northern *campos rupestres*, only *Melocactus violaceus* occurring in sand-dunes of the *Mata atlântica* and *M. ernestii* being occasional on rocks above *brejo* forest (cf. Porembski *et al.* 1998: 116).

Karyotype information has been published in two papers by Das *et al.* (1998a & b), but is not included below, since no permanently preserved voucher materials were cited and considerable doubt must exist concerning the identities of the taxa studied, which were of unstated origin from a living collection (the unreliable identity of most living plants offered from commercial sources is well known and compounded by the widespread occurrence of man-made hybrids). A more recent study (Resende 2002, Assis *et al.* 2003), based on wild-collected material identified by Marlon Machado, indicates that at least two of the species exist in both diploid and tetraploid races and that tetraploidy is the norm amongst the 11 Brazilian taxa counted. Such studies, if expanded, may be valuable to enable an understanding of why some sympatric taxa produce hybrids, while others do not.

Various species occur sympatrically and sometimes form hybrid swarms. The key below does not attempt to account for plants of hybrid origin, which include (cf. Taylor 1991a): No. 1a × No. 10, 1b × 14, 1b × 15, 2a × 10, 2a × 11, 2a × 13, 8 × 11, 9a × 11, 11 × 14, 13 × 14, 14 × 15.

For all taxa, fruit, seed and edaphic data are essential for precise identification. Plates 56.1–64.4.

1. Fruit entirely white, white but very pale pink at apex, or pale lilac-pink to pink and only 10–20 mm long 2
1. Fruit red or pinkish magenta, at least at apex, 10–45 mm long 13

2. Stem lacking mucilage; fruit length 1.5–2.0 × diam. 3
2. Stem with at least some mucilage in the green cortical tissues or highly mucilaginous; fruit length 2 or more × diam. 4

3. Lowermost radial spine markedly longer than longest central spine; flower pinkish magenta; seed to 1.35 mm, testa-cells strongly convex (granite/gneiss inselbergs & *lajedos*, cent.-S Bahia) . . . . **5. deinacanthus**
3. Lowermost radial spine ± equal to or shorter than longest central spine; flower red, at least without; seed 1.35–1.75 mm, testa-cells almost flat (limestone, N Minas Gerais, W & cent.-S Bahia) . . . . **6. levitestatus**

4. Fruit white, or white and very pale pink only at apex, or pale pink and seed with almost flat testa-cells 5
4. Fruit lilac-pink to pink; seed with testa-cells strongly convex at end opposite hilum 8

5. Central spine(s) > 20 × 1.5 mm (inland Bahia, limestone) 6
5. Central spine lacking or < 20 × 1.5 mm (coastal sand dunes)
. . . . **16c. violaceus** subsp. **margaritaceus**

6. Perianth-segments 0.7–1.7 mm wide; fruit to 6 mm diam. 7
6. Perianth-segments 1.4–2.2 mm wide; fruit 7–9.5 mm diam. . . . . **9. pachyacanthus**

7. Stem glaucous, at least when young; cephalium apex with brownish tufts of wool, bristles not or scarcely exserted; seeds (1.3–)1.4–1.75 mm . . . . **7. azureus**
7. Stem never glaucous; cephalium apex lacking brownish wool tufts, bristles usually well-exserted; seeds 1.05–1.30 mm . . . . **8. ferreophilus**

8. Flowers cleistogamous (N & SE Pernambuco & E Paraíba) . . . . **12. lanssensianus**
8. Flowers opening prior to fruit development 9

9. Stem pale bluish waxy-glaucous, at least when young 10
9. Stem always plain green 11

10. Stem 11–16 cm diam., pith chlorophyllous; ribs 8–12; areoles 13–20 mm apart on the ribs
. . . . **14. concinnus**
10. Stem to 25 cm diam., pith white; ribs 10–22; areoles 20–40 mm apart on the ribs . . **11. zehntneri**

11. Spines (3–)4–6 per areole, largest 1.2–2.0 mm thick (cent. & S Bahia, 900–1500 m)
. . . . **15. paucispinus**
11. Spines > 6 per areole or only 0.5–1.0 mm thick 12

12. Radial spines 5–10(–11), almost straight, 0.5–1.5 mm thick (*restinga* and sandy places inland, 0–1100 m) . . . . **16. violaceus**
12. Radial spines (6–)7–11, ± curved, to 2.5 mm thick (inland, especially in the *caatinga*)
. . . . **11. zehntneri**

13. Fruit to 16 × 7 mm, dark red to base; cephalium white-woolly throughout, bristles hidden (Bahia: Mun. Ourolândia, Mun. Morro do Chapéu & Mun. Barra) . . . **13. glaucescens**
13. Fruit to 45 × 12 mm, reddish to magenta at apex, paler below; cephalium with conspicuous bristles or white-woolly only at apex 14

14. Ribs ± rounded in cross-section (but edge often acute) or very low; stems depressed-globose to taller than broad 15
14. Ribs triangular in cross-section; stems broader than tall 18

15. Lowermost radial spine recurved at apex, to 35 mm, central spine 1, to 22 mm; stem ± hemispheric; ribs very low (Bahia: Mun. Vitória da Conquista) . . . . **4. conoideus**
15. Lowermost radial spine straight or outcurved at apex, or > 35 mm, central spines > 1, > 22 mm, or stem and ribs not as above 16

16. Lowermost radial spine > 40 mm and < 1.5 mm thick, or stem taller than broad or lacking mucilage and/or flowers with pinkish, long-exserted stigma-lobes; ribs 9–16 17
16. Lowermost radial spine < 40 mm or > 1.5 mm thick; stem depressed to globose, with mucilage in the green cortical tissues; stigma-lobes white, scarcely exserted; ribs 8–10
. . . . **3. bahiensis (inconcinnus)**

17. Spines 9–14 per areole, lowermost radial 40–80 mm; stem to 15 × 18 cm . . . . **1. oreas**
17. Spines (11–)14–21 per areole, lowermost radial 50–150 mm; stem to 45 × 35 cm . . . . **2. ernestii**

18. Stem light greyish blue-green or quite glaucous; flower with c. 23 perianth-segments visible from above; spines usually < 40 mm (S & E Bahia) . . . . **10. salvadorensis**
18. Stem pale to dark green; flower with 25–33 perianth-segments visible from above; spines to 60 mm (Pernambuco, Bahia & N Minas Gerais) . . . . **3. bahiensis**

MELOCACTUS OREAS Group (Nos 1–4)

Characteristic of exposed outcrops of gneiss/granite and sandstones or crystalline rocks/gravels (especially quartz) between low shrubs, only very rarely occurring on limestone.

1. Melocactus oreas *Miq.*, Monogr. Melocacti: 113 (1840). Type: 'Habitat circa Bahiam' [Salvador] (assumed not to have been preserved). Neotype (Taylor 1991a: 24–25, illus.): Bahia, Mun. Itatim, *Zappi* 181A (K, photo.).

Cactus oreas (Miq.) Britton & Rose, Cact. 3: 227 (1922).
Melocactus rubrisaetosus Buining *et al.* in Succulenta 56: 161 (1977). *M. oreas* subsp. *rubrisaetosus* (Buining *et al.*) P. J. Braun in Bradleya 6: 95 (1988). *M. oreas* var. *rubrisaetosus* (Buining *et al.*) Rizzini, Melocactus no Brasil. 52–53 (1982), nom. inval. (Art. 33.2). Holotype: Bahia, Mun. Milagres, *Horst* 137 (U 531282?).
M. oreas var. *submunitis* Rizzini, tom. cit.: 54, fig. 22 (1982), nom. inval. (Art. 37.1).
[*M. oreas* subsp. *bahiensis* (Britton & Rose) Rizzini, tom. cit.: 52–53 (1982), excl. typ. *M. oreas* var. *bahiensis* (Britton & Rose) Rizzini, l.c., fig. 23 (1982), excl. typ.]
[*M. conoideus sensu* Rizzini, tom. cit.: figs 22 & 34, above right (1982) non Buining & Brederoo (1973).]

Stem depressed-globose to elongate, 8.5–15.0(–35.0) × 10–18 cm, mid- to dark green; ribs 10–16, rather variable in height, to 40 mm broad near stem base, always ± rounded in cross-section, but edge often acute; areoles 6–9 × 4–6 mm, 10–18 mm apart on the ribs. Spines 9–14, horn-coloured, pale brown or dull reddish brown, acicular, central(s) 1–4, lower 27–45 mm, radials 8–11, lowermost 40–80 mm, slender and flexuous, rarely stout and stiff, outcurved or straight. Cephalium to 12 × 4–8 cm, bristles dark red-brown. Flowers light to dark pink-magenta, 17–22 × (7–)9–10 mm. Fruit red at apex, magenta to pinkish below, fading to almost or quite white at base, elongate clavate, 14–28 × 5–9 mm. Seed nearly globose, 0.85–1.20 × 0.8–1.15 mm, testa-cells convex.

CONSERVATION STATUS. Least Concern [LC]; known from > 10 locations, but requires regular monitoring in view of its limited range.

This endemic NE Brazilian species is divisible into the following subspecies:

1. Ribs 10–13; stem usually depressed (N Chapada Diamantina, Bahia, 700–1000 m) . . . . **1b.** subsp. **cremnophilus**
1. Ribs 12–16; stem depressed to elongate (E Bahia < 500 m) . . . . **1a.** subsp. **oreas**

1a. subsp. **oreas**

Stem subglobose to elongate, 10–15(–35) × 10–18 cm; ribs (11–)12–16, to 15–25 mm high; areoles 9 × 4–6 mm. Spines 9–14, central(s) 1–4, lower 35–45 mm, radials 8–10, lowermost to 40–80 mm. Chromosome number: 2n = 44 (*Assis et al.* 2003).

Eastern *caatinga* element: on more or less exposed, horizontal, granite/gneiss *lajedos* and arenitic rocks in *caatinga*, within the drainage of the Rio Paraguaçu, at 300–500 m, eastern Bahia. Endemic to Bahia. Map 42.

BAHIA: Mun. Macajuba, 2.5 km N of town towards Baixa Grande, 9 Feb. 1991, *Taylor et al.* 1605A (K, photos); Mun. Ipirá, 42.5 km E of Baixa Grande, 9 Feb. 1991, *Taylor et al.* 1609C (K, photos); Mun. Rui Barbosa, quarry 1.5 km SW of town by road to Morro das Flores, 7 Feb. 1991, *Taylor et al.* 1565A (K, photo); Mun. Itaberaba, 9.5 km from Rui Barbosa on road to village of Alagoas, 7 Feb. 1991, *Taylor et al.* 1578C (K, photo); Mun. Iaçu, 5.5 km W of the town, 7 Feb. 1991, *Taylor et al.* 1580A (K, photos); Mun. Itatim, 7 km N of Milagres beside road BR 116, 27 June 1989, *Zappi* 181A (K, photo); Mun. Milagres, *Horst* 137 (U 531282, ZSS), *K. Preston-Mafham* s.n. (K, photo); Mun. Nova Itarana, 34 km S of Milagres on road BR 116, 27 July 1989, *Zappi* 180A (K, photo).

CONSERVATION STATUS. Least Concern [LC] (1); PD=1, EI=1, GD=2. Short-list score (1×4) = 4. Least Concern at present, but further modification of its habitats (eg. quarrying) and collection (sold for medicinal purposes) is likely and it requires monitoring in view of its relatively restricted distribution.

The name *M. oreas* Miq. is here employed (subsp. *oreas*) for a plant having a relatively limited, lowland distribution and stems with up to 16 ribs, the high rib-count being the only detail in Miquel's original description that positively excludes its application to the more common and widespread *M. ernestii* Vaupel, which has up to 13(–15) ribs only and ranges into higher and more humid areas (including outcrops in *brejo* forest), but is otherwise much more variable.

Subspecies *oreas* has been observed sympatric with *M. ernestii* and *M. salvadorensis* (all 3 were found together at Iaçu). It will occasionally hybridize with the latter. It may also hybridize or intergrade with *M. bahiensis* in north-eastern Bahia.

1b. subsp. **cremnophilus** (*Buining & Brederoo*) *P. J. Braun* in Bradleya 6: 95 (1988). Holotype: Bahia, Mun. Morro do Chapéu, *Horst* 223 (ZSS, apparently never deposited; U 531296, lecto. designated here).

Melocactus cremnophilus Buining & Brederoo in Cact. Succ. J. (US) 44: 3, with illus. (1972).

Like subsp. *oreas*, but stem globose to strongly depressed-globose, usually rather small, 8.5–12.0 × 13.5–14.5 cm; ribs 10–13, often very low; areoles to c. 6 × 5 mm. Spines 9–13, central(s) 1–4, 27–40 × 1 mm, lower slightly upcurved, radials 8–9, straight to slightly outcurved, lowermost to 55–70 × 1 mm, more strongly outcurved. Chromosome number: 2n = 44 (Assis *et al.* 2003).

Caatinga/Northern *campo rupestre* (Chapada Diamantina) element: on ± exposed, usually horizontal, crystalline/sandstone and granitic rocks, 700–1000 m, northern part of the Chapada Diamantina, Bahia. Endemic to Bahia. Map 42.

BAHIA: Mun. Jacobina, 21 km W of Jacobina on road BR 324 to Lajes (Lages), 16 July 1989, *Zappi* 131B (K, frs in spirit, photos); Mun. Morro do Chapéu, c. 8 km NW of town on road to Várzea Nova, *M. Machado* (obs.); l.c., c. 5 km SE of town, 2 Aug. 2002, *Taylor* (K, photos); l.c., 19 km E of town beside road BA 052, 24 Dec. 1988, *Taylor & Zappi* in *Harley* 27385 (K, SPF, CEPEC), *Horst* 223 (U 531296); Mun. Seabra, Serra da Água de Rega, 25 km N of Seabra on road to Água de Rega, near waterfall, 25 Feb. 1971, *Harley* s.n. (K, photo); l.c., 11.5 km W of Seabra on road BR 242, 18 July 1989, *Zappi* 143B (K, photo).

CONSERVATION STATUS. Least Concern [LC] (1); area of occupancy estimated to be < 2000 km²; PD=1, EI=1, GD=1. Short-list score (1×3) = 3. No serious deterioration to its habitats is known to be occurring. The habitat to the east and north-east of Morro do Chapéu is crossed by roads and may be visited by tourists, but the rock outcrops involved are sufficiently extensive to ensure its survival.

Subspecies *cremnophilus* is isolated from subsp. *oreas* by a zone of Atlantic Forest on the eastern flank of the Chapada Diamantina, Bahia. It has been found sympatric with *M. bahiensis*, *M. concinnus* and *M. paucispinus*, and will hybridize with the last two taxa. The population previously identified as belonging to this subspecies from Pernambuco (Mun. Caetés, 4 km NW of the town, then 0.5 km NE towards a fazenda, 13 Feb. 1991, *Taylor & Zappi* 1627B, K, photos) requires further study (cf. Taylor 1991a).

2. Melocactus ernestii *Vaupel* in Monatsschr. Deutsch. Kakteen-Ges. 30: 8 (1920); Buining in Krainz, Die Kakteen, Lfg 58 (1974). Type: Bahia, Mun. Barra da Estiva (?), *Ule* (photos, see Vaupel, l.c. and Ule 1908).

M. oreas Miq. subsp. *ernestii* (Vaupel) P. J. Braun in Bradleya 6: 95 (1988).

M. erythracanthus Buining & Brederoo in Cact. Succ. J. (US) 45: 223 (1973). *M. oreas* f. *erythracanthus* (Buining & Brederoo) P. J. Braun in Bradleya 6: 95 (1988). Holotype: Bahia, Mun. Morro do Chapéu, *Horst* 220 (U 531270).

M. azulensis Buining, Brederoo & S. Theun. in Kakt. and. Sukk. 28: 154 (1977). *M. oreas* f. *azulensis* (Buining *et al.*) P. J. Braun in Bradleya 6: 95 (1988). Holotype: Minas Gerais, Mun. Pedra Azul, *Horst* 168 (U 531281?).

M. longispinus Buining, Brederoo & S. Theun. in Succulenta 56(6): 137, with illus. on journal cover (1977). *M. oreas* var. *longispinus* (Buining *et al.*) P. J. Braun in Bradleya 6: 95 (1988). Holotype: Bahia, Mun. Iaçu, *Horst* 435 (U, not found Apr. 1989; K, lecto. designated here).

M. nitidus F. Ritter, Kakt. Südamer. 1: 139 (1979). Holotype: Minas Gerais, Mun. Itaobim, *Ritter* 1210 (U 531257).

M. interpositus F. Ritter, Kakt. Südamer. 1: 140 (1979). Holotype: Bahia, Mun. Iaçu, *Ritter* 1207 (U 531258).

[*M. oreas sensu auctt. pro parte et* Buining in Krainz, Die Kakteen, Lfg. 55–56 (1973), non Miq. (1840).]

Stem with or without mucilage, subglobose to shortly cylindric, 9–45 × 7–22(–35) cm, light yellow-green to dark green; ribs 9–13(–14), to c. 30 mm high and 70 mm broad near stem base, ± rounded in cross-section but sometimes with an acute edge; areoles to 13 × 9.5 mm, to 28 mm apart. Spines (11–)14–21, red and yellow banded or dull reddish to brownish, centrals (3–)4–8, lower somewhat upcurved or straight, to 32–90 × 1.3 mm, radials 7–13, lowermost to 45–150 × 1.3 mm, straight or outcurved. Cephalium to 18 × 8 cm, composed of pale to bright pinkish red bristles, sometimes covered by white wool at apex. Flower 19.5–29.0 × 9–18 mm; stigma-lobes white to reddish, sometimes exserted. Fruit deep pink to red at apex, rather variable in size, 14–45 × 5–12 mm. Seed 1.05–1.35 × 0.8–1.1 mm, testa-cells convex.

CONSERVATION STATUS. Least Concern [LC].

This species is divisible into two subspecies:

1. Stigma-lobes not or scarcely exserted, white; green stem tissues mucilaginous, ribs 10–13(–15); central spines 1–4(–6) (NE Brazil & NE Minas Gerais) . . . . **2a.** subsp. **ernestii**
1. Stigma-lobes exserted, often pinkish red; stem lacking mucilage or with traces in the green cortical tissues, ribs 9–11; central spines 4–8 (S Bahia & N Minas Gerais: between the Rio São Francisco and 42°W) . . . . **2b.** subsp. **longicarpus**

2a. subsp. **ernestii**

Stem mucilaginous at least in the green cortical tissues, subglobose to shortly cylindric, 10–45 × 14–22(–35) cm; ribs 10–13(–15), to c. 30 mm high and 40–70 mm broad near stem base; areoles to 10 × 9 mm. Spines

(11–)14–18, centrals usually 4(–6), lower somewhat upcurved, to 40–90 × 1.3 mm, radials 7–13, lowermost to 50–150 × 1.3 mm, outcurved. Flower 19.5–25.0 × 9–15 mm; stigma-lobes white, not exserted. Chromosome numbers: 2n = 22, 44 (Assis *et al.* 2003).

Eastern *caatinga* element: locally dominant on exposed crystalline/sandstone rocks and especially gneissic inselbergs, including outcrops in *brejo* forest, 50–1190 m, southern Paraíba, eastern Pernambuco, western Alagoas, western Sergipe, Bahia (E of 42°W) and north-eastern Minas Gerais. Reported from Ceará and Rio Grande do Norte by Luetzelburg (1925–26) and likely to occur there. Endemic to Eastern Brazil. Map 42.

PARAÍBA: Mun. Esperança, 16 Nov. 1920, *Luetzelburg* 15000 (M); Mun. Pocinhos, July 1921, *Luetzelburg* 12602 (M); between Soledade and Campina Grande, Nov. 1920, *Luetzelburg* 51, 57B (M); Mun. Maturéia, Serra de Teixeira, Pico do Jabre, 1994, *M.F. Agra et al.* (JPB).

PERNAMBUCO: Mun. Sertânia, Serra do Pinheiro, reported by Andrade-Lima (1966: 1458, as *M. oreas*); l.c., 13 km W of Arcoverde, 12 Feb. 1991, *Taylor & Zappi* 1624 (K, ZSS, HRCB, UFP); Mun. Brejo da Madre de Deus, 9 Feb. 1993, *F.A.R. Santos* 65 (HUEFS); Mun. Pesqueira, E of Cimbres, 'alto da serra' on road between Pesqueira and border with Paraíba state, 11 Feb. 1995, *Taylor* (K, photos); Mun. Caruaru, 9 km NE of town, 8°14'21"S, 35°54'52"W, IPA reserve, Dec. 1999, *Zappi* (K, photos); Mun. Alagoinha, 1992, *F.A.R. Santos* 25 (HUEFS); l.c., 4.5 km SW of the town on road to Garanhuns, 13 Feb. 1991, *Taylor & Zappi* 1626B (K, photos); Mun. Caetés, 4 km NW of town, 13 Feb. 1991 (observed growing with *M. bahiensis*, see *Taylor & Zappi* 1627).

ALAGOAS: reported from Piranhas etc. by Luetzelburg (1925–26, 1: 107).

SERGIPE: Mun. São Francisco de Canindé, reported as '*M. oreas*' by Rizzini (1982: 53, figs 8 & 19).

BAHIA: Mun. Jaguarari, Flamengo, *Ritter* 1210 (SGO 122126, *fide* Eggli *et al.* 1995: 493); Mun. Paulo Afonso, rock beside waterfall of the Rio São Francisco, 1919, *Luetzelburg* 92 (M); Mun. Morro do Chapéu, 21 km W of Morro do Chapéu beside road BA 052, 25 Dec. 1988, *Taylor & Zappi* in *Harley* 27394 (K, SPF, CEPEC), 1971, *Horst* 220 (U 531270); l.c., c. 2 km W of Ventura, 24 Dec. 1988, *Taylor & Zappi* in *Harley* 27388 (K, SPF, CEPEC); Mun. Candeal, beside road BR 324, 20.5 km NW of Tanquinho, 25 Apr. 1992, *Taylor & Zappi* (K, photo); Mun. Ipirá, 30 km E of Ipirá along road BA 052 to Feira de Santana, 2 Apr. 1976, *Davidse et al.* 11832 (SP, MO), 9 Feb. 1991, *Taylor et al.* 1611D (K, photos); Mun. Feira de Santana, 74.5 km E of Ipirá on road BA 052, 9 Feb. 1991, *Taylor et al.* 1611E (K, photos); l.c., N of town, Serra de São José, Nov. 2001, *Zappi* (K, photos); Mun. Cachoeira/São Félix, Barragem de Bananeiras, 12°32'S, 39°5'W, May 1980, *Iscardino et al.* in Grupo Pedra do Cavalo 7 (ALCB, RB); Mun. Iaçu, 1964, *Ritter* s.n. (U 531258); l.c., 5.5 km W, 7 Feb. 1991, *Taylor et al.* 1580B (K, photos); Mun. Itatim, Morro das Tocas, ápice, 12°43'S, 39°42'W, 31 Aug. 1996, *F. França* 1791, 1792, 1793, 1794 (HUEFS, K, photos) & *F. França* s.n. (K, photos); l.c., 12°47'S, 39°47'W, *L.P. de Queiroz & T.S.N. Sena* 3193 (K, HUEFS); Mun. Santa Teresinha, Serra da Jibóia, 7 Aug. 2002, *Taylor* (K, photos); Mun. Marcionílio Sousa, Machado Portella, reported by Buining (1974b); Mun. Maracás, 6–7 km E of Porto Alegre, immediately N of the Rio de Contas, 4 Feb. 1991, *Taylor et al.* 1552A (K, photos); Mun. Barra da Estiva, 'vom Rio de Contas' [Rio Sincorá, *cf.* Ule (1908)], photos in Vaupel, l.c. (1920); Mun. Iramaia, 38 km from Contendas do Sincorá towards Maracás, 4 Feb. 1991, *Taylor et al.* 1550A (K, photos); Mun. Dom Basílio, 21 km SE of Livramento do Brumado, 13°48'S, 41°47'W, 21 Mar. 1991, *G.P. Lewis* s.n. (K, photos); Mun. Ituaçu, c. 2 km NE of the town, 19 Aug. 1988, *Eggli* s.n. (ZSS, photo); Mun. Jequié, 41.5 km E of Porto Alegre, 4 Feb. 1991, *Taylor et al.* 1555A (K, photos); l.c., c. 30 km E of town, along the Rio de Contas, reported by Buining (1974b); Mun. Anajé, 11 km SE of town, 19 Dec. 1988, *Taylor & Zappi* in *Harley* 25594 (K, SPF, CEPEC); l.c., 14 km from Vitória da Conquista, 11 Apr. 1980, *Chalet* 32 (B).

MINAS GERAIS: NE Minas Gerais, Mun. Pedra Azul, near the town, *Horst* 168 (U 531281); Mun. Itaobim, 1 km W of town, 18 Nov. 1988, *Taylor & Zappi* in *Harley* 25527 (K, SPF); l.c., 8 km from road BR 116 towards Itinga, 9 Apr. 1983, *Martinelli & Leuenberger* 9228 (B).

CONSERVATION STATUS. Least Concern [LC] (1); PD=1, EI=2, GD=2. Short-list score (1×5) = 5.

This subspecies can be found sympatric with *M. oreas, M. bahiensis, M. salvadorensis, M. zehntneri, M. glaucescens* and *M. concinnus*. It can be distinguished from the first-named with difficulty, when they are not growing sympatrically, mainly because subsp. *ernestii* remains a highly variable and complex taxon. It requires much further study, which should also embrace *M. oreas, M. bahiensis* and *M. conoideus*.

Although disputed by Braun & Esteves Pereira (2002: 132), the following may represent a name for hybrids between *M. ernestii* and *M. zehntneri*, which have been observed in the municípios of Poção/Jataúba (*F. A. R. Santos* 5 & 6, PEUFR, K, photos) and Alagoinha (*Taylor & Zappi*, K, photos), Pernambuco: *M. horridus* Werderm. in Notizbl. Bot. Gart. Berlin-Dahlem 12: 227 (1934). Holotype: Pernambuco, Mun. Serra Talhada, 1932, *Werdermann* 2934a (B, in spirit; K, photos). It has yet to be recollected at the type locality, but the plants observed match the type well. Although the ranges of *M. ernestii* and *M. zehntneri* overlap, they are usually separated ecologically, *M. ernestii* preferring more humid rocks at higher altitudes.

2b. subsp. **longicarpus** (*Buining & Brederoo*) *N. P. Taylor* in Bradleya 9: 26 (1991). Holotype: Minas Gerais, Mun. Porteirinha, *Horst* 149 (U 531269).

Melocactus longicarpus Buining & Brederoo in Cact. Succ. J. (US) 46: 191 (1974). *M. deinacanthus* subsp. *longicarpus* (Buining & Brederoo) P. J. Braun in Bradleya 6: 94 (1988).

M. florschuetzianus Buining & Brederoo in Ashingtonia 2: 25 (1975). *M. deinacanthus* subsp. *florschuetzianus* (Buining & Brederoo) P. J. Braun in Bradleya 6: 94 (1988). Holotype: Minas Gerais, Mun. Francisco Sá / Grão Mogol, *Horst* 148 (U 531286).

M. mulequensis Buining & Brederoo in Succulenta 55: 46 (1976). *M. deinacanthus* f. *mulequensis* (Buining & Brederoo) P. J. Braun in Bradleya 6: 94 (1988). Holotype: Bahia, Mun. Rio do Antônio (?), *Horst* 122 (U 531299).

M. montanus F. Ritter, Kakt. Südamer. 1: 141 (1979). Holotype: Bahia, Mun. Urandi, *Ritter* 1358 (U 531254).

M. neomontanus Van Heek & Hovens in Succulenta 63: 78, illus. (1984). Holotype: Bahia, *van Heek & Hovens* 81-135 (U, not found Apr. 1989). Lectotype (designated here): van Heek & Hovens, l.c., illus.

Stem lacking mucilage or with a little in the green cortical tissues, epidermis light to dark green, occasionally somewhat glaucous, 9–20 × 7–20 cm; ribs 9–11(–12), to 25–50 mm broad near stem base; areoles to 13 × 9.5 mm. Spines (13–)14–21, sometimes very thinly overlaid with grey, nearly straight, but sometimes strongly curved to hooked at apex in seedlings, centrals (3–)4–8, lower to 32–53 × 1–1.8 mm, radials (8–)10–13, lowermost to 45–90 × 0.8–1.8 mm. Cephalium bristly or densely white-woolly at apex. Flowers often large, 22–29 × 9–18 mm, with whitish, pink or reddish, ± exserted stigma-lobes. Fruit pink to red at apex, elongate-clavate, 16–45 × 6.5–12.0 mm.

Southern *caatinga* element: on gneissic inselbergs and in other rocky places in *caatinga*, c. 450–950 m, between the Rio São Francisco and 42°W in southern Bahia, and on west side of the Serra do Espinhaço in northern Minas Gerais. Endemic to the core area of Eastern Brazil. Map 42.

BAHIA: Mun. Guanambi, 7 km W of Guanambi beside road BR 030, 20 July 1989, *Zappi* 152 (K, SPF); Mun. Rio do Antônio, 43.5 km W of Brumado on road BR 030, 2 Feb. 1991, *Taylor et al.* 1523B (K, photos), l.c. (?), 1972, *Horst* 122 (U 531299); Mun. Urandi, 3 km S of town, above railroad, 22 July 1989, *Zappi* 160 (K, SPF).

MINAS GERAIS: Mun. Espinosa, 8 km S of town on road BR 122 to Mato Verde, 22 July 1989, *Zappi* 163 (K, SPF); Mun. Monte Azul, c. 8 km E of the town, Serra Geral, 'Serra Gillette', 28 Jan. 1991, *Taylor et al.* 1460 (K, ZSS, HRCB, BHCB); Mun. Porteirinha, 8 km S of town on road to Riacho dos Machados, 7 Nov. 1988, *Taylor & Zappi* in *Harley* 25514 (K, SPF); Mun. Francisco Sá / Grão Mogol, vicinity of Barrocão, 950 m, *Horst* 148 (U 531286).

CONSERVATION STATUS. Least Concern [LC] (1); PD=1, EI=1, GD=2. Short-list score (1×4) = 4. Least Concern at present, but habitats are being disturbed in certain parts of its range; needs to be monitored.

Subspecies *longicarpus* is sympatric with *M. salvadorensis* in Mun. Rio do Antônio, Bahia.

3. Melocactus bahiensis *(Britton & Rose) Luetzelb.*, Estud. bot. Nordéste 3: 111 (1926). Type: Bahia, Mun. Marcionílio Sousa, *Rose & Russell* 19935 (US, lecto. designated here; NY, lectopara.).

Cactus bahiensis Britton & Rose., Cact. 3: 234 (1922). *Melocactus oreas* subsp. *bahiensis* (Britton & Rose) Rizzini, Melocactus no Brasil: 52–53 (1982), tantum quoad typ. *M. oreas* var. *bahiensis* (Britton & Rose) Rizzini, l.c. (1982), tantum quoad typ.
? *M. brederooianus* Buining in Succulenta 51: 28 (1972). *M. inconcinnus* var. *brederooianus* (Buining) F. Ritter, Kakt. Südamer. 1: 137 (1979). Holotype: Bahia, Mun. Senhor do Bomfim, *Buining* 1001 (U 531252).
M. acispinosus Buining & Brederoo in Krainz, Die Kakteen, Lfg 62 (1975), excl. photograph of plant apex (= *M. azureus* Buining & Brederoo). Holotype: Bahia, Mun. Jacobina, *Horst* 258A (U 531308).
? *M. inconcinnus* Buining & Brederoo in Kakt. and. Sukk. 26: 193 (1975). Holotype: Bahia, Mun. Brumado (?), *Buining* 1003 (U 531295).

Stem globose, depressed-globose or pyramidal, 9.5–21 × 11–21 cm, pale to dark green, with mucilage at least in the green cortical tissues; ribs 8–14, very low or up to 30 mm high, very variable in shape; areoles 6–14 × 4.5–8.0 mm, to 24 mm apart on the ribs. Spines 9–16, mostly straight, central(s) 1–4, 17–50 mm, radials (7–)8–12, lowermost to 22–60 mm. Cephalium usually small, rarely more than 5 × 6.5–8.5 cm. Flower pinkish magenta, 20–23 × 10–12.5 mm. Fruit reddish to magenta at apex, paler below, 17–25 × 6.5–9.0(–10.0) mm. Seed 1.05–1.35 × 0.85–1.30 mm, testa-cells convex.

CONSERVATION STATUS. Least Concern [LC].

Until recently the name *M. bahiensis* was commonly misapplied in Brazilian literature to the more widely ranging *M. zehntneri*. This misuse can be attributed to Werdermann, whose incorrect determinations of herbarium material preserved at Recife (IPA) seem to have misled two generations of botanists and ecologists.

The true *M. bahiensis* is highly variable and its taxonomy is poorly understood, as are its relationships with other species in the M. OREAS Group. The following infraspecific taxa are provisionally recognized:

1. Central spine(s) 1–4, the lower and largest usually > 25 mm, lowermost radial spine 24–60 mm . . . 2
1. Central spine 1, to 25 mm, lowermost radial spine 22–32 mm (N Bahia)
. . . . **3a(ii).** subsp. **bahiensis** f. **acispinosus**

2. Ribs rounded to somewhat acute but scarcely triangular in cross-section, or lowermost radial spine > 40 × 1.5 mm (Pernambuco & NE to S Bahia) 3
2. Ribs acute and triangular in cross-section; lowermost radial spine to 40 × 1.5 mm (S edge of Bahia to central Minas Gerais: Diamantina) . . . . **3b.** subsp. **amethystinus**

3. Ribs c. 8–10, sharply acute at edge, to 60 mm diam.; areoles 8–14 mm; spines to 2 mm diam. (S Bahia) . . . . **3a(iii).** subsp. **bahiensis** f. **inconcinnus**
3. Ribs 10–13, ± rounded, to 45 mm diam.; areoles to c. 8 mm; spines to c. 1 mm diam.
. . . . **3a(i).** subsp. **bahiensis** f. **bahiensis**

3a. subsp. **bahiensis**

Stem globose to depressed globose or pyramidal, 9.5–21 × 11–21 cm; ribs 8–13, high and acute to low and rounded, to 60 mm broad; areoles to 8–14 × 6.5–8.0 mm (to only 6 × 4.5 mm in f. *acispinosus*). Spines

9–16, pale reddish brown overlaid with grey, central(s) 1–4, lower 17–50 × 1–2 mm, porrect, straight or upcurved, radials (7–)8–12, lowermost 22–60 × 1–2 mm. Seed 1.05–1.20 × 0.8–1.05 mm (to 1.35 × 1.3 mm in f. *acispinosus*).

Eastern *caatinga*/Northern *campo rupestre* element: on more or less exposed crystalline rock formations (quartzitic-arenitic, granite/gneiss), quartz gravel etc., rarely on limestone, *campo rupestre / caatinga*, 380–1300 m, northern and eastern Pernambuco, to c. 14°50'S in Bahia. Endemic to the core area within North-eastern Brazil. Map 43.

PERNAMBUCO: Mun. Triunfo, Lajedo do Piripiri, 1140 m, 22 Nov. 1992, *F.A.R. Santos* 59 (HUEFS, PEUFR); Mun. Taquaritinga do Norte, road BR 104, c. 3 km S of turning to the town, 8 Feb. 1993, *F.A.R. Santos* 63 (PEUFR 13225, K, photo); Mun. Serra Talhada, 'Serra da Carnaubeira' [Serra do Morcego ou da Carnaúba?], on rock at top of the Serra, 22 May 1971, *Heringer, D. Andrade-Lima et al.* 842 (IPA 19729, UB); Mun. Caetés, 4 km NW of the town, then 0.5 km NE towards a fazenda, 13 Feb. 1991, *Taylor & Zappi* 1627 (K, ZSS, HRCB, PEUFR).

BAHIA: Mun. Senhor do Bomfim, Andorinha, *Buining* s.n. (U 531252); Mun. Monte Santo, above Monte Santo by path to santuário, 14 July 1989, *Zappi* 123 (K, SPF); l.c., near village of Pedra Vermelha, 14 July 1989, *Zappi* 123A (K, fruit in spirit, photos); Mun. Itiúba/Cansanção, 15 km E of Itiúba on road to Cansanção, 19 Feb. 1974, *Erskine* 91, cult. R.B.G. Kew, 1980 (K, photo); Mun. Jacobina, 2 km E of Jacobina, valley slope N of river, 16 July 1989, *Zappi* 131 (K, SPF), 1972, *Horst* 258A ['258'] (U 531308, ZSS AA-20-255); (?) Mun. Santa Bárbara, 43.5 km from Feira de Santana on road BR 116 to Serrinha, 13 July 1989, *Zappi* 119 (K, SPF); Mun. Rui Barbosa, Serra de Rui Barbosa, *Taylor et al.* 1582 (K, ZSS, HRCB, CEPEC); Mun. Marcionílio Sousa, Machado Portella, on tops of nearly barren hills, 1915, *Rose & Russell* 19935 (US, NY), *Horst* 388 (U 531261); Mun. Rio de Contas, between Rio de Contas and the Pico das Almas, Fazenda Brumadinho, 27 Nov. 1988, *Taylor in Harley* 25553 (K, SPF, CEPEC); l.c., between Faz. Brumadinho and Faz. Silvino, near the Pico das Almas, 1300 m, Dec. 1988, *Taylor* (K, photos); Mun. Paramirim, between Caraíbas and Paramirim, 30 Nov. 1988, *Taylor in Harley* 27029A (K, frs in spirit, photos); Mun. Livramento do Brumado, just N of Livramento, rocks by waterfall of Rio Brumado, 13°38'S, 41°50'W, Jan. 1974, *Harley et al.* 15345 (K, U, CEPEC, IPA 22990); Mun. Jussiape, 25 km from Rio de Contas towards the town, 25 Nov. 1988, *Taylor & Zappi in Harley* 25549 (K, SPF, CEPEC); Mun. Brumado (?), Rio São João, between Brumado and Livramento, 1973, *Buining* 1003 (U 531295); Mun. Ituaçu, c. 2 km SW of the town, exposed limestone, 18 Aug. 1988, *Eggli* s.n. (ZSS, photo); Mun. Tanhaçu, E of Sussuarana, Serra Escura, 18 April 2003, *Taylor et al.* (K, photos); Mun. Vitória da Conquista, Pradoso, Gameleira, 17 April 2003, *Taylor et al.* (K, photos).

CONSERVATION STATUS. Least Concern [LC] (1); PD=1, EI=1, GD=2. Short-list score (1×4) = 4.

Two distinctive local Bahian taxa of this complex are distinguished in the key above. Forma *acispinosus* (Buining & Brederoo) N. P. Taylor (1991a: 28) is represented by the collections cited from the municípios of Senhor do Bomfim, Itiúba and Jacobina, and f. *inconcinnus* (Buining & Brederoo) N. P. Taylor (1991a: 30) by those from Paramirim, Livramento do Brumado, Brumado and Ituaçu. However, recently (April 2003), the last-named has been found by the authors sympatric with typical *M. bahiensis* f. *bahiensis* near Vitória da Conquista and east of Sussuarana (Serra Escura) and would appear to be a distinct species, distinguishable by the characters given in the key, above.

Subspecies *bahiensis* can be found sympatric with *M. oreas*, *M. ernestii*, *M. zehntneri*, *M. concinnus* and *M. violaceus* subsp. *ritteri*.

3b. subsp. **amethystinus** (*Buining & Brederoo*) *N. P. Taylor* in Bradleya 9: 30 (1991). Holotype: Bahia, Mun. Caitité, *Horst* 270 (U 531291).

Melocactus amethystinus Buining & Brederoo in Krainz, Die Kakteen, Lfg. 50–51 (1972).

M. *lensselinkianus* Buining & Brederoo in Succulenta 53: 68 (1974). Holotype: Minas Gerais, Mun. Itaobim, *Horst* 381 (U 531275).

M. *griseoloviridis* Buining & Brederoo in Kakt. and. Sukk. 25: 98 (1974) ('*grisoleoviridis*'). Holotype: Minas Gerais, Mun. Itamarandiba, *Horst* 405 (U 531274).

M. *glauxianus* Brederoo & C. A. L. Bercht in Succulenta 63: 55, illus. (1984). Holotype: Minas Gerais, *Horst* 382 (U, not found Apr. 1989). Lectotype (designated here): Brederoo & Bercht, l.c., illus.

M. *ammotrophus* Buining, Brederoo & C. A. L. Bercht in Succulenta 63: 33 (1984). Holotype: Minas Gerais, Mun. Grão Mogol, *Horst* 353 (U 531306).

Stem depressed-globose to pyramidal, 10–13 × 13–19 cm; ribs 9–14, usually triangular in cross-section, edge acute; areoles 6–10 × 5–7 mm. Spines 9–13, all ± straight, brown to reddish or horn yellow, thinly overlaid with grey, central(s) 1–4, lower 15–33 × 1–1.8 mm, radials 8–10, lowermost 24–40 × 1.2–1.5 mm. Seed 1.1–1.3 × 0.95–1.15 mm. Chromosome number: 2n = 44 (Assis *et al*. 2003).

Eastern *caatinga*/Northern *campo rupestre* element: under and between shrubs on mainly crystalline (quarzitic/arenitic) rock formations in the Serra do Espinhaço region and on exposed granite/gneiss further east, *campo rupestre/caatinga*, 300–1000 m, southern Bahia to central Minas Gerais. Endemic to the core area of Eastern Brazil. Map 43.

BAHIA: Mun. Caetité, Brejinho das Ametistas, 26 July 1989, *Zappi* 175 (K, SPF), *Horst* 270 (U 531291); Mun. Licínio de Almeida, 10 June 2002, *G. Charles* (obs.); l.c. (?), E of Urandi, reported by Braun (1988a: 208–209, as 'HU 535'); Mun. Presidente Jânio Quadros, reported by Rizzini (1982: 75, fig. 39 above left).

MINAS GERAIS: Mun. Taiobeiras, between Taiobeiras and Curral de Dentro, by bridge over Rio Atoleiro, 17 Oct. 1988, *Taylor & Zappi in Harley* 25148 (K, SPF); Mun. (?), c. 15 km W of Curral de Dentro on road BR 251, 13 Aug. 1988, *Eggli* s.n. (ZSS, photo), 31 Jan. 1991, *Taylor et al*. 1516C (K, photos); Mun. Salinas, Fruta de Leite, reported by Vandecaveye & Keirse in Succulenta 63: 213 (1984); Mun. Medina, reported by Rizzini (1982: 74); Mun. Grão Mogol, Córrego da Bonita, 16°35'S, 42°54'W, 29 May 1988, *Zappi et al*. in CFCR 12095 (SPF 50304); Mun. Cristália, S bank of Rio Itacambiruçú near Grão Mogol, 15 Oct. 1988, *Taylor & Zappi in Harley* 25073 (K, SPF); Mun. Coronel Murta, 5 km SE of town, 21 Feb. 1988, *Supthut* 8860 (ZSS); Mun. Itinga, c. 5 km E of town, S of road to Itaobim, 19 Nov. 1988, *Taylor & Zappi in Harley* 25532A (K, frs in spirit, photo); Mun. Itaobim, c. 1 km W of town on N side of Rio Jequitinhonha, 18 Nov. 1988, *Taylor & Zappi in Harley* 25526 (K, SPF), 1972, *Horst* 381 (U 531275); Mun. Itinga, 1999, *Machado et al*. (obs.); Mun. Araçuaí, reported by Rizzini (1982: 74); Mun. Itacambira, N of the town, 1987, *Braun* 844 (ZSS); Mun. Itamarandiba, near Itamarandiba, 1972, *Horst* 405 (U 531274); Mun. Diamantina, Mercês, *Horst & Uebelmann* HU 174 (Uebelmann, mss; Eerkens 1983: 277).

CONSERVATION STATUS. Least Concern [LC] (1); PD=1, EI=1, GD=2. Short-list score (1×4) = 4.

This subspecies is sometimes sympatric with *M. ernestii*, *M. zehntneri* and *M. concinnus*, but hybrids between them have not been observed.

4. Melocactus conoideus *Buining & Brederoo* in Krainz, Die Kakteen, Lfg. 55–56 (1973). Holotype: Bahia, Mun. Vitória da Conquista, *Horst* 183 (U 531247).

VERNACULAR NAME. Cabeça-de-frade-do-Periperi.

Stem mucilaginous in the green cortical tissues, strongly depressed-globose to hemispheric, to 10 × 17 cm; ribs 11–15, very low and rounded, to 2.5 × 4 cm; areoles 6.5–7.5 × 6.5 mm. Spines dark brown overlaid with grey, central 1, 20–22 × 1.5 mm, radials 8–11, straight to slightly recurved, lowermost 20–35 × 1.5–1.7 mm, recurved near apex. Flower pinkish magenta, c. 22 × 10 mm. Fruit lilac-magenta, 18–21 × 5–6 mm. Seed 1.05–1.25 × 0.9–1.05 mm, testa-cells strongly convex at periphery. Chromosome number: 2n = 44 (Assis *et al*. 2003).

CACTI OF EASTERN BRAZIL   **377**
MELOCACTUS

Eastern *caatinga*/Northern *campo rupestre* element: under and between shrubs in quartz gravel, *campo sujo/cerrado de altitude*, 900–1050 m, Serra do Periperi, Mun. Vitória da Conquista, south-eastern Bahia. Endemic to Bahia. Map 43.

BAHIA: Mun. Vitória da Conquista, Morro do Cruzeiro above Vitória da Conquista, 26 July 1989, *Zappi* 179 (K, SPF), *Horst* 183 (K, in spirit [ex cult. Kew, Accn. No. 413-83.04910], U 531247, ZSS AA-20-73); l.c., NW of the above, 16 April 2003, *Taylor et al.* (K, photos).

CONSERVATION STATUS. Critically Endangered [CR B1ab(iii, iv, v) + 2ab(iii, iv, v)] (4); extent of occurrence < 100 km² and area of occupancy estimated to be < 10 km²; PD=2, EI=1, GD=1. Short-list score (4×4) = 16. Regrettably, survival of *M. conoideus* in the wild continues to be severely threatened by extraction of the quartz gravel in which it grows, and it was formerly impacted by commercial collection for the European horticultural market (Taylor 1992b). Since June 1992 it has been listed on Appendix I of C.I.T.E.S. In 1989, *M. conoideus* appeared close to extinction at the type locality in the Serra do Periperi above Vitória da Conquista, BA, where it is sympatric with the widespread *M. concinnus*. However, from the original locality other adjacent areas of similar habitat could be seen and subsequently Brazilian cactus enthusiasts have confirmed that healthy extensions of the population exist, some of these including 1000s of individuals. These areas are more distant from the BR 116 highway and one is now officially protected, although the extraction of gravel and wood continues throughout. Staff at the Universidade Estadual do Sudoeste da Bahia (Vitória da Conquista) have successfully propagated a substantial number of plants from wild seed and hope to reintroduce the species into a securely protected area of its original habitat in the near future. One encouraging piece of news is that it appears *M. conoideus* is capable of recolonizing areas from which gravel has been extracted and may even be able to take advantage of the reduced competition that such disturbance temporarily creates. A substantial grant to fence the officially protected area was made available in 2003 by the British Cactus & Succulent Society. This should improve the chances of the species' survival considerably.

The strongly depressed habit of this species is, as in the case of *M. paucispinus*, assumed to be an adaptation to minimize the effects of fire. It is very close to forms of *M. bahiensis* and may ultimately be best treated as a subspecies of it, once the latter has been better studied in the field.

A plant that could represent a southern subspecies of *M. conoideus* has recently been found in Mun. Itinga, Minas Gerais, by Inês Ribeiro de Andrade. Its habitat, at 1160 m altitude, is comparable and it differs from its Bahian relative principally in having stouter, strongly recurved spines.

MELOCACTUS DEINACANTHUS Group (No. 5)
Restricted to gneiss outcrops in central-southern Bahia.

5. Melocactus deinacanthus *Buining & Brederoo* in Kakt. and. Sukk. 24: 217 (1973). Holotype: Bahia, *Horst* 153 (U 531251).

Stem not mucilaginous, globose to elongate, 15–35 × 12–25 cm; ribs 10–12, acute, to 40 × 60 mm broad near stem base; areoles to 15 × 10 mm, to 33 mm apart on the ribs. Spines 15–21, reddish brown, centrals 4–7, lower to 53 × 3 mm, radials 11–14, lowermost to 80 × 2.5 mm, somewhat recurved. Cephalium to 25 × 9 cm, with exposed bristles at apex. Flowers pinkish magenta, to 26 × 11.5 mm; stigma-lobes whitish, not exserted. Fruit entirely white, rarely pinkish, shortly clavate, 12–22 × 6–12 mm. Seed 1.15–1.35 × 1.05–1.25 mm, scarcely narrowed or broadest at the hilum, testa-cells strongly convex.

Southern Rio São Francisco *caatinga* element: on gneissic inselbergs and *lajedos* in *caatinga* east of the Rio São Francisco, 500–540 m, central-southern Bahia. Endemic to Bahia. Map 44.

BAHIA: Mun. Bom Jesus da Lapa, Juá, 21 Apr. 1992, *Hatschbach & E. Barbosa* 56810 (MBM); l.c., 34.5 km SE of Bom Jesus on road BR 430 to Riacho de Santana, south side of road, Morro da Barriguda, 20 July 1989, *Zappi* 150 (K, SPF); l.c., 22 km from Riacho de Santana, 16 Apr. 1983, *Leuenberger et al.* 3075 (B, CEPEC); l.c., 1971, *Horst* 153 (U 531251); l.c., 2–10 km N of Juá towards Baixada Grande and Favelândia, incl. Faz. Barauninha, Jan. 2003, *M. Machado* (K, photos).

CONSERVATION STATUS. Critically Endangered [CR B1ab(iii) + 2ab(iii)] (4); area of occupancy estimated to be < 10 km², PD=3, EI=2, GD=1. Short-list score (4×6) = 24. It seems to be known from only c. 5 sites within 10 km of each other and it may be hybridising with *M. zehntneri* at one of these (M. Machado, pers. comm.). Its rarity and vulnerability to commercial exploitation has caused it to be placed on Appendix I of C.I.T.E.S., since June 1992. Part of the original locality is now a quarry.

M. deinacanthus is readily distinguished from all other members of the genus by the combination of shortly clavate, pure white fruits and uniquely shaped seeds, which are very broad at the hilum and with strongly convex testa-cells. It appears to have a very limited distribution east of the Rio São Francisco and is presently known from only the locality cited above, where it is a co-dominant element of the rupicolous vegetation.

MELOCACTUS LEVITESTATUS Group (No. 6)

Restricted to outcrops of *Bambuí* limestone in central-northern Minas Gerais and western Bahia.

6. **Melocactus levitestatus** *Buining & Brederoo* in Cact. Succ. J. (US) 45: 271 (1973). Holotype: Bahia, Mun. Bom Jesus da Lapa, *Horst* 397 (U 531285).

M. diersianus Buining & Brederoo in Kakt. and. Sukk. 26: 169 (1975). Holotype: Minas Gerais, Mun. Bocaiúva, *Horst* 404 (U 531287).

M. securituberculatus Buining & Brederoo in Cact. Succ. J. (US) 48: 38 (1976). *M. levitestatus* f. *securituberculatus* (Buining & Brederoo) P. J. Braun & E. Esteves Pereira in Schumannia 3: 188 (2002). Holotype: Bahia, Mun. Iuiú, *Horst* 446 (U 531298).

M. warasii E. Pereira & Bueneker in Bradea 2(30): 213, with illus. (1977). Holotype: Bahia, Mun. Bom Jesus da Lapa, *R. Buenecker* (HB 1827, in spirit, †). Lectotype (designated here): Pereira & Bueneker, l.c., illus.

M. rubrispinus F. Ritter, Kakt. Südamer. 1: 135 (1979). *M. diersianus* f. *rubrispinus* (F. Ritter) P. J. Braun in Bradleya 6: 94 (1988). Holotype: Minas Gerais, Mun. Bocaiúva, *Ritter* 1330 (U 531267).

M. uebelmannii P. J. Braun in Kakt. and. Sukk. 36: 232 (1985). Holotype: Bahia, Mun. Brejolândia, *Horst* 528 (ZSS).

[*M.* aff. *zehntneri sensu* Andrade-Lima in Revista Brasil. Biol. 37: 182–191, fig. 7 (1977).]

[*M. azureus sensu* Rizzini, Melocactus no Brasil: 60, fig. 12 above (1982) non Buining & Brederoo (1971).]

Stem without mucilage, light grey-green to dark green, or sometimes rather glaucous and with reddened edges to the ribs, slightly depressed or globose to cylindric, 15–68 × 14–30 cm; ribs 9–12(–15), to 5 × 8 cm; areoles to 11 × 10 mm, with abundant creamy wool at first, to 30–42 mm apart on the ribs. Spines 8–15, brownish red thinly overlaid with grey, central(s) 1–4(–6), 17–33 × 1–2.5 mm, mostly ascending, lower up- or decurved towards apex, radials 7–10, all strongly recurved, lower 3 21–33 × 1.5–2.0 mm. Cephalium to 18 × 7–12 cm. Flower entirely red or deep magenta within, red without, 20–27 × 6–9 mm, expanding fully or the inner perianth-segments remaining ± erect and forming a tube around the exserted stigma-lobes. Fruit pure white or faintly pinkish at apex, shortly clavate, 12–20(–24) × 7–12 mm. Seed 1.35–1.75 × (1.2–)1.3–1.65 mm, testa-cells flat and smooth.

Southern Rio São Francisco *caatinga* element: usually on elevated outcrops of *Bambuí* limestone amidst high *caatinga* forest, c. 450–780 m, western and central-southern Bahia and central-northern Minas

Gerais. Endemic to the core area of Eastern Brazil. Map 44.

BAHIA: Mun. Vanderlei, near Vanderlei, 25 km from road BR 242, 19 July 1989, *Zappi* 149 (K, SPF); Mun. Brejolândia, 1985, *Horst* 528 (ZSS AA-58-25/26/27, AV-20-72); Mun. Santana, 28 km S of the town, Faz. São Geraldo, path to Gruta do Padre, 15 Jan. 1991, *Taylor et al.* 1424A (K, photos); l.c., c. 8 km N of Porto Novo, Jan. 1991, *Eggli* (K, photo); Mun. Bom Jesus da Lapa, W of Rio São Francisco, 1978, *Horst* 456 (ZSS AV-20-36), 1972, *Horst* 397 (U 531285, ZSS AV-20-89/132); Mun. Iuiú, 2 km S of town, 21 July 1989, *Zappi* 155 (K, SPF), 1974, *Horst* 446 (U 531298).

MINAS GERAIS: Mun. (?) Manga, 'Mocambinho, Morro Solto', 15°6'S, 43°59'W, 18 Mar. 1998, *I. Pimenta* (photos); l.c., 'Serra Azul' [between 'Mocambinho' and Jaíba], 18 Mar. 1998, *I. Pimenta* (K, photos); Mun. Itacarambi/Januária, vale do Rio Peruaçu, illustrated in Costa *et al.* (1998: Fig. 29); Mun. Januária, c. 15 km SE of the Rio São Francisco on road BR 135, 11 Aug. 1988, *Eggli* s.n. (ZSS, photo); Mun. Varzelândia, c. 10 km N of the town, 12 Aug. 1988, *Eggli* s.n. (ZSS, photo); Mun. Capitão Enéas, reported by Rizzini (1982: 60, fig. 24 above); Mun. Montes Claros, 1964, *Ritter* s.n. (U 531250); Mun. Bocaiúva, Eng. Dolabela, limestone outcrop in sugarcane plantation near road BR 135, 4 Nov. 1988, *Taylor & Zappi in Harley* 25505 (K, SPF); Mun. Monjolos, between Rodeador and Conselheiro Mata, 18°17'27"S, 44°2'W, Aug. 2003, *M. Machado* (obs.).

CONSERVATION STATUS. Least Concern [LC] (1); PD=3, EI=1, GD=2. Short-list score (1×6) = 6. Least Concern at present, but in need of monitoring in view of the potential threat from limestone quarrying, though this is unlikely to affect the plant over the length of its considerable range.

MELOCACTUS AZUREUS Group (Nos 7–9)

Restricted to limestone outcrops (*Bambuí*) in northern Bahia.

7. Melocactus azureus Buining & Brederoo in Kakt. and. Sukk. 22: 101 (1971). Holotype: Bahia, [Mun. Juçara / 'Juassara' *fide* Uebelmann (1996)], 1968, *Horst* 256 (U, not found Apr. 1989). Lectotype (Taylor 1991a): colour photograph in Buining & Brederoo, l.c.

M. krainzianus Buining & Brederoo in Krainz, Die Kakteen, Lfg 62 (1975), excl. photo of plant apex (= *M. bahiensis* (Britton & Rose) Luetzelb.). *M. azureus* var. *krainzianus* (Buining & Brederoo) P. J. Braun in Bradleya 6: 94 (1988). Holotype: Bahia, Mun. Irecê, *Horst* 264 (U 531300).

Stem mucilaginous in the chlorophyllous cortical tissues, depressed-globose to shortly cylindric, usually intensely glaucous, less so when very old, 13–26 × 14–19 cm; ribs (9–)10, triangular in cross-section, to 35 × 40 mm; areoles to 11 × 7.5 mm, to 28–35 mm apart on the ribs. Spines 9–13, blackish or dark brown to reddish overlaid with grey, some hooked in seedlings, all somewhat recurved to decurved or almost straight, lower central 21–43 × 1.5–2.0 mm, radials 8–10, lowermost 21–40 × 1–2 mm. Cephalium to c. 12 × 7–9 cm, composed of reddish bristles and white wool, apex with brownish tufts of wool, bristles not or scarcely exserted. Flowers pinkish magenta, 19–23 × 8–11.5 mm. Fruit entirely white, slightly pinkish at apex or pale pink throughout, to 17 × 6 mm. Seed (1.3–)1.4–1.75 × 1.2–1.5 mm, testa-cells flat and smooth.

Northern Rio São Francisco *caatinga* element: on flat, exposed *Bambuí* limestone in *caatinga* at c. 450–750 m, in the region of Irecê, drainage of the Rio Jacaré and Rio Verde, central-northern Bahia. Endemic to Bahia. Map 44.

BAHIA: Mun. São Gabriel, Sítio Arqueológico, 10°57'54"S, 41°41'38"W, 27 Aug. 2002, *A. Cardoso* (photos); Mun. Itaguaçu da Bahia, road BA 052, just S of the Rio Verde, 14 July 2000, *G. Charles* (photos); Mun. Juçara, near Juçara, 25.5 km from road BA 052, 17 July 1989, *Zappi* 139 (K, SPF); Mun. Irecê, 9 km W of cross-roads with road BA 052, 18 June 1994, *L.P. de Queiroz* 3996 (HUEFS, K); l.c., beside road to Seabra, 1 km from traffic island immediately S of Irecê, 17 July 1989, *Zappi* 140A (K, fls in spirit); 12 km E of Irecê, S of road BA 052 to Morro do Chapéu, 25 Dec. 1988, *Taylor & Zappi in Harley* 27396 (K, SPF, CEPEC), 1972, *Horst* 264 (U 531300).

CONSERVATION STATUS. Endangered [EN B1ab(i, ii, iii, iv) + 2ab(i, ii, iii, iv)] (3); extent of occurrence < 2000 km² and area of occupancy < 50 km²; PD=2, EI=1, GD=2. Short-list score (3×5) = 15. Endangered by actual or potential habitat destruction (agricultural development) at its few known localities and not included within any protected area.

This species was observed sympatric with *M. pachyacanthus* subsp. *viridis*, q.v., but it is doubtful whether this habitat still holds plants of either species. Near the type locality of *M. azureus* there is a large population of plants that on their vegetative appearance seem to be somewhat intermediate between this species and *M. pachyacanthus* (G. Charles, photos).

8. Melocactus ferreophilus *Buining & Brederoo* in Krainz, Die Kakteen, Lfg 52 (1973). Holotype: Bahia, Mun. Barro Alto / Souto Soares, *Horst* 217 (U 531260).

M. azureus subsp. *ferreophilus* (Buining & Brederoo) N.P. Taylor in Bradleya 9: 40 (1991).

Like *M. azureus*, but usually cylindric, dark to grey-green, 15–45 × 13–19 cm; areoles to 13 × 11 mm. Spines 11–15, lower central to 30–53 × 2–2.2 mm, radials 7–11, lowermost to 40–53 × 2 mm, often strongly decurved towards its apex. Cephalium lacking brownish woolly tufts at apex, scarcely woolly in old plants, bristles conspicuous, ± well-exserted. Seed 1.05–1.30 × 0.95–1.10 mm.

Northern Rio São Francisco *caatinga* element: on raised, exposed, karstic *Bambuí* limestone in *caatinga* at c. 520–850 m, in the drainage of the Rio Jacaré (Rio Vereda do Romão Gramacho) and tributaries, between São Gabriel (APA de Gruta dos Brejões) and Mulungo do Morro, central Bahia. Endemic to Bahia. Map 44.
BAHIA: Mun. São Gabriel, Sítio Arqueológico, 10°57'54"S, 41°41'38"W, 27 Aug. 2002, *A. Cardoso* (photos); l.c., APA de Gruta dos Brejões, escarpment above Rio Jacaré, 4 Aug. 2002, *Taylor* (K, photos); Mun. Morro do Chapéu, 2.5 km E of América Dourada, south bank of Rio Jacaré (Rio Vereda do Romão Gramacho), N of road BA 052, 17 July 1989, *Zappi* 138 (K, SPF), 13 Jan. 1991, *Taylor et al.* 1408B (K, frs in spirit, photos); Mun. Barro Alto, Gameleira [near Lagoa do Boi], *Horst* 217 (U 531260, ZSS TP-20-236); Mun. Mulungu do Morro, *Horst & Uebelmann* 606 (ZSS TP-58-86).

CONSERVATION STATUS. Endangered [EN B1ab(iii) + 2ab(iii)] (3); area of occupancy < 50 km². PD=2, EI=1, GD=1. Short-list score (3×4) = 12. Endangered from potential limestone quarrying, actual modification of the environment surrounding its habitat and in view of its very limited known area of occupancy at 5 localities.

This taxon is here elevated to specific rank following the discovery by Aloísio Cardoso (Administrador, APA de Gruta dos Brejões, BA) of the first population cited above, where it grows sympatric with *M. azureus* without evidence of intermediates or hybrids. At the second and third localities cited this species was found sympatric at the edge of its habitat with *M. zehntneri* and there was some evidence of introgression between them. Its epithet was given in the mistaken belief that the limestone upon which it occurs was an iron-rich rock of volcanic origin.

9. Melocactus pachyacanthus *Buining & Brederoo* in Kakt. and. Sukk. 27(1): 1, with illus. (1975) [vol. for 1976 but Heft 1 publ. Dec. 1975]. Holotype: Bahia, Mun. Ourolândia, *Horst* 407 (U 531288).

Stem mucilaginous, depressed-globose to ovoid, pale green to grey-green, glaucous when young, 15–30 × 20 cm; ribs 9–11, low, to 25–37 × 50–65 mm; areoles to 13–15 × 9–11 mm, to 25 mm apart on the ribs. Spines 10–12, all very stout, reddish brown overlaid with grey, central(s) 1–3, lower 28–46 × 2–3(–4) mm, porrect or slightly ascending, radials 8–9, straight to slightly recurved, lower 3 25–49 × 2–2.5 mm. Cephalium to 30 × 10 cm, with dull pinkish red bristles, wool conspicuous or sparse. Flowers reddish to deep pinkish magenta, 22–25 × 7–10 mm, sometimes only half-expanded or scarcely exserted from cephalium. Fruit whitish or rather pale pink becoming very pale in lower half, 16.5–20.0 × 7–9.5 mm, somewhat flattened. Seed 1.2–1.4 × 0.95–1.15 mm, testa-cells flat and smooth, some elongate.

CONSERVATION STATUS. Endangered [EN B2ab(i, ii, iii, iv)]; area of occupancy estimated to be < 100 km² and numbers of plants at 5 of the c. 10 known locations are undergoing severe decline or may have already been lost.

An endemic Bahian species divisible into two subspecies:

1. Stem globose to elongate-ovoid, strongly glaucous except when old; cephalium with bristles and wool visible . . . . **9a.** subsp. **pachyacanthus**
1. Stem depressed-globose, plain green or slightly glaucous when young; cephalium of very dense red bristles only . . . . **9b.** subsp. **viridis**

9a. subsp. **pachyacanthus**

Stem glaucous. Cephalium to c. 9 cm diam., with bristles and conspicuous wool.

Northern Rio São Francisco *caatinga* element: on flat, ± exposed *Bambuí* limestone in *caatinga* at 400–645 m, near the Rios Salitre and Jacaré, northern Bahia. Endemic to Bahia. Map 44.
BAHIA: Mun. Juazeiro, 19 km W from Maçaroca on road to Fazenda Lagoa do Angico, 8 Jan. 1991, *Taylor* (K, photo); Mun. Juazeiro/Campo Formoso, 2.5 km W of Curral Velho, 9 Jan. 1991, *Taylor et al.* 1386 (K, ZSS, HRCB, CEPEC); Mun. Ourolândia, 65 km W of Jacobina, 1972, *Horst* 407 (U 531288); Mun. Morro do Chapéu, Área de Proteção Ambiental (APA) de Gruta dos Brejões, 4 Aug. 2002, *Taylor* (K, photos); Mun. São Gabriel, 57 km from Umburanas on road to Gameleira, 14 km NE of the latter, 13 July 2000, *G. Charles* (photos).

CONSERVATION STATUS. Endangered [EN B1ab(i, ii, iii, iv)] (3); area of occupancy = < 500 km²; PD=1, EI=1, GD=2. Short-list score (3×4) = 12. Endangered by ongoing habitat destruction (urban/agricultural development, quarrying) at 3 out of its 5 known, widely separated localities, but the largest population at the APA Gruta dos Brejões is safe and numbers a few 1000s of mature individuals and seedlings.

This subspecies is sometimes partially sympatric with *M. zehntneri*, with which it may exchange genes, but can be distinguished by its darker coloured flowers, smooth seeds and stronger spination.

9b. subsp. **viridis** N. P. *Taylor* in Bradleya 9: 40, pl. 16 (1991). Holotype: Bahia, Mun. América Dourada, *Taylor & Zappi* in *Harley* 27400 (CEPEC; SPF, K, isos.).

Stem never or only slightly glaucous when young. Cephalium c. 10 cm diam. or wider, with very dense bristles and little or no wool.

Northern Rio São Francisco *caatinga* element: on flat, ± exposed *Bambuí* limestone in *caatinga* at 700–800 m, in the region of Irecê, Gruta dos Brejões and to the west of Várzea Nova, central-northern Bahia. Endemic to Bahia. Map 44.
BAHIA: Mun. Morro do Chapéu, 10°57'53"S, 41°23'57"W, Sep. 2003, *A. Cardoso* (photos); l.c., E bank of the Rio Salitre, nr 'lagedo bordado', N of the village of Brejão (Formosa), W of Várzea Nova and c.

8 km from Icó, 5 Aug. 2002, *Taylor* (K, photos); Mun. Irecê, beside road BA 432 to Seabra, 1 km beyond traffic island S of Irecê, 17 July 1989, *Zappi* 140 (K, SPF); Mun. América Dourada, c. 17 km S of road BA 052, 2 km E of Belo Campo, 26 Dec. 1988, *Taylor & Zappi in Harley* 27400 (K, SPF, CEPEC).

CONSERVATION STATUS. Critically Endangered [CR B2ab(i, ii, iii, iv); C2a(ii); D] (4); area of occupancy < 10 km². PD=1, EI=1, GD=1. Short-list score (4×3) = 12. Almost certainly extirpated by habitat destruction (agricultural development) at two of its localities, apart from the second cited above, where a total of less than 50 individuals are known to survive. It has recently been reported that additional populations occur on top of limestone escarpments along the course of the Rio Salitre (M. Machado, *in litt.*).

Near Irecê, Bahia, subsp. *viridis* has been found sympatric with *M. azureus*, but can be distinguished by its greener epidermis, stouter spination, massive cephalium of very dense bristles and scant wool, less expanded flowers with much broader perianth-segments, larger, more pinkish fruits and smaller seeds. The habitats of these taxa in this region have mostly been destroyed by agricultural development. The second population cited above is somewhat morphologically intermediate with subsp. *pachyacanthus*.

MELOCACTUS VIOLACEUS Group (Nos 10–16)

Species Nos 13, 15 & 16 are restricted to sand or very sandy/gravelly substrates amongst low, sparse shrubs and No. 14 is also commonly found on similar materials, but more often associated with taller or denser vegetation. Nos 10 & 12 are most commonly found on crystalline rocks, gneiss etc. or very stony substrates, whereas No. 11 is found on any of the above as well as on limestone and soils derived from it.

10. **Melocactus salvadorensis** *Werderm*. in Notizbl. Bot. Gart. Berlin-Dahlem 12: 228 (1934). Type (syntypes): Bahia, Mun. Ipirá (Camisão) and Mun. São Félix/Muritiba, Bananeiras, 1932, *Werdermann* 3391, 3392 (both B†). Neotype (Taylor 1991a): Bahia, Bananeiras, *Horst* 301 (U 531290), illustrated in Krainz, Die Kakteen, Lfg 54 (1973).

[*M. macrodiscus sensu* Rizzini, Melocactus no Brasil: fig. 32 below (1982) non Werderm. (1932).]
[*M. inconcinnus sensu* F. Ritter, Kakt. Südamer. 1: 136–137, Abb. 78 (1979) non Buining & Brederoo (1975).]

Stem with mucilage in the green cortical tissues, pyramidal-globose to depressed-globose, greyish blue-green or quite glaucous, 12–20 × 12–25 cm; ribs 8–14, triangular-acute, 25–30 × 30–50 mm; areoles to 7–8.5 × 7–8 mm, 20–30 mm apart on the ribs. Spines horn yellow to reddish brown, overlaid with grey at first, central(s) 1–4, 15–30 × 1.5–2.5 mm, radials 7–10, lowermost 20–46 × 1.8–2.2 mm. Cephalium to 15 × 6–10 cm, composed of dense reddish bristles and sparse grey-white wool. Flowers pinkish magenta, opening c. 4–5 p.m., to 25 × 12 mm, but sometimes scarcely exserted from cephalium and less expanded. Fruit deep lilac-magenta, clavate, c. 17 × 6.5–9.0 mm. Seed 1.1–1.3 × 0.9–1.15 mm, some testa-cells markedly elongated, others strongly convex, especially at periphery. Chromosome number: 2n = 44 (Assis *et al.* 2003).

Eastern *caatinga* element: usually on or adjacent to exposed gneiss/granitic rocks/inselbergs and in stony soil of the *caatinga*, low elevations to 1040 m, within the Rio Paraguaçu (Rio Jacuípe) and Rio de Contas (Rio Gavião) drainage systems, eastern and southern Bahia. Endemic to Bahia. Map 45.
BAHIA: Mun. Riachão do Jacuípe, near road BR 324, 9 km NW of bridge over Rio Jacuípe, 25 Apr. 1992, *Taylor & Zappi* (K, photo); Mun. Candeal, beside road BR 324, 20.5 km NW of Tanquinho, 25 Apr. 1992, *Taylor & Zappi* (K, photos); Mun. Ipirá, 42.5 km E of Baixa Grande on road BA 052, 9 Feb. 1991, *Taylor et al.* 1609D (K, photos); Mun. São Félix/Muritiba/Conceição da Feira, Bananeiras, 1971,

Horst 301 (U 531290); Mun. Boa Vista do Tupim, c. 14 km da balsa para travessia do Rio Paraguaçú para João Amaro, na estrada para Boa Vista do Tupim, 27 Apr. 1994, *L.P. de Queiroz & Nascimento* 3886 (HUEFS, K); Mun. Iaçu, 5.5 km W of the town, 7 Feb. 1991, *Taylor et al.* 1580C (K, photos); Mun. Itatim, Morro do Agenor, 12°42'S, 39°46'W, 1 Sep. 1996, *F. França* 1815 (HUEFS, K, photo); l.c., 7 km N of Milagres beside road BR 116 to Feira de Santana, 27 July 1989, *Zappi* 181 (K, SPF); Mun. Rio de Contas, 8 km S of Jussiape (Juçiape) beside road to town of Rio de Contas, just W of the Rio de Contas, 22 Dec. 1988, *Taylor & Zappi in Harley* 25595 (K, SPF, CEPEC); Mun. Iramaia, 34 km from Contendas do Sincorá towards Maracás, 4 Feb. 1991, *Taylor et al.* 1549A (K, photo); Mun. Maracás, 6–7 km E of Porto Alegre, 4 Feb. 1991, *Taylor et al.* 1553 (K, ZSS, HRCB, CEPEC); Mun. Jequié, c. 1 km S of Jequié beside road BR 116, 0.5 km S of the Rio de Contas, 27 July 1989, *Zappi* 180 (K, SPF); l.c., 10 km S, 21 Aug. 1996, *Hatschbach* 63224 (MBM, K); Mun. Aracatu, 25 km along road BR 030 towards Sussuarana, 3 Feb. 1991, *Taylor et al.* 1542A (K, photos); Mun. Tanhaçu, 54 km E of Brumado along road BR 030 towards Sussuarana, 3 Feb. 1991, *Taylor et al.* 1545A (ZSS, photo); Mun. Rio do Antônio, 36 km from Brumado on road BR 030 to Caetité, 14 Apr. 1983, *Leuenberger et al.* 3064 (B, CEPEC), 2 Feb. 1991, *Taylor et al.* 1523A (K, fr. in spirit, photos); Mun. Brumado, 7 km N of Brumado beside road to Livramento, 22 Nov. 1988, *Taylor & Zappi in Harley* 25536 (K, SPF, CEPEC); Mun. Jacaraci, 2 km NW of Jacaraci, *Horst & Uebelmann* 535 *pro parte* (ZSS AV-20-37); Mun. Condeúba and Mun. Piripá, reported by Rizzini (1982: 71, figs 36 & 37); Mun. Cordeiros, 15°4'S, 41°53.5'W, Sep. 2003, *M. Machado* (obs.).

CONSERVATION STATUS. Least Concern [LC] (1); PD=2, EI=1, GD=1. Short-list score (1×4) = 4. Least Concern at present in view of its extensive range and number of locations, but its habitat continues to decline.

This species can be found sympatric with *M. oreas*, *M. ernestii* (both subspp.), *M. zehntneri* and, probably, *M. bahiensis* and *M. concinnus*. It occasionally hybridizes with *M. oreas* and *M. ernestii* subsp. *ernestii*. It is easily confused with *M. zehntneri*, but has much darker, magenta fruit. Its pollination biology has been studied by Raw (1996).

11. Melocactus zehntneri *(Britton & Rose) Luetzelb.*, Estud. bot. Nordéste 3: 111 (1926). Type: Bahia, Mun. Juazeiro, *Rose & Russell* 19728 (US, lectotype designated here; NY, lectopara.).

Cactus zehntneri Britton & Rose, Cact. 3: 236, with illus. of type, fig. 248 (1922).
Melocactus macrodiscus Werderm., Blühende Kakteen (Repert. Spec. Nov. Regni Veg. Sonderbeih. C), Lfg 12 (1932). Holotype: Bahia, Mun. Caitité, *Werdermann* s.n. (B†). Lectotype (Taylor 1991a): Werdermann, l.c. tab. 47 (1932).
M. curvicornis Buining & Brederoo in Kakt. and. Sukk. 23: 33 (1972). *M. zehntneri* var. *curvicornis* (Buining & Brederoo) P. J. Braun in Bradleya 6: 98 (1988). Holotype: Bahia, Mun. Macaúbas, *Horst* 128 (U 531266).
M. giganteus Buining & Brederoo in Cact. Succ. J. (US) 45: 227 (1973). Holotype: Bahia, Mun. Gentio do Ouro, *Horst* 266 (U 531294).
M. helvolilanatus Buining & Brederoo in Succulenta 55: 261 (1976). Holotype: Piauí, Santo Antônio de Lisboa, *Horst* 444 (U 531289).
M. canescens F. Ritter, Kakt. Südamer. 1: 134 (1979). *M. zehntneri* subsp. *canescens* (F. Ritter) P. J. Braun in Bradleya 6: 98 (1988). Holotype: Bahia, Mun. Tanhaçu, *Ritter* 1333 (U 531248).
M. canescens var. *montealtoi* F. Ritter, l.c. (1979). Holotype: Bahia, Mun. Palmas de Monte Alto, *Ritter* 1437 (U 531249).
M. zehntneri var. *viridis* F. Ritter, l.c., 132 (1979). Holotype: Piauí, Dom Expedito Lopes, *Ritter* 1206a (U 531268).
M. zehntneri var. *ananas* Rizzini, Melocactus no Brasil: 64 (1982), nom. inval. (Art. 37.1).
M. arcuatispinus Brederoo & Eerkens in Succulenta 62: 97, with illus. (1983). *M. zehntneri* subsp. *arcuatispinus* (Brederoo & Eerkens) P. J. Braun & E. Esteves Pereira in Schumannia 3: 188 (2002). Holotype: Bahia, Mun. (?), *Horst* 424 (U, not found Apr. 1989).

M. douradaensis Hovens & W. Strecker in ibid. 63: 3, with illus. (1984). Holotype: Bahia, Mun. América Dourada, *Hovens et al.* 81/172 (U, not found Apr. 1989).

M. saxicola L. Diers & E. Esteves Pereira in Kakt. and. Sukk. 35: 196, with illus. (1984). Holotype: Bahia, Mun. Barreiras, *E. Esteves Pereira* 119 (KOELN ['Succulentarium'], n.v.; UFG, iso.).

[*M. bahiensis sensu* Werderm., Bras. Säulenkakt.: 40 (1933) et auctt. mult. bras., non (Britton & Rose) Luetzelb. (1926).]

[*M. salvadorensis sensu* Rizzini, Melocactus no Brasil: 72, fig. 36 (1982) non Werderm.]

Stem highly mucilaginous, at least in the non-chlorophyllous tissues, dark to grey- or light green, often glaucous, sometimes strongly so when growing on limestone, but sometimes plain dark green (when on sand), hemispheric to cylindric, very variable in shape, 11–48 × (9–)13–25 cm; ribs 10–22, 22–35 × 22–57 mm, with a sharply acute edge; areoles to 12 × 9 mm, set into notches and c. 20–40 mm apart on the ribs. Spines 8–13(–15), horn yellow, brown or dull pink, but usually thickly overlaid with grey except at the dark tips, at least some hooked at apex in seedlings, centrals (0–)1–2(–4), upcurved, 15–40(–45) × 1.3–3.0 mm, radials 7–11, weakly to strongly recurved, lower 1–3 largest, 19–39(–45) × 1.5–2.5 mm. Cephalium to 11(–30) × 6–10 cm, composed of fine, dense, pale pinkish red bristles and sparse to abundant, white to creamy wool. Flowers pale to deep pink, not at all to well-exserted from cephalium, sometimes with the tube exserted also, ± expanded, to 25 × 4–13 mm. Fruit very pale (rarely almost white) to deep lilac-pink, rarely magenta (SE Pernambuco), 12–20 × 5–9(–10) mm. Seed 1.0–1.4 × 0.8–1.35 mm, testa-cells strongly convex at end opposite hilum. Chromosome numbers: 2n = 22, 44 (Assis *et al.* 2003).

Widespread *caatinga* / Northern *campo rupestre* element: in soil or sand and on rocks of various types, including limestone, gneiss/granite (inselbergs), sandstones, quartzitic and other crystalline formations, in the *caatinga* (rarely on rocks in *cerrado*, W Bahia, or in *campo rupestre*, S Bahia), c. 200–1000 m, northern Piauí, northern Ceará and Rio Grande do Norte to southern Bahia. Endemic to North-eastern Brazil. Map 45.

PIAUÍ: N Piauí, Mun. Buriti dos Lopes, 5 km NE of town on road BR 343, rocks by the Rio Pirangi, 19 Feb. 1995, *Taylor* (K, photos); Mun. Cocal, c. 17 km E of town, 19 Feb. 1995, *Taylor* (K, photos); SE Piauí, Mun. Dom Expedito Lopes, Gaturiano, *Ritter* 1206a (U 531268); Mun. Santo Antônio de Lisboa, *Horst* 444 (U 531289); Mun. Simões, foot of Chapada do Araripe on road from Marcolândia, 7°36'S, 40°45'W, 5 Apr. 2000, *E.A. Rocha et al.* (K, photos); Mun. Jaicós, 28 km S of town on road BR 407, 6 Apr. 2000, *E.A. Rocha et al.* (obs.); Mun. Canto do Buriti, c. 15 km S of town, partly exposed granite, 5 Feb. 1990, *Zappi* 215 (HRCB, ZSS); Mun. Paulistana, 25 km S of Paulistana, 1982, *Horst* 519 (ZSS AV-20-48).

CEARÁ: Mun. Tauá, 6 km along road to Marrecas/Cocotá from junction with road BR 020, S of Tauá, 30 Jan. 1990, *Zappi* 209 (HRCB, K, photos).

RIO GRANDE DO NORTE: Mun. Açu, 1978, *Horst* 480 (ZSS); Mun. Tangará, road BR 226, W of town towards Santa Cruz, 6°14'S, 35°51'W, 1 Apr. 2000, *E.A. Rocha et al.* (K, photos).

PARAÍBA: Mun. Pocinhos, Parque das Pedras, 28 June 1998, *E.A. Rocha* 420 (JPB, PEUFR); Mun. Taperoá, Nov. 1920, *Luetzelburg* 54 (M); Mun. Campina Grande, Nov. 1920, *Luetzelburg* 14001 (M); Mun. Itaporanga, Serra Água Branca, 7–10 Jan. 1994, *M.F. Agra et al.* 2559 (JPB); Mun. São João do Cariri, c. 1–2 km from the town, on road BR 412, 29 June 1998, *E.A. Rocha* 426 (JPB); Mun. Sumé, 29–30 June 1994, *M.F. Agra et al.* 2827 (JPB); Mun. Monteiro, s.d., *Luetzelburg* 86 (M), Nov. 1920, *Luetzelburg* 12321 (M).

PERNAMBUCO: Mun. Araripina, 15 Feb. 1990, *Zappi* 226B (HRCB); Mun. Taquaritinga do Norte, 8 Feb. 1993, *F.A.R. Santos* 63 (HUEFS); Mun. Ouricuri, Faz. Barrinha, 12 July 1984, *G. Costa Lima* 19 (IPA 42153); Mun. Salgueiro, 1.5 km N of city, granite outcrop, 17 Feb. 1990, *Zappi* 227 (HRCB, ZSS, K, photos); Mun. Serra Talhada, serra N of city, exposed granite, 13 Feb. 1990, *Zappi* 224 & 224A (HRCB, ZSS, K, photos); Mun. Brejo da Madre de Deus, Faz. Nova, 23 Jan. 1934, *B. Pickel* 3503 (IPA 4478), Jan. 1938, *V. Sobrino* s.n. (IPA 657); Mun. Ibimirim, 18 Nov. 1991, *F.A.R. Santos* 41 (HUEFS); Mun. Alagoinha, 9 Dec. 1991, *F.A.R. Santos* 20 (HUEFS, PEUFR 12412, K, photo); l.c., Faz. Cajueiro Seca, 6 km from town, near the quarry, 31 Mar. 1998, *E.A. Rocha & F. Lins* 360 (IPA); Mun. Afrânio, road BR 407, 5 km NW of Rajada, 8°47'S, 40°52'W, 6 Apr. 2000, *E.A. Rocha et al.* (obs.); Mun. Floresta, Serra de São Gonçalo, 23 May 1971, *Heringer, Andrade-Lima et al.* 864 (IPA 19753); l.c., road PE 360, N of the Serra Negra, 8°33'S, 38°1'W, 8 Apr. 2000, *E.A. Rocha et al.* (K, photos); Mun. Garanhuns, reported by Rizzini

(1982: 68, fig. 30); Mun. Santa Maria da Boa Vista, 18 Nov. 1992, *F.A.R. Santos* 47 (PEUFR, HUEFS); Mun. Petrolina, 19 Nov. 1992, *F.A.R. Santos* 50 (HUEFS); l.c., 20 km from Petrolina towards Recife, granite serra, 13 Aug. 1966, *Hunt* 6494 (K, photo); Mun. Nova Petrolândia, 12 km NE of the turning to the old town on road BR 110, 12 Feb. 1991 (observed with *Pilosocereus tuberculatus, Taylor & Zappi* 1621).
ALAGOAS: Mun. Ouro Branco, *fide* Uebelmann (1996): HU 475; Mun. Piranhas, Oct. 1993, *R. de Lyra-Lemos* s.n. (MAC); Mun. Dois Riachos and Mun. Santana do Ipanema, reported by Rizzini (1982: 67, fig. 29 left).
SERGIPE: Mun. Canindé de São Francisco, 23 Apr. 1981, *M. Fonseca et al.* 481 (ASE 896); Mun. Porto da Folha, 12.5 km from Monte Alegre de Sergipe towards Paulo Afonso (Bahia), 11 Feb. 1991, *Taylor & Zappi* 1617C (K, photo); Mun. Nossa Senhora da Glória and Mun. Tobias Barreto, reported by Rizzini (1982: 67, figs 28 above & 29 right).
BAHIA: W Bahia, Mun. Barreiras, NW of the city, Serra da Bandeira, cliff edges, road to new airport and near Telebahia tower, 18 Jan. 1991, *Taylor et al.* 1441 (K, ZSS, HRCB, CEPEC) & 1441A (K, HRCB); N Bahia, Mun. Glória, Brejo do Burgo, 27 Aug. 1995, *F.P. Bandeira* 267 (HUEFS); Mun. Remanso, 11 km E of town, 9°33'S, 42°W, 7 Apr. 2000, *E.A. Rocha et al.* (K, photos); l.c., c. 30 km from Remanso towards Casa Nova, 10 Feb. 1990, *Zappi* 220B (K, seeds); Mun. Juazeiro, 6.5 km along dirt road to Rodeodouro from road BA 210, 10 Jan. 1991, *Taylor et al.* 1389 (K, ZSS, HRCB, CEPEC); Mun. Paulo Afonso, reported by Rizzini (1982: 67, fig. 34 below left); Mun. Sento Sé, 10 km E of turning to Bazuá on road BA 210, Jan. 1991, *Taylor et al.* 1390A (K, photo); Mun. Curaçá, 4 km SE of Poço de Fora, 7 Jan. 1991, *Taylor et al.* 1368A (ZSS, photo); Mun. Uauá, 12.5 km N towards Patamuté, 6 Jan. 1991, *Taylor et al.* s.n. (ZSS, photo); Mun. Campo Formoso, 2.5 km W of Curral Velho, 9 Jan. 1991, *Taylor et al.* 1386C (K, photo); Mun. Jaguarari, 10 km N of Flamengo on road BR 407 to Juazeiro, 15 July 1989, *Zappi* 125A (K, frs in spirit, photo); Mun. Xique-Xique, 26 km SE of the town on road BA 052, 13 Jan. 1991, *Taylor et al.* (obs.); Mun. Queimadas, 3 June 2000, *C.T.S. Andrade* (photos); Mun. Jacobina, 10 km from Lajes (Lages) on road to Catinga do Moura, on limestone, 16 July 1989, *Zappi* 133A (K, photo); Mun. Gentio do Ouro, 16 June 1994, *L.P. de Queiroz* 3951 (HUEFS, K); l.c., S of Xique-Xique, Serra de Santo Inácio, *Horst* 266 (U 531294); Mun. Morro do Chapéu, 2.5 km E of América Dourada, south (east) bank of Rio Jacaré (Rio Vereda do Romão Gramacho), N of road BA 052, exposed *Bambuí* limestone, 17 July 1989, *Zappi* 138A (K, SPF); Mun. Valente, 6 km N of Valente beside road to Santa Luz, on fine quartz sand, 14 July 1989, *Zappi* 121 (K, SPF); Mun. Araci, 14.5 km N towards Tucano, 5 Jan. 1991, *Taylor et al.* 1354A (K, photos); Mun. Ibotirama, few km E of the town, 1986, *P.J. Braun* 604 (ZSS); S Bahia, Mun. Bom Jesus da Lapa, c. 32 km NE of Bom Jesus da Lapa, just beyond Caldeirão, 43°13'W, 13°10'S, *caatinga* with damp sandy area, 18 Apr. 1980, *Harley* 21525 (K, CEPEC); Mun. Macaúbas, *Horst* 128 (U 531266, 531280); Mun. Paramirim, 43 km W of Livramento do Brumado beside road to Paramirim, 4 km from Caraíbas, 28 Nov. 1988, *Taylor in Harley* 25555 (K, SPF, CEPEC); Mun. Livramento do Brumado, 12 km S of Livramento beside road to Brumado, 23 Nov. 1988, *Taylor & Zappi in Harley* 25545A (K, fls in spirit); Mun. Dom Basílio, 21 km S of Livramento do Brumado, 21 Mar. 1991, *G.P. Lewis* s.n. (K, photos); Mun. Brumado, reported by Rizzini (1982: 67, fig. 28 below); Mun. Aracatu, 25 km E along road BR 030 towards Sussuarana, 3 Feb. 1991, *Taylor et al.* 1542B (K, photo); Mun. Tanhaçú, Ourives, 1964, *Ritter* s.n. (U 531248); Mun. Palmas de Monte Alto, 1964, *Ritter* s.n. (U 531249); Mun. Caetité, near Brejinho das Ametistas, c. 1000 m, 26 July 1989, *Zappi* 177 (K, SPF).

CONSERVATION STATUS. Least Concern [LC] (1); PD=2, EI=1, GD=2. Short-list score (1×5) = 5. Least Concern, with an enormous distribution over much of NE Brazil, but exploited in many parts of its range as cattle fodder, for cactus candy and even for horticultural plantings, as seen in Pernambuco (Taylor 1991a) and elsewhere.

With the exception of one disjunct occurrence in the *cerrado* of western Bahia (near Barreiras), the range of *M. zehntneri* corresponds very closely to the limits of the *caatinga*. It is absent, however, from the *caatingas* of northern Minas Gerais and adjacent southernmost Bahia, and is replaced by the related and similar *M. salvadorensis* (q.v.) in the dry valleys of the Rio Paraguaçu and Rio de Contas in eastern Bahia (E of 41°W). It can be found sympatric with *M. ernestii*, *M. bahiensis*, *M. ferreophilus*, *M. pachyacanthus*, *M. salvadorensis*, *M. lanssensianus* (*fide* Braun) and *M. concinnus*. Braun & Esteves Pereira (2002: 79) indicate that *Melocactus* occurs in Maranhão, which could relate to *M. zehntneri*, but this remains to be backed up by published evidence.

The name *M. horridus* Werderm. is here applied to hybrids between *M. zehntneri* and *M. ernestii* (q.v.). It does not appear to be a synonym of *M. zehntneri*, as suggested by Braun & Esteves Pereira (2002: 132). Introgression between *M. zehntneri* and *M. concinnus* is evident at São Rafael and Tareco, Mun. Morro do Chapéu (M. Machado & Taylor, obs.).

M. zehntneri varies greatly in stem size, shape and colour, in spine number, length, thickness and colour, and in the degree to which its flowers are exserted from the cephalium and thus able to expand. Its pollination biology has been studied by Locatelli & Machado (1999a).

12. Melocactus lanssensianus *P. J. Braun* in Succulenta 65: 26, 61–63 (1986); P. J. Braun & E. Esteves Pereira in Schumannia 3: 135, Abb. 150 (2002). Holotype: Pernambuco, Mun. Caetés, 1977, *Horst* 474 (ZSS).

Stem highly mucilaginous, depressed-globose, to 8 × 14 cm, bluish grey-green; ribs c. 12, acute, 25–30 mm high, to 40 mm wide near stem base; areoles circular, 5–6 mm diam., c. 13 mm apart on the ribs. Spines 8–9, pinkish grey to beige, yellowish to brownish at apex, stiff, central 1, upcurved, to 30–35 × 1.5–2.0 mm, radials 7–8, recurved, lowermost longest, to 35–40 × 1.5 mm. Cephalium to 2.5 × 7 cm, with pale red to salmon-coloured bristles. Flowers apparently cleistogamous, drying up without expanding to form 6 mm long floral remains attached to the fruit. Fruit pink, c. 17 × 6 mm, containing 18–32 seeds. Seed 1.1 × 0.9 mm, testa-cells strongly convex.

Eastern *caatinga* element: on exposed granitic outcrops of serras in *caatinga*, c. 900 m, Mun. Caetés, south-eastern Pernambuco. Map 45.
PERNAMBUCO: Mun. Caetés, *Horst* 474 (ZSS AV-20-33/44, AA-58-55, TP-58-35).

CONSERVATION STATUS. Data Deficient [DD]; (material from the type locality is cultivated at the Royal Botanic Gardens, Kew).

At its type locality said by Braun, l.c., to occur sympatrically with *M. zehntneri*, but neither species could be observed by the present authors in the município cited above during field studies in 1991. However, it is not the only cleistogamous taxon of this relationship known from the region — the following cleistogamous entities may also belong here, the first being most similar to *Horst* 474 (see Map 45):

PARAÍBA: Mun. Tacima, near border of Mun. Caiçara, inselberg known as 'Pão de Açucar', 6°36'S, 35°28'W, 1 Apr. 2000, *E.A. Rocha et al.* (K, photos) — like *M. zehntneri*, but stem rather small and spines longer and more slender; Plate 61.
PERNAMBUCO: Mun. Serra Talhada, rocky mountain slope c. 1 km north of the town, 13 Feb. 1990, *Zappi* 223 (HRCB, ZSS, K) — not dissimilar from *M. bahiensis* in some respects and possibly worthy of description as a new taxon (see Taylor 1991a: pl. 14).

13. Melocactus glaucescens *Buining & Brederoo* in Cact. Succ. J. (US) 44: 159, with illus. (1972). Holotype: Bahia, Mun. Morro do Chapéu, [W of town], *Horst* 219 (U, not found Apr. 1989). Lectotype (Taylor 1991a): Buining & Brederoo, l.c. fig. 2.

[*M. pruinosus sensu* P. J. Braun in Succulenta 67: 114–117 (1988) and in Bradleya 6: 95 (1988) pro parte, non Werderm. (1934).]

Stem highly mucilaginous, depressed-globose to pyramidal, intensely light blue-glaucous at first, light grey-green in age, 13–18 × 12.5–24.0 cm; ribs 7–15, to 40 mm high and 60 mm broad near stem base; areoles to 8 × 5 mm, to 21 mm apart on the ribs. Spines 6–10, brown thickly overlaid with grey, dark brown to blackish at tip, centrals (0–)1–2, lower 11–20 × 0.8–1.7 mm, ascending and upcurved, radials 5–8, nearly straight to somewhat recurved, lower 3 11–55 × 1–1.8 mm. Cephalium to 10 × 6–7.5 cm, bristles hidden beneath dense, creamy white wool. Flowers lilac-magenta, c. 25 × 15.5 mm. Fruit entirely deep magenta-red, 9.5–16.0 × 5–7 mm, terete or somewhat flattened. Seed 1.1–1.35 × 0.9–1.15 mm, testa-cells strongly convex. Chromosome number: 2n = 44 (Assis *et al.* 2003).

Caatinga/Northern *campo rupestre* (Chapada Diamantina) element: in the open and between low shrubs of the *caatinga* on very sandy soil between small stones or in pure sand, sometimes occuring on adjacent, flat sandstone rocks, 650–1000 m, Chapada Diamantina, Bahia. Endemic to Bahia. Map 46.
BAHIA: Mun. Ourolândia, N of Olho d'água do Fagundes, 10°54'41"S, 41°15'4"W, Aug. 2003, *M. Machado* (obs.); Mun. Morro do Chapéu, NE of Gruta dos Brejões, between 'Vermelho' and 'Mulungu da Gruta', 4 Aug. 2002, *Taylor* (K, photos); l.c., north of town towards Várzea Nova, W bank of the Rio Salitre, c. 8 km from Icó, N of Brejão (Formosa), 'lagedo bordado', 5 Aug. 2002, *Taylor* (K, photos); l.c., 12 km N of Tareco, along the Rio Jacaré, after village of 'Canabrava', 'Cavalo Morto', 10 April 2003, *M. Machado* (obs.); l.c., 20–21 km W of Morro do Chapéu near road BA 052 to Irecê, 25 Dec. 1988, *Taylor & Zappi* in *Harley* 27393 (K, SPF, CEPEC), 3 Aug. 2002, *Taylor* (K, photos); c. 25 km W of Morro do Chapéu, 1 km S of road BA 052 towards Cafarnaum, slope E of road, 26 Dec. 1988, *Taylor & Zappi* in *Harley* 27397 (K, SPF, CEPEC).

CONSERVATION STATUS. Critically Endangered [CR B1ab(iii) + 2ab(iii)] (4); area of occupancy estimated to be < 10 km²; PD=2, EI=1, GD=2. Short-list score (4×5) = 20. In view of its rarity and the threats from commercial collection *M. glaucescens* has been placed on Appendix I of C.I.T.E.S. since June 1992. A local reserve to protect this and other rare species, incorporating the locality of *Harley* 27393 (above), has recently been implemented, but plants can still be damaged by fire. Its conservation status may soon need to be revised in view of the new discoveries made by Marlon Machado and colleagues over the past 4 years.

On account of its striking white-woolly cephalium, lilac-magenta flowers and small, deep magenta-red fruits, this is one of the most distinctive species and cannot be confused with any other member of the genus. The plant differs significantly at each of its sites of occurrence. At the northernmost site in Mun. Morro do Chapéu the plants have much longer spines (to 55 mm), whereas at the third site (listed above) the stems are smaller and have consistently fewer ribs (7–9).

M. glaucescens is restricted in distribution, being mainly known from 6 small areas north and west of Morro do Chapéu, Bahia, the latter two south-western sites being close to a main road (BA 052) and therefore easily accessed. At these two sites there is also evidence of hybridization with contiguous populations of other members of the genus and it is disturbing that 'pure' *M. glaucescens* is rapidly disappearing or restricted to very small areas of fine sand. At the fifth locality cited above (*Harley* 27393) *M. glaucescens* hybridizes with *M. ernestii* and the product of these two very different taxa can be identified as *M. × albicephalus* Buining & Brederoo in Krainz, Die Kakteen, Lfg 52 (1973). At the final locality (*Harley* 27397) *M. glaucescens* had formed a hybrid swarm with *M. concinnus*, although this may not be the only taxon involved (Taylor 1991a: plate 17, below). According to Marlon Machado (pers. comm.) pure *M. glaucescens* can no longer be found at this site. In the limited area of pure white sand at the previous locality, where the plant is found in pure form, it is at risk from fires set by the local farmer and various badly damaged specimens were seen (2002). At its new northern sites, which are c. 80 and 40 km distant from the type locality, there are no other *Melocactus* taxa close enough for gene exchange to occur and the species seems more secure from other forms of disturbance.

Further investigation may reveal yet more localities for this exceptionally beautiful species and, remarkably, Alex Braga, a contact of Marlon Machado, says he has recently found the species in the dunes of the Rio São Francisco, in Mun. Barra, on the west bank of the river, far from the Chapada Diamantina.

14. Melocactus concinnus *Buining & Brederoo* in Kakt. and. Sukk. 23: 5–7 (1972). Holotype: Bahia, Mun. Seabra, *Horst* 214 (U 531262).

? M. pruinosus Werderm. in Notizbl. Bot. Gart. Berlin-Dahlem 12: 228 (1934). Type: Bahia, Mun. Morro do Chapéu, *Werdermann* 3285 (B†).
M. pruinosus var. *concinnus* (Buining & Brederoo) P. J. Braun in Bradleya 6: 95 (1988).
M. axiniphorus Buining & Brederoo in Succulenta 55: 193 (1976). *M. concinnus* subsp. *axiniphorus* (Buin. & Brederoo) P. J. Braun & E. Esteves Pereira in Schumannia 3: 188 (2002). Holotype: Bahia, Mun. Vitória da Conquista, *Horst* 450 (U 531307).
M. robustispinus Buining, Brederoo & S. Theun. in ibid. 56: 116–119 (1977). *M. zehntneri* subsp. *robustispinus* (Buining *et al.*) P. J. Braun in Bradleya 6: 98 (1988). Holotype: Minas Gerais, Mun. Mato Verde, *Horst* 403 (U 531293).
[*M. macrodiscus* var. *macrodiscus* sensu F. Ritter, Kakt. Südamer. 1: 133 (1979) non Werderm. (1932).]

Stem highly mucilaginous, with green, chlorophyllous pith, grey-green and glaucous, at least when young, intensely so in seedlings, depressed-globose, 8–12(–13) × 11–16 cm; ribs 8–12, to 20(–30) mm high and 35–60 mm broad near stem base, often laterally creased; areoles to 6.5 × 6 mm, 13–20 mm apart on the ribs. Spines 6–9, red then blackish when developing, later horn yellow or pale reddish brown to brown, thickly overlaid with grey except at the dark tips, central (0–)1, 10–19 × 1 mm, upcurved, radials 6–8, slightly to strongly recurved, lower 1–3 largest, 15–26(–28) × 1–2 mm, sometimes hooked at apex. Cephalium to 5.5 × 4–9 cm, composed of fine, dense, pale pinkish red bristles and creamy wool. Flowers deep pink, scarcely to well-exserted from cephalium, 20–23 × 6–12 mm. Fruit pale lilac-pink to pink, 13–18(–22.5) × 5–8.5 mm, terete or somewhat flattened. Seed 1.05–1.35 × 0.95–1.3 mm, testa-cells strongly convex. Chromosome number: 2n = 44 (Assis *et al.* 2003).

Eastern *caatinga*/Northern *campo rupestre* element: in the open or beneath low to tall shrubs and trees in *caatinga, carrasco, cerrado de altitude* and *campo rupestre*, in stony soil, quartz sand or gravel, or between crystalline rocks, 550–1300 m, Chapada Diamantina, Serra do Espinhaço and Serra do Periperi (Vitória da Conquista), central Bahia to northern Minas Gerais. Endemic to core area of Eastern Brazil. Map 46.
BAHIA: Mun. Morro do Chapéu, at Tareco and São Rafael, 25 Dec. 1988 & 5 Aug. 2002, *Taylor* (obs.); l.c., 19 km W of town on road BA 052 to Irecê, dense *caatinga*, 25 Dec. 1988, *Taylor & Zappi* in *Harley* 27389 (K, SPF, CEPEC); c. 25 km W of Morro do Chapéu, 1 km S of road BA 052 *en route* to Cafarnaum, E of road, crystalline rocks and gravel, 26 Dec. 1988, *Taylor & Zappi* in *Harley* 27398 (K, SPF, CEPEC); l.c., c. 2 km S of town, terreno de M. Machado, 3 Aug. 2002, *Taylor* (K, photos); Mun. Seabra, 29 km N of town on road to Água de Rega, 27 Feb. 1971, *Irwin et al.* 31221 (K, UB); l.c., 11.5 km W of Seabra on road BR 242 to Ibotirama, *caatinga* on granite with sand between the rocks, 18 July 1989, *Zappi* 143 (K, SPF); Mun. Piatã, 12 km S of Boninal, *Taylor & Zappi* in *Harley* 25596 (K, SPF, CEPEC); Mun. Abaíra, between Catolés and Abaíra, 13°18'S, 43°49'W, 31 Jan. 1992, *Pirani et al.* in *Harley* 51396 (SPF, CEPEC, HUEFS, K); Mun. Érico Cardoso [Água Quente], 2 hours' walk W of 'Campo Queiroz', Serra do Pau Queimado (region of Pico das Almas), gravel from mine workings, 11 Dec. 1988, *Taylor* in *Harley* 25563 (K, SPF, CEPEC); Mun. Jussiape (Juçiape), 25 km from Rio de Contas

Plate 46. 46.1 *Pilosocereus magnificus*. NE Minas Gerais, *Taylor & Zappi* 763 (NT). **46.2** *P. pachycladus* subsp. *pachycladus*. Bahia, *Taylor et al.* 1406 (UE). **46.3** *P. machrisii*. W Bahia, *Taylor et al.* 1440, cult. Kew (K). **46.4** *P. aurisetus* subsp. *aurisetus*. Minas Gerais, S of Diamantina, 2002 (GC).

Plate 47. **47.1** *Pilosocereus aurisetus* subsp. *aurilanatus*. Minas Gerais, Joaquim Felício, 2002 (MM). **47.2** *P. aurisetus* subsp. *aurisetus* (*densilanatus*). Minas Gerais, S of Itamarandiba, 2002 (GC). **47.3** *Ibid.,* in fruit. Minas Gerais, *Zappi & Prado* in CFCR 11825 (DZ). **47.4** *P. aureispinus*. Bahia, *Taylor et al.* 1422 (NT). **47.5** *P. multicostatus*. NE Minas Gerais, *Harley et al.* 25188 (NT).

Plate 48. 48.1 *Pilosocereus multicostatus*. NE Minas Gerais, *Taylor et al.* 1517 (NT). **48.2** *P. piauhyensis*, flower. Piauí, *Zappi* 217 (NT). **48.3** *P. chrysostele*. N Pernambuco, *Zappi* 222 (NT).

Plate 49. 49.1. *Pilosocereus densiareolatus*. W Bahia, Porto Novo, 2002 (MM). **49.2** *P. parvus*. Bahia, *Horst & Uebelmann* HU 203, cult. G. Charles (GC). **49.3** *P. bohlei*. Type locality, 2000 (GC). **49.4** *Micranthocereus violaciflorus*. Minas Gerais, *Taylor et al.* 1514 (NT).

Plate 50. 50.1 *Micranthocereus albicephalus*. Bahia, N of Licínio de Almeida, 2002 (GC). **50.2** *Ibid*. Minas Gerais, *Harley et al*. 25519 (NT). **50.3** *M. purpureus*. Bahia, *Harley et al*. 27429 (NT). **50.4** *M. auriazureus*. Minas Gerais, Grão Mogol, 2002 (GC).

Plate 51. 51.1 *Micranthocereus auriazureus*. Minas Gerais, Grão Mogol, 2002 (GC). **51.2** *M. streckeri*. Type locality, 2002 (MM). **51.3** *Ibid*. Ibid. (MM). **51.4** *M. polyanthus*. S Bahia, Brejinho das Ametistas, 2002 (GC).

Plate 52. **52.1** *Micranthocereus flaviflorus, sens. lat.* Bahia, SE of Morro do Chapéu, 2002 (NT). **52.2** *M. flaviflorus.* Bahia, *Zappi* 135 (NT). **52.3** *Ibid.,* in fruit. Ibid. (NT). **52.4** *M. dolichospermaticus.* W Bahia, Porto Novo, 2002 (GC).

Plate 53. **53.1** *Coleocephalocereus buxbaumianus* subsp. *buxbaumianus*. Minas Gerais, *Taylor & Zappi* 771 (NT). **53.2** *C. buxbaumianus* subsp. *flavisetus*. E Minas Gerais, *Taylor & Zappi* 754 (NT). **53.3** *C. fluminensis* subsp. *decumbens*. NE Minas Gerais, S of Padre Paraíso, 2002 (GC).

Plate 54. 54.1 *Coleocephalocereus fluminensis* subsp. *decumbens*. NE Minas Gerais, S of Padre Paraíso, 2002 (GC).
54.2 *C. fluminensis* subsp. *fluminensis*. Espírito Santo, *Taylor & Zappi* 786 (NT). **54.3** *C. pluricostatus*. E Minas
Gerais, *Taylor & Zappi* 776 (NT). **54.4** *C. goebelianus*. Bahia, Mun. Maracás, 2002 (MM).

Plate 55. 55.1 *Coleocephalocereus aureus*. NE Minas Gerais, Pedra Azul, 2002 (MM). **55.2** *Ibid*. NE Minas Gerais, Mun. Salinas, 2002 (GC). **55.3** *C. purpureus*, NE Minas Gerais, *Harley et al.* 25532 (NT). **55.4** *Ibid*., in fruit. Ibid. (NT).

Plate 56. 56.1 *Melocactus oreas* subsp. *oreas*. Bahia, *Taylor et al*. 1578C (NT). **56.2** *M. oreas* subsp. *cremnophilus*. Bahia, SE of Morro do Chapéu, 2002 (NT). **56.3** *M. ernestii* subsp. *ernestii*. E Bahia, Jequié, 2002 (GC). **56.4** *M. ernestii* subsp. *longicarpus*. Minas Gerais, *Harley et al*. 25514 (NT).

Plate 57. **57.1** *Melocactus bahiensis* subsp. *bahiensis*. Bahia, nr Vitória da Conquista, 2003 (NT). **57.2** *M. bahiensis* subsp. *amethystinus*. Minas Gerais, *Harley et al.* 25073 (NT). **57.3** *M. conoideus*. Type locality, 2002 (MM). **57.4** *M. bahiensis (M. inconcinnus)*. Bahia, *Harley et al.* 27029A (NT).

Plate 58. 58.1 *Melocactus deinacanthus*. Bahia, type locality, 2003 (MM). 58.2. *Ibid. Ibid., Zappi* 150 (NT). **58.3** *Ibid*. Bahia, new locality 8–10 km from type, 2003 (MM).

Plate 59. 59.1 *Melocactus levitestatus.* W Bahia, Porto Novo, 2002 (GC). **59.2** *M. azureus.* N Bahia, Mun. Itaguaçu da Bahia, 2000 (GC). **59.3** *Ibid.* N Bahia, *Harley et al.* 27396 (NT). **59.4** *M. ferreophilus.* N Bahia, *Taylor et al.* 1408B (NT).

Plate 60. 60.1 *Melocactus pachyacanthus* subsp. *pachyacanthus*. Bahia, APA Gruta dos Brejões, 2002 (NT). **60.2** *M. pachyacanthus* subsp. *viridis*. Type collection (NT). **60.3** *M. salvadorensis*. Bahia, *Harley et al.* 25536 (NT).

Plate 61. 61.1 *Melocactus zehntneri*. Bahia, *Zappi* 177 (NT). **61.2** *Ibid.* Bahia, *Harley et al.* 25555 (NT). **61.3** *M.* cf. *lanssensianus*. Paraíba, Tacima (ER).

Plate 62. 62.1 *Melocactus glaucescens*. Bahia, NE of Gruta dos Brejões, 2002 (NT). **62.2** *Ibid*. Bahia, W bank of the Rio Salitre, 'lagedo bordado', 2002 (NT). **62.3** *Ibid*. Bahia, W of Morro do Chapéu, *Harley et al.* 27393 (NT). **62.4** *Ibid*. Ibid., *Harley et al.* 27397 (NT).

Plate 63. 63.1 *Melocactus concinnus*. Bahia, Morro do Chapéu, 2002 (NT). **63.2** *M. paucispinus*. Ibid. (NT).

Plate 64. 64.1 *Melocactus concinnus* × *M. paucispinus*. Bahia, Morro do Chapéu, 2002 (NT). **64.2** *M. violaceus* subsp. *violaceus*. Paraíba, NE of Pedras de Fogo, 2000 (NT). **64.3** *M. violaceus* subsp. *ritteri*. Bahia, *Taylor et al.* 1583 (NT). **64.4** *M. violaceus* subsp. *margaritaceus*. Bahia, Itapoã, 2002 (GC).

Plate 65. 65.1 *Harrisia adscendens,* in bud. Bahia, *Taylor et al.* 1388 (NT). **65.2** *Ibid.*, in fruit. Bahia, S of Feira de Santana, 1989 (NT).

Plate 66. 66.1 *Leocereus bahiensis*, in fruit. Bahia, W bank of the Rio Salitre, 'lagedo bordado', 2002 (NT). **66.2** *Ibid.,* flower sectioned. Bahia, *Harley et al.* 25552 (NT). **66.3** *Facheiroa ulei*. Bahia, near Santo Inácio, 2000 (GC). **66.4** *Ibid*. Ibid. (GC).

Plate 67. 67.1 *Facheiroa cephaliomelana* subsp. *cephaliomelana*. Cult. Univ. Bonn (WR). **67.2** *F. cephaliomelana* subsp. *estevesii*. S Bahia, Iuiú, 2002 (GC). **67.3** *F. squamosa*. Type locality, 1990 (NT). **67.4** *Ibid*. N Bahia, *Taylor et al*. 1372 (UE).

Plate 68. 68.1 *Espostoopsis dybowskii*. N Bahia, Jaguarari, 2000 (GC). **68.2** *Ibid*. S Bahia, *Taylor et al*. 1551 (NT).
68.3 *Arthrocereus melanurus* subsp. *melanurus*. Minas Gerais, *Horst & Uebelmann* HU 594, cult. G. Charles (GC).
68.4 *A. melanurus* subsp. *magnus*. Type locality, Aug. 2003 (NT).

Plate 69. 69.1 *Arthrocereus melanurus* subsp. *odorus*. Minas Gerais, *Horst & Uebelmann* HU 1555, cult. G. Charles (GC). **69.2** *A. rondonianus*. Minas Gerais, Joaquim Felício, 2002 (GC). **69.3** *Ibid*. Loc. cit., *Horst & Uebelmann* HU 145, cult. Univ. Heidelberg (WR).

Plate 70. 70.1 *Arthrocereus glaziovii*. Minas Gerais, *Horst & Uebelmann* HU 330, cult. G. Charles (GC). **70.2** *Ibid.*, in bud. Minas Gerais, *Harley et al.* 25500 (NT). **70.3** *Ibid.* Ibid. (NT).

Plate 71. 71.1 *Discocactus zehntneri* subsp. *zehntneri*. N Bahia, Serra do Francisco, S of Piçarrão, 2000 (GC).
71.2 *D. zehntneri* subsp. *boomianus*. Bahia, W of Morro do Chapéu, 2002 (NT). **71.3** *Ibid. (D. horstiorum)*. Bahia,
Serra do Curral Feio, 2000 (GC). **71.4** *D. bahiensis*. Bahia, Rodeadouro, 2000 (NT). **71.5** *Ibid*. Bahia, *Horst &*
Uebelmann HU 438, cult. G. Charles (GC). **71.6** *Ibid*. Bahia, São Rafael, 2000 (GC).

Plate 72. **72.1** *Discocactus heptacanthus* subsp. *catingicola*. Bahia, *Harley* 25558 (NT). **72.2** *D. placentiformis*, with spent flowers and fire-scorched stem. Minas Gerais, *Harley et al.* 24988 (NT). **72.3** *Ibid.*, in fruit. Minas Gerais, *Taylor et al.* 1450 (NT).

Plate 73. 73.1 *Discocactus pseudoinsignis*. Minas Gerais, Grão Mogol, 2002 (GC). **73.2** *D. horstii*. Ibid., 1981 (WR).

Plate 74. 74.1 *Uebelmannia buiningii.* Type locality, 2002 (GC). **74.2** *U. gummifera (meninensis).* Minas Gerais, Pedra Menina, 2002 (GC).

Plate 75. 75.1 *Uebelmannia gummifera*. Minas Gerais, SW of Itamarandiba, 2002 (GC). **75.2** *Ibid*. Minas Gerais, Penha da França, 2002 (GC). **75.3** *Ibid*. Minas Gerais, Tromba D'Anta, 2002 (MM).

Plate 76. 76.1 *Uebelmannia pectinifera* subsp. *pectinifera*. Minas Gerais, W of Mendanha, 2002 (GC). **76.2** *Ibid*. (green form). Minas Gerais, Inhaí, 2002 (MM). **76.3** *U. pectinifera* subsp. *flavispina*. Minas Gerais, N of Datas, 2002 (GC).

Plate 77. *Uebelmannia pectinifera* subsp. *horrida*. Minas Gerais, Mun. Bocaiúva, 2002 (GC).

on road to town, 25 Nov. 1988, *Taylor & Zappi* in *Harley* 25550 (K, SPF, CEPEC); Mun. Ituaçu, 13°48'S, 41°16'W, 22 June 1987, *L.P. de Queiroz* 1632A (HUEFS); l.c., c. 2 km NE of the town, 18 Aug. 1988, *Eggli* s.n. (ZSS, photo); Mun. Caetité, Brejinho das Ametistas, *campo rupestre* on quartzitic rocks, 26 July 1989, *Zappi* 174 (K, SPF); Mun. Licínio de Almeida, *Braun* 402 (ZSS TP58-116); Mun. Jacaraci, *Horst* 468 (ZSS AV-20-32); Mun. Vitória da Conquista, Morro do Cruzeiro above Vitória da Conquista, low shrub vegetation on quartz gravel, 26 July 1989, *Zappi* 178 (K, SPF), *Horst* 450 (U 531307); Mun. Cordeiros, c. 15°4'S, 41°53.5'W, Sep. 2003, *M. Machado* (obs.).

MINAS GERAIS: Mun. Monte Azul, Serra Geral, c. 9 km E of the town by path to village of Gerais, 28 Jan. 1991, *Taylor et al.* 1474 (K, ZSS, HRCB, BHCB); Mun. Mato Verde, 12 km E of Mato Verde on road to Santo Antônio do Retiro, *cerrado de altitude* and *campo rupestre* on quartz sand and gravel, 9 Nov. 1988, *Taylor & Zappi* in *Harley* 25517 (K, SPF), 22 July 1989, *Zappi* 165A (K, fls in spirit, photos), Aug. 1972, *Horst* 403 (U 531293, K); Mun. Grão Mogol, Serra da Bocaina, 48.5 km W of bridge over Rio Vacaria on road BR 251, 31 Jan. 1991, *Taylor et al.* 1515A (K, photo); Mun. Turmalina, Peixe Cru, by the Rio Jequitinhonha, July 1991, *M.G. Carvalho* (K, photo).

CONSERVATION STATUS. Least Concern [LC] (1); PD=2, EI=1, GD=2. Short-list score (1×5) = 5. Least Concern, but some populations affected by burning of vegetation started by man (*campos rupestres*) and the southernmost population is likely to be submerged under an hydro-electric dam-lake.

When found sympatric or contiguous with populations of other species, *M. concinnus* not infrequently forms hybrid swarms. These include allied species, such as *M. glaucescens*, *M. zehntneri* and *M. paucispinus*, as well as the unrelated *M. oreas* subsp. *cremnophilus*. It is also found with *M. bahiensis*, *M. conoideus* and *M. salvadorensis*.

15. Melocactus paucispinus *Heimen & R. J. Paul* in Kakt. and. Sukk. 34: 227–229, with illus. (1983). Holotype: Bahia, Mun. Seabra, *Heimen et al.* 81/149 (KOELN, n.v.).

Stem highly mucilaginous, strongly depressed, hemispheric or disc-shaped, usually partly buried in the sand, light grey-green, never glaucous even when young, 7–11.5 × 15–20 cm; ribs 9–10, to 2.5–4.0 cm high and 3.7–5.0 cm broad near stem base; areoles c. 3.5–7.0 × 2.5–5.5 mm, to 24 mm apart on the ribs. Spines (3–)4–6, recurved, pale grey with dark brown tips, all radial, lowermost largest, 20–32 × 1.2–2.0 mm, uppermost 1–3 much smaller, to 10 × 1 mm. Cephalium to 3–6 × 7–8 cm, composed of dense, fine, pinkish red bristles and creamy white wool. Flowers deep pink, (18–)24–26 × (9–)12.5–14.5 mm. Fruit pale lilac-pink, (14–)16.5–19.0 × 5.5–7.5 mm. Seed 1.3–1.6 × 1.2–1.5 mm, testa-cells strongly convex. Chromosome number: 2n = 44 (Assis *et al.* 2003).

Northern *campo rupestre* element: in sand or more rarely quartz gravel, *caatinga/cerrado de altitude*, c. 900–1500 m, in the Chapada Diamantina and northern Serra do Espinhaço, northern to southern Bahia. Endemic to Bahia. Map 46.

BAHIA: Mun. Morro do Chapéu, W of town, S of road BA 052 towards 'Pedra Duas Irmãs', c. 11°33'40"S, 41°17'50"W, June 2002, *M. Machado* (K, photos); l.c., E & S of town, in various sites within the area delimited by 11°33–37'S & 41°0–10'W, 2 Aug. 2002, *Taylor* (K, photos); Mun. Seabra, 28.5 km W of Seabra near road BR 242 to Ibotirama, 18 July 1989, *Zappi* 144 (K, SPF); Mun. Piatã, between Piatã and Inúbia, *R. Harley* (K, photos); Mun. Abaíra, 9 km N of Catolés, 'caminho de Ribeirão de Baixo a Piatã, gerais entre Serra do Atalho e a Serra da Tromba', 13°14'S, 41°55'W, 10 July 1995, *L.P. de Queiroz et al.* 4420 (HUEFS, K); Mun. Érico Cardoso [Água Quente], Pico das Almas, valley near Serra do Pau Queimado, near old road between Mato Grosso do Norte and Água Quente, quartz gravel, 23 Dec. 1988, *Hind* in *Harley* 27430 (K, SPF, CEPEC); Mun. Rio de Contas, Pico das Almas, between Rio de Contas and Faz. Brumadinho, 0.5 km before Faz. Brumadinho, 27 Nov. 1988, *Taylor* in *Harley* 25554 (K, SPF, CEPEC); Mun. (?), E of Urandi, reported by Braun (1988a: 208, 171, illus. 'HU 536').

CONSERVATION STATUS. Endangered [EN B2ab(v)] (3); extent of occurrence > 10,000 km², but area of occupancy assumed to be < 50 km²; PD=2, EI=1, GD=1. Short-list score (3×4) = 12. In view of its erratic distribution, rarity and desirability to collectors, this species has been listed in Appendix I of C.I.T.E.S. since June 1992. Only the population to the E & S of Morro do Chapéu is known to be large, the remaining 6 being much smaller (some with < 50 individuals) and severely fragmented/disjunct.

The strongly depressed stem of *M. paucispinus* seems to be an adaptation to minimize damage caused by fire, which periodically sweeps through its *cerrado/campo rupestre* habitat. By remaining partly buried in sand and exposing only the upper half of its flattened stem, it benefits from the cooler air drawn in at ground level as the fire passes, though the edges of its ribs may still get scorched. This adaptation is identical to that displayed by members of the ecologically comparable genus *Discocactus*, with which juvenile plants of *M. paucispinus* are readily confounded. Besides hummingbirds, a small lizard, c. 5 cm long, was seen visiting the flowers at Morro do Chapéu.

A population photographed by Ray Harley (K, photos), between Piatã and Inúbia (Mun. Piatã, Bahia), appears to represent plants showing introgression with *M. concinnus*, with bluish grey-green epidermis, 11 ribs and mostly 5, well-developed spines per areole. Introgression has also been observed by Taylor (K, photos) in a population south of Morro do Chapéu, in which the hybrid plants were larger than either parent and strongly resembled *M. zehntneri*. Similar plants were seen south of Rio de Contas. It also hybridizes with *M. oreas* subsp. *cremnophilus*.

16. Melocactus violaceus *Pfeiff.* in Allg. Gartenzeitung 3: 313 (1835); Enum. Cact.: 45–46 (1837). Type (see Schumann 1897–98): Brazil, cult. Schelhase (B†). Neotype (Taylor 1991a): Brazil, Rio de Janeiro, *D. Sucre* 9186 (RB 192529).

? Cactus melocactoides Hoffmanns., Verz. Pfl.-Kult. Nachtr. 3: 24 (1826). Type not known to have been preserved; see Taylor (1980: 67). *? Melocactus melocactoides* (Hoffmanns.) DC., Prodr. 3: 461 (1828).
M. depressus Hook. in Bot. Mag. 65: tab. 3691 (1838). *M. melocactoides* var. *depressus* (Hook.) Rizzini, Melocactus no Brasil: 82 (1982). *M. melocactoides* f. *depressus* (Hook.) Rizzini, tom. cit. 82, 84, fig. 49 (1982). Type: Pernambuco, *Gardner* (apparently not preserved). Lectotype (Taylor 1991a): Hooker, l.c., tab. 3691 (1838).
M. melocactoides var. *violaceus* (Pfeiff.) Rizzini, tom. cit., 81, 83 (1982).
M. violaceus subsp. *natalensis* P. J. Braun & E. Esteves Pereira in Cact. Succ. J. (US) 69: 71–74 (1997). Type: Rio Grande do Norte, *Esteves Pereira* 340 (UFG 17657, n.v.). *M. melocactoides* var. *natalensis* Rizzini, tom. cit. (1982), nom. inval. (Art. 37.1).

Stem highly mucilaginous, dark green, never glaucous even when young, usually broader than tall, often very small, 5–18(–20) × 6–17(–20) cm; ribs 8–15; areoles to 5 × 5 mm, 6–18 mm apart on the ribs. Spines 5–10(–12), brownish but thickly overlaid with grey except at the dark brown to blackish tips, very slender, 0.5–1.5 mm thick, centrals absent or 1, to 19 mm, ascending and often slightly upcurved, radials almost straight or slightly recurved, lower 3–5 largest and nearly equal, 14–24(–26) mm. Cephalium to 5.5(–11) × 3.7–8.5 cm, with abundant white wool and sparse to numerous, ± exserted, pale red, very fine bristles. Flowers deep pink, 15–25 × 6–13.5 mm, scarcely to well-exserted from the cephalium. Fruit pale pink to lilac-pink or white, 12.5–19.0 × 5.5–7.5 mm. Seed 1.2–1.5 × 1–1.4 mm, testa-cells strongly convex.

CONSERVATION STATUS. Vulnerable [VU A3c]; see assessments of subspecies below.

Two of the 3 subspecies recognized here are endemic to the core area of Eastern Brazil:

1. Fruit lilac- to pale pink 2
1. Fruit white to very pale pink (NE Bahia, Sergipe & Alagoas) . . . . . . **16c.** subsp. **margaritaceus**

2. Flower to 25 × 13.5 mm; spines 6–12; ribs 9–15 (NE Minas Gerais; and coastal regions
   of E Brazil up to 35 km inland, from Rio Grande do Norte to Rio de Janeiro) . . . .
   . . . . **16a.** subsp. **violaceus**
2. Flower c. 18–22 × 7–10 mm; spines 5–6; ribs 8–10 (Bahia inland: Jacobina & Rui Barbosa)
   . . . . **16b.** subsp. **ritteri**

16a. subsp. **violaceus**

Stem depressed-globose, hemispheric or disc-shaped, 6–16 cm diam.; ribs 9–15, to 10–25 × 40 mm broad near stem base; areoles 6–18 mm apart on the ribs. Spines 6–12. Cephalium to 5.5(–7.0) × 3.7–8.5 cm. Flower ± exserted from cephalium, with 17–25 perianth-segments visible from above.

Widespread humid forest (*Mata atlântica/restinga*) and Northern *campo rupestre* (*cerrado*) element: between shrubs in sand of the coastal *restinga*, riverine sand-dunes and similar habitats further inland, 0–150 m (in sandy *cerrado de altitude* at 1100 m, NE Minas Gerais only), Rio Grande do Norte to Rio de Janeiro, but apparently rather discontinuous, perhaps for lack of suitable habitats and due to extensive habitat destruction. Map 46.
RIO GRANDE DO NORTE: Mun. Genipabu, reported by Braun & Esteves Pereira (2002: 133, Abb. 146); Mun. Natal, c. 1902, *F.H. Rodier Heath* s.n., cult. R.B.G. Kew, 1908 (K), cf. also Cactus J. (Croyden) 6: 63–65, figs 1 & 2 (1938).
PARAÍBA: Mun. Mamanguape/Mun. Rio Tinto, Reserva de Guaribas, Aug. 2001, *J. Jardim* (obs.); Mun. Santa Rita, *M.F. Agra* (obs.); l.c. (?), plants on sale in market in João Pessoa, assumed to be from Santa Rita site, 1992, *McRobb* (K, photo ref. BZ 92-151); Mun. Pedras de Fogo, beside sugarcane plantation E of dirt road leading N from Destilaria Giasa, 7°20'S, 35°4'W, c. 150 m, 31 Mar. 2000, *E.A. Rocha et al.* (K, photos) — c. 30 km from the coast.
PERNAMBUCO: Mun. Jaboatão dos Guararapes, Prazeres (Estação), Oct. 1932, *B. Pickel* 3224 (IPA 4479); Mun. Ipojuca, *restinga* near baía de Suape, 1978, *Andrade-Lima & Medeiros-Costa* 241 (IPA 22705); l.c., Porto das Galinhas, 1988, reported by Braun & Esteves Pereira (2002: 21).
SERGIPE: between Penedo (AL) and Pindoba, c. 30–35 km from the ocean, c. 6 km from the Rio São Francisco, 100 m, 1978, *Horst* 482 (ZSS AV-20-24/41/94).
BAHIA: Mun. Maraú, Jan. 2000, *E.A. Rocha* (CEPEC); Mun. Uruçuca/Ilhéus, west of Serra Grande, 'Campo Cheiroso', Aug. 2001, *Zappi* (obs.); Mun. Canavieiras, *A.M. de Carvalho* (living plant seen at Ilhéus, Feb. 1991); Mun. Santa Cruz Cabrália, 2–6 km S of Santa Cruz Cabrália beside road BR 367, 31 July 1989, *Zappi* 185 (K, SPF); Mun. Porto Seguro, c. 11 km N of Porto Seguro on road BR 367 to Santa Cruz Cabrália, beneath shrublets of *Chamaecrista, Cuphea* and Melastomataceae, 30 July 1989, *Zappi* 183 (K, SPF); Mun. Prado, near Hotel Praia do Prado, 23 Sep. 1989, *M.C. Vianna* 1994 (GUA).
MINAS GERAIS: Mun. Jequitinhonha, c. 47 km S of Pedra Azul on road to Jequitinhonha, 'Serra da Areia' [Scrra da Sapucaia?], 1100 m, 20 Oct. 1988, *Taylor & Zappi in Harley* 25265 (K, SPF).
ESPÍRITO SANTO: Mun. Itaúnas, 15 May 1987, *G.P. Lewis* (K, photo); Mun. Linhares, Reserva CVRD, near Parajú road, 19°6'–18'S, 39°45'–40°19'W, reported by Figueira et al. (1994: 295 & fig. 1); Mun. Guarapari, KM 29–30 on road ES 060, Lagoa Vermelha, 21 May 1990, *O.J. Pereira* 2124 & *Zappi* (HRCB).

CONSERVATION STATUS. Vulnerable [VU A3c] (2); PD=1, EI=1, GD=2. Short-list score (2×4) = 8. Vulnerable from habitat modification by development for the tourist industry and for pineapple and sugarcane plantations. In Paraíba, the locality at Santa Rita has already been destroyed by a pineapple plantation, while that north-east of Pedras de Fogo has all but disappeared due to the cultivation of sugarcane. However, it may be protected in the northern part of this state in the Guaribas reserve (see above). Bahian populations are similarly affected, including habitat destruction due to the planting of coconut groves (eg. at 'Campo Cheiroso', Uruçuca/Ilhéus).

16b. subsp. **ritteri** *N. P. Taylor* in Bradleya 9: 57 (1991). Holotype: Bahia, Mun. Rui Barbosa, *Ritter* 1209a (U 531256).

Melocactus macrodiscus var. *minor* Ritter, Kakt. Südamer. 1: 133 (1979). Type: as above.

Stem depressed-globose, 8–13.5 cm diam.; ribs 8–10, to c. 15 × 50 mm; areoles 10–14 mm apart on the ribs. Spines 5–6. Cephalium to c. 5 × 5 cm. Flower with up to 24 perianth-segments visible from above, sometimes scarcely exserted from cephalium.

Eastern *caatinga* / Northern *campo rupestre* element: between *Vellozia* shrubs in fine quartz sand or gravel, *campo rupestre*, c. 450–860 m, near Jacobina and above Rui Barbosa, central-eastern Bahia. Endemic to Bahia. Map 46.
BAHIA: Mun. Jacobina, 2 km E of Jacobina town centre, valley slope N of Rio Itapicurumirim, 16 July 1989, *Zappi* 130 (K, SPF); Mun. Rui Barbosa, 1964, *Ritter* s.n. (U 531256); l.c., Serra de Rui Barbosa, 8 Feb. 1991, *Taylor et al.* 1583 (K, ZSS, HRCB, CEPEC).

CONSERVATION STATUS. Critically Endangered [CR B2ab(iii, v)] (4); area of occupancy < 10 km²; PD=1, EI=1, GD=1. Short-list score (4×3) = 12. Known from only two small locations, both near to towns, giving a projected decline in habitat and numbers of individuals (through collection).

This very local taxon occurs sympatrically with different forms of *M. bahiensis* at both localities cited above. It is very similar to the following, but has pink fruits like subsp. *violaceus*.

16c. subsp. **margaritaceus** *N. P. Taylor*, l.c., pl. 19 (1991). Holotype: Sergipe, Mun. Santo Amaro das Brotas, *Rizzini & Mattos* (RB 215018).

? Melocactus pentacentrus Lemaire, Cact. Nov. Gen. Sp.: 108 (1839). Type: Bahia [= Salvador], not known to have been preserved.
M. ellemeetii Miquel in Nederl. Kruidk. Arch. 4(3): 336, tab. 1 (1858) [a separate print from Miquel filed in the herbarium at BR is annotated 1857]. Type: Bahia, Salvador, a living plant in the garden of Ellemeet. Lectotype (designated here): Miquel, l.c., tab. 1 (1858).
M. margaritaceus Rizzini, Melocactus no Brasil: 76, figs 42–45 (1982), nom. inval. (Art. 37.1).
M. margaritaceus var. *disciformis* Rizzini, tom. cit.: 78 (1982), nom. inval. (Art. 43.1).
M. margaritaceus var. *salvadoranus* Rizzini, tom. cit.: 78 (1982), nom. inval. (Art. 43.1).
[*M. depressus* sensu Schumann, Gesamtb. Kakt. Nachtr.: 129 (1903) pro parte.]

Stem depressed-globose, rarely shortly cylindric, 5–15(–25) × 7–13(–20) cm; ribs 8–15, to 20 × 40 mm; areoles 7–10 mm apart on the ribs. Spines 5–10. Cephalium to 5(–11) × 3.7–6.5 cm. Flower exserted from cephalium, with c. 19 perianth-segments visible from above. Chromosome number: 2n = 22 (Assis *et al.* 2003).

Northern humid/subhumid forest (*restinga*) element: on coastal dunes of fine, white sand and inland (Serra de Itabaiana, SE), near sea level to c. 400 m, Alagoas, Sergipe and eastern Bahia (south to Salvador). Endemic to the core area within North-eastern Brazil. Map 46.
ALAGOAS: (?) Mun. Maragogi, coast N of the town, 1988, *Braun* 1006 (ZSS); Mun. Marechal Deodoro, 16 km S of Maceió, 9°47'S, 35°52'W, 2 Feb. 1982, *J.H. Kirkbride jr* 4620 (MAC, UB); l.c., *restinga* S of Ponta do Cavalo Ruço, 14 Feb. 1995, *Taylor & Zappi* (K, photos); Mun. Feliz Deserto, 30 Sep. 1981, *R.F. Rocha* 145 (MAC).
SERGIPE: Mun. Itabaiana, on rocks and sand of east slope (to c. 400 m) and at base of Serra da Itabaiana, 3 Aug. 1998, *Taylor* (ASE, K, photos); Mun. Santo Amaro das Brotas, 10 Mar. 1981, *Rizzini & Mattos*, cult. Jard. Bot. Rio de Janeiro (RB 215018); Mun. São Cristóvão, reported by Rizzini (1982: 77).
BAHIA: Mun. Camaçari, near Arembepe, Rod. Linha Verde, 18 Aug. 1995, *Hatschbach et al.* 63363

(MBM, K); Mun. Lauro de Freitas, 8 km N of Itapoã, Mar. 1989, *Supthut* 89-72 (ZSS); Mun. Salvador, 1 km N of praia of Itapoã, Lagoa de Abaeté, 13 July 1989, *Zappi* 115 (K, SPF); l.c., Dunas de Itapoã, 12°56'S, 38°21'W, 13 Mar. 1996, *M. Gomes* 12 (ALCB).

CONSERVATION STATUS. Vulnerable [VU A3c] (2); PD=1, EI=1, GD=2. Short-list score (2×4) = 8. Vulnerable from habitat modification by the rapidly expanding tourist industry and agricultural development (sugar cane, coconut), especially along the extensive coastal area in Bahia served by the 'Linha Verde' road and along the coast of NE Alagoas. Away from the coast it is most abundant and currently best protected in the state ecological reserve at the Serra de Itabaiana, Sergipe, managed by the local office of IBAMA. A decline of the populations of this taxon by at least 30% can be projected within 20 years (c. 3 generations).

If this taxon should one day prove to be worthy of specific rank, as Rizzini believed, then the earliest, clearly typified species name for it is *M. ellemeetii* Miquel, l.c. (1858). However, at least one of its northern populations has flowers and very pale pink fruits, which are somewhat intermediate with subsp. *violaceus*, further weakening the differences between them.

Tribe **TRICHOCEREEAE** F. Buxbaum

As noted earlier, the distinction between this tribe and Cereeae Salm-Dyck is currently unclear when analysed using DNA gene-sequence data (Wallace unpubl., Soffiatti unpubl.). Their traditional circumscriptions, based primarily on the presence or absence of hair-spines/spines on the pericarpel and flower-tube, is maintained here, although *Espostoopsis* F. Buxb. (Trichocereeae) and some species of *Cipocereus* Ritter (Cereeae) represent exceptions in each case. In the gene sequence study by Soffiatti (2002, unpubl.) the enigmatic *Uebelmannia* is placed amongst the Brazilian Cereeae, in a position close to *Cipocereus*, but here it is left at the end of the Cereeae/Trichocereeae pending further DNA data.

24. **HARRISIA** Britton

in Bull. Torr. Bot. Club. 35: 561 (1908). Type: *H. gracilis* (Mill.) Britton (*Cereus gracilis* Mill.).

Including *Eriocereus* (A. Berger) Riccob. (1909); *Harrisia* subg. *Eriocereus* (A. Berger) Britton & Rose (1920).

A genus of c. 10 species, with a disjunct distribution between the Caribbean (Subg. *Harrisia*, fruits indehiscent) and central South America (Subg. *Eriocereus*, fruits dehiscent). Two species are native of Brazil: *H. balansae* (K. Schum.) N. P. Taylor & Zappi (Mato Grosso & Mato Grosso do Sul, first recorded by Hoehne 1915; '*H. pomanensis*' misapplied by Braun & Esteves Pereira 2002: 127, Abb. 140) and that treated here, which is endemic to the Nordeste and isolated from its nearest congeners by some 1800 km. In its gross morphology it is most similar to the east Andean *H. pomanensis* (F.A.C. Weber) Britton

& Rose (syn. *Eriocereus tarijensis* F. Ritter; Argentina & Bolivia). However, gene sequence data (Wallace 1997: 11) indicate that *H. adscendens* is sister taxon to the Caribbean Subg. *Harrisia*, linking the two subgenera and indicating the path of radiation of the genus from its presumed origin in the eastern Andes of Bolivia, where it is believed to have common ancestry with *Samaipaticereus* Cárdenas. Plate 65.

1. Harrisia adscendens *(Gürke) Britton & Rose*, Cact. 2: 155 (1920). Type: Brazil, Bahia, 'Tambury', Oct. 1906, *Ule* 7072 (B, lecto. designated here; HBG, lectopara. [K, photos ex B]).

Cereus adscendens Gürke in Monatsschr. Kakt.-Kunde 18: 66 (1908). *Eriocereus adscendens* (Gürke) A. Berger, Kakteen: 341 (1929).

VERNACULAR NAMES. Rabo-de-raposa, Passa-prá-lá.

Shrubby, to 2 m or more, free-standing or semi-scandent and partly pendulous, sometimes forming dense masses, with basitonic or mesotonic branching, ± segmented. Juvenile growth of markedly different appearance to the adult, the ribs scarcely visible and bearing very small areoles and fine spines. Adult stems erect to decumbent, 2–7 cm diam.; ribs 6–10, low, rounded, tuberculate at the areoles, sinuses sinuous, marked by a dark line (as in *H. pomanensis*); epidermis pale grey-green, sometimes reddish. Areoles 6–10 mm diam., 25–50 mm apart (much closer in juvenile forms), with white, greyish or brownish, early deciduous hairs, subtended by red scale leaves when first emerging from apical meristem. Central spines 1–3, 15–30 mm, radials 4–7, 3–15 mm. Flower-bearing region not differentiated; young flower-buds covered in golden hair-spines to 15 mm long, flowers nocturnal, 15–22 × 12–17 cm; pericarpel greenish, covered in tubercles bearing areoles with bristles and triangular bract-scales; flower-tube 5–7 cm, infundibuliform, 1–2 cm diam., bearing scattered areoles with triangular bract-scales and hairs to 15 mm; perianth-segments to 6 cm, outer segments reflexed, lanceolate to linear, fleshy, dark red with green shades, inner segments spreading, lanceolate and fimbriate, delicate, white; stamens exerted in relation to the perianth-segments, curved, anthers linear; style 13–20 × 0.3–0.4 cm, stigma-lobes 12–15, exserted; ovary locule ovoid to oblong in longitudinal section. Fruit globose, to 10 cm diam., dehiscent by splitting irregularly, flower remnants deciduous; pericarp greenish then bright orange-red, covered in podaria bearing woolly areoles; funicular pulp white. Seeds c. 3–4 mm, cochleariform, black, shiny; testa-cells flat, smooth.

Widespread central-southern *caatinga* element: common along roadsides, often in farm hedges, amongst semi-open vegetation or scrambling over rocks (inselbergs), *caatinga-agreste,* especially on soils containing clay, c. 50–700 m, from north-western (Xique-Xique), central-northern and -eastern Bahia (drainage of Rio Paraguaçu), from 13°S, northwards to southern Ceará and southern Paraíba (c. 7°S). Endemic to the core area within North-eastern Brazil. Map 47.

PIAUÍ: reported by Luetzelburg (1925–26, 3: 111), but see comments below.

CEARÁ: S Ceará, Mun. Barro, road CE 292, between Cuncas and border with PB, 7°6'S, 38°42'W, 4 Apr. 2000, *E.A. Rocha et al.* (K, photos); Mun. Milagres, 2 km E of town on road BR 116, 4 Apr. 2000, *E.A. Rocha et al.* (obs.); Mun. Mauriti, 'Mamalu', road CE 296, shortly after the border with PB, 7°32'S, 38°40'W, 3 Apr. 2000, *E.A. Rocha et al.* (K, photos); Mun. Jardim, 1.5 km S of town towards Cedro (PE), 4 Apr. 2000, *E.A. Rocha et al.* (obs.).

PARAÍBA: Mun. Itaporanga, Serra Água Branca, 7–10 Jan. 1994, *M.F. Agra et al.* 2515, 2561 & 2562 (JPB); Mun. Desterro, 1998, *E.A. Rocha* (JPB); Mun. Conceição, planted in the town ex Serra do Cardoso, 3 Apr. 2000, *E.A. Rocha et al.* (obs.); Mun (?), between Sumé and Monteiro, 22 Feb. 1962, *Castellanos* 23269 (GUA); Mun. Monteiro, very common around the town, 11 Feb. 1995, *Taylor* (obs.).

PERNAMBUCO: Mun. Araripina, behind hotel 'Pousada do Araripe', 15 Feb. 1990, *Zappi* 226A (K, photos); Mun. Exu, 3 km S of town, 7°32'S, 39°43'W, 5 Apr. 2000, *E.A. Rocha et al.* (obs.); Mun. Serrita, Fazenda

Pirapora, 7°48'S, 39°14'W, 4 Apr. 2000, *E.A. Rocha et al.* (obs.); Mun. Triunfo, 8 Dec. 1991, *F.A.R. Santos* 15 (PEUFR, HUEFS); Mun. Serra Talhada, Estação experimental do IPA, 7°59'S, 38°19'16"W, 18 Jan. 1996, *M.L. Gomes* 125 (IPA); Mun. Parnamirim, road BR 122, Poço do Fumo, 8 Dec. 1971, *Andrade-Lima et al.* 1219 (IPA, PEUFR); l.c., 18 km from the town, road BR 316, 0.5 km after the Rio Favela, 11 Feb. 1993, *F.A.R. Santos* 71 (ALCB, HUEFS, PEUFR); Mun. Custódia, Rio Salgado, road BR 232, 9 Dec. 1991, *F.A.R. Santos* 19 (PEUFR, HUEFS); Mun. Sertânia, Moderna, 8 Apr. 2000, *E.A. Rocha et al.* (obs.); Mun. Arcoverde, W of town, 8 Apr. 2000, *E.A. Rocha et al.* (obs.); Mun. Brejo da Madre de Deus, Fazenda Nova, 23 Jan. 1934, *Pickel* 3500 (IPA); Mun. Pesqueira, Ipanema, 12 Jan. 1956, *Andrade-Lima* 56-2488 (IPA); Mun. Alagoinha, 13/14 Feb. 1992, *F.A.R. Santos* 21, 24, 27 (PEUFR, HUEFS); Mun. Afrânio, 5 km NW of Rajada on road BR 407, 8°47'S, 40°52'W, 6 Apr. 2000, *E.A. Rocha et al.* (K, photos); Mun. Belém do São Francisco, 12 km E of Cabrobó on road BR 428, 8 Apr. 2000, *E.A. Rocha et al.* (obs.); Mun. Floresta, 8°36'S, 38°27'W, 8 Apr. 2000, *E.A. Rocha et al.* (obs.); Mun. Inajá, Reserva Biol. Serra Negra, 9 Dec. 1995, *A. Laurênio et al.* 265 (K, PEUFR); Mun. Petrolina, Dec. 1914, *Luetzelburg* 729 (M), l.c., 5 km norte da CPATSA, 16 Aug. 1983, *Fotius* 3552 (IPA); l.c., road W of Petrolina towards village of Tapera, immediately N of the Rio São Francisco, 9°27'S, 40°41'W, 7 Apr. 2000, *E.A. Rocha et al.* (obs.).

ALAGOAS: Mun. Arapiraca, near Craíbas, Tingui, 22 Nov. 1983, *Staviski* 669 (MAC).

SERGIPE: Mun. Porto da Folha, 9.5 km from Monte Alegre de Sergipe towards Poço Redondo, 12 Feb. 1991, *Taylor & Zappi* (obs.); Mun. Nossa Senhora da Glória, Faz. Olhos D'agua, 21 Dec. 1981, *G. Viana* 331 (ASE 2006); Mun. Frei Paulo, 5 km após o povoado de Mocambo, 26 Mar. 1981, *M. Fonseca et al.* 450 & 451 (ASE 852 & 853).

BAHIA: NW Bahia, Mun. Barra, dunes by the Rio São Francisco, 10°47'16"S, 42°46'22"W, 23 June 1996, *Hind* s.n. (K, photos); N & NE Bahia, Mun. Juazeiro, near Rodeadouro, 10 Jan. 1991, *Taylor et al.* 1388 (HRCB, K, ZSS, CEPEC); Mun. Uauá, 1.7 km ESE of town towards Bendengó, 6 Jan. 1991, *Taylor et al.* 1362 (HRCB, K, ZSS, CEPEC); Mun. Jaguarari, Barrinha, 8 June 1915, *Rose & Russell* s.n. (US); Mun. Monte Santo, 14 July 1989, *Taylor & Zappi* (obs.); Mun. Itiúba, road to Filadélfia, 10°39'S, 39°44'W, 11 May 2002, *Nascimento & Nunes* 103 (HUEFS); Mun. Queimadas, 3 June 2000, *C.T.S. Andrade* (photos); Mun. Jacobina, 'margem esquerda do Rio Jacuípe', 21 Jan. 1985, *Passos* 25 (BAH 6022); l.c., 27–28 km WNW of town towards Lajes, 12 Jan. 1991, *Taylor et al.* (obs.); Mun. Morro do Chapéu, NE of Gruta dos Brejões, 4 Aug. 2002, *Taylor* (obs.); l.c., W of Icó, Brejão (Formosa), 'Lagedo Bordado', 5 Aug. 2002, *Taylor* (obs.); l.c., 26 km E of town, near Ventura, 11°42'S, 40°58'W, 24 Dec. 1988, *Taylor & Zappi in Harley* 27387 (K, SPF, CEPEC); Mun. Miguel Calmon, 30 km S of Jacobina, 26 Apr. 1992, *Taylor & Zappi* (obs.); Mun. São José do Jacuípe, 20 km SE of Capim Grosso, 6 km NW of bridge over the Rio Jacuípe, 25 Apr. 1992, *Taylor & Zappi* (obs.); Mun. Teofilândia, 2.5 km S of town, 16 km N of Serrinha towards Araci, 5 Jan. 1991, *Taylor et al.* 1349 (HRCB, K, ZSS, CEPEC); Mun. Riachão do Jacuípe, 33 km NW of Tanquinho, 25 Apr. 1992, *Taylor & Zappi* (K, photo); l.c., 24.5 km NW of Tanquinho, 25 Apr. 1992, *Taylor & Zappi* (obs.); Mun. Lamarão, 15 July 1959, *Castellanos* 22441 (R); Mun. Tanquinho, 8 July 1994, *F.A.R. Santos* 93 (HUEFS); Mun. Ipirá, 24 km E of town on road BA 052, 9 Feb. 1992, *Taylor et al.* (K, photos); Mun. Iaçu, 5.5 km W of town towards Marcionílio Souza, 7 Feb. 1991, *Taylor et al.* 1580 (HRCB, K, ZSS, CEPEC); Mun. Rafael Jambeiro, Km 16 on road BR 242, 29 Sep. 1973, *A.L. Costa* s.n. (ALCB); Mun. Castro Alves, May & Oct. 1972, *G.C.P. Pinto* s.n. (ALCB 02903, 02905); Mun. Itatim, 1997, *F. França* s.n. (K, photo.); Mun. Milagres, 10 km towards Itaberaba and then 2 km towards the Fazendas Morros and Antônio Romeu, 27 Jan. 1973, *I. & G. Gottsberger* 11-27173 (K); Mun. Brejões, 3 km N of the junction to the town on road BR 116, Aug. 2001, *Zappi* (obs.); unlocalized, 'Mina Poço, Mina Angico', 27 Mar. 1966, *Castellanos* 25943 (HB).

CONSERVATION STATUS. Least Concern [LC] (1); PD=2, EI=1, GD=1. Short-list score (1×4) = 4.

Although reported from Piauí by Luetzelburg (1925–26, 3: 111), this species has not been seen there by the authors. However, its occurrence at Araripina, Pernambuco is close to the border with that state, although it is not known if suitable edaphic conditions exist (ie. soils with clay content). Braun & Esteves Pereira (2002: 79, 128) indicate that it occurs in Rio Grande do Norte and western Minas Gerais, but concrete evidence is awaited and their distribution map (l.c. 78) is contradictory. If it does occur in Minas Gerais, it will be important to determine whether it is truly native or introduced there.

The places in which this species is commonly found today suggest that its dispersal and establishment is assisted by man and his animals. Its fruits are edible and it is often seen in farmhouse hedges.

25. LEOCEREUS Britton & Rose

Cact. 2: 108 (1920). Type: *L. bahiensis* Britton & Rose.

An isolated genus comprising a single species endemic to the core area of Eastern Brazil. Its placement in Trichocereeae is provisional and awaits confirmation via DNA gene sequence data.

Other species referred to *Leocereus* by Britton & Rose (l.c.) belong in *Arthrocereus*, except for *Cereus oligolepis* Vaupel (Britton & Rose, l.c., 225), which is a *Pilosocereus* (Zappi 1994). *L. squamosus* (Gürke) Werderm. (1933) is *Facheiroa squamosa* (Gürke) P. J. Braun & E. Esteves Pereira. For *L. paulensis* Speg., see *Coleocephalocereus fluminensis* (Zappi & Taylor 1992b). Plates 66.1–66.2.

1. Leocereus bahiensis *Britton & Rose*, Cact. 2: 108 (1920); N. P. Taylor & Zappi in Bradleya 8: 107–108 (1990). Type: Bahia, Mun. Sento Sé / Xique-Xique, Serra do Tinga ['Tingga'], *Zehntner* 266 (US, lecto. designated here; NY, K, lectoparas.).

Cereus bahiensis (Britton & Rose) Luetzelb., Estud. Bot. Nordéste 3: 111 (1926).
Leocereus urandianus F. Ritter, Kakt. Südamer. 1: 222 (1979). *L. bahiensis* var. *urandianus* (F. Ritter) P. J. Braun & E. Esteves Pereira in Kakt. and. Sukk. 41: 205 (1990). *L. bahiensis* subsp. *urandianus* (F. Ritter) P. J. Braun & E. Esteves Pereira in Succulenta 74: 132 (1990). Holotype: Bahia, Mun. Urandi, *Ritter* 1231 (U).
L. estevesii P. J. Braun in Kakt. and. Sukk. 41: 204 (1990). Holotype: SW Piauí, Mun. Corrente/Gilbués, *E. Esteves Pereira* 207 (UFG 12380; ZSS, iso.).
L. bahiensis var. *barreirensis* P. J. Braun & E. Esteves Pereira in ibid. 41: 205 (1990). *L. bahiensis* subsp. *barreirensis* (P. J. Braun & E. Esteves Pereira) P. J. Braun & E. Esteves Pereira in Succulenta 74: 132 (1995). Holotype: Bahia, Mun. Barreiras, *E. Esteves Pereira* 118 (UFG 12360; ZSS, iso.).
L. bahiensis var. *exiguospinus* P. J. Braun & E. Esteves Pereira in Kakt. and. Sukk. 41: 205 (1990). *L. bahiensis* subsp. *exiguospinus* (P. J. Braun & E. Esteves Pereira) P. J. Braun & E. Esteves Pereira in Succulenta 74: 132 (1995). Holotype: Bahia, Mun. (?), 'Canabrava', *E. Esteves Pereira* 135 (UFG [not found Jan. 1991]; ZSS, iso.).
L. bahiensis var. *robustispinus* P. J. Braun & E. Esteves Pereira in Kakt. and. Sukk. 41: 205 (1990). *L. bahiensis* subsp. *robustispinus* (P. J. Braun & E. Esteves Pereira) P. J. Braun & E. Esteves Pereira in Succulenta 74: 132 (1995). Holotype: Bahia, Mun. Gentio do Ouro, *P. J. Braun* 702 (ZSS).

Erect or semi-scandent, to 2(–3) m high, with one or more seldom-branched stems arising from a woody rootstock; stems cylindrical, very woody, lacking mucilage, bright or olive-green, to 200 × 1–2.5 cm; ribs (10–)12–20, rounded, obtuse, low, c. 1–2 × 2–4.5 mm, sinuses sinuate; areoles circular, 1–2.8 mm, 4–7 mm apart. Spines 8–21, not clearly differentiated into centrals and radials, slender acicular, yellowish to dark brownish or blackish, central(s) 1–9, ± erect, 6–20 mm, radials 7–12, adpressed, 3–5 mm. Flowers subapical, nocturnal, tubular, 40–73 × 20–34 mm, white; pericarpel and tube to c. 50 mm, green, clothed with 2–3 × 0.8–1.0 mm, brownish, triangular-acuminate bract-scales, pericarpel c. 9–10 × 6.5–7.5 mm, with 4 mm, porrect, pale spines in the bract-scale axils, the tube with dark hairs and bristles to 12 mm; nectar-chamber 12–17 × 8–12 mm, narrowed to 6–8.5 mm at base and to 7–11 mm above, apex of tube 12–18 mm diam.; perianth-segments 4–17 × 1.8–4.5 mm, outer series patent to reflexed, green to dull brownish red, oblong-apiculate, inner series suberect, pure white, lanceolate; lowermost stamens with thickened filaments protecting the nectar-chamber, 15–18 mm, anthers cream, 1.6–2.5 mm; style 30–40 × 0.7–1.0 mm, stigma-lobes 6–11, white, 1.5–4.5 × 0.6 mm. Fruit globose to ovoid, 23–36 × 16–32 mm, red or purplish, pericarp very shiny, indehiscent, at maturity the spine-clusters becoming detached, floral remains persisting, erect, funicular pulp magenta. Seeds 1.2–1.8 × 1–1.2 mm, black, glossy, hilum oblique, testa-cells flat to slightly convex, with intercellular pits (Barthlott & Hunt 2000: pl. 27.6–8).

Widespread central-southern *caatinga* element: growing between and through shrubs, rocky and sandy places in *'caatinga de altitude'*, *campo rupestre* and their ecotone, and amongst rocks and cliffs in *cerrado* (W Bahia), 550–1500 m, central-northern Minas Gerais to northern Bahia (especially in the Serra do Espinhaço and Chapada Diamantina), westwards to southern Piauí and in the Chapadão da Bahia (Espigão Mestre) towards the border with Goiás. On present evidence, endemic to Eastern Brazil. Map 47.

PIAUÍ: Mun. Canto do Buriti, *fide* Uebelmann (1996): HU 663; Mun. Corrente/Gilbués, *E. Esteves Pereira* 207 (UFG 12380, ZSS).

BAHIA: W Bahia, Mun. Roda Velha, *fide* Uebelmann (1996): HU 950; Mun. Barreiras, Serra da Bandeira, edge of cliff overlooking Barreiras with Telebahia installations, 18 Jan. 1991, *Taylor et al.* 1436 (K, ZSS, HRCB, CEPEC); l.c. (?), Apr. 1981, *E. Esteves Pereira* 118 (UFG, ZSS); N Bahia, Mun. Sento Sé, 10°7'S, 41°25'W & 10°19'S, 41°24'W, 11 July 2000, *G. Charles* (photos); l.c., Serra do Tinga ['Tingga'], Aug. 1912, *Zehntner* 266 (US, NY, K), *Zehntner* in IFOCS 2040 (M); Mun. Jaguarari, Barrinha, 7–8 June 1915, *Rose & Russell* 19820 (K); Mun. Gentio do Ouro, nr Gameleira, July 1986, *Braun* 702 (ZSS); Mun. Morro do Chapéu, W of Icó, Brejão (Formosa), 'Lagedo Bordado', 5 Aug. 2002, *Taylor* (K, photos); l.c., c. 20 & 25 km W of Morro do Chapéu, S of road BA 052, 26 Dec. 1988, *Taylor & Zappi* (obs.); Mun. Cafarnaum, 12 Nov. 1956, *E. Pereira* 2152 (RB); Mun. Barra do Mendes/Seabra, c. 20 km W of Seabra, 1986, *Braun* 721 (ZSS); Mun. Ibitiara, Serra Malhada, c. 52 km W of Seabra on road BR 242, 14 Jan. 1991, *Taylor et al.* 1416 (K, ZSS, HRCB, CEPEC); Mun. Seabra, 6 km E of Seabra beside road BR 242, 25 Aug. 1988, *Eggli* 1283 (ZSS); S Bahia, Mun. Piatã/Boninal, 12 km S of Boninal on road to Piatã, 12°48'S, 41°48'W, 22 Dec. 1988, *Taylor & Zappi* in *Harley* 25596A (SPF, K, photos); Mun. Érico Cardoso [Água Quente], NW slopes of Pico das Almas, 13°29'S, 42°W, 11 Dec. 1988, *Taylor & Hind* in *Harley* 25552B (K, photos); Mun. Rio de Contas, 3 km from the town towards Livramento, 13°36'S, 41°50'W, 23 Nov. 1988, *Taylor & Zappi* in *Harley* 25540 (K, SPF, CEPEC); Mun. Ituaçu, c. 2 km NE of the town, 18 Aug. 1988, *Eggli* s.n. (ZSS, photo), *Ritter* 1470 (U); Mun. Urandi, 1964, *Ritter* 1231 (U).

MINAS GERAIS: Mun. Monte Azul, mountains c. 9 km E of the town, 15°10'S, 42°48'W, 28 Jan. 1991, *Taylor et al.* 1473 (K, ZSS, HRCB); Mun. Mato Verde, 12 km from Mato Verde on road to S. Antônio do Retiro, 15°23'S, 42°45'W, 9 Nov. 1988, *Taylor & Zappi* in *Harley* 25521 (K, SPF).

CONSERVATION STATUS. Least Concern [LC] (1); PD=4, EI=1, GD=2. Short-list score (1×7) = 7.

Braun, l.c. and Braun & Esteves Pereira (2002: 79, 85, Abb. 95) report the synonymous *L. estevesii* from southern Maranhão, but no corroborating material has been seen by the present authors.

26. **FACHEIROA** Britton & Rose

Cact. 2: 173 (1920); P. J. Braun & E. Esteves Pereira (1986–89). Type: *Facheiroa pubiflora* Britton & Rose (= *Facheiroa ulei* (Gürke) Werderm.).

Including *Zehntnerella* Britton & Rose (1920); *Facheiroa* subg. *Zehntnerella* (Britton & Rose) P. J. Braun & E. Esteves Pereira (1986).

Columnar, erect, treelike, sometimes shrubby, to 8 m; central cylinder woody, solid. Branches 2.4–7.0 cm diam., erect, not constricted, epidermis dull to bright green, not glaucous; ribs 15–20, low, crenate, sinuses straight. Areoles with felt and long hairs; spines weak, flexible to brittle, not very pungent, central spines porrect to deflexed, longer than the numerous radials. Flower-bearing region of stem lateral, subapical or flowers randomly disposed, when differentiated sometimes forming a cephalium that transforms the structure of the stem. Flowers nocturnal, to 5 × 2.5 cm; pericarpel and tube covered in triangular bract-scales subtending tufts of hair; tube cylindric, stout; sometimes with a tuft of hairs between the nectar-chamber and the filament bases; innermost filaments curved towards the style; perianth-segments short, triangular to lanceolate or spathulate, innermost white, pinkish or purplish. Fruit globose to depressed-globose or pear-shaped, indehiscent, covered in scales and hairs, flower remnant drying brownish, erect. Seeds c. 1–1.5 mm, cochleariform; testa-cells flat or convex; cuticular folds present or absent.

A genus of 3 allopatric species, endemic within the core area of Eastern Brazil (Rio São Francisco drainage), mainly in *caatinga* vegetation, and of uncertain relationship within the Trichocereeae (but cf. *Yungasocereus* Ritter (1980) and *Vatricania* Backeberg (1950), both from the eastern edge of the Andes in Bolivia, the latter genus nowadays sometimes included within *Espostoa* Britton & Rose). Buxbaum (1959) published *Espostoa* subg. *Facheiroa* (Britton & Rose) F. Buxbaum. Plates 66.3–67.4.

1. Flower-bearing part of stem strongly modified, rarely only poorly differentiated 2
1. Flower-bearing part of stem not at all modified (cent.-S Bahia northwards, non-calcareous rocks)
. . . . **3. F. squamosa**

2. Flowers to 3 × 2.8 cm; perianth-segments sometimes pinkish (*Bambuí* limestone outcrops of SW Bahia and cent.-N Minas Gerais) . . . . **2. F. cephaliomelana**
2. Flowers 4–4.7 × 2 cm; perianth-segments white (cent.-N Bahia, non-calcareous rocks) . . . **1. F. ulei**

1. Facheiroa ulei *(Gürke) Werderm.*, Bras. Säulenkakt.: 113 (1933); P. J. Braun & E. Esteves Pereira in Kakt. and. Sukk. 38: 26, Abb. 1 (1987); ibid. 39: 64 (1988). Holotype: Brazil, Bahia, Mun. Gentio do Ouro, Serra de Santo Inácio, 1907, *Ule* 12 (B, n.v.).

Cephalocereus ulei Gürke in Monatsschr. Kakt.-Kunde 18: 85 (1908). *Espostoa ulei* (Gürke) Buxb. in Oesterr. Bot. Zeitschr. 106: 155 (1959), nom. inval. (Art. 33.2).
Facheiroa pubiflora Britton & Rose, Cact. 2: 173 (1920) ('*publiflora*'). Type: Brazil, Bahia, Mun. Xique-Xique, Serra da Cana-brava, Oct. 1917, *Zehntner* in *Rose* s.n. (US, lecto. designated here; K, lectopara.)

VERNACULAR NAME. Facheiro-preto.

Treelike or often shrubby, to 6 m or more, much branched. Stems erect, to 7 cm diam.; ribs 18–20, 8–9 × 7–8 mm, epidermis dark green to grey-green. Areoles 4–5 mm, 10–12 mm apart, long hairs grey. Spines flexible, brown, central spines 2–3, to 18 mm, radials 13–15, 3–15 mm. Flower-bearing region strongly differentiated, lateral, sunken into stem, with greyish wool and bristles to 10 mm. Flowers 4–4.7 × 1.7–2.0 cm; pericarpel and tube wine-coloured, covered in lanceolate to triangular, 2–5 mm bract-scales, bearing tufts of short, reddish to golden brown hairs; flower tube cylindric, to 30 × 22 mm; perianth-segments 8–10 × 3.5–5.0 mm, lanceolate-spathulate, whitish; stamens included in relation to the perianth-segments; nectar-chamber 10 × 12 mm; style 27 mm, stigma-lobes 10–15, exserted; ovary locule saucer-shaped in longitudinal section, 3 × 8 mm. Fruit depressed-globose to pear-shaped, 3.5–6.0 × 4 cm. Seeds c. 1 × 0.7 mm, broadly oval, brown (Barthlott & Hunt 2000: pl. 29.2–3).

Northern Rio São Francisco *caatinga* element: in *caatinga* at c. 500 m or more, north-western edges of the Chapada Diamantina, from the region of Xique-Xique to the Serra da Chapada, central-northern Bahia. Endemic to Bahia. Map 48.

BAHIA: Mun. Xique-Xique/Itaguaçu da Bahia, Serra da Cana-brava, Oct. 1917, *Zehntner* in *Rose* s.n. (K, US); Mun. Xique-Xique/Gentio do Ouro, 11°6'S, 42°43'W, 24 June 1996, *Guedes et al.* in PCD 3025 (K, ALCB); Mun. Gentio do Ouro, Serra do Santo Inácio, 19 July 1972, *Horst & Uebelmann* 265 (U); Mun. Brotas de Macaúbas / Barra do Mendes, S of Minas de Espírito Santo and Serra da Chapada, reported by Braun & Esteves Pereira, l.c. (1988).

CONSERVATION STATUS. Data Deficient [DD]. No population data and indications of threats to its habitats are currently available for this species.

This taxon has not been studied in the field by the authors and little is known about its ecology and relationship with other members of the genus, although it is clearly a distinct species.

2. Facheiroa cephaliomelana *Buining & Brederoo* in Kakt. and. Sukk. 26: 121–124 (1975). Holotype: Brazil, Bahia, Santa Maria da Vitória, 25 July 1974, *Horst & Uebelmann* 447 (U).

F. *pilosa* F. Ritter, Kakt. Südamer. 1: 219 (1979). Holotype: Brazil, Minas Gerais, Mun. Januária, 1959, *Ritter* 1000 (U).
F. *tenebrosa* P. J. Braun & E. Esteves Pereira in Kakt. and. Sukk. 39: 126–131 (1988). Holotype: Brazil, Bahia, W of the Rio São Francisco [Mun. Cocos], *E. Esteves Pereira* 183 (KOELN ['Succulentarium'], n.v.; ZSS, UFG, K, isos.).

Treelike or usually shrubby, sometimes to 4 m or more, much branched. Stems erect, 2.4–6.0 cm diam.; ribs 18–32, 4–8 × 4–9 mm; epidermis green to grey-green. Areoles 1.5–3.0 mm, 1–8 mm apart, long hairs grey to brownish. Spines flexible, brown or golden, central spine(s) 1–5, to 40 mm, radials 10–12, 3–15 mm. Flower-bearing region differentiated, lateral, with brownish to greyish or white wool and bristles to 10 mm. Flowers 2.8–3.5 × 1.9–2.8 cm; pericarpel and tube wine-coloured, covered in lanceolate to triangular, 4.5–6.0 mm bract-scales, bearing tufts of short or long, white to golden brown hairs in their axils; flower-tube 19–20 × 18 mm; perianth-segments 8–10 × 3.5–5.0 mm, lanceolate-spathulate, pink; stamens included in relation to the perianth-segments; nectar-chamber 5–6 × 10–13 mm; style 23–27 mm, stigma-lobes 10–12; ovary locule saucer-shaped in longitudinal section, 3 × 8–10 mm. Fruit depressed-globose to turbinate, 2.5 × 1.8–2.5 cm. Seeds c. 1.2–1.5 mm, cochleariform, brown, dull or shiny.

CONSERVATION STATUS. Vulnerable [VU B2ab(iii)]; area of occupancy estimated to be < 2000 km² and decline projected in extent of habitat at its < 10 known locations (see below).

The distribution of *F. cephaliomelana,* as broadly circumscribed here, is paralleled or somewhat exceeded by that of other similarly variable cactus taxa exclusive to *Bambuí* limestone outcrops from the same region, eg. *Tacinga saxatilis, Pilosocereus densiareolatus* and *Melocactus levitestatus*. It is divisible into two subspecies, the first of which is somewhat heterogeneous and may merit further subdivision (cf. Braun & Esteves Pereira 2002: 120, Abb. 133):

1. Cephalium somewhat sunken into stem, conspicuous (SW Bahia & N Minas Gerais) . . . .
   . . . . **2a.** subsp. **cephaliomelana**
1. Cephalium superficial or only weakly developed (cent.-S Bahia, E of the Rio São
   Francisco) . . . . **2b.** subsp. **estevesii**

2a. subsp. **cephaliomelana**

VERNACULAR NAMES. Facheiro-preto, Xique-xique-preto.

Southern Rio São Francisco *caatinga* element: locally co-dominant on outcrops of raised *Bambuí* limestone within *caatinga/cerradão*, 460–750 m, south-western Bahia (W of the Rio São Francisco) and central-northern Minas Gerais. Endemic to the core area of Eastern Brazil. Map 48.
BAHIA: Mun. Santa Maria da Vitória, 25 July 1974, *Horst & Uebelmann* 447 (U); Mun. Serra do Ramalho, Serra do Ramalho, 13°31'S, 43°45'W, 2001, *J. Jardim* 3486 (CEPEC); l.c. (?), 'Mun. Bom Jesus da Lapa Faz. Serra Solta', 17 July 1975, *Andrade-Lima* 75-8203 (IPA); Mun. Cocos, July 1984, *Esteves Pereira* 183 (UFG, K).
MINAS GERAIS: Mun. Januária, 1959, *Ritter* 1000 (U); l.c., 3 km W on road MG 479 to Pandeiros, 11 Aug. 1988, *Eggli* 1135 (ZSS), l.c., 5 km towards Serra dos Araras, 5 Sep. 1985, *Horst & Uebelmann* 718 (K); Mun. Varzelândia, W of town, 12 Aug. 1988, *Eggli* 1147 (ZSS).

CONSERVATION STATUS. Vulnerable [VU B1ab(iii)] (2); extent of occurrence = 7554 km^2; PD=3, EI=2, GD=2. Short-list score (2×7) = 14. Relatively rare, with only c.5 populations currently known, and restricted to isolated limestone outcrops, which may be quarried in future leading to rapid decline in some local populations.

This taxon has not been studied in the field by the present authors, but the characteristics of the populations treated here have been recorded in some detail by Braun & Esteves Pereira (1986–89).

2b. subsp. **estevesii** *(P. J. Braun) N. P. Taylor & Zappi* in Cact. Consensus Initiatives 3: 7 (1997). Holotype: Brazil, Bahia, E of the Rio São Francisco, [E of Malhada, Iuiú], July 1984, *E. Esteves Pereira* 186 (KOELN ['Succulentarium'], n.v.; UFG, K, isos.).

Facheiroa estevesii P. J. Braun in Kakt. and. Sukk. 37: 74–79 (1986).

VERNACULAR NAMES. Facheiro-da-serra, Xique-xique-amarelo.

Southern Rio São Francisco *caatinga* element: on outcrops of raised *Bambuí* limestone amidst high *caatinga* forest, 600–650 m, central-southern Bahia (E of the Rio São Francisco). Endemic to the locality cited below. Map 48.
BAHIA: Mun. Iuiú, July 1984, *E. Esteves Pereira* 186 (UFG, K); l.c., 2 km S of Iuiú, July 1989, *Taylor & Zappi* (K, photos), 28 Aug. 1988, *Eggli* 1315 (ZSS).

CONSERVATION STATUS. Vulnerable [VU D2] (2); area of occupancy < 20 km^2; PD=3, EI=1, GD=1. Short-list score (2×5) = 10. Restricted to a single, albeit extensive locality, near a town, where there is the potential for limestone quarrying.

3. **Facheiroa squamosa** *(Gürke) P. J. Braun & E. Esteves Pereira* in Kakt. and. Sukk. 40: 202 (1989). Holotype: Brazil, Piauí, Serra Branca [Mun. São Raimundo Nonato], Jan. 1907, *Ule* 10 (B; K, SPF photos ex B).

Cereus squamosus Gürke in Monatsschr. Kakt.-Kunde 18: 70 (1908). *Leocereus squamosus* (Gürke) Werderm., Bras. Säulenkakt.: 98 (1933).
Zehntnerella squamulosa Britton & Rose, Cact. 2: 177 (1920). Type: Brazil, Bahia, Juazeiro, Serra do Atoleiro, 4 June 1915, *Rose & Russell* 19760 (US, lecto. designated here; NY, lectopara.).

Z. polygona F. Ritter, Kakt. Südamer. 1: 214 (1979). *Facheiroa squamosa* var. *polygona* (F. Ritter) P. J. Braun & E. Esteves Pereira in Kakt. and. Sukk. 40: 202 (1989). *F. squamosa* subsp. *polygona* (F. Ritter) P. J. Braun & E. Esteves Pereira in Succulenta 74: 130 (1995). Holotype: Brazil, Bahia, Flamengo, 1963, *Ritter* 1228 (U).

Zehntnerella chaetacantha F. Ritter, Kakt. Südamer. 1: 215 (1979). *Facheiroa chaetacantha* (F. Ritter) P. J. Braun & E. Esteves Pereira in Kakt. and. Sukk. 40: 202 (1989). Holotype: Bahia, Guanambi, 1964, *Ritter* 1229 (U).

Zehntnerella chaetacantha var. *montealtoi* F. Ritter, Kakt. Südamer. 1: 215–216 (1979). *Facheiroa chaetacantha* var. *montealtoi* (F. Ritter) P. J. Braun & E. Esteves Pereira in Kakt. and. Sukk. 40: 202 (1989). Holotype: Brazil, Bahia, Palmas de Monte Alto, 1964, *Ritter* 1229a (U).

F. deinacanthus Rauh ex Innes & Glass, Illustrated Encyclopaedia of Cacti: 112 (1991), nom. inval. (Arts 36 & 37).

VERNACULAR NAMES. Facheiro, Facheiro-preto.

Shrubby to treelike, to 6 m or more, much branched. Stems erect, old branches weak, becoming curved, 4–8 cm diam.; ribs 13–24, 3–5 × 3–4 mm; epidermis green to grey-green. Areoles to 2 mm, 4–7 mm apart, long hairs grey to brownish. Spines flexible, brown, central spines 1–8, to 30 mm, radials 8–18, 3–15 mm. Flower-bearing region not differentiated, flowers lateral, subapical or randomly distributed. Flowers 3–5 × 2–3 cm; pericarpel and tube green, covered in lanceolate to triangular, green 6–8 mm bract-scales, bearing tufts of short, white to golden or reddish brown hairs; flower-tube 20–40 × 17–22 mm, cylindric or infundibuliform; perianth-segments 8–9 × 3–4 mm, lanceolate spathulate, white; stamens included in relation to the perianth-segments; a tuft of hairs between the nectar-chamber and the filaments bases appearing randomly; nectar-chamber 8–9 × 8 mm; style 23–27 mm, stigma-lobes 10–12; ovary locule saucer-shaped in longitudinal section, 2–4 × 8–10 mm. Fruit globose to depressed-globose, 2–3.5 cm diam. Seeds c. 1–1.2 mm, broadly oval, dark reddish brown, dull (Barthlott & Hunt 2000: pl. 29.5–6).

Rio São Francisco *caatinga* element: locally co-dominant with other arborescent cacti on non-calcareous (mostly granite/gneiss) inselbergs or *lajedos* and in very stony ground of the *caatinga* (rarely on arenitic rocks in *cerrado* or at the margins of *campo rupestre*), 390–1020 m, south-eastern Piauí, western Pernambuco and northern Bahia, and disjunctly in central to southern Bahia, eastwards to the watershed with the Rio de Contas drainage system. Endemic to the core area within North-eastern Brazil. Map 48.

PIAUÍ: Mun. São Raimundo Nonato, Serra de Capivara, Feb. 1990, *Taylor & Zappi* (obs.); l.c., Serra Branca, Jan. 1907, *Ule* 10 (K, SPF [photos ex B]), Feb. 1990, *Taylor & Zappi* (K, photos).

PERNAMBUCO: Mun. Petrolina, road BR 122 [428], Serra [de] Santa, 18 Nov. 1992, *F.A.R. Santos* 49 (HUEFS, PEUFR); l.c., 20 km NE of Petrolina on road to Recife, 13 Aug. 1966, *Hunt* 6491 (K).

BAHIA: N Bahia, Mun. Juazeiro, 14 km E of town towards Curaçá, 7 Jan. 1991, *Taylor et al.* 1372 (K, HRCB, ZSS, CEPEC); l.c., Serra do Atoleiro, 4 June 1915, *Rose & Russell* 19760 (US, NY); Mun. Casa Nova, Santana, Crista da Serra, 4 Jan. 1984, *Fotius* 3672 (IPA); Mun. Sento Sé, 13.5 km S of main road to Juazeiro in the direction of Cabeluda, 11 Jan. 1991, *Taylor et al.* 1395 (K, HRCB, ZSS, CEPEC); l.c., 10°7'S, 41°25'W, 11 July 2000, *G. Charles* (photos); Mun. Sento Sé / Campo Formoso, Serra São Francisco, reported by Buining & Brederoo (1974b: 257, fig. 8); Mun. Jaguarari, Flamengo, 1963, *Ritter* 1228 (U); cent. & S Bahia, Mun. Ibotirama, 14 km E of town on road BR 242, 18 July 1989, *Taylor & Zappi* (obs.); l.c. (?), 12°13'18"S, 43°8'48"W, 2000, *F.M. Sene* J78 'sp. B' (K, photos); Mun. Ibitiara, Serra Malhada, 52 km W of Seabra on road to Ibotirama, 1020 m, 14 Jan. 1991, *Taylor et al.* 1418 (K, HRCB, ZSS, CEPEC); Mun. 'Seabra', July 1987, *Esteves Pereira* 238 (UFG); Mun. Rio do Pires, Ibiajara, *fide* Uebelmann (1996): HU 268; Mun. Macaúbas, *fide* Uebelmann (1996): HU 127; Mun. Paramirim, near Rio Paramirim, 13°26'S, 42°15'W, 28 Nov. 1988, *Taylor in Harley* 25557 (K, SPF, CECEC), ibid., between Caraíbas and Paramirim, north of the road from Livramento, 30 Nov. 1988, *Taylor* (K, photos); Mun. Riacho de Santana, 22 km from Bom Jesus da Lapa, 16 Apr. 1983, *Leuenberger et al.* 3074 (CEPEC, B), 20 July 1989, *Taylor & Zappi* (obs.); Mun. Rio de Contas, southern limits of município bordering on Mun. Brumado, 13°49'39"S, 41°34'25"W, 'Boa Sentença', 26 Jan. 2001, *Harley & Giulietti* (K, photos); Mun. Guanambi, 1964, *Ritter* 1229 (U); Mun. Palmas de Monte Alto, 1964, *Ritter* 1229a (U), 21 July 1989, *Taylor & Zappi* (obs.).

CONSERVATION STATUS. Least Concern [LC] (1); PD=3, EI=2, GD=2. Short-list score (1×7) = 7.

A rather variable species in terms of stem size, rib number, flower size and flower-tube indumentum colour/abundance etc., the variation being only partly correlated on a regional basis. The characters utilized by Braun & Esteves Pereira to maintain the southern-ranging variant, *F. chaetacantha,* as a separate species do not seem to be consistent on the basis of the materials studied here. The distribution of *F. squamosa sens. lat.* is markedly disjunct, being interrupted in central-northern Bahia (drainages of Rio Salitre and Rio Jacaré) due to the presence of limestone derived substrates, which it appears to avoid, and in the north-western part of the Chapada Diamantina, where it is replaced by *F. ulei.*

27. ESPOSTOOPSIS F. Buxbaum

in Krainz, Die Kakteen, Lfg. 38–39 (July 1968). Type and only species: *E. dybowskii* (Roland-Gosselin) Buxb. (*Cereus dybowskii* Gosselin).

A monotypic genus endemic to the *caatinga* of Bahia and possibly related to *Espostoa* Britton & Rose (*sens. str.*), from the central Andes of southern Ecuador and northern Peru. It differs from *Espostoa* in its naked pericarpel and tube, in which it strongly resembles some cephalium-bearing members of tribe Cereeae (cf. *Micranthocereus* Backeb., *sens. lat.*), but in other respects it is extremely similar to the Andean genus and could easily be mistaken for it when in sterile condition. *Gerocephalus* F. Ritter (see synonymy below) is an illegitimate generic name based on *Cereus dybowskii,* but published one month later than *Espostoopsis* Buxb. Plate 68.1–68.2.

1. Espostoopsis dybowskii *(Gosselin) Buxb.,* l.c. (1968). Type: Brazil, N Bahia, Itumirim, *Dybowski* (living material), *Casabianca* (fls & frs), apparently not extant (cf. Kiesling 1986). Neotype (designated here): Brazil, Bahia, Mun. Jaguarari, 38 km N of Senhor do Bonfim on road BR 407, 15 July 1989, *Zappi* 125 (SPF; HRCB, ZSS isoneotypi; K, photos).

Cereus dybowskii Gosselin in Bull. Soc. Bot. France 55: 695 (1908). *Cephalocereus dybowskii* (Gosselin) Britton & Rose, Cact. 2: 30 (1920). *Espostoa dybowskii* (Gosselin) Frič in Kreuzinger, Verzeichnis: 6 (1935). *Austrocephalocereus dybowskii* (Gosselin) Backeb. in Cact. Succ. J. (US) 23: 149 (1951). *Gerocephalus dybowskii* (Gosselin) F. Ritter in Kakt. and. Sukk. 19: 156 (Aug. 1968). *Coleocephalocereus dybowskii* (Gosselin) F. Brandt in Kakt. Orchid. Rundschau 6: 124 (1981). *Micranthocereus dybowskii* (Gosselin) D. Herbel in Kakt. and. Sukk. 49: 275 (1998), nom. inval. (Art. 33).

VERNACULAR NAMES. Cabeça-de-velho, Homem-velho, Cabeça-branca, Mandacaru-cabeça-branca, Mandacaru-de-penacho ['perracho'].

Columnar, erect, shrubby to almost treelike, to 5 m, trunk to 80 cm diam. Stems erect, 4–8 cm diam., often braking in the region of the cephalium; vascular cylinder woody, solid, very hard, with scattered perforations, cortex scarcely mucilaginous; ribs 24–33, 2 × 5–6 mm, strongly tuberculate, sinuses straight; epidermis light green with numerous pale stomata; areoles 2–4 mm, to 6 mm apart, with abundant, long, white to grey hairs covering the stem. Spines flexible to brittle, golden or reddish brown, central spines 1–2, to 2(–2.5) cm, radials 10–15, 0.3–1.5 cm. Flower-bearing region a strongly differentiated, lateral cephalium sunken into the stem and markedly protruding from its surface, with abundant, very compact white wool and bristles to 2.5 cm, the flowers developing in an irregular manner (not in order of age of cephalium areoles). Flowers 4–4.2 × 3.5–3.8 cm, all parts turning orange at the slightest touch, then blackening; pericarpel 7 × 10 mm, pale green with minute and scarcely visible bract-scales; tube to 30 × 22 mm, cylindric, white to pale pink, naked, with broad bract-scales only at apex; perianth-segments 6–7 × 3 mm, lanceolate, acuminate, white; stamens included in relation to the perianth-segments; nectar-chamber 12 × 6–8 mm; style 28–35 mm, stigma-lobes c. 15–16, exserted; ovary locule 6 × 3–4 mm, hemiglobose in longitudinal section. Fruit depressed-globose, 20 × 22 mm, slightly compressed laterally, flower remnants deeply sunken into apex and carrying with them hairs from the cephalium; pericarpel wrinkled, olive-green, pale pink or reddish, becoming exposed at cephalium surface or ± buried. Seeds c. 1.5 mm, cochleariform, black, shiny (Barthlott & Hunt 2000: pl. 50.1).

Eastern *caatinga* element: dominant or co-dominant on gneiss/granite inselbergs or quartzitic rock outcrops and in the surrounding *caatinga*, 300–750 m, central-northern and eastern Bahia (markedly disjunct). Endemic to Bahia. Map 48.

BAHIA: cent.-N Bahia, Mun. (?), 'Juazeiro', 20 Oct. 1967, *A.P. Duarte 10589* (RB); Mun. Jaguarari, 38 km N of Senhor do Bonfim on road BR 407, 15 July 1989, *Zappi 125* (SPF, HRCB, ZSS; K, photos); l.c., 86 km S of Juazeiro, between Flamengo and Juacema, 1 km W of road BR 407, 8 Jan. 1991, *Taylor et al. 1384* (K, HRCB, ZSS, CEPEC); l.c., Itumirim/Flamengo, 8 Feb. 1970, *A.L. Costa s.n.* (ALCB, in spirit); ibid., Barrinha, 7–8 June 1915, *Rose & Russell 19785* (US, NY); E Bahia, Mun. Maracás, 6–7 km E of Porto Alegre towards Jequié, 4 Feb. 1991, *Taylor et al. 1551* (K, HRCB, ZSS, CEPEC); Mun. Jequié, E of Porto Alegre on dirt road connecting with road BR 330, gneissic inselberg, 4 Feb. 1991, *Taylor et al.* (obs.); Mun. Tanhaçu, nr Sussuarana, Serra Escura, Sep. 2001, *Avaldo de Oliveira Soares Filho* (photos), 19 April 2003, *Taylor et al.* (K, photos).

CONSERVATION STATUS. Endangered [EN B2ab(iii)] (3); area of occupancy < 100 km², PD=4, EI=2, GD=2. Short-list score (3×8) = 24. All populations of limited extent, the northern one affected by clearance of the *caatinga* in some places, the central of the three southern ones recently burnt. Creation of reserves to accommodate this and associated species at one or more of its southern locations in the Rio de Contas valley (between Sussuarana and Jequié) is recommended.

The distribution of this isolated species is markedly disjunct between northern and eastern Bahia (Rio de Contas valley), the southern form showing some differences (notably smaller stems and more exserted olive-brown fruits) that could justify subspecific status.

The better-known northern population, whence the type came, covers an extensive area within the município of Jaguarari and in some parts of this region the plant dominates the vegetation, forming impenetrable groves around and upon quartzitic outcrops. It may range beyond this area into the neighbouring municípios of Sento Sé, Campo Formoso, Senhor do Bomfim, Itiúba and further east to Jeremoabo, as implied by Andrade-Lima (1989: 6) and Luetzelburg (1925–26, 3: 69), but, despite being so conspicuous, there are no collections or other reports to confirm its presence in these other localities and in northern Bahia it has not been seen outside Mun. Jaguarari by the present authors.

The southern populations are found on either gneissic inselbergs or quartzite, but are associated with a quite different suite of *caatinga* Cactaceae. Its disjunct distribution and the differences between the plants from the two areas implies that the species may be a relict in decline.

28. **ARTHROCEREUS** A. Berger

Kakteen: 337 (1929), *nom. cons.*; F. Knuth, Den Nye Kaktusbog: 111 (1930). Type: Monatsschr. Kakt.-Kunde 28: illus. opp. p. 62 (1918), as '*Cereus damazioi*', *typ. cons.* (= *A. glaziovii* (K. Schum.) N. P. Taylor & Zappi).

Literature: Taylor (1991d, 1992a); Brummitt (1994).

Shrubby or solitary, 1–2(–5) m tall, basally branched, sometimes with a thickened, corky rootstock (adaptation to withstand fire?). Ribs 9 or more, low. Areoles close together, with felt and long hairs. Spines golden or ferrugineous, translucent, weakly differentiated into centrals and radials, centrals ascending or porrect, pungent, radials numerous, adpressed, bristle-like. Flower-bearing part of stem not differentiated. Flowers salverform, nocturnal, white, or pink and remaining open in the early morning in *A. rondonianus* (secondary bird-pollination syndrome?); pericarpel globose, it and the tube bearing few to many bract-scales with hair-spines in their axils; tube elongate, cylindric to infundibuliform, sometimes ± S-shaped; perianth-segments mostly narrow; inner segments white to lilac-pink; stamens usually inserted in 2 series, when distinguishable the outer series in a ring at the mouth of the tube; ovary locule hemicircular or rounded in longitudinal section. Fruit indehiscent, globose to obpyriform, with bract-scales and areoles bearing long hair-spines or bristles, funicular pulp white. Seeds ovoid, 1.2–2.0 mm.

An interesting genus endemic to Brazil, with 3 species in Eastern Brazil (*campos rupestres* of central to SW Minas Gerais) and a fourth, *A. spinosissimus* (Buining & Brederoo) F. Ritter, which has flowers with striking purple stamens rather similar to No. 2 below, geographically isolated in Mato Grosso (Chapada dos Guimarães — Map 13). This disjunct taxon has been separated into subgenus *Chapadocereus* P. J. Braun & E. Esteves Pereira (1995: 82), but its similarities to *A. melanurus* and *A. rondonianus* would seem to indicate that subdivision of the genus on geographical lines is untenable. The genus is assumed to be related to *Echinopsis* Zucc. (*sens. lat.*), but differs in its pollen (Leuenberger 1976), indehiscent fruits and unusual habit form.

In the past the genus *Arthrocereus* has been used in a broader sense than now, including species presently referred to *Echinopsis*, ie. *E. mirabilis* Speg. (Argentina), and *Pygmaeocereus* H. Johnson & Backeb. (= *Echinopsis*), ie. *P. bylesianus* Backeb. (Peru).

The name of the genus was approved for conservation with a new type, as above, in 1993 (XV Int. Bot. Congress, Tokyo), following the discovery that the former type, *Cereus microsphaericus* K. Schum. (and its illegitimate, homotypic synonym, *C. damazioi* Weingart), was misapplied by Berger and is actually identifiable with a species of *Schlumbergera* Lem. (Rhipsalideae). However, more recently, Eggli & Nyffeler (1996) have cast doubt on whether the generic name can be accepted as validly published by Berger (1929), although this has been refuted by Doweld & Greuter (2001). Plates 68.3–70.3.

1. Inner perianth-segments pink, stamens purple (Serra do Cabral & Conselheiro Mata) . . . . **2. A. rondonianus**
1. Inner perianth-segments and stamens white or cream (Serra do Espinhaço and Serra da Mantiqueira) 2

2. Flower-tube ± naked, with only few bract-scales and hairs; stems usually decumbent to ascending, often segmented, sometimes very short to nearly spherical (usually on *canga* formation) . . . . **3. A. glaziovii**

2. Flower-tube with conspicuous bract-scales and hairs; stems ± erect, not segmented (on
   substrates other than *canga*) .... **1. A. melanurus**

1. Arthrocereus melanurus *(K. Schum.) L. Diers, P. J. Braun & E. Esteves Pereira* in
Kakt. and. Sukk. 38: 312–315 (1987). Type (syntypes): 'Brasilia', (?) Minas Gerais, *Sello*
1000 (B†; an extant isosyntype at MO resembles *A. glaziovii*); Serra de São João del Rey
[Serra do Lenheiro], *Glaziou* s.n. (living plant, assumed not to have been preserved, but
illustrated by Schumann, l.c. infra, tab. 39, 1890). If an extant duplicate of the Glaziou
collection cannot be found, this name will require conservation, or its application may
have to change in line with the Sello isosyntype at MO (cf. ICBN Art. 9.10).

Cereus melanurus K. Schum. in Martius, Fl. bras. 4(2): 200, tab. 39 (1890). *Leocereus melanurus* (K. Schum.)
   Britton & Rose, Cact. 2: 109 (1920).
A. mello-barretoi Backeb. & Voll. in Arch. Jard. Bot. Rio de Janeiro 9: 157 (1949, publ. 1950). *A. melanurus*
   subsp. *mello-barretoi* (Backeb. & Voll) P. J. Braun & E. Esteves Pereira in Succulenta 74: 82 (1995).
   Holotype: Brazil, Minas Gerais, Serra do Lenheiro [São João del Rei], *Mello-Barreto* s.n. (RB 65044)
   — this is the type locality of *A. melanurus*!
A. melanurus var. *estevesii* L. Diers & P. J. Braun in Kakt. and. Sukk. 39(5): 100–105 (1988). *A. melanurus*
   subsp. *estevesii* (L. Diers & P. J. Braun) P. J. Braun & E. Esteves Pereira in Succulenta 74: 82 (1995).
   Holotype: Brazil, Minas Gerais, São João del Rei, *Horst et al.* 1 (KOELN, n.v.; K, iso.; ZSS iso.,
   numbered as *Braun* 458) — this is the type locality of *A. melanurus*!

VERNACULAR NAME. Sabugo-do-capeta.

Shrubby or single-stemmed, 0.4–3.0(–5.0) m tall, generally with a corky, thickened rootstock. Branches
erect, not segmented; epidermis bright green; ribs 9–19. Areoles to 3 mm diam., 4–6 mm apart, felt
greyish, basal areoles with indeterminate growth. Flowers nocturnal, sweetly scented; pericarpel and tube
± covered in triangular bract-scales with brownish woolly hairs; perianth-segments linear, acuminate,
fleshy, pale green with brownish shades, strongly reflexed, touching the outside of the flower tube at full
anthesis, inner segments linear-lanceolate, slender, delicate, cream or whitish, erect; stamens in 2 series;
stigma-lobes exserted. Seed: Barthlott & Hunt (2000: pl. 38.1–2).

CONSERVATION STATUS. Vulnerable [VU B2ab(iii)]; area of occupancy estimated to be < 2000 km²,
number of known locations < 10 and some of these subject to anthropogenic alterations (see below).

The following subspecies are recognized:

1. Flowers 9–11.5 cm; inner perianth-segments 23–40 mm (SW Minas Gerais) . . **1a.** subsp. **melanurus**
1. Flowers 10–16.5 cm; inner perianth-segments 40–55 mm (Cent. & S Minas Gerais) 2

2. Plants with few or no basal branches, reaching 2 m or more, stout rootstock not visibly
   developed; ribs 12–19; perianth-segments spathulate, rounded-apiculate (Serra do Ibitipoca)
   .... **1b.** subsp. **magnus**
2. Plants shrubby, branching freely at base, usually not > 1 m, stems slender; rootstock well-
   developed; ribs 9–12; perianth-segments tapering more evenly to apex (N of Diamantina
   & Serra do Cipó) .... **1c.** subsp. **odorus**

1a. subsp. **melanurus**

Shrubby, to 80 cm tall, with a corky, thickened rootstock. Branches to 4.5 cm diam., tapering to 1.5 cm,
erect; ribs 10–17, 3 × 5 mm. Spines yellow to reddish brown, central spines 5–7, stout, porrect, to 3.2
cm, radials numerous, thin to bristle-like and flexible, 4–5 mm. Flowers 9.2–11.5 × 6.5–8.0 cm;

pericarpel and tube completely covered in triangular bract-scales with densely hairy areoles and yellow, brown or ferrugineous stiff hairs to 1.5 cm; tube 5–6 × 1 cm at its narrowest point; outer perianth-segments to 23–40 × 4–6 mm, inner segments to 32 × 3.5–5.0 mm; the innermost stamens inserted at 2.5 cm above pericarpel, with long filaments, the outermost series at the apex of tube, filaments 10 mm, anthers to 1.9 mm; ovary locule to 7 mm diam.; style 6–7.7 cm, stigma-lobes 12–13, exserted, 8–10 × 0.4 mm, slender. Fruit rounded, c. 3.0 cm diam., pericarp green, densely covered in bract-scales and trichomes. Seeds c. 1.1 mm, testa-cells convex, with cuticular folds.

South-eastern *campo rupestre* element: in sandy, gravelly and rocky places in *campo rupestre*, 600–1000 m, drainage of the Rio Grande, south-western Minas Gerais. Endemic to Minas Gerais. Map 49.
MINAS GERAIS: Mun. Alpinópolis, Furnas, 1994, *N.M. Dinez* (K, photo); Mun. São João del Rei, Serra do Lenheiro, 1936, *Mello-Barreto s.n.* (RB 65044); Mun. Tiradentes/São João del Rei, Serra de São José, 24 Feb. 1987, *Zappi et al.* in CFCR 10338 (SPF), 11 Nov. 1988, *Taylor & Zappi* in Harley 25524 (SPF, K), s.d., *Horst et al.* 1 & *Horst* 594 (K), *Braun et al.* 458 (ZSS), 3 Nov. 1990, *E. Tameirão Neto* 594 (K, BHCB); Mun. Itutinga, between Lavras and São João del Rei, 56 km from the latter, 13 Oct. 1992, *V.C. Souza & C. Sakuragui* 2105 (ESA, SPF); Mun. São Tomé das Letras, 29 Sep. 1968, *L. Monteiro* 116 & *M.C. Vianna* 335 (GUA), 19 Dec. 1971, *L. Monteiro s.n.* (GUA).

CONSERVATION STATUS. Vulnerable [VU B1ab(iii)] (2); extent of occurrence = 9669 km²; PD=3, EI=1, GD=1. Short-list score (2×5) = 10. The type locality is very close to an expanding town and other sites may be subject to mining activities.

It would not be surprising if this taxon were to be found in NE São Paulo.

1b. subsp. **magnus** *N. P. Taylor & Zappi* in Cact. Consensus Initiatives 3: 7 (1997). Holotype: Brazil, Minas Gerais, Mun. Lima Duarte, Parque Estadual do Ibitipoca, 27 July 1991, *Zappi et al.* 262 (SPF; CESJ, HRCB, K isos.).

Erect, single-stemmed or somewhat branched at base or above, to 3(–5) m. Stems stout, 3.5–5.0(–8.0) cm diam.; ribs 12–19, 3 × 7 mm. Spines gold to golden brown, central spines to 2–3.5 cm, radials 8–10 mm. Flowers c. 16.5 × 12.0 cm, pericarpel c. 1.5 × 2.0 cm; nectar-chamber and flower-tube together 10 cm, greenish, expanding from 1.5 cm diam. at base to 2.5 cm diam. at apex, bearing greenish bract-scales to 2 × 1 mm, with 15 mm brownish hair-spines in their axils, the interior of the tube (throat) with faint reddish brown markings; perianth-segments spathulate, the outermost 30–45 × 5–10 mm, green with brownish midrib, inner segments to 55 × 20 mm, pure white, apex rounded-apiculate; stamens in 2 well-marked series, the outermost (throat circle) to c. 35 mm, the inner to 120 mm, bearing cream-coloured anthers to 3 × 1 mm; style c. 13.5 cm × 1.5 mm, greenish white; stigma-lobes c. 13, to 10 × 1 mm, cream. Fruit to 45 × 50 mm; pericarp green with orange tinge, areoles hairy. Seeds 1.9–2.0 mm.

South-eastern *campo rupestre* element: amongst sand and rocks in *campo rupestre*, 1200–1700 m, Serra do Ibitipoca, southern Minas Gerais. Endemic to the core area within Minas Gerais. Map 49.
MINAS GERAIS: Mun. Lima Duarte, Parque Estadual do Ibitipoca, 27 July 1991, *Zappi et al.* 262 (SPF, CESJ, HRCB, K), ibid., cult. R.B.G. Kew, 6 Aug. 2001 (K, flower in spirit, photos), l.c., *H.C. Souza s.n.*, (K, photo), 1 Jan. 1982, *A. Vilaça* 193 (GUA), s.d., anon. in *F. Sene s.n.* (K, photos), 3–5 Aug. 2003, *Taylor et al.* (K, photos); l.c., 17 km from Lima Duarte on road to the Parque Estadual, 21 Sep. 1994, *Splett* 616 (K); l.c. (?), 'Juiz de Fora', Oct. 1969, *Krieger* 7379, 7525 & 7527 (CESJ).

CONSERVATION STATUS. Near Threatened [NT] (1); area of occupancy estimated to be < 50 km²; PD=3, EI=1, GD=1. Short-list score (1×5) = 5. Population occurs within a protected area (Parque Estadual do Ibitipoca), the management quality of which will determine whether this plant becomes threatened in future. At present the park is well managed and the plant appears safe.

Distinct from other races of this species for its much greater size and broader, markedly spathulate-apiculate perianth-segments. The first illustration of this giant subspecies in the above-cited habitat is that in Leme & Marigo (1993: 68). It develops the densest colonies and largest specimens in sandy flat areas, where there is less competition.

The plant recently illustrated as from Ibitipoca by Braun & Esteves (2003: 23,F) may be incorrectly captioned — it appears to be the pale-spined form of *A. rondonianus* from Diamantina.

1c. subsp. **odorus** *(F. Ritter) N. P. Taylor & Zappi* in Cact. Consensus Initiatives 3: 7 (1997). Holotype: Brazil, Minas Gerais, Serra do Cipó, 1964, *Ritter* 1354 (U).

? Cereus parvisetus Otto ex Pfeiff., Enum. cact.: 79 (1837); K. Schum., Gesamtb. Kakt.: 67 (1897). Type: Brazil [Minas Gerais, Serra da Lapa, *Riedel, fide* Schumann, l.c.], no longer extant.
Arthrocereus odorus F. Ritter, Kakt. Südamer. 1: 225 (1979).

Shrubby, to 1 m tall, branching at base, with a corky, thickened rootstock. Branches to 2–3.3 cm diam., erect or hanging from cliffs; ribs 9–12, 2–3 × 5 mm. Spines yellow to reddish, central spines 6–7, stout, the largest porrect, to 4(–6) cm, radials numerous, thin to bristle-like and flexible, 4–5 mm. Flowers 10–16 × 8–10 cm; pericarpel and tube partially covered in triangular bract-scales with pale to pinkish brown hairs to 15 mm in their axils; tube 4–8 cm × 8–9 mm at its narrowest point, ridged; outer perianth-segments 40 × 5–7 mm; inner segments 40–50 × 6–9 mm; the innermost stamens inserted at 4 cm above pericarpel, with long filaments, the outermost series at the apex of tube, filaments 15–18 mm, anthers to 2.2 mm; ovary locule 8 mm diam.; style 10–13 cm, stigma-lobes 12–13, 10 mm, slender. Fruit rounded to ovate, 3–4 cm diam., pericarp green, covered in bract-scales and pale brown trichomes. Seeds c. 1.2–1.8 mm, testa-cells convex, with cuticular folds.

South-eastern *campo rupestre* element: amongst rocks in *campo rupestre*, 600–900 m, central-southern Minas Gerais. Endemic to the core area within Minas Gerais. Map 49.
MINAS GERAIS: Mun. Diamantina, Mendanha, *Uebelmann* HU 1555, cult. G. Charles (photo), l.c., illustrated in Herm *et al.* (2001: 91); Mun. Santana do Riacho, path to the Serra das Bandeirinhas, near Cachoeira da Farofa, 13 Sep. 1987, *Zappi et al.* in CFSC 10419; Mun (?) Santana do Riacho, [W foot of the] Serra do Cipó, 1964, *Ritter* 1354 (U); l.c., Mun. Jaboticatubas, along the Rio Cipó, 28 Oct. 1988, *Taylor & Zappi* in *Harley* 25430 (K, SPF), 20 Nov. 1989, *Zappi et al.* 193 (HRCB, SPF), 1997, *Ribeiro de Andrade & Pimentel Mendonça* (K, photo).

CONSERVATION STATUS. Vulnerable [VU D2] (2); area of occupancy estimated to be < 20 km2; PD=3, EI=1, GD=1. Short-list score (2×5) = 10. Certainly known from only 3 localities in two distant and very limited areas within the Serra do Espinhaço.

2. Arthrocereus rondonianus *Backeb. & Voll* in Cact. Succ. J. (US) 23: 120 (1951). Type: Brazil, Minas Gerais, Diamantina, 700 m (presumed not to have been preserved). Lectotype (designated here): Blätt. Kakteenforsch. 1935(4): [unpaged] (1935), illus. 'Arthrocereus rondonianus Bckbg. et Voll n.sp.'; the same illustration also in Arch. Jard. Bot. Rio Janeiro 9: 158, fig. 1 (1949, publ. 1950) and Backeberg, Die Cact. 4: 2110, Abb. 1990 (1960).

Shrubby or single-stemmed, to 1 m tall or more, with a corky, thickened rootstock; stems 1.5–3.0(–4.0) cm diam., erect, not segmented unless damaged; epidermis bright green; ribs 13–18, 3 × 4 mm. Areoles to 3 mm diam., 5–7(–10) mm apart, felt greyish. Spines yellow to pale brown (or whitish in cultivation), central spines 6–7, stout, the largest porrect, to 7 cm, radials numerous, thin to bristle-like and flexible, 5–10 mm. Flowers nocturnal, remaining open until early morning, 10–13 × 9 cm, not scented; pericarpel and tube with scattered triangular bract-scales and pale to pinkish brown hairs to 1.5 cm; tube 6–7 × 0.8 cm at its narrowest point, often S-shaped and ridged, the limb being directed horizontally (erect in other species); outer perianth-segments to 50 × 4 mm, linear-lanceolate, fleshy, pale green with pinkish shades or pale pink, strongly reflexed, inner segments to 40 × 6–7 mm, lanceolate, delicate, lilac-pink, erect to spreading; stamens in two series, the innermost inserted at 4 cm above pericarpel, with long filaments, the outermost series at the apex of tube, filaments 15–18 mm, anthers purple; style 8–12 cm, stigma-lobes 10–11, exserted, whitish. Fruit rounded to ovoid, 4 × 3 cm; pericarp purplish to brownish, with bract-scales, bristles and hairs. Seeds c. 2 mm, testa-cells convex, with cuticular folds.

South-eastern *campo rupestre* element: between rocks and in bushy places, *campo rupestre*, c. 700–1200 m, Serra do Cabral and west slope of Serra do Espinhaço, central Minas Gerais. Endemic to the core area within Minas Gerais. Map 49.

MINAS GERAIS: Mun. Joaquim Felício, *Ritter* 1355 (SGO 124966, *fide* Eggli *et al.* 1995: 533); Mun. Buenópolis, Serra do Cabral, 13 Oct. 1988, *Taylor & Zappi* in *Harley* 24989 & 24994 (K, SPF), *Horst & Uebelmann* 145 (ZSS); Mun. Diamantina, Conselheiro Mata, s.d., *A. Hofacker* AH261, cult. M. Winberg (colour photo on Winberg website, June 2001); l.c. (?), '80 km W of Diamantina', 1983, *Braun* 439 (ZSS); l.c. (?), 'Diamantina', without collector, cult. Univ. Calif. Bot. Gard., Berkeley, 5 Apr. 1961 (US 2830736); Mun. Monjolos, 8 June 2002, *G. Charles* (photos); l.c., nr Rodeador, 8 June 2002, *G. Charles* (photos).

CONSERVATION STATUS. Vulnerable [VU D2] (2); area of occupancy < 100 km²; PD=3, EI=1, GD=1. Short-list score (2×5) = 10. Known to have a very limited range, there being only c. 60 km separating the two small areas recorded above (Serra do Cabral and W slope of Serra do Espinhaço), which represent about 5 localities or perhaps just two partially fragmented populations, separated by the lowlands of the Rio Curimataí and Rio Pardo.

The lilac-pink flowers of this species are spectacular, but the colour is of variable intensity.

3. Arthrocereus glaziovii *(K. Schum.) N. P. Taylor & Zappi* in Bradleya 9: 84–85 (1991). Holotype: Brazil, Minas Gerais, Pico d'Itabira do Campo, 20 Dec. 1888, *Glaziou* s.n. (B [K, SPF, photos ex B]).

Cereus glaziovii K. Schum. in Martius, Fl. bras. 4(2): 200, tab. 39 (1890). *Leocereus glaziovii* (K. Schum.) Britton & Rose, Cact. 2: 109 (1920). *Trichocereus glaziovii* (K. Schum.) Werderm., Bras. Säulenkakt.: 94 (1933).

Trichocereus campos-portoi Werderm., ibid. 94–95 (1933). *Arthrocereus campos-portoi* (Werderm.) Backeb. in Backeb. & Knuth, Kaktus-ABC: 211 (1935, publ. 1936). Holotype: Brazil, Minas Gerais, Serra do Curral [Mun. Belo Horizonte], June 1932, *Werdermann* 3994 (B, K [photo ex B]).

A. itabiriticola L. Diers, P. J. Braun & E. Esteves Pereira in Kakt. and. Sukk. 38: 312–315 (1987). Holotype: Brazil, Minas Gerais, Mun. Itabirito, Pico do Itabirito [formerly Pico do Itabira do Campo], *Horst & Uebelmann* HU 330 (KOELN, n.v.; ZSS, iso.) — this is the type locality of *A. glaziovii*!

? *A. damazioi* P.V. Heath in Calyx 2(2): 65 (1992), nom. inval. (Art. 32). Based on an invalid name of Schumann (1903).

[*A. microsphaericus* sensu (K. Schum.) A. Berger, Kakteen: 337 (1929) et auctt., non *Cereus microsphaericus* K. Schum. (= *Schlumbergera microsphaerica* (K. Schum.) Hövel).]

[*Cereus damazioi* sensu Quehl in Monatsschr. Kakt.-Kunde 28: illustr. opp. p. 62 (1918) et auctt., non Weingart, nom. illeg.]

[*Trichocereus damazioi* sensu (Weingart) Werderm., Bras. Säulenkakt.: 94 (1933), non *Cereus damazioi* Weingart, nom. illeg.]

Procumbent or erect, shrubby, 20–30 cm tall, stems rarely to 60 cm or more. Branches 3–4(–4.5) cm diam., erect or decumbent, generally segmented, sometimes losing the growing point and the segments becoming very short or almost spherical; epidermis dark green; ribs 10–16, 2–3 × 4–6 mm. Areoles to 2.5

mm diam., 4 mm apart, felt greyish. Spines yellow to dark reddish brown, central spines 4–8, stout, the largest porrect, to 4 cm, radials numerous, thin to bristle-like and flexible, to 8 mm. Flowers nocturnal, 8–15 × 5–8.5 cm, sweetly scented; pericarpel and tube bearing few triangular bract-scales with hairy areoles and few, pale hairs; tube 60–75 × 6–8 mm at its narrowest point, ridged; outer perianth-segments to 30 × 6 mm, linear, acuminate, fleshy, pale brownish pink or greenish, strongly reflexed; inner segments to 18 × 5 mm, linear, acute, slender, delicate, cream or whitish, erect; stamens in two poorly distinct series, the innermost inserted at 2.5–3.5 cm above pericarpel, with long filaments, the outermost series at the apex of tube, filaments 5–10 mm, anthers to 1.5 mm; ovary locule 8–9 mm diam.; style 5–9 cm, stigma-lobes 7–12, exserted, 6 mm, slender. Fruit depressed-globose to obpyriform, 2–3 cm diam.; pericarp reddish brown, naked or nearly so. Seeds c. 1.5–1.8(–2.2) mm, testa-cells convex, with cuticular folds.

South-eastern *campo rupestre* element: on iron-rich rock (*canga*) in *campo rupestre*, to 1300–1750 m, east and south of Belo Horizonte, central-southern Minas Gerais. Endemic to the core area within Minas Gerais. Map 49.

MINAS GERAIS: Mun. Belo Horizonte, Serra do Curral, 22 Dec. 1996, *M.F. de Vasconcelos* in BHCB 37421 (BHCB, K); l.c., border with Mun. Nova Lima, 13 Jan. 1998, *I. Ribeiro de Andrade* (K, photo); Mun. Caeté, Serra da Piedade, *Mello-Barreto* 2213 (BHMG), 19 Jan. 1971, *Irwin et al.* 28758 (NY), *M. Braga* s.n. (BHCB 7512), 3 Nov. 1988, *Taylor & Zappi* in *Harley* 25500 (SPF, K), 12 Dec. 1990, *Taylor & Zappi* 750 (HRCB, K, ZSS, BHCB); Mun. Nova Lima, Serra da Moeda, 1989, *L. Martens* 104 (SPF, HRCB); Mun. Itabirito, Pico do Itabirito [Itabira, Itabira do Campo], Dec. 1888, *Glaziou* s.n. (K, SPF, photos ex B), s.d., *Horst & Uebelmann* HU 330 (ZSS), 23 July 1966, *L. Emygdio de Mello F.* 2209 & *A. Andrade* 2113 (R), 11 Sep. 1993, *W. Antunes Teixeira* s.n. (K, BHCB); Mun. Ouro Preto, reported by Schumann (1903: 38), but awaiting confirmation; without locality or date, *Sello* 1000 (MO) [isosyntype for *A. melanurus,* q.v.].

CONSERVATION STATUS. Endangered [EN B1ab(i, ii, iii, iv, v) + 2ab(i, ii, iii, iv, v)] (3); extent of (historical) occurrence = 672 km^2; PD=3, EI=1, GD=2. Short-list score (3×6) = 18. Many of its former habitats have been eliminated through the mining of iron ore and its isolated populations are sometimes very small (< 100 individuals). Less than 5 surviving localities are known.

Further field studies are needed to evaluate the status of the erect and prostrate forms of this species.

29. DISCOCACTUS Pfeiff.

in Allg. Gartenzeitung 5: 241 (1837). Type: *D. insignis* Pfeiff. (= *D. placentiformis* (Lehm.) K. Schum.).

VERNACULAR NAMES. Roseta-do-diabo, Coroa-do-diabo, Coroa-de-frade, Cabeça-de-frade.

Stems depressed-globose, often half buried in the ground, non-mucilaginous, lacking woody tissues, very small or to c. 10 × 25 cm, remaining solitary or sprouting offsets; ribs 9–26, well-defined or broken up into tubercles; areoles small to large but soon becoming glabrous, indeterminate growth absent. Spines relatively few, usually 3–17 per areole, sometimes very stout or minute. At maturity stem apex transformed into a white-woolly and often bristly cephalium, bearing flowers and fruit from its deeply sunken apex. Flowers salverform, nocturnal, sweetly scented (hawkmoth syndrome), 1–6 or more rapidly developing at the same time; pericarpel ovoid, usually quite naked, tube (including the nectar-chamber) elongate, bearing leafy bract-scales glabrous in their axils; perianth-segments narrowly oblanceolate, the outermost often greenish, brownish or even deep pink, inner series white, rarely pale pinkish; stamens numerous, inserted along the inside of the tube, and at its apex, on short filaments, sometimes the lowermost inserted above a whorl of hairs which protect the nectar-chamber; style long or short, stigma-lobes c. 5–6, slender, whitish. Fruit clavate, usually naked, dehiscing by longitudinal splits in the pericarp, red, orange-yellow, greenish or white, pulp scanty, but very attractive to ants. Seeds 0.8–2.0 × 0.8–2.2 mm, hat-shaped, black, very shiny, testa tuberculate, the tubercles sometimes elongated and pointed.

A genus of 7 or more, very closely related species, the majority in Eastern Brazil, including Maranhão (*fide* Braun & Esteves Pereira 2002: 79), Ceará, Piauí, Pernambuco (*fide* Uebelmann 1996), Bahia and Minas Gerais. Of these, 5 are endemic and mostly either patchy in distribution or extremely local, rare and often in danger of extinction (the whole genus has accordingly been placed in Appendix I of C.I.T.E.S. since 1992). However, the complex species, *D. heptacanthus* (with a subspecies ranging into NW Minas Gerais, W & S-cent. Bahia & SW Piauí), ranges as far as north-eastern Paraguay and eastern Bolivia, through Mato Grosso do Sul, Goiás and Mato Grosso. It is curious that, except for a single population of *D. heptacanthus* subsp. *catingicola* from Mun. Paramirim, the genus has not been recorded from central and southern Bahia (east of the Rio São Francisco), although suitable habitats exist, eg. in the Chapada Diamantina. Instead, such habitats are characterized by similarly adapted *Melocactus* species (eg. *M. paucispinus*, *M. conoideus*, *M. concinnus*, *M. violaceus*).

As the generic name implies, the plants are disc-shaped, and over much of its range *Discocactus* is found in habitats through which fire passes regularly (notably the *cerrado* and associated *campo rupestre*). This depressed, ground-hugging habit protects the plant against the worst effects of burning, since the region a few centimetres above the ground is usually much cooler from air being drawn in as the fire passes. Three of the rarer, northern species, however, occur in the *caatinga* and apparently are not normally subjected to fire.

Discocactus is clearly very closely related to the much larger genus, *Gymnocalycium* Pfeiff. ex Mittler (1844), which replaces it to the south and south-west, in Southern Brazil, Paraguay, Argentina and eastern Bolivia. *Gymnocalycium* differs mainly in lacking a cephalium, in having broader, diurnal, brightly coloured flowers and in displaying a greater diversity in seed-morphology. It may prove to be paraphyletic in respect of *Discocactus* and the latter name has priority, potentially raising the fraught question of which name should be preferred or conserved, although they do not necessarily need to be combined and are certainly not confusable. Plates 71–73.2.

1. Radial spines 3–8 per areole, often > 1.5 mm thick; stem solitary or offsetting when damaged by fire, > 11 cm diam. 2
1. Radial spines > 8 per areole, to 1.5 mm thick; stems freely offsetting, or solitary and < 11 cm diam. 5

2. Exposed part of vegetative stem with only 3 areoles visible per rib; ribs not broken up by tubercles (N Bahia to Piauí and Ceará) . . . . **2. bahiensis**
2. Exposed part of vegetative stem with > 3 areoles per rib and/or ribs tuberculate with deep sinuses between successive areoles 3

3. Seed 1.5–2.0 mm; fruit white or greenish, rarely pinkish red at apex; ribs 9–26, ± tuberculate; spines often flattened, > 1.5 mm thick 4
3. Seed 1–1.4 mm; fruit yellow-orange at apex; ribs 12–13, scarcely tuberculate; spines ± terete, to 1.5 mm thick (N Minas Gerais: near Grão Mogol, sand) . . . . **5. pseudoinsignis**

4. Ribs 9–26, acute-edged, with shallow sinuses between areoles on the same rib (Minas Gerais: E of Rio São Francisco) . . . . **4. placentiformis**

4. Ribs 10–12, composed of rounded tubercles with deep, acute sinuses between areoles on the same rib (SW Piauí, W & cent.-S Bahia & NW Minas Gerais)
.... **3. heptacanthus** subsp. **catingicola**

5. Ribs not tuberculate; stem solitary, dark purplish green to brownish; spines minute, adpressed, claw-like (N Minas Gerais: near Grão Mogol, in pure quartz gravel) .... **6. horstii**
5. Ribs ± tuberculate; stem usually offsetting, light green or spines not as above (N Bahia) 6

6. Tubercles strongly developed, spiralled, obscuring the ribs .... **1. zehntneri**
6. Tubercles weakly developed, arranged in clearly defined ± vertical ribs .... **2. bahiensis**

1. Discocactus zehntneri *Britton & Rose*, Cact. 3: 218 (1922); N. P. Taylor in Cact. Succ. J. Gr. Brit. 43: 40 (1981). Type: Bahia, Mun. Sento Sé, *Zehntner* in *Rose & Russell* 19779 (US, lecto. designated here).

Echinocactus zehntneri (Britton & Rose) Luetzelb., Estud. Bot. Nordéste 3: 69 (1926).
D. albispinus Buining & Brederoo in Cact. Succ. J. (US) 46: 252–257 (1974). *D. zehntneri* f. *albispinus* (Buining & Brederoo) Riha in Kaktusy 19: 26 (1983). *D. zehntneri* var. *albispinus* (Buining & Brederoo) P. J. Braun in Succulenta 69: 215 (1990). *D. zehntneri* subsp. *albispinus* (Buining & Brederoo) P. J. Braun & E. Esteves Pereira in Kakt. and. Sukk. 46: 64 (1995). Holotype: Bahia, Mun. Sento Sé / Campo Formoso, *Horst* 390 (U).

Solitary or strongly caespitose; stem 3–7 × 2–12 cm, light to dark green; ribs 12–21, broken up into ± spiralled tubercles; areoles to c. 15 mm apart on the ribs. Spines creamy white to grey, interwoven and ± hiding the stem, centrals 0–2, radials 9–18. Cephalium to 3.5 × 4.5 cm, white-woolly, with 7–60 mm, erect, dark bristles at its margin. Flowers 33–77 × 35–55 mm, greenish in bud; outer perianth-segments 19–28 × 3.5–6.0 mm, green to yellowish, inner segments 17–25 × 4–6 mm, white. Fruit commonly bright red, or white to greenish, 18–46 × 6–9 mm. Seed 0.8–2.0 × 0.8–2.2 mm, testa densely covered by elongate tubercles.

CONSERVATION STATUS. Vulnerable [VU B1ab(iii, v)]; extent of occurrence estimated to be < 20000 km², known from < 10 locations and continuing decline projected (see below).

This species is divisible into two subspecies:

1a. subsp. **zehntneri**

Solitary or sparingly offsetting, stems globose to elongate; ribs c. 12–13, tubercles 10 × 15–20 × 15–20 mm; areoles c. 9–10 × 5–6 mm. Central spines 0–2, to 70 × 2 mm, radials c. 11, to 25–75 × 1.5 mm. Fruit red.

Northern Rio São Francisco *caatinga* element: on exposed rocks and gravelly soil in *caatinga*, c. 530 m, north of the Chapada Diamantina, Bahia. Endemic to Bahia. Map 50.
BAHIA: Mun. Sento Sé, *Zehntner* in *Rose & Russell* 19779 (US), 29 km SE of the [old] town, nr 'Brejo Grande', July 1974, *Horst* 441 (U, ZSS); Mun. Sento Sé/Campo Formoso, Serra São Francisco, S of Piçarrão, *Horst* 390 (U, ZSS).

CONSERVATION STATUS. Data Deficient [DD]. Part of its former habitat is believed to have been submerged beneath the Represa (Lago) de Sobradinho leaving, perhaps, only a single locality intact (that of the type of the synonymous *D. albispinus,* which was observed by G. Charles in July 2000). Detailed knowledge of its extent of occurrence and area of occupancy is currently lacking. However, it is likely to be Vulnerable (VU).

1b. subsp. **boomianus** (*Buining & Brederoo*) *N. P. Taylor & Zappi* in Bradleya 9: 86 (1991). Holotype: Bahia, Mun. Morro do Chapéu, *Horst* 222 (U).

Discocactus boomianus Buining & Brederoo in Succulenta 50: 26 (1971). *D. zehntneri* var. *boomianus* (Buining & Brederoo) P. J. Braun in ibid. 69: 218 (1990).

D. araneispinus Buining *et al.* in Succulenta 56: 258 (1977); Buining, Gen. Discocactus: 39 (1980). *D. zehntneri* var. *araneispinus* (Buining *et al.*) P. J. Braun in Succulenta 69: 215 (1990). *D. zehntneri* subsp. *araneispinus* (Buining *et al.*) P. J. Braun & E. Esteves Pereira in Kakt. and. Sukk. 46: 64 (1995). Holotype: Bahia, Mun. Sento Sé, *Horst* 440 (U).

? *D. buenekeri* W.R. Abraham in Kakt. and. Sukk. 38: 284, with illus. (1987); P. J. Braun in Succulenta 69: 219–221 (1990). *D. zehntneri* subsp. *buenekeri* (W.R. Abraham) P. J. Braun & E. Esteves Pereira in Kakt. and. Sukk. 44: 64 (1993). Holotype: N Bahia, sine loc., *CWRA* 27 (KOELN [Succulentarium?], n.v.).

D. zehntneri var. *horstiorum* P. J. Braun in ibid. 69: 218 (1990). *D. zehntneri* subsp. *horstiorum* (P. J. Braun) P. J. Braun & E. Esteves Pereira in ibid. 46: 64 (1995). Holotype: Bahia, *K.I. Horst* 667 (ZSS; K, iso.).

Offsetting freely, stems depressed, sometimes ± buried when growing in sand; ribs 13–21, tubercles 3–11 × 2–13 × 2–13 mm; areoles 3–7 × 1.5–3 mm. Central spine 0–1, to 35 × 1 mm, but to 60 mm beneath the cephalium, radials 4–30 × < 1 mm. Fruit red or white (in the same population). Seed: Barthlott & Hunt (2000: pl. 44.2).

Northern *campo rupestre* (Chapada Diamantina) element: on exposed arenitic rocks often with an accumulation of gravel or in pure quartz sand, *caatinga/campo rupestre*, c. 700–1000 m, northern Bahia. Endemic. Map 50.

BAHIA: Mun. Sento Sé, E part of Serra do Mimoso, N of Limoeiro, *Horst* 440 (U, ZSS); l.c., serra near Minas do Mimoso [Serra do Curral Feio, bordering on Mun. Umburanas], 1988, *K.I. Horst* 667 (K, ZSS); Mun. Morro do Chapéu, 21–22 km W of town on road BA 052, 11°28'S, 41°20'W, 25 Dec. 1988, *Taylor & Zappi* in *Harley* 27395 (K, SPF, CEPEC), 23 July 1988, *Eggli* 1274 (ZSS), 3 Aug. 2002, *Taylor* (K, photos); l.c. (?), cult. ZSS, *Horst & Uebelmann* HU 222 (K).

CONSERVATION STATUS. Vulnerable [VU D2] (2); PD=1, EI=1, GD=2. Short-list score (2×4) = 8. Habitats fragmented and subject to collection. There are at least 4 locations for this plant at the southern end of its range to the west of Morro do Chapéu. Of these, 3 are accessible by road. According to Marlon Machado, the locations vary in size from 3.5 to 150 ha with variable densities of individuals, numbers ranging from 50,000 to at least 150,000.

Abraham's *D. buenekeri* remains inadequately known in view of the characteristic secrecy surrounding the location of its wild habitat, but it may be sufficiently distinct to merit treatment as a third subspecies of *D. zehntneri*.

2. Discocactus bahiensis *Britton & Rose*, Cact. 3: 220 (1922); Buining, Gen. Discocactus: 123–129 (1980); Hofacker in Kakt. and. Sukk. 53: 200–205 (2002). Holotype: Bahia, *Zehntner* in *Rose* 19783 (US).

Echinocactus bahiensis (Britton & Rose) Luetzelb., Estud. Bot. Nordéste 3: 69 (1926).

Discocactus subviridigriseus Buining *et al.* in Succulenta 56: 262 (1977); Buining, Gen. Discocactus: 115 (1980). *D. bahiensis* subsp. *subviridigriseus* (Buining *et al.*) P. J. Braun & E. Esteves Pereira in Kakt. and. Sukk. 44: 63 (1993). Holotype: Bahia, Mun. Juazeiro, *Horst* 438 (U).

D. bahiensis subsp. *gracilis* P. J. Braun & E. Esteves Pereira in ibid. 52: 286–290, illus. (2001). Holotype: Bahia, Vereda do Romão Gramacho (Rio Jacaré), São Rafael, s.d., *Braun & Esteves Pereira* s.n. (UFG 22432, n.v.).

[*D. placentiformis* sensu Britton & Rose, Cact. 3: 220, fig. 233 (1922) non (Lehm.) K. Schum.]

VERNACULAR NAME. Frade-de-cavalo.

Solitary or freely offsetting in age, 4–7 × 5.5–18.0 cm, grey- to light or yellow-green; ribs 10–15, to 30 × 20 mm, sometimes composed of tubercles to c. 7–10 × 15–30 × 15 mm; areoles oval, c. 6–10 × 3–8 mm, 10–23 mm apart, sometimes only 3 visible per rib above ground. Spines 5–11, all radial, strongly recurved, sometimes hooked at apex, in some forms interwoven with those of adjacent areoles, pale yellow to pinkish brown at first, later greyish or dark brown to blackish, to 35(–45) × 0.5–2.5(–3.0) mm, the uppermost 1–3 in each areole very small when present. Cephalium low, to 47 mm diam., with soft whitish wool, bristles when present hairlike, to 30 mm, yellow or brownish. Flowers 40–72 × 30–54 mm; outer perianth-segments green, yellow or pinkish brown, inner segments 8–20 × 5–7 mm, white. Fruit greenish white or red, 25–50 × 8–10 mm. Seed 1.2–1.8 × 1.5–1.9 mm, testa with sparse to dense patches of pointed tubercles.

Northern *caatinga* element: on exposed, gravelly river terraces amongst limestone or iron-stained quartzite, and seasonally inundated river flood plain, under and between jurema-preta (*Mimosa tenuiflora*) and carnaúba (*Copernicia prunifera*) within the *caatinga*, 380–700 m, Rio São Francisco drainage of northern Bahia (probably in adjacent Pernambuco), Ceará and north-western Piauí (Rio Canindé). Map 50.

CEARÁ: unlocalized, photograph communicated by Prof. Mattos, Univ. Fed. Ceará, Fortaleza (K).
PIAUÍ: Mun. Francisco Ayres, near the Rio Canindé, *Horst & Uebelmann* 943 (ZSS).
PERNAMBUCO: vaguely reported from near Santa Maria da Boa Vista, *Uebelmann* (mss).
BAHIA: N Bahia, Mun. Sento Sé, reported from S of the original town by Buining (1980: 121, photo); Mun. Juazeiro, near the town, 2–6 June 1915, *Rose & Russell* 19742, 19783 (US); l.c., SSW of town, 0.5 km S of Rodeadouro, 10 Jan. 1991, *Taylor et al.* 1387 (K, ZSS, HRCB, CEPEC), *Horst* 438 (U, ZSS), 9°28'S, 40°33'W, 7 Apr. 2000, *E.A. Rocha et al.* (K, photos); l.c., E of town, 2–6 June 1915, *Rose & Russell* 19764 (US); l.c., [11 km W of] Juremal towards Curral Velho, *Horst & Uebelmann* 633 (ZSS T02127 & TP58-114/163/164); Mun. Campo Formoso, c. 60 km W of Maçaroca (Massaroca), W of Abreus, W bank of the Rio Salitre, *Horst* 437 (U, ZSS); ibid., reported from Vargem Grande, Laje, Panelas, Delfino and [Faz.] Vargem do Sal ['Sol'] by Braun & Esteves Pereira, l.c. (2001); Mun. Itaguaçu da Bahia, Serra Azul, reported by Braun & Esteves Pereira, l.c. (2001); Mun. Ourolândia (Ouro Branco), reported by Braun & Esteves Pereira, l.c. (2001); Mun. Jacobina, Caatinga do Moura, reported by Braun & Esteves Pereira, l.c. (2001); Mun. Morro do Chapéu, E of Rio Jacaré, São Rafael, 5 Aug. 2002, *Taylor* (K, photos).

CONSERVATION STATUS. Endangered [EN B2ab(i, ii, iii, iv, v)] (3); area of occupancy < 500 km²; PD=2, EI=1, GD=2. Short-list score (3×5) = 15. Part of its range was eliminated by permanent inundation from the Represa de Sobradinho (BA/PE) in the 1970s and the remainder has been heavily impacted by agriculture and road/house construction during the past 10 years. Some of its sites are accessible by road and have been visited by hobbyist collectors.

The distribution of this taxon seems rather disjunct towards its northern limits, but remains inadequately known at present. It is related to the preceding species, occasionally bearing red fruits (cf. *Rose & Russell* 19742) and both are assumed to have been derived from within the *D. heptacanthus* complex, sharing similarities with the geographically close *D. heptacanthus* subsp. *catingicola* (q.v.), especially via the dwarf, southern form called *D. bahiensis* subsp. *gracilis* by Braun & Esteves Pereira, l.c. (2001) — see last locality cited above. Hofacker (2002) has recently illustrated the diversity of populations in Bahia.

The form found in the vicinity of Juazeiro, which was described as *D. subviridigriseus* Buining *et al.*, is connected to typical *D. bahiensis* by a series of populations from northern Bahia, comprising *Horst* 437 (*D. bahiensis* sensu Buining), the actual type of the species (*Rose & Russell* 19783), *Horst & Uebelmann* 633 and others from the extensive area covered by Mun. Campo Formoso. The '*D. subviridigriseus*' form extends into Piauí and Ceará and may be expected to occur in Pernambuco. It seems to be an ecotype of river flood plains in the *caatinga*, where in northern Bahia, at least, it is associated with semi-barren habitats characterized by bushes of jurema-preta (*Mimosa tenuiflora* (Willd.) Poir.) and carnaúba palms, and is evidently, or was formerly, at times subjected to temporary inundation.

3. Discocactus heptacanthus *(Barb.Rodr.) Britton & Rose*, Cact. 3: 218 (1922); Taylor in Cact. Succ. J. Gr. Brit. 43: 38 (1981). Type: Brazil, Mato Grosso, nr Cuiabá, *Rodrigues* (†, see Britton & Rose, l.c.). Lectotype (designated here): Rodrigues, Pl. Mato Grosso: tab. 11 (1898).

Malacocarpus heptacanthus Barb. Rodr., Pl. Mato Grosso: 29, tab. 11 (1898).

CONSERVATION STATUS. Data Deficient [DD]. The status of this species, comprising many populations found outside the area treated here, remains to be investigated, but it cannot be assumed to be Least Concern in view of the extensive alterations that are known to have already occurred to the *cerrado* biome.

Only the following subspecies is found in Eastern Brazil:

3a. subsp. **catingicola** *(Buining & Brederoo) N. P. Taylor & Zappi* in Cact. Consensus Initiatives 3: 7 (1997). Holotype: W Bahia, Mun. São Desidério, *Horst 392* (U).

Discocactus catingicola Buining & Brederoo in Kakt. and. Sukk. 25: 265–267 (1974).
D. griseus Buining & Brederoo in Succulenta 54: 185–190 (1975). *D. catingicola* var. *griseus* (Buining & Brederoo) P. J. Braun & E. Esteves Pereira in Kakt. and. Sukk. 44: 112 (1993). *D. catingicola* subsp. *griseus* (Buining & Brederoo) P. J. Braun & E. Esteves Pereira in ibid. 46: 64 (1995). Holotype: Minas Gerais, [Mun. João Pinheiro], Serra dos Alegres, 850 m, s.d., *Horst & Uebelmann* HU 343 (U; ZSS, iso.).
D. rapirhizus Buining & Brederoo in Ashingtonia 2: 44–47 (1975). *D. catingicola* subsp. *rapirhizus* (Buining& Brederoo) P. J. Braun & E. Esteves Pereira in Kakt. and. Sukk. 44: 112 (1993). Holotype: Goiás, surroundings of Posse, 19 June 1974, *Buining & Horst* HU 200 (U).
D. spinosior Buining et al. in Succulenta 56: 261 (1977); Buining, Gen. Discocactus: 146 (1980). Holotype: W Bahia, Mun. Barreiras, *Horst* 205A (U? [labelled with number only]).
D. nigrisaetosus Buining et al. in Succulenta 56: 260 (1977); Buining, tom. cit.: 129 (1980). *D. catingicola* var. *nigrisaetosus* (Buining et al.) P. J. Braun & E. Esteves Pereira in Kakt. and. Sukk. 44: 112 (1993). Holotype: W Bahia, Mun. Santana, *Horst* HU 448 (U? [labelled with number only]).
D. piauiensis P. J. Braun & E. Esteves Pereira in Kakt. and. Sukk. 46: 57–62, with figs (1995). Holotype: S Piauí, 600 m, *E. Esteves Pereira* 114 (UFG 14761, n.v.; B?, iso., n.v.), l.c., 1993, *Braun* 1652, 1655 (B, paratypi, n.v.).

Solitary, 4–9 × 11–20 cm, plain green; ribs 10–12, completely broken into tubercles, which are 15–20 × 22–40 × 15–30 mm; areoles positioned slightly below and sunken into tubercle apex, oval, 6–8 × 4–5 mm, 15–30 mm apart, 4–5 visible above ground on each rib. Central spine 0(–1), to 20 mm, radials 5–8, weakly recurved, pale yellow then grey to brownish in age, to 36 × 1.25–2.50 mm. Cephalium 3–7.5 cm diam., with white wool and 35–45 mm, yellow to dark brownish bristles. Flowers 50–60 × 40–60 mm, light brown to olive-green in bud; outer perianth-segments 22–24 × 7–8, white to greenish without, inner segments 18–20 × 5–7, white. Fruit white or faintly pinkish, 40–45 × 8–13 mm. Seed 1.5–2.0 × 1.5–2.0 mm, with irregularly spaced, elongate, pointed tubercles (Barthlott & Hunt 2000: pl. 44.3–4).

Western *cerrado* element: on exposed gravel or sand, *cerrado* and *cerrado-caatinga* ecotone, 450–700 m, south-western Piauí, western and central-southern Bahia and north-western Minas Gerais; to Goiás & Tocantins. Map 50.

PIAUÍ: SW Piauí, Mun. Bom Jesus, 30 km S of town, 1982, *Horst & Uebelmann* 523 (ZSS).
BAHIA: W Bahia, Mun. Barreiras, near old airport, *Horst* 205A (U, ZSS, K); Mun. São Desidério, west of Sítio Grande at waterfall, *Horst* 392 (U); Mun. Santana, near Porto Novo, *Horst* 448 (U, ZSS); l.c. (?), *Braun* 341 (ZSS); Mun. Santa Maria da Vitória, airport, *Horst & Uebelmann* 760 (ZSS, K); Mun. Coribe, 8 km south of Coribe, *Horst & Uebelmann* 760A (ZSS); cent.-S Bahia, Mun. Paramirim, N bank of rio Paramirim, 28 Nov. 1988, *Taylor* in *Harley* 25558 (K, SPF, CEPEC).
MINAS GERAIS: Mun. Januária, Pandeiros, *Horst & Uebelmann* 720 (ZSS).

CONSERVATION STATUS. Vulnerable [VU B2ab(iii)] (2); area of occupancy estimated to be < 2000 km²; PD=1, EI=1, GD=2. Short-list score (2×4) = 8. Vulnerable due to the fragmented nature of its distribution and small population size. In need of regular monitoring, since its *cerrado* and *caatinga* habitats are undergoing much destructive change.

Despite Buining's use of the epithet *catingicola,* this subspecies is more typical of the *cerrado* and probably occurs in the *caatinga* only at its eastern limits. The collections cited from south-western Piauí and central-southern Bahia (*Harley et al.* 25558, which is not precisely localized for conservation reasons) are somewhat intermediate with *D. bahiensis.*

4. Discocactus placentiformis *(Lehm.) K. Schum.* in Engler & Prantl, Pflanzenfam. 3(6a): 190 (1894). Type: 'Brasilia meridionali', apparently not preserved. Neotype (designated here): Lehmann in Nov. Act. Nat. Cur. 16(1): tab. 16 (1832) (iconotype).

Cactus placentiformis Lehm., Sem. hort. bot. Hamburg: 17 (1826) and in Nov. Act. Nat. Cur. 16(1): 318–319 (1832). *Echinocactus placentiformis* (Lehm.) K. Schum. in Martius, Fl. Bras. 4(2): 246 (1890). *Melocactus besleri* Link & Otto in Verh. Ver. Beförd. Gartenb. 3: 420 (1827), nom. illeg. (Art. 52.1). *Discocactus lehmannii* Pfeiff. in Nov. Act. Nat. Cur. 19(1): 120 (1839), nom. illeg. (Art. 52.1).
D. insignis Pfeiff. in Allg. Gartenz. 5: 241 (1837). Type: a living plant in the collection of Schelhase, not preserved at the time of publication. Lectotype (designated here): Pfeiffer in Nov. Act. Nat. Cur. 19(1): tab. 15 (1839) — assumed to be an illustration prepared prior to the publication of the name in 1837.
D. linkii Pfeiff., l.c. (1839). Based on *Melocactus besleri* Link & Otto, l.c. (1827), typ. excl. Lectotype (designated here): Link & Otto, l.c., tab. 21 (1827). *D. besleri* F.A.C. Weber in Bois, Dict. hort.: 450 (1896). Type: as above.
? *D. alteolens* Lem. ex A. Dietr. in Allg. Gartenzeitung 14: 202 (1846). *D. placentiformis* var. *alteolens* (Lem. ex A. Dietr.) P. J. Braun & E. Esteves Pereira in Kakt. and. Sukk. 44: 105 (1993). *D. placentiformis* subsp. *alteolens* (Lem. ex A. Dietr.) P. J. Braun & E. Esteves Pereira in ibid. 46: 64 (1995). *Echinocactus alteolens* (Lem. ex A. Dietr.) K. Schum. in Martius, Fl. bras. 4(2): 246 (1890). *Cactus alteolens* (Lem. ex A. Dietr.) O. Kuntze in Deutsch. Bot. Monatsschr. 21: 173 (1903). Type: not known to have been preserved.
D. tricornis Monv. ex Pfeiff., Abbild. Beschr. Cact. 2: tab. 28 (1850). Type: not known to have been preserved. Lectotype (designated here): Pfeiffer, l.c., tab. 28 (1850).
D. pugionacanthus Buining *et al.* in Succulenta 56: 260–261 (1977); Buining, Gen. Discocactus: 61 (1980). *D. placentiformis* var. *pugionacanthus* (Buining *et al.*) P. J. Braun & E. Esteves Pereira in Kakt. and. Sukk. 44: 105 (1993). *D. placentiformis* subsp. *pugionacanthus* (Buining *et al.*) P. J. Braun & E. Esteves Pereira in ibid. 46: 64 (1995). Holotype: Minas Gerais, Mun. Grão Mogol (?), *Horst* 462 ['275'] (U).
D. pulvinicapitatus Buining & Brederoo in Buining, tom. cit.: 100 (1980). *D. latispinus* subsp. *pulvinicapitatus* (Buining & Brederoo) P. J. Braun & E. Esteves Pereira in Kakt. and. Sukk. 44: 105 (1993). Holotype: Minas Gerais, Mun. Bocaiúva, *Horst* 425 (U? [fl. labelled with number only]).
D. latispinus Buining *et al.* in Succulenta 56: 259 (1977); Buining, tom. cit.: 95 (1980). Holotype: Minas Gerais, Serra do Cabral, *Horst* 146 (U).

D. *multicolorispinus* P. J. Braun & Brederoo in Kakt. and. Sukk. 32: 54–59, with illus. (1981). *D. placentiformis* subsp. *multicolorispinus* (P. J. Braun & Brederoo) P. J. Braun & E. Esteves Pereira in Kakt. and. Sukk. 44: 105 (1993). Holotype: Minas Gerais, Serra de Minas [collector?] (U, n.v.; ZSS, iso.). The type plant is illustrated in Cact. Succ. J. (US) 51: 17, fig. 27 (1979).

D. *crystallophilus* L. Diers & E. Esteves Pereira in ibid. 32: 258–262, with illus. (1981). Holotype: Minas Gerais, Mun. Augusto de Lima, *E. Esteves Pereira* E-84 (KOELN ['Succulentarium'], n.v.; ZSS, iso.).

D. *pseudolatispinus* L. Diers & E. Esteves Pereira in ibid. 38: 242–247 (1987). *D. latispinus* subsp. *pseudolatispinus* (L. Diers & E. Esteves Pereira) P. J. Braun & E. Esteves Pereira in Kakt. and. Sukk. 44: 105 (1993). Holotype: Minas Gerais, Mun. Claro dos Poções, *E. Esteves Pereira* 111 (KOELN, n.v.; UFG, iso.).

Very variable; stem solitary unless damaged by fire, 3.5–10.0 × 12–25 cm, light to dark green; ribs 9–26, well-defined, but weakly to strongly tuberculate, with shallow sinuses between successive tubercles, to 30 mm high and 40 mm wide near stem base, the tubercles 8–15 × 20–30 × 12–25 mm; areoles circular to oval, 4–9(–15) × 3–8(–12) mm, somewhat sunken, 15–25 mm apart on the ribs, 3–7 visible per rib above ground level. Spines 3–8(–10), whitish, dull yellow, brownish, pinkish or reddish, later grey, central spine 0(–1), to 20 mm, lower 3–5 radials to 40(–45) × 4 mm, mostly flattened and slightly recurved, sometimes splitting longitudinally, other spines when present much smaller. Cephalium 1–7 × 4–11 cm, white-woolly, with or without 30–45 mm, brownish bristles. Flowers 45–85 × 40–80 mm, pale green, brown or deep pink in bud; outer perianth-segments 24–35 × 3–7 mm, inner segments 15–30 × 2–5 mm, white or pale pinkish. Fruit 30–50 × 5–15 mm, white to pink or reddish at apex. Seed 1.5–1.9 × 1.4–2.0 mm, testa with irregularly spaced, short to elongate, pointed tubercles.

Widespread South-eastern *campo rupestre/cerrado* element: on arenitic rocks, quartz sand and gravel, *cerrado/campo rupestre*, rarely within the southern limits of the *caatinga*, 550–1275 m, east of the Rio São Francisco, central and northern Minas Gerais. Endemic to the core area within Minas Gerais. Map 50.

MINAS GERAIS: Mun. Manga (?), 'Mocambinho', 15°6'S, 43°59'W, [*caatinga* region!], Mar. 1998, *I. Pimenta* (K, photo); Mun. Grão Mogol, 48.5 km west of bridge over Rio Vacaria on road BR 251, Serra da Bocaina, pass at telecommunications tower, 31 Jan. 1991, *Taylor et al.* 1512 (K, ZSS, HRCB, BHCB); Mun. Claro dos Poções, 67 km from Montes Claros on road BR 365, June 1979, *E. Esteves Pereira* 111 (UFG), 1986, *P.J. Braun* 707 (ZSS); Mun. Jequitaí, 91.5 km NE of Pirapora on road BR 365, 27 Jan. 1991, *Taylor et al.* 1450 (K, ZSS, HRCB, BHCB); Mun. Francisco Dumont, c. 2 km from the town towards Jequitaí, 7 Aug. 1988, *Eggli* s.n. (ZSS, photo); Mun. Bociaúva, W slope of Serra do Espinhaço, near Sítio, Faz. Laginha, *Horst & Uebelmann* 548 (ZSS, K); l.c., Faz. Olho d'Agua, 1978, *Horst & Uebelmann* 461 (ZSS); near Rio Jequitaí, Faz. Timboré, *Horst & Uebelmann* 425 (U, ZSS, K); Mun. Buenópolis, Serra do Cabral, 6 & 7 km from the town on the road to Barreira da Lapa, 17°53'S, 44°15'W, 13 Oct. 1988, *Taylor & Zappi in Harley* 24988, 24993 (K, SPF); Mun. Augusto de Lima, above Santa Barbara industrial site, 31 Aug. 1985, *Horst & Uebelmann* 356 (K, ZSS); ibid., near road BR 135, 4 km from the border of Mun. Corinto, 6 Aug. 1988, *Eggli* 1096 (ZSS, photo); Mun. Diamantina, near Senador Mourão, 1985, *Horst & Uebelmann* 707 (ZSS T00227, K); ibid., c. 15 km NE of Diamantina on road to Mendanha, 30 Jan. 1969, *Irwin et al.* 22884 (K); ibid., between Diamantina and Biribiri, 18°11'S, 43°36'W, 8 Mar. 1995, *Splett* 808 (UB); ibid., estrada para o vilarejo de Três Barras, 15 Apr. 1987, *Zappi in CFCR* 10540 (SPF); ibid., estrada para Conselheiro Mata, KM 165, 18 July 1987, *Zappi et al. in CFCR* 11297 (SPF); 1 km from Conselheiro Mata, *Horst & Uebelmann* 542 (ZSS); Mun. Couto de Magalhães de Minas, near the town, 3 Aug. 1988, *Eggli* 1056 (ZSS), *Horst & Uebelmann* 232 (ZSS TP58-46, K); Mun. Datas, 5.5 km N of Datas, 17 Feb. 1988, *Supthut* 8835 (ZSS); Mun. Gouveia, just east of Ponte de Paraúna (E of Pres. Juscelino), reported by Buining (1980: 74); Mun. Sete Lagoas, Lagoa Grande, 16 Oct. 1959, *Heringer* 7228 (UB); Mun. Ouro Branco, Serra do Ouro Branco, illustrated in Martius, Flora brasiliensis 1 (1, Tabulae physiognomicae), tab. XLVI (1855).

CONSERVATION STATUS. Vulnerable (2) [VU B2ab(iii, v)]; extent of occurrence > 50,000 km², but area of occupancy almost certainly < 2000 km²; PD=2, EI=1, GD=2. Short-list score (2×5) = 10. Vulnerable, since most populations are small and isolated from one another. Especially affected by unsustainable exploitation in the production of cactus candy, when located close to towns and villages.

Typical *D. placentiformis* (Lehm.) K. Schum. is assumed to be the plant found originally about the towns of Diamantina and Ouro Branco, Minas Gerais. These were visited before 1827 by botanists such as Riedel (see Urban 1906: 90), who was cited as the collector of the type of the synonymous *D. linkii* Pfeiff. (*Melocactus besleri* Link & Otto *pro parte*), based on material contemporary with that of *Cactus placentiformis* Lehm.

The large-stemmed, many-ribbed, thick-spined form of this species (syn. *D. pulvinicapitatus, D. latispinus, D. pseudolatispinus*), from the western slopes of the Serra do Espinhaço, and from the Serra do Cabral and northwards (municípios Claro dos Poções, Jequitaí, Bocaiúva, Francisco Dumont & Buenópolis), is distinctive and may be worthy of recognition as a subspecies. It is connected to typical forms of the species by populations found near the western edge of Mun. Diamantina (syn. *D. multicolorispinus*). The form from the north-eastern population (Mun. Grão Mogol, syn. *D. pugionacanthus*) is also distinctive for its ± strongly tuberculate stem and could be mistaken for *D. heptacanthus*. Some plants from the region north of Diamantina, eg. *Horst & Uebelmann* 232 cited above, superficially resemble *D. pseudoinsignis* (see below) in their spination.

5. Discocactus pseudoinsignis *N. P. Taylor & Zappi* in Bradleya 9: 86 (1991). Holotype: Minas Gerais, Mun. Cristália, *Zappi et al.* in CFCR 12045 (SPF).

[*D. insignis* sensu Buining, Gen. Discocactus: 81–82 (1980) non Pfeiff. (1837).]

Solitary, 7–9 × 12–21 cm, light to dark green; ribs 12–13, almost straight and even, not tuberculate, 15–25 mm high, 20–40 mm broad; areoles 6 × 4–5 mm, 10–30 mm apart and 5–6 visible per rib above ground. Spines 5–8(–9), ± terete or at least isodiametric, grey to blackish, central 0–1, 10–30 × 0.5–1.5 mm, radials 5–7(–8), lower 3 c. 25–42 × 1–1.5 mm, straight or variously curved, others much smaller. Cephalium to 5 × 10 cm, white-woolly, usually with exserted, to 40 mm, dark brownish bristles. Flowers c. 75 × 60 mm, pale brownish olive-green in bud; outer perianth-segments c. 30 × 4 mm, inner segments c. 22 × 2 mm, white. Fruit 32–45 × 5–9 mm, orange-yellow to reddish at apex. Seed 1–1.4 × 1.0 mm, regularly tuberculate.

South-eastern *campo rupestre* (Grão Mogol) element: in pure quartz sand or sand between arenitic rocks, *campo rupestre*, 700–1000 m, Mun. Cristália and Mun. Grão Mogol, northern Minas Gerais. Endemic to the core area within Minas Gerais. Map 50.
MINAS GERAIS: Mun. Grão Mogol, 1972/1974, *Horst* 347 (U, K, ZSS); l.c., 10 km NW of the town, 22 Oct. 1978, *Hatschbach* 41364 (MBM); l.c., Vale do Ribeirão das Mortes, 4 Sep. 1986, *I. Cordeiro & R. Mello-Silva* in CFCR 10088 (SPF); l.c., Córrego Escurona, 13 May 1990, *Zappi et al.* in CFCR 12901 (SPF, HRCB); l.c., Várzeas Escuras, 6 Nov. 1997, *F. Fernande* (K, photo); Mun. Cristália, 'margem direita do Rio Itacambiruçu, encosta da Serra das Cabras', 28 May 1988, *Zappi et al.* in CFCR 12045 (SPF, holo.).

CONSERVATION STATUS. Endangered [EN B1ab(iii) + 2ab(iii)] (3); extent of occurrence = 89 km² (possibly under-recorded); PD=2, EI=1, GD=1. Short-list score (3×4) = 12. Populations may be affected by habitat modification, including projected inundation by an hydro-electric dam-lake, but are offered some protection within the Parque Estadual da Serra do Barão, Grão Mogol.

Following the discovery of this species in the early 1970s, Buining (1980) misidentified it as *D. insignis* Pfeiff., a name previously and correctly referred to the synonymy of *D. placentiformis* (Lehm.) K. Schum. by earlier authors (see Taylor 1981: 40). Pfeiffer's 19th Century description calls for a plant with only 10 ribs (consistently 12–13 in *D.*

pseudoinsignis) and, together with his illustration of the type, clearly indicates that the bract-scales on the flower-tube and outer perianth-segments of the flower were deep pink, which is a feature of some forms of *D. placentiformis*, but not of *D. pseudoinsignis* (outer segments pale brownish olive-green).

D. pseudoinsignis is similar to the variable *D. placentiformis* and falls within the geographical range of the latter (which is recorded from the northern part of Mun. Grão Mogol, Serra da Bocaina), but can be distinguished by its non-tuberculate ribs, slender spines, distinctively coloured outer perianth-segments and fruit apex, and smaller seeds. Its closest relative is probably the following, with which it is partly sympatric.

6. Discocactus horstii *Buining & Brederoo* in Krainz, Die Kakteen, Lfg 52 (1973); Buining, Gen. Discocactus: 106–110 (1980). Holotype: Minas Gerais, Mun. Grão Mogol, *Horst* 360 (U).

Solitary, to 2 × 6 cm, dark brown- to purplish green; ribs 15–22, straight and narrow, 6–8 mm high, c. 4 mm wide at the rounded edge, and to 10 mm apart, not tuberculate; areoles oval, c. 1.5 × 1.0 mm, very slightly sunken into the ribs, to 5 mm apart, 4–6 visible per rib above ground level. Spines 9–11, all radial, pectinate, tightly adpressed to stem, claw-like, brown with a grey coating, to 3.5 × 0.75 mm. Cephalium to 1.5 × 2 cm, white-woolly, with erect, dark brown, 20 mm bristles. Flowers solitary or paired, 60–75 × 60 mm, pale yellow-brown in bud; tube only 4–5 mm diam. at narrowest point; perianth-segments white, outer 30–35 × 3.5–8.0 mm, inner 20–24 × 6–9 mm. Fruit 30 × 4 mm, white. Seed 1–1.1 × 0.9–1.0 mm, tuberculate (Barthlott & Hunt 2000: pl. 44.1).

South-eastern *campo rupestre* (Grão Mogol) element: in quartz gravel and sand beneath shrubs in *campo rupestre*, c. 1000 m, Serra do Barão, northern Minas Gerais. Endemic to the core area within Minas Gerais. Map 50.
MINAS GERAIS: Mun. Grão Mogol, *Horst* 360 (U), 1987, *P.J. Braun* 851 (ZSS).

CONSERVATION STATUS. Endangered [EN B1ab(iii) + 2ab(iii)] (3); extent of occurrence < 100 km²; PD=2, EI=1, GD=1. Short-list score (3×4) = 12. Now known from two adjacent populations, one heavily impacted by collectors in the recent past and by ongoing habitat modification through quartz extraction. A protected area has recently been established (Parque Estadual da Serra do Barão) and should result in increased security, the authorities being well aware of the interest in the plant (M. Machado, *in litt.*, 20.05.2000).

A dwarf neotenic ally of *D. pseudoinsignis*, and perhaps the most remarkable of all Brazilian cacti. Heavily collected in the early 1970s for the European horticultural market (Buining 1974a: 70) and formerly regarded as 'Critically Endangered'. *D. woutersianus* Brederoo & Broek in Succulenta 59: 203 (1980) was said to have been based on material of *Horst* 360, but according to Riha in Kaktusy 26: 59 (1990) it is identifiable as the hybrid *D. horstii* × *D. pseudoinsignis* ['*D. insignis*'] originating in cultivation and now also reported from the wild (Uebelmann 1996: HU 1497).

The name *D. subnudus* Britton & Rose (1922: 217) was based on a photograph of a badly damaged plant said to emanate from the coast of Bahia. No *Discocactus* has subsequently been reported from coastal Brazil and the provenance of this plant and its identity remain doubtful. Luetzelburg (1925–26, 3: 69, 111) combined this epithet as *Echinocactus subnudus*.

30. UEBELMANNIA Buining

in Succulenta 46: 159–163 (1967). Type: *U. gummifera* (Backeb. & Voll) Buining (*Parodia gummifera* Backeb. & Voll).

Globose to elongate, usually single-stemmed, 5–100 cm; vascular cylinder not woody, outer cortex generally with gummiferous cells (disintegrating and fusing vertically into ducts in *U. gummifera*) and positioned within or close to the ribs. Seedlings globose, sometimes developing beneath the surface of the substrate. Stems erect to decumbent, 5–17 cm diam., epidermis bright to grey-green, whitish, reddish or dark brown, distinctly roughened; hypodermis with dense accumulations of pectin; ribs c. 13–40, broken or not into tubercles. Areoles with felt and white or brownish long hairs; spines hard, golden to greyish, central spines porrect, radials sometimes absent. Flower-bearing region of stem apical, not differentiated. Flowers diurnal, yellow or greenish, sometimes ageing to pale orange or pinkish, 0.8–2.5 × 0.6–3.0 cm; pericarpel and tube with bract-scales, hairs and eventually bristles; tube rotate; perianth-segments triangular to lanceolate, spreading to reflexed. Fruit narrowly turbinate, few-seeded, flower remnant drying whitish, erect; pericarpel with bract-scales and some glabrescent hairs, red; funicular pulp translucent. Seeds 1.1–2.4 mm, cochleariform, brown to black, with the hilum-micropylar region depressed; testa-cells flat or convex; cuticular folds absent.

Literature: Nyffeler (1997, 1998); Schulz & Machado (2000).

A remarkable, taxonomically isolated genus of 3 species, endemic to a relatively small region of central Minas Gerais (*campos rupestres, sensu lato*) and since 1992 placed in Appendix I of C.I.T.E.S. in view of its rarity and the potential risks to its survival from commercial and private collection in the wild. The distribution map for this genus (Map 51) is partly based on data kindly supplied as GPS co-ordinates by Marlon Machado, which are not given here in order to protect these rare plants from unscrupulous collectors.

It is assumed that the flowers are adapted for visits by hymenoptera, as observed by Schulz & Machado (2000), but Heek & Strecker (1995) have also noted hummingbirds visiting the flowers of *U. gummifera*. Plates 74.1–77.1.

1. Ribs not broken up into tubercles; flowers to 18 × 10 mm; fruit reddish to deep pink, conspicuous; seeds with flat testa-cells (Subg. *Leopoldohorstia*) . . . . **3. pectinifera**
1. Areoles borne on pronounced tubercles; flowers > 20 × 15 mm; fruit yellowish or greenish, inconspicuous; seeds with convex testa-cells (Subg. *Uebelmannia*) 2

2. Ribs 15–22; stem cortex with mucilage cells but lacking ducts . . . . **1. buiningii**
2. Ribs 22–42; stem cortex with vertically arranged mucilage ducts . . . . **2. gummifera**

Subg. *Uebelmannia*
See Key, above. Found only on quartz sands and gravels, or on rocks adjacent to these substrates.

1. Uebelmannia buiningii Donald in Nat. Cact. Succ. J. (UK) 23: 2–3 (1968); P. J. Braun & E. Esteves Pereira in Kakt. and. Sukk. 39: 2–3 (1988); van Heek & Strecker in ibid. 45: 234–235, illus. (1994). Holotype: Brazil, Minas Gerais, Mun. Itamarandiba, Serra Negra, Tromba d'Anta, 1966, 'c. 1000 m', Horst HU 141 (U; ZSS, iso.).

Globose to subcylindric, solitary, to 12 × 8 cm; without gummiferous ducts under the epidermis. Epidermis greenish red to brownish, rough. Ribs to 16–18(–22), strongly tuberculate; tubercles conical, deflexed, to 7.5 mm high and 6 mm broad, c. 5 mm apart. Areoles on top of the tubercles, 2 mm diam., with sparse greyish wool when young, later glabrescent. Spines yellow-brown, black tipped, soon becoming whitish, curved upwards, flattened on upper surface, central spines 4, to 15 mm, radials 0–4, to 5 mm. Flower to 27 × 25 mm; tube c. 9 mm, with bract-scales, white hairs and brownish bristles; perianth-segments deep yellow; style 14 mm; stigma-lobes 6–7. Fruit 5–6 × 4–5 mm, yellowish. Seed 1.2–1.4 × 0.8–0.9 mm, cochleariform to reniform, black; testa-cells convex (Barthlott & Hunt 2000: pl. 44.7–8).

South-eastern *campo rupestre* (Serra Negra) element: on slabs of quartzitic rock amongst gravel, *campo rupestre*, c. 1200 m, Serra Negra, Minas Gerais. Endemic to the core area within Minas Gerais. Map 51. **MINAS GERAIS:** Mun. Itamarandiba, 24 Aug. 1985, *Horst & Uebelmann* 141 (ZSS, K); Serra Mata Virgem, south slope, 4 Aug. 1988, *Eggli* 1074 (ZSS).

CONSERVATION STATUS. Critically Endangered [CR C2a] (4); PD=3, EI=1, GD=1. Short-list score (4×5) = 20. According to Braun & Esteves Pereira (l.c.) and Schulz & Machado (2000) this species is on the verge of extinction and is scarce in cultivation, where it has proved difficult. Affected by collection of plants and seed, fire and trampling by cattle.

Marlon Machado (pers. comm.) believes there are good reasons for continuing to accept this taxon as a species distinct from the following, but it is currently too rare to permit a proper survey of its described anatomical differences to be carried out.

2. Uebelmannia gummifera *(Backeb. & Voll) Buining* in Succulenta 46: 159–160 (1967). Holotype: Brazil, Minas Gerais, Serra do Ambrósio [Mun. Rio Vermelho], 1938, *Mello-Barreto* s.n. (RB 64065).

Parodia gummifera Backeb. & Voll. in Arq. Jard. bot. Rio de Janeiro 9: 158 (1949, publ. 1950).
Uebelmannia meninensis Buining in Kakt. and. Sukk. 19(8): 151–152 (1969); W. van Heek *et al.* in ibid. 46: 9–12 (1995). *U. gummifera* subsp. *meninensis* (Buining) P. J. Braun & E. Esteves Pereira in Succulenta 74: 226 (1995). Holotype: Brazil, Minas Gerais, [Mun. Rio Vermelho], Pedra Menina, *Horst* HU 108 (U).
U. meninensis var. *rubra* Buining in Krainz, Die Kakteen, Lfg. 55–56 (1975). *U. gummifera* var. *rubra* (Buining) P. J. Braun & E. Esteves Pereira in Succulenta 74: 226 (1995). Holotype: Brazil, Minas Gerais, Serra Negra, 1100–1200 m, *Horst* HU 406 (U).

VERNACULAR NAMES. Coroa-de-frade, cabeça-de-frade.

Globose to cylindric, elongate, solitary, 5–30(–40) × 6–15 cm; with gummiferous ducts under the epidermis. Epidermis greenish red to brownish, rough. Ribs to c. 30–40, strongly tuberculate; tubercles conical, deflexed, to 4.5 mm high and 4–5 mm broad, c. 4 mm apart. Areoles on top of the tubercles, 2 mm diam., when young with sparse greyish wool, later glabrescent. Spines greyish or whitish, porrect or curved upwards, flattened on upper surface, central spines 2–6, 10–25 mm long, radials 0–4, to 8 mm. Flower 20–25 × 18–30 mm; tube c. 10 mm, with triangular bract-scales and white hairs; perianth-segments deep yellow; style 12 mm; stigma-lobes 7–8. Fruit 6–8 × 4–6 mm, yellowish or greenish. Seed 1.1–1.3 × 0.8–0.9 mm, black or dark reddish brown; testa-cells convex.

South-eastern *campo rupestre* (Serra Negra/Serra do Ambrósio) element: in quartz sand, *campo rupestre*, 900–1600 m, Serra Negra, Minas Gerais. Endemic to the core area within Minas Gerais. Map 51. **MINAS GERAIS:** Mun. Itamarandiba, above Penha de França, 24 Aug. 1985, *Horst & Uebelmann* 282 (ZSS, K), 3 Aug. 1988, *Eggli* 1059 (ZSS); 25 km SW of town, 4 Aug. 1988, *Eggli* 1062 (ZSS); c. 20 km SSW of town, Serra d'Anta, 4 Aug. 1988, *Eggli* 1069 (ZSS); Mun. Rio Vermelho, Pedra Menina, 22 Aug. 1985, *Horst & Uebelmann* 406 (ZSS, K); l.c., c. 9 km SSE, (?) Rio Mundo Velho, 22 Aug. 1985, *Horst & Uebelmann* 108 (U, ZSS, K), 5 Aug. 1988, *Eggli* 1086, 1088 (ZSS); l.c., Serra do Ambrósio, 1938, *Mello-Barreto* s.n. (RB); l.c., Faz. do Sr José Batista, 15 July 1984, *A.M. Giulietti et al.* in CFCR 4511 (SPF, SP), 6 Mar. 1988, *Zappi & Prado* in CFCR 11821 (SPF).

CONSERVATION STATUS. Vulnerable [VU B1ab(iii, v)] (2); extent of occurrence = 198 km²; PD=3, EI=1, GD=2. Short-list score (2×6) = 12. Populations with often large or very large numbers of individuals, but overall range of species very limited and some habitats affected by charcoal production and collection of plants and seed. Local morphological variation significant.

The plant described as *U. meninensis* may be worthy of recognition as a subspecies (*fide* M. Machado), but has not been studied adequately in habitat by the authors.

Subg. *Leopoldohorstia* P. J. Braun & E. Esteves Pereira in Succulenta 74: 134 (1995). See Key, above. Found on or beside crystalline rocks. Type and only species:

3. Uebelmannia pectinifera *Buining* in Nat. Cact. Succ. J. (UK) 22: 86–87 (1967). Holotype: Brazil, Minas Gerais, Mun. Couto de Magalhães de Minas, Feb. 1966, *Horst* HU 106 (U).

U. pectinifera var. *pseudopectinifera* Buining in Kakt. and. Sukk. 23(5): 125 (1972). Holotype: Brazil, Minas Gerais, SE of Datas, towards Serro, *Horst* HU 280 (U).
U. pectinifera var. *multicostata* Buining & Brederoo in Krainz, Die Kakteen, Lfg 62 (1975). Holotype: Brazil, Minas Gerais, 10 km E of Mendanha towards the Rio Jequitinhonha, *Horst* HU 362 (U).
'U. ammotrophus' hort., *nom. nud.*

VERNACULAR NAME. Quiabo-da-lapa.

Globose to fusiform or cylindric, elongate, solitary or aggregated, 10–50(–100) × 10–17 cm; with gummiferous ducts under the epidermis. Epidermis green to grey-green, with or without waxy plates, rough. Ribs 13–29, straight, not tuberculate, to 5 mm high and 5–7 mm broad. Areoles 2 mm diam., very condensed, to 3 mm apart, with sparse greyish or brownish wool when young, later glabrescent. Spines greyish, whitish or yellowish brown, porrect or spreading, straight, flattened on upper surface, central spines 2–6, 10–40 mm, radials 0–3, spreading. Flower 8–16 × 6–10 mm; tube to c. 8 mm, with triangular bract-scales, white hairs and bristles; perianth-segments pale yellow, tinged reddish or greenish; style 6–7 mm; stigma-lobes 7–8. Fruit 15–25 × 6–8 mm, pericarp with few glabrescent or slightly woolly bract-scales, reddish to bright pinkish. Seed 1.7–2.4 × 1.3–1.6 mm, cochleariform, keeled, brown, shiny, smooth; testa cells flat.

CONSERVATION STATUS. Vulnerable [VU B1ab(iii, v)]; extent of occurrence < 20000 km². Although it has a total of at least 18 distinct locations/populations, the severe fragmentation of these, their generally very small size, the decline in the quality of habitat and ongoing collection of adult individuals warrants an assessment of Vulnerable.

A variable species, comprising a complex of numerous local forms, the following circumscription into subspecies possibly representing a convenient over-simplification of the situation in nature (cf. Schulz & Machado 2000):

1. Plants with grey-green, white-scaly epidermis; ribs 13–20(–26) . . . . **3a.** subsp. **pectinifera**
1. Plants with green epidermis; white scaly plates absent; ribs (16–)18–29 2

2. Spines yellow, ascending, organized in rows following the edge of the rib; plants to 50 cm
. . . . **3b.** subsp. **flavispina**
2. Spines greyish, spreading, not organized in rows; plants to 100 cm . . . . **3c.** subsp. **horrida**

3a. subsp. **pectinifera**

Plants 20–85 cm tall, with grey-green, white-scaly epidermis; ribs 13–20(–26). Spines greyish or whitish, ascendent, organized in rows following the edge of the ribs, central spines 2–4, 10–25 mm, radials 0–3.

South-eastern *campo rupestre* (Diamantina) element: crystalline rocks in *campo rupestre*, c. 650–1250 m, Serra do Espinhaço, north-east and south of Diamantina, Minas Gerais. Endemic to the core area within Minas Gerais. Map 51.
MINAS GERAIS: Mun. Couto de Magalhães de Minas, Feb. 1966, *Horst* HU 106 (U, ZSS, K); Mun. Presidente Kubitschek, near junction with road BR 259, 25 July 1998, *Hatschbach et al.* 68223 (K); l.c., Trinta Réis, Apr. 1986, *Zappi* (obs.); SE of Datas, towards Serro, *Horst* HU 280 (ZSS, K); Mun. Diamantina, 29 Aug. 1985, *Horst & Uebelmann* 362 (ZSS, K); l.c., 10 km E of Mendanha, *Horst & Uebelmann* 362 (U); l.c., Inhaí, reported by van Heek & Strecker (1994: 135).

CONSERVATION STATUS. Vulnerable [VU B1ab(iii, v)] (2); extent of occurrence = 765 km²; PD=3, EI=1, GD=2. Short-list score (2×6) = 12. Endangered from collection of plants and seed throughout its limited range, where its 9 known populations are mostly small (area of occupancy < 500 km²).

The form reported from Inhaí seems somewhat morphologically and geographically intermediate between all three subspecies recognized here.

3b. subsp. **flavispina** *(Buining & Brederoo) P. J. Braun & E. Esteves Pereira* in Succulenta 74: 135 (1995). Holotype: Brazil, Minas Gerais, Barão do Guaçuí, 1280 m, *Horst & Uebelmann* 361 (U).

Uebelmannia flavispina Buining & Brederoo in Succulenta 52: 9–10 (1973).
'*U. pectinifera* var. *crebrispina*' Strecker in Kakt. and. Sukk. 45: 34–35, illus. (1994), nom. nud. (*Horst & Uebelmann* 642).
'*U. warasii*' F. Ritter, nom. nud.

Plants to 50 cm tall, with green epidermis; ribs (16–)21–29. Spines yellow, ascendent, organized in rows following the edge of the rib, central spines 2–4, 10–25 mm.

South-eastern *campo rupestre* (Diamantina) element: crystalline rocks in *campo rupestre*, c. 1200–1350 m, Serra do Espinhaço, west of Diamantina, Minas Gerais. Endemic to the core area within Minas Gerais. Map 51.
MINAS GERAIS: Mun. Diamantina, 8 km on the road to Conselheiro Mata, 18°15'S, 43°44'W, 30 Oct. 1988, *Taylor & Zappi in Harley* 25454 (K, SPF), 17 Apr. 1987, *Zappi et al. in CFCR* 10593 (SPF), 29 July 1988, *Horst & Uebelmann* 854 (ZSS, K); [N of] Barão do Guaçuí, *Horst & Uebelmann* 361 (U, ZSS, K), 20 Aug. 1985, *Horst & Uebelmann* 642 (ZSS, K).

CONSERVATION STATUS. Vulnerable [VU B1ab(iii, v)] (2); extent of occurrence = 131 km²; PD=3, EI=1, GD=2. Short-list score (2×6) = 12. Subjected to the collection of plants and seed throughout its very restricted range, comprising c. 6 populations of variable size, but mostly rather small.

3c. subsp. **horrida** *(P. J. Braun) P. J. Braun & E. Esteves Pereira* in Succulenta 74: 135 (1995). Holotype: Brazil, Minas Gerais, Mun. Bocaiúva, near Sítio, 1982, *Horst & Uebelmann* 550 (ZSS; K, iso.).

Uebelmannia pectinifera var. *horrida* P. J. Braun in Kakt. and. Sukk. 35: 264–266 (1984).

Plants to 100 cm tall, with green epidermis; ribs 23–27. Spines greyish, spreading, not organized in rows; central spines 3–6, 20–40 mm.

South-eastern *campo rupestre* (Rio São Francisco drainage) element: on sandstone outcrops on the western slopes of the Serra do Espinhaço ('Serra Mineira'), c. 700–850 m, Mun. Bocaiúva, Minas Gerais. Endemic to the core area within Minas Gerais. Map 51.
MINAS GERAIS: Mun. Bocaiúva, NE of Engenheiro Dolabela, near Sítio, 1982, *Horst & Uebelmann* 550 (ZSS, K), 8 Aug. 1988, *Eggli* 1124 (ZSS).

CONSERVATION STATUS. Vulnerable [VU D2] (2); PD=3, EI=1, GD=1. Short-list score (2×5) = 10. Known from only a single small group of localities, close to agricultural activities and highly sought-after by collectors, but currently not experiencing marked decline (Schulz & Machado 2000).

Apparently disjunct from the remainder of the genus, but the region in between is very poorly known.

The untypifiable name *Uebelmannia centeteria* (Lehm. ex Pfeiff.) Schnabel (*Echinocactus centeterius* Lehm. ex Pfeiff., Enum. Cact.: 65. 1837) is doubtfully referred to the Chilean species, *Eriosyce curvispina* (Colla) Kattermann, by Hunt *et al.* (1994: 146). It was stated by Pfeiffer, l.c., to have originated from Minas Gerais, but this is assumed to be an error.

6. BIBLIOGRAPHY AND ACKNOWLEDGEMENTS

6.1 LITERATURE CITED OR CONSULTED

Ab'Sáber, A. N. (1974). O domínio morfoclimático semi-árido das caatingas brasileiras. *Geomorphologia (São Paulo)* 43: 1–39.

Agra, M. de F. (1996). *Plantas da medicina popular dos Cariris Velhos, Paraíba, Brasil.* Editora União, João Pessoa. Pp. 125.

Anderson, E. F. (1999). The IOS Cactaceae Working Party: A Major Contribution of the IOS. *I.O.S. Bull.* 7: 18.

Andrade-Lima, D. de (1950). Catálogo do Herbário da Escola Superior de Agricultura da Tapera. *Bol. Secr. Agric. (Recife)* 17: 279–281.

—— (1966). Cactaceae de Pernambuco (Cactaceous plants of Pernambuco). *Anais do IX Congresso Int. Pastagens (São Paulo)* 2: 1453–1458.

—— (1975). A vegetação da bacia do Rio Grande, Bahia. Nota preliminar. *Revta Brasil. Biol.* 35: 223–232.

—— (1977). A flora de áreas erodidas de calcário Bambuí, em Bom Jesus da Lapa, Bahia. *Ibid.* 37: 179–194.

—— (1981). The caatingas dominium [*sic*]. *Revta Brasil. Bot.* 4: 149–163.

—— (1989). *Plantas das Caatingas.* Academia Brasileira de Ciências, Rio de Janeiro. Pp. 243.

Assis, J.G.A., Oliveira, A.L.P.C., Resende, S.V., Senra, J.F.V. & Machado, M. (2003). Chromosome numbers in Brazilian *Melocactus* (Cactaceae). *Bradleya* 21: 1–6.

Azevedo, A. de, ed. (1972). *Brasil, a terra e o homem,* ed. 2, 1. Companhia Editora Nacional, São Paulo.

Backeberg, C. (1938). *Blätter für Kakteenforschung* 1938(6): [unpaged].

—— (1942). Cactaceae Lindley. *Jahrb. Deutsch. Kakteen Ges.* 1941(2): 1–80 (publ. 1942).

—— (1950). Nova genera et subgenera. *Cact. Succ. J. (US)* 22: 153–154.

—— (1958–62). *Die Cactaceae. Handbuch der Kakteenkunde,* 1–6. VEB Gustav Fischer Verlag, Jena.

—— (1966). *Das Kakteenlexikon. Enumeratio diagnostica Cactacearum.* Ibid.

—— & Knuth, F. M. (1936). *Kaktus-ABC.* Gyldendalske Boghandel-Nordisk Forlag, Copenhagen. Pp. 432 (1935, publ. 1936).

Bandeira, R. L. (1995). *Chapada Diamantina, história, riquezas e encantos.* Onavlis Editora, Salvador, Bahia.

Barthlott, W. (1976). In: Löve, A., IOPB chromosome reports, LIV. *Taxon* 25: 644–645.

—— (1977). *Kakteen*. Belser, Stuttgart.

—— (1979). *Cacti*. Stanley Thornes, Cheltenham, U.K.

—— (1987). New names in Rhipsalidinae (Cactaceae). *Bradleya* 5: 97–100.

—— (1988). Über die systematischen Gliederung der Cactaceae. *Beitr. Biol. Pflanz.* 63: 17–40.

—— & Hunt, D. R. (1993). Cactaceae. In: Kubitsky, K., *The Families and Genera of Vascular Plants,* 2: 161–197. Springer Verlag, Berlin etc.

—— & —— (2000). Seed diversity in the Cactaceae subfamily Cactoideae. *Succulent Pl. Res.* 5: 1–173.

——, Poremski, S., Kluge, M., Hopke, J. & Schmidt, L. (1997). *Selenicereus wittii* (Cactaceae): an epiphyte adapted to Amazonian Igapó inundation forests. *Pl. Syst. Evol.* 206: 175–185.

—— & Rauh, W. (1975). Notes on the morphology, palynology and evolution of the genus *Schlumbergera*. *Cact. Succ. J. (US), Yearbook* 1975: 5–21.

—— & Taylor, N. P. (1995). Notes towards a monograph of Rhipsalideae (Cactaceae). *Bradleya* 13: 43–79.

—— —— (1996). *Hatiora* — die Osterkakteen und ihre Verwandten. *Kakt. and. Sukk.* 47: 73–77.

—— & Voit, G. (1979). Mikromorphologie der Samenschalen und Taxonomie der Cactaceae. *Pl. Syst. Evol.* 132: 205–229.

Bauer, R. (2003). A synopsis of the tribe Hylocereeae F. Buxb. *Cact. Systematics Initiatives* 17: 3–64.

Benson, L. (1969). The cacti of the United States and Canada — new names and nomenclatural combinations. *Cact. Succ. J. (US)* 41: 124–128.

—— (1982). *The Cacti of the United States and Canada*. Stanford Univ. Press, California. Pp. 1044.

Berger, A. (1905). A systematic revision of the genus *Cereus*. *Ann. Rep. Missouri Bot. Gard.* 16: 57–86.

—— (1920). Einiges über *Rhipsalis*. *Monatsschr. Kakt.-Kunde* 30: 1–4.

—— (1926). *Die Entwicklungslinien der Kakteen*. Gustav Fischer, Jena.

—— (1929). *Kakteen*. Ulmer Verlag, Stuttgart.

Bohle, B. (2000). In der Chapada Diamantina, Brasilien. *Kakt. and. Sukk.* 51: 43–48.

Boyle, T. H. & Idnurm, A. (2001). Physiology and genetics of self-incompatibility in *Echinopsis chamaecereus* (Cactaceae). *Sex Plant Reprod.* 13: 323–327.

Bradley, R. (1716–1727). *The History of Succulent Plants*. W. Mears, London.

Brandão, M. (2000). Caatinga. In: Mendonça, M. P. & Lins, L. V., *Lista Vermelha das Espécies Ameaçadas de Extinção da Flora de Minas Gerais*. Fundação Biodiversitas, Fundação Zoo-Botânica de Belo Horizonte, Minas Gerais, Brazil. Pp. 160.

Brasil (1998). *Primeiro Relatório Nacional para a Convenção sobre Diversidade Biológica. Brasil*. Ministério do Meio Ambiente, dos Recursos Hídricos e da Amazônia Legal, Brasília, D.F., Brazil. Pp. 283.

—— (2000). *Política Nacional de Biodiversidade. Roteiro de consulta para elaboração de uma proposta*. Ibid., ibid. Pp. 48.

Braun, P. J. (1986). *Melocactus lanssensianus* P. J. Braun species nova — een nieuwe soort uit Pernambuco, Brazilië. *Succulenta* 65: 29.

—— (1988a). Eindrücke aus dem nordostbrasilienischen Sertão. *Kakt. and. Sukk.* 39: 169–171.

—— (1988b). On the taxonomy of Brazilian Cereeae (Cactaceae). *Bradleya* 6: 85–99.

—— (1990). Arquipélago de Fernando de Noronha (Brasilien) — Eindrücke von einer Reise zu einem der letzten Inselparadiese und zum Standort von *Cereus insularis* Hemsley. *Kakt. and. Sukk.* 41: 254–258.

—— & Esteves Pereira, E. (1986–89). Revision der Gattung *Facheiroa* Britton et Rose. *Ibid.* 37: 56, 74–79; 38: 26–33, 82–85, 184–187; 39: 64–68, 126–131; 40: 198–203, 298–301.

—— —— (1988). *Uebelmannia buiningii* Donald — eine bald ausgerottete Art. *Ibid.* 39: 2–3.

—— —— (1989). The *Opuntia inamoena* complex in Brazil, I. *Cact. Succ. J. (US)* 61: 268–273.

—— —— (1990). *Siccobaccatus* Braun et Esteves — Een nieuw cactus geslacht uit Brasilië. *Succulenta* 69: 3–8.

—— —— (1991a). *Micranthocereus* Backeberg subgenus *Austrocephalocereus* (Backcbcrg) Braun & Esteves. Een nieuw ondergeslacht (Cactaceae) uit Brazilië. *Ibid.* 70: 62–67.

—— —— (1991b). *Arrojadoa dinae* Buining & Brederoo var. *nana* Braun & Esteves. Eine neue *Arrojadoa*-Sippe aus Minas Gerais / Brasilien. *Kakt. and. Sukk.* 42: 190–195.

—— —— (1992). *Pilosocereus rizzoianus* Braun & Esteves Pereira, a new species of Cactaceae from Goiás, Brazil. *Cact. Succ. J. (US)* 64: 148–155.

—— —— (1995a). The *Opuntia inamoena* complex in Brazil, VI. *Ibid.* 67: 111.

—— —— (1995b). Nieuwe combinaties en namen voor Cactussen uit Brazilië, Bolivia en Paraquay. *Succulenta* 74: 81–85, 130–135, 226.

—— —— (1999a). *Pilosocereus chrysostele* (Vaupel) Byles & Rowley subsp. *cearensis* P. J. Braun & Esteves — A new subspecies of *Pilosocereus* from the state of Ceará, Brazil. *Brit. Cact. Succ. J.* 17: 21–27.

—— —— (1999b). Eine neue Kakteensippe aus Bahia, Brasilien. *Pilosocereus densiareolatus* subsp. *brunneolanatus* P. J. Braun & Esteves. *Kakt. and. Sukk.* 50: 283–289.

—— —— (2002). Kakteen und andere Sukkulenten in Brasilien. *Schumannia* 3: 1–235.

—— —— (2003). Brazil and its Columnar Cacti — 70 years after Werdermann. *Kaktusy* (CZ) 39 (Special 2002/1): 1–47.

Britton, N. L. & Rose, J. N. (1919–1923). *The Cactaceae*, 1–4. Carnegie Institution, Washington, D.C. (Vol. 1, 1919, vol. 2, 1920, vol. 3, 1922, vol. 4, 1923.)

Brummitt, R. K. (1994). Report of the Committee for Spermatophyta: 41. *Taxon* 43: 271–277.

—— (1996). Ibid.: 44. *Ibid.* 45: 677–678.

—— & Taylor, N. P. (1990). To correct or not to correct? *Ibid.* 39: 298–306.

Buining, A. F. H. (1967). Een nieuw geslacht der Cactaceae uit Minas Gerais. *Succulenta* 46: 159–163.

—— (1974a). *Discocactus* in Brazil. *I.O.S. Bull.* 3(3): 67–72.

—— (1974b). *Melocactus ernestii* Vaupel. In: Krainz, H., *Die Kakteen*, Lfg 58.

—— (1975a). *Melocactus florschützianus* [sic] Buin. et Bred. spec. nova. *Ashingtonia* 2: 25–27.

—— (1975b). *Austrocephalocereus purpureus* (Gürke) non Backeberg. *Ibid.* 28–29.

—— (1980). *The genus* Discocactus *Pfeiffer.* Buining-fonds, Succulenta, The Netherlands.

—— & Brederoo (1974a). *Pseudopilocereus superfloccosus* Buin. & Brederoo sp. nova. *Cact. Succ. J. (US)* 46: 60–63, 96.

—— & —— (1974b). A new *Discocactus* species from Brazil. *Ibid.* 252–257.

Burnham, R. J. (1995). A new species of winged fruit from the Miocene of Ecuador: *Tipuana ecuatoriana* (Leguminosae). *Amer. J. Bot.* 82: 1599–1608.

—— & Graham, A. (1999). The history of neotropical vegetation: new developments and status. *Ann. Missouri Bot. Gard.* 86: 546–589.

Buxbaum, F. (1950–54). *Morphology of cacti*. Parts I–III. Abbey Garden Press, Pasadena.

—— (1958). The phylogenetic division of the subfamily Cereoideae [*sic*], Cactaceae. *Madroño* 14: 177–206.

—— (1959). Die behaartblütigen Cephalienträger Südamerikas. *Oesterr. Bot. Zeitschr.* 106: 138–158.

—— (1962). Das phylogenetische System der Cactaceae. In: Krainz, H., *Die Kakteen*, Lfg. 21.

—— (1968a). Die Entwicklungslinien der Tribus Cereeae Britt. et Rose emend. F. Buxb. (Cactaceae, Cactoideae), 1. *Beitr. Biol. Pflanzen* 44: 215–276.

—— (1968b). Ibid., 2. Versuch einer phylogenetischen Gliederung innerhalb der Tribus. *Ibid.* 389–433.

—— (1971). *Buiningia*. In: Krainz, H., *Die Kakteen*, Lfg. 46–47.

—— (1972). *Coleocephalocereus luetzelburgii*. *Ibid.,* Lfg. 48–49.

—— (1975). Provisorium einer Neugliederung der Tribus Notocacteae (C VI). *Ibid.,* Lfg. 60.

Byles, R. S. & Rowley, G. D. (1957). *Pilosocereus* Byl. & Rowl. nom. gen. nov. (Cactaceae). *Cact. Succ. J. Gr. Brit.* 19: 66–67, 69.

Cavalcante Bernardes, L. M. (1951). Notas sobre o clima da bacia do São Francisco. *Revta Bras. Geogr.* 13: 473–479.

Costa, C. M. R., Hermann, G., Martins, C. S., Lins, L. V. & Lamas, I. R. (1998). *Biodiversidade em Minas Gerais: um atlas para sua conservação*. Biodiversitas, Belo Horizonte. Pp. 94.

Das, A. B., Mohanty, S. & Das, P. (1998a). Variation in Karyotype and 4C DNA content in six species of *Melocactus* of the family Cactaceae. *Cytologia* 63: 9–16.

—— —— —— (1998b). Interspecific variation in Nuclear DNA content and chromosome analysis in *Melocactus*. *Ibid.* 63: 239–247.

Davis, S. D., Heywood, V. H., Herrera-MacBryde, O., Villa-Lobos, J., & Hamilton, A. C. (1997*). Centres of Plant Diversity. A guide and Strategy for their Conservation,* 3. *The Americas*. The World Wide Fund For Nature (WWF) and IUCN – The World Conservation Union, IUCN Publications Unit, Cambridge, U.K.

Dickie, J. B., Ellis, R. H., Kraak, H. L., Ryder, K. & Tompsett, P. B. (1990). Temperature and seed storage longevity. *Ann. Bot.* 65: 197–204.

Dobat, K. & Piekert-Holle, T. (1985). *Blüten und Fledermäuse:* 238–243. Senckenbergischen Naturforschenden Gesellschaft, Frankfurt.

Domingues, A. J. P. (1973). Relevo. Pp. 11–27 in Instituto Brasileiro de Geografia, *"Novo" Paisagens do Brasil*, ed. 2 (Biblioteca Geográfica Brasileira, Serie D, publ. no. 2). Ministério do Planejamento e Coordenação Geral, Rio de Janeiro.

Doweld, A. & Greuter, W. (2001). Nomenclatural notes on *Notocactus* and on Alwyn Berger's "Kakteen". *Taxon* 50: 879–885.

Eerkens, G. (1983). Braziliaanse Melocactussen met een HU nummer. *Succulenta* 62: 275–278.

—— (2000). *Melocactus* species HU 534. *Ibid*. 79: 274–275.

Eggli, U., Muñoz Schick, M. & Leuenberger, B. E. (1995). Cactaceae of South America: The Ritter Collections. *Englera* 16: 1–646.

—— & Taylor, N., eds (1991). *List of Cactaceae names from Repertorium Plantarum Succulentarum (1950–1990)*. Royal Botanic Gardens, Kew, Richmond, U.K. Pp. 222.

Egler, W. (1951). Contribuição ao estudo da caatinga Pernambucana. *Revta Bras. Geogr.* 13: 577–590.

Eiten, G. (1983). *Classificação da vegetação do Brasil*. Conselho Nacional de Desenvolvimento Científico e Tecnológico (CNPq), Brasília.

Endler, J. & Buxbaum, F. (1974). *Die Pflanzenfamilie der Kakteen*, ed. 3. A. Philler, Minden.

Esteves Pereira, E. (1997). *Pierrebraunia* Esteves, a new genus of Cactaceae from Brazil. *Cact. Succ. J. (US)* 69: 296–302.

Farjon, A. & Page, C. N., comps (1999). *Conifers — Status Survey and Conservation Action Plan*. IUCN/SSC Conifer Specialist Group, IUCN, Gland, Switzerland and Cambridge, U.K.

Figueira, J. E. C., Vasconcellos-Neto, J., Garcia, M. A. & Teixeira de Souza, A. L. (1993). O cactus e o lagarto. *Ciência Hoje* (BR) 15(89): 12–13.

—— —— —— —— (1994). Saurocory in *Melocactus violaceus* (Cactaceae). *Biotropica* 26: 295–301.

Funk, V. A. & Brooks, D. R. (1990). Phylogenetic Systematics as the Basis of Comparative Biology. *Smithsonian contributions to botany,* 73. Pp. 45.

Gibson, A. C. & Horak, K. E. (1978). Systematic anatomy and phylogeny of Mexican columnar cacti. *Ann. Missouri Bot. Gard*. 65: 999–1057 (1977, publ. 1978).

Giulietti, A. M. & Pirani, J. R. (1988). Patterns of geographic distribution of some plant species from the Espinhaço range, Minas Gerais and Bahia, Brazil. In: Heyer, W. R. & Vanzolini, P. E., eds, *Proceedings of a Workshop on Neotropical Distribution Patterns:* 39–69. Academia Brasileira de Ciências, Rio de Janeiro.

Glaziou, A. F. M. (1909). Plantae Brasiliae centralis a Glaziou lectae. *Mem. Soc. Bot. France* 1(3): 325–327.

Golding, J. S. & Smith, P. P. (2001). A 13-point strategy to meet conservation challenges. *Taxon* 50: 475–477.

Greuter, W., McNeill, J., Barrie, F. R., Burdet, H. M., Demoulin, V., Filgueiras, T. S., Nicolson, D. H., Silva, P. C., Skog, J. E., Trehane, P., Turland, N. J. & Hawksworth, D. L., eds (2000). International Code of Botanical Nomenclature (Saint Louis Code). *Regnum Vegetabile,* 138.

Gürke, M. (1908). Neue Kakteen-Arten aus Brasilien. *Monatsschr. Kakt.-Kunde* 18: 52–57.

Gusmão, M. (1999). O sertão virou pó. *Veja* (BR) 32(35): 122–125 (publ. 1 Sep. 1999).

Gutman, F., Bar-Zvi, D., Nerd, A. & Mizrahi, Y. (2001). Molecular typing of *Cereus peruvianus* clones and their genetic relationships with other *Cereus* species evaluated by RAPD analysis. *J. Hort. Sci. Biotech.* 76: 709–713.

Harley, R. M. (1988). Evolution and distribution of *Eriope* (Labiatae), and its relatives, in Brazil. In: Heyer, W. R. & Vanzolini, P. E., eds, *Proceedings of a Workshop on Neotropical Distribution Patterns:* 71–120. Academia Brasileira de Ciências, Rio de Janeiro.

—— & Mayo, S. J. (1980). *Towards a Checklist of the flora of Bahia.* Royal Botanic Gardens, Kew, Richmond, U.K. Pp. 250.

—— & Simmons, N. A. (1986). *Florula of Mucugê.* Royal Botanic Gardens, Kew, Richmond, U.K. Pp. 228.

Heath, P. V. (1992). The type of *Monvillea* Britton & Rose (Cactaceae). *Taxon* 41: 85–87.

—— (1994). New combinations in *Pseudopilocereus* Buxbaum. *Calyx* 4: 140–141.

Heek, W. van & Strecker, W. (1994). An den Standorten von *Uebelmannia,* 4. *Uebelmannia pectinifera* Buining var. *multicostata* Buining & Brederoo (HU 362). *Kakt. and. Sukk.* 45: 134–136.

—— —— (1995). *Uebelmannia gummifera* (Backeberg & Voll) Buining HU 282 und HU 859. *Ibid.* 46: 121–123.

—— —— (2002). Am Fundort von *Micranthocereus streckeri. Ibid.* 53: 113–116.

Herm, K., Hofacker, A., Charles, G., Heek, W. van, Bohle, B., Strecker, W. & Heimen, G. (2001). *Kakteen in Brasilien / Cacti in Brazil.* Published by the authors (ISBN 3-00-007573-9). Pp. 176.

Herschkovitz, M. A. & Zimmer, E. A. (1997). On the evolutionary origin of the cacti. *Taxon* 46: 217–232.

Hoehne, F. C. (1915). Annexo 5, História Natural: Botânica 6: 55–56. In: *Commissão de Linhas Telegráphicas Estratégicas de Matto Grosso ao Amazonas.* Rio de Janeiro.

Hofacker, A. (2001). *Pilosocereus bohlei* Hofacker spec. nov. – eine neue Art aus Bahia. *Kakt. and. Sukk.* 52: 253–257, illus.

—— (2002). Der Formenkreis um *Discocactus bahiensis* Britton & Rose in Bahia, Brasilien. *Ibid.* 53: 200–205, illus.

—— & Braun, P. J. (1998). Staubfäden in zwei Reihen: Die Gattung *Brasilicereus* Backeberg. *Ibid.* 49: 265–268, illus.

Hooker, J. D. (1899). *Cereus paxtonianus. Curtis's Bot. Mag.* 125: tab. 7648.

Hooker, W. J. (1838). *Melocactus depressus. Ibid.* 65: tab. 3691.

Howard, R. (1989). *Flora of the Lesser Antilles, Leeward and Windward Islands,* 5: 398–422. Arnold Arboretum, Harvard Univ., Jamaica Plain, Massachusetts.

—— & Touw, M. (1982). *Opuntia* species in the Lesser Antilles. *Cact. Succ. J. (US)* 54: 170–179.

Hunt, D. R. (1967). Cactaceae. In: Hutchinson, J., *Families and Genera of Flowering Plants*, ed. 2. Oxford.

—— (1969). A synopsis of *Schlumbergera* Lem. (Cactaceae). *Kew Bull.* 23: 255–263.

—— (1984). The Cactaceae of Plumier's Botanicum Americanum. *Bradleya* 2: 39–64.

—— (1988). New and unfamiliar names for use in the European Garden Flora: Addenda and Corrigenda. *Ibid.* 6: 100.

—— (1989). Notes on *Selenicereus* (A. Berger) Britton & Rose and *Aporocactus* Lemaire (Cactaceae-Hylocereinae). *Ibid.* 7: 89–96.

—— (1991). Stabilization of Names in Succulent Plants. Pp. 151–155, in: Hawksworth, D. L., Improving the Stability of Names: needs and options. *Regnum Vegetabile*, 123. Koeltz Scientific Books, Königstein.

—— (1992a). *Selenicereus*. A further note on the 'Pitaya'. In: Hunt, D. R. & Taylor, N. P., eds, Notes on miscellaneous genera of Cactaceae (2). *Bradleya* 10: 32.

—— (1992b). *CITES Cactaceae Checklist*. Royal Botanic Gardens, Kew, Richmond, U.K.

—— (1999a). *Ibid.*, ed. 2. Ibid.

—— (1999b). The opuntioids: all sections or all genera? *Cact. Consensus Initiatives* 8: 3–6.

—— (2000). Classification of subfam. Cactoideae. *Cact. Systematics Initiatives* 9: 3.

—— & Taylor, N. P., eds (1986). The genera of the Cactaceae: towards a new consensus. *Bradleya* 4: 65–78.

—— —— (1990). The genera of Cactaceae: progress towards consensus. *Ibid.* 8: 85–107.

—— ——, eds (1991). Notes on miscellaneous genera of Cactaceae. *Ibid.* 9: 81–92.

—— ——, eds (1992). Notes on miscellaneous genera of Cactaceae (2). *Ibid.* 10: 17–32.

—— —— & Zappi, D. (1994). Appendix IV: Documentation of accepted names and synonyms. In: Kattermann, F., *Eriosyce* (Cactaceae), The genus revised and amplified. *Succulent Pl. Res.* 1: 131–169.

Ibisch, P. L., Boegner, A., Nieder, J. & Barthlott, W. (1996). How diverse are neotropical epiphytes? An analysis based on the "Catalogue of the Flowering Plants and Gymnosperms of Peru". *Ecotropica* 2: 13–28.

Innes, C. & Glass, C. (1991). *Illustrated Encyclopaedia of Cacti*. Quarto Publishing, London. Pp. 320.

IUCN (2001). *IUCN Red List Categories: Version 3.1*. IUCN, Gland, Switzerland and Cambridge, UK. Pp. 23.

Judd, W. S., Campbell, C. S., Kellogg, E. A. & Stevens, P. F. (1999). *Plant Systematics. A phylogenetic approach*. Sinauer Associates, Sunderland, Massachusetts. Pp. 464.

Kiesling, R. (1986). Tipos de Cactáceas Sudamericanas en herbarios extranjeros. *Bol. Soc. Argent. Bot.* 24: 381–386.

—— (1990). *Cactus de la Patagonia*. Reprinted with English summary from *Flora Patagónica* 5: 218–243 (1988).

—— (1994). *Monvillea kroenleinii*, a new species from Paraguay. *Cact. Succ. J. (US)* 66: 157–165.

—— (1999). Origin, Evolution and Distribution of *Opuntia ficus-indica*. *XVI International Botanical Congress Abstracts:* 36.

King, L. C. (1956). A Geomorfologia do Brasil Oriental. *Revta Bras. Geogr.* 18: 147–265.

Krook, R. & Mottram, R. (2001). *Opuntia* Index. Part 7: Nomenclatural note and P–Q. *Bradleya* 19: 91–116.

Labouret, J. (1853). *Monographie des Cactées*. Dusacq, Librarie Agricole de la Maison Rustique, Paris.

Lemaire, C. C. (1838). *Cactearum aliquot novarum ac insuetarum in Horto Monvilliano cultarum accurata descriptio*. Levrault, Paris. Pp. 40.

—— (1862). Histoire et Révision du Gènre *Pilocereus*. *Rev. Hort.* (Paris) 34: 426–430.

Leme, E. M. C. & Marigo, L. C. (1993). *Bromeliads in the Brazilian wilderness*. Marigo Comunicação Visual, Rio de Janeiro.

Leuenberger, B. E. (1976). Die Pollenmorphologie der Cactaceae und ihre Bedeutung für die Systematik. *Diss. Bot.* 31: 1–321.

—— (1978). Type specimens of Cactaceae in the Berlin-Dahlem Herbarium. *Cact. Succ. J. Gr. Brit.* 40: 101–104.

—— (1986). *Pereskia* (Cactaceae). *Mem. New York Bot. Gard.* 41: 1–140.

—— (1991). Interpretation of *Cactus ficus-indica* L. and *Opuntia ficus-indica* (L.) Miller (Cactaceae). *Taxon* 40: 621–627.

—— (1993). Interpretation of *Cactus opuntia* L., *Opuntia vulgaris* Mill., and *O. humifusa* (Rafin.) Rafin. (Cactaceae). *Ibid.* 42: 419–429.

—— (1997). Cactaceae. In: Görts van Rijn, A. R. A. & Jansen-Jacobs, M. J., eds, *Flora of the Guianas*, 31(18). Royal Botanic Gardens, Kew, Richmond, U.K.

—— (2002). The South American *Opuntia* ser. *Armatae* (= *O.* ser. *Elatae*) (Cactaceae). *Bot. Jahrb. Syst.* 123: 413–439.

Lewis, G. P. (1987). *Legumes of Bahia*. Royal Botanic Gardens, Kew, Richmond, U.K. Pp. 369.

—— (1998). *Caesalpinia. A revision of the Poincianella-Erythrostemon Group*. Ibid. Pp. 233.

Linder, H. P. (1995). Setting Conservation Priorities: The Importance of Endemism and Phylogeny in the Southern African Orchid Genus *Herschelia*. *Conserv. Biol.* 9: 585–595.

Linke, A. (1858). *Nachtrag zum Verzeichniss der Cacteen-Sammlung bei August Linke*. Berlin.

Locatelli, E. & Machado, I. C. (1999a). Comparative Study of the Floral Biology in Two Ornithophilous Species of Cactaceae: *Melocactus zehntneri* and *Opuntia palmadora*. *Bradleya* 17: 75–85.

—— & —— (1999b). Floral Biology of *Cereus fernambucensis*: a sphingophilous cactus of restinga. *Ibid.* 17: 86–94.

—— —— & Medeiros, P. (1997). Floral Biology and Bat Pollination in *Pilosocereus catingicola* (Cactaceae) in Northeastern Brazil. *Ibid.* 15: 28–34.

Luetzelburg, P. von (1925–26). *Estudo Botânico do Nordéste*, 1–3. Inspetoria Federal de Obras Contra as Secas, Rio de Janeiro (dated 1922–23, but vol. 1 publ. 1925, vols 2–3 publ. 1926).

Madsen, J. E. (1989). 45. Cactaceae. In: Harling, G. & Andersson, L., *Flora of Ecuador*, 35. Pp. 79.

Maffei, M., Meregalli, M. & Scannerini, S. (1997). Chemotaxonomic significance of surface wax *n*-alkanes in the Cactaceae. *Biochem. Syst. Ecol.* 25: 241–253.

Martius, C. P. F. von (1832). Beschreibung einiger neuen Nopaleen. *Nova Acta Leop.* 16: 321–362.

—— (1841). *Flora brasiliensis*, 1(1). Oldenbourg, Munich.

Mauseth, J. D. (1996). Comparative Anatomy of Tribes Cereeae and Browningieae (Cactaceae). *Bradleya* 14: 66–81.

McMillan, A. J. S. & Horobin, J. (1995). Christmas Cacti: The genus *Schlumbergera* and its hybrids. *Succulent Pl. Res.*, 4.

Menezes, A. I. (1949). *Flora da Bahia*. Companhia Ed. Nacional, São Paulo.

Mittler, L. (1844). *Taschenbuch für Cactusliebhaber*, 2. Schreck, Leipzig.

Moser, G. (1985). *Kakteen. Adolfo Maria Friedrich und sein schönes Paraguay*. G. Moser, Kufstein, Austria.

Myers, N., Mittermeier, R. A., Mittermeier, C. G., Fonseca, G. A. B. da & Kent, J. (2000). Biodiversity hotspots for conservation priorities. *Nature* 403: 853–858.

Nassar, J. M., Hamricks, J. L. & Fleming, T. H. (2002). Genetic variation and population structure of the mixed-mating cactus, *Melocactus curvispinus* (Cactaceae). *Heredity* 87: 69–79.

Nimer, E. (1973). Clima. Pp. 33–49 in Instituto Brasileiro de Geográfia, *"Novo" Paisagens do Brasil*, ed. 2 (Biblioteca Geográfica Brasileira, Serie D, publ. no. 2). Ministério do Planejamento e Coordenação Geral, Rio de Janeiro.

Nixon, K. C. & Wheeler, Q. D. (1990). An amplification of the phylogenetic species concept. *Cladistics* 6: 211–223.

Nyffeler, R. (1997). Stem anatomy of *Uebelmannia* (Cactaceae) — with special reference to *Uebelmannia gummifera*. *Bot. Acta* 110: 489–495.

—— (1998). The genus *Uebelmannia* Buining (Cactaceae: Cactoideae). *Bot. Jahrb. Syst.* 120: 145–163.

—— (1999). The Evolution of Stem Characters in the Cacti. *I.O.S. Bull.* 7: 27.

—— (2000a). A Sketch of the Cactus Tribe[s] — What do the Molecules Tell Us? *I.O.S. Bull.* 8: 25.

—— (2000b). Molecular Systematics and the Polyphyly of the Tribe Notocacteae (Cactaceae). *Ibid.* 8: 25–26.

—— (2000c). Should *Pfeiffera* be resurrected? *Cact. Systematics Initiatives* 10: 10–11.

—— (2001). What about the tribe Notocacteae? *Ibid.* 12: 25–27.

—— (2002). Phylogenetic relationships in the cactus family (Cactaceae) based on evidence from trnK/matK and trnL-trnF sequences. *Amer. J. Bot.* 89: 312–326.

—— & Eggli, U. (1996). Berger's "Kakteen" — the end of a nomenclatural nightmare in sight? *Taxon* 45: 301–304.

Oldfield, S., comp. (1997). *Cactus and Succulent Plants — Status Survey and Conservation Action Plan*. IUCN/SSC Cactus and Succulent Specialist Group. IUCN, Gland, Switzerland and Cambridge, U.K. Pp. 10 + 212.

O'Leary, M. C. & Boyle, T. H. (1998). Segregation distortion at isozyme locus *Lap-1* in *Schlumbergera* (Cactaceae) is caused by linkage with the gametophytic self-incompatibility (*S*) locus. *J. Heredity* 89(3): 206–210.

Oliveira-Filho, A. T. de (1993). Gradient analysis of an area of coastal vegetation in the state of Paraíba, northeastern Brazil. *Edinb. J. Bot.* 50: 217–236.

—— & Ratter, J. A. (1995). A study of the origin of central Brazilian forests by the analysis of plant species distribution patterns. *Ibid.* 52: 141–194.

Pennington, R. T., Prado, D. E. & Pendry, C. A. (2000). Neotropical seasonally dry forests and Quaternary vegetation changes. *J. Biogeography* 27: 261–273.

Petit, S. (1999). The effectiveness of two bat species as pollinators of two species of columnar cacti on Curaçao. *Haseltonia* 6: 22–31 (1998, publ. 1999).

Pfeiffer, L. (1837). *Enumeratio diagnostica cactearum hucusque cognitarum:* 43–46. L. Oehmigke, Berlin.

Pickel, B. (1938). Catálogo do Herbário da Escola Superior de Agricultura em Tapera (Pernambuco). *Bol. Mus. Nac. (Rio de Janeiro)* 13: 63–132.

Pinkava, D. J. (2002). On the evolution of the continental North American Opuntioideae (Cactaceae). *Succulent Pl. Res.* 6: 59–98.

Piso, W. (1648): Piso, W. & Marcgrave, G. (1648). *Historia naturalis Brasiliae .....* Frans Hack, Leiden & Ludovic Elzevier, Amsterdam.

Porembski, S., Martinelli, G., Ohlemüller, R. & Barthlott, W. (1998). Diversity and ecology of saxicolous vegetation mats on inselbergs in the Brazilian Atlantic rainforest. *Diversity and Distributions* 4: 107–119.

Prado, D. E. (1991). *A critical Evaluation of the Floristic Links between Chaco and Caatingas Vegetation in South America.* PhD thesis, University of St Andrews, Scotland. [Unpublished.]

—— & Gibbs, P. E. (1993). Patterns of species distributions in the dry seasonal forests of South America. *Ann. Missouri Bot. Gard.* 80: 902–927.

Prance, G. T. & Mori, S. A. (1982). In Memoriam, Dárdano de Andrade-Lima (1919–1981). *Brittonia* 34: 351–354.

Raw, A. (1996). Territories of the ruby-topaz hummingbird *Chrysolampis mosquitus* at flowers of the "turk's-cap" cactus, *Melocactus salvadorensis* in the dry caatinga of northeastern Brazil. *Revta Bras. Biol.* 56: 581–584.

Reid, W. V. (1998). Biodiversity hotspots. *Trends Ecol. Evol.* 13: 275–280.

Renvoize, S. A. (1984). *Grasses of Bahia.* Royal Botanic Gardens, Kew, Richmond, U.K. Pp. 301.

Resende, S. V. (2002). *Estudos Citogenéticos em Melocactus (Cactaceae).* Depto de Biologia Geral, Instituto de Biologia, Universidade Federal da Bahia, Salvador. Pp. 39. [Unpublished.]

Riccobono, V. (1909). Studii sulle Cattee del R. Orto Botanico di Palermo. *Boll. Ort. Bot. Palermo* 8: 215–266.

Ritter, F. (1968). Die Cephalien tragenden Kakteen Brasiliens. *Kakt. and. Sukk.* 19: 87–96, 119–123, 140, 156–162.

—— (1979). *Kakteen in Südamerika,* 1. F. Ritter Selbstverlag, Spangenberg.

—— (1980). *Ibid.,* 2. Ibid.

Rizzini, C. T. (1979). *Tratado de Fitogeografia do Brasil.* HUCITEC, Universidade de São Paulo, São Paulo.

—— (1982). *Melocactus no Brasil.* Instituto Brasileiro de Defesa Florestal (IBDF), Rio de Janeiro.

—— (1986). Sobre a Cactácea dendróide do calcário de Minas Gerais. *Rev. Brasil. Biol.* 46: 781–784.

—— & Mattos-Filho, A. (1992). Contribuição ao conhecimento das floras do nordeste de Minas Gerais e da Bahia Mediterrânea. Jardim Botânico do Rio de Janeiro. *Estudos e Contribuições,* 9. Pp. 95.

Rocha, E. A. & Agra, M. F. (2002). Flora do Pico do Jabre, Paraíba, Brasil: Cactaceae Juss. *Acta bot. bras.* 16: 15–21.

Rodal, M. J. N., Sales, M. F. de, Mayo, S. J. (1998). *Florestas Serranas de Pernambuco. Localização e Conservação dos Remanescentes dos Brejos de Altitude.* Universidade Federal Rural de Pernambuco, Recife, Brazil. Pp. 30.

Ross, R. (1981). Chromosome counts, cytology and reproduction in the Cactaceae. *Amer. J. Bot.* 68: 463–470.

Rowley, G. D. (1992). *Didiereaceae. 'Cacti of the Old World'.* British Cactus & Succulent Society, Richmond, U.K.

Ruiz, A. M. S., Soriano, P. J., Cavelier, J. & Cadena, A. (1997). Relaciones mutualísticas entre el murciélago *Glossophaga longirostris* y las cactáceas columnares en la zona árida de La Tatacoa, Colombia. *Biotropica* 29: 469–479.

Sales, M. F. de, Mayo, S. J. & Rodal, M. J. N. (1998). *Plantas Vasculares das Florestas Serranas de Pernambuco. Um Checklist da Flora Ameaçada dos Brejos de Altitude, Pernambuco, Brasil.* Universidade Federal Rural de Pernambuco, Recife, Brazil. Pp. 130.

Santos, F. A. R., Watanabe, H. M. & Alves, J. L. de H. (1997). Pollen morphology of some Cactaceae of North-Eastern Brazil. *Bradleya* 15: 84–97.

Savolainen, V., Fay, M. F., Bank, M. van der, Powell, M., Albach, D. C., Weston, P., Backlund, A., Johnson, S. A. & Chase, M. W. (2000). Phylogeny of the eudicots: a nearly complete familial analysis based on *rbc*L gene sequences. *Kew Bull.* 55: 257–309.

Scheinvar, L. (1985). Cactáceas. In: Reitz, R., *Flora Ilustrada Catarinense,* I. Itajaí, Santa Catarina, Brazil.

—— (1988). *Cactus triangularis* Linne, *Cereus ocamponis* Salm-Dyck, *Cactus triangularis* Vellozo, *Cereus undatus* Haworth és *Cereus trigonus* Haworth. *Kaktusz Világ* (Hungary) 1988(4): 63–69.

Schlindwein, C. & Wittmann, D. (1997). Stamen movements in flowers of *Opuntia* (Cactaceae) favour oligolectic pollinators. *Pl. Syst. Evol.* 204: 179–193.

Schulz, R. & Machado, M. (2000). *Uebelmannia* and their environment. Schulz Publishing, Teesdale, Australia. Pp. 160.

Schumann, K. M. (1890). Cactaceae. In: Martius, C. P. F. von, *Flora brasiliensis,* 4(2).

—— (1894). Cactaceae. In: Engler, A. & Prantl, K., *Das Pflanzenfamilien,* 3(6a): 156–205.

—— (1897–98). *Gesamtbeschreibung der Kakteen.* J. Neumann, Neudamm.

—— (1903). *Ibid., Nachträge 1898 bis 1902.* Ibid.

Stafleu, F. & Cowan, R. S. (1983). Taxonomic Literature, ed. 2, 4. *Regnum Vegetabile,* 110.

Stannard, B. E., ed. (1995). *Flora of the Pico das Almas, Chapada Diamantina, Bahia*. Royal Botanic Gardens, Kew, Richmond, U.K.

Stuppy, W. (2002). Seed characters and generic classification of the Opuntioideae. *Succulent Pl. Res.* 6: 25–58.

—— & Huber, H. (1991). Samenmerkmale und Gattungsgliederung der Opuntioideae. *Kakt. and. Sukk.* 42: 122–126, 144–147, 165–167, 208–211.

Supthut, D. (1999). A short Chronicle of 50 Years of IOS. *I.O.S. Bull.* 7: 6–16.

Taylor, N. P. (1980). Notes on the genus *Melocactus* (1): E Brazil. *Cact. Succ. J. Gr. Brit.* 42: 63–70.

—— (1981). Reconsolidation of *Discocactus* Pfeiff. *Ibid.* 43: 37–40.

—— (1982). Additional notes on E Brazilian *Melocactus*. *Ibid.* 44: 7–8.

—— (1991a). The genus *Melocactus* in Central and South America. *Bradleya* 9: 1–80.

—— (1991b). *Cereus, Micranthocereus*. In: Hunt, D. R. & Taylor, N. P., eds, Notes on miscellaneous genera of Cactaceae. *Ibid.* 9: 85–86, 89.

—— (1991c). [Submission to the Republic of Brazil's CITES Management Authority (IBAMA), via the CITES Secretariat, Lausanne, Switzerland, proposing the inclusion of *Discocactus, Uebelmannia* and 4 species of *Melocactus* in Appendix I of the Convention on International Trade in Endangered Species of fauna & flora.]

—— (1991d). Proposal to conserve 5401a *Arthrocereus* A. Berger (Cactaceae) with a new type. *Taxon* 40: 660–662.

—— (1992a). *Arthrocereus, Cereus*. In: Hunt, D. R. & Taylor, N. P., eds, Notes on miscellaneous genera of Cactaceae (2). *Bradleya* 10: 17, 20–25.

—— (1992b). Plants in peril, 17. *Melocactus conoideus*. *Kew Magazine* 9: 138–142.

—— (1994). Proposal to conserve *Cactus cruciformis* Vell. (Cactaceae). *Taxon* 43: 469–470.

—— (1995a). *Hylocereus*. In: Rowley, G. D., ed., The Sessé & Mociño cactus plates – A postscript. *Bradleya* 13: 119–120.

—— (1995b). Validation of *Monvillea kroenleinii* Kiesling as *Cereus kroenleinii* and a note on extension of its range. *Kew Bull.* 50: 819–820.

—— (1999). *Rhipsalis puniceodiscus*. Cactaceae. *Curtis's Bot. Mag.* 16: 29–33.

—— (2000). *Taxonomy and Phytogeography of the Cactaceae of Eastern Brazil*. PhD thesis. Life Sciences, The Open University, UK. Pp. 398+16. [Unpublished.]

—— (mss. [1996]). Cladistic analysis of 25 taxa representative of major elements of Rhipsalideae (Cactaceae), based on 29 morphological characters and utilising PAUP3 with Adams consensus tree.

—— & Eggli, U. (1991). *Stephanocereus*. In: Hunt, D. R. & Taylor, N. P., eds, Notes on miscellaneous genera of Cactaceae. *Bradleya* 9: 91.

—— Stuppy, W. & Barthlott, W. (2002). Realignment and revision of the Opuntioideae of Eastern Brazil. *Succulent Pl. Res.* 6: 99–132.

—— & Zappi, D. C. (1989). An alternative view of generic delimitation and relationships in tribe Cereeae (Cactaceae). *Bradleya* 7: 13–40.

—— —— (1990). Notes on *Leocereus* Britton & Rose. *Ibid.* 8: 108.

—— —— (1991). *Arthrocereus, Discocactus*. In: Hunt, D. R. & Taylor, N. P., eds, Notes on miscellaneous genera of Cactaceae. *Ibid.* 9: 84–86.

———— ———— (1992a). Cactaceae do Vale do Rio Jequitinhonha, Minas Gerais. *Acta Bot. Bras.* 51: 63–69 (1991, publ. 1992).

———— ———— (1992b). *Leocereus.* In: Hunt, D. R. & Taylor, N. P., eds, Notes on miscellaneous genera of Cactaceae (2). *Bradleya* 10: 25–26.

———— ———— (1992c). Proposal to conserve *Cereus jamacaru* DC. (Cactaceae) with a new type. *Taxon* 41: 590–591.

———— ———— (1995). Appendix II: Herbarium records. In: McMillan, A. J. S. & Horobin, J. F., Christmas Cacti. The genus *Schlumbergera* and its hybrids. *Succulent Pl. Res.* 4: 80–81.

———— ———— (1996). Two species of *Arrojadoa* from Eastern Brazil. *Curtis's Bot. Mag.* 13: 70–78.

———— ———— (1997). Nomenclatural adjustments and novelties in Brazilian Cactaceae. *Cact. Consensus Initiatives* 3: 7–8.

———— ———— & Eggli, U. (1992). *Pseudoacanthocereus.* In: Hunt, D. R. & Taylor, N. P., eds, Notes on miscellaneous genera of Cactaceae (2). *Bradleya* 10: 28–31, pl. V.

Tricart, J. (1985). Evidence of Upper Pleistocene dry climates in northern South America. Chapter 10, pp. 197–217, in: Douglas, I. & Spencer, T., *Environmental change and tropical geomorphology.* Allen & Unwin, London.

Uebelmann, W. J. (1996). *Horst & Uebelmann Feldnummernliste.* W. J. Uebelmann, Zufikon, Switzerland. Pp. 204.

———— & Braun, P. J. (1984). *25 Jahre HU Horst–Uebelmann. Beschriebene Arten und Feldnummern.* Flora-Buchhandel, Titisee-Neustadt, Germany. Pp. 64.

Ule, E. (1908). Kautschukgewinnung und Kautschukhandel in Bahia. *Notizbl. Bot. Gart. Berlin* 5(41a): Tab. IV [map].

Urban, I. (1906). Vitae itineraque collectorum botanicorum. In: Martius, C. P. F. von, *Flora brasiliensis* 1(1): 1–154.

Velloso, A. L., Sampiao, E. V. S. B. & Pareyn, F. G. C., eds (2002). *Ecorregiões Propostas para o Bioma Caatinga.* Associação Plantas do Nordeste & Instituto de Conservação Ambiental, The Nature Conservancy do Brasil, Recife. Pp. 75.

Wallace, R. S. (1995). Molecular systematic study of the Cactaceae: Using chloroplast DNA variation to elucidate Cactus phylogeny. *Bradleya* 13: 1–12.

———— (1997). The phylogenetic position of *Mediocactus hahnianus. Cact. Consensus Initiatives* 4: 11–12.

———— & Cota, J. H. (1996). An intron loss in the chloroplast gene *rpo*C1 supports a monophyletic origin for the subfamily Cactoideae of the Cactaceae. *Curr. Genet.* 29: 275–281.

———— & Dickie, S. L. (2002). Taxonomic implications of chloroplast DNA sequence variation in subfam. Opuntioideae (Cactaceae). *Succulent Pl. Res.* 6: 9–24.

———— & Gibson, A. C. (2002). Evolution and Systematics. Pp. 1–21. In: Nobel, P. S., *Cacti. Biology and Uses.* Univ. California Press, Berkeley etc.

Walters, S. M., Alexander, J. C. M., Brady, A., Brickell, C. D., Cullen, J., Green, P. S., Heywood, V. H., Matthews, V. A., Robson, N. K. B., Yeo, P. F. & Knees, S. G. (1989). *The European Garden Flora,* 3. Cambridge Univ. Press, Cambridge, U.K. etc.

Warming, J. E. B. (1908). *Lagoa Santa*. Trans. A. Löfgren. In: Warming, J. E. B. & Ferri, M. G. (1973), *q.v.*

—— & Ferri, M. G. (1973). *Lagoa Santa & A Vegetação dos Cerrados Brasileiros*. Editora Univ. São Paulo and Itatiaia Editora, Belo Horizonte.

WCMC (1992). *Protected Areas of the World: A review of national systems*, 4. Neartic and Neotropical. IUCN, Gland, Switzerland and Cambridge, U.K.

Werdermann, E. (1932). *Cephalocereus Lehmannianus* Werd. etc. In: Fedde, F., *Rep. Sp. Nov., Sonder-Beih. C*, Lfg. 11.

—— (1933). *Brasilien und seine Säulenkakteen*. J. Neumann, Neudamm.

—— (1934). Cactaceae novae. *Notizbl. Bot. Gart. Berlin* 12: 228–229.

—— (1942). *Brazil and its Columnar Cacti*. Abbey Garden Press, Pasadena. [Werdermann (1933), trans. R. W. Kelly; pp. 84–120 trans. & ed. L. Croizat.]

Whitehead, P. J. P. & Boeseman, M. (1989). *A portrait of Dutch 17th century Brazil. Animals, plants and people by the artists of Johan Maurits of Nassau*. North-Holland Publishing Company. Amsterdam / Oxford / New York.

Wurdack, J. (1970). Erroneous data in Glaziou collections of Melastomataceae. *Taxon* 19: 911–913.

Yang, X.-Y. (1999). *Storage, germination and characterisation of Cactaceae seed*. PhD thesis, Jilin Agricultural Univ., P. R. China. [Unpublished.]

Zappi, D. C. (1989). *A família Cactaceae nos Campos Rupestres da Cadeia do Espinhaço, Minas Gerais*. MSc thesis, Univ. São Paulo. [Unpublished.]

—— (1991). Flora da Serra do Cipó, Minas Gerais: Cactaceae. *Bol. Bot. Univ. São Paulo* 12: 43–59 (1990, publ. 1991).

—— (1992). *Revisão taxonómica do gênero Pilosocereus (Cactaceae) no Brasil*. PhD thesis, Univ. São Paulo. [Unpublished; subsequently updated and published as the following:]

—— (1994). *Pilosocereus* (Cactaceae). The genus in Brazil. *Succulent Pl. Res.* 3: 1–160.

—— & Taylor, N. P. (1991). *Cipocereus*. In: Hunt, D. R. & Taylor, N. P., eds, Notes on miscellaneous genera of Cactaceae. *Bradleya* 9: 86.

—— —— (1994). Cactaceae del Brasile orientale / Eastern Brazilian Cacti. *Piante Grasse* 13(4), suppl.: 65–78 (1993, publ. 1994).

6.2 ACKNOWLEDGEMENTS & PHOTO CREDIT CODES

Any study in systematics and biogeography of this size inevitably relies on the help and kindness of many individuals. These acknowledgements cannot hope to be comprehensive and thus, at the outset, the authors wish to apologize to anyone who feels under-valued by the omission of their name or organization from the following list.

To many colleagues, both current, former and retired, at the Royal Botanic Gardens, Kew, we owe more than a mention: Dr DAVID HUNT (Honorary Research Fellow and Cactaceae specialist), Dr RAY HARLEY (shared field knowledge & photos), Dr ALAN PATON (NT's PhD supervisor), Dr GWIL LEWIS (for taking the trouble to photograph cacti on his travels and for help with literature), Dr EIMEAR NIC LUGHADHA (for varied advice), Dr NICHOLAS HIND (for cactus collections and photographs), Dr DICK BRUMMITT (for much nomenclatural advice), Dr WOLFGANG STUPPY (for collaboration on Opuntioideae and IUCN Red Listing), Dr MIKE MAUNDER (conservation literature), STEVE RUDDY (for help with software on countless occasions), JUSTIN MOAT (Herbarium GIS Unit, for help with mapping), PHIL GRIFFITHS (field companion), Dr CHARLIE BUTTERWORTH and ROBYN COWAN (for molecular phylogenies), Dr ILIA LEITCH (for obtaining cytogenetics literature), Dr SIMON MAYO (for helpful discussions), MARILYN WARD (illustrations research, Kew Library), ANDREW MCROBB (Photographic Unit, Information Services Dept) and Dr BOB ALLKIN (CNIP, for logistical support in Brazil).

The following Brazilian friends are sincerely thanked for facilitating field excursions and herbarium visits (some generously providing accommodation in their own homes) and/or for supplying information and photographs (institutions in which they are based identified by *Index Herbariorum* codes): Prof. EMERSON ROCHA (JPB, PEUFR and now Univ. Santa Cruz, Ilhéus, Bahia) & Profa MARIA DE FÁTIMA AGRA (JPB); the late Dr ANDRÉ DE CARVALHO & JOMAR JARDIM (CEPEC, the latter now at HUEFS); MARLON MACHADO (HUEFS, for much valuable information, photographs and assistance) and also his collaborators from Vitória da Conquista, Bahia, namely RAYMUNDO REIS JUNIOR & JOSÉ CARLOS FALCON (both members of the Associação Conquistense de Orquidófilos), AVALDO DE OLIVEIRA SOARES FILHO & MARCELLO MOREIRA (Univ. Estadual do Sudoeste da Bahia) and RITA AUGUSTO (Cruz das Almas, BA, who acted as guide at the Serra da Jibóia, Mun. Santa Teresinha); HÉLIO BOUDET FERNANDES (MBML); Dr FRANCISCO DE ASSIS DOS SANTOS (SP, now HUEFS); Profa MARIA LUIZA SANTOS (Aracaju, Sergipe, for facilitating access to ASE and the Serra da Itabaiana); Dra ANA MARIA GIULIETTI (SPF, now HUEFS, DZ's PhD supervisor), Dra NANUZA LUIZA DE MENEZES (SPF, 'que maravilha !'), Dr J. R. PIRANI (SPF, DZ's mentor); JOÃO BATISTA (field companion) & Dra MANOELA DA SILVA (MG); Dr OBERDAN PEREIRA (Univ. Fed. Vitória, ES); Dr LUCIANO PAGANUCCI QUEIROZ, Dr EDUARDO BORBA, CÁSSIA TATIANA ANDRADE, Dra LIGIA FUNCH & postgraduate students (HUEFS); ALOÍSIO CARDOSO (Administrador, APA de Gruta dos Brejões, BA); Irmão DELMAR (Presidente, Grupo Ambientalista Morrense, Morro do Chapéu, BA); MÍRIAM PIMENTEL MENDONÇA, INÊS RIBEIRO DE ANDRADE, MARIA GUADALUPE CARVALHO & JULIANA ORDONES REGO (Fundação Zoo-Botânica de Belo Horizonte); Padre CÉLIO M. DELL'AMORE (Director,

RPPN do Caraça, Catas Altas, Minas Gerais); JULIO ANTÔNIO LOMBARDI & JOÃO RENATO STEHMANN (BHCB); Dra MARIA DO CARMO E. AMARAL, Dr VOLKER BITTRICH & LIDYANNE AONA (UEC); PATRÍCIA SOFFIATTI & RAFAELA FORZZA (SPF, the latter now at RB); Dra LÚCIA ROSSI & E. L. CATHARINO (SP); Dr FABIO DE MELO SENE (Univ. São Paulo, Ribeirão Preto); Dr GERT HATSCHBACH (MBM); LÚCIA HELENA DE OLIVEIRA (IBAMA, Brasília) and ROSÂNGELA LYRA-LEMOS (MAC). The Brazilian government's agencies, CNPq and IBAMA, who approved collecting permits and special visas in 1988, 1990 & 1999 (EX-14/88, 181/90-DEVIS), with IBAMA also part funding NT's participation in a Brazilian ecosystems conservation symposium in 1997, are gratefully acknowledged for their support.

European and North American friends and acquaintances whose help has been invaluable include Dr URS EGGLI (ZSS, field companion, photographer and bibliographer), Prof. Dr WILHELM BARTHLOTT (BONN, collaborator on Rhipsalideae and seed-morphology, many photographs), Dr STEFAN POREMBSKI (ROST, inselberg cacti), Dr BEAT LEUENBERGER (B, photographs of types & botanists), Dr MATS HJERTSON (supplier of bibliography and field companion), Dr RALF BAUER (epiphytes), ANDREAS HOFACKER and acquaintances (Böblingen, Germany; photographs and discussions), Dr TOBY PENNINGTON (E, biogeography discussions and references), Dr PETER GIBBS (STA, NT's PhD external examiner), Prof. Dr PAUL MAAS (U, for facilitating study visits), Dr ROBERT WALLACE (AMES, for information on gene sequence phylogenies), Dr DAN NICOLSON (US), the late Dr JOHN WURDACK (US), MYRON KIMNACH (HNT), Dr W. WAYT THOMAS (NY) and the directors/curators of other herbaria listed in Chapter 1.

U.K. cactophiles, GRAHAM CHARLES, PAUL KLAASSEN, TERRY HEWITT (Holly Gate Nursery) and ROY MOTTRAM (Whitestone Gardens) kindly assisted with plant material and/or information/photographs (especially G. Charles!), while BRITISH AIRWAYS' 'Assisting Conservation' programme provided free flights to NT for travel to Brazil and Europe on two occasions, enabling participation in conservation planning, presentation of papers and receipt of feedback.

PHOTO CREDIT CODES

B = © courtesy of the Botanic Garden and Botanical Museum Berlin-Dahlem, WR = Werner Rauh† (courtesy of Wilhelm Barthlott), NT = Nigel Taylor, DZ = Daniela Zappi, GC = Graham Charles, WB = Wilhelm Barthlott, UE = Urs Eggli, K = © Trustees of the Royal Botanic Gardens Kew, AJ = Andrea L. Jones, MM = Marlon Machado, RB = Ralf Bauer, ER = Emerson Rocha, RH = Ray Harley, AH = Andreas Hofacker, KH = Konrad Herm, WvH = Werner van Heek.

7. GLOSSARY AND INDEXES

7.1 GLOSSARY OF SPECIALIZED BOTANICAL TERMS (CACTACEAE)

A few specialized terms used by botanists to describe cactaceous structures are defined below. Some other botanical terms are explained where they are first used and, therefore, are not repeated here. For the terminology used to describe vegetation types, see Chapter 3.2.

areoles are the felted cushions (actually highly telescoped short shoots) found on various parts of the cactus plant, bearing the spines (= modified leaves), trichomatous hairs etc. Their meristems, when active, can give rise to new shoots and flower-buds.

aril (funicular envelope) in the sense used here is the bony and sometimes fibrous covering of the true seed developed in Opuntioideae, being a highly specialized development of the funicle and hence correctly described as an aril, although it is very different from so-called structures (or *strophioles*) in other parts of the Cactus Family (Barthlott & Hunt 2000: 24).

brachyblast leaves (Pereskioideae) are those produced *by* the *areoles* (see above), as opposed to those subtending areoles on the main extension axes (long shoots).

bract-scales are the minute to conspicuous scalelike appendages seen on the pericarpel and flower-tube below (ie. anterior to) the perianth-segments and often intergrading with the latter. They may be naked in their axils (eg. most Cereeae) or subtend areoles bearing trichomes, spines etc.

cephalium is a ± modified part of the stem, whether apical or lateral, whence the flowers and fruits are borne (cf. Barthlott & Hunt 1993: 164). It is in effect a kind of inflorescence structure in which the areoles may be enlarged or reduced relative to those in the vegetative part, and often compressed together, bearing abundant trichomes and/or dense spines/bristles, distinguishing the fertile part of the stem from the purely vegetative. It may be either a chlorophyllous or non-chlorophyllous part of the stem, whose cross-section in the case of lateral *cephalia* (plural) may remain normal (terete) or deformed, as in the so-called 'sunken' cephalia.

cereoid cacti are those with usually erect, leafless, ± cylindrical stems, as in *Cereus*, which can be contrasted with the globular to depressed stems of, eg., *Melocactus* or *Discocactus*.

epilithic is used in this book in contrast to *'epiphytic'* to describe cacti found growing upon rocks or rock surfaces when the plants in question are representative of taxa whose life-form is commonly that of an epiphyte.

flower-tube is the hollow or partially hollow structure above the pericarpel which comprises fused floral and receptacular tissues; the latter on the exterior, often bearing bract-scales; the former within and subtending the perianth-segments at its apex.

funicular pulp is the term used for the either solid or semi-liquid pulp found within the pericarp of cactus fruits, surrounding the seeds, which is derived from the ovule funicles.

glochids are normally short, strongly barbed, easily detached, specialized spines produced by the areoles of Opuntioideae and distinct from the generally much larger normal spines, which are also barbed in this subfamily.

hair-spines are specialized true cactus spines (ie. multicellular structures representing modified leaves) with a soft and often woolly, hairlike quality, often found on flowers of Trichocereeae. True areolar hairs are normally single-celled trichomes, short and felt-like, but sometimes longer and woolly (see *long hairs*).

hypertrophic spines are those developed in an exaggerated manner from areoles of indeterminate growth at the base of stems in certain rock-dwelling cacti, especially *Coleocephalocereus* and *Micranthocereus* spp.

long hairs are fine, elongate, woolly hairs produced by the areoles of a variety of Cereeae species and distinguishable from the short felt-like areolar trichomes and thicker hair-spines (see above)

pericarp is the fruit wall formed by the fusion of stem (receptacular) and floral tissues (see below). The visible exterior is the stem component and may bear bract-scales, areoles, spines etc., or be almost or quite naked.

pericarpel is the structure comprising the lower part of the specialized stem or receptacle into which the ovary of the inverted cactus flower is sunken.

podaria (sing. *podarium*) are the swellings often subtending areoles that represent the points of attachment of leaves or bracts that have been lost, or almost lost, in the course of evolution of the highly succulent habit.

stem-segments refers to stems where the seasonal growth is begun and terminated at a constriction, or marked by the development of an entirely new growth segment. This is best seen in Rhipsalideae (*Rhipsalis, Hatiora, Schlumbergera*), but also occurs, for example, in the cereoid taxa bearing terminal cephalia, where the subsequent vegetative phase grows 'through' the cephalium (*Stephanocereus leucostele, Arrojadoa* spp.).

GLOSSÁRIO DE TERMOS BOTÂNICOS (CACTACEAE)

Alguns termos especializados são atualmente utilizados por botânicos para descrever estruturas típicas de Cactaceae. A seguir, são apresentadas as traduções para a língua portuguesa e respectivas interpretações.

areoles — *aréolas* são estruturas parecidas com pequenas almofadas felpudas, consistindo em ramos extremamente encurtados encontrados em várias partes das plantas, dando origem aos espinhos, tricomas, cerdas, e, no caso de apresentarem meristemas ativos no seu interior, podem originar novos ramos e botões florais.

aril (funicular envelope) — *arilo (envelope funicular)* no sentido aplicado aqui trata-se da cobertura óssea e às vezes fibrosa cobrindo a semente em Opuntioideae, tratando-se de uma estrutura desenvolvida a partir de uma modificação extrema do funículo e portanto corretamente descrita como um arilo, apesar de mostrar-se bastante diferente das estruturas chamadas arilos (ou estrofíolos) em outros membros da família Cactaceae (Barthlott & Hunt 2000: 4)

brachyblast leaves — *folhas do braquiblasto* (Pereskioideae) são aquelas produzidas pelas aréolas (ver acima), contrastando com as folhas que subtendem aréolas nos eixos principais de extensão (ramos longos ou 'long shoots').

bract-scales — *brácteas escamiformes* são apêndices diminutos até bem visíveis encontrados no pericarpelo e no tubo-floral abaixo dos segmentos do perianto e muitas vezes intergradando-se com os mesmos. Essas brácteas podem apresentar axilas nuas (maioria dos membros da tribo Cereeae) ou subtender aréolas apresentando tricomas, espinhos, etc.

cephalium — *cefálio* é uma parte mais ou menos modificada do ramo, podendo ser apical ou lateral, a partir da qual surgem flores e frutos (Barthlott & Hunt 1993: 164). Trata-se de uma estrutura do tipo inflorescência, na qual as aréolas podem apresentar-se maiores ou reduzidas com respeito àquelas da parte vegetativa, e muitas vezes comprimidas, comportando tricomas ou espinhos e cerdas abundantes, distinguindo a parte fértil daquela puramente vegetativa. Pode tratar-se de uma parte tanto fotossintetizante como não fotossintetizante do ramo, sendo que uma secção transversal de um dito cefálio lateral pode permanecer cilíndrico ou apresentar-se deformado, no caso de um cefálio definido como 'aprofundado'.

cereoid — *cereóide* é o adjetivo utilizado para caracterizar aqueles cactos com ramos normalmente eretos, sem folhas, mais ou menos cilíndricos, como no caso de *Cereus*, contrastando com aqueles que apresentam ramos globosos ou depressos, como por exemplo *Melocactus* ou *Discocactus*.

epilithic — *epilítico*, utilizado neste livro em contraste com 'epifitico' ou 'epífita', é relativo a cactos encontrados crescendo sobre rochas ou superfícies rochosas, quando o tipo de planta em questão é mais comumente encontrado crescendo como epífita.

flower-tube — *tubo-floral* trata-se da estrutura oca ou partialmente oca acima do pericarpelo, constituída de tecidos florais e receptaculares adnados; externamente apresentando brácteas-escamiformes e internamente subtendendo segmentos do perianto na porção apical.

funicular pulp — *polpa funicular* trata-se da polpa sólida ou líquida encontrada no interior do pericarpo de frutos de Cactaceae, e que envolve as sementes, sendo derivada do funículo dos óvulos.

glochids — *gloquídeos* são normalmente espinhos curtos, com os lados fortemente serreados e que se destacam facilmente, presentes nas aréolas das espécies da subfamília Opuntioideae e distintos dos espinhos normais, que são muito maiores, mas também serreados na mesma subfamília.

hair-spines — *tricomas-capiliformes* são espinhos verdadeiros especializados, ou seja, estruturas multicelulares representando folhas modificadas, com textura macia e muitas vezes lanosa, assemelhando-se a fios de cabelo, encontrados muitas vezes em flores de membros da tribo Trichocereeae. Tricomas areolares são normalmente unicelulares, curtos e com aspecto aveludado, mas podem às vezes apresentar-se mais longos e lanosos (ver 'tricomas lanosos' – 'long hairs').

hypertrophic spines — *espinhos hipertróficos* são aqueles desenvolvidos de maneira exagerada a partir de aréolas de crescimento indeterminado presentes na base dos ramos em certos cactos que crescem sobre rochas, especialmente *Coleocephalocereus* e *Micranthocereus* spp.

long hairs — *tricomas lanosos* são tricomas finos, longos, lanosos, produzidos pelas aréolas de muitas espécies de Cereeae, distintos dos tricomas curtos das aréolas e dos 'tricomas capiliformes' – 'hair-spines' (ver acima).

pericarp — *pericarpo* é a parede do fruto formada pela fusão do ramo (receptáculo) e dos tecidos florais (ver 'pericarpel' — 'pericarpelo' abaixo). A parte visível é o exterior do ramo e pode apresentar brácteas-escamiformes, aréolas, espinhos, costelas, etc., ou ser quase ou completamente liso.

pericarpel — *pericarpelo* trata-se da estrutura composta pela parte inferior do ramo especializado ou receptáculo no interior do qual o ovário invertido da flor de Cactaceae encontra-se imerso.

podaria (sing. *podarium*) — *podários* trata-se dos espessamentos encontrados subtendendo aréolas e que representam o ponto de conexão de folhas ou brácteas que sofreram redução ou perda no curso da evolução de ramos com hábito extremamente suculento e fotossintetizante.

stem-segment — *segmento do ramo* refere-se aos ramos onde o crescimento estacional inicia-se e finaliza em uma constricção, ou é marcado pelo desenvolvimento do crescimento de um segmento inteiramente novo. Este fenômeno pode ser observado na tribo Rhipsalideae, mas também ocorre em cactos cereóides dotados de cefálio terminal (*Stephanocereus leucostele*, *Arrojadoa* spp.).

7.2 INDEX TO BOTANICAL NAMES AND EPITHETS

Accepted names (or their epithets) are indicated in **bold** type, synonyms, inadequately known taxa, orthographic variants and invalid names by *italics*. Synonyms based on types collected outside the Eastern Brazil area, in adjacent regions of Brazil, are also included. Other names of taxa from outside the region or those that are only doubtfully represented, which are mentioned in discussion or cited as the types of generic names, are not usually included, unless given in plain Roman type. Autonyms representing accepted names of included taxa are also listed. The accepted names of introduced or widely cultivated taxa are followed by an asterisk (★). Authority citations are omitted except for misapplied names and homonyms (the citation, if given in abbreviated form being of the author who misapplied the name or published the homonym in question). Entries are referenced by the number of the accepted generic name followed, after a colon, by the species number and, if applicable, the letter(s) indicating the accepted infraspecific classification. Binomials referenced with 'see' to the generic number only are discussed under the given genus heading in Chapter 5; similarly, those followed by a genus:species number. Hybrid taxa are indicated by means of a multiplication sign (✕) connecting the species numbers, eg. **Tacinga ✕ quipa** (**T. palmadora** ✕ **T. inamoena**) is indicated as 3:5✕3:6. A few names listed here do not appear as formally cited synonyms in chapter 5, especially some archaic and 'nude' (description-less) names in *Lepismium* and *Uebelmannia*.

7.3 INDEX TO VERNACULAR NAMES

As all users of vernacular plant names should be aware, the application of such nomenclature is very haphazard, so that some plants have many different names, while many different species are known by the same. Yet other, common species, eg. *Brasilicereus phaeacanthus*, have no recorded vernacular, while sometimes rare, narrow endemic taxa possess a distinctive local name. Some commonly applied vernaculars are used more or less indescriminately for a wide variety of cacti, so that, eg., the name 'Mandacaru' (commonly *Cereus jamacaru*) can be used to identify almost any Cactaceae in Brazil; similarly 'Facheiro' and sometimes 'Xique-xique' (strictly speaking = *Pilosocereus gounellei*) are used for almost all columnar or cylindrical-stemmed cacti in the Nordeste, while 'Quipá', 'Palma' and 'Palmatória' are applied to anything vaguely opuntioid in appearance. There are countless spelling variants, requiring the user to look for other possible ways of spelling a name when consulting the index below. It is also important to note that the following list is probably far from complete:

7.4 APPENDICES I & II

APPENDIX 1

Table 7.1

Data on the scientific discovery and description of the native Cactaceae of Eastern Brazil

TAXA (IN SYSTEMATIC ORDER) *Key: † = endemic to total area; ‡ = endemic to core area*	Earliest illustration or herbarium record as native in Eastern Brazil; state code; author ref./*collector(s)*	Date of earliest correctly applied valid Latin name; origin of type; author(s)
Pereskia aculeata	1896; MG; *Silveira*	1753; ex cult.; Linnaeus
P. grandifolia subsp. *grandifolia*	1815/1816; ES/RJ; *Wied-Neuwied*	1819; Rio de Janeiro; Haworth
P. grandifolia subsp. *violacea* ‡	1922; MG; *Santos*	1986; Minas Gerais; Leuenberger
P. bahiensis ‡	1906; BA; *Ule*	1908; Bahia; Gürke
P. stenantha ‡	1912; BA; *Zehntner*	1979; Bahia; Ritter
P. aureiflora ‡	1950; BA; *Pinto*	1979; Minas Gerais; Ritter
Quiabentia zehntneri ‡	1912; BA; *Zehntner*	1919; Bahia; Britton & Rose
Tacinga funalis ‡	Before 1915; BA; *Zehntner*	1919; Bahia; Britton & Rose
T. braunii ‡	Before 1890; MG; *anon.* in *Glaziou*	1989; Minas Gerais; Esteves Pereira
T. (Opuntia) werneri ‡	1915; BA; *Rose & Russell*	1992; Bahia; Eggli
T. (Opuntia) palmadora ‡	Before 1645; NE Brazil; see Whitehead & Boeseman (1989: tt. 89b, 99b)	1919; Bahia; Britton & Rose
T. (Opuntia) saxatilis subsp. *saxatilis*	1959; MG; *Ritter*	1979; Minas Gerais; Ritter
T. (Opuntia) saxatilis subsp. *estevesii* ‡	1984; BA; *Esteves Pereira*	1990; Bahia; Braun
T. (Opuntia) inamoena †	Before 1890; MG; *anon.* in *Glaziou*	1890; Minas Gerais; Schumann
Brasiliopuntia (Opuntia) brasiliensis	Before 1645; NE Brazil; see Whitehead & Boeseman (1989: tt. 3a, 7a, 51b)	1814; NE Brazil; Willdenow
Opuntia monacantha	1959; MG; *Ritter*	1819; 'Barbados'; Haworth
Hylocereus setaceus	Before 1867; MG; Warming (1908)	1828; Brazil; Salm-Dyck
Epiphyllum phyllanthus	[data not available]	1753; ex cult.; Linnaeus
Pseudoacanthocereus brasiliensis ‡	1912; BA; *Zehntner*	1920; Bahia; Britton & Rose
Lepismium houlletianum	1978; MG; *Martinelli*	1858; Brazil; Lemaire
L. warmingianum	Before 1867; MG; Warming (1908)	1890; Minas Gerais; Schumann
L. cruciforme	1944; MG; *Heringer*	1827; ex cult.; Haworth
Rhipsalis russellii	1915; BA; *Rose & Russell*	1923; Bahia; Britton & Rose
R. elliptica	1959; MG; *Castellanos*	1890; São Paulo; Schumann
R. oblonga	1917; ES; *Luetzelburg*	1918; Rio de Janeiro; Löfgren
R. crispata	1970, PE, *Andrade-Lima*	1830; Brazil; Haworth

R. floccosa subsp. *floccosa*	1915; BA; *Rose & Russell*	1837; Brazil; Pfeiffer
R. floccosa subsp. *oreophila* ‡	1964; MG; *Ritter*	1979; Minas Gerais; Ritter
R. floccosa subsp. *pulvinigera*	1915; MG; *Hoehne*	1889; Brazil; Lindberg
R. paradoxa subsp. *septentrionalis* ‡	Before 1966; BA; *Martins*	1995; Bahia; Taylor & Barthlott
R. pacheco-leonis subsp. *catenulata*	1986; ES; *Rauh & Kautsky*	1992; Rio de Janeiro; Kimnach
R. cereoides	1986; ES; *Rauh & Kautsky*	1936; Rio de Janeiro; Backeberg & Voll
R. sulcata ‡	1986; ES; *Rauh & Kautsky*	1898; ex cult.; Weber
R. lindbergiana	Before 1823; MG; *Saint-Hilaire*	1890; Rio de Janeiro; Schumann
R. teres	1898; MG; *Jaguaribe*	1829; Rio de Janeiro; Vellozo
R. baccifera subsp. *baccifera*	1945; CE; *Cutler*	1770–1777; ex cult.; Mueller
R. baccifera subsp. *hileiabaiana* ‡	Before 1857; BA; *Blanchet*	1995; Bahia; Taylor & Barthlott
R. pulchra	1970; MG; *Krieger*	1915; Rio de Janeiro; Löfgren
R. burchellii	1986; ES; *Rauh*	1923; São Paulo; Britton & Rose
R. juengeri	1991; MG; *Zappi*	1995; ex cult.; Barthlott & Taylor
R. clavata	1986; ES; *Rauh & Kautskyi*	1892; Rio de Janeiro; Weber
R. cereuscula	1893; MG; *Silveira*	1830; Brazil; Haworth
R. pilocarpa	1986; ES; *Rauh & Kautsky*	1903; São Paulo; Löfgren
R. hoelleri ‡	1987; ES; *Orssich*	1995; Espírito Santo; Barthlott & Taylor
Hatiora salicornioides	1892; MG; *Ule*	1819; Rio de Janeiro; Haworth
H. cylindrica	1989; BA; *Augusto*	1915; Rio de Janeiro; Britton & Rose
Schlumbergera kautskyi ‡	1986; ES; *Kautsky*	1991; Espírito Santo; Horobin & McMillan
S. microsphaerica	1941; MG; *Brade*	1890; Rio de Janeiro; Schumann
S. opuntioides	1991; MG; *Zappi*	1905; Rio de Janeiro; Löfgren & Dusén
Brasilicereus phaeacanthus ‡	1906; BA; *Ule*	1908; Bahia; Gürke
B. markgrafii ‡	1938; MG; *Markgraf et al.*	1950; Minas Gerais; Backeberg & Voll
Cereus mirabella	1964; MG; *Ritter*	1979; Minas Gerais; Ritter
C. albicaulis ‡	1915; BA; *Rose & Russell*	1920; Bahia; Britton & Rose
C. fernambucensis subsp. *fernambucensis*	Before 1645; NE Brazil; see (1989: tt. 45, 93, 97)	1839; Pernambuco; Lemaire Whitehead & Boeseman
C. fernambucensis subsp. *sericifer*	1917; ES; *Luetzelburg*	1979; Rio de Janeiro; Ritter
C. insularis †	Before 1877; F. de Noronha; *Moseley*	1884; F. de Noronha; Hemsley
C. jamacaru subsp. *jamacaru* †?	Before 1645; NE Brazil; see Whitehead & Boeseman (1989: tt. 70, 78b, 89b, 99b)	1828; NE Brazil; De Candolle
C. jamacaru subsp. *calcirupicola* ‡	Before 1867; MG; *Warming* (1908)	1979; Minas Gerais; Ritter
C. hildmannianus	Before 1890; MG; *Glaziou* (1909)	1890; SE Brazil; Schumann
Cipocereus laniflorus ‡	1987; MG; *Zappi & Scatena*	1997; Minas Gerais; Taylor & Zappi

C. crassisepalus ‡	1959; MG; *Ritter*	1973; Minas Gerais; Buining & Brederoo
C. bradei ‡	Before 1935; MG; *Brade* ?	1942; Minas Gerais; Backeberg & Voll
C. minensis subsp. *leiocarpus* ‡	1934; MG; *Brade*	2004; Minas Gerais; Taylor & Zappi
C. minensis subsp. *minensis* ‡	1932; MG; *Werdermann*	1933; Minas Gerais; Werdermann
C. pusilliflorus ‡	1964; MG; *Ritter*	1979; Minas Gerais; Ritter
Stephanocereus leucostele ‡	1906; BA; *Ule*	1908; Bahia; Gürke
S. luetzelburgii ‡	1913; BA; *Luetzelburg*	1923; Bahia; Vaupel
Arrojadoa bahiensis ‡	1981; BA; *Furlan et al.*	1993; Bahia; Braun & Esteves Pereira
A. dinae subsp. *dinae* ‡	1964; BA; *Ritter*	1973; Bahia; Buining & Brederoo
A. dinae subsp. *eriocaulis* ‡	1972; MG; *Horst*	1973; Minas Gerais; Buining & Brederoo
A. penicillata ‡	1906; BA; *Ule*	1908; Bahia; Gürke
A. rhodantha ‡	1907; PI; *Ule*	1908; Bahia; Gürke
Pilosocereus tuberculatus ‡	1932; PE; *Werdermann*	1933; Pernambuco; Werdermann
P. gounellei subsp. *gounellei* †	Before 1645; NE Brazil; see Whitehead & Boeseman (1989: tab. 80)	1897; Pernambuco; Weber
P. gounellei subsp. *zehntneri* ‡	1917; BA; *Zehntner*	1920; Bahia; Britton & Rose
P. catingicola subsp. *catingicola* ‡	c. 1907; BA; *Ule*	1908; Bahia; Gürke
P. catingicola subsp. *salvadorensis* †	1915; BA; *Rose & Russell*	1933; Bahia; Werdermann
P. azulensis ‡	1968; BA/MG; *Castellanos*	1997; Minas Gerais; Taylor & Zappi
P. arrabidae	1973; ES; *Horst & Uebelmann*	1862; Rio de Janeiro; Lemaire
P. brasiliensis subsp. *brasiliensis*	1990; ES; *O.J. Pereira*	1920; Rio de Janeiro; Britton & Rose
P. brasiliensis subsp. *ruschianus* ‡	1917; ES; *Luetzelburg*	1980; Espírito Santo; Buining & Brederoo
P. flavipulvinatus †	1935; CE; *Drouet*	1979; Piauí; Buining & Brederoo
P. pentaedrophorus subsp. *pentaedr.* ‡	Before 1853; BA; *Morel*	1858; Bahia; Cels
P. pentaedrophorus subsp. *robustus* ‡	1972; BA; *Horst & Uebelmann*	1994; Bahia; Zappi
P. glaucochrous ‡	1932; BA; *Werdermann*	1933; Bahia; Werdermann
P. floccosus subsp. *floccosus* ‡	Before 1867; MG; *Warming* (1908)	1950; Minas Gerais; Backeberg & Voll
P. floccosus subsp. *quadricostatus* ‡	1965; MG; *Ritter*	1979; Minas Gerais; Ritter
P. fulvilanatus subsp. *fulvilanatus* ‡	1968; MG; *Horst*	1973; Minas Gerais; Buining & Brederoo
P. fulvilanatus subsp. *rosae* ‡	1982; MG; *Horst & Uebelmann*	1984; Minas Gerais; Braun
P. pachycladus subsp. *pachycladus* ‡	1907; BA; Ule (1908)	1975; Bahia; Buining & Brederoo
P. pachycl. subsp. *pernambucoensis* †	1915; BA; *Rose & Russell*	1979; Pernambuco; Ritter
P. magnificus ‡	1964; MG; *Ritter*	1972; Minas Gerais; Buining & Brederoo

P. machrisii	Before 1975; BA; *Horst & Uebelmann*	1957; Goiás; Dawson
P. aurisetus subsp. *aurisetus* ‡	Before 1862; MG; *anon.*	1933; Minas Gerais; Werdermann
P. aurisetus subsp. *aurilanatus* ‡	1964; MG; *Ritter*	1979; Minas Gerais; Ritter
P. aureispinus ‡	c. 1972; BA; *Horst & Uebelmann*	1974; Bahia; Buining & Brederoo
P. multicostatus ‡	1965; MG; *Ritter*	1979; Minas Gerais; Ritter
P. piauhyensis †	1907; PI; *Ule*	1908; Piauí; Gürke
P. chrysostele †	1920; PB/PE; *Luetzelburg*	1923; Paraíba/Pernambuco; Vaupel
P. densiareolatus ‡	1936; MG; *Duarte*	1979; Minas Gerais; Ritter
Micranthocereus violaciflorus ‡	1968; MG; *Buining*	1969; Minas Gerais; Buining
M. albicephalus ‡	1964; MG; *Ritter*	1979; Minas Gerais; Buining & Brederoo
M. purpureus ‡	c. 1906; BA; *Ule*	1908; Bahia; Gürke
M. auriazureus ‡	1971; MG; *Buining*	1973; Minas Gerais; Buining & Brederoo
M. streckeri ‡	1985; BA; *Van Heek & Van Criekinge*	1986; Bahia; Van Heek & Van Criekinge
M. polyanthus ‡	1932; BA; *Werdermann*	1932; Bahia; Werdermann
M. flaviflorus ‡	1912; BA; *Zehntner*	1974; Bahia; Buining & Brederoo
M. dolichospermaticus ‡	1972; BA; *Horst & Uebelmann*	1974; Bahia; Buining & Brederoo
Coleocephaloc. buxbaum. subsp. *b.* ‡	1972; MG; *Buining & Horst*	1974; Minas Gerais; Buining
C. buxbaumianus subsp. *flavisetus* ‡	1965; MG; *Ritter*	1979; Minas Gerais; Ritter
C. fluminensis subsp. *fluminensis*	1972; MG; *Horst & Uebelmann*	1838; Rio de Janeiro; Miquel
C. fluminensis subsp. *decumbens* ‡	1965; MG; *Ritter*	1968; Minas Gerais; Ritter
C. pluricostatus ‡	1917; MG/ES; *Luetzelburg*	1971; Minas Gerais; Buining & Brederoo
C. goebelianus ‡	Before 1920; BA; *Zehntner*	1923; Bahia; Vaupel
C. aureus ‡	1964; MG; *Ritter*	1968; Minas Gerais; Ritter
C. purpureus ‡	1972; MG; *Horst & Uebelmann*	1973; Minas Gerais; Buining & Brederoo
Melocactus oreas subsp. *oreas* ‡	Before 1840; BA; *anon.*	1840; Bahia; Miquel
M. oreas subsp. *cremnophilus* ‡	1971; BA; *Harley*	1972; Bahia; Buining & Brederoo
M. ernestii subsp. *ernestii* †	c. 1906; BA; *Ule*	1920; Bahia; Vaupel
M. ernestii subsp. *longicarpus* ‡	1964; BA; *Ritter*	1974; Minas Gerais; Buining & Brederoo
M. bahiensis subsp. *bahiensis* ‡	1915; BA; *Rose & Russell*	1922; Bahia; Britton & Rose
M. bahiensis subsp. *amethystinus* ‡	1959; MG; *Ritter*	1972; Bahia; Buining & Brederoo
M. conoideus ‡	1972; BA; *Horst*	1973; Bahia; Buining & Brederoo
M. deinacanthus ‡	1971; BA; *Horst*	1973; Bahia; Buining & Brederoo
M. levitestatus ‡	1964; MG; *Ritter*	1973; Bahia; Buining & Brederoo
M. azureus ‡	1968; BA; *Horst*	1971; Bahia; Buining & Brederoo
M. ferreophilus ‡	1967; BA; *Horst*	1973; Bahia; Buining & Brederoo
M. pachyacanthus subsp. *pachyac.* ‡	1972; BA; *Horst*	1975; Bahia; Buining & Brederoo
M. pachyacanthus subsp. *viridis* ‡	1988; BA; *Taylor & Zappi*	1991; Bahia; Taylor

M. salvadorensis ‡	1932; BA; *Werdermann*	1934; Bahia; Werdermann
M. zehntneri †	1915; BA; *Rose & Russell*	1922; Bahia; Britton & Rose
M. lanssensianus ‡	1977; PE; *Horst*	1986; Pernambuco; Braun
M. glaucescens ‡	1967; BA; *Horst*	1972; Bahia; Buining & Brederoo
M. concinnus ‡	1964; BA; *Ritter*	1972; Bahia; Buining & Brederoo
M. paucispinus ‡	1981; BA; *Heimen et al.*	1983; Bahia; Heimen & Paul
M. violaceus s.l.	Before 1645; NE Brazil; see Whitehead & Boeseman (1989: tab. 48c)	1835; Brazil; Pfeiffer
M. violaceus subsp. *violaceus*	1837; PE; *Gardner*	as above
M. violaceus subsp. *ritteri* ‡	1964; BA; *Ritter*	1979; Bahia; Ritter
M. violaceus subsp. *margaritaceus* ‡	Before 1858; BA; *anon.*	1857/58; Bahia; Miquel
Harrisia adscendens ‡	Before 1645?; NE Brazil; see Whitehead & Boeseman (1989: tt. 89b, 99b)?	1908; Bahia; Gürke
Leocereus bahiensis ‡	1912; BA; *Zehntner*	1920; Bahia; Britton & Rose
Facheiroa ulei ‡	1907; BA; *Ule*	1908; Bahia; Gürke
F. cephaliomelana subsp. *cephaliom.* ‡	1959; MG; *Ritter*	1975; Bahia; Buining & Brederoo
F. cephaliomelana subsp. *estevesii* ‡	1984; BA; *E. Esteves Pereira*	1986; Bahia; Braun
F. squamosa ‡	1907; PI; *Ule*	1908; Piauí; Gürke
Espostoopsis dybowskii ‡	Before 1908; BA; *Dybowski*	1908; Bahia; Gosselin
Arthrocereus melanurus subsp. *melan.*	Before 1890; MG; *Glaziou*	1890; Minas Gerais; Schumann
A. melanurus subsp. *magnus* ‡	1969; MG; *Krieger*	1997; Minas Gerais; Taylor & Zappi
A. melanurus subsp. *odorus* ‡	1964; MG; *Ritter*	1979; Minas Gerais; Ritter
A. rondonianus ‡	Before 1935; MG; *Brade* ?	1951; Minas Gerais; Backeberg & Voll
A. glaziovii ‡	Before 1832; MG; *Sello*	1890; Minas Gerais; Schumann
Discocactus zehntneri subsp. *zehntn.* ‡	c. 1915; BA; *Zehntner*	1922; Bahia; Britton & Rose
D. zehntneri subsp. *boomianus* ‡	1967; BA; *Horst*	1971; Bahia; Buining & Brederoo
D. bahiensis †	c. 1915; BA; *Zehntner*	1922; Bahia; Britton & Rose
D. heptacanthus subsp. *catingicola*	1972; BA; *Horst*	1974; Bahia; Buining & Brederoo
D. placentiformis ‡	Before 1826; MG; *Riedel* ?	1826; Brazil; Lehmann
D. pseudoinsignis ‡	1972; MG; *Horst*	1991; Minas Gerais; Taylor & Zappi
D. horstii ‡	1972; MG; *Horst*	1973; Minas Gerais; Buining & Brederoo
Uebelmannia buiningii ‡	1966; MG; *Horst*	1968; Minas Gerais; Donald
U. gummifera ‡	1938; MG; *Mello-Barreto*	1950; Minas Gerais; Backeberg & Voll
U. pectinifera subsp. *pectinifera* ‡	1966; MG; *Horst*	1967; Minas Gerais; Buining
U. pectinifera subsp. *flavispina* ‡	1972; MG; *Horst & Uebelmann*	1973; Minas Gerais; Buining & Brederoo
U. pectinifera subsp. *horrida* ‡	1982; MG; *Horst & Uebelmann*	1984; Minas Gerais; Braun

APPENDIX 2

Checklists of Cactaceae from adjacent Brazilian Regions

The states for species of restricted occurrence are indicated by means of 2-letter codes, explained below; otherwise a listed taxon is known or expected to occur in all states included in the Region, those with a dagger (†) being endemic to the region or area indicated in the heading to the list. The most up-to-date published sources on the Cactaceae native to Brazil as a whole are Hunt (1999a) and Braun & Esteves Pereira (2002, 2003).

Northern Brazil

Comprising the states/territories of Acre (AC), Amazonas (AM), Rondônia (RO), Roraima (RR), Amapá (AP), Pará (PA) and Tocantins (TO). This region has only 16 species in total:

Cereus hexagonus (syn. *C. perlucens*, *Pilocereus*
  perlucens)
C. sp. nov.† (TO: Mun. Palmas, limestone)*
Discocactus heptacanthus subsp. *catingicola* (TO)
Disocactus amazonicus (*Wittia amazonica*,
  Wittiocactus amazonicus) (AM)
Epiphyllum phyllanthus
Melocactus estevesii† (RR)
M. neryi (AM,RR,RO)
M. smithii (syn. *M. roraimensis*, *Echinocactus
  amazonicus*?) (RR)

Micranthocereus (*Siccobaccatus*) *estevesii* (TO)
Pilosocereus flexibilispinus† (TO)
P. oligolepis (RR)
P. machrisii (*sens. lat.*) (PA,TO)
Pseudorhipsalis ramulosa (AC)
Rhipsalis baccifera subsp. *baccifera*
Hylocereus cf. *setaceus* (PA,RR)
Selenicereus wittii (*Strophocactus wittii*) (AM)

Central-western Brazil

Comprising the states/districts of Mato Grosso (MT), Mato Grosso do Sul (MS), Goiás (GO) and Distrito Federal (DF). This region has a total of c. 36 species:

Arrojadoa sp. nov (?), cf. *A. dinae*† (GO)
Arthrocereus spinosissimus† (MT)
Brasiliopuntia brasiliensis (MT,MS)
Cereus adelmarii† (MT)
C. bicolor
C. kroenleinii (MT,MS)
C. lanosus (MS)?
C. pierre-braunianus (GO)
C. saddianus† (MT)
C. spegazzinii (syn. *Monvillea spegazzinii*) (MS)
C. mirabella (GO,DF)?
Cleistocactus baumannii subsp. *horstii* (MT,MS)
Discocactus heptacanthus sens. *lat.*

D. ferricola† (MS)
Echinopsis calochlora (MS,MT)
E. rhodotricha (MS)
Epiphyllum phyllanthus
Frailea cataphracta (syn. *F. matoana*) (MS)
Gymnocalycium anisitsii (MS)
G. marsoneri (MS)
Harrisia balansae (syn. *H. guelichii*, *Eriocereus
  guelichii*, *Cereus balansae*; 'Harrisia pomanensis'
  misapplied) (MS)
Hylocereus setaceus (MS,MT)
Lepismium cruciforme (MS)
Micranthocereus (*Siccobaccatus*) *estevesii* (GO)

---

* Seen in the nursery of the Jardim Botânico de Brasília, July 2000. Rootstock said not to be tuberous; stems creeping over rocks, elongate cylindric, 2 cm diam., plain green, not glaucous; ribs 4–6(–8), low; spines brownish with pale tips, centrals 1(–4), to 12 mm, radials 6, shorter. Flowers and fruit unknown.

Opuntia anacantha (MS)
O. cf. *quimilo* (MS)
O. roborensis (MS)
Pereskia aculeata (GO)
P. sacharosa (MS,MT)
Praecereus euchlorus subsp. *euchlorus* (syn. *Monvillea alticostata*, 'M. cavendishii') (MS)

P. saxicola (*Cereus ritteri*) (MS)
Pilosocereus diersianus† (GO)
P. machrisii, sens. lat.
P. parvus (GO)
P. vilaboensis (syn. *P. rizzoanus*) † (GO)
Rhipsalis russellii (GO,MT)

Southern Brazil and the parts of South-eastern Brazil not included within E Brazil

Comprising the states of Minas Gerais (MG, in part), Rio de Janeiro (RJ, in part), São Paulo (SP), Paraná (PR), Santa Catarina (SC) and Rio Grande do Sul (RS). This area has a total of c. 103 species:

Cereus aethiops (RS)
C. hildmannianus (sens. lat.)
C. mirabella (MG)
Discocactus heptacanthus (MG)
Echinopsis calochlora (RS)
E. eyriesii (RS)
E. rhodotricha (RS)
Epiphyllum phyllanthus
Frailea castanea (RS)
F. cataphracta (RS)
F. curvispina† (RS)
F. gracillima, sens. lat. (RS)
F. mammifera (RS)
F. perumbilicata† (RS)
F. phaeodisca (RS)
F. pumila (RS)
F. pygmaea, sens. lat. (RS)
Gymnocalycium denudatum (RS)
G. horstii† (RS)
G. netrelianum (RS)
Hatiora cylindrica (MG?,RJ,SP?)
H. epiphylloides† (SP,RJ)
H. gaertneri† (PR,SC,RS)
H. herminiae† (SP)
H. rosea† (PR,SC,RS)
H. salicornioides (PR,SP,RJ,MG)
Hylocereus setaceus (PR,SP,RJ,MG)
Lepismium cruciforme
L. houlletianum
L. lumbricoides
L. warmingianum
Micranthocereus (Siccobaccatus) estevesii (MG)
Opuntia monacantha (syn. *O. arechavaletae*)
Parodia alacriportana† (RS,SC)
P. arnostiana† (RS)
P. buiningii (RS)

P. carambeiensis (PR,SC)
P. concinna, sens. lat. (RS)
P. crassigibba† (RS)
P. erinacea (RS)
P. fusca (RS)
P. haselbergii† (RS,SC)
P. herteri (RS)
P. horstii† (RS)
P. langsdorfii (RS)
P. leninghausii (RS)
P. linkii (RS,SC,PR)
P. magnifica† (RS)
P. mammulosa (RS)
P. muricata (RS)
P. neoarechavaletae (RS)
P. neohorstii† (RS)
P. nothominuscula† (RS)
P. nothorauschii (RS)
P. ottonis (RS,SC,PR)
P. oxycostata (RS)
P. rudibuenekeri† (RS)
P. rutilans (RS)
P. scopa (RS)
P. tenuicylindrica† (RS)
P. warasii (RS)
Pereskia aculeata
P. grandifolia (SP)
P. nemorosa (RS)
Pilosocereus albisummus† (MG)
P. brasiliensis (RJ)
P. machrisii (MG,SP)
Praecereus euchlorus subsp. *euchlorus* (SP,PR)
Rhipsalis agudoensis† (RS)
R. baccifera subsp. *shaferi* (SP)
R. burchellii (SP,PR)
R. campos-portoana

R. cereoides (RJ)

R. cereuscula

R. clavata (SP,RJ)

R. crispata (SP,RJ,SC)

R. dissimilis† (SP,PR)

R. elliptica

R. ewaldiana† (RJ)

R. floccosa (sspp. *floccosa* & *pulvinigera*)

R. grandiflora† (RJ,SP,PR)

R. lindbergiana (RJ,SP)

R. mesembryanthemoides† (RJ)

R. neves-armondii

R. oblonga (RJ,SP)

R. olivifera† (SP,RJ)

R. ormindoi† (RJ)

R. pacheco-leonis (RJ)

R. pachyptera†

R. paradoxa (subsp. *paradoxa*†)

R. pentaptera† (RJ)

R. pilocarpa

R. pulchra (MG,SP,RJ)

R. puniceodiscus†

R. russellii (PR)

R. teres

R. trigona† (SP,PR,SC)

Schlumbergera microsphaerica (RJ)

S. opuntioides (SP,RJ)

S. orssichiana† (SP,RJ)

S. russelliana† (RJ)

S. truncata† (RJ)

Tacinga saxatilis subsp. *saxatilis* (MG).